Steel Structures

SERIES IN CIVIL ENGINEERING

STEEL STRUCTURES
Design and Behavior

Second Edition

Charles G. Salmon
John E. Johnson
University of Wisconsin-Madison

HARPER & ROW, PUBLISHERS, New York
Cambridge, Hagerstown, Philadelphia, San Francisco
London, Mexico City, São Paulo, Sydney

17

Sponsoring Editor: Charles Dresser
Project Editor: Pamela Landau
Designer: Emily Harste
Production Manager: Marion A. Palen
Compositor: The Universities Press Limited
Printer and Binder: Halliday Lithograph Corporation
Art Studio: J & R Technical Services, Inc.

STEEL STRUCTURES: Design and Behavior, 2d edition

Copyright © 1980 by Charles G. Salmon and John E. Johnson

Library of Congress Cataloging in Publication Data
Salmon, Charles G
  Steel structures.
  (Series in civil engineering)
  Includes bibliographies and index.
  1. Building, Iron and steel. 2. Steel, Structural.
I. Johnson, John Edwin, 1931- joint author.
II. Title.
TA684.S24 1980   624′.1821   79-20924
ISBN 0-06-045694-9

CONTENTS

Chapter 6 - Compression Members **251**

Preface

The publication of the second edition reflects the continuing changes occurring in design requirements for structural steel. Design of structural steel members has developed over the last 75 years from a simple approach involving a few basic properties of steel and elementary mathematics to a sophisticated treatment demanding a thorough knowledge of structural and material behavior. Present design practice utilizes knowledge of mechanics of materials, structural analysis, and particularly, structural stability, in combination with nationally recognized design rules for safety. The most widely used design rules are those of the American Institute of Steel Construction (AISC), given in *Specification for the Design, Fabrication and Erection of Structural Steel for Buildings,* referred to hereafter as the AISC Specification.

The specific occurrences dictating this second edition are the publication of the 1978 AISC Specification (effective November 1, 1978) with Commentary, and the publication by AISC of the handbook, *Manual of Steel Construction,* 8th Edition, 1980. Steel members and components are selected from this handbook, referred to hereafter as the AISC Manual.

The second edition follows the same philosophical approach that has gained wide acceptance of users since the first edition was published in 1971. In the second edition, effort has been made to present in a logical manner the theoretical

background needed for developing and explaining design requirements, particularly those of the 1978 AISC Specification. Beginning with coverage of background material including references to pertinent research, the development of specific AISC Specification formulas is followed by a generous number of design examples explaining in detail the process of selecting minimum weight members to satisfy given conditions.

Considerable emphasis has been put on presenting for the beginning, as well as the advanced student the necessary elastic and inelastic stability concepts, the understanding of which is deemed essential to properly apply the AISC Specification requirements. This treatment is incorporated into the chapters in such a way that the reader may either study in detail the stability concepts in logical sequence, or omit or postpone study of sections containing the detailed development, merely accepting qualitative explanation and proceeding directly to the design.

In the 1978 AISC Specification, the traditional working stress method, focusing on working (service) loads and elastically computed stresses, remains the primary design philosophy and is Part 1 of the Specification. The strength design philosophy using factored service loads and "ultimate" strength is reflected in Part 2 where the rules for plastic design appear. Plastic design is a special case of the strength design philosophy wherein the "ultimate" strength must be the plastic strength. Throughout the text the theory and background material, being common to both design philosophies, have been integrated. The specific AISC design provisions and illustrative examples are, however, treated in separate sections within the chapters so that one may treat either the working stress or the plastic design portions separately.

The second edition contains an introduction to metrication using SI units as an addition to the primary use of U.S. customary units. Although the 1978 AISC Specification does not contain SI units, some use of SI units is incorporated throughout the text. AISC Specification formulas have their SI equivalent (conversions are practical ones made by the authors) given as a footnote on the text page containing the U.S. customary version. Tables and diagrams have both U.S. customary and SI units. Many problems at the ends of chapters contain the given numerical data of the problem repeated in SI in parenthesis at the end of the problem statement.

Depending on the proficiency required of the student, this textbook may provide material for two courses of three or four semester-credit hours each. It is suggested that the beginning course in steel structures for undergraduate students might contain the material of Chapters 1 through 7, 9, 10, 12, and 16, except Sections 6.4, 6.6, 6.12 to 6.18, 7.9 to 7.10, 9.3, 9.4, 9.10 to 9.12, and 12.6 and 12.7. The second course would review some of the same topics as in the first course, but much more rapidly, emphasizing items omitted in the first course. In addition, the remaining chapters—namely Chapter 8 on torsion, Chapter 11 on plate girders, Chapter 13 on connections, Chapter 14 on braced and unbraced frames, and Chapter 15 on frame design—are suggested for inclusion.

The reader will need to have ready access to the AISC Manual* throughout the study of the text, particularly when working with the examples. However, it is not the objective of this text that the reader become proficient in the routine use of tables; the tables serve only as a guide to obtaining experience with variation of design parameters and as an aid in arriving at good design. The AISC Specification and Commentary are contained in the AISC Manual and are therefore not included in the textbook, except for various individual provisions quoted where they are explained.

This second edition has used in all examples the new wide-flange and structural tee profiles that have been the standard sections rolled by the principal mills since about September 1, 1978. The new profiles are those whose dimensions are given by American Society for Testing and Materials (ASTM), *Standard Specification for General Requirements for Rolled Steel Plates, Shapes, Sheet Piling, and Bars for Structural Use* (ANSI/ASTM A6-77b), 1977. The new series contains 187 wide-flange (W) shapes, including 111 from the previous series with slight dimensional modifications along with 76 new wide-flange shapes; 81 wide-flange shapes from the previous series were dropped.

In the numerical examples, the selection of members was done in anticipation of the 8th Edition of the AISC Manual using special design aids published by various steel producers and by the authors. It is intended that the reader use the 1980 AISC Manual; and topical references to that manual correctly correspond to the latest edition. Appendix Tables A3 and A4 giving lateral-torsional buckling design information have been corrected in this second edition for the 1979 AISC Specification and the new 1978 section profiles.

Special features to be found in the second edition are: (a) comprehensive treatment of design of I-shaped members subject to torsion (Chapter 8), including a simplified practical method; (b) detailed treatment of plate girder theory (Chapter 11) and a comprehensive design example of a two-span continuous girder utilizing two different grades of steel; (c) extensive treatment of connections (Chapter 13), including significant discussion and illustration of the design of components, in addition to completely revised treatment of high-strength bolts (Chapter 4) and welds (Chapter 5) as fastening devices; (d) special treatment of bracing for beams and columns in Chapter 9 and for rigid frames in Chapter 14; and (e) detailed presentation of stability concepts for braced and unbraced rigid frames, using stiffness and flexibility coefficients with the objective of explaining the effective length concept for compression members in frames.

The authors are indebted to students, colleagues, and other users of the first

---

Manual of Steel Construction, 8th Edition. Since nearly continuous reference is made to the AISC Manual (which also contains the 1978 Specification and Commentary), the reader will find it desirable to secure a copy of it from the American Institute of Steel Construction, Chicago, Illinois. The AISC Specification and Commentary is also available as a separate paper cover document.

Conversion Factors

Some Conversion Factors, between US Customary and SI Metric Units, Useful in Structural Steel Design

	To Convert	To	Multiply by
Forces	kip force	kN	4.448
	lb	N	4.448
	kN	kip	0.2248
Stresses	ksi	MPa (i.e., N/mm²)	
			6.895
	psi	MPa	0.006895
	MPa	ksi	0.1450
	MPa	psi	145.0
Moments	ft·kip	kN·m	1.356
	kN·m	ft·kip	0.7376
Uniform Loading	kip/ft	kN/m	14.59
	kN/m	kip/ft	0.06852
	kip/ft²	kN/m²	47.88
	psf	N/m²	47.88
	kN/m²	kip/ft²	0.02039

For proper use of SI, see *Standard for Metric Practice* (ASTM E380-76), American Society for Testing and Materials, Philadelphia, 1976. Also see *Standard Practice for the Use of Metric (SI) Units in Building Design and Construction (Committee E-6 Supplement to E380)* (ANSI/ASTM E621-78), American Society for Testing and Materials, Philadelphia, 1978.

Basis of Conversions (ASTM E380): 1 in. = 25.4 mm; 1 lb force = 4.448 221 615 260 5 newtons.

Basic SI units relating to structural steel design:

Quantity	Unit	Symbol
length	metre	m
mass	kilogram	kg
time	second	s

Derived SI units relating to structural steel design:

Quantity	Unit	Symbol	Formula
force	newton	N	kg·m/s²
pressure, stress	pascal	Pa	N/m²
energy, or work	joule	J	N·m

edition who have suggested improvements of wording, identified errors, and recommended items for inclusion or deletion. These suggestions have been carefully considered and included in this complete revision wherever possible. Particularly, the authors are appreciative of the many suggestions made by Professors J. C. Smith of North Carolina State University and Leland S. Riggs of Georgia Institute of Technology, the discussions and advice from Dr. Raymond H. R. Tide of AISC and Dr. Joseph A. Yura of the University of Texas at Austin relating to laterally unbraced beams, the suggestions of Dr. Chai Hong Yoo of Marquette University relating to torsion, and the cooperation of AISC through Frank W. Stockwell, Frederick Palmer, Raymond H. R. Tide, and Robert Lorenz.

Users of the second edition are urged to communicate with the authors regarding all aspects of this book, particularly on identification of errors and suggestions for improvement.

The senior author gives special credit to his wife Bette on her continued patience and encouragement, without which the task of revision would not have been completed.

Charles G. Salmon
John E. Johnson

1
Introduction

1.1 STRUCTURAL DESIGN

Structural design may be defined as a mixture of art and science, combining the experienced engineer's intuitive feeling for the behavior of a structure with a sound knowledge of the principles of statics, dynamics, mechanics of materials, and structural analysis, to produce a safe economical structure which will serve its intended purpose.

Until about 1850, structural design was largely an art relying on intuition to determine the size and arrangement of the structural elements. Early man-made structures essentially conformed to those which could also be observed in nature; such as beams and arches. As the principles governing the behavior of structures and structural materials have become better understood, design procedures have become more scientific.

Computations involving scientific principles should serve as a *guide* to decision making and not be followed blindly. The art or intuitive ability of the experienced engineer is utilized to make the decisions, guided by the computational results.

1350 ft World Trade Center, New York. (Photo by C. G. Salmon)

1.2 PRINCIPLES OF DESIGN

Design is a process by which an optimum solution is obtained. In this text the concern is with the design of structures—in particular, *steel* structures. In any design, certain criteria must be established to evaluate whether or not an optimum has been achieved. For a structure, typical criteria may be (a) minimum cost; (b) minimum weight; (c) minimum construction time; (d) minimum labor; (e) minimum cost of manufacture of owner's products; (f) maximum efficiency of operation to owner. Usually several criteria are involved, each of which may require weighting. Observing the above possible criteria, it may be apparent that setting clearly measurable criteria (such as weight and cost) for establishing an optimum frequently will be difficult, and perhaps impossible. In most practical situations the evaluation must be qualitative.

If a specific objective criterion can be expressed mathematically, then optimization techniques may be employed to obtain a maximum or

minimum for the objective function. Optimization procedures and techniques comprise an entire subject that is outside the scope of this text. The criterion of minimum weight is emphasized throughout, under the general assumption that minimum material represents minimum cost. Other subjective criteria must be kept in mind, even though the integration of behavioral principles with design of structural steel elements in this text utilizes only simple objective criteria, such as weight or cost.

Design Procedure

The design procedure may be considered as composed of two parts—functional design and structural framework design. Functional design is the design which insures that the intended results are achieved such as (a) providing adequate working areas and clearances; (b) providing for ventilation and/or air conditioning; (c) adequate transportation facilities, such as elevators, stairways, and cranes or materials handling equipment; (d) adequate lighting; and (e) exhibiting architectural attractiveness.

The structural framework design is the selection of the arrangement and sizes of structural elements so that service loads may be safely carried.

The iterative design procedure may be outlined as follows:

1. *Planning.* Establishment of the functions for which the structure must serve. Set criteria against which to measure the resulting design for being an optimum.
2. *Preliminary structural configuration.* Arrangement of the elements to serve the functions in step 1.
3. *Establishment of the loads* to be carried.
4. *Preliminary member selection.* Based on the decisions of steps 1, 2, and 3 selection of the member sizes to satisfy an objective criterion, such as least weight or cost.
5. *Analysis.* Structural analysis to ascertain whether members selected are safe, but not excessively so. This would include checking of all strength and stability factors for members and their connections.
6. *Evaluation.* Are all requirements satisfied and is the result optimum? Compare the result with the predetermined criteria.
7. *Redesign.* Repetition of any part of the sequence 1 through 6 found necessary or desirable as a result of evaluation. Steps 1 through 6 represent an iterative process. Usually in this text only steps 3 through 6 will be subject to this iteration since the structural configuration and external loading will be prescribed.
8. *Final decision.* The determination of whether or not an optimum design has been achieved.

1.3 HISTORICAL BACKGROUND OF STEEL STRUCTURES

Metal as a structural material began with cast iron, used on a 100-ft (30-m) arch span which was built in England in 1777–1779 [1].* A number of cast-iron bridges were built during the period 1780–1820, mostly arch-shaped with main girders consisting of individual cast-iron pieces forming bars or trusses. Cast iron was also used for chain links on suspension bridges until about 1840.

Wrought iron began replacing cast iron soon after 1840, the earliest important example being the Brittania Bridge over Menai Straits in Wales which was built in 1846–1850. This was a tubular girder bridge having spans 230–460–460–230 ft (70–140–140–70 m), which was made from wrought-iron plates and angles.

The process of rolling various shapes was developing as cast iron and wrought iron received wider usage. Bars were rolled on an industrial scale beginning about 1780. The rolling of rails began about 1820 and was extended to I-shapes by the 1870s.

The developments of the Bessemer process (1855), the introduction of a basic liner in the Bessemer converter (1870), and the open-hearth furnace brought widespread use of iron ore products in building material. Since 1890 steel has replaced wrought iron as the principal metallic building material. Currently (1979), steels are available for structural uses which have yield stresses varying from 24,000 to 100,000 pounds per square inch, psi (165 to 690 megapascals†, MPa). The various steels, their uses and their properties are discussed in Chapter 2.

1.4 LOADS

The accurate determination of the loads to which a structure or structural element will be subjected is not always predictable. Even if the loads are well known at one location in a structure, the distribution of load from element to element throughout the structure usually requires assumptions and approximations. Some of the most common kinds of loads are discussed in the following sections.

Dead Load

Dead load is a fixed position gravity service load, so called because it acts continuously toward the earth when the structure is in service. The weight of the structure is considered dead load, as well as attachments to the structure such as pipes, electrical conduit, air-conditioning and heating ducts, lighting fixtures, floor covering, roof covering, and suspended ceilings; that is, all items that remain throughout the life of the structure.

Dead loads are usually known accurately but not until the design has

---

* Numbers in brackets refer to the Selected References at the end of the chapter.
† MPa, megapascals, are equivalent to Newtons per square millimeter, N/mm², in SI units.

been completed. Under steps 3, 4, and 5 of the design procedure discussed in Sec. 1.2, the weight of the structure or structural element must be estimated, preliminary section selected, weight recomputed, and member selection revised if necessary. The dead load of attachments is usually known with reasonable accuracy prior to the design.

Live Load

Gravity loads acting when the structure is in service, but varying in magnitude and location, are termed *live loads*. Examples of live loads are human occupants, furniture, movable equipment, vehicles, and stored goods. Some live loads may be practically permanent, others may be highly transient. Because of the unknown nature of the weight, location, and density of live load items, realistic magnitudes and the positions of such loads are very difficult to determine.

Because of the public concern for adequate safety, live loads to be taken as service loads in design are usually prescribed by state and local building codes. These loads are generally empirical and conservative, based on experience and accepted practice rather than accurately computed values. Wherever local codes do not apply, or do not exist, the provisions from one of several regional and national building codes may be used. One such widely recognized code is the The American National Standard Building Code of the American National Standards Institute (ANSI) [2], from which some typical live loads are presented in Table 1.4.1.

Live load when applied to the structure should be positioned to give maximum effect, including partial loading, alternate span loading, or full loading as may be necessary. The simplified assumption of full uniform loading everywhere should be used only when it agrees with reality or is an appropriate approximation. The probability of having the prescribed loading uniformly applied over an entire floor, as well as over all other floors of a building simultaneously, is almost nonexistent. Most codes recognize this by allowing for some percentage reduction from full loading. For instance, for live loads of 100 psf or less the ANSI–1972 [2] allows the design live load on any member supporting 150 sq ft (14 sq m) or more to be reduced at the rate of 0.08% per sq ft (0.86% per sq m) of area supported by the member, except in public assembly areas, garages, and roofs. The greatest benefit from such a reduction will occur in the columns of a tall building where the total area tributary to columns over ten or twenty stories may be very large. The ANSI–1972 limits the reduction to

$$\text{Max } R = 23\left(1 + \frac{D}{L}\right) \leq 60 \tag{1.4.1}$$

where R = reduction in percent, D = dead load in psf, and L = live load in psf.

Table 1.4.1 Typical Uniformly Distributed Live Loads
(Adapted from Ref. 2)

Occupancy or Use	Live Load	
	lb/ft^2	MPa^*

1. Hotel guest rooms
   School class rooms
   Private apartments
   Hospital private rooms
	40	1900

2. Offices	50	2400
3. Assembly halls, fixed seat Library reading rooms	60	2900
4. Corridors	80	3800

5. Theater aisles, and lobbies
   Assembly halls, movable seats
   Office building lobbies
   Main floor, retail stores
   Dining rooms and restaurants
	100	4800

6. Storage warehouse, light Manufacturing	125	6000
7. Library stack rooms	150	7200
8. Storage warehouse, heavy Sidewalks, driveways, subject to trucking	250	12000

* SI values are approximate conversions.

Highway Live Loads

Highway vehicle loading in the United States has been standardized by the American Association of State Highway and Transportation Officials (AASHTO) [3] into standard truck loads and lane loads that approximate a series of trucks. There are two systems, designated H and HS, that are identified by the number of axles per truck. The H system has two axles, whereas the HS system has three axles per truck. Altogether there are five classes of loading: H10, H15, H20, HS15, and HS20. The loading is shown in Fig. 1.4.1.

In designing a given bridge, either one equivalent truck loading is applied to the entire structure, *or* the equivalent lane loading is applied. When the lane loading is used, the uniform portion is distributed over as much of the span or spans as will cause the maximum effect. In addition, the one concentrated load is positioned for the greatest effect. On continuous structures, in determining maximum negative moment at the support only, an additional concentrated load must be used in a span other than the position of the first one. The load distribution across the width of a bridge to its various supporting members is taken in accordance with semiempirical rules that depend on the type of bridge deck and supporting structure.

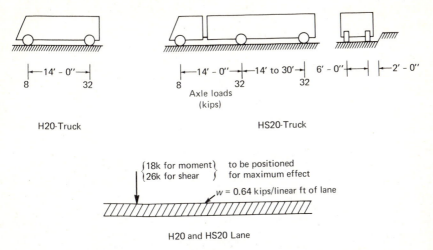

Fig. 1.4.1 AASHTO–1977 Highway H20 and HS20 loadings (for H15 and HS15 use 75 percent of H20 and HS20, and for H10 use 50 percent of H20). (1 kip = 4.45 kN)

The single truck loading provides the effect of a heavy concentrated load and usually governs on relatively short spans. The uniform lane load is to simulate a line of traffic, and the added concentrated load is to account for the possibility of one extra heavy vehicle in the line of traffic. These loads have been used with no apparent difficulty since 1944, before which time a line of trucks was actually used for the loading. On the interstate system of highways a military loading is also used that consists of two 24–kip (107-kN) axle loads spaced 4 ft (1.2 m) apart.

Railroad bridges are designed to carry a similar semiempirical loading known as the Cooper E72 train, consisting of a series of concentrated loads a fixed distance apart followed by uniform loading. This loading is prescribed by the American Railway Engineering Association (AREA) [4].

Impact

The term *impact* as ordinarily used in structural design refers to the dynamic effect of a suddenly applied load. In the building of a structure the materials are added slowly; people entering a building are also considered a gradual loading. Dead loads are static loads; i.e., they have no effect other than their weights. Live loads may be either static or they may have a dynamic effect. Persons and furniture would be treated as static live load, but cranes and various types of machinery also have dynamic effects.

Consider the spring-mass system of Fig. 1.4.2a where the spring may be thought of as analogous to an elastic beam. When load is gradually

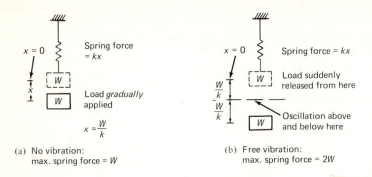

(a) No vibration:
    max. spring force = W

(b) Free vibration:
    max. spring force = $2W$

Fig. 1.4.2 Comparison of static and dynamic loading.

applied (i.e., static loading) the mass (weight) deflects an amount x and the load on the spring (beam) is equal to the weight W. In Fig. 1.4.2b the load is suddenly applied (dynamic loading) and the maximum deflection is $2x$; i.e., the maximum load on the spring (beam) is $2W$. In this case the mass vibrates in simple harmonic motion with its neutral position equal to its static deflected position. In real structures the harmonic (vibratory) motion is damped out (reduced to zero) very rapidly. Once the motion has stopped the force remaining in the spring is the weight W. To account for the increased force during the time the member is in motion a load equal to twice the static load W should be used—add 100% of the static load to represent the dynamic effect. This is called a 100% impact factor.

Any live load that can have a dynamic effect should be increased by an impact factor. While a dynamic analysis of a structure could be made to accurately determine these effects, such a procedure is usually too complex or too costly in ordinary design. Thus empirical formulas and impact factors are usually used. In cases where the dynamic effect is small (say where impact would be less than about 20%) it is ordinarily accounted for by using a conservative (higher) value for the specified live load. The dynamic effects of persons in buildings and of slow-moving vehicles in parking garages are examples where ordinary design live load is conservative and no explicit impact factor is usually added.

For highway bridge design, however, impact is always to be considered. AASHTO–1977 [3] prescribes empirically that the impact factor expressed as a portion of live load is

$$I = \frac{50}{L+125} \leq 0.30 \qquad (1.4.2)$$

In Eq. 1.4.2, L (expressed in feet) is the length of the portion of the span that is loaded to give the maximum effect on the member. Since vehicles travel directly on the superstructure, all parts of it are subjected to vibration and must be designed to include impact. The substructure, including all portions not rigidly attached to the superstructure such as abutments, retaining walls, and piers, are assumed to have adequate damping or be sufficiently remote from the application point of the dynamic load so that impact is not to be considered. Again, conservative static loads may account for the smaller dynamic effects.

In buildings it is principally in the design of supports for cranes and heavy machinery that impact is explicitly considered. The American Institute of Steel Construction (AISC) Specification [5] (AISC–1.3.3) states that if not otherwise specified, the impact percentage shall be:

For supports of elevators	100%
For cab operated traveling crane support girders and their connections	25%
For pendant operated traveling crane support girders and their connections	10%
For supports of light machinery, shaft or motor driven, not less than	20%
For supports of reciprocating machinery or power driven units, not less than	50%
For hangers supporting floors and balconies	33%

In the design of crane runway beams (see Fig. 1.4.3) and their connections, the horizontal forces caused by moving crane trolleys must be considered. AISC–1.3.4 prescribes using "20 percent of the sum of the weights of the lifted load and of the crane trolley (but exclusive of other parts of the crane)." It also states that "The force shall be assumed to be applied at the top of the rails, acting in either direction normal to the runway rails, and shall be distributed with due regard for lateral stiffness of the structure supporting the rails."

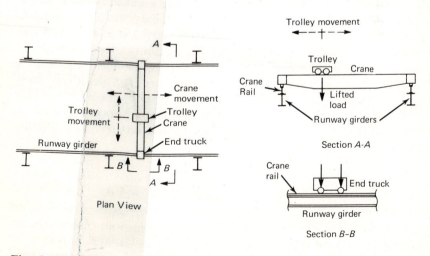

Fig. 1.4.3 Crane arrangement, showing movements that contribute impact loading.

In addition, due to acceleration and deceleration of the entire crane, a longitudinal force is transmitted to the runway girder through friction of the end truck wheels with the crane rail. AISC–1.3.4 says "The longitudinal force shall, if not otherwise specified, be taken as 10 percent of the maximum wheel loads of the crane applied at the top of the rail."

Snow Load

The live loading for which roofs are designed is either totally or primarily a snow load. Since snow has a variable specific gravity, even if one knows the depth of snow for which design is to be made, the load per unit area of roof is at best only a guess.

The best procedure for establishing snow load for design is to follow ANSI–1972 [2]. This Code uses a map of the United States giving isolines of ground snow corresponding to a 50-year mean recurrence interval for use in designing most permanent structures. The ground snow is then multiplied by a coefficient that includes the effect of roof slope, wind exposure, nonuniform accumulation on pitched or curved roofs, multiple series roofs, and multilevel roofs and roof areas adjacent to projections on a roof level.

It is apparent that the steeper the roof the less snow can accumulate. Also partial snow loading should be considered, in addition to full loading, if it is believed such loading can occur and would cause maximum stresses. Wind may also act on a structure that is carrying snow load. It is certainly unlikely, however, that maximum snow and wind loads would act simultaneously.

In general, the basic snow load used in design varies from 30 to 40 psf (1400 to 1900 MPa) in the northern and eastern states to 20 psf (960 MPa) or less in the southern states. Flat roofs in normally warm climates should be designed for 20 psf (960 MPa) even when such accumulation of snow may seem doubtful. This loading may be thought of as due to people gathered on such a roof. Furthermore, though wind is frequently ignored as a vertical force on a roof, nevertheless it may cause such an effect. For these reasons a 20 psf (960 MPa) minimum loading, even though it may not always be snow, is reasonable. Local codes, actual weather conditions, or ANSI–1972 [2] should be used when designing for snow.

Wind Load

All structures are subject to wind load, but it is usually only those more than three or four stories high, other than long bridges, for which special consideration of wind is required.

On any typical building of rectangular plan and elevation, wind exerts pressure on the windward side and suction on the leeward side, as well as either uplift or downward pressure on the roof. For most ordinary situations vertical roof loading from wind is neglected on the assumption that snow loading will require a greater strength than wind loading. Furthermore, the total lateral wind load, windward and leeward effect, is assumed to be applied to the windward face of the building.

In accordance with Bernoulli's theorem for an ideal fluid striking an object, the increase in static pressure equals the decrease in dynamic pressure, or

$$q = \tfrac{1}{2}\rho V^2 \qquad (1.4.3)$$

where q is the dynamic pressure on the object, ρ is the mass density of air (specific weight $w = 0.07651$ pcf at sea level and 15°C), and V is the wind velocity. In terms of velocity V in miles per hour the dynamic pressure q (psf) would be

$$q = \frac{1}{2}\left(\frac{0.07651}{32.2}\right)\left(\frac{5280\,V}{3600}\right)^2 = 0.0026\,V^2 \qquad (1.4.4)^*$$

In common design procedures for usual types of buildings the dynamic pressure q is converted into equivalent static pressure p, which may be expressed [16]

$$p = qC_eC_gC_p \qquad (1.4.5)$$

where C_e is an exposure factor that varies from 1.0 (for 0–40-ft height) to 2.0 (for 740–1200-ft height); C_g is a gust factor, such as 2.0 for structural members and 2.5 for small elements including cladding; and C_p is a shape factor for the building as a whole. Excellent details of application of wind loading to structures is available in ANSI–1972 [2] and in the National Building Code of Canada [16].

The commonly used wind pressure of 20 psf, as specified by many building codes, corresponds to a velocity of 88 miles per hour (mph) from Eq. 1.4.4. For an exposure factor C_e of 1.0, a gust factor C_g of 2.0, and a shape factor C_p of 1.3 for an airtight building, corresponds to a 55-mph wind for an equivalent static pressure p of 20 psf from Eq. 1.4.5. For all buildings with nonplanar surfaces, plane surfaces inclined to the wind, or having significant openings, careful examination of wind forces should be made using, for example, ANSI–1972 [2], or the National Building Code of Canada [16]. For more extensive study of wind loads and effects, see the work of the Task Committee on Wind Forces [6].

Earthquake Load

An earthquake consists of horizontal and vertical ground motions, with the vertical motion usually having much the smaller magnitude. Since the

---

* In SI units,

$$q = 0.63\,V^2, \quad \text{for } q \text{ in MPa and } V \text{ in m/sec}$$

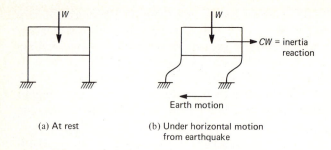

(a) At rest

(b) Under horizontal motion from earthquake

Fig. 1.4.4 Force developed by earthquake.

horizontal motion of the ground causes the most significant effect it is that effect which is usually thought of as earthquake load. When the ground under an object (structure) having a certain mass suddenly moves, the inertia of the mass tends to resist the movement, as shown in Fig. 1.4.4. A shear force is developed between the ground and the mass. Most building codes having earthquake provisions require the designer to consider a lateral force CW that is usually empirically prescribed. The dynamics of earthquake action on structures is outside the scope of this text, but the designer is referred to Refs. 7, 8, and 9.

One of the most commonly used codes is that of the Structural Engineers Association of California (SEAOC) [10]. In its 1967 recommendations the base shear force to be designed for is

$$V = KCW \tag{1.4.6}$$

where V = base shear to represent the dynamic effect of the inertia force

W = weight of building

$C = 0.05/\sqrt[3]{T}$, the seismic coefficient, equivalent to the maximum acceleration in terms of acceleration due to gravity

T = natural period of the structure, i.e., time for one cycle of vibration

K = coefficient varying from 0.67 to 3.0, indicating capacity of the members to absorb plastic deformation (low values indicate high ductility)

The total lateral load is recommended [10] to be distributed in accordance with the following:

$$F_n = \frac{W_n h_n}{\sum Wh} V \tag{1.4.7}$$

where F_n = lateral at the nth-floor level

W_n = weight at the nth-floor level

h_n = height above ground of the nth-floor level

$\sum Wh$ = total sum of Wh for all floor levels

If the natural period T cannot be determined by a rational means from technical data, it may be assumed to be

$$T = 0.05H/\sqrt{D} \qquad (1.4.8)$$

where H = height of the building above its base
$\quad\quad D$ = dimension of the building in the direction parallel to the applied forces (D to be in the same units as H)

The foregoing discussion based on the SEAOC is presented to show how earthquake loads may be treated empirically by code procedures.

1.5 TYPES OF STRUCTURAL STEEL MEMBERS

As discussed in Sec. 1.2 the function of a structure is the principal factor determining the structural configuration. Using the structural configuration along with the design loads, individual elements, or components, are selected to properly support and transmit loads throughout the structure. Steel members are selected from among the standard rolled shapes adopted by the American Institute of Steel Construction (also given by American Society for Testing and Materials (ASTM) A6 Specification). Of course, welding permits combining plates and/or other rolled shapes to obtain any shape the designer or architect may require.

Typical rolled shapes, the dimensions for which are found in the AISC Manual [11], are shown in Fig. 1.5.1. The AISC Manual will be frequently referred to throughout this text and should be available to the reader. The most commonly used section is the wide-flange shape (Fig. 1.5.1a) which is formed by hot rolling in the steel mill. The wide-flange

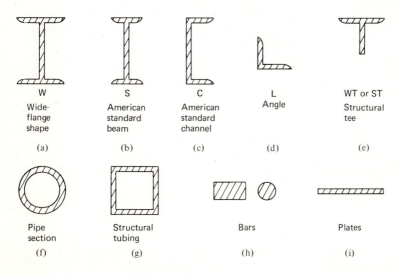

W	S	C	L	WT or ST
Wide-flange shape	American standard beam	American standard channel	Angle	Structural tee
(a)	(b)	(c)	(d)	(e)

Pipe section	Structural tubing	Bars	Plates
(f)	(g)	(h)	(i)

Fig. 1.5.1 Standard rolled shapes.

shape is designated by the nominal depth and the weight per foot, such as a W18×97 which is nominally 18 in. deep (actual depth = 18.59 in. according to AISC Manual) and weighs 97 pounds per foot. (In SI units the W18×97 section could be designated W460×142, meaning nominally 460 mm deep and having a mass of 142 kg/m.) Two sets of dimensions are found in the AISC Manual, one set stated in decimals for the designer to use in computations, and another set expressed in fractions ($\frac{1}{16}$ in. as the smallest increment) for the detailer to use on plans and shop drawings. Rolled W shapes are also designated by ANSI/ASTM A6 [20] in accordance with web thickness as Groups I through V, with the thinnest web sections in Group I.

The American Standard beam (Fig. 1.5.1b) commonly called the I-beam, has relatively narrow and sloping flanges and a thick web compared to the wide-flange shape. Use of most I-beams has become relatively uncommon in recent years because of excessive material in the web and relative lack of lateral stiffness due to the narrow flanges.

The channel (Fig. 1.5.1c) and angle (Fig. 1.5.1d) are commonly used either alone or in combination with other sections. The channel is designated, for example, as C12×20.7, a nominal 12-in. deep channel having a weight of 20.7 pounds per foot. Angles are designated by their leg length (long leg first) and thickness, such as, L6×4×$\frac{3}{8}$.

The structural tee (Fig. 1.5.1e) is made by cutting wide-flange or I-beams in half and is commonly used for chord members in trusses. The tee is designated, for example, as WT5×44, where the 5 is the nominal depth and 44 is the weight in pounds per foot; this tee being cut from a W10×88.

Pipe sections (Fig. 1.5.1f) are designated "standard," "extra strong," and "double-extra strong" in accordance with the thickness and are also nominally prescribed by diameter; thus 10-in.-diam double-extra strong is an example of a particular pipe size.

Structural tubing (Fig. 1.5.1g) is used where pleasing architectural

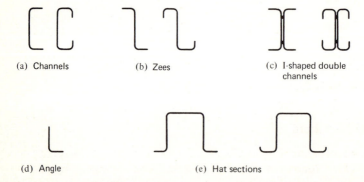

(a) Channels   (b) Zees   (c) I-shaped double channels

(d) Angle   (e) Hat sections

Fig. 1.5.2 Some cold-formed shapes.

appearance is desired with exposed steel. Tubing is designated by outside dimensions and thickness, such as structural tubing, $8 \times 6 \times \frac{1}{4}$.

The sections shown in Fig. 1.5.1 are all hot-rolled; that is, they are formed from hot billet steel (blocks of steel) by passing through rolls numerous times to obtain the final shapes.

Many other shapes are cold-formed from plate material having a thickness not exceeding 1 in., as shown in Fig. 1.5.2.

Regarding size and designation of cold-formed steel members, there are no truly standard shapes even though the properties of many common shapes are given in the *Cold-Formed Steel Design Manual* [12]. Various manufacturers produce many proprietary shapes.

Tension Members

The tension member occurs commonly as a chord member in a truss, as diagonal bracing in many types of structures, as direct support for balconies, as cables in suspended roof systems, and as suspension bridge main cables and suspenders that support the road-way. Typical cross sections of tension members are shown in Fig. 1.5.3, and their design (except for special factors relating to suspension-type cable supported structures) is treated in Chapter 3.

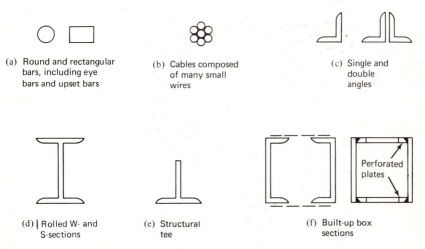

(a) Round and rectangular bars, including eye bars and upset bars

(b) Cables composed of many small wires

(c) Single and double angles

(d) | Rolled W- and S-sections

(e) Structural tee

(f) Built-up box sections

Perforated plates

Fig. 1.5.3 Typical tension members.

Compression Members

Since compression member strength is a function of the cross-sectional shape (radius of gyration), the area is generally spread out as much as is practical. Chord members in trusses, and many interior columns in buildings are examples of members subject to axial compression. Even under the most ideal condition, pure axial compression is not attainable; so, design for "axial" loading assumes the effect of any small simultaneous bending may be neglected. Typical cross sections of compression members are shown in Fig. 1.5.4 and their behavior and design are treated in Chapter 6.

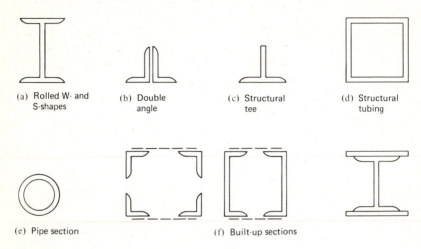

(a) Rolled W- and S-shapes   (b) Double angle   (c) Structural tee   (d) Structural tubing

(e) Pipe section   (f) Built-up sections

Fig. 1.5.4 Typical compression members.

Beams

Beams are members subjected to transverse loading and are most efficient when their area is distributed so as to be located at the greatest practical distance from the neutral axis. The most common beam sections are the wide-flange (W) and I-beams (S) (Fig. 1.5.5a), as well as smaller rolled I-shaped sections designated as "miscellaneous shapes" (M).

    For deeper and thinner-webbed sections than can economically be rolled, welded I-shaped sections (Fig. 1.5.5b) are used, including stiffened plate girders.

    For moderate spans carrying light loads, open-web "joists" are often used (Fig. 1.5.5c). These are parallel chord truss-type members used for the support of floors and roofs. The steel may be hot-rolled or cold-formed. Such joists are designated "J-Series" when the material used has a yield strength* of 36 ksi (248 MPa) and "H-Series" when the chord yield strength is 50 ksi (345 MPa). When longer and more heavily loaded joists are required, those designated "Longspan Steel Joists LJ- and LH-Series" or "Deep Longspan Steel Joists DLJ- and DLH-Series" are used. The LJ- and LH-Series are designed using 36 ksi (248 MPa) yield strength, and the DLJ- and DLH-Series are designed using material

* Refer to Sec. 2.5 for definition.

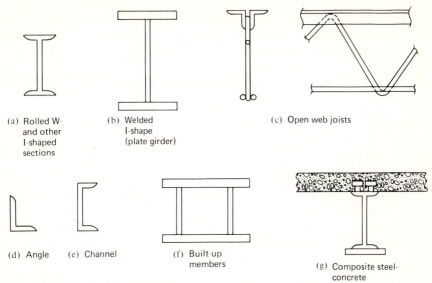

(a) Rolled W-
    and other
    I-shaped
    sections

(b) Welded
    I-shape
    (plate girder)

(c) Open web joists

(d) Angle

(e) Channel

(f) Built-up
    members

(g) Composite steel-
    concrete

Fig. 1.5.5 Typical beam members.

having a yield strength at least 36 ksi (248 MPa) but not greater than 50 ksi (355 MPa). All of these joists are designed according to specifications adopted by the Steel Joist Institute [13] and endorsed by AISC.

For beams (known as lintels) carrying loads across window and door openings, angles are frequently used; and for beams (known as girts) in wall panels, channels are frequently used.

Bending and Axial Load

When simultaneous action of tension or compression along with bending occurs, a combined stress problem arises and the type of member used will be dependent on the type of stress that predominates. A member subjected to axial compression and bending is usually referred to as a *beam-column*, the behavior and design of which is dealt with in Chapter 13.

The aforementioned illustration of types of members to resist various kinds of stress is intended only to show common and representative types of members and not to be all inclusive.

1.6 STEEL STRUCTURES

Structures may be divided into three general categories: (a) framed structures, where elements may consist of tension members, columns, beams, and members under combined bending and axial load; (b) shell-type structures, where axial stresses predominate, and (c) suspension-type structures, where axial tension predominates the principal support system.

Framed Structures

Most typical building construction is in this category. The multistory building usually consists of beams and columns, either rigidly connected or having simple end connections along with diagonal bracing to provide stability. While the multistory building is three-dimensional, when designed for rigid joints it usually has much greater stiffness in one direction than another so that it may reasonably be treated as a series of plane frames. However, if the framing is such that the behavior of the members in one plane substantially influences the behavior in another plane, the frame must be treated as a three-dimensional space frame.

Industrial buildings and special one-story buildings such as churches, schools, and arenas, generally are either wholly or partly framed structures. Particularly the roof system may be a series of plane trusses (see Fig. 1.6.1), a space truss (see Fig. 1.6.2), a dome (see Fig. 1.6.3), or it may be part of a flat or gabled one-story rigid frame.

Fig. 1.6.1 Floor joists (plane trusses) and steel decking.

Fig. 1.6.2 Space truss roof for Upjohn Company Office Building, Kalamazoo, Michigan. Courtesy Whitehead and Kales Company, Detroit.

Bridges are mostly framed structures, such as beams and plate girders (see Fig. 1.6.4), or trusses, usually continuous (see Fig. 1.6.5).

Most of this text is devoted to behavior and design of elements in framed structures.

Shell-type Structures

In this type of structure the shell serves a use function in addition to participation in carrying loads. One common type where the main stress is tension is the containment vessel used to store liquids (for both high and low temperatures), of which the elevated water tank is a notable example. Storage bins, tanks, and the hulls of ships are other examples. On many shell-type structures, a framed structure may be used in conjunction with the shell.

On walls and flat roofs the "skin" elements may be in compression while they act together with a framework. The aircraft body is another such example.

Shell-type structures are usually designed by a specialist and are not within the scope of this text.

Fig. 1.6.3 Dome roof, Brown University auditorium. Courtesy Bethlehem Steel Corporation.

Fig. 1.6.4 Continuous orthotropic plate girder, Poplar Street Bridge, St. Louis, Missouri. Courtesy Bethlehem Steel Corporation.

Fig. 1.6.5 Continuous truss bridge. Outerbridge Crossing, Staten Island, New York. (Photo by C. G. Salmon)

Fig. 1.6.6 Cable-suspended roof for Madison Square Garden Sports and Entertainment Center, New York. Courtesy Bethlehem Steel Corporation.

Fig. 1.6.7 Newport Bridge, Rhode Island. Courtesy Bethlehem Steel Corporation.

Suspension-type Structures

In the suspension-type structure tension cables are major supporting elements. A roof may be cable supported, as shown in Fig. 1.6.6. Probably the most common structure of this type is the suspension bridge, as shown in Fig. 1.6.7. Usually a subsystem of the structure consists of a framed structure, as in the stiffening truss for the suspension bridge. Since the tension element is the most efficient way of carrying load, structures utilizing this concept are coming into increasing use.

Many unusual structures utilizing various combinations of framed, shell-type, and suspension-type structures have been built. However, the typical designer must principally understand the design and behavior of framed structures.

1.7 SPECIFICATIONS AND BUILDING CODES

Structural steel design of buildings in the United States is principally based on the specifications of the American Institute of Steel Construction (AISC) [5]. AISC is comprised of steel fabricator and manufacturing companies, as well as individuals interested in steel design and research. The AISC Specification is the result of the combined judgment of researchers and practicing engineers. The research efforts have been synthesized into practical design procedures to provide a safe, economical structure. The advent of the digital computer into design practice has generally made feasible more elaborate specification rules, a trend that began with the adoption of the 1961 AISC Specification and has continued with the introduction of the 1978 specification. Throughout this text unless otherwise indicated, specification sections referred to by number are from the 1978 edition.

A specification containing a set of rules is intended to insure safety; however, the designer must understand the behavior for which the rule applies, otherwise an absurd, a grossly conservative, and sometimes unsafe design may result. The authors contend that it is virtually impossible to write rules that fully apply to every situation. Behavioral understanding must come first; application of rules then follows. No matter what set of rules is applicable, *the designer has the ultimate responsibility for a safe structure.*

A specification when adopted by AISC is actually a set of recommendations put forth by a highly respected group of experts in the field of steel research and design. Only when governmental bodies, such as city, state, and federal agencies, who have legal responsibility for public safety, adopt or incorporate a specification such as AISC–1978 into their building codes does it become legally official.

The design of steel bridges is generally in accordance with specifications of the American Association of State Highway and Transportation Officials (AASHTO) [3]. This becomes a legal set of rules since it has been adopted by the States (usually the state highway departments have this responsibility). The 1977 edition is the most recent one, but yearly supplements with changes are usually approved.

Railroad bridges are designed in accordance with the specifications adopted by the American Railway Engineering Association (AREA) [4]. In this case the railroads have the responsibility for safety and through their own organization adopt the rules to insure safe designs.

The term *building code* is sometimes used synonomously with specifications. More correctly a building code is a broadly based document, either a legal document such as a state or local building code, or a document widely recognized even though not legal which covers the same wide range of topics as the state or local building code. Building codes generally treat all facets relating to safety, such as structural design, architectural details, fire protection, heating and air conditioning, plumbing and sanitation, and lighting. On the other hand, specifications frequently refer to rules set forth by the architect or engineer that pertain to only one particular building while under construction. Building codes also ordinarily prescribe standard loads for which the structure is to be designed, as discussed in Sec. 1.4.

The reader should not be disturbed by the interchangeable use of building code and specification, but should clearly understand that which is legally required for design and that which could be thought of as recommended practice.

1.8 PHILOSOPHIES OF DESIGN

Two philosophies of design are in current use. The *working stress design* philosophy has been the principal one used during the past 90 years. According to this philosophy, a structural element is designed so that unit stresses computed under the action of working, or service, loads do not exceed predesignated allowable values. These allowable stresses are pre-scribed by a building code or specification (such as the 1978 AISC Specification [5]) to provide a factor of safety against attainment of some limiting stress, such as the minimum specified yield stress or the stress at which buckling occurs. The computed stresses are well within the elastic range; i.e., stresses are proportional to strains. In terms of a beam, for example, the safety criterion in working stress design may be expressed

$$\left[f_b = \frac{Mc}{I} \right] \le \left[F_b = \frac{F_y}{\text{FS}} \left(\text{or } \frac{F_{cr}}{\text{FS}} \right) \right] \qquad (1.8.1)$$

where f_b is the unit stress at the extreme fiber of the beam cross section, caused by the maximum service load moment M and computed under the assumption that the beam is elastic; c is the distance from the neutral axis to the extreme fiber; and I is the moment of inertia of the beam section. The allowable stress F_b is obtained by dividing the limiting stress, such as the yield stress F_y, or the buckling stress F_{cr}, by a factor of safety FS.

The other design philosophy may generally be referred to as *limit states design*. This relatively recent term includes the methods commonly referred to as "ultimate strength design," "strength design," "plastic design," "load factor design," "limit design," and more recently, "load and resistance factor design (LRFD)." *Limit states* is a general term meaning "those conditions of a building structure in which the building ceases to fulfill the function for which it was designed" [16]. Those states can be divided into the categories of *strength* and *serviceability*. Strength (i.e., safety) limit states are ductile maximum strength (commonly called plastic strength), buckling, fatigue, fracture, overturning, and sliding. Serviceability limit states are those concerned with occupancy of the building, such as deflection, vibration, permanent deformation, and crack-ing. In limit states design, the strength or safety-related limit states are dealt with by applying factors to the loadings, focusing attention on the failure modes (limit states) by making comparisons for safety at the limit state condition, rather than in the service load range as is done for working stress design. In terms of a beam, for example, the safety criterion in limit states design may be expressed

$$M(\text{FS}) \le M_u \qquad (1.8.2)$$

where M is the maximum service load moment, which is increased by multiplying it by the factor FS for safety. The factored moment should cause the beam to reach a strength-related limit state. M_u is the actual limit state strength that is achievable.

AISC—Working Stress Design

The principal method of the AISC Specification [5] is working stress design (sometimes called allowable stress design). The focus is on service load conditions (i.e., unit stresses assuming an elastic structure) when satisfying the safety requirement (adequate strength) for the structure. The allowable stresses prescribed by the Specification for stresses at service load are determined, however, by the strength capable of being achieved if the structure is overloaded. When the section is ductile and buckling does not occur, strains greater than that at which yielding first occurs can be accommodated by the section. Such ductile inelastic behavior *may permit* higher loads to be carried than those possible if the structure had remained entirely elastic. In the case of yielding throughout the depth of a beam, the upper bound for moment strength is defined as *plastic strength.*

In the working stress method, the allowable stress is adjusted upward whenever the plastic strength is the true limit state. When the true limit state is instability (buckling) or some other behavior that might prevent achievement of the strain at onset of yielding, the allowable stress is adjusted downward.

Serviceability requirements such as deflection are always investigated at service load conditions.

AISC—Plastic Design

Part 2 of the AISC Specification [5] called *plastic design* is a special case of limit states design, wherein the limit state for strength *must* be achievement of plastic strength. This precludes having limit states based on instability, fatigue, or brittle fracture. In plastic design, the inherent ductility of steel is recognized and utilized in the design of statically indeterminate structures, such as continuous beams and rigid frames. Achievement of plastic strength at one location in a statically indeterminate structure may not constitute achievement of maximum strength *for the structure.* After one location reaches its plastic strength, additional load is carried in different proportion throughout the structure until a second location of plastic strength is achieved. Once the structure has no further ability to carry an increased load it is said to have reached a "collapse mechanism."

Even when plastic design is used for the strength requirement, serviceability requirements such as deflection are investigated at service load conditions.

Load and Resistance Factor Design (LRFD)

In the past few years the general limit states approach has been moving toward acceptance by AISC. Termed LRFD, this general approach is the result of work by an Advisory Task Force under the direction of T. V. Galambos. Papers by Pinkham and Hansell [17], Galambos and Ravindra [18, 21], and Wiesner [19] present the current thinking.

The proposed format [17] for the limit states is

$$\phi R_n \geq \gamma_0 \sum \gamma_i Q_i, \qquad i = \text{(DL), (LL), W, S,} \ldots \qquad (1.8.3)$$

where the left side is the nominal strength R_n multiplied by an under-capacity factor ϕ, a number less than 1.0 to account for resistance uncertainties. The right side sums the products of load effects Q_i and the overload factors γ_i. The resulting sum of products is multiplied by an analysis factor γ_0, a number larger than 1.0 to account for uncertainties in structural analysis. For example, commonly a three-dimensional rigid frame is analyzed as a two-dimensional system. Connections are frequently treated as either simple (hinged) or rigid (fixed) when actually they are something in between. The subscript i indicates load types, such as dead load (DL), live load (LL), wind (W), and snow (S). For comparison with conventional design philosophies, the ϕ factor may be put into the denominator on the right side of the equation and the combined safety factor may be determined.

Historical Development of the AISC Design Philosophy

Maximum strength was the earliest basis for design because the load a member could carry when it failed could be easily measured by experiment. A knowledge of magnitude and distribution of internal stresses was not required. With interest in and understanding of the elastic methods of analysis in the early 1900s, the elastic-based working stress design was adopted almost universally by specifications as the best for design. Since steel is an elastic material up to the yield stress (actually a well-defined yield point for most structural steels), it seemed ideally suited for the method. As better understanding of the actual behavior of steel structures was gained, particularly under loads at the limit states, adjustments in the method were continuously made, resulting in the 1978 AISC Specification working stress design which is largely adjusted to reflect behavior at the limit states.

The design of connectors and connections in working stress design has always been based on maximum strength rather than service load behavior, since to do otherwise would have required a highly complex analysis. Column design also has been based on maximum strength since the earliest specifications. With the adoption of the 1961 AISC Specification, design of beams became based on maximum strength (or buckling), and the design of plate girders was completely revised to be based on maximum strength.

The plastic design method was first approved by AISC in 1958 and has now (1979) become widely used. Undoubtedly, in the near future

AISC will adopt a general limit states method, referred to above as *load and resistance factor design* (LRFD), to include plastic design. AASHTO [3] already has such a method (known as *load factor design*) permitted as an alternate to working stress design.

1.9 FACTOR OF SAFETY

Structures and structural members must always be designed to carry some reserve load beyond that which is expected under normal use. The reserve capacity is provided to account primarily for the possibility of overload; in addition, however, it also accounts for the possibility of understrength. Deviations in the dimensions of rolled sections, even though within accepted tolerances, can result in understrength. Occasionally a steel section will have a yield strength slightly below the minimum specified value; thus giving a reduced strength. Overloads can arise from changing the use for which a particular structure was designed, from underestimation of the effect of loads by oversimplifications in calculation procedures, and from variations in erection procedures. Normally, the possible drastic change of use is not explicitly considered nor intended to be covered by the safety provision; however, erection procedures known to cause particular stress conditions must be explicitly taken into account.

The safety that is required for designs is really a combination of economics and statistics. Obviously it will not be economically feasible to design a structure so that it will be impossible for it to fail—that is, it is not feasible to design so that there is zero probability of failure. The load factor, or safety factor, is intended to limit the probability below a certain reasonable level.

The apparently simple task of defining what is meant by the term "factor of safety" was a principal goal of the ASCE Task Committee on Factors of Safety during the period 1956–1966. The Final Report [14] of the Task Committee summarizes concepts necessary to a full understanding of structural safety and its relationship to theory of probability. The Final Report [14] indicates "the Committee has not been successful in its efforts to resolve the 'factor of safety' question, it is the belief ... that the probability approach deserves considerable more study than it has received." More recently, safety requirements for structures based on probability have been proposed [15], which are intended "to enhance realism and improve consistency in the code treatment of uncertainty of both the measurable and the 'professional' kinds." Measurable uncertainty refers to such things as statistical differences in material strengths and strengths related to dimensional tolerances. "Professional" kinds of uncertainty refers to things seldom found in statistical observations, such as incomplete knowledge of structural performance, or situations in which it is not economically feasible to apply an "exact" analysis.

Most building codes have not identified the various factors that are involved in determining the safety requirements. One may state that the minimum resistance must exceed the maximum applied load by some prescribed amount. Suppose the actual load exceeds the service load by an amount ΔS, and the actual resistance is less than the computed resistance by an amount ΔR. A structure that is just adequate would have

$$R - \Delta R = S + \Delta S$$
$$R(1 - \Delta R/R) = S(1 + \Delta S/S) \tag{1.9.1}$$

The degree of safety, or safety factor, would be the ratio of the nominal strength to nominal design load, R/S; or

$$\text{FS} = \frac{R}{S} = \frac{1 + \Delta S/S}{1 - \Delta R/R} \tag{1.9.2}$$

Equation 1.9.2 illustrates the effect of overload ($\Delta S/S$) and undercapacity ($\Delta R/R$); however, it does not identify all the factors that may contribute to either. If one assumes that the occasional overload ($\Delta S/S$) may be 40% greater than its mean value, and that an occasional understrength ($\Delta R/R$) may be 15% less than its mean value, then

$$\text{FS} = \frac{1 + 0.4}{1 - 0.15} = \frac{1.4}{0.85} = 1.65$$

Obviously even if the percentage variations are correct they only relate to a certain probability of the variation occurring. It does *not* mean there is zero probability of having greater variations.

A treatment of statistics and probability is outside the scope of the text, but the reader is referred to Refs. 14 and 15 for extended treatment.

The 1978 AISC uses FS = 1.67 as the basic value for the working stress method, and uses FS = 1.7 as the value for plastic design; i.e., essentially the same value for both. Dividing capacities by 1.67 gives a multiplier of 0.6 (a convenient multiple) in the working stress method. In plastic design, the loads are multiplied by the factor 1.7; a convenient value.

SELECTED REFERENCES

1. Hans Straub, *A History of Civil Engineering*. Cambridge, Mass.: M.I.T. Press, 1964 (pp. 173–180).
2. *American National Standard Building Code Requirements for Minimum Design Loads in Buildings and Other Structures*. New York: American Standards Institute, ANSI A58.1–1972.
3. *Standard Specifications for Highway Bridges*, American Association of State Highway and Transportation Officials, 12th ed., Washington, D.C., 1977 (also 1978–79 Interim Provisions).
4. "Specifications for Steel Railway Bridges." Chicago, Ill.: American Railway Engineering Association, 1965.

5. *Specification for the Design Fabrication and Erection of Structural Steel for Buildings.* New York: American Institute of Steel Construction, 1978.
6. "Wind Forces on Structures," Task Committee on Wind Forces, Committee on Loads and Stresses, Structural Division, ASCE, Preliminary Reports, *Journal of Structural Division,* ASCE, 84, ST4 (July 1958); and Final Report, *Transactions,* ASCE, 126, pt. II (1961), 1124–1198.
7. "Lateral Forces of Earthquake and Wind," Joint Committee of San Francisco Section, ASCE, and Structural Engineers Association of Northern California, *Transactions,* ASCE, 117 (1952), 716–780. (Includes extensive bibliography.)
8. John M. Biggs, *Introduction to Structural Dynamics.* New York: McGraw-Hill Book Company, Inc., 1964, Chap. 6.
9. C. H. Norris et al., *Structural Design for Dynamic Loads.* New York: McGraw-Hill Book Company, Inc., 1959, Chaps. 16–18.
10. *Recommended Lateral Force Requirements and Commentary.* San Francisco: Seismology Committee, Structural Engineers Association of California, 1967.
11. *Manual of Steel Construction,* 8th ed., Chicago: American Institute of Steel Construction, Inc., 1980.
12. *Cold-Formed Steel Design Manual.* New York: American Iron and Steel Institute, 1977 (see part V, Charts and Tables).
13. *Standard Specifications Load Tables and Weight Tables.* Richmond, Va.: Steel Joist Institute, 1978.
14. Alfred M. Freudenthal, Jewell M. Garrelts, and Masanobu Shinozuka, "The Analysis of Structural Safety," *Journal of Structural Division,* ASCE, 92, ST1 (February 1966), 267–325.
15. C. Allin Cornell, "A Probability-Based Structural Code," *ACI Journal, Proceedings,* 66, December 1969, 974–985.
16. *National Building Code of Canada.* Ottawa: Associate Committee on the National Building Code, National Research Council of Canada, 1977.
17. C. W. Pinkham and W. C. Hansell, "An Introduction to Load and Resistance Factor Design for Steel Buildings," *Engineering Journal,* AISC, 15, 1 (First Quarter 1978), 2–7.
18. Theodore V. Galambos and M. K. Ravindra, "Proposed Criteria for Load and Resistance Factor Design," *Engineering Journal,* AISC, 15, 1 (First Quarter 1978), 8–17.
19. Kenneth B. Wiesner, "LRFD Design Office Study," *Engineering Journal,* AISC, 15, 1 (First Quarter 1978), 18–25.
20. *Standard Specification for General Requirements for Rolled Steel Plates, Shapes, Sheet Piling, and Bars for Structural Use,* ANSI/ASTM A6-78. Philadelphia, Pa: American Society for Testing and Materials, 1978. Also adopted by the American National Standards Institute.
21. Mayasandra K. Ravindra and Theodore V. Galambos, "Load and Resistance Factor Design for Steel," *Journal of Structural Division,* ASCE, 104, ST9 (September 1978), 1337–1353.

2
Steels and Properties

2.1 STRUCTURAL STEELS

During most of the period from the introduction of structural steel as a major building material until about 1960, the steel used was classified as a carbon steel with the ASTM (American Society for Testing and Materials) designation A7, and had a minimum specified yield stress of 33 ksi. Most designers merely referred to "steel" without further identification, and the AISC specification prescribed allowable stresses and procedures only for the A7 type of steel. Other structural steels, such as a special corrosion resistant low alloy steel (A242) and a more readily weldable steel (A373), were available but they were rarely used in buildings. Bridge design made occasional use of these other steels.

Today (1979) the many steels available to the designer permit him to increase the strength of the material in highly stressed regions rather than greatly increase the size of members. The designer can decide whether maximum rigidity or least weight is the more desirable attribute. Corrosion resistance, hence elimination of frequent painting, may be a highly important factor. Some steels now oxidize to form a dense protective coating that prevents further oxidation (corrosion), acquiring a pleasing even-textured dark red-brown appearance. Since painting is not required,

Steel framework using multi-grades of steel. IBM Building, Pittsburgh. (Courtesy of United States Steel Corporation)

it may be economical to use these weathering steels even though the initial cost is somewhat higher than traditional carbon steels.

Certain steels provide better weldability than others; some are more suitable than others for pressure vessels, either at temperatures well above or well below room temperatures.

Structural steels are referred to by ASTM designations, and also by many proprietary names. For design purposes the yield stress in tension is the quantity that specifications, such as AISC, use as the material property variable to establish allowable unit stresses under various types of member loading. The term *yield stress* is used to include either "yield point," the well-defined deviation from perfect elasticity exhibited by most of the common structural steels; or "yield strength," the unit stress at a certain offset strain for steels having no well-defined yield point. In 1979 steels are readily available with yield stresses from 24 to 100 ksi (165 to 689 MPa).

Table 2.1.1 Steels Used for Buildings and Bridges

ASTM† Designation	Grade (if any)	F_y Minimum Yield Stress (ksi)	(MPa)	F_u Tensile Strength (ksi)	(MPa)	Maximum Thickness for Plates (in.)	A6 Groups* for Shapes	Common Usage
A36		32	220	58–80	400–550	Over 8	—	General structural purposes; bolted and welded, mainly for buildings
		36	250	58–80	400–550	To 8	All	
A53	A	30	210	48	330			Welded and seamless pipe
	B	35	240	60	415			
A242		42	290	63	435	Over $1\frac{1}{2}$ to 4	4, 5	Welded and bolted bridge construction where corrosion resistance is desired; essentially superseded by A709, Grade 50W
		46	315	67	460	Over $\frac{3}{4}$ to $1\frac{1}{2}$	3	
		50	345	70	485	To $\frac{3}{4}$	1, 2	
A440		42	290	63	435	Over $1\frac{1}{2}$ to 4	4, 5	Bolted construction; essentially superseded by A572 for buildings and A709 for bridges
		46	315	67	460	Over $\frac{3}{4}$ to $1\frac{1}{2}$	3	
		50	345	70	485	To $\frac{3}{4}$	1, 2	
A441		40	275	60	415	Over 4 to 8	—	Welded construction; largely superseded by A572 for buildings and A709 for bridges
		42	290	63	435	Over $1\frac{1}{2}$ to 4	4, 5	
		46	315	67	460	Over $\frac{3}{4}$ to $1\frac{1}{2}$	3	
		50	345	70	485	To $\frac{3}{4}$	1, 2	

Designation	Grade	Min. yield (ksi)	Min. yield (MPa)	Tensile (ksi)	Tensile (MPa)	Thickness/size (in.)		Application
A500	A	33	228	45	310	Round		Cold-formed welded and seamless round and shaped tubing for general structural purposes
	B	42	290	58	400			
	C	46	317	62	427			
	A	39	269	45	310	Shaped		
	B	46	317	58	400			
	C	50	345	62	427			
A501		36	248	58	400			Hot-formed welded and seamless round and shaped tubing for general structural purposes
A514		90	620	100–130	690–895	Over $2\frac{1}{2}$ to 6		Alloy steel plates for welded construction; superseded by A709 for bridges
		100	690	110–130	760–895	To $2\frac{1}{2}$		
A529		42	290	60–85	414–586	To $\frac{1}{2}$	1	Pre-engineered rigid frames
A570	A	25	170	45	310			Cold-formed sections
	B	30	210	49	340			
	C	33	230	52	360			
	D	40	280	55	380			
	E	42	290	58	400			
A572	42	42	290	60	415	To 6	All	Welded and bolted construction for buildings; welded bridges in Grades 42, and 50 only; essentially superseded by A709, Grade 50 for bridges
	50	50	345	65	450	To 2	All	
	60	60	415	75	520	To $1\frac{1}{4}$	1–2	
	65	65	450	80	550	To $1\frac{1}{4}$	1	

Table 2.1.1 (Continued)

ASTM† Designation	Grade (if any)	F_y Minimum Yield Stress (ksi)	(MPa)	F_u Tensile Strength (ksi)	(MPa)	Maximum Thickness for Plates (in.)	A6 Groups* for Shapes	Common Usage
A588		42	290	63	435	Over 5 to 8		Weathering steel for welded and bolted construction; essentially superseded by A709, Grade 50W for bridges
		46	315	67	460	Over 4 to 5	All	
		50	345	70	485	To 4		
A606		45	310	65	450	(Hot-rolled cut lengths only)		Hot- and cold-rolled sheet and strip steel available in coils or cut lengths, used for cold-formed sections
		50	345	70	480			
A607	45	45	310	60	410			Hot-rolled and cold-rolled sheet and strip steel in coils or cut lengths, used in cold-formed sections
	50	50	345	65	450			
	55	55	380	70	480			
	60	60	415	75	520			
	65	65	450	80	550			
	70	70	485	85	590			
A611	A	25	170	42	290			Cold-rolled sheet steel for cold-formed sections
	B	30	205	45	310			
	C	33	230	48	330			
	D	40	275	52	360			
	E	80	550	82	570			

		Min yield (ksi)	(MPa)	Tensile (ksi)	(MPa)	Thickness (in)	Group	
A618	I	50	345	70	483			Hot-formed welded and seamless tubing for general structural purposes
	II	50	345	70	483			
	III	50	345	65	448			
A709	36	32	220	58	400	Over 8		Bridge construction: Grade 36 is approximately the same as A36; Grade 50 as A441; Grade 50W as A588; and Grade 100 as A514
	36	36	250	58–80	400–550	To 8	All	
	50	50	345	65	450	To 2	1–4	
	50W	50	345	70	485	To 4	All	
	100 & 100W	90	635	100–130	700–915	Over 2½ to 4		
	100 & 100W	100	700	110–130	775–915	To 2½		

* Structural rolled shapes (W, M, S, HP, C, MC, and L) are grouped according to size for tensile property classification by ANSI/ASTM A6 (see Ref. 1). These Groups are numbered 1 through 5. Included are all rolled flanged sections having at least one dimension of the cross section 3 in. (75 mm), or greater. The Groups are established approximately according to the web thickness that corresponds to the maximum thickness for plates, with the thinnest webs in Group 1 and the thickest in Group 5. For the specific sections included in each Group, the reader is referred to ANSI/ASTM A6, or the AISC Manual.

† All steels listed are approved under the 1978 AISC Specification, except A440, A570, Grades A, B, and C, A611, and A709.

Steels for structural use in hot-rolled applications may be classified as *carbon steels, high-strength low-allow steels*, and *alloy steels*. The general requirements for such steels are covered under ANSI/ASTM A6 Specification [1]. Table 2.1.1 lists the common steels, their minimum yield stresses, and tensile strengths.

Carbon Steels

Carbon steel is the term applied to steels containing the following maximum percentages of elements other than iron: (a) carbon, 1.7, (b) manganese, 1.65, (c) silicon, 0.60, and (d) copper, 0.60. Carbon and manganese are the main elements to increase strength over that of pure iron. The category includes material from ingot iron containing essentially no carbon to cast iron which has at least 1.7% carbon. These steels are divided into four categories: low carbon (less than 0.15%); mild carbon (0.15–0.29%); medium carbon (0.30–0.59%); and high carbon (0.60–1.70%). Structural carbon steels are in the mild-carbon category; a steel such as A36 has maximum carbon varying from 0.25 to 0.29% depending on thickness. These structural carbon steels exhibit marked yield points as shown in curve (a) of Fig. 2.1.1. Increased percentage of carbon raises the yield stress but reduces ductility, making welding more difficult. Satisfactory economical welding without preheat, postheat, or special welding electrodes can usually be accomplished when carbon content does not exceed 0.30%. Over the years, the various carbon steels have had more restrictive limitations placed on their carbon content as the need for good weldability has increased. The carbon steels of Table 2.1.1 are A36, A53, A500, A501, A529, A570, A611, and A709, Grade 36.

High-Strength Low-Alloy Steels

This category includes steels having yield stresses from 40 to 70 ksi (275 to 480 MPa), exhibiting the well-defined yield points shown in curve (b) of Fig. 2.1.1, the same as shown by carbon steels. The addition to carbon steels of small amounts of alloy elements such as chromium, columbium, copper, manganese, molybdenum, nickel, phosphorus, vanadium or zirconium, improve some of the mechanical properties. Whereas carbon steels gain their strength by increasing carbon content, the alloy elements create increased strength from a fine rather than coarse microstructure obtained during cooling of the steel. High-strength low-alloy steels are used in the as-rolled or normalized condition; i.e., no heat treating is used.

The high-strength low-alloy steels of Table 2.1.1 are A242, A441, A572, A588, A606, A607, A618, and A709, Grades 50 and 50W. A440 is a high-strength steel having properties similar to the high-strength low-alloy steels but its increased strength is obtained by using higher carbon and manganese content than for A36 steel.

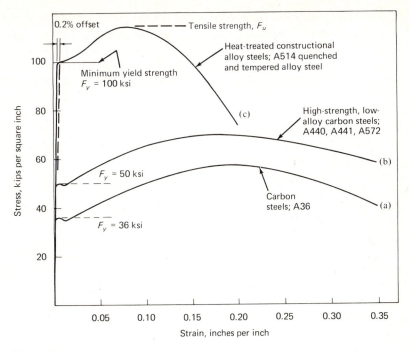

Fig. 2.1.1 Typical stress-strain curves.

Alloy Steels

Low-alloy steels may be quenched and tempered to obtain yield strengths of 80 to 110 ksi (550 to 760 MPa). Yield strength is usually defined as the stress at 0.2% offset strain, since these steels do not exhibit a well-defined yield point. A typical stress-strain curve is shown in Fig. 2.1.1, curve (c). These steels are weldable with proper procedures, and ordinarily require no additional heat treatment after they have been welded. For special uses, stress relieving may occasionally be required. Some carbon steels, such as certain pressure vessel steels, may be quenched and tempered to give yield strengths in the 80 ksi (550 MPa) range, but most steels of this strength are low-alloy steels. These low-alloy steels generally have a maximum carbon content of about 0.20% in order to limit the hardness of any coarse-grain microstructure (martensite) that may form during heat treating or welding, thus reducing the danger of cracking.

The heat treatment consists of quenching (rapid cooling with water or oil from at least 1650°F (900°C) to about 300–400°F); then tempering by reheating to at least 1150°F (620°C) and allowing to cool. Tempering, even though reducing the strength and hardness somewhat from the quenched material, greatly improves the toughness and ductility. Reduction in strength and hardness with increasing temperature is somewhat counteracted by occurrence of a secondary hardening, resulting from precipitation of fine columbium, titanium, or vanadium carbides. This precipitation begins at about 950°F (510°C) and accelerates up to about 1250°F (680°C). Tempering at or near 1250°F to get maximum benefit from precipitating carbides may result in entering the transformation zone, thus producing the weaker microstructure that would have been obtained without quenching and tempering.

In summary, the quenching produces martensite, a very hard, strong, and brittle microstructure; reheating reduces the strength and hardness somewhat while increasing the toughness and ductility. For more detailed information concerning the metallurgy of the quenching and tempering process, the reader is referred to the *Welding Handbook* [2]. The quenched and tempered alloy steels of Table 2.1.1 are A514 and A709, Grades 100 and 100W.

2.2 FASTENER STEELS

The detailed treatment of threaded fasteners appears in Chapter 4. A brief description of the materials used for bolts appears in the following paragraphs.

A307, Low-Carbon Steel Externally and Internally Threaded Standard Fasteners

This material is used for what are commonly referred to as "machine bolts." These are usually used only for temporary installations. Included are Grade A bolts for general applications, which have a *minimum* tensile strength of 60 ksi (415 MPa); and Grade B bolts for flanged joints in piping systems where one or both flanges are cast iron. The Grade B bolts have a *maximum* tensile strength limitation of 100 ksi (700 MPa). No well-defined yield point is exhibited by these bolts, and no minimum yield strength (for instance, 0.2% offset strength) is specified.

A325, High-Strength Bolts for Structural Steel Joints Including Suitable Nuts and Plain Hardened Washers

This quenched and tempered medium carbon steel is used for bolts commonly known as "high-strength structural bolts," or high-strength bolts. This material has maximum carbon of 0.30%. It is heat-treated by quenching and then by reheating (tempering) to a temperature of at least 800°F. This steel behaves in a tension test more similarly to the heat-treated low-alloy steels than to carbon steel. It has an ultimate tensile strength of 105 ksi (733 MPa) ($1\frac{1}{8}$ to $1\frac{1}{2}$ in.-diam bolts) to 120 ksi (838 MPa) ($\frac{1}{2}$ to 1-in.-diam bolts). Its yield strength, measured at 0.2% offset, is prescribed at 81 ksi (566 MPa) minimum for $1\frac{1}{8}$ to $1\frac{1}{2}$-in.-diam bolts, and 92 ksi (643 MPa) for bolts $\frac{1}{2}$ to 1 in. diam (see Table 4.1.1).

A449, Quenched and Tempered Steel Bolts and Studs

These bolts have tensile strengths and yield stresses (strength at 0.2% offset) the same as A325 for bolts $1\frac{1}{2}$ in. diam and smaller; however, they have the regular hexagon head and longer thread length of A307 bolts. They are also available in diameters up to 3 in. The AISC Specification permits use of A449 bolts only for certain structural joints requiring diameters exceeding $1\frac{1}{2}$ in. and for high-strength anchor bolts and threaded rods.

A490, Quenched and Tempered Alloy Bolts for Structural Steel Joints

This material has carbon content that may range up to 0.53% for $1\frac{1}{2}$ in.-diam bolts, and has alloying elements in amounts similar to the A514 steels. After quenching in oil the material is tempered by reheating to at least 900°F. The minimum yield strength, obtained by 0.2% offset, ranges from 115 ksi (803 MPa) (over $2\frac{1}{2}$ in. to 4 in. diam) to 130 ksi (908 MPa) (for $2\frac{1}{2}$ in. diam and under).

2.3 WELD ELECTRODE AND FILLER MATERIAL

The detailed treatment of welding and welded connections appears in Chapter 5. The electrodes used in shielded metal arc welding (SMAW) (see Sec. 5.2) also serve as the filler material and are covered by AWS A5.1 and A5.5 Specifications [3]. Such consumable electrodes are classified E60XX, E70XX, E80XX, E90XX, E100XX, and E110XX. The "E" denotes electrode. The first two digits indicate the tensile strength in ksi; thus the tensile strength ranges from 60 to 110 ksi (414 to 760 MPa). The "X's" represent numbers indicating the usage of the electrode.

For submerged arc welding (SAW) (see Sec. 5.2), the electrodes which also serve as filler material are covered by AWS A5.17 and A5.23, and are designated F6X-EXXX, F7X-EXXX, F8X-EXXX, F9X-EXXX, F10X-EXXX, and F11X-EXXX. The "F" designates a granular flux material that shields the weld as it is made. The first of the two digits following the "F" indicate the tensile strength (6 means 60 ksi), while the

Table 2.3.1 Electrodes Used for Welding*

	Process						
Shielded Metal arc Welding (SMAW) AWS A5.1 and A5.5	Submerged Arc Welding (SAW) AWS A5.17 and A5.23	Gas Metal Arc Welding (GMAW) AWS A5.18	Flux Cored Arc Welding (FCAW) AWS A5.20	Minimum Yield Stress		Minimum Tensile Strength	
				(ksi)	(MPa)	(ksi)	(MPa)
E60XX				50	345	67	460
	F6X–EXXX			50	345	62–80	425–550
			E60T–X	50	345	62	425
E70XX				57	395	70	485
	F7X–EXXX			60	415	70–95	485–655
		E70S–X	E70–X	60	415	72	495
E80XX				67	460	72	495
	F8X–EXXX			68	470	80–100	550–690
		Grade E80S		65	450	80	550
			Grade E80T	68	470	80–95	550–655
E100XX				87	600	100	690
	F10X–EXXX			88	605	100–130	690–895
		Grade E100S		90	620	100	690
			Grade E100T	88	605	100–115	690–790
E110XX				97	670	110	760
	F11X–EXXX			98	675	110–130	760–895
		Grade E110S		98	675	110	760
			Grade E110T	98	675	110–125	760–860

* Filler metal requirements given by AWS D1.1–79, Table 4.1.1 to match the various structural steels.

second digit gives the Charpy V-notch impact strength. The "E" and the other X's represent numbers relating to the use. For gas metal arc welding (GMAW) (see Sec. 5.2) and flux cored arc welding (FCAW) (see Sec. 5.2) the electrodes are designated E70S-X and E70T-X, respectively. The number 70 is the tensile strength in ksi. The yield stresses and tensile strengths of the commonly used electrodes are given in Table 2.3.1.

2.4 STRESS-STRAIN BEHAVIOR (TENSION TEST) AT ATMOSPHERIC TEMPERATURES

Typical stress-strain curves for tension are shown in Fig. 2.1.1 for the three categories of steel already discussed: carbon, high-strength low-alloy, and heat-treated high-strength low-alloy. The same behavior occurs in compression when support is provided so as to preclude buckling. The portion of each of the stress-strain curves of Fig. 2.1.1 that is utilized in ordinary design is shown enlarged in Fig. 2.4.1.

The stress-strain curves of Fig. 2.1.1 are determined using a unit stress obtained by dividing the load by the original cross-sectional area of the specimen, and the strain (inches per inch) is obtained as the elongation divided by the original length. Such curves are known as *engineering*

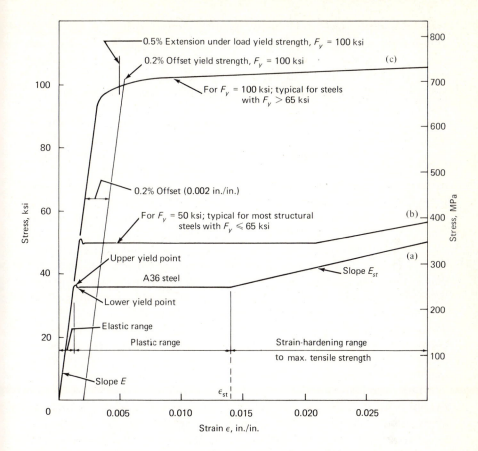

Fig. 2.4.1 Enlarged typical stress-strain curves for different yield stresses.

stress-strain curves and rise to a maximum stress level (known as the tensile strength) and then fall off with increasing strain until they terminate as the specimen breaks. Insofar as the material itself is concerned the unit stress continues to rise until failure occurs. The so-called *true-stress/true-strain* curve is obtained by using the actual cross-section even after necking down begins and using the instantaneous incremental strain. The engineering stress-strain curve permits the practical use of the curve to determine the maximum *load* that can be carried (ultimate tensile strength).

Stress-strain curves (as per Fig. 2.4.1) show a straight line relationship up to a point known as the *proportional limit*, which essentially coincides with the yield point for most structural steels with yield points not exceeding 65 ksi (450 MPa). For the quenched and tempered low-alloy steels the deviation from a straight line occurs gradually, as in curve (c), Fig. 2.4.1. Since the term yield point is not appropriate to curve (c), *yield strength* is used for the stress at an offset strain of 0.2%; or alternatively, a 0.5% extension under load, as shown in Fig. 2.4.1. *Yield stress* is the general term to include both the unit stress at a yield point, when such exists; and a unit stress corresponding to a specified strain for material showing a gradual nonlinear stress-strain behavior.

The ratio of stress to strain in the initial straight line region is known as the modulus of elasticity, or Young's modulus, *E*, which for structural steels may be taken approximately as 29,000 ksi (200,000 MPa). In the straight-line region loading and unloading results in no permanent deformation; hence it is the *elastic range*. The service load unit stress in steel design is always intended to be safely within the proportional limit, even though in order to ascertain safety factors against failure or excessive deformation, knowledge is required of the stress-strain behavior up to a strain about 15 to 20 times the maximum elastic strain.

For steels exhibiting yield points, as curves (a) and (b) of Fig. 2.4.1, the large strain for which essentially constant stress exists is known as the *plastic range*. The plastic design method consciously uses this range for determining plastic strength (which is usually assumed to be ultimate or maximum strength). The higher strength steels typified by curve (c) of Fig. 2.4.1 also have a region that might be called the plastic range; however, even in this zone the stress is continuously increasing as strain increases instead of remaining constant, so that as yet (1979) the plastic strength methods are not applied to these steels.

For strains greater than 15 to 20 times the maximum elastic strain the stress again increases but with a much flatter slope than the original elastic slope. This increase in strength is called *strain hardening*, the strain-hardening range continuing up to tensile strength. The slope of the stress-strain curve is known as the strain-hardening modulus, E_{st}. Average values for this modulus and the strain, ϵ_{st} at which it begins have been determined [4] for two steels as: A36 steel, $E_{st} = 900$ ksi (6200 MPa) at $\epsilon_{st} = 0.014$ in. per in.; and for A441, $E_{st} = 700$ ksi (4800 MPa) at $\epsilon_{st} = 0.021$ in. per in. General use of the strain-hardening range is not made in design, but certain of the buckling limitations are conservatively derived to preclude buckling even at strains well beyond onset of strain hardening.

The stress-strain curve also indicates the *ductility*. Ductility is defined as the amount of permanent strain (i.e., strain exceeding proportional limit) up to the point of fracture. Measurement of ductility is obtained from the tension test by determining the percent elongation (comparing final and original cross-sectional areas) of the specimen. Ductility is important because it permits yielding locally due to high stresses and thus allows the stress distribution to change. Design procedures based on ultimate strength behavior require large inherent ductility, particularly for treatment of stresses near holes or abrupt change in member shape, as well as for design of connections.

2.5 TOUGHNESS AND RESILIENCE

Toughness and resilience are measures of the ability of a metal to absorb mechanical energy. For uniaxial stress these quantities are obtainable from the tension test (engineering stress-strain) curves, such as those of Fig. 2.1.1.

Resilience relates to the elastic energy absorption of the material. Sometimes referred to as *modulus of resilience,* resilience is the amount of elastic energy able to be absorbed by a unit volume of material loaded in tension; i.e., it equals the area under the stress-strain diagram up to the yield stress.

Toughness relates to the total energy, both elastic and inelastic, able to be absorbed by a unit volume of material before it fractures. For uniaxial tension, it is the area under the tension stress-strain curve out to the fracture point where the diagram terminates. This area is sometimes called the *modulus of toughness.* Because all parts of the tensile specimen do not deform the maximum amount, the area only gives an approximate value for the metal's toughness. To illustrate the magnitude of these quantities for some typical steels, Ref. 4 gives the following values:

	Resilience		Toughness	
Steels	$(in.-lb/in.^3)$	$(kN \cdot m/m^3)$	$(in.-lb/in.^3)$	$(kN \cdot m/m^3)$
Carbon				
(A36 with $F_y = 36$ ksi)	22	152	12,000	82,700
High-Strength Low-Alloy				
(A441 with $F_y = 50$ ksi)	43	296	15,000	103,000
Quenched and Tempered Carbon				
($F_y = 70$ to 80 ksi)	110	758	18,000	124,000
Quenched and Tempered Low Alloy				
(A514 with $F_y = 100$ ksi)	170	1170	19,000	131,000

The values for the A36, A441, and A514 agree closely with values computed from the curves of Fig. 2.1.1.

Since uniaxial tension rarely exists in real structures, particularly in the region of connections, a more practical index of toughness is used based on the more complex stress condition (probably triaxial) below the root of a notch. *Notch toughness* is the term used to describe the resistance of a metal to the start and propagation of a crack at the base of a standard notch. Notch toughness is most commonly measured by the Charpy V-notch test. This test uses a small rectangular simply supported beam with a V-notch at midlength. The bar is fractured by a blow from a swinging pendulum. The amount of energy absorbed is calculated from the height the pendulum raises after breaking the specimen.

The Charpy V-notch test is widely used to determine the transition temperature from brittle to ductile behavior. For different temperatures the fracture energy absorption is determined and plotted as in Fig. 2.5.1a.

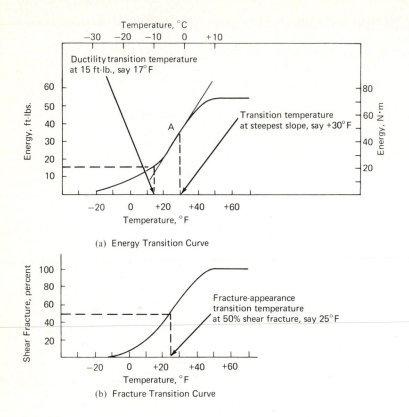

Fig. 2.5.1 Transition temperature curves for carbon steel obtained from Charpy V-notch impact tests. (Adapted from Ref. 4)

The temperature at the point where the slope is steepest (point A of Fig. 2.5.1a) is the transition temperature. Since brittleness and ductility are qualitative terms, the various structural steels have different degrees of ductility which are important only when related to the task they must perform. Thus for the steels used in ordinary structures an arbitrary amount of energy absorption is often required, such as 15 ft-lb, and the temperature at which the energy equals that value is the *ductility transition temperature*, commonly called just transition temperature.

The various alloying elements used to increase the strength of steel all have the effect of increasing the ductility transition temperature; i.e., adversely affecting notch toughness. The few elements that improve notch toughness either (a) result in high shrinkage (due to removing undissolved oxygen because of a high affinity for it) during the process of finishing a heat of steel (hence, lower quantities yielded); (b) are themselves too expensive, or (c) decrease weldability. Elements such as carbon, vanadium, and nitrogen increase the transition temperature about 5 to 6°F per 1000 psi (3.5 to 4°C per 10 N/mm^2) increase in yield stress. Certain combinations of vanadium, nitrogen, and columbium, as specified in A572 steel have been found to increase transition temperature only about 4°F per 1000 psi (2.7°C per 10 N/mm²) increase in yield stress.

When reduced transition temperatures are necessary, such as for pressure vessels at −50°F, heat treatment such as quenching and tempering is required.

An alternative use of the Charpy V-notch test involves examining the appearance of the fracture surface. The percentage of the surface that appears to have fractured by shear is plotted against temperature, as in Fig. 2.5.lb. The shear fracture portion gives a fine fibrous appearance, while the remainder appears brittle or crystalline. The temperature at which the shear fracture portion is 50 percent may be called the *fracture-appearance transition temperature* [4].

2.6 YIELD STRENGTH FOR MULTIAXIAL STATES OF STRESS

Only when the load-carrying member is subject to uniaxial tensile stress can the properties from the tension test be expected to be identical with those of the structural member. One may forget that yielding in a real structure is usually *not* the well-defined behavior observed in the tension test. Yielding is commonly assumed to be achieved when any one component of stress reaches the uniaxial value F_y.

For all states of stress other than uniaxial, a definition of yielding is needed. These definitions, and there are frequently several for a given state of stress, are called *yield conditions* (or theories of failure) and are equations of interaction between the stresses acting.

Energy-of-Distortion (Huber-vonMises-Hencky) Yield Criterion

This most commonly accepted theory gives the uniaxial yield stress in terms of the three principal stresses. The yield criterion* may be stated

$$\sigma_y^2 = \tfrac{1}{2}[(\sigma_1 - \sigma_2)^2 + (\sigma_2 - \sigma_3)^2 + (\sigma_3 - \sigma_1)^2] \qquad (2.6.1)$$

where σ_1, σ_2, σ_3 are the tensile or compressive stresses that act in the three principal directions; i.e., the stresses that act in the three mutually perpendicular planes of zero shear, and σ_y is the "yield stress" that may be compared with the uniaxial value F_y.

For most structural design situations, one of the principal stresses is either zero or small enough to be neglected; hence Eq. 2.6.1 reduces to the following for the case of plane stress (all stresses considered are acting in a plane)

$$\sigma_y^2 = \sigma_1^2 + \sigma_2^2 - \sigma_1\sigma_2 \qquad (2.6.2)$$

* See Fred B. Seely and James O. Smith, *Advanced Mechanics of Materials*. 2nd ed. (New York: John Wiley & Sons, Inc., 1952), pp. 76–91.

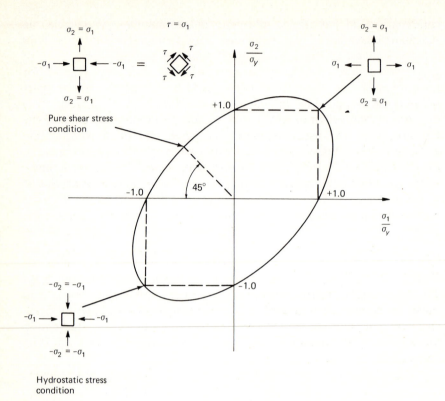

Fig. 2.6.1 Huber-von Mises-Hencky energy-of-distortion yield criterion for plane stress.

When stresses on thin plates are involved, the principal stress acting transverse to the plane of the plate is usually zero (at least to first-order approximation). Flexural stresses on beams assume zero principal stress perpendicular to the plane of bending. Furthermore, structural shapes are all comprised of thin plate elements, so that each is subject to Eq. 2.6.2. The plane stress yield criterion, Eq. 2.6.2, is the one used throughout the remaining chapters where needed, and is illustrated in Fig. 2.6.1.

Shear Yield Stress

The yield point for pure shear can be determined from a stress-strain curve with shear loading, or if the multiaxial yield criterion is known, that relationship can be used. Pure shear occurs on 45° planes to the principal planes when $\sigma_2 = -\sigma_1$, and the shear stress $\tau = \sigma_1$. Substitution of $\sigma_2 = -\sigma_1$ into Eq. 2.6.2 gives

$$\sigma_y^2 = \sigma_1^2 + \sigma_1^2 - \sigma_1(-\sigma_1) = 3\sigma_1^2 \tag{2.6.3}$$

$$\sigma_1 = \tau = \sigma_y/\sqrt{3} = \text{shear yield} \tag{2.6.4}$$

which indicates that the yield condition for shear stress acting alone equals $\sigma_y/\sqrt{3}$.

Poisson's Ratio, μ

When stress is applied in one direction, strains are induced not only in the direction of applied stress but also in the other two mutually perpendicular directions. The usual value of μ used is that obtained from the uniaxial stress condition, where it is the ratio of the transverse strain to longitudinal strain under load. For structural steels, Poisson's ratio is approximately 0.3 in the elastic range where the material is compressible and approaches 0.5 when in the plastic range where the material is essentially incompressible (i.e., constant resistance no matter what the strain).

Shear Modulus of Elasticity

Loading in pure shear produces a stress-strain curve with a straight line portion whose slope represents the shear modulus of elasticity. If Poisson's ratio μ and the tension-compression modulus of elasticity E are known, the shear modulus G is defined by the theory of elasticity as

$$G = \frac{E}{2(1+\mu)} \tag{2.6.5}$$

which for structural steel is just over 11,000 ksi (75,800 MPa).

2.7 HIGH TEMPERATURE BEHAVIOR

The design of structures to serve under atmospheric temperature rarely involves concern about high temperature behavior. Knowledge of such behavior is desirable when specifying welding procedures, and is necessary when concerned with the effects of fire.

When temperatures exceed about 200°F (93°C) the stress-strain curve begins to become nonlinear, gradually eliminating the well-defined yield point. The modulus of elasticity, yield strength, and tensile strength all reduce as temperature increases. The range from 800 to 1000°F (430 to 540°C) is where the rate of decrease is maximum. While each steel, because of its different chemistry and microstructure, behaves somewhat differently, the general relationships are shown in Fig. 2.7.1. Steels having relatively high percentages of carbon, such as A36 and A440, exhibit "strain aging" in the range 300 to 700°F (150 to 370°C). This is evidenced by a relative rise in yield point and tensile strength in that temperature range over what is shown as average in Fig. 2.7.1a and b. Tensile strength may rise to about 10% above that at room temperature and yield point may recover to about its room temperature value when

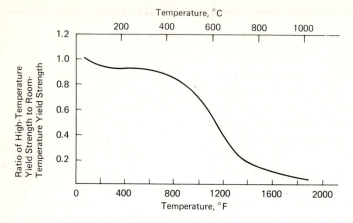

(a) Average Effect of Temperature on Yield Strength

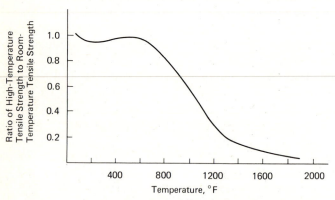

(b) Average Effect of Temperature on Tensile Strength

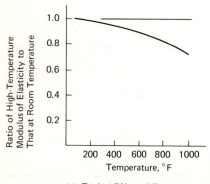

(c) Typical Effect of Temperature on Modulus of Elasticity

Fig. 2.7.1 Typical effects of high temperature on stress-strain curve properties of structural steels. (Adapted from Ref. 4)

the temperature reaches 500 to 600°F (260 to 320°C). Strain aging results in decreased ductility.

The modulus of elasticity decrease is moderate up to 1000°F (540°C); thereafter it decreases rapidly. More importantly, when temperatures above about 500 to 600°F (260 to 320°C) exist, steels exhibit deformation which increases with increasing time under load, a phenomenon known as *creep*. Creep is well known in concrete structures; and its effect in steel, which does not occur at atmospheric temperatures, increases with increasing temperature.

Other high temperature effects are (a) improved notch impact resistance up to about 150 to 200°F (65 to 95°C), as discussed in Sec. 2.5; (b) increased brittleness due to metallurgical changes, such as carbide precipitation discussed in Sec. 2.1, begins to occur at about 950°F (510°C); and (c) corrosion resistance of structural steels increases for temperatures up to about 1000°F (540°C). Most steels are used in applications below 1000°F, and some heat treated steels should be kept below about 800°F (430°C).

2.8 COLD WORK AND STRAIN HARDENING

After the strain $\epsilon_y = F_y/E_s$ at first yield has been exceeded appreciably and the specimen is unloaded, reloading may give a stress-strain relationship differing from that observed during the initial loading. Elastic loading and unloading results in no residual strain; however, initial loading beyond the yield point such as to point A of Fig. 2.8.1 results in unloading to a strain at point B. A permanent set OB has occurred. The ductility capacity has been reduced from a strain OF to the strain BF. Reloading exhibits behavior as if the stress-strain origin were at point B; the plastic zone prior to strain hardening is also reduced.

When loading has occurred until point C is reached, unloading follows the dashed line to point D; i.e., the origin for a new loading is now point D. The length of the line CD is greater, indicating that the yield point has increased. The increased yield point is referred to as a strain hardening effect; the ductility remaining when loading from point D is severely reduced from its original value prior to the initial loading. The process of loading beyond the elastic range to cause a change in available ductility, when done at atmospheric temperature, is known as *cold work*. Since real structures are not loaded in uniaxial tension-compression the cold work effect is much more complex and any theoretical study of it is outside the scope of the text.

When structural shapes are made by cold-forming from plates at atmospheric temperature, inelastic deformations occur at the bends. Cold working into the strain hardening range at the bend locations increases the yield strength, which design specifications may permit taking into

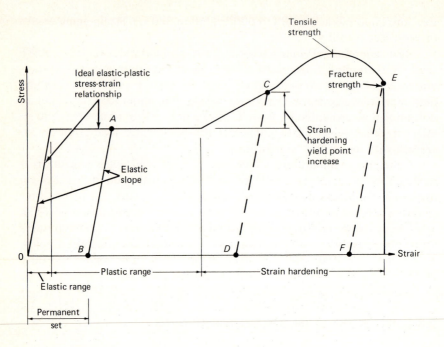

Fig. 2.8.1 Effects of straining beyond the elastic range.

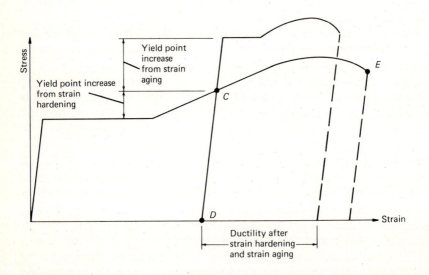

Fig. 2.8.2 Effect of strain aging after straining into strain-hardening range and unloading.

account. The Cold-Formed Steel Design Specification [5] has such provisions.

Based on the previous discussion it would at first appear that the increase in strength is obtained at the expense of ductility, and with the loss of the original well-defined yield point and its associated constant stress plastic range. Upon unloading and after a period of time, the steel will have acquired different properties from those represented by points *D*, *C*, and *E* of Fig. 2.8.1 by a phenomenon known as *strain aging*. Strain aging, as shown in Fig. 2.8.2, produces an additional increase in yield point, restores a plastic zone of constant stress, and gives a new strain hardening zone at an elevated stress. The original shape of the stress-strain diagram is restored but the ductility is reduced. The new stress-strain diagram may be used as if it were the original for analyzing cold-formed sections, as long as the ductility that remains is sufficient. The corner regions of cold-formed shapes generally would not require high ductility for rotational strain about the axis of the bend.

Stress relieving by annealing will eliminate the effects of cold work should that be desired. Annealing involves heating to a temperature above transformation range and allowing to slowly cool; a recrystallization occurs to restore the original properties.

2.9 BRITTLE FRACTURE

As has been discussed in several sections, steel that is ordinarily ductile can become brittle under various conditions. The designer must understand the causes in order to preclude brittle fracture.

Rolfe [6] has provided an excellent summary of fracture and fatigue control for structural engineers. He defines *brittle fracture* as "a type of catastrophic failure that occurs without prior plastic deformation and at extremely high speeds." Fracture behavior is affected by temperature, loading rate, stress level, flaw size, plate thickness or constraint, joint geometry, and workmanship.

Effect of Temperature

Notch toughness, as determined by the Charpy impact transition temperature curves (see Sec. 2.5), is an indication of the susceptibility to brittle fracture. Temperature is a vital factor in several ways: (a) the value below which notch toughness is inadequate; (b) in the 600 to 800°F (320 to 430°C) range causes formation of brittle microstructure; and (c) over 1000°F (540°C) causes precipitation of carbides of alloying elements to give more brittle microstructure. The other temperature factors have already been discussed in earlier sections.

Effect of Multiaxial Stress

The complex stress condition found in usual structures, particularly at joints, is another major factor affecting brittleness. The *Primer on Brittle Fracture* [7] has provided an excellent rational presentation of this and forms the basis for what follows. The engineering stress-strain curve is for uniaxial stress; prior to fracture a necking down occurs, as shown in Fig. 2.9.1a. If biaxial lateral loading as shown in Fig. 2.9.1b could be applied, "plastic behavior can be suppressed to the point where the bar would break in a brittle manner with no elongation and no reduction in area." The fracture stress based on the unreduced cross-sectional area would be the same high value as that based on the necked-down cross section in the uniaxial tension case. The unit stress would be far above the nominal maximum tensile strength of the engineering stress-strain curve, which is always computed on the basis of original cross section. This is a further extension of the yield criterion (or failure criterion) concept discussed in Sec. 2.6.

Also the effects of notches have been alluded to in the discussion of notch toughness in Sec. 2.5. The notch serves somewhat the same purpose as the theoretical triaxial loading of Fig. 2.9.1, in that it restrains plastic flow which otherwise would occur and thus at some higher stress may likely fail in a brittle manner. Figure 2.9.2 shows the effect of a notch in a tensile test specimen. The cross-sectional area at the base of the notch corresponds to the area of the original specimen of Fig. 2.9.1b. The reduced section tries to become narrower as the axial tension increases,

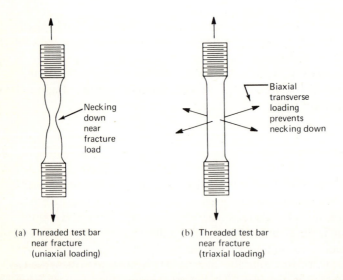

(a) Threaded test bar near fracture (uniaxial loading)

(b) Threaded test bar near fracture (triaxial loading)

Fig. 2.9.1 Uniaxial and triaxial loading.

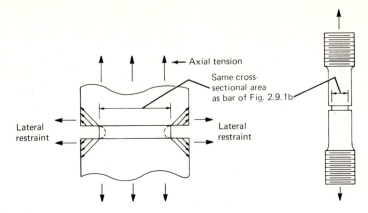

Fig. 2.9.2 Effect of notch on uniaxial tension test.

but is resisted by the diagonal pull that develops in the corners, as shown in Fig. 2.9.2. The test bar will fail at high stress by brittle fracture.

Notches can occur in real structures by use of unfilleted corners in design or from improperly made welds that may crack. Such occurrences can lead to brittleness. Notches and cracked welds can, however, be minimized by good design and welding procedures.

Unusual configurations and changes in section should be made gradually so the stress flow lines are not required to make abrupt changes. Whenever the complexity is such as to give rise to three-dimensional stresses, the tendency for brittleness increases. Castings, for instance, have the reputation for brittleness. Primarily this is because of the built-in three-dimensional continuity.

Multiaxial Stress Induced by Welding

In general, welding creates a built-in continuity that gives rise to biaxial and triaxial stress and strain conditions, which result in brittle behavior. To illustrate, consider the loaded simply supported beam of Fig. 2.9.3, which in turn supports a plate in tension. Due to flexure, the bottom flange of the beam is in tension; therefore, the stress at point A is uniaxial tension (neglecting the small effects of beam width and attachment of flange to web). Application of the tension plate using angles and bolts puts the flange bolts and the angles essentially in uniaxial tension and the bolt which passes through the suspender plate in shear, so that there is no appreciable effect on the stress at point A. In other words, the stress conditions in the connection of Fig. 2.9.3a are approximately uniaxial in nature.

Next, consider the tensile suspender plate welded to the tension flange of the beam, as in Fig. 2.9.3b. The stress at point A is now biaxial because of the direct attachment to the flange at that point. The weld

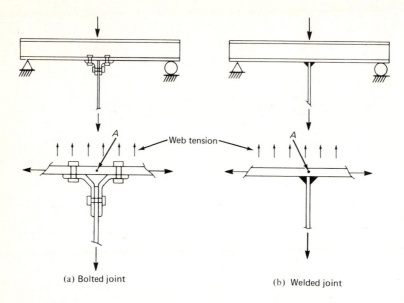

(a) Bolted joint   (b) Welded joint

Fig. 2.9.3 Comparison of stress conditions in bolted and welded joint.

region, therefore, is subject to triaxial stress; biaxial from the directly applied loads, plus the resistance to deformation along the axis of the welds resulting from continuous attachment (Poisson's ratio effect). The design of welded joints should consider the possibilities of brittleness due to three-dimensional stressing. The subject of *lamellar tearing* is treated in Sec. 2.10.

Effect of Thickness

As discussed in Sec. 2.6, if plane stress exists, such as with thin plates where stress in the direction transverse to the plane of the plate may be disregarded, the third dimensional effect is eliminated. For thick plates, because of the three-dimensional effects, the tendency for brittleness increases. From the manufacturing process, thick plates also tend to be more brittle than thin ones; (a) the slower cooling rate gives rise to a coarser microstructure and (b) the higher carbon content that is necessary to obtain the same yield strength for thick sections as obtained by additional hot working in thin sections, also produces a more brittle material.

Effect of Dynamic Loading

The stress-strain properties referred to so far have been for static loading slowly applied. More rapid loading, such as that of forge drop hammers, earthquake, or nuclear blast changes the stress-strain properties. Ordinarily, the increased strain rate from dynamic loading increases the yield point, tensile strength, and ductility. At temperatures about 600°F (320°C) there is a moderate decrease in strength. Some increased brittleness has been noted with high strain rate, but it seems principally associated with other factors already discussed, such as notches where stress concentrations exist and the temperature effect on toughness. The more important factor relating to dynamic load application is not that a rapid increasing strain rate occurs, but that it is combined with a rapid *decreasing* strain rate. The effect of stress *variation* is discussed in the section on fatigue.

Table 2.9.1, from Ref. 7, provides a list of factors "to help determine whether or not the risk of brittle fracture is serious and requires special design considerations."

Table 2.9.1 The Element of Risk: Factors to Analyze in Estimating Seriousness of Brittle Fracture (from Ref. 7)

---

1. What is the minimum anticipated service temperature? The lower the temperature, the greater the susceptibility to brittle fracture.
2. Are tension stresses involved? Brittle fracture can occur only under condition of tensile stress.
3. How thick is the material? The thicker the steel, the greater the susceptibility to brittle fracture.
4. Is there three-dimensional continuity? Three-dimensional continuity tends to restrain the steel from yielding and increases susceptibility to brittle fracture.
5. Are notches present? The presence of sharp notches increases susceptibility to brittle fracture.
6. Are multiaxial stress conditions likely to occur? Multiaxial stresses will tend to restrain yielding and increase susceptibility to brittle fracture.
7. Is loading applied at a high rate? The higher the rate of loading, the greater susceptibility to brittle fracture.
8. Is there a changing rate of stress? Brittle fracture occurs only under conditions of increasing rate of stress.
9. Is welding involved? Weld cracks can act as severe notches.

2.10 LAMELLAR TEARING

Lamellar tearing is a form of brittle fracture occurring as a separation in the base material of a highly restrained welded joint caused by "thru-thickness" strains induced by weld metal shrinkage. When welds are made in highly restrained welded joints, the localized strains due to weld metal shrinkage can be several times larger than yield point strains. Since the stresses due to service loads are well below the yield stress, the strains due to such loads are not believed to initiate or propagate lamellar tears.

The subject of lamellar tearing has received considerable attention

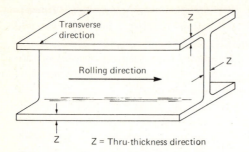

Fig. 2.10.1 Definition of direction terminology. (From Ref. 8)

during the early 1970s, resulting in a tendency for structual engineers to blame lamellar tearing for many brittle fractures. The AISC has provided an excellent summary of the phenomenon [8]. Thornton [9] has provided design and supervision procedures to minimize lamellar tearing.

As a result of the hot rolling operation in manufacture, steel sections have different properties in the direction parallel to rolling (see Fig. 2.10.1), in the transverse direction, and in the "thru-thickness" direction.

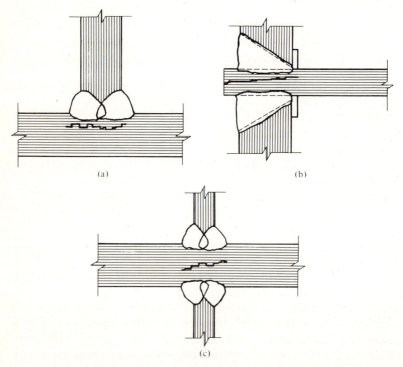

Fig. 2.10.2 Joints showing typical lamellar tears resulting from shrinkage of large welds in thick material under high restraint. (From Ref. 8)

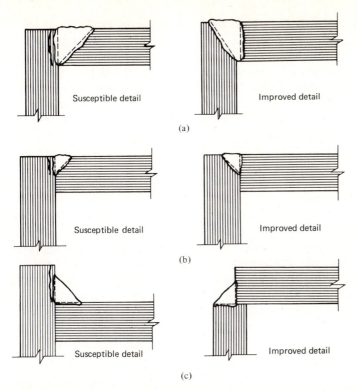

Fig. 2.10.3 Susceptibility to lamellar tearing can be reduced by careful detailing of welded connections. (From Ref. 8)

In the elastic range, both the rolling and transverse directions exhibit similar behavior, with the elastic limit for the transverse direction being only slightly below that for the rolling direction. The ductility (strain capability), however, in the "thru-thickness" direction may be well below that for the rolling direction.

Generally, I-shaped steel sections are adequately ductile when loaded either parallel or transverse to the rolling direction. They will deform locally to strains greater than the yield strain (F_y/E_s), carrying load with some of the material acting at the yield stress and bringing adjacent material into participation if added strength is needed. When, however, the strain is localized for instance in the "thru-thickness" direction at one thick flange of a section, a restrained situation exists because the strain cannot redistribute from the flange through the web to the opposite flange. The large localized "thru-thickness" strain may exceed the yield point strain, causing decohesion and leading to a lamellar tear.

Figure 2.10.2 illustrates the relationship of a lamellar tear to a welded joint. The condition of connection restraint is not related to continuity as referred to by structural engineers in the analysis of a statically indeterminate rigid frame. The restraint potentially giving rise to lamellar tearing is internal joint restraint that inhibits the large unit strains resulting from weld shrinkage. Referring to Fig. 2.10.3, when the weld shrinkage occurs in the "thru-thickness" direction, the material being connected becomes susceptible to lamellar tearing. The weld detail should be made so that weld shrinkage occurs in the rolling direction.

2.11 FATIGUE STRENGTH

Repeated loading and unloading even if the yield point is never exceeded may result in eventual failure. Such a phenomenon is known as *fatigue*. While fatigue may be observed even if all conditions are ideal, i.e., excellent notch toughness, no stress concentrations from holes or notches, uniaxial stress condition, ductile microstructure, etc., adverse conditions affecting ductility and the existence of multiaxial stress conditions greatly reduce fatigue strength. Actually, the factors discussed throughout this chapter are all interrelated.

Consider the possible stress cycles of Fig. 2.11.1. The most extreme variation is the full reversal; zero to maximum tension, unload to zero,

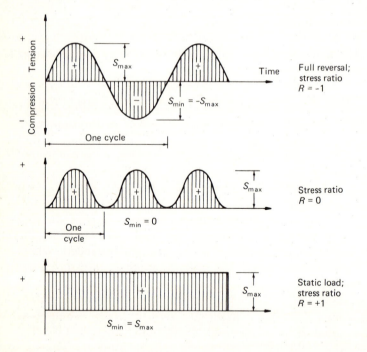

Fig. 2.11.1 Types of stress cycles, showing extreme range of stress ratio from $R = +1$ (nonfatigue condition) to $R = -1$ (stress reversal).

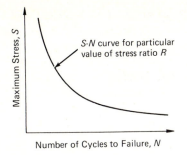

Fig. 2.11.2 Maximum stress S obtainable for various number of cycles N of loading (plotted to linear scale).

load in compression to same numerical stress as in tension, unload to zero. The ratio of S_{max} (tension) to S_{min} (compression) is said to be $R = -1$, which results in the lowest fatigue strength. The least extreme, of course, is static load without variation; with the stress ratio $R = +1$. When the maximum stress S_{max} is plotted against the number of cycles to which it was subjected before failure occurred, a curve such as in Fig. 2.11.2 results. Knowing the maximum number of cycles to which the structure will be subjected, along with the stress ratio, the fatigue strength can be determined. When plotted to log-log scale, as in Fig. 2.11.3, the curves closely approximate straight lines. When a curve reaches a constant stress that is independent of the number of cycles of loading, the corresponding stress is referred to as the *fatigue limit* or *endurance limit*. This usually occurs at about 2 million cycles of loading. At the lower end, there is usually little strength reduction for fewer than 100,000 cycles. Since most buildings are subject to 100,000 cycles of loading or less, fatigue ordinarily is not considered. Highway bridges usually are expected to have more than 100,000 cycles of loading, so that fatigue is an important consideration in their design.

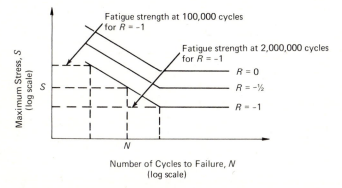

Fig. 2.11.3 Typical S–N curves (approximation by straight lines) for several stress ratios plotted to log–log scale.

The sloping portion of Fig. 2.11.3 may be expressed [10] as

$$F_n = S(N/n)^k \qquad (2.11.1)$$

where F_n = fatigue strength computed for failure at n cycles
S = stress which produced failure at N cycles
k = slope of the best-fit straight line representing the data

As shown in Fig. 2.11.3 each stress ratio requires a different $S–N$ curve. Usually a diagram known as a *Goodman diagram* is used to summarize the results for the various types of stress cycles. Figure 2.11.4 shows such a diagram from Ref. 10.

Frequently for design purposes a modified Goodman diagram is used on which the effects of several different life cycles may be presented, as shown in Fig. 2.11.5. Stress Category B from 1978 AISC Appendix B (Table B3) has been used for illustration. The maximum tensile stress has been taken as 30 ksi ($0.60F_y$ for $F_y = 50$ ksi). For static tension ($R = +1$) the upper limit represented by the horizontal line at $0.60F_y$ governs. For stress ratios (R) between $+\frac{1}{2}$ and $+1$, fatigue has no effect, i.e., so long as the minimum stress is not less than one-half the maximum and is of the same sign.

When the stress ratio is less than $+\frac{1}{2}$, some reduction for fatigue is required when the number of cycles N exceeds 2 million. The sloping line intersecting the $R = 0$ line at 16 ksi indicates a stress variation of 16 ksi is permitted when $N > 2,000,000$. Notice that full reversal ($R = -1$) would allow the stress to vary from 8 ksi in tension to 8 ksi in compression. For any given stress ratio R, one may graphically obtain the maximum stress permitted during the cycle by means of a modified Goodman diagram.

According to the Appendix B of the AISC Specification some

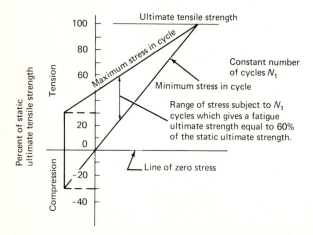

Fig. 2.11.4 Typical Goodman diagram showing effect of various stress ratios on fatigue strength for N_1 cycles of loading. (Adapted from Ref. 10)

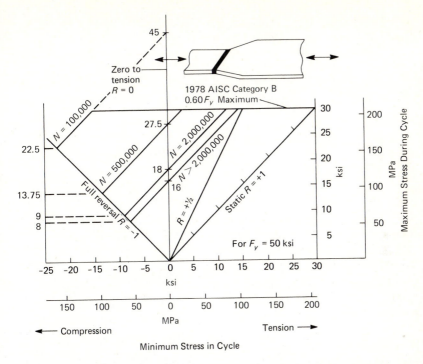

Fig. 2.11.5 Modified Goodman diagram, giving 1978 AISC allowable stresses for base metal and weld metal at full joint penetration groove welded splices at transitions in width and thickness.

reduction in strength due to fatigue may be expected under certain stress ratios and types of loading when the number of expected cycles exceeds 20,000 (approximately 2 applications per day for 25 years).

The mechanism of fatigue is still not entirely understood, but it is known to be closely related to the factors relating to ductility. Welding, in particular, may have a dramatic effect on fatigue strength. For a more detailed treatment the reader is referred to *Fatigue of Welded Structures* [10], or the *Welding Handbook* [2] (pp. 186–201).

2.12 CORROSION RESISTANCE AND WEATHERING STEELS

Since the earliest uses of steel, one of the important drawbacks was that painting was required to prevent the deterioration of the metal by corrosion (rusting). The lower-strength carbon steels were inexpensive but very vulnerable to corrosion. Corrosion resistance may be improved by the addition of copper as an alloy element. However, copper-bearing carbon steel is too expensive for general use.

High-strength low-alloy steels have several times [11] the corrosion resistance of structural carbon steel, with or without the addition of

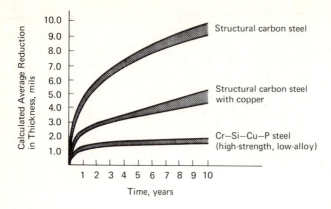

Fig. 2.12.1 Comparative corrosion of steels in an industrial atmosphere. Shaded areas indicate range for individual specimens. (Adapted from Ref. 11)

copper, as shown in Fig. 2.12.1. The high-strength low-alloy steels do not pit as severely as carbon steels and the rust that forms becomes a protective coating to prevent further deterioration. With certain alloy elements the high-strength low-alloy steel will develop an oxide protective coating that is pleasing in appearance and is described as follows:*
"It is a very dense corrosion—actually a deeply colored brown, red, purple.... It has a texture and color which cannot be reproduced artificially—a character only nature can give, as with stone, marble, and granite." When steels are to be unpainted and left exposed they are called weathering steels.

As might be expected, the corrosion properties of any steel, including the weathering steels, are dependent on the chemical composition, the degree of pollution in the atmosphere, and the frequency of wetting and drying of the steel.

Since its first major use in 1958, for the Administrative Center for Deere & Company in Moline, Illinois, the use of weathering steel has received considerable attention. At first such steels were specified under ASTM A242 which as previously discussed is very general, allowing a wide variation in chemistry.

With the adoption of A588 steel in 1969, and A709 in 1975, A242 is now essentially obsolete. A588 is generally used for weathering steel in buildings and A709 Grades 50W and 100W for weathering steel in bridges (see Table 2.1.1).

Fabrication and erection of weathering steel requires care. Unsightly gouges, scratches, and dents should be avoided. Painting, even for identification, should be minimized, since all marks must be removed after the erection is completed. Scale and discoloration from welding also

---

* *Architectural Record*, August 1962.

must be removed. The extra expense resulting from fabrication and erection is offset by the elimination of painting at intervals during the life of the structure.

To a large extent the need for fireproofing of steel in buildings has slowed the use of exposed steel. Two examples of innovative solutions involve (1) keeping the exposed steel members entirely outside the enclosed portion of a building which could contain a fire; and (2) filling exposed hollow steel columns with chemically treated water which will act as a heat sink to keep the temperature of the steel down should the columns be subject to fire [12].

SELECTED REFERENCES

1. *Standard Specification for General Requirements for Delivery of Rolled Steel Plates, Shapes, Sheet Piling, and Bars for Structural Use* (ANSI/ASTM A6-78). Philadelphia, Pa.: American Society for Testing and Materials, 1978. Also adopted by the American National Standards Institute.
2. *Welding Handbook*, 7th ed., Vol. 1, *Fundamentals of Welding*, Miami, Fla.: American Welding Society, 1977.
3. *Structural Welding Code (AWS D1.1–79)*. Miami, Fla.: American Welding Society, Inc., 1979.
4. R. L. Brockenbrough and B. G. Johnston, *Steel Design Manual*, Pittsburgh, Pa.: U.S. Steel Corporation, 1968, Chap. 1.
5. *Specification for the Design of Cold-Formed Steel Structural Members*. New York: American Iron and Steel Institute, 1968, with Addendum No. 1(November 19, 1970) and Addendum No. 2 (February 4, 1977).
6. S. T. Rolfe, "Fracture and Fatigue Control in Steel Structures," *Engineering Journal*, AISC, 14, 1(First Quarter 1977), 2–15.
7. *A Primer on Brittle Fracture*, Booklet 1960–A, Steel Design File, Bethlehem Steel Corp., Bethlehem, Pa.
8. "Commentary on Highly Restrained Welded Connections," *Engineering Journal*, AISC, 10, 3(Third Quarter 1973), 61–73.
9. Charles H. Thornton, "Quality Control in Design and Supervision Can Eliminate Lamellar Tearing," *Engineering Journal*, AISC, 10, 4(Fourth Quarter 1973), 112–116.
10. W. H. Munse and LaMotte Grover, *Fatigue of Welded Structures*, New York: Welding Research Council, 1964.
11. C. P. Larrabee, "Corrosion Resistance of High-Strength Low-Alloy Steels as Influenced by Composition and Environment," *Corrosion Magazine*, 9, 8(August 1953), 259–271.
12. "Weathering Steels Become Loadbearing," *Progressive Architecture* (September 1967).

3
Tension Members

3.1 INTRODUCTION

Tension members are encountered in most steel structures. They occur as principal structural members in bridge and roof trusses, in truss structures such as transmission towers and wind bracing systems in multi-storied buildings. They frequently appear as secondary members, being used as tie rods to stiffen a trussed floor system or to provide intermediate support for a wall girt system. Tension members may consist of a single structural shape or they may be built up from a number of structural shapes. The cross sections of some typical tension members are shown in Fig. 3.1.1.

In general, the use of single structural shapes is more economical than the built-up sections. However, built-up members may be required when (a) the tensile capacity of a single rolled section is not sufficient, (b) the slenderness ratio (the ratio of the unbraced length L to the minimum radius of gyration r) does not provide sufficient rigidity, (c) the effect of bending combined with the tensile behavior requires a larger lateral stiffness, (d) unusual connection details require a particular cross section, or (e) esthetics control.

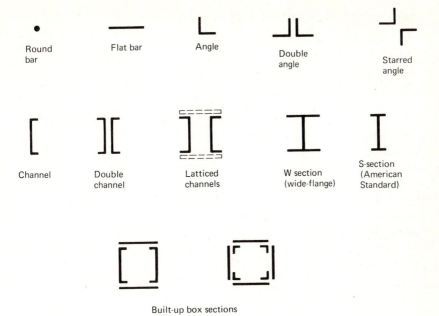

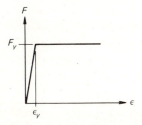

Built-up box sections

Fig. 3.1.1   Cross section of typical tension members.

3.2 STRENGTH AS A DESIGN CRITERION

The design of a tension member is one of the simplest and most straightforward problems in structural engineering. Since stability is only of secondary concern, the problem of designing a tension member is basically one of providing a member with a sufficient cross-sectional area to resist the applied loads with an adequate factor of safety against failure.

Usual design procedures (although nominally the working stress method using stresses at service load level) are actually based on ultimate strength. A tension member without holes (such as with welded connections) achieves its strength when all fibers of the cross section have yielded (Fig. 3.2.1); i.e., the tensile stress distribution is uniform at

Fig. 3.2.1   Stress-strain curve for steel.

ultimate strength. The strength may be expressed as

$$T_u = F_y A_g \tag{3.2.1}$$

where A_g is the gross cross-sectional area.

For tension members having holes, such as for rivets or bolts, or for threads cut on rods, the reduced cross section is referred to as the *net area*. Holes or threads in a member cause stress concentrations (nonuniform stresses); for example, a hole in a plate will give rise to a stress distribution at service load as shown in Fig. 3.2.2a. Theory of elasticity shows that tensile stress adjacent to a hole will be about three times the average stress on the net area. However, as each fiber reaches yield strain its stress then becomes a constant F_y, with deformation continuing with increasing load until finally all fibers have achieved or exceeded yield strain (Fig. 3.2.2b).

The strength of a tension member having holes or threads may be expressed as (Fig. 3.2.2b)

$$T_u = F_y A_n \tag{3.2.2}$$

(where A_n is the net area of the cross section. The safe service load T may then be obtained by dividing the strength by a factor of safety (FS). Thus

$$T = \frac{F_y A_n}{FS} = F_t A_n \tag{3.2.3}$$

where F_t may be termed the allowable stress for working (service) conditions.

A suitable value for FS may be established by using probabilistic methods or from historical precedent. The latter has been traditional for steel structures design where the basic FS for a ductile mode of failure (such as in a tension member) is 1.67 under the AISC Specification. A rationale as described in Sec. 1.9 may be used as justification along with a history of satisfactory performance. From Eq. 3.2.3, the allowable stress

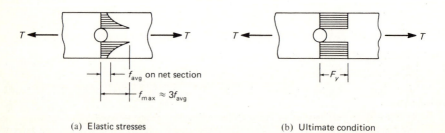

(a) Elastic stresses   (b) Ultimate condition

Fig. 3.2.2 Stress distribution with holes present.

F_t would be $F_y/1.67$, or $0.60F_y$. For many years prior to 1978, AISC has used Eq. 3.2.3 as the basis for design of all tension members.

Although yielding on the net section (Fig. 3.2.2b) *may* constitute maximum strength, tests [1] have indicated that yielding on the gross section away from the holes will frequently occur prior to reaching the ultimate tensile stress (usually referred to as tensile strength) F_u (Fig. 2.1.1) on the net section. In other words, actual failure may occur when yield stress is reached on the gross section. If the designer could be certain that gross section yielding would control, the design would be simpler since A_n would not have to be computed. The requirement may be stated

$$\phi F_u A_n \geq F_y A_g \tag{3.2.4}$$

where ϕ is a capacity reduction factor to account for the variability in the connection strength. If the gross area controls the design of the *member*, then $F_y A_g$ may be divided by the basic FS (1.67) for members to obtain the safe service load capacity; thus

$$\frac{\phi F_u A_n}{1.67} \geq 0.60 F_y A_g \tag{3.2.5}$$

The left side of the equation representing *connection* strength (or strength in the vicinity of a connection) should provide a factor of safety of 2.0 consistent with the value used for connector design (fasteners, Chapter 4 and welds, Chapter 5). The net section service load capacity will be $0.50 F_u A_n$ when $\phi = 0.85$.

Thus AISC–1.5.1.1 is based on Eq. 3.2.5 but uses an *effective net area* A_e instead of actual net area A_n. For other than pin-connected members AISC requires

$$f_a = \frac{T}{A_g} \leq 0.60 F_y \tag{3.2.6}$$

and

$$f_a = \frac{T}{A_e} \leq 0.50 F_u \tag{3.2.7}$$

The effective net area A_e may be less than the actual net area so as to include the effect of stress concentrations and eccentricity of loading not otherwise taken into account (see Sec. 3.5). The tensile strength F_u may be obtained from Table 2.1.1 (or from AISC–Appendix Table 2) for the various steels.

A summary of the allowable stresses for axial tension in members (tension on fasteners and welds is treated elsewhere) is given in Table 3.2.1.

Table 3.2.1 Allowable Stresses on Tension Members

AISC–1.5.1.1–1978

  For *other* than pin-connected members:

    $F_t = 0.60F_y$ on gross area

    $F_t = 0.50F_u$ on effective net area

  For pin-connected members:

    $F_t = 0.45F_y$ on net area

  For threaded rods of steels approved in AISC–1.4.1:

    $F_t = 0.33F_u$ on major diameter*

      (for static loading only)

AASHTO–1.7.1–1977

    $F_t = 0.55F_y$ on *net* area

    $F_t = 0.46F_u$ on net area

  (except where fatigue limitations require lower value)

---

* Major diameter is that measured to the *outer* projections of the threads (essentially gross area A_D as given in standard dimensions for threaded fasteners).

3.3 NET AREA

Whenever a tension member is to be fastened by means of bolts or rivets, holes must be provided at the connection. As a result, the member cross-sectional area at the connection is reduced and the allowable tensile load in the member *may* also be reduced depending on the size and location of the holes.

Several methods are used to cut holes. The most common and least expensive method is to punch *standard* holes $\frac{1}{16}$ in. (1.6 mm) larger than the diameter of the rivet or bolt. During the punching operation the metal at the edge of the hole is damaged. This is accounted for in design by assuming that the extent of the damage is limited to a radial distance of $\frac{1}{32}$ in. (0.8 mm) around the hole. Therefore the total width to be deducted (AISC–1.14.4) is to be taken as the nominal dimension of the *hole* normal to the direction of applied stress plus $\frac{1}{16}$ in. (1.6 mm). For fasteners in standard holes this is equivalent to the fastener diameter plus $\frac{1}{8}$ in. (3.2 mm).

A second method of cutting holes consists of subpunching them $\frac{3}{16}$ in. (4.8 mm) diam undersize and then reaming the holes to the finished size after the pieces being joined are assembled. This method is more expensive than that of punching standard holes but does offer the advantage of accurate alignment.

A third method consists of drilling holes to a diameter of the bolt or rivet plus $\frac{1}{32}$ in. (0.8 mm). This method is used to join thick pieces, and is the most expensive of the common methods.

When greater latitude is needed in meeting dimensional tolerances during erection, larger than *standard* holes can be used with high-strength bolts larger than $\frac{5}{8}$ in. diam without adversely affecting the performance. In Chapter 4, high-strength bolted connections are treated involving the use of oversize, short-slotted, and long-slotted holes. The maximum hole size for such holes is given in Table 4.7.2.

Example 3.3.1

What is the net area A_n for the tension member shown in Fig. 3.3.1?

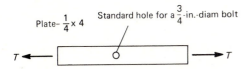

Fig. 3.3.1   Tension member for Example 3.3.1.

SOLUTION
$$A_g = 4(0.25) = 1.0 \text{ sq in.}$$

Width to be deducted for hole $= \frac{3}{4} + \frac{1}{8} = \frac{7}{8}$ in.

$$A_n = A_g - (\text{width for hole})(\text{thickness of plate})$$
$$= 1.0 - 0.875(0.25) = 0.78 \text{ sq in.}$$

3.4 EFFECT OF STAGGERED HOLES ON NET AREA

Whenever there is more than one hole and the holes are *not* lined up transverse to the loading direction, more than one potential failure line may exist. The controlling failure line is that which gives the minimum net area.

In Fig. 3.4.1a the failure line is along the section $A{-}B$. In Fig. 3.4.1b showing two lines of staggered holes, the failure line might be through one hole (section $A{-}B$) or it might be along a diagonal path, $A{-}C$. At first glance one might think section $A{-}B$ is critical since the path $A{-}B$ is obviously shorter than path $A{-}C$. However, from path $A{-}B$, only one hole would be deducted while two holes would have to be deducted from path $A{-}C$. In order to determine the controlling section, both paths $A{-}B$ and $A{-}C$ must be investigated. Accurate checking of strength along path $A{-}C$ is complex. However, a simplified empirical relationship proposed by Cochrane [2] has been adopted by AISC–1.14.2 to account for the difference between the path $A{-}C$ and the path $A{-}B$ expressed as a

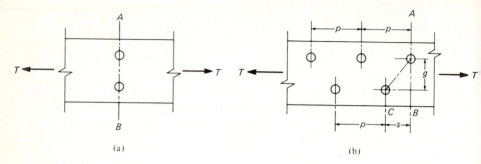

Fig. 3.4.1 Paths of failure on net section.

length correction,

$$\frac{s^2}{4g}$$

where s is the stagger, or spacing of adjacent holes parallel to the loading direction (see Fig. 3.4.1), and g the gage distance transverse to the loading direction. Thus the net lengths of paths $A-B$ and $A-C$ would be

Net length of $A-B =$ length of $(A-B) -$ (width of hole $+ \frac{1}{16}$ in.)

Net length of $A-C =$ length of $(A-B) - 2$(width of hole $+ \frac{1}{16}$ in.) $+ \dfrac{s^2}{4g}$.

The minimum net area would then be determined from the minimum net length multiplied by the thickness of the plate.

In the years since Cochrane proposed the simple $s^2/4g$ expression many investigators have proposed other rules [3–6] but none of them give significantly better results and all are more complicated.

Consistent with the general trend toward using design approaches that relate to the ultimate strength, the work of Bijlaard [7] and others [8–10] has provided limit analysis theories to obtain net area in tension. These theories do not deviate from the $s^2/4g$ method by more than 10 to 15 percent.

The reader is referred to McGuire [11] for a more complete coverage of this subject of net section through staggered lines of fasteners.

Example 3.4.1

Determine the minimum net area of the plate shown in Fig. 3.4.2, assuming $\frac{15}{16}$-in.-diam holes are located as shown.

SOLUTION

According to AISC–1.14.2 and 1.14.4, the width used in deducting for holes is the hole diameter plus $\frac{1}{16}$ in., and the staggered length correction is $s^2/4g$.

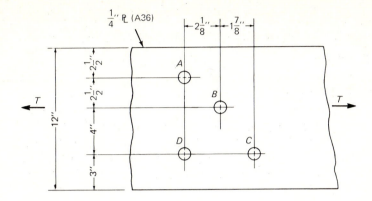

Fig. 3.4.2 Example 3.4.1.

Path AD:

$$\left[12 - 2\left(\frac{15}{16} + \frac{1}{16}\right)\right]0.25 = 2.50 \text{ sq in.}$$

Path ABD:

$$\left[12 - 3\left(\frac{15}{16} + \frac{1}{16}\right) + \frac{(2.125)^2}{4(2.5)} + \frac{(2.125)^2}{4(4)}\right]0.25 = 2.43 \text{ sq in.}$$

Path ABC:

$$\left[12 - 3\left(\frac{15}{16} + \frac{1}{16}\right) + \frac{(2.125)^2}{4(2.5)} + \frac{(1.875)^2}{4(4)}\right]0.25 = 2.42 \text{ sq in.}$$

(controls)

Angles

When holes are staggered on two legs of an angle, the gage length g for use in the $s^2/4g$ expression is obtained by using a length between the centers of the holes measured along the centerline of the angle thickness i.e., the distance A–B in Fig. 3.4.3. Thus the gage distance g is

$$g = g_a - \frac{t}{2} + g_b - \frac{t}{2} = g_a + g_b - t \tag{3.4.1}$$

Every rolled angle has a standard value for the location of holes (i.e., gage distances g_a and g_b) depending on the length of the leg. Table 3.4.1 shows *usual gages* for angles as listed in the AISC Manual. Unless special requirements dictate using other than usual gages, the higher fabrication costs resulting from using special gage values cannot be justified.

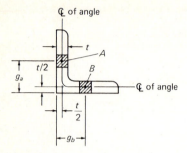

Fig. 3.4.3 Gage distances for an angle.

Example 3.4.2

Determine the net area A_n for the angle given in Fig. 3.4.4 if $\frac{15}{16}$-in.-diam holes are used.

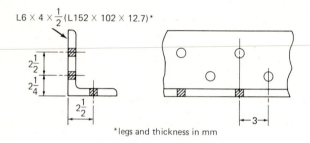

L6 × 4 × $\frac{1}{2}$ (L152 × 102 × 12.7)*

*legs and thickness in mm

Fig. 3.4.4 Example 3.4.2.

SOLUTION

For net area calculation the angle may be visualized as being flattened into a plate as shown in Fig. 3.4.5.

$$A_n = A_g - Dt + \frac{s^2}{4g}t$$

where D is the width to be deducted for the hole.

Path AC:

$$4.75 - 2\left(\frac{15}{16} + \frac{1}{16}\right)0.5 = 3.75 \text{ sq in.}$$

Path ABC:

$$4.75 - 3\left(\frac{15}{16} + \frac{1}{16}\right)0.5 + \left[\frac{(3)^2}{4(2.5)} + \frac{(3)^2}{4(4.25)}\right]0.5 = 3.96 \text{ sq in.}$$

Since the smallest A_n is 3.75 sq in., that value governs.

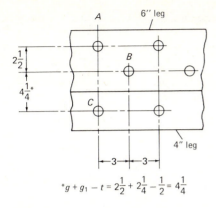

$$^*g + g_1 - t = 2\frac{1}{2} + 2\frac{1}{4} - \frac{1}{2} = 4\frac{1}{4}$$

Fig. 3.4.5   Angle for Example 3.4.2 with legs shown "flattened" into one plane.

Table 3.4.1   Usual Gages for Angles, Inches (from AISC Manual)

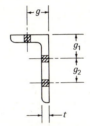

Leg	8	7	6	5	4	$3\frac{1}{2}$	3	$2\frac{1}{2}$	2	$1\frac{3}{4}$	$1\frac{1}{2}$	$1\frac{3}{8}$	$1\frac{1}{4}$	1
g	$4\frac{1}{2}$	4	$3\frac{1}{2}$	3	$2\frac{1}{2}$	2	$1\frac{3}{4}$	$1\frac{3}{8}$	$1\frac{1}{8}$	1	$\frac{7}{8}$	$\frac{7}{8}$	$\frac{3}{4}$	$\frac{5}{8}$
g_1	3	$2\frac{1}{2}$	$2\frac{1}{4}$	2										
g_2	3	3	$2\frac{1}{2}$	$1\frac{3}{4}$										

3.5 EFFECTIVE NET AREA

The net area as computed in Secs. 3.3 and 3.4 gives the reduced section that resists tension but still may not correctly reflect the strength. This is particularly true when the tension member has a profile consisting of elements not in a common plane and where the tensile load is transmitted at the end of the member by connection to some but not all of the elements. An angle section having connection to one leg only is an example of such a situation. For such cases the tensile force is not uniformly distributed over the net area. To account for the nonuniformity, AISC–1.14.2 provides for an *effective net area* A_e equal to $C_t A_n$ where C_t is a reduction coefficient. By using the effective net area the nonuniformity of stress is accounted for in a simplified manner.

For short tension members consisting of splice or gusset plates, where the elements of the cross section lie essentially in a common plane,

Table 3.5.1 Effective Net Area, A_e (adapted from AISC–1.14.2.2 and 1.14.2.3)

Types of Members	Minimum Number of Fasteners per Line	Special Requirements	Effective Net Area, A_e
(a) Full length tension members having *all* cross-sectional elements connected to transmit the tensile force	1	None	A_n*
(b) Short tension member fittings, such as splice plates, gusset plates, or beam-to-column fittings	1	None	A_n but not exceeding $0.85A_g$
(c) W, M, or S rolled shapes	3	$\dfrac{\text{flange width}}{\text{section depth}} \geq \dfrac{2}{3}$ connection is to flange or flanges	$0.90A_n$
(d) Structural tees cut from sections meeting requirements of (c) above	3		$0.90A_n$
(e) W, M, or S shapes not meeting the conditions of (c), and other shapes, including built-up sections, having unconnected segments not in the plane of the loading	3	None	$0.85A_n$
(f) All shapes in (c), (d), or (e)	2	None	$0.75A_n$

* Actual net area computed according to AISC–1.14.2.1 and 1.14.4.

the effective net area is taken (AISC–1.14.2.3) equal to A_n, but the effective net area may not be taken greater than 85% of the gross area A_g. Tests [1] have shown that when any holes are present in such connections, there will be at least a 15% reduction in strength from that computed using gross area. Table 3.5.1 summarizes the AISC requirements for effective net area.

3.6 TENSION RODS

A common and simple tension member is the threaded rod. Such rods are usually secondary members where the design stress is small, such as (a) sag rods to help support purlins in industrial buildings (Fig. 3.6.1a);

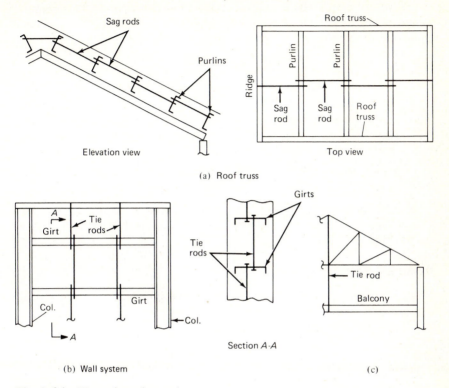

Fig. 3.6.1 Uses of tension rods.

(b) vertical ties to help support girts in industrial building walls (Fig. 3.6.1b); (c) hangers, such as tie rods supporting a balcony (Fig. 3.6.1c); and (d) tie rods to resist the thrust of an arch.

Tie rods are frequently used with an initial tension as diagonal wind bracing in walls, roofs, and towers. The initial tension effectively adds to the stiffness and reduces deflection and vibrational motion which tends to cause fatigue failures in the connections. Such initial tension can be obtained by designing the member something on the order of $\frac{1}{16}$ in. short for a 20-ft length.

Example 3.6.1

Determine the size of a threaded round steel tension rod to carry 9 kips using A36 steel having $F_y = 36$ ksi.

SOLUTION

From Table 3.2.1 the allowable stress on threaded rods is

$$F_t = 0.33F_u$$

based on gross area A_D using the major thread diameter. The minimum tensile strength F_u for A36 steel is 58 ksi from Table 2.1.1 (or AISC–Appendix Table 2).

The gross area A_D required is

$$A_D = \frac{T}{0.33F_u} = \frac{9}{0.33(58)} = 0.47 \text{ sq in.}$$

Information on standard threaded rods is to be found in the section, "Threaded Fasteners," of the AISC Manual, where the gross area A_D based on the major diameter measured to the outer extremity of the threads is to be used as the basis for design.

Use $\frac{7}{8}$-in.-diam rod (9 threads per inch) ($A_D = 0.601$ sq in.).

Note: AISC–1.15.1 requires "connections carrying calculated stresses, except for lacing, sag bars, and girts, shall be designed to support not less than 6 kips."

Example 3.6.2

Design sag rods to support the purlins of the industrial building roof of Fig. 3.6.2. Sag rods are spaced at $\frac{1}{3}$ points between roof trusses, which are

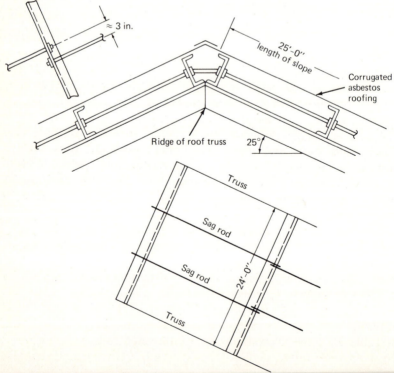

Fig. 3.6.2 Details for Example 3.6.2.

spaced 24 ft apart. Assume $F_t = 0.33F_u = 0.33(58) = 19.1$ ksi. Use 20 psf snow load.

SOLUTION

(a) Loads: Roofing—corrugated asbestos = 3 psf

Purlin (assume given)   = 3.5 psf

Since purlins must have been designated first, their weight is known.

Snow 20 psf (cos 25°)   = 18.1 psf

Total (per sq ft of roof area)   = 24.6 psf

Snow load customarily is prescribed as having a certain intensity in pounds per square foot (psf) of horizontal projection. Generally 20 psf is considered a minimum, with 30 to 40 psf (1.44 to 1.92 kN/m^2) being used in some northern areas (see Sec. 1.4).

(b) Sag rods are expected to carry the component of the load acting parallel to the roof.

$$w \text{ (parallel to roof)} = 24.6 \sin 25°$$

$$= 24.6(0.423) = 10.4 \text{ psf}$$

(c) Choose rod size.

Load carried by one rod:

$$T = w \text{ (tributary area)} = 0.0104(25)(8) = 2.08 \text{ kips}$$

$$\text{required } A_D = \frac{T}{F_t} = \frac{2.08}{19.1} = 0.11 \text{ sq in.}$$

Select from AISC Manual Table, "Threaded Fasteners, Screw Threads," a $\frac{3}{8}$-in.-diam threaded rod having $A_D = 0.110$ sq in.

Use $\frac{3}{8}$-in.-diam rod.

3.7 STIFFNESS AS A DESIGN CRITERION

Even though stability is not a criterion in the design of tension members, it is still necessary to limit their length in order to prevent a member from becoming too flexible. Tension members that are too long may sag excessively due to their own weight. In addition, they may also vibrate when subjected to wind forces as in an open truss or when supporting vibrating equipment such as fans or compressors.

In order to reduce the problems associated with excessive deflections and vibrations a stiffness criterion was established. This criterion is based on the slenderness ratio, L/r, of a member where L is the length and r the least radius of gyration ($r = \sqrt{I/A}$). The accepted maximum slenderness ratios for tension members (see AISC–1.8.4 and AASHTO–1.7.11) are:

	AISC	AASHTO
For main members	240	200
For lateral bracing and other secondary members	300	240
For members subject to stress reversal	—	140

In applying the stiffness criterion to tension members, the higher slenderness ratio based on the two principal axes must be used. A symmetrical member may have two different radii of gyration, and for nonsymmetrical members one must consider the weakest principal axis. When a tension member is built up from a number of sections, the radius of gyration must be computed using the moment of inertia I and the cross-sectional area A. The value for r will be with respect to the same axis as that of the moment of inertia used.

Example 3.7.1

Determine the maximum length permitted by the AISC Specification for a tension member whose cross section is a flat bar 1×6.

SOLUTION
Determine the least radius of gyration, which may be shown to be a function of the lateral dimension of the member only.

$$I = bt^3/12; \qquad A = bt$$

$$r = \sqrt{\frac{I}{A}} = \sqrt{\frac{bt^3/12}{bt}} = t\sqrt{\frac{1}{12}} = 0.288t$$

Thus the least r occurs with respect to the least lateral dimension:

$$r_{min} = 0.288(1) = 0.288 \text{ in.}$$

$$\frac{L}{r} = 240 = \frac{L}{0.288t}; \qquad L = 69 \text{ in.}$$

Example 3.7.2

Determine the maximum unsupported length permitted by the AISC Specification for the cross sections indicated in Fig. 3.7.1.

SOLUTION
The values for r are taken from the AISC Manual.

(a) For the C12×20.7, $r_x = 4.61$ in., $r_y = 0.80$ in.

$$\text{Max} \frac{L}{r} = 240 \qquad \text{(AISC–1.8.4)}$$

$$\text{Max } L = (\text{Min } r)(240) = 0.80(240) = 192 \text{ in.}$$

(b) For the $L5 \times 5 \times \frac{1}{2}$, $r_x = r_y = 1.54$ in., $r_z = 0.98$ in.

$$\text{Max } L = (\text{Min } r)(240) = 0.98(240) = 235 \text{ in.}$$

(c) For the double angle member, $2 - L6 \times 4 \times \frac{3}{8}$, $r_x = 1.93$ in., $r_y = 1.50$ in.

$$\text{Max } L = (\text{Min } r)(240) = 1.50(240) = 360 \text{ in.}$$

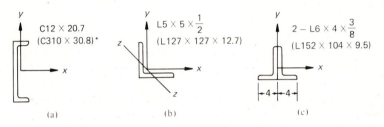

*In SI units, 310-mm nominal depth and mass of 30.8 kg/m.

Fig. 3.7.1   Sections for Example 3.7.2.

3.8 LOAD TRANSFER AT CONNECTIONS

Normally the holes that are to be dealt with in tension members are those for rivets or bolts to transfer load from the tension member into another member.

Although the detailed treatment of fasteners and their behavior is in Chapter 4, the basic assumption is that each equal size fastener transfers an equal share of the load whenever the fasteners are arranged symmetrically with respect to the centroidal axis of the tension member. The following example is to illustrate the idea and its relationship to net area calculations.

Example 3.8.1

Calculate the governing net area for plate A of the single lap joint in Fig. 3.8.1 and show free-body diagrams of portions of plate A with sections taken through each line of holes. Assume that plate B has adequate net area and does not control the capacity T.

SOLUTION

The full tensile force T in plate A acts on section 1–1 of Fig. 3.8.1. Examination of other sections in plate A to the left of section 1–1 will involve *less than* 100% of T acting, since part of that force will have already been transferred from plate A to plate B. At section 4–4, 100% of T must now be acting in plate B while only 20% of T acts in plate A.

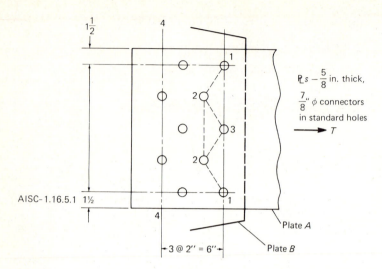

Fig. 3.8.1 Single lap connection for Example 3.8.1.

Since there must be zero force acting on the end of plate A a short distance to the left of section 4–4, the force T must have been entirely transferred to plate B over the distance from sections 1–1 to 4–4. The free bodies of the various segments are shown in Fig. 3.8.2.

$$\text{Deduction for 1 hole} = \text{Diam of hole} + \tfrac{1}{16} \text{ in.}$$

$$= \text{Diam of fastener} + \tfrac{1}{8} \text{ in.}$$

$$\text{for } \textit{standard} \text{ hole}$$

$$= \tfrac{7}{8} + \tfrac{1}{8} = 1 \text{ in.}$$

Net area (section 1–1) $= \tfrac{5}{8}(15 - 3) = 7.50$ sq in. on which 100% of T acts (Fig. 3.8.2d).

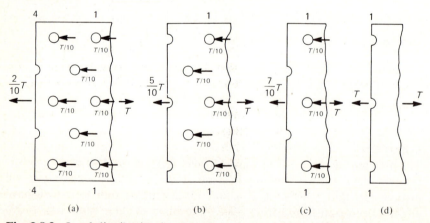

Fig. 3.8.2 Load distribution in plate A.

Net area (staggered path 1–2–3–2–1):

$$= \frac{5}{8}\left[15 - 5(1) + 4\frac{(2)^2}{4(3)}\right] = 7.08 \text{ sq in.}$$

$$\underbrace{}_{s^2/4g}$$

on which 100% of T also acts.

Net area (staggered path 1–2–2–1):

$$= \frac{5}{8}\left[15 - 4 + 2\frac{(2)^2}{4(3)}\right] = 7.29 \text{ sq in.}$$

on which 0.9 of T is presumed to act. One connector has already transferred its share (0.10) of the load prior to reaching section 1–2–2–1. The 7.29 sq in. with 0.9T acting would compare with 7.29/0.9 = 8.10 sq in. with T acting. A comparison of 7.50, 7.08, and 8.10 shows that section 1–2–3–2–1 governs; then $A_n = 7.08$ sq in.

3.9 EXAMPLES—AISC PROCEDURE

Example 3.9.1

Determine the service load capacity in tension based on effective net area for an L6×4×$\frac{1}{2}$ (L152×102×12.7) of A572 Grade 50 steel connected with $\frac{7}{8}$-in.-diam bolts in standard holes as shown in Fig. 3.9.1.

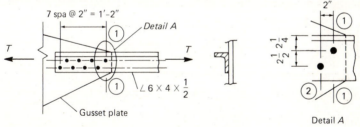

Fig. 3.9.1 Tension member for Example 3.9.1.

SOLUTION

The angle tension member is connected to a gusset plate, typical of truss joints. The gusset plate is the plate at the intersection of members to which they are connected.

The maximum net area strength will be based on section 1–1 with one hole deducted, or on the staggered section 1–2 through two holes, the governing one of which will have 100% of T acting on it.

For section 1–1,

$$A_n = A_g - 1 \text{ hole}$$

$$= 4.75 - \left(\frac{7}{8} + \frac{1}{8}\right)0.50 = 4.25 \text{ sq in.}$$

For section 1–2,

$$A_n = A_g - 2 \text{ holes} + (s^2/4g)t$$

$$= 4.75 - 2\left(\frac{7}{8} + \frac{1}{8}\right)0.50 + \frac{(2)^2}{4(2.5)}(0.50) = 3.95 \text{ sq in.}$$

Since the angle does not have both legs connected to transfer the tensile force, the effective net area is less than the computed net area, accounting for the eccentricity at the connection. From AISC–1.14.2, as summarized in Table 3.5.1 [case (e)],

$$A_e = 0.85A_n = 0.85(3.95) = 3.36 \text{ sq in.}$$

The allowable capacity is the smaller of

$$T = 0.60F_yA_g = 30(4.75) = 143 \text{ kips}$$

or

$$T = 0.50F_uA_e = 0.50(65)(3.36) = 109 \text{ kips}$$

Thus the allowable capacity T is 109 kips.

Example 3.9.2

Determine the service load capacity based on effective net area for the $\frac{5}{8}$-in. splice plate A of Fig. 3.8.1 if the plate material is A36 steel.

SOLUTION

The net area A_n was computed in Example 3.8.1 to be 7.08 sq in. (4570 mm²). This plate connection has no eccentricity through the connection since the tension member elements lie essentially in a common plane. According to AISC–1.14.2.3 [case (b) of Table 3.5.1],

$$A_e = A_n \le 0.85A_g$$

For this example,

$$A_n = 7.08 \text{ sq in.} < [0.85(0.625)15 = 7.97 \text{ sq in.}]$$

Thus

$$A_e = 7.08 \text{ sq in.}$$

$$T = 0.60F_yA_g = 22(0.625)(15) = 206 \text{ kips}$$

or

$$T = 0.50F_uA_e = 0.50(58)(7.08) = 205 \text{ kips}$$

The smaller value of 205 kips controls.

The following example illustrates a rational procedure for combining (1) the strength requirement and (2) the stiffness requirement, to select standard rolled shapes containing holes.

Example 3.9.3

A tension diagonal member for a roof truss is to be selected of A572 Grade 50 steel. The axial tension is 60 kips and the member is 12 ft long. Assume $\frac{7}{8}$-in.-diam bolts will be located on a single gage line in standard holes.
(a) Select the lightest single angle member.
(b) Select the lightest double angle member having legs separated by $\frac{1}{4}$ in. back-to-back.

SOLUTION
For angle tension members the allowable stress requirements are

$$\text{Required } A_e = \frac{T}{0.50F_u} = \frac{60}{0.5(65)} = 1.85 \text{ sq in.}$$

$$\text{Required } A_g = \frac{T}{0.60F_y} = \frac{60}{30} = 2.00 \text{ sq in.}$$

Since the effective area is $0.85A_n$ according to case (e) of Table 3.5.1,

$$\text{Required } A_n = \frac{A_e}{0.85} = \frac{1.85}{0.85} = 2.17 \text{ sq in.}$$

The net area requirement obviously controls since it exceeds the gross area requirement.

Also, the minimum r to satisfy AISC–1.8.4 may be established,

$$\text{Min } r = \frac{L}{240} = \frac{12(12)}{240} = 0.6 \text{ in.}$$

(a) Select single angle member. The required gross area in each case depends on the area deducted for one hole, which in turn depends on the thickness. The following tabular procedure may be found useful in making the selection:

Standard Thickness t	Reduction for One Hole	Required Gross Area	Choices from AISC Manual Single Angle Properties
1/4 in.	0.250*	2.42‡	L6×4×$\frac{1}{4}$, $A = 2.44$, $r = 0.89$†
5/16	0.313	2.48	L5×3$\frac{1}{2}$×$\frac{5}{16}$, $A = 2.56$, $r = 0.77$
3/8	0.375	2.55	
7/16	0.438	2.61	
1/2	0.500	2.67	

* $(\frac{7}{8}+\frac{1}{8})0.25 = 0.25$ sq in.
† *Note:* Min $r = r_z$ for single angles
‡ Required $A_g = 2.17 + 0.25 = 2.42$.

In searching for possible choices, the $L6 \times 4 \times \frac{1}{4}$ is the lightest with at least the required area. The angles $L4 \times 4 \times \frac{5}{16}$ and $L5 \times 3 \times \frac{5}{16}$, having an area $A_g = 2.40$ sq in., would both be slightly more than 3% understrength.

Use $L6 \times 4 \times \frac{1}{4}$ ($L152 \times 102 \times 6.4$) single angle member.

(b) Select double angle member. For this type of section two holes must be deducted. Selection should be made from the double angle properties in AISC Manual.

Standard Thickness t	Reduction for Two Holes	Required Gross Area	Choices from AISC Manual Double Angle Properties
1/4	0.500	2.67	$L3 \times 2\frac{1}{2} \times \frac{1}{4}$, $A = 2.63$, $r = 0.95$
5/16	0.626	2.80	$L3 \times 2 \times \frac{5}{16}$, $A = 2.93$, $r = 0.77$
3/8	0.750	2.92	

Use $2 - L3 \times 2\frac{1}{2} \times \frac{1}{4}$ ($L76 \times 64 \times 6.4$) with long legs back-to-back. The 1.5% understrength is generally considered acceptable.

SELECTED REFERENCES

1. John W. Fisher and John H. A. Struik, *Guide to Design Criteria for Bolted and Riveted Joints.* New York: John Wiley & Sons, Inc., 1974, Chapter 5.
2. V. H. Cochrane, "Rules for Rivet Hole Deductions in Tension Members," *Engineering News-Record*, 89 (Nov. 16, 1922), 847–848.
3. W. M. Wilson, Discussion of "Tension Tests of Large Rivet Joints," *Transactions*, ASCE, 105 (1942), p. 1268.
4. W. M. Wilson, W. H. Munse, and M. A. Cayci, "A Study of the Practical Efficiency under Static Loading of Riveted Joints Connecting Plates," U. of Illinois Engg. Experiment Station Bulletin 402, 1952.
5. F. W. Schutz, "Effective Net Section of Riveted Joints," Proc. Second Illinois Structural Engg. Conf., November 1952.
6. "Here's a Better Way to Design Splices," *Engineering News-Record*, 150, Part I (Jan. 8, 1953), 41.
7. P. P. Bijlaard, Discussion of "Investigation and Limit Analysis of Net Area in Tension," *Transactions*, ASCE, 120 (1955), 1156–1163.
8. G. W. Brady and D. C. Drucker, "Investigation and Limit Analysis of Net Area in Tension," *Transactions*, ASCE, 120 (1955), 1133–1154.
9. W. H. Munse and E. Chesson, Jr., "Riveted and Bolted Joints: Net Section Design," *Journal of Structural Division*, ASCE, 89, ST2 (February 1963), 107–126.
10. E. Chesson and W. H. Munse, "Behavior of Riveted Connections in Truss-Type Members," *Journal of Structural Division*, ASCE, 83, ST1 (January 1957).
11. William McGuire, *Steel Structures*, Englewood Cliffs, N.J.: Prentice-Hall, Inc., 1968 (pp. 310–328).

PROBLEMS*

Note: For all problems assume fastener strength is adequate and does not control. Use AISC Specifications and consider all bolt holes as standard unless otherwise indicated. Values of tensile strength F_u for the various steels are given in Table 2.1.1.

3.1. Determine the allowable tensile load on an $L6 \times 4 \times \frac{3}{4}$ ($L152 \times 102 \times 19.0$) using (a) A36 steel, (b) A572 Grade 50 steel. Assume welded connections so there are no holes.

3.2. Determine the allowable tensile load on the angle in Prob. 3.1 if a single gage line of $\frac{7}{8}$-in.-diam bolts is used in the 4 in. leg and a double gage line of $\frac{7}{8}$-in.-diam bolts is used in the 6-in. leg. Assume no stagger of bolts in the lines, and that all bolts participate in carrying the tensile load.

3.3. Determine the allowable tensile load on an A36 steel plate $\frac{1}{4}$ in. $\times$ 12 in. having a single line of holes parallel to the direction of loading. $\frac{7}{8}$-in.-diam bolts are to be used.

3.4. Find the effective net area A_e for the splice plate shown in the accompanying figure and determine the maximum value for T if A36 is specified and the holes are $\frac{13}{16}$-in.-diam. (20-mm-diam)

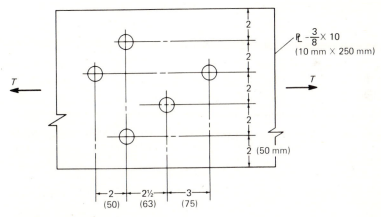

Probs. 3.4 & 3.5

3.5. Repeat Prob. 3.4 using A572 Grade 60 steel and $\frac{15}{16}$-in.-diam. holes. (24-mm-diam)

3.6. Select a pair of angles to support a tensile load of 90 kips using A36 steel. Assume a $\frac{3}{8}$-in. gusset plate between the angles at the connection, and that the connection is to be welded. ($T = 400$ kN; 10-mm gusset plate)

---

* Many problems may be solved as either a problem stated in U.S. customary units or as a problem in SI units using the numerical data in parenthesis at the end of the statement. The conversions are only approximate in order to avoid having the given data imply a degree of accuracy greater in SI than in U.S. customary units.

3.7. Repeat Prob. 3.6 using A572 Grade 60 steel.

3.8. Select a single angle of A36 steel to support a tensile load of 55 kips assuming a single gage line of $\frac{3}{4}$-in.-diam bolts, and at least 3 bolts are to be used. ($T = 250$ kN; 20-mm-diam bolts)

3.9. Repeat Prob. 3.8 using A572 Grade 50 steel.

3.10. Select a standard threaded rod to carry a tensile force T of 10 kips. Use A572 Grade 50 steel. ($T = 44$ kN)

3.11. Select a standard threaded rod to carry a tensile force T of 6 kips. Use A36 steel. ($T = 27$ kN)

3.12. Design sag rods to support the purlins of an industrial building roof. Sag rods are placed at $\frac{1}{3}$ points between roof trusses, which are spaced 30 ft apart. Assume roofing and purlin weight is 9 psf of roof surface. Use standard threaded rods and A36 steel. The snow load to be carried is 20, 30, or 40 psf of horizontal projection, whichever is appropriate for your locale.

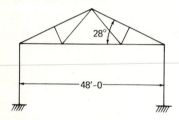

Prob. 3.12

3.13. Repeat Prob. 3.8 if the length of this main member is 20 ft. (6 m)

3.14. Repeat Prob. 3.6 if the length of this main member is 30 ft. (11 m)

3.15. Determine the maximum allowable tensile load for a single C15 × 33.9 fastened to a $\frac{1}{2}$-in. gusset plate. Use A36 steel and assume holes are for $\frac{5}{8}$-in.-diam bolts. (C380 × 50.4, 13-mm gusset, 16-mm-diam bolts)

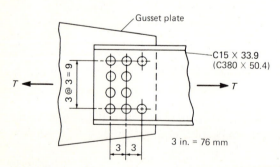

Prob. 3.15

3.16. Repeat Prob. 3.15 using an MC18 × 42.7. (MC460 × 63.5)

3.17. Determine the tensile load permitted by AISC for a pair of L6×4×$\frac{3}{8}$ angles with the standard gage distances shown in Table 3.4.1. Use A36 steel and $\frac{3}{4}$-in.-diam bolts. The force T is transmitted to the gusset plate by the fasteners on lines A and B, assume only open holes in the outstanding legs. (L152×102×9.5; 19-mm-diam bolts)

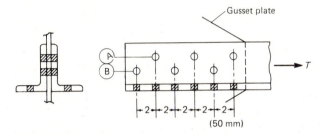

Prob. 3.17

3.18. Repeat Prob. 3.17 using a pair of L8×6×$\frac{3}{4}$ angles and $\frac{7}{8}$-in.-diam bolts.

3.19. Given the splice shown:
    (a) Determine the maximum capacity T based on the A36 steel plates having holes arranged as shown.
    (b) What value of s should be specified to provide the maximum capacity T as computed in part (a), if the final design is to have $s_1 = s_2 = s$?

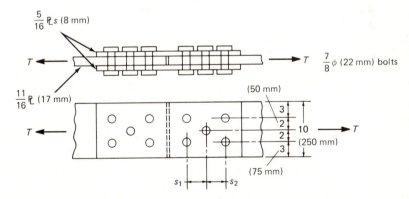

Prob. 3.19

3.20. An L5×3$\frac{1}{2}$×$\frac{1}{2}$ angle is to carry a load of 90 kips with the shortest length of connection using two gage lines of bolts in the 5-in. leg. What is the minimum acceptable stagger, theoretical and specified ($\frac{1}{2}$-in. multiples), using A572 Grade 50 steel? (L127×89×12.7; $T = 400$ kN; 12 mm multiples)

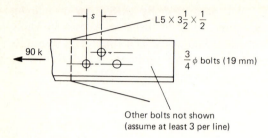

Other bolts not shown
(assume at least 3 per line)

Prob. 3.20

3.21. An $L5 \times 3\frac{1}{2} \times \frac{1}{2}$ angle is to carry a load of 80 kips. Using one gage line of holes for $\frac{7}{8}$-in.-diam bolts in each leg, what would be the minimum stagger s required to accomplish this? Consider the load to be transferred by bolts in the 5-in. leg, while in the $3\frac{1}{2}$-in. leg the holes may be considered open ones (i.e., not to transmit the tensile load). Use A36 steel. ($L127 \times 89 \times 12.7$; $T = 350$ kN; 22-mm-diam bolts)

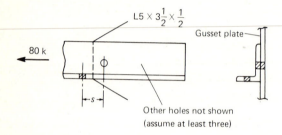

Other holes not shown
(assume at least three)

Prob. 3.21

3.22. What is the minimum value of s that could be used on the angle of the accompanying figure such that the maximum tensile force T may be carried?

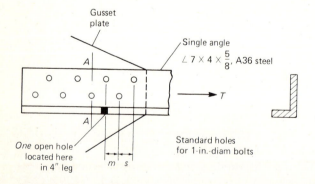

Probs. 3.22, 3.23 & 3.24

3.23. Assuming that s for the angle of Prob. 3.22 is made large enough that a staggered path involving s will not govern the failure mode, determine the minimum distance m required so that the open hole in the 4-in. leg will not reduce the strength below its maximum possible value.

3.24. Assuming the distances s and m are large enough so that staggered paths involving either s or m will not govern the strength of the angle, compute the nominal stress on the controlling section passing through the hole at section A–A.

3.25. Design an eyebar to carry 100 kips, using flame-cut A572 Grade 50 steel plate. (Refer to AISC–1.14.5) ($T = 450$ kN)

4
Structural Fasteners

4.1 TYPES OF FASTENERS

Every structure is an assemblage of individual parts or members that must be fastened together, usually at the member ends, by some means. Welding is one such means and is treated in Chapter 5. The other is to use fasteners, such as rivets or bolts. This chapter is primarily concerned with bolting; in particular, high-strength bolts. High-strength bolts have for the most part replaced rivets as the principal means of making nonwelded structural connections. However, for completeness, a brief description of the other fasteners, including rivets and unfinished machine bolts, is given.

High-Strength Bolts

The two basic types of high-strength bolts are designated by ASTM as A325 and A490, the material properties of which are discussed in Sec. 2.2 and summarized in Table 4.1.1. These bolts are heavy hexagon-head bolts, used with heavy semifinished hexagon nuts, as shown in Fig. 4.1.1b. The threaded portion is shorter than for bolts in nonstructural applications, and may be cut or rolled. A325 bolts are of heat-treated *medium*

Table 4.1.1 Properties of Bolts

ANSI/ASTM Designation	Bolt Diameter in. (mm)	Proof Load,[a] Length Measurement[b] Method, ksi (MPa)	Proof Load,[a] Yield Strength[c] Method, ksi (MPa)	Minimum Tensile Strength, ksi (MPa)
A307[d], low-carbon steel Grades A and B	$\frac{1}{4}$ to 4 (6.35 to 104)	—	—	60
A325[e], high-strength steel Types 1, 2, and 3	$\frac{1}{2}$ to 1 (12.7 to 25.4)	85 (585)	92 (635)	120 (825)
Types 1, 2, and 3	$1\frac{1}{8}$ to $1\frac{1}{2}$ (28.6 to 38.1)	74 (510)	81 (560)	105 (725)
A449[f], high-strength steel (Note: AISC permits use only for bolts larger than $1\frac{1}{2}$ in. and for threaded rods and anchor bolts)	$\frac{1}{4}$ to 1 (6.35 to 25.4)	85 (585)	92 (635)	120 (825)
	$1\frac{1}{8}$ to $1\frac{1}{2}$ (28.6 to 38.1)	74 (510)	81 (560)	105 (725)
	$1\frac{3}{4}$ to 3 (6.35 to 76.2)	55 (380)	58 (400)	90 (620)
A490[g], quenched and tempered alloy steel	$\frac{1}{2}$ to $1\frac{1}{2}$ (12.7 to 38.1)	120 (825)	130 (895)	150 (1035)

[a] Actual proof load and tensile load obtained by multiplying given stress value by the tensile stress area A_s; $A_s = 0.7854 \, [D - (0.9743/n)]^2$, where A_s = stress area in square inches, D = nominal diameter of bolt in inches, and n = number of threads per inch.
[b] 0.5% extension under load.
[c] 0.2% offset value.
[d] ANSI/ASTM A307–78.
[e] ANSI/ASTM A325–78a
[f] ANSI/ASTM A449–78a.
[g] ANSI/ASTM A490–78.

carbon steel having an approximate yield strength of 81 to 92 ksi (558 to 634 MPa) depending on diameter. A490 bolts are also heat-treated but are of *alloy* steel having an approximate yield strength of 115 to 130 ksi (793 to 896 MPa) depending on diameter. A449 bolts are occasionally used when diameters over $1\frac{1}{2}$ in. up to 3 in. are needed, and also for anchor bolts and threaded rods.

High-strength bolts range in diameter from $\frac{1}{2}$ to $1\frac{1}{2}$ in. (3 in. for A449). The most common diameters used in building construction are $\frac{3}{4}$ in. and $\frac{7}{8}$ in., whereas the most common sizes in bridge design are $\frac{7}{8}$ in. and 1 in.

High-strength bolts are tightened to develop a specified tensile stress in them which results in a predictable clamping force on the joint. The

a) Rivet   b) High-strength   c) High-strength inter-
                hexagon head bolt   ference-body bolt

Fig. 4.1.1 Types of fasteners.

actual transfer of service loads through a joint is therefore due to the friction developed in the pieces being joined. Joints containing high-strength bolts are designed either as *friction type*, where high slip resistance is desired; or as *bearing type*, where high slip resistance is unnecessary.

Rivets

For many years rivets were the accepted means of connecting members but in recent years have become virtually obsolete in the United States. Undriven rivets are formed from bar steel, a cylindrical shaft with a head formed on one end, as shown in Fig. 4.1.1a. Rivet steel is a mild carbon steel designated by ASTM as A502 Grade 1 ($F_y = 28$ ksi) (190 MPa) and Grade 2 ($F_y = 38$ ksi) (260 MPa), with the minimum specified yield strengths based on bar stock as rolled. The forming of undriven rivets and the driving of rivets cause changes in the mechanical properties.

The method of installation is essentially that of heating the rivet to a light cherry-red color, inserting it into a hole and then applying pressure to the preformed head while at the same time squeezing the plain end of the rivet to form a rounded head. During this process the shank of the rivet completely or nearly fills the hole into which it had been inserted. Upon cooling, the rivet shrinks, thereby providing a clamping force. However, the amount of clamping produced by the cooling of the rivet varies from rivet to rivet and therefore cannot be counted on in design calculations. Rivets may also be installed cold but then they do not develop the clamping force since they do not shrink after driving.

Unfinished Bolts

These bolts are made from low-carbon steel, designated as ASTM A307, and are the least expensive type of bolt. They may not, however, produce the least expensive connection since many more may be required in a particular connection. Their primary use is in light structures, secondary or bracing members, platforms, catwalks, purlins, girts, small trusses, and similar applications in which the loads are primarily small and static in nature. Such bolts are also used as temporary fitting up connectors in cases where high-strength bolts, rivets, or welding may be the permanent means of connection. Unfinished bolts are sometimes called common, machine, or rough bolts and may come with square heads and square nuts.

Turned Bolts

These practically obsolete bolts are machined from hexagonal stock to much closer tolerances (about $\frac{1}{50}$ in.) than unfinished bolts. This type of bolt was primarily used in connections which required close-fitting bolts in drilled holes, such as in riveted construction where it was not possible to drive satisfactory rivets. They are sometimes useful in aligning mechanical equipment and structural members which require precise positioning. They are now (1979) rarely if ever used in ordinary structural connections, since high-strength bolts are better and cheaper.

Ribbed Bolts

These bolts of ordinary rivet steel which have a rounded head and raised ribs parallel to the shank were used for many years as an alternative to rivets. The actual diameter of a given size of ribbed bolt is slightly larger than the hole into which it is driven. In driving a ribbed bolt, the bolt actually cuts into the edges around the hole producing a relatively tight fit. This type of bolt was particularly useful in bearing connections and in connections that had stress reversals.

A modern variation of the ribbed bolt is the *interference-body bolt* shown in Fig. 4.1.1c which is of A325 bolt steel and instead of longitudinal ribs has serrations around the shank as well as parallel to the shank. Because of the serrations around the shank through the ribs, this bolt is often called an *interrupted-rib* bolt. Ribbed bolts were also difficult to drive when several layers of plates were to be connected. The current high-strength A325 interference-body bolt may also be more difficult to insert through several plates; however, it is used when tight fit of the bolt in the hole is desired and it permits tightening the nut without the simultaneous holding of the bolt head as may be required with smooth loose-fitting ordinary A325 bolts.

4.2 HISTORICAL BACKGROUND OF HIGH-STRENGTH BOLTS

The first experiments indicating the possibility of using high-strength bolts in steel-framed construction were reported by Batho and Bateman [1] in 1934. Batho and Bateman concluded that bolts with a minimum yield strength of 54 ksi (372 MPa) could be relied on to prevent slippage of the connected parts. Follow-up tests by Wilson and Thomas [2] substantiated the earlier work by reporting that high-strength bolts smaller in diameter than the holes in which they were inserted had fatigue strengths equal to that of well-driven rivets provided that the bolts were sufficiently pretensioned.

The next major step occurred in 1947 with the formation of the Research Council on Riveted and Bolted Structural Joints. This organization began by using and extrapolating information from studies of riveted joints; in particular, the extensive annotated Bibliography by De Jonge [3], completed in 1945, was used. From this beginning, the Research Council has continued to organize and sponsor research on high-strength bolted connections, and issue specifications at intervals on the basis of research findings.

In 1948 the American Railway Engineering Association (AREA) also became interested and initiated studies on the use of high-strength bolts in railroad bridge maintenance. In the same year the Association of American Railroads initiated a number of field test installations confirming the adequacy of connections made with high-strength bolts.

By 1950 the concept of using high-strength bolts and a summary of research and behavior was presented [4] to practicing engineers and the steel-fabrication industry. As the next step, in 1951 the Research Council published its first specifications, permitting the replacement of rivets with bolts on a one-to-one basis. It was conservatively assumed that friction transfer of the load was necessary in all joints under service load conditions. The factor of safety against slip was established at a high enough level so that good fatigue resistance (i.e., no slip under varying stress or stress reversal consisting of many load cycles) was provided in every joint, similar to or better than that shown by riveted joints.

In 1954 a revision was made in the specifications to include the use of flat washers on 1:20 sloping surfaces and to allow the use of impact wrenches for installing high-strength bolts. Also, the 1954 revision permitted the surfaces in contact to be painted when the bolts were to create a *bearing-type* connection; i.e., when the ultimate strength of the connection was to be based on the bolt in bearing against the side of the hole.

In 1956 W. H. Munse [5] summarized bolt behavior and concluded that if high-strength bolts are to be as efficient and economical as possible, an initial tension as high as practicable must be induced in the bolts. By 1960 much additional research justified increasing the minimum bolt tension, recognized that the *bearing-type* connection was ordinarily an acceptable substitute for a riveted connection, and accepted that the connection having its design strength based on slip resistance, known as a *friction-type* connection, may only be necessary when direct tension acts on the bolts or when fatigue conditions are important. Also, a turn-of-the-nut procedure was introduced as an alternative to the torque wrench method for tightening the high-strength bolts. The previous requirement of two washers was reduced to require only one under the element (head or nut) being turned, if the turn-of-the-nut method of installation was used.

In 1962 the Research Council again liberalized the Specifications, completely deleting the requirement for washers [6]. In 1964, the A490 bolt was included in the specification for use with high-strength steels, while in 1966 the required initial tension for the A490 bolts was reduced.

The most recent revision by the Research Council [7] was probably the most significant change since 1951. Based on the *Guide to Design Criteria for Bolted and Riveted Joints* by Fisher and Struik [8] (hereafter referred to as the *Guide*), the changes include increased allowable capacities, explicit recognition of other than standard size holes for fasteners, and changes in the philosophy of design for friction-type and bearing-type connections. The changes have the objective of providing a more uniform reliability in connection performance for all types of loading and joint arrangements. This entire chapter involves new concepts and requirements relating to the 1976 Research Council Specifications [7] that are also included in the 1978 AISC Specification.

4.3 CAUSES OF RIVET OBSOLESCENCE

Riveting is a method of connecting a joint by inserting ductile metal pins into holes in the pieces being joined and forming a head at each end to prevent the joint from coming apart. Typical types of rivets are shown in Fig. 4.3.1 (see also Fig. 4.1.1a). The principal causes for rivet obsolescence have been the advent of high-strength bolts and development of welding techniques. The development of welding procedures is discussed in Chapter 5 and the development of high-strength bolting forms the basis for this chapter. However, in addition to the above-mentioned causes, a number of inherent disadvantages have hastened the obsolescence of structural riveting, especially field riveting.

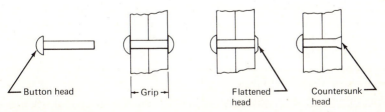

Button head   Grip   Flattened head   Countersunk head

Fig. 4.3.1   Types of rivets.

Riveting crews require 4 or 5 men each of whom must be experienced. Such experienced crews have not always been available in all sections of the country. On the other hand, the crews required for high-strength bolts do not need to be highly skilled and are readily available. Inspection procedures in riveted connections are difficult and unfortunately even the most experienced riveting crews require rigid inspection. Cutting out and replacing bad rivets is an expensive procedure. Even the preheating immediately prior to driving is critical in developing the necessary tightness after cooling.

Some additional disadvantages of rivets are the ever-present danger of fire and the high level of noise caused by driving the rivets. The level of sound produced by riveting can be very disruptive to business and has caused many expensive lawsuits.

Historically rivets had a higher fatigue strength than unfinished bolts because the rivets more or less completely filled the holes as a result of the driving and head-forming procedure. High fatigue strength is particularly important to structures in which there is a cycling of stresses or stress reversals as in the case of highway and railroad bridges.

The principal factor that delayed immediate acceptance of high-strength bolts was the cost of the material. High-strength A325 bolts are about three times the cost of rivets. In addition, the hardened washers added cost. In the early 1950s the reduced labor cost for installing bolts did not offset the higher bolt material cost. Once the washers could be reduced to one or eliminated, and the greater strength of a bolt over that of a rivet could be utilized in design, high-strength bolts became economical. Now (1979) with ever higher labor costs and connection design generally requiring fewer bolts than would be required for rivets, the economy is clearly with high-strength bolts.

Shop fabrication using rivets persisted long after field erection use of high-strength bolts was standard for nonwelded connections. Since extensive labor was involved in the shop fabrication of the holes little extra labor was needed for riveting. By about 1970, welding procedures had advanced so that shop fabrication is nearly always welded. In addition, properly welded joints are simpler and better.

4.4 DETAILS OF HIGH-STRENGTH BOLTS AND INSTALLATION PROCEDURES

Both the most commonly used A325 bolt and the occasionally used A490 bolt are heavy hexagon head bolts, identified by the ASTM designation and the manufacturer's symbol marked on the top of the head as shown in Fig. 4.4.1. Both have heavy hexagon nuts, containing standard markings along with the manufacturer's symbol on one of the faces.

Heavy hex bolts have shorter threaded portions than other standard bolts; this reduces the probability of having threads occurring where

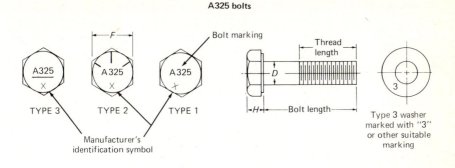

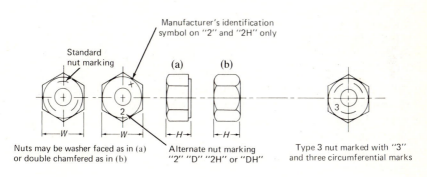

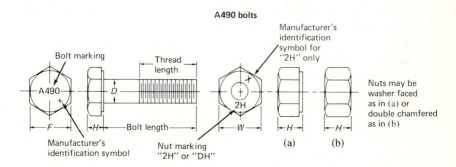

Fig. 4.4.1   Control dimensions for A325 and A490 bolts. (From Ref. 7)

maximum strength is required across the shank of the bolt. The control dimensions, thread lengths, and standard identifications for A325 and A490 bolts are given in Fig. 4.4.1 and Table 4.4.1.

The A325 and A490 bolts are available as Types 1, 2, or 3. The A325 Type 1 is the medium-carbon steel bolt that has been available for many years and is the one that would be supplied if not otherwise specified. Similarly for A490, Type 1 is the ordinary alloy steel bolt. Type 2 for both A325 and A490 is a low-carbon martensite steel alternative to

Table 4.4.1 A325 and A490 Bolt Dimensions (from Ref. 7)

Nominal bolt size, D (in.)	Bolt Dimensions (in.)			Nut Dimensions (in.)	
	Heavy Hex Structural Bolts			Heavy Hex Nuts	
	Width across flats, F	Height, H	Thread length	Width across flats, W	Height, H
1/2	7/8	5/16	1	7/8	31/64
5/8	1 1/16	25/64	1 1/4	1 1/16	39/64
3/4	1 1/4	15/32	1 3/8	1 1/4	47/64
7/8	1 7/16	35/64	1 1/2	1 7/16	55/64
1	1 5/8	39/64	1 3/4	1 5/8	63/64
1 1/8	1 13/16	11/16	2	1 13/16	1 7/64
1 1/4	2	25/32	2	2	1 7/32
1 3/8	2 3/16	27/32	2 1/4	2 3/16	1 11/32
1 1/2	2 3/8	15/16	2 1/4	2 3/8	1 15/32

Type 1 for atmospheric temperature applications. For elevated temperature applications Type 1 must be used. Type 3 for both A325 and A490 is a weathering steel bolt having corrosion resistance comparable to A588 weathering steel (see Table 2.1.1).

Occasionally, ASTM A449 bolts are used where diameters larger than $1\frac{1}{2}$ in. are required.

A variation of the standard A325 and A490 used for bearing connections is the *interference-body* bolt described in Sec. 4.1. These bolts have a button head and may be used with the standard high-strength nuts or with self-locking nuts. The interference-body type bolts have particular applications where high bearing capacity is desired together with stress reversals or vibratory loads.

Proof Load and Bolt Tension

The primary requirement when installing high-strength bolts is to provide a sufficient *pretension* force. The pretension should be as high as possible without chancing permanent deformation or failure of the bolt. Bolt material exhibits a stress-strain (load-deformation) behavior that has no well-defined yield point, as shown in Fig. 4.4.2. Instead of directly using a yield stress, a so-called *proof load* is used. The *proof load* is the load obtained by multiplying the tensile stress area* by a yield stress established by using either a 0.2% offset strain or a 0.5% extension under load (see Sec. 2.4). ASTM tabulates this proof load for each diameter fastener using, for example, for $\frac{1}{2}$ to 1-in.-diam bolts a strain offset value of 92 ksi (634 MPa) and a length measurement value of 85 ksi (586 MPa). The proof load stress is approximately a minimum of

---

* See footnote to Table 4.1.1.

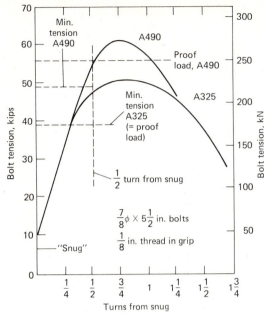

Fig. 4.4.2 Typical load vs nut rotation relationships for A325 and A490 bolts. (From Ref. 8)

70% and 80% of the minimum tensile strengths for A325 and A490 bolts, respectively.

Since the early 1950s the minimum required pretension equals the proof load for A325 bolts. Using the turn-of-the-nut installation method (as discussed later in this section) no difficulty is encountered in obtaining proof load for these bolts with $\frac{1}{2}$ turn of the nut from snug position, as shown in Fig. 4.4.2. With the A490 bolt, however, the $\frac{1}{2}$ turn from snug may not achieve the proof load. Also for long bolts, more than $\frac{1}{2}$ turn from snug will be required to achieve the same tension as for short bolts.

Thus AISC–1.23.5 requires a pretension equal to 70% of minimum tensile strength, as given in Table 4.4.2. This equals the proof load for A325 bolts and about 85 to 90% of proof load for A490 bolts.

The magnitude of pretension that is desirable and necessary has been the subject of considerable study by researchers [9–11].

Installation Techniques

The three general techniques for obtaining the required pretension indicated in Table 4.4.2 are the *calibrated wrench* method, the *turn-of-the-nut* method, and the *direct tension indicator* method.

The calibrated wrench method includes the use of manual torque wrenches and power wrenches adjusted to stall at a specified torque

Table 4.4.2 Minimum Bolt Tension*
(AISC–Table 1.23.5)

| Bolt Size | | A325 Bolts | | A490 Bolts | |
(in.)	(mm)	(kips)	(kN)	(kips)	(kN)
1/2	12.7	12	53	15	67
5/8	15.9	19	85	24	107
3/4	19.1	28	125	35	156
7/8	22.2	39	173	49	218
1	25.4	51	227	64	285
1 1/8	28.6	56	249	80	356
1 1/4	31.8	71	316	102	454
1 3/8	34.9	85	378	121	538
1 1/2	38.1	103	458	148	658

* Equal to 70% of specified minimum tensile strengths
of bolts, rounded off to the nearest kip.

value. Variations in bolt tension produced by a given torque have been found [12–14] to be as high as ±30% with an average variation of ±10%. The Research Council [8] has therefore recommended that torque or calibrated wrenches be set to produce a bolt tension at least 5% in excess of the values indicated in Table 4.4.2.

Beginning in the early 1950s and continuing into the 1960s the turn-of-the-nut method was developed whereby the pretensioning force in the bolt is obtained by a *specified rotation* of the nut from an initially *snug tight* position which causes a specific amount of strain in the bolt. According to the Research Council [8], a nut is considered to be "snug tight" after "a few impacts of an impact wrench or the full effort of a man using an ordinary spud wrench." Although snugness or initial tightness can vary due to the condition of the surfaces being tightened, this variation is not significant as can be seen from Fig. 4.4.3 taken from Fisher, Ramseier, and Beedle [15]. The clamping force of 48.6 kips (216 kN) corresponding to half a turn-of-the-nut is seen to occur sufficiently far along the flat portion of the tension curve to make any variation in the snug tension insignificant to the overall behavior of the bolt.

One may wonder whether there is danger of having inadequate reserve strength if the pretension exceeds the proof load; i.e., when it approaches 90% of tensile strength. Figure 4.4.4 from Rumpf and Fisher [9] shows the effect of various turns of the nut with the margin of safety indicated. If the calibrated wrench method is used, *strength* is the critical factor, with the typical safety margin shown in Fig. 4.4.4. The possibility of overtorquing the bolts with power wrenches is not considered a problem since such overtorquing usually fractures the bolts and they are

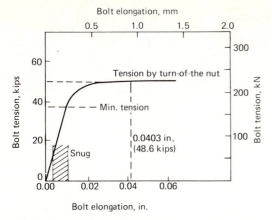

Fig. 4.4.3 Bolt elongations in a typical test joint. (From Ref. 15)

replaced during installation. In the turn-of-the-nut method, *deformation* is the critical factor with the typical safety margin shown in Fig. 4.4.4. For either installation process one can expect a minimum of $2\frac{1}{4}$ turns from snug to fracture. When the turn-of-the-nut method is used and bolts are tensioned using $\frac{1}{8}$ turn increments, frequently as many as four turns may be obtained from snug to fracture. The turn-of-the-nut method is the cheapest, is more reliable, and is generally the preferred method. The approved nut rotations as given by the Research Council [7] are indicated in Table 4.4.3.

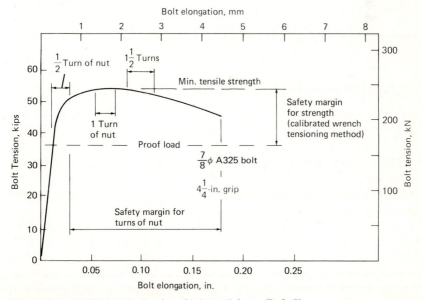

Fig. 4.4.4 A325 bolt behavior. (Adapted from Ref. 9)

Table 4.4.3 Nut Rotation* from Snug Tight Condition (from Ref. 7)

Bolt Length (as measured from underside of head to extreme end of point)	Disposition of Outer Faces of Bolted Parts		
	Both faces normal to bolt axis	One face normal to bolt axis and other face sloped not more than 1:20 (bevel washer not used)	Both faces sloped not more than 1:20 from normal to bolt axis (bevel washers not used)
Up to and including 4 diameters	$\frac{1}{3}$ turn	$\frac{1}{2}$ turn	$\frac{2}{3}$ turn
Over 4 diameters but not exceeding 8 diameters	$\frac{1}{2}$ turn	$\frac{2}{3}$ turn	$\frac{5}{6}$ turn
Over 8 diameters but not exceeding 12 diameters†	$\frac{2}{3}$ turn	$\frac{5}{6}$ turn	1 turn

* Nut rotation is relative to bolt, regardless of the element (nut or bolt) being turned. For bolts installed by $\frac{1}{2}$ turn and less, the tolerance should be plus or minus 30°; for bolts installed by $\frac{2}{3}$ turn and more, the tolerance should be plus or minus 45°.
† No research work has been performed by the Council to establish the turn-of-the-nut procedure when bolt lengths exceed 12 diameters. Therefore the required rotation must be determined by actual tests in a suitable tension device simulating the actual conditions.

The third and most recent method for establishing bolt tension is by the *direct tension indicator.* This device is a hardened washer with a series of protrusions on one face. The washer is inserted between the bolt head and the gripped material with the protrusions bearing against the underside of the bolt leaving a gap maintained by the protrusions. Upon tightening the bolt, the protrusions are flattened and the gap is reduced. The bolt tension is determined by measuring the remaining gap, which for properly tensioned bolts will be about 0.015 in. (0.38 mm) or less [16].

4.5 LOAD TRANSFER BY STRUCTURAL FASTENERS

Loads are transferred from one member to another by means of the connection between them. A few typical connections are shown in Fig. 4.5.1.

The simplest device for transferring load from one steel piece to another is with a pin (a cylindrical piece of steel) inserted in holes that are aligned in the two pieces as shown in Fig. 4.5.2. The cotter pins shown would prevent the pin from sliding out. Load would be transferred by bearing of the shank of the pin against the side of the hole. From the free bodies of the pin it may be noted the transfer between piece A and B is actually made via shear on the pin (the slight rotation of the pin due to unbalanced moment would be relatively negligible). There would be negligible friction between pieces A and B. The earliest steel structures, particularly trusses, were actually connected by pins.

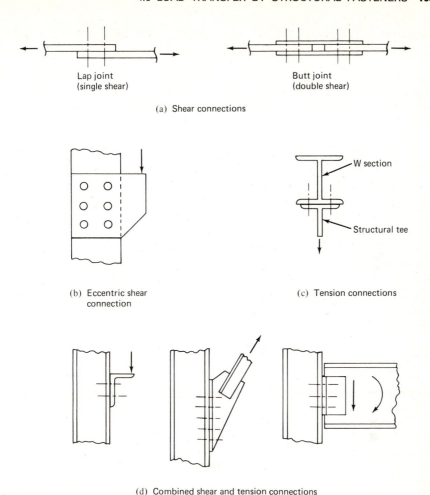

(a) Shear connections

(b) Eccentric shear connection

(c) Tension connections

(d) Combined shear and tension connections

Fig. 4.5.1 Typical bolted connections.

A riveted connection behaves differently. Cooling of the rivet tends to shorten it, but since the underside of the head of the rivet bears against the pieces being joined and restrains the shortening, a tensile force is induced in the rivet. For equilibrium, the tensile force in the rivet is balanced by a compression in the pieces being joined; in effect there is a clamping force on the joint. Free-body diagrams showing these forces are in Fig. 4.5.3. Transfer of load between plates A and B occurs through a combination of frictional resistance (μT) and bearing (B) of the shank against the side of the hole. The riveting process involves considerable variability, so that the initial tension in rivets is variable, giving a variable friction force (coefficient of friction μ times clamping force T). The design process historically has neglected the frictional force.

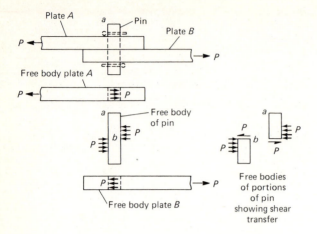

Fig. 4.5.2   Transfer of load in pin connection.

The high-strength bolt behaves in a manner similar to the rivet. The bolt, however, is installed by one of the methods discussed in Sec. 4.4 to have a specified initial tension. Since the specified tension is large enough to give a frictional force capable of transferring the entire load, bearing of the bolt shank against the side of the hole will generally not occur under service load. The free-body diagrams for the transfer of load in a high-strength bolted connection are shown in Fig. 4.5.4.

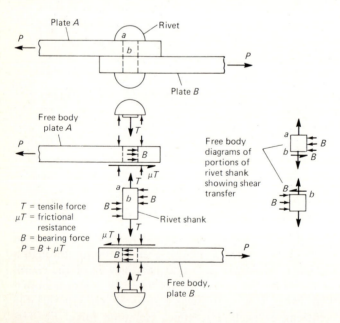

Fig. 4.5.3   Transfer of load in riveted connection.

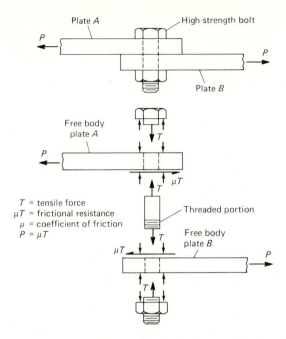

Fig. 4.5.4 Transfer of load in high-strength bolted connection.

Actual connections consist of more than one fastener (sometimes a large number of fasteners). This means that as long as friction alone can transfer the load, all fasteners participate equally in transmitting the load (assuming they are the same size and material). However, after overload, once the frictional resistance is inadequate to transfer the load, bearing against the side of the hole will occur. When failure of the connection is imminent, the frictional force will not greatly affect the failure mode. Instead, the plate strength and the bolt tensile and shear strengths will control the connection strength. The possible failure modes are shown in Fig. 4.5.5.

Nominal Stresses

For many years, the design of connections has been based on the behavior when failure is imminent rather than on the behavior at service load, *even though the calculations are made using service loads.* Essentially the modes indicated in Fig. 4.5.5 are used to compute unit stresses. Since the behavior assumed for the calculation does not occur when the connection is behaving elastically, such stresses are *not* real stresses but instead merely serve as criteria of safety. Such stresses used in design calculations are known as *nominal stresses.* These nominal stresses were used in Chapter 3 when calculating the stress on the net section through holes in members.

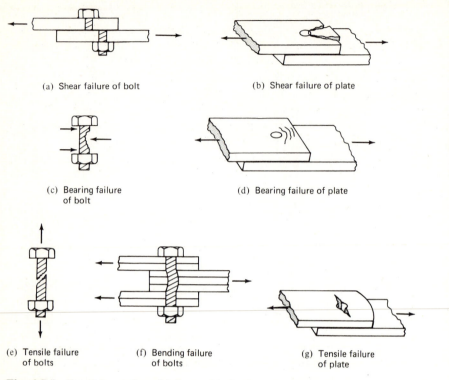

(a) Shear failure of bolt

(b) Shear failure of plate

(c) Bearing failure of bolt

(d) Bearing failure of plate

(e) Tensile failure of bolts

(f) Bending failure of bolts

(g) Tensile failure of plate

Fig. 4.5.5 Possible modes of failure of bolted connections.

The nominal stress approach used in design focuses on the individual fastener capacity. This means all fasteners of the same size and material are treated as having equal strength in transmission of loads. Thus when, for instance, five fasteners are acting in a line to transmit load in a tension lap joint, as in Fig. 4.5.6, each fastener participating equally transmits $\frac{1}{5}$ of the load. When friction is overcome and the fasteners act in bearing the plate deformation between fasteners would not actually be the same. The nominal stress assumption would mean rigid plates, otherwise the actual deformation on the fasteners could not be the same.

For design purposes, the nominal shear stress f_v and the nominal

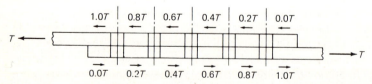

Fig. 4.5.6 Assumed forces in a row of 5 bolts for a lapped plate tension connection.

tensile stress f_t in the bolts are computed on the nominal bolt cross-sectional area. Assuming P is the load to be carried by one fastener, the equations become

$$f_v = \frac{P}{m\left(\dfrac{\pi D^2}{4}\right)} \qquad\qquad (4.5.1)$$

$$f_t = \frac{P}{\dfrac{\pi D^2}{4}} \qquad\qquad (4.5.2)$$

where P = load per bolt

D = nominal diameter of bolt

m = number of shear planes participating [usually one (*single shear*) or two (*double shear*) as in Fig. 4.5.1a].

The nominal bearing stress f_p is computed on the basis of the nominal diameter and the thickness of the plate; thus

$$f_p = \frac{P}{Dt} \qquad\qquad (4.5.3)$$

where t = thickness of plate.

In summary, the primary reasons *why nominal stresses differ from actual stresses* are:

1. Friction resistance to slip is neglected.
2. Deformation of plates in neglected.
3. Tensile stress concentrations at holes are neglected.
4. Shearing deformation of connectors is assumed proportional to shearing stress.
5. Shearing stress is assumed uniform over connector cross section.
6. Bearing stress is assumed uniform over a nominal contact surface equal to connector diameter times plate thickness.
7. Bending of connectors is neglected.

Thus it may be apparent that even if desired, the actual stresses and the force distribution at service load level in a bolted connection would be difficult to obtain. Furthermore, in addition to items 1 through 7, the actual stresses are dependent on many factors often beyond the control of the designer, such as poor alignment of holes, loose or unequal tightening, eccentric loading not anticipated, and poor construction.

In view of the above discussion, it is fortuitous that the excellent ductility inherent to steel eliminates the necessity of complex stress computations for design. Prior to reaching ultimate capacity, the bolts and plates are capable of deforming plastically an amount sufficient to redistribute any unequal forces. As failure of the connection is approached, the assumptions used for nominal stresses at service load are reasonably correct. The result is that the designer is able to proportion connections by determining the loads acting on each bolt using elementary mechanics and nominal stresses, with the knowledge that even though service load stresses are unknown the connection will have an adequate factor of safety with regard to strength.

4.6 BEARING-TYPE CONNECTIONS

Two categories of connections are provided for in the AISC Specification: *friction-type* when no slip at service load is desired, and *bearing-type* when prevention of slip is not considered necessary. In the friction-type, the serviceability condition of adequate slip-resistance must be provided, in addition to adequate strength of the connection. For the bearing-type, adequate strength of the connection is the sole criterion. Under the current AISC Specification, high-strength bolts are required to be installed in the same manner to obtain the same pretension, whether the connection is to be friction-type or bearing-type. Performance at service load is generally identical; that is, service loads are transmitted by friction between the connected pieces. Any difference in performance is entirely due to the difference in factor of safety against slip. Detailed treatment of friction-type connections appears in Sec. 4.7.

The bearing-type connection is used when it is not important if slip occurs under an occasional overload to bring the bolt shank into bearing against the side of the hole. For any subsequent loading, load is transferred by friction in combination with bearing on the plate. As long as the loading is static and does not reverse direction such slip will occur only once; thereafter the bolt is already bearing against the material at the side of the hole.

Shear Strength

In general, the allowable nominal shear stress F_v for either the friction-type or the bearing-type connection may be expressed

$$F_v = \beta_1 \beta_2 \beta_3 (\text{Basic } F_v) \qquad (4.6.1)$$

where β_1, β_2, and β_3 are serviceability factors relating to slip-resistant connections (i.e., friction-type connections). The Basic F_v is given by AISC for bearing-type connections as 30 ksi (207 MPa) for A325 bolts and 40 ksi (276 MPa) for A490 bolts, with the stress acting on the effective cross-sectional area of the bolt. When no threads are in the shear plane, the full area A_b through the bolt shank is effective. When threads are in the shear plane the tensile stress area is effective. In order to permit use of the nominal bolt area A_b for all calculations, the Basic F_v is reduced in the ratio of the tensile stress area to the nominal bolt area (approximately 0.7). Thus when threads are in the shear plane, the Basic

F_v is 21 ksi (0.70 of 30 ksi) for A325 and 28 ksi (0.70 of 40 ksi) for A490 bolts.

For bearing-type connections where slip is not important, the factors β_1, β_2, and β_3 are taken as 1.0. For slip-resistant (friction-type) connections the values for β are discussed in Sec. 4.7.

For many years, the design philosophy was to provide a balanced design, wherein the shear strength of the bolt was equated to the tensile strength of the plate material based on its net area. Thus when an allowable tensile stress of $0.60F_y$ on net area of plate material was equated to a nominal shear stress of 22 ksi (1969 AISC Spec.) for A325 bolts, the following result was obtained:

$$0.60F_y A_n = 22A_b$$

For A36 steel, $0.60F_y = 22$ ksi giving the ratio $A_n/A_b = 1.0$. However, one may note that with $A_n/A_b = 1.0$ the actual FS for a shear failure of the bolt is

$$\text{FS} = \frac{\text{shear strength}}{F_v} = \frac{0.62(\text{tensile strength})}{F_v}$$

$$= \frac{0.62(120)}{22} = 3.4$$

and the actual FS for a failure on net area is

$$\text{FS} = \frac{\text{tensile strength}}{0.60F_y} = \frac{58}{22} = 2.6$$

When $A_n/A_b = 1.31$ (i.e., 3.4/2.6) the actual strengths would be equal but would correspond to an FS greater than 3.

Figures 4.6.1 and 4.6.2 illustrate the factor of safety actually provided when various shear stresses are used for A325 and A490 bolts. Even when the 1969 AISC allowable values of 22 ksi for A325 and 32 ksi for A490 bolts were used the factor of safety varied from above 3 for short joints to about 2 for long joints. The factor of safety about 2 has given satisfactory performance for these longer joints, and is a comparable value used generally in the design of connections. Increasing the basic values of F_v to 30 ksi and 40 ksi for A325 and A490 bolts, respectively, has reduced the factor of safety closer to 2.0 for short joints and will be at least 2.0 up to about a 50-in. joint length. AISC–1.5.2.1 requires the allowable shear values to be reduced 20% in bearing-type connections when the length exceeds 50 in. between extreme fasteners measured parallel to the line of axial force. A summary of the allowable shear stress F_v values for bearing-type connections is given in the right-hand column of Table 4.6.1. Note that the designations A325–N and A325–X are used to indicate the situations where threads are *not* excluded (N) from the shear planes and where threads are excluded (X) from shear planes.

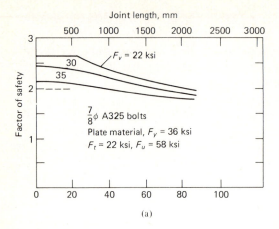

(a)

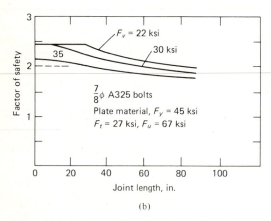

(b)

Fig. 4.6.1 Factor of safety for joints fastened by A325 bolts using various nominal shear stress values. (From Ref. 8)

Bearing Strength

In addition to adequate tensile capacity of the plate at its critical net section and adequate shear capacity of the fastener, the bearing strength of the plate material must be adequate if failure is to be prevented.

Once slip has occurred the bolt shank bears against the side of the hole. If the resistance of the plate is inadequate the hole will elongate or the fastener may actually tear through the end of the plate, as shown in Figs. 4.5.5b and d.

The end distance required to prevent the plate from splitting out may be established by equating the shear strength of the plate material to the load transmitted by the end bolt. Referring to Fig. 4.6.3, the actual tearing would occur along lines 1–1 and 2–2. As a lower bound for

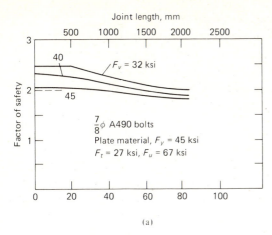

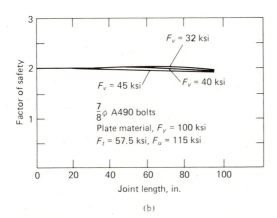

Fig. 4.6.2 Factor of safety for joints fastened by A490 bolts using various nominal shear stress values. (From Ref. 8)

strength α could be taken as zero, giving the strength as

$$P = 2t\left(L_e - \frac{D}{2}\right)\tau_u^P \qquad (4.6.2)$$

where τ_u^P = shear strength of plate material
≈ 0.70 of tensile strength F_u

Thus

$$P = 2t\left(L_e - \frac{D}{2}\right)(0.70)F_u \qquad (4.6.3)$$

The load on the fastener in terms of bearing stress f_p (Eq. 4.5.3) is

$$P = f_p D t \qquad (4.6.4)$$

Table 4.6.1 Allowable Shear Stress F_v on Fasteners
(Stress given in ksi and (MPa) to be applied to nominal bolt cross-sectional area) (from AISC–1.5.2.1)

Description of Fasteners	Friction-type Connections*			Bearing-type Connections
	Standard Size Holes	Oversize and Short-slotted Holes	Long-slotted Holes	
(1) A502, Grade 1, hot-driven rivets				17.5† (121)
(2) A502, Grade 2, hot-driven rivets				22.0† (152)
(3) A307 bolts				10.0 (69)
(4) Threaded parts meeting requirements of AISC–1.4.1 and A449 bolts when threads are *not* excluded from shear plane				$0.17F_u$
(5) Threaded parts meeting requirements of AISC–1.4.1 and A449 bolts when threads are excluded from shear plane				$0.22F_u$
(6) A325 bolts, when threading is *not* excluded from the shear planes	17.5 (121)	15.0 (103) A325–F	12.5 (86)	21.0† (145) A325–N
(7) A325 bolts, when threading is excluded from the shear planes	17.5 (121)	15.0 (103) A325–F	12.5 (86)	30.0† (207) A325–X
(8) A490 bolts, when threading is *not* excluded from the shear planes	22.0 (152)	19.0 (131) A490–F	16.0 (110)	28.0† (193) A490–N
(9) A490 bolts, when threading is excluded from the shear planes	22.0 (152)	19.0 (131) A490–F	16.0 (110)	40.0† (276) A490–X

† Values are to be reduced 20% when length of connection between extreme fasteners, measured parallel to the line of axial force, exceeds 50 in.
* Applicable for contact surfaces with *clean mill scale* (Class A) condition (see also Table 4.7.3 for other classes of surface condition).

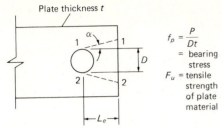

Fig. 4.6.3 Bearing strength related to end distance.

Equating Eqs. 4.6.3 and 4.6.4 gives

$$f_p Dt = 2t\left(L_e - \frac{D}{2}\right)(0.70)F_u \tag{4.6.5}$$

To prevent a tear-out failure the L_e/D must be at least the value obtained from Eq. 4.6.5,

$$\frac{L_e}{D} \geq \left[0.5 + 0.714\frac{f_p}{F_u}\right] \tag{4.6.6}$$

which is shown in Fig. 4.6.4. Equation 4.6.6 may be approximated by the following simpler expression:

$$\frac{L_e}{D} \geq \frac{f_p}{F_u} \tag{4.6.7}$$

Since a minimum factor of safety of 2 is normally used in connection design, Eqs. 4.6.6 and 4.6.7 will have FS = 2 if $2f_p$ replaces f_p; thus

$$\frac{L_e}{D} \geq \left[0.5 + 1.43\frac{f_p}{F_u}\right] \tag{4.6.8}$$

or

$$\frac{L_e}{D} \geq \frac{2f_p}{F_u} \tag{4.6.9}$$

AISC–1.5.1.5.3, 1.16.4, and 1.16.5 cover the bearing strength requirements. The basic provision is Eq. 4.6.9, wherein if $f_p = P/(Dt)$, the minimum end distance L_e is obtained:

$$L_e \geq \frac{2P}{F_u t} \tag{4.6.10}$$

where L_e is the distance from the center of a fastener to the nearest edge of an adjacent fastener or to the edge of the connected part toward which the force is directed. From this definition the L_e relates to spacing of fasteners in the line of stress as well as the end distance.

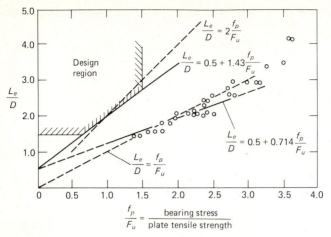

Fig. 4.6.4 Comparison of test results and design requirements for minimum end distance to prevent bearing or tear-out failure. (Adapted from Ref. 8)

Table 4.6.2 Minimum Edge Distances (center of standard hole† to edge of connected part) (adapted from AISC–Table 1.16.5.1)

Nominal Rivet or Bolt Diameter		Minimum Edge Distance			
		At Sheared Edges		At Rolled Edges of Plates, Shapes or Bars, or Gas Cut Edges*	
(in.)	(mm)	(in.)	(mm)	(in.)	(mm)
1/2	12.7	7/8	22.2	3/4	19.1
5/8	15.9	1 1/8	28.6	7/8	22.2
3/4	19.1	1 1/4	31.8	1	25.4
7/8	22.2	1 1/2	38.1	1 1/8	28.6
1	25.4	1 3/4	44.4	1 1/4	31.8
1 1/8	28.6	2	50.8	1 1/2	38.1
1 1/4	31.8	2 1/4	57.2	1 5/8	41.3
Over 1 1/4	Over 31.8	1 3/4×diam		1 1/4×diam	

* All edge distances in this column may be reduced 1/8 in. (3.2 mm) when the hole is at a point where stress does not exceed 25% of the maximum allowed stress in the element.
† When oversize or slotted holes are used, see AISC–1.16.5.4.

Figure 4.6.4 provides a comparison between test results and Eqs. 4.6.6 and 4.6.7. When the factor of safety of 2.0 is applied, the diagonal side of the design region is defined. For L_e/D less than about 3, the tear-out condition controls; when greater than 3 the failure mode changes to one in which material deformation occurs resulting in elongation of the hole. To prevent excessive hole deformation, AISC–1.5.1.5.3 gives the limitation on bearing nominal stress as

$$f_p \leq 1.5F_u \tag{4.6.11}$$

which corresponds to $L_e/D = 2.65$ and forms the righthand boundary of the design region in Fig. 4.6.4. The lower boundary is minimum $L_e/D = 1.5$, which corresponds approximately to traditional practice. Minimum end distances from AISC–1.16.5 are given in Table 4.6.2 (see also AISC–Table 1.16.5.1). Minimum center-to-center distances for standard holes are given by AISC–1.16.4.1 as $2\frac{2}{3}$ times bolt diameter. In the line of transmitted force, the distance from the center of a fastener to the edge of the next adjacent hole must satisfy Eq. 4.6.10 also; thus occasionally the minimum of $2\frac{2}{3}D$ will not be sufficient.

Example 4.6.1

Determine the tensile capacity of the bearing-type connection in Fig. 4.6.5 if (a) the bolt threads are excluded from the shear plane and (b) if the bolt threads are included in the shear plane. Use AISC Spec. with $\frac{7}{8}$-in.-diam A325 bolts and A572 Grade 50 plate material with standard holes.

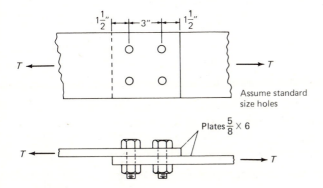

Fig. 4.6.5 Example 4.6.1.

SOLUTION

(a) Threads excluded from shear plane. Consider first the tensile capacity of the plates (Chapter 3).

$$A_n = [6 - 2(\tfrac{7}{8} + \tfrac{1}{8})]0.625 = 2.50 \text{ sq in.}$$
$$A_g = 6(0.625) = 3.75 \text{ sq in.}$$
$$A_e = A_n = 2.50 \text{ sq in.}$$
$$T = 0.60F_y A_g \quad \text{or} \quad 0.50F_u A_e, \quad \text{whichever is smaller}$$
$$= 30(3.75) = 113 \text{ kips} \quad \text{or} \quad 0.5(65)(2.50) = 81.3 \text{ kips}$$
$$\text{(controls)}$$

The action of the entire connection is a shear transfer of load between the two plates. The plane of contact may be thought of as the shear plane. When there is a single plane of contact involved in the load transfer, it is referred to as "single shear."

The allowable capacity R_{SS} per bolt in single shear for A325–X (see Table 4.6.1) is

$$R_{SS} = F_v A = 30(0.6013) = 18.04 \text{ kips}$$

giving for the total capacity for the connection

$$T = 4R_{SS} = 4(18.04) = 72.2 \text{ kips}$$

Bearing strength must also be checked.

The allowable bearing stress F_p from AISC–1.5.1.5.3 is

$$F_p = 1.5F_u = 1.5(65) = 97.5 \text{ ksi}$$

The allowable capacity R_B per bolt in bearing is

$$R_B = F_p Dt = 97.5(\tfrac{7}{8})(\tfrac{5}{8}) = 53.3 \text{ kips}$$

$$R_{SS} < R_B \qquad \therefore R_{SS} = 18.04 \text{ kips controls}$$

The end distance from the center of a standard hole to the edge of the plate may not be less than

$$L_e \geq \left[\frac{2P}{F_u t} = \frac{2(18.04)}{65(0.625)} = 0.89 \text{ in.} \right]$$

The 1.5 in. provided exceeds the minimum required 0.89 in. and also satisfies the $1\tfrac{1}{2}$ in. minimum for sheared edges given by Table 4.6.2.

Since the shear capacity ($T = 72.2$ kips) is lower than the plate tensile capacity ($T = 81.3$ kips), then

$$T = 72.2 \text{ kips} \quad \text{(for A325–X)}$$

(b) Threads included in shear plane. The tensile capacity based on single shear for A325–N (Table 4.6.1) with $F_v = 21.0$ ksi is

$$T = 4(21.0)0.6013 = 50.5 \text{ kips}$$

Nominal shear stress again governs since there is no change in net section tensile capacity or in the capacity based on bearing.

Example 4.6.2

Determine the number of $\tfrac{3}{4}$-in.-diam A325 bolts required to develop the full strength of the A572 Grade 65 steel plates in Fig. 4.6.6 (a portion of a double lap splice connection) for a bearing-type connection with threads excluded from the shear planes. Use AISC Specification and assume a double row of bolts with standard size holes.

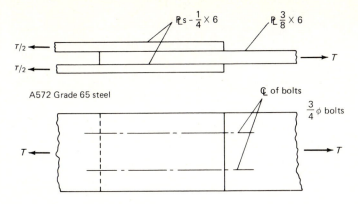

Fig. 4.6.6 Example 4.6.2.

SOLUTION

It is determined by inspection that the cross-sectional area of the center plate is less than the sum of the areas of the two outer plates. Therefore in this case one need only check the capacity of the center plate.

$$A_n = [6 - 2(\tfrac{3}{4} + \tfrac{1}{8})]0.375 = 1.59 \text{ sq in.}$$

$$\text{Max } A_n = 0.85\, A_{\text{gross}} = 0.85(6)0.375 = 1.91 \text{ sq in.}$$

$$A_e = A_n = 1.59 \text{ sq in.}$$

$$T = 0.50 F_u A_e = 0.5(80)1.59 = 63.6 \text{ kips}$$

but not more than

$$0.60 F_y A_g = 0.60(65)(6)(0.375) = 87.8 \text{ kips}$$

$$T = 63.6 \text{ kips}\quad \text{(tensile capacity of plates)}$$

In this case the action of the entire connection may be thought of as a shear transfer between plates occurring at the *two* planes of contact. When connectors are positioned so that they cross two shear planes of contact, it is called "double shear." *Double shear* is a symmetrically loaded situation so far as the shear planes and directions of shear transfer are concerned, whereas the single-shear case is unsymmetrical (Fig. 4.6.5).

Allowable shear capacity per bolt (double shear) for A325–X (Table 4.6.1) is

$$R_{\text{DS}} = 2(\text{area})\, F_v = 2(0.4418)(30) = 26.5 \text{ kips}\quad \text{(controls)}$$

The allowable capacity per bolt based on bearing strength is

$$R_B = F_p D t = 1.5 F_u D t = 1.5(80)(\tfrac{3}{4})(\tfrac{3}{8}) = 33.8 \text{ kips}$$

$$\text{Number of bolts} = \frac{63.6}{26.5} = 2.4$$

Use 4—$\frac{3}{4}$-in.-diam A325 bolts (A325–X).
The end distance required would be

$$L_e \geq \frac{2P}{F_u t}$$

The load per fastener is $63.6/4 = 15.9$ kips.

$$L_e \geq \left[\frac{2(15.9)}{80(0.375)} = 1.06 \text{ in.} \right]$$

Table 4.6.1 controls, requiring $L_e \geq 1.25$ in.

Example 4.6.3

Determine the number of $\frac{3}{4}$-in.-diam A325 bolts in standard size holes required to develop the full strength of the plates in Fig. 4.6.7 if A36 steel is used. Assume the portion of a double lap splice is a bearing-type connection with the threads excluded from the shear plane and a double row of bolts. Use AISC Specification.

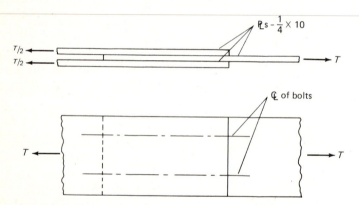

Fig. 4.6.7 Example 4.6.3.

SOLUTION

In this example, the center plate will control the tensile capacity.

$$A_n = [10 - 2(\tfrac{3}{4} + \tfrac{1}{8})]0.25 = 2.06 \text{ sq in.}$$

$$\text{Max } A_n = 0.85 A_g = 0.85(2.50) = 2.13 \text{ sq in.}$$

$$\therefore A_e = A_n = 2.06 \text{ sq in.}$$

$$T = 0.50 F_u A_e = 0.50(58)2.06 = 59.7 \text{ kips}$$

or

$$T = 0.60 F_y A_g = 22(2.50) = 55.0 \text{ kips} \quad \text{(controls)}$$

For shear, using A325–X bolts,

$$R_{DS} = 2(30.0)(0.4418) = 26.5 \text{ kips}$$

For bearing,

$$R_B = 1.5 F_u D t = 1.5(58)(\tfrac{3}{4})(0.25) = 16.3 \text{ kips}$$

The number of bolts required is

$$n = \frac{T}{R_B} = \frac{55.0}{16.3} = 3.4$$

Use 4—$\tfrac{3}{4}$-in.-diam A325–X bolts, in two rows.
  The minimum end distance L_e is

$$P = \frac{55.0}{4} = 13.75 \text{ kips/bolt}$$

$$L_e \geq \left[\frac{2P}{F_u t} = \frac{2(13.75)}{58(0.25)} = 1.90 \text{ in.} \right]$$

Since this exceeds the value from Table 4.6.1, this larger value governs.
Use 2 in. end distance.

Summary of Design Considerations for High-Strength Bolted Joints Subjected to Axial Tension or Compression

The following summary is intended to combine factors dealt with in this section with those in Chapter 3 so that strength along with other limitations of design specifications will be considered in the design of axially loaded joints. The elements of a good design are:

1. Adequate *net area and gross area* to carry tensile load (see Chapter 3).
2. Adequate *gross area* to carry compressive load (see Chapter 6).
3. Adequate number of bolts to provide shear strength.
4. Adequate number of bolts, or use of thick enough plates, so that bearing strength is adequate (normally of concern only for bearing-type connections).
5. Adequate *end distance* so that connectors cannot shear out (see Table 4.6.1 and the requirements of AISC–1.16.5).
6. Maximum edge distance not exceeded, so that "dishing" or curling of the edge will not occur (see AISC–1.16.6).
7. Reasonable spacing between connectors, measured center-to-center, so that tearing of plates cannot occur and so that wrenches can easily be applied for bolt tightening. AISC–1.16.4.1 states that the minimum spacing is $2\tfrac{2}{3}$ times bolt diameter, but preferably three bolt diameters. Commonly a 3-in. spacing is used for $\tfrac{3}{4}$-in. and $\tfrac{7}{8}$-in.-diam bolts. Provisions of AISC–1.16.4.2 also apply to center-to-center spacing but rarely govern.

8. Long lines of connectors avoided; use compact joints if feasible. Compact joints result when connector spacing both transverse and parallel to the direction of stress is approximately 5 times the diameter of the bolts, and the length of the joint does not exceed about 5 pitches in the direction of stress.

The above summary relates essentially to connection *strength* and applies equally for both bearing-type and friction-type connections. The *additional* serviceability requirement of slip-resistance for friction-type connections is treated in the next section.

4.7 FRICTION-TYPE CONNECTIONS

When slip-resistance at service load is desired, the friction-type connection should be used. In developing the rationale for the AISC provisions relating to friction-type connections, it is helpful to study the behavior of high-strength bolts under load, as shown in Fig. 4.5.4. The pretension force (T) in the bolt equals the clamping force between the pieces being joined. The resistance to shear is a frictional force μT, where μ is the coefficient of friction.

The coefficient of friction, or more properly the slip coefficient, depends on the surface condition of the pieces being joined, with factors such as mill scale, oil, paint, or special surface treatments determining the value of μ.

A number of investigators [17–22] have evaluated the slip coefficient experimentally, using as a definition of the maximum frictional resistance that occurring when "the friction bond is definitely broken and the two surfaces slip with respect to one another by a relatively large amount" [21]. The range of μ has been found to be from about 0.2 to 0.6 depending on the surface condition of the pieces being joined [8].

Example 4.7.1

Determine the amount of force P required to cause slip of a $\frac{7}{8}$-in.-diam A325 bolt loaded as in Fig. 4.7.1, if the slip coefficient μ is 0.34 (a typical value for the usual "clean mill scale" surface condition). Using the force P, compute the nominal shear stress, $f_v = P/A_b$.

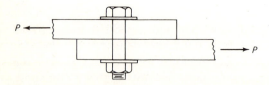

Fig. 4.7.1   Example 4.7.1.

SOLUTION

Using the pretension load from Table 4.4.2,

$$T_i = 39 \text{ kips}$$

$$P = \mu T_i = 0.34(39) = 13.3 \text{ kips}$$

Since the overall action of the connection is a shearing effect, the nominal shear stress f_v in the $\frac{7}{8}$-in.-diam high-strength bolt at the load causing slip to begin is

$$f_v = \frac{P}{A_b} = \frac{13.3}{0.6013} = 22.1 \text{ ksi}$$

AISC–1.5.2 (see also Table 4.6.1) uses the nominal shear stress approach rather than directly use the slip coefficient; thus an allowable nominal shear stress F_v of 17.5 ksi is given for A325 bolts in friction-type connections using standard holes.

Example 4.7.2

Determine the factor of safety against slip of a $\frac{7}{8}$-in.-diam A325 bolt in a standard hole for the situation shown in Fig. 4.7.1.

SOLUTION

The factor of safety provided against slip (for $\mu = 0.34$) is

$$FS = \frac{\text{Frictional shear resistance}}{\text{Allowable force } P} = \frac{\mu T_i}{F_v A_b}$$

which for $\frac{7}{8}$-in.-diam A325 bolts in friction-type connections (A325–F) is

$$FS = \frac{0.34(39)}{17.5(0.6013)} = 1.26$$

For other bolt sizes, the variation in the factor of safety against slip may be observed from Table 4.7.1. The clamping force is the initial tension from Table 4.4.2.

Using static loading conditions the factor of safety against slip as determined in Example 4.7.2 represents the margin against a once in the life of the structure movement such as a shock or severe wind loading. Slip is a serviceability requirement and does not constitute failure; therefore a lower value for the factor of safety such as 1.2 (compared with the basic 1.67) is acceptable.

In general, strength and serviceability can both be treated in design by the nominal shear stress approach. As indicated in the discussion of bearing-type connections, Eq. 4.6.1 can give the allowable nominal shear stress for both slip-resistant and not slip-resistant connections; i.e.,

$$F_v = \beta_1 \beta_2 \beta_3 \text{ (Basic } F_v) \qquad [4.6.1]$$

Table 4.7.1 Factor of Safety against Slip, Friction-Type Connection (Coefficient of Slip = 0.34)

Bolt Diameter (in.)	Nominal Area (sq. in.)	(mm²)	Factor of Safety A325	A490
5/8	0.3068	197.9	1.20	1.21
3/4	0.4418	285.0	1.23	1.22
7/8	0.6013	387.9	1.26	1.26
1	0.7854	506.7	1.26	1.26
1 1/8	0.9940	641.3	1.09	1.24
1 1/4	1.2272	791.7	1.12	1.28

Table 4.7.2 Maximum Size (Nominal) for Oversize and Slotted Holes (adapted from Commentary to Ref. 7)

Nominal Bolt Size (in.)	(mm)	Oversize Holes* (in.)	(mm)	Short-slotted Holes† (in.)	(mm)	Long-slotted Holes‡ (in.)	(mm)
5/8	15.9	13/16	20.6	11/16×7/8	17.5×22.2	11/16×1 9/16	17.5×39.7
3/4	19.1	15/16	23.8	13/16×1	20.6×25.4	13/16×1 7/8	20.6×47.6
7/8	22.2	1 1/16	27.0	15/16×1 1/8	23.8×28.6	15/16×2 3/16	23.8×55.6
1	25.4	1 1/4	31.8	1 1/16×1 5/16	27.0×33.3	1 1/16×2 1/2	27.0×63.5
1 1/8	28.6	1 7/16	36.5	1 3/16×1 1/2	30.2×38.1	1 3/16×2 13/16	30.2×71.4
1 1/4	31.8	1 9/16	39.7	1 5/16×1 5/8	33.3×41.3	1 5/16×3 1/8	33.3×79.4
1 3/8	34.9	1 11/16	42.9	1 7/16×1 3/4	36.5×44.5	1 7/16×3 7/16	36.5×87.3
1 1/2	38.1	1 13/16	46.0	1 9/16×1 7/8	39.7×47.6	1 9/16×3 3/4	39.7×95.3

* Nominal diameters up to 3/16 in. (4.76 mm) larger than bolts 7/8 in. (22.2 mm) and less in diameter; 1/4 in. (6.35 mm) larger than 1 in. (25.4 mm) diam bolts; and 5/16 in. (7.94 mm) larger than bolts 1 1/8 in. (38.6 mm) and greater in diameter.
† Nominally 1/16 in. (1.59 mm) wider than bolt diameter and having a length which does not exceed the oversize diameter (footnote *) by more than 1/16 in. (1.59 mm).
‡ Nominally 1/16 in. (1.59 mm) wider than bolt diameter and having a length longer than for short-slotted holes but not exceeding 2 1/2 times the bolt diameter.

The allowables given for bearing-type connections (Table 4.6.1) are the "Basic F_v" values for Eq. 4.6.1. The "Basic F_v" values assume adequate strength to provide an FS of at least 2.0.

When slip-resistance is desired, the β values are involved; β_1 is a factor relating to the probability of slip; β_2 is a factor related to method of installation; and β_3 is a fabrication factor, particularly relating to size of holes.

To establish an appropriate value for β_1, Fisher and Struik in their *Guide* [8] compared β_1 for various surface conditions for both the A325 and A490 bolts using a 5% and a 10% probability of slip. For example, for A325 bolts and a *clean mill scale* (Class A) surface condition, β_1 is 0.59 for a 5% probability of slip and 0.68 for a 10% probability [8]. The probability of slip occurring is a more precise way of treating slip than is the factor of safety approach since both the clamping force and the slip coefficient have considerable variation in actual situations.

A value for β_2 recommended by the *Guide* [8] is 1.0 for the turn-of-the-nut method and 0.85 for the calibrated wrench method, indicating the higher reliability of obtaining the specified tension when using the turn-of-the-nut method. No reference was made to the load indicator method, but it would appear to be a highly reliable method, justifying $\beta_2 = 1.0$.

The factor β_3 relating to size of holes was recommended by the *Guide* [8] to be 1.0 for standard holes and 0.70 for oversize and slotted holes.

Using Eq. 4.6.1 with recommended values from the *Guide* [8], an allowable shear stress F_v may be established. For example, A325 bolts in standard holes ($\beta_3 = 1.0$) having clean mill scale surface condition (Class A) using a 10% probability of slip ($\beta_1 = 0.68$), and installed by the calibrated wrench method ($\beta_2 = 0.85$), would give for Eq. 4.6.1

$$F_v = \beta_1\beta_2\beta_3 \,(\text{Basic } F_v) = 0.68(0.85)(1.0)30 = 17.3 \text{ ksi} \quad (119 \text{ MPa})$$

This compares favorably with the AISC allowable ($F_v = 17.5$ ksi) for this same condition. In effect, AISC assumes a 10% probability of slip and the calibrated wrench value of $\beta_2 = 0.85$ for all friction-type connection allowable values. No difference in performance related to installation method is recognized.

The values allowable for the Class A (clean mill scale) surface condition are given in Table 4.6.1 (AISC–Table 1.5.2.1). The values given for oversize and short-slotted holes are 85% and those for long-slotted holes are 70% of the values for standard holes. The values used prior to the 1978 AISC Specification correspond to the situation of oversize holes.

The larger than standard holes, whose maximum dimensions are given in Table 4.7.2, are used when greater latitude is needed to meet dimensional tolerances during erection. Allen and Fisher [23] have shown that such larger holes can be used with $\frac{5}{8}$ in. and larger diameter high-strength bolts without adversely affecting the performance. Details of this are reported in the *Guide* [8].

If the designer specifies special treatment (Classes B through I) of the contact surfaces in a *friction-type* connection, the allowable nominal shear stress of Table 4.7.3 may be used. These values were also established statistically to correspond to about a 10% probability of slip [8].

In addition to consideration of slip, the actual strength of the connection after slip must be adequate to carry twice the service load (i.e., FS = 2.0 for strength); the usual strength requirement for connections. Thus for friction-type connections the capacities used in design may not exceed the capacities what would be permitted if the connection were

Table 4.7.3 Allowable Shear Stresses,* Based on Surface Conditions of Bolted Parts, for Friction-Type Shear Connections (from AISC–Appendix Table E1)

Class	Surface Condition of Bolted Parts	Standard Holes		Oversize Holes and Short-slotted Holes		Long-slotted Holes	
		A325	A490	A325	A490	A325	A490
	F_v (ksi)						
A	Clean mill scale	17.5	22.0	15.0	19.0	12.5	16.0
B	Blast-cleaned carbon & low alloy steel	27.5	34.5	23.5	29.5	19.5	24.0
C	Blast-cleaned quenched & tempered steel	19.0	23.5	16.0	20.0	13.5	16.5
D	Hot-dip galvanized and roughened†	21.5	27.0	18.5	23.0	15.0	19.0
E	Blast-cleaned, organic zinc rich paint†	21.0	26.0	18.0	22.0	14.5	18.0
F	Blast-cleaned, inorganic zinc rich paint	29.5	37.0	25.0	31.5	20.5	26.0
G	Blast-cleaned, metallized with zinc	29.5	37.0	25.0	31.5	20.5	26.0
H	Blast-cleaned, metallized with aluminum	30.0	37.5	25.5	32.0	21.0	26.5
I	Vinyl wash	16.5	20.5	14.0	17.5	11.5	14.5
	F_v (MPa) (See Note‡)						
A	Clean mill scale	121	152	103	131	86	110
B	Blast-cleaned carbon & low alloy steel	190	238	162	203	134	165
C	Blast-cleaned quenched & tempered steel	131	162	110	138	93	114
D	Hot-dip galvanized and roughened†	148	186	128	159	103	131
E	Blast-cleaned, organic zinc rich paint†	145	179	124	152	100	124
F	Blast-cleaned, inorganic zinc rich paint	203	255	172	217	141	179
G	Blast-cleaned, metallized with zinc	203	255	172	217	141	179
H	Blast-cleaned, metallized with aluminum	207	259	176	221	145	183
I	Vinyl wash	114	141	97	121	79	100

* Values from this table are applicable *only* when they do not exceed the lowest appropriate allowable working stresses for *bearing-type* connections, taking into account the position of threads relative to shear planes and, if required, the 20% reduction due to joint length (see Table 4.6.1).

† Not recommended if more than half of the load is due to gravity and frame deformation caused by joint slip into bearing cannot be tolerated.

‡ Allowables given in MPa are the AISC values multiplied by 6.8947. AISC–1978 does not give allowable values in units other than ksi.

considered bearing-type. This can be better understood by referring to Eq. 4.6.1, where the product $\beta_1\beta_2\beta_3$ might exceed 1.0, implying a higher value for F_v than "Basic F_v". When the "Basic F_v" is 30 ksi for A325 bolts or 40 ksi for A490 bolts as it is when no threads are in the shear plane, the allowable shear stress F_v for friction-type connections can never exceed "Basic F_v" for *any* class of surface condition given in Table 4.7.2. However, when threads are *not* excluded from the shear planes, the friction-type allowables for some classes of surface condition do exceed "Basic F_v."

The only place the AISC Specification refers to the requirement that the friction-type connection strength may not exceed the strength as a bearing-type connection is by a footnote to Appendix Table E1, which itself is only referred to by footnote 5 to Table 1.5.2.1. The Table E1 footnote states that values for friction-type connections are applicable "*only* when they do not exceed the lowest appropriate allowable working stress for *bearing-type* connections." Past practice did *not* require an examination of strength (i.e., review as a bearing-type connection) *in addition to* the design as a friction-type connection because previously friction-type connection allowable values were always lower than those for bearing-type connections.

Example 4.7.3

Determine the tensile capacity of the connection previously investigated in Example 4.6.1 as a bearing-type connection (Fig. 4.6.5); however, consider it as a friction-type connection with the usual clean mill scale (Class A) surface condition. Use $\frac{7}{8}$-in.-diam A325 bolts in standard holes with A572 Grade 50 plate material. Use AISC Spec.

SOLUTION
The plate capacity is as determined in Example 4.6.1,

$$T = 81.3 \text{ ksi} \quad (\text{based on } 0.50 F_u A_e)$$

From Table 4.6.1, the allowable capacity per bolt in single shear is

$$R_{SS} = F_v A_b = 17.5(0.6013) = 10.52 \text{ kips}$$
$$T = 4 R_{SS} = 4(10.52) = 42.1 \text{ kips}$$

Thus the capacity is controlled by the bolt strength; $T = 42.1$ kips.

Since the load is expected to be transferred by friction, bearing strength is not expected to control for the Class A surface condition. However, AISC–1.5.1.5.3 as well as AISC–1.16.4 and 1.16.5 for spacing, end, and edge distances apply for both bearing-type and friction-type connections. These requirements were examined in Example 4.6.1 and there would be no change if the connection is friction-type.

Example 4.7.4

Redesign the connection for Fig. 4.6.6 as a friction-type connection using oversize holes with a Class A surface condition. Use $\frac{3}{4}$-in.-diam A325 bolts in a double row, with AISC Spec.

SOLUTION
The allowable tensile load based on plate strength was determined in Example 4.6.2 to be 63.6 kips.

The allowable capacity per bolt based on shear (double shear) is, using Table 4.6.1,

$$R_{DS} = F_v A_b = 15.0(2)0.4418 = 13.25 \text{ kips}$$

$$\text{Number of bolts} = \frac{63.6}{15.25} = 4.8$$

Use 6—$\frac{3}{4}$-in.-diam bolts (A325–F).

The bearing strength ($R_B = 33.8$ kips per bolt) is not controlling. The spacing and edge distances must conform to AISC–1.16.4 and 1.16.5.

Example 4.7.5

Redesign the connection of Fig. 4.6.6 as a friction-type connection with standard holes for Class B (blast-cleaned carbon and low alloy steel) surface condition. Use $\frac{3}{4}$-in.-diam A325 bolts in a double row, with AISC Spec. Assume there may be threads in the shear planes.

SOLUTION

The allowable tensile load based on plate strength was determined in Example 4.6.2 to be 63.6 kips.

The allowable capacity per bolt based on shear (double shear), using Table 4.7.3, is

$$R_{DS} = F_v A_b = 27.5(2)0.4418 = 24.3 \text{ kips}$$

If this connection were bearing-type (with threads in the shear plane) the allowable would be (Table 4.6.1)

$$R_{DS} = F_v A_b = 21.0(2)0.4418 = 18.5 \text{ kips}$$

This latter value is based on strength rather than factor of safety against slip; thus the value permitted is the smaller (18.5 vs 24.3) one.

$$R_{DS} = 18.5 \text{ kips}$$

$$\text{Number of bolts} = \frac{63.6}{18.5} = 3.4$$

Use 4—$\frac{3}{4}$-in.-diam bolts (A325–F).

In this case, the Class B frictional resistance was permitted to increase the bolt capacity above the 13.25 kips for Class A but not as high as the 24.3 kips based on slip for Class B. Strength based on shear strength of the bolt through the threaded portion restricted the capacity to 18.5 kips. If no threads were in the shear plane, the strength value would increase to 26.5 kips (based on 30 ksi) so that the slip related value of 24.3 kips would control.

4.8 ECCENTRIC SHEAR

When the load P is applied on a line of action that does not pass through the center of a bolt group, there will be an eccentric loading effect, such as in Fig. 4.8.1. A load P at an eccentricity e, as shown in Fig. 4.8.2, is statically equivalent to a moment P times e plus a concentric load P both acting on the connection. Since both the moment and the concentric load contribute shear effects on the bolt group, the situation is referred to as *eccentric shear.*

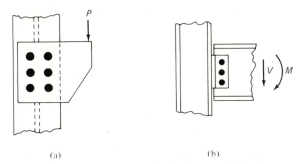

(a)   (b)

Fig. 4.8.1   Typical eccentric shear connections.

Only in the past few years has significant experimental work [8, 24, 25] been done to evaluate the strength of such joints. Essentially, three general approaches are available to the designer: (1) the traditional elastic (vector) analysis assuming no friction with the plates rigid and the fasteners elastic; (2) an empirical modification of the elastic (vector) analysis using the same assumptions as in (1) but with an empirically reduced eccentricity e; and (3) an ultimate strength analysis (plastic analysis) wherein it is assumed the eccentrically loaded fastener group rotates about an instantaneous center of rotation and the deformation at each fastener is proportional to its distance from the center of rotation.

Since the AISC Specification prescribes the allowable capacity for a fastener but does not specify the means of analysis, any of these approaches may be used by the designer. Each of these approaches is dealt with in detail in the remainder of this section.

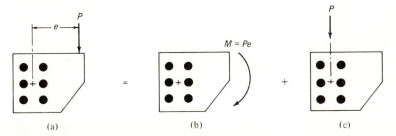

(a)   (b)   (c)

Fig. 4.8.2   Combined moment and direct shear.

Traditional Elastic (Vector) Analysis

For many years eccentrically loaded fastener groups have been analyzed by considering the fastener group areas as an elastic cross section subjected to direct shear and torsion. The stresses resulting are nominal in the sense that they have stress units (say, psi) and provide a guide to safety but are not real stresses because the service loads are actually carried by friction. This elastic analysis method has been used because it makes use of simple mechanics of materials concepts and has been found to be a conservative procedure.

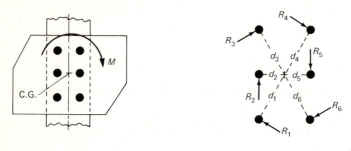

(a) Connection

(b) Forces on connectors

Fig. 4.8.3 Pure moment connection.

To develop the equations for use in this procedure, consider first the connection acted upon by the moment M, as shown in Fig. 4.8.3. Neglecting friction between the plates, the moment equals the sum of the forces shown in Fig. 4.8.3b times their distances to the centroid (C.G.) of the fastener areas:

$$M = R_1 d_1 + R_2 d_2 + \ldots + R_6 d_6 = \sum Rd \qquad (4.8.1)$$

Assuming all the fasteners to be of the same cross-sectional area A the unit stresses are

$$f_1 = \frac{R_1}{A}; \qquad f_2 = \frac{R_2}{A} \ldots f_6 = \frac{R_6}{A} \qquad (4.8.2)$$

Also, since the deformation of each fastener is proportional to its distance d from the assumed center of twist and the fasteners are considered elastic, the stress in each fastener is also proportional to its distance d from the centroid,

$$\frac{f_1}{d_1} = \frac{f_2}{d_2} = \ldots \frac{f_6}{d_6} \qquad (4.8.3)$$

Rewriting the stresses in terms of f_1 and d_1,

$$f_1 = \frac{f_1 d_1}{d_1}; \qquad f_2 = \frac{f_1 d_2}{d_1} \dots f_6 = \frac{f_1 d_6}{d_1} \tag{4.8.4}$$

Substituting Eqs. 4.8.2 and 4.8.4 into Eq. 4.8.1 gives

$$M = \frac{f_1 d_1^2}{d_1} A + \frac{f_1 d_2^2}{d_1} A + \dots + \frac{f_1 d_6^2}{d_1} A$$

$$= \frac{f_1}{d_1} A [d_1^2 + d_2^2 + d_3^2 + \dots + d_6^2] \tag{4.8.5a}$$

$$= \frac{f_1}{d_1} \sum A d^2 \tag{4.8.5b}$$

The stress in fastener 1 is therefore

$$f_1 = \frac{M d_1}{\sum A d^2} \tag{4.8.6a}$$

and by similar reasoning, stress on the other fasteners are

$$f_2 = \frac{M d_2}{\sum A d^2}; \qquad f_3 = \frac{M d_3}{\sum A d^2}; \qquad \dots f_6 = \frac{M d_6}{\sum A d^2} \tag{4.8.6b}$$

or in general,

$$f = \frac{M d}{\sum A d^2} \tag{4.8.7}$$

which is the same as the familiar mechanics of materials formula for torsion on a circular shaft, Tr/J, which is also discussed in Sec. 8.2. The twisting moment, $T = M$; the radius from the center of rotation to the point at which stress is computed, $r = d$; and the polar moment of inertia, $J = \sum A d^2$.

In computing the stresses on the fasteners it is usually more convenient to resolve the distances and forces into their horizontal and vertical components as shown in Fig. 4.8.4. The horizontal and vertical components of the stress, f, are f_x and f_y, respectively, and similarly the components of the distance d are x and y. From Fig. 4.8.4,

$$f_x = f \frac{y}{d} \quad \text{and} \quad f_y = f \frac{x}{d} \tag{4.8.8}$$

Substituting Eq. 4.8.8 into Eq. 4.8.7 gives

$$f_x = \frac{M y}{\sum A d^2} \quad \text{and} \quad f_y = \frac{M x}{\sum A d^2} \tag{4.8.9}$$

Assuming that all the connectors have the same area, and noting that

$$d^2 = x^2 + y^2$$

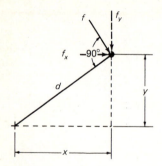

Fig. 4.8.4 Horizontal and vertical components of the stress f.

the vertical and horizontal components of the stresses may be written

$$f_x = \frac{My}{A(\sum x^2 + \sum y^2)} \qquad (4.8.10a)$$

$$f_y = \frac{Mx}{A(\sum x^2 + \sum y^2)} \qquad (4.8.10b)$$

By taking the vector sum of f_x and f_y, the total stress f on the connector becomes

$$f = \sqrt{f_x^2 + f_y^2} \qquad (4.8.11)$$

In order to compute the total shear stresses in an eccentric connection such as shown in Fig. 4.8.2a, the direct shear stresses f_s of Fig. 4.8.2c must be added to the pure moment condition in Fig. 4.8.2b. Assuming the vertical force is equally shared by all the connectors, the direct shear stress f_s is

$$f_s = \frac{P}{\sum A} \qquad (4.8.12)$$

The total resultant stress f then becomes

$$f = \sqrt{(f_y + f_s)^2 + f_x^2} \qquad (4.8.13)$$

Example 4.8.1

Using the elastic (vector) method determine if the eccentric shear connection shown in Fig. 4.8.5a can be used as either a friction-type or bearing-type connection when $\frac{7}{8}$-in.-diam A325 bolts are used. Assume the threads are excluded from the shearing plane and neglect the stress in the plate.

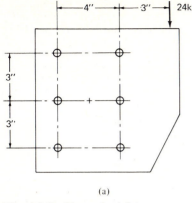

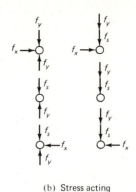

(a)

(b) Stress acting on connector

Fig. 4.8.5 Example 4.8.1.

SOLUTION

From Fig. 4.8.5 it can be seen that the upper and lower right fasteners are the most highly stressed. Since these two fasteners are equally stressed only one need be investigated; check upper-right fastener. The eccentricity e, as measured from the centroid (assumed center of rotation), is

$$e = 3 + 2 = 5 \text{ in.}$$

$$M = 24(5) = 120 \text{ in.-kips}$$

$$\sum x^2 + \sum y^2 = 6(2)^2 + 4(3)^2 = 60 \text{ sq in.}$$

$$f_x = \frac{My}{A(\sum x^2 + \sum y^2)} = \frac{120(3)}{0.601(60)} = 9.98 \text{ ksi} \rightarrow$$

$$f_y = \frac{Mx}{A(\sum x^2 + \sum y^2)} = \frac{120(2)}{0.601(60)} = 6.65 \text{ ksi} \downarrow$$

$$f_s = \frac{P}{\sum A} = \frac{24}{6(0.601)} = 6.65 \text{ ksi} \downarrow$$

$$f = \sqrt{(6.65 + 6.65)^2 + (9.98)^2} = 16.6 \text{ ksi}$$

This is less than $F_v = 17.5$ ksi for the friction-type connection using standard holes; therefore it is acceptable for either a friction-type or bearing-type connection ($F_v = 30.0$ ksi).

Example 4.8.2

Determine the maximum nominal shear stress in the top right fastener of the connection shown in Fig. 4.8.6 if 1-in.-diam fasteners are used. Use the elastic (vector) method.

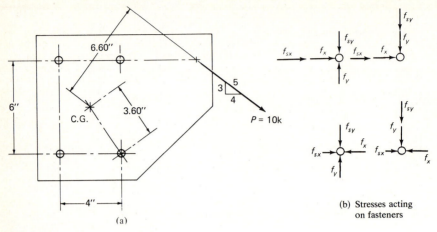

(b) Stresses acting on fasteners

(a)

Fig. 4.8.6 Example 4.8.2.

SOLUTION

$$e = 6.60 \text{ in}$$

$$M = 10(6.60) = 66.0 \text{ in.-kips}$$

$$\sum Ad^2 = 4(0.785)(3.60)^2 = 40.7 \text{ in.}^4$$

$$f_x = \frac{My}{\sum Ad^2} = \frac{66.0(3)}{40.7} = 4.86 \text{ ksi} \rightarrow$$

$$f_y = \frac{Mx}{\sum Ad^2} = \frac{66.0(2)}{40.7} = 3.24 \text{ ksi} \downarrow$$

$$f_{sx} = \frac{P_x}{\sum A} = \frac{0.8(10)}{4(0.785)} = 2.55 \text{ ksi} \rightarrow$$

$$f_{sy} = \frac{P_y}{\sum A} = \frac{0.6(10)}{4(0.785)} = 1.91 \text{ ksi} \downarrow$$

$$f = \sqrt{(f_y + f_{sy})^2 + (f_x + f_{sx})^2}$$
$$= \sqrt{(3.24 + 1.91)^2 + (4.86 + 2.55)^2} = 9.02 \text{ ksi}$$

(Maximum nominal shear stress)

Note this is the only fastener of the four where the two x-components (4.86 and 2.55) and the two y-components (3.24 and 1.91) are both algebraically additive.

Elastic (Vector) Analysis Using Reduced Effective Eccentricity

For many years, designers have used the nominal stress elastic (vector) analysis developed earlier in this section without any apparent difficulty.

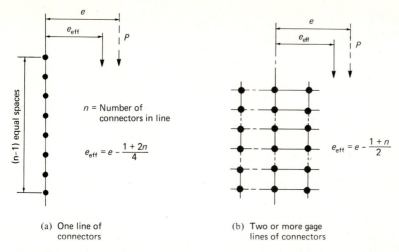

(a) One line of
connectors

(b) Two or more gage
lines of connectors

Fig. 4.8.7 1970 AISC Manual recommendation for effective eccentricity.

The conservatism of the procedure was verified in 1964 by tests sponsored by AISC [26]. As a result the recommendation given in the 1970 AISC Manual (though *not* a specification requirement) was to use a reduced effective eccentricity of load in accordance with the following:

1. For fasteners equally spaced on a single gage line (Fig. 4.8.7a)

$$e_{\text{eff}} = e - \frac{1+2n}{4}$$
(4.8.14)

   where n = number of fasteners in one line.
2. For fasteners equally spaced on two or more gage lines,

$$e_{\text{eff}} = e - \frac{1+n}{2}$$
(4.8.15)

As shown later in illustrating the ultimate strength analysis method, the use of the effective eccentricity with the higher allowable shear stresses for fasteners in the 1978 AISC Specification gives connection capacities too high for the desired 2.5 factor of safety for such connections. Thus *the effective eccentricity should not be used with the 1978 AISC allowable shear stresses for high-strength bolts.*

Example 4.8.3

Repeat Example 4.8.1 using the reduced effective eccentricity of Eq. 4.8.15 and determine the percentage reduction in the computed nominal stress on the most highly stressed fastener (see Fig. 4.8.5).

SOLUTION

The reduced effective eccentricity is

$$e_{\text{eff}} = 5 - \frac{1+3}{2} = 3 \text{ in.}$$

$$M = 24(3) = 72 \text{ in.-kips}$$

From the stress equations, it is apparent the flexural stress is directly proportional to M; then the maximum stress is

$$f_x = 5.99 \text{ ksi}, \qquad f_y = 3.99 \text{ ksi}, \qquad f_s = 6.65 \text{ ksi}$$

Thus $f = 12.2$ ksi, a 27% reduction from the elastic (vector) analysis using the full eccentricity.

It should be noted that on many simple connections that have small eccentricity of load, it has long been the practice to ignore such eccentricity. This concept of a reduced effective eccentricity is in agreement with that practice.

Ultimate Strength Analysis

This method, also called plastic analysis, currently is recognized by the *Guide* [8] as that giving the factor of safety (2.5) most consistent with that used in other types of connection design.

In this method it is recognized that the eccentric load P causes a rotation as well as translation effect on the fastener group. The translation and rotation can be reduced to a pure rotation about a point defined as the instantaneous center of rotation (see Fig. 4.8.8).

The requirements for equilibrium are as follows:

$$\sum F_H = 0; \qquad \sum_{i=1}^{n} R_i \sin \theta_i - P \sin \delta = 0 \qquad \text{(4.8.16)}$$

$$\sum F_V = 0; \qquad \sum_{i=1}^{n} R_i \cos \theta_i - P \cos \delta = 0 \qquad \text{(4.8.17)}$$

$$\sum M = 0; \qquad \sum_{i=1}^{n} d_i R_i - P(e + r_0) = 0 \qquad \text{(4.8.18)}$$

Actually the concept of instantaneous center is identical to the elastic (vector) method if the resistance R_i is proportional to the deformation (i.e., stress is proportional to strain). For either the elastic or the strength method, the deformation is proportional to the distance d_i from the instantaneous center of rotation.

For the strength analysis, two approaches have been used [8]. For the *bearing-type* connection, slip is neglected so that the *deformation* of each fastener is proportional to its distance from the instantaneous center. The resistance of each fastener is related to its deformation according to its

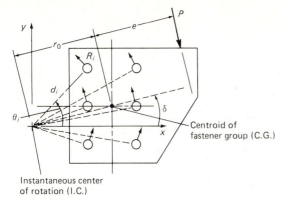

Fig. 4.8.8 Instantaneous center of rotation.

load-deformation relationship. An expression proposed by Fisher [30] and used by Crawford and Kulak [24] for this load (R) -deformation (Δ) response is

$$R_i = R_{ult}(1 - e^{-10\Delta})^{0.55} \tag{4.8.19}$$

where $R_{ult} = \tau_u A_b$. The coefficients "10" and "0.55" were experimentally determined, and the maximum Δ at failure was about 0.35 in. [24]. Note that e in Eq. 4.8.19 is the Naperian base (2.718) and not the eccentricity. For A325 bolts, the ultimate shear strength τ_u is approximately 62% of the tensile strength (120 ksi minimum). The experimental work relating to Eq. 4.8.19 used bolts loaded symmetrically (that is, in double shear); however, the general strength method could use *any* appropriate load (R) -deformation (Δ) relationship, not necessarily Eq. 4.8.19.

For the *friction-type* connection, a modified strength method approach may be used wherein the resistance R_i from each fastener is the same, say equal to R_s. With frictional load transmission, the frictional resistance offered by each fastener is essentially the same.

Example 4.8.4

Illustrate the general strength method by determining the ultimate load P_u that may be applied to the fastener group of Fig. 4.8.5. Use Eq. 4.8.19 as the load-deformation expression, and assume that the maximum deformation Δ_{max} at failure is 0.35 in.

SOLUTION

For $\frac{7}{8}$-in.-diam A325 bolts, Eq. 4.8.19 becomes

$$R_i = 0.62(120)(0.6013)(1 - e^{-10\Delta})^{0.55}$$
$$= 44.74(1 - e^{-10\Delta})^{0.55} \tag{a}$$

The load is applied in the y-direction; therefore $\delta = 0$ (Fig. 4.8.8). Using y_i/d_i for $\sin \theta_i$ and x_i/d_i for $\cos \theta_i$, Eqs. 4.8.16 through 4.8.18 become

$$\sum R_i \frac{y_i}{d_i} = 0 \qquad \text{(b)}$$

$$\sum R_i \frac{x_i}{d_i} = P_u \qquad \text{(c)}$$

$$\sum R_i d_i = P_u(e + r_0) \qquad \text{(d)}$$

Also, a basic deformation assumption is

$$\Delta_i = \frac{d_i}{d_{max}} \Delta_{max} = \frac{d_i}{d_{max}} (0.35) \qquad \text{(e)}$$

(a) Since an iterative process will be required to solve Eqs. (b), (c), and (d), let the first trial $r_o = 3$ in. (see Fig. 4.8.9).

Fasteners	x_i	y_i	d_i	Δ_i	R_i	$\dfrac{R_i x_i}{d_i}$	$R_i d_i$
1	1	3	3.162	0.19	40.9	12.94	129.4
2	1	0	1.0	0.06	28.9	28.88	28.9
3	1	−3	3.162	0.19	40.9	12.94	129.4
4	5	3	5.831	0.35	44.0	37.72	256.5
5	5	0	5.0	0.30	43.5	43.50	217.5
6	5	−3	5.831	0.35	44.0	37.72	256.5
						173.7	1018.2

Eq. (c): $P_u = 174$ kips

Eq. (d): $P_u = \dfrac{1018.2}{8} = 127$ kips

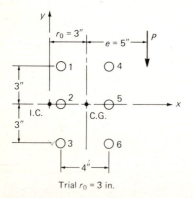

Trial $r_0 = 3$ in.

Fig. 4.8.9 Strength method for Example 4.8.4.

Since the values are not identical, further trials are required. After several trials:

Try $r_o = 2.1$ in.

Fasteners	x_i	y_i	d_i	Δ_i	R_i	$\dfrac{R_i x_i}{d_i}$	$R_i d_i$
1	0.1	3	3.002	0.207	41.5	1.38	124.7
2	0.1	0	0.10	0.007	10.2	10.2	10.3
3	0.1	−3	3.002	0.207	41.5	1.38	124.7
4	4.1	3	5.080	0.35	44.0	35.51	223.5
5	4.1	0	4.100	0.282	43.2	43.25	177.3
6	4.1	−3	5.080	0.35	44.0	35.51	223.5
						127.2	884.0

$$P_u = 127 \text{ kips}$$

$$P_u = \frac{884.1}{7.1} = 124.5 \text{ kips}$$

Close enough. Say $P_u = 125$ kips.

Recent studies [24, 25, 27–29] have indicated that an ultimate strength (plastic) analysis is the most rational approach to the strength of eccentric shear connections. The elastic analysis using full eccentricity (with the lower 1969 AISC allowable stresses for fasteners) was overly conservative, giving factors of safety for strength averaging 4.5 for connections tested [29]. Using the effective eccentricity (with the same 1969 AISC stresses) the factor of safety for these tests was reduced to about 3.2. Using the plastic analysis, the factor of safety is about 2.5 for the tests reported [8]. Using the higher 1978 AISC allowable fastener stresses the elastic analysis with full eccentricity still tends to be conservative (though erratic), giving factors of safety between 2.5 and 3.0. However, the effective eccentricity with the elastic method and 1978 AISC fastener stresses gives an FS of only about 2.0. *The effective eccentricity should not be used with the 1978 AISC fastener stresses.*

If a factor of safety of 2.5 is applied to the ultimate load obtained in the last example, the safe load P would be $125/2.5 = 50$ kips.

For slip-resistant connections (i.e., friction-type) the resistance R_i may be considered constant (say, R_s) for all fasteners (assuming the same size in the same kind of holes).

Example 4.8.5

Repeat Example 4.8.4 (Fig. 4.8.5) using $R_i = R_s$ as for a slip-resistant connection using the ultimate strength approach.

SOLUTION

For $R_i = R_s$ and $\delta = 0$, Eqs. 4.8.16 through 4.8.18 become

$$R_s \sum \frac{y_i}{d_i} = 0 \qquad \text{(a)}$$

$$R_s \sum \frac{x_i}{d_i} - P = 0 \qquad \text{(b)}$$

$$R_s \sum d_i - P(e + r_o) = 0 \qquad \text{(c)}$$

Try $r_o = 2$ in. Referring to Fig. 4.8.9,

Fastener	x_i	y_i	d_i	$\dfrac{x_i}{d_i}$
1	0	3	3.0	0
2	0	0	0	0
3	0	-3	3.0	0
4	4	3	5.0	0.8
5	4	0	4.0	1.0
6	4	-3	5.0	0.8
			20.0	2.6

From Eq. (b),

$$P = R_s \sum \frac{x_i}{d_i} = 2.6 R_s$$

From Eq. (c),

$$P = \frac{R_s \sum d_i}{e + r_o} = \frac{R_s(20)}{7} = 2.86 R_s$$

For this assumption, fastener No. 2 is at the center of rotation and therefore is not involved in Eq. (c). Also, fastener No. 2 is assumed to have *no* contribution to Eq. (b). When r_o is assumed slightly larger than 2.0, Eq. (b) gives $P = 3.6 R_s$ because $x_i/d_i = 1.0$ for fastener No. 2. When r_o is assumed slightly smaller than 2.0, Eq. (b) gives $P = 1.6 R_s$ because $x_i/d_i = -1$ for the same fastener. Thus the value $P = 2.86 R_s$ from Eq. (c) is accepted as the answer (i.e., $r_o = 2.0$ in.). If a factor of safety is applied, R_s may be interpreted as the allowable value for a friction-type connection and P the safe applied load. For $\frac{7}{8}$-in.-diam fasteners, using $F_v = 17.5$ ksi, gives for single shear,

$$R_{ss} = 17.5(0.6013) = 10.5 \text{ kips}$$

and the service load capacity would be

$$P = 2.86(10.5) = 30.0 \text{ kips}$$

Recommended Practical Method for Design

The method recommended by the *Guide* [8] is essentially the ultimate strength method. Instead of using a trial-and-error analysis, coefficients have been developed [8, 24] from a computerized version of the trial-and-error method. Little difference (generally less than 5 percent) was found between the strength analysis results for the friction-type and bearing-type connections.

Thus the *Guide* [8] recommends the same procedure for both friction-type and bearing-type connections, as follows:

$$P = CmA_b F_v \qquad (4.8.20)$$

where m = number of shear planes
$\quad A_b$ = nominal bolt area
$\quad F_v$ = allowable values from Table 4.6.1 (AISC–1.5.2.1)
$\quad C$ = fastener group coefficient
$\quad\quad = \alpha' I^\beta$ (equation approximating the tabular values given by the *Guide* [8])
$\quad I = I_x + I_y$, using $A_b = 1$
$$\alpha' \approx 0.0104 + \frac{0.625}{e} + \frac{4.719}{e^2} - \frac{6.750}{e^3}$$
$\quad\quad$ (for one line of fasteners)
$$\alpha' \approx 0.0125 + \frac{0.814}{e} + \frac{5.550}{e^2} - \frac{8.220}{e^3}$$
$\quad\quad$ (for two lines of fasteners)
$$\beta \approx 0.645 - \frac{0.129}{e} - \frac{3.85}{e^2} + \frac{7.43}{e^3}$$
$\quad\quad$ (for one line of fasteners)
$$\beta \approx 0.651 - \frac{0.183}{e} - \frac{3.13}{e^2} + \frac{6.25}{e^3}$$
$\quad\quad$ (for two lines of fasteners)

Example 4.8.6

Compare the results of the various methods of treating the eccentric shear connection, specifically for the connection of Fig. 4.8.5. Determine the allowable capacity P for (1) friction-type connection and (2) bearing-type connection with no threads in the shear plane. Use $\frac{7}{8}$-in.-diam A325 bolts.

SOLUTION
$\quad$ (a) Elastic analysis: (Example 4.8.1)

$$f_v = 16.6 \text{ ksi} \quad \text{for} \quad P = 24 \text{ kips}$$

For $F_v = 17.5$ ksi with friction-type connections:

$$P = 24(17.5)/16.6 = 25.3 \text{ kips}$$

For $F_v = 30$ ksi with bearing-type connection:

$$P = 24(30)/16.6 = 43.4 \text{ kips}$$

(b) Effective eccentricity: (Example 4.8.3)

$$f_v = 12.2 \text{ ksi}$$
$$P = 34.4 \text{ kips}   \text{(friction-type)}$$
$$P = 59.0 \text{ kips}   \text{(bearing-type)}$$

(c) Ultimate strength analysis: (Examples 4.8.5 and 4.8.6)

$$P = 50 \text{ kips}   \text{(bearing-type using FS} = 2.5)$$
$$P = 30.0 \text{ kips}   \text{(friction-type)}$$

(d) Recommended procedure of *Guide* [8]:

$$I_x + I_y = 4(3)^2 + 6(2)^2 = 60 \text{ in.}^2$$

$$\alpha' = 0.0125 + \frac{0.814}{5} + \frac{5.550}{(5)^2} - \frac{8.220}{(5)^3} = 0.3315$$

$$\beta = 0.651 - \frac{0.183}{5} - \frac{3.13}{(5)^2} + \frac{6.25}{(5)^3} = 0.5392$$

$$C = \alpha' I^\beta = 0.3315(60)^{0.5392} = 3.015$$

This formula value is somewhat high (as expected) compared with the computerized strength analysis value $C = 2.72$ given by the table in the *Guide* [8] (Table 13.2, p. 224); however, the table is for a gage of 3 in. between fastener lines whereas in this example the gage is 4 in.

$$P = 3.015(0.601)17.5 = 31.7 \text{ kips}   \text{(friction-type)}$$
$$P = 3.015(0.601)30 = 54.4 \text{ kips}   \text{(bearing-type)}$$

Comparisons

	Friction-type		Bearing-type	
Procedure	Load P (kips)	Relative* Value	Load P (kips)	Relative* Value
1. Elastic (vector)	25.3	0.84	43.4	0.87
2. Effective eccentricity	34.4	1.15	59.0	1.18
3. Ultimate strength	30.0	1.00	50.0	1.00
4. *Guide* [8] recommended method using approx. equations for coefficient C	31.7	1.06	54.4	1.09
5. *Guide* [8] method using table value for C obtained by computerized strength analysis for 3-in. gage between fastener lines	28.6	0.95	49.1	0.98

* Assumes the ultimate strength analysis procedure (No. 3) is the "best."

Design Formula for Moment on Single Line of Fasteners

The AISC Manual in its section, "Eccentric Loads on Fastener Groups," gives a series of tables that allow the designer to determine the number of connectors required for a given load and eccentricity. In unusual cases for which the tables do not apply, or when they may not be readily available, it is desirable to have a simple alternative method to use. The following development from Shedd* provides a useful simple formula.

Consider a single line of equally spaced fasteners subjected to moment alone, as shown in Fig. 4.8.10. Since with uniform spacing the resistance of the fasteners is uniform from top to bottom, and according to Eq. 4.8.7, $f = Md/(\sum Ad^2)$, the stress distribution varies linearly as shown in Fig. 4.8.10.

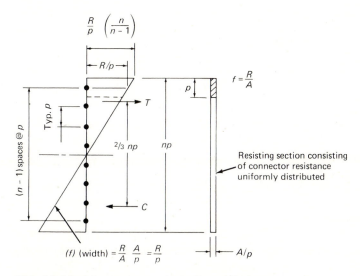

Fig. 4.8.10 Moment on a single line of fasteners.

Assuming R is the force, fA, in the outermost fastener, and that it represents the accumulation of stress that would occur on a rectangular resisting section over the height p, one may designate the average load per inch of height at the outermost fastener as R/p.

Using similar triangles the load per inch at the extreme fiber may be determined,

$$\text{Extreme fiber value} = \frac{R}{p}\left(\frac{n}{n-1}\right) \tag{4.8.21}$$

The tensile force is the area of the triangle represented by the force per

---

* Thomas C. Shedd, *Structural Design in Steel* (John Wiley & Sons, New York 1934), p. 287.

inch diagram,

$$T = \frac{1}{2}\left(\frac{np}{2}\right)\left(\frac{R}{p}\right)\left(\frac{n}{n-1}\right) = \frac{Rn^2}{4(n-1)} \tag{4.8.22}$$

The internal resisting moment is

$$M = T(\tfrac{2}{3} np) \tag{4.8.23}$$

Substitution of Eq. 4.8.22 into Eq. 4.8.23 gives

$$M = \frac{Rn^2}{4(n-1)}\left(\frac{2}{3} np\right) = \frac{Rn^3 p}{6(n-1)} \tag{4.8.24}$$

Solving Eq. 4.8.24 for n^2, one obtains

$$n = \sqrt{\frac{6M}{Rp}\left(\frac{n-1}{n}\right)} \tag{4.8.25}$$

which as a first approximation becomes

$$n = \sqrt{\frac{6M}{Rp}} \tag{4.8.26}$$

which is suggested for design use.

Since Eq. 4.8.26 is for moment alone acting on a single row of fasteners, the numerical value for R to be used in it should be adjusted to account for direct shear and for more than one row of fasteners. It is suggested to use a reduced effective R for the direct-shear effect and use an increased effective R for the effect of lateral spread. For lateral spread use a multiplier on R of 1.0 for one line up to about 2.0 for a square array of connectors.

More complicated formulas have been developed to compute the maximum stress or force on a connector, but no direct solution for the number of connectors or the number of rows is possible from such equations.

Example 4.8.7

Determine the required number of $\frac{7}{8}$-in.-diam A325 bolts for one vertical line of fasteners Ⓐ–Ⓐ in the bracket shown in Fig. 4.8.11. Assume it to be a bearing-type connection with threads included in the shear planes (A325–N).

SOLUTION

(a) Using allowable nominal stresses determine the maximum allowable load per fastener

$$F_v = 21 \text{ ksi} \quad (\text{A325–N})$$
$$R_{DS} = 21(0.6013)(2) = 25.3 \text{ kips} \quad (\text{controls})$$

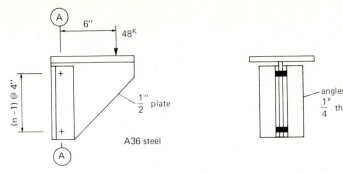

Fig. 4.8.11 Example 4.8.7.

It is a case of double shear, since two shear planes are resisting failure of each fastener

$$F_p = 1.5F_u = 1.5(58) = 87 \text{ ksi}$$

$$R_B = 87(\tfrac{7}{8})(0.5) = 38.1 \text{ kips}$$

(b) Estimate the number of bolts required using Eq. 4.8.26.

$$n = \sqrt{\frac{6M}{Rp}} = \sqrt{\frac{6(48)(6)}{25.3(4)}} = 4.1$$

The full eccentricity has been used as recommended by the authors for use with the elastic method and the R value has *not* been adjusted for the direct shear effect; try 4 fasteners.

(c) Check adequacy of 4 bolts using elastic method with full eccentricity.

$$f_s = \frac{P}{\sum A} = \frac{P}{An}$$

or multiplying by A, the force per connector is obtained:

$$f_s A = \frac{P}{n} = \frac{48}{4} = 12.0 \text{ kips} \downarrow$$

the moment component is

$$f_x = \frac{My}{A(\sum x^2 + \sum y^2)}$$

$$f_x A = \frac{My}{\sum x^2 + \sum y^2}$$

$$\sum x^2 + \sum y^2 = 2[(2)^2 + (6)^2] = 80 \text{ in.}^2$$

$$f_x A = \frac{48(6)6}{80} = 21.6 \text{ kips} \rightarrow$$

$$\text{Actual } R = \sqrt{(12.0)^2 + (21.6)^2} = 24.7 \text{ kips} < 25.3 \text{ kips} \qquad \text{OK}$$

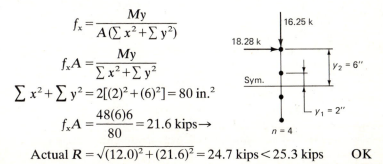

It may be noted the edge distance measured in the direction of the resultant force must satisfy Eq. 4.6.10 (AISC–1.16.5.1) requiring $L_e \geq 2P/(F_u t)$.

Use $4—\frac{7}{8}$-in.-diam A325–N bolts @ 4-in. pitch.

Example 4.8.8

Determine the required number of $\frac{3}{4}$-in.-diam A325 bolts in standard holes for the bracket plate of Fig. 4.8.12, assuming 4 vertical rows. Use a friction-type connection with clean mill scale (Class A) surface condition.

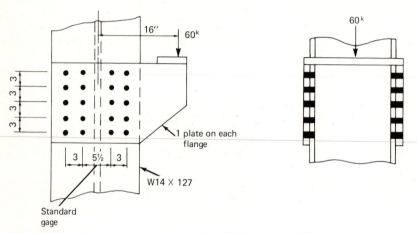

Fig. 4.8.12 Eccentric shear connection of Example 4.8.8.

SOLUTION

(a) General design method using elastic (vector) method of analysis. Choose a vertical pitch of 3 in., which is a commonly used value. The maximum allowable load per connector is

$$R_{SS} \text{ (single shear)} = 17.5(0.4418) = 7.73 \text{ kips}$$

Bearing need not be considered for friction-type connections with Class A surface condition.

Use the full eccentricity,

$$n = \sqrt{\frac{6M}{Rp}} = \sqrt{\frac{6(30/4)(16)}{7.73(3)}} = 5.6$$

In the above equation, the load per plate is 30 kips, and the load per line of fasteners is 30/4, which must be used since Eq. 4.8.26 applies to one line of fasteners. No adjustment in R was made either for direct shear or for several lines of fasteners.

Check 5 bolts per row by elastic method:

$$f_s A = \frac{P}{n} = \frac{30}{20} = 1.50 \text{ kips } \downarrow$$

$$\sum x^2 + \sum y^2 = 10[(2.75)^2 + (5.75)^2] + 8[(3)^2 + (6)^2] = 766 \text{ in.}^2$$

$$f_x A = \frac{My}{\sum x^2 + \sum y^2} = \frac{30(16)6}{766} = 3.76 \text{ kips} \rightarrow$$

$$f_y A = \frac{Mx}{\sum x^2 + \sum y^2} = \frac{30(16)5.75}{766} = 3.60 \text{ kips } \downarrow$$

$$\text{Actual } R = \sqrt{(1.50 + 3.60)^2 + (3.76)^2} = 6.34 \text{ kips} < 7.73 \qquad \text{OK}$$

Use 5—$\frac{3}{4}$-in.-diam A325 bolts per row.

4.9 FASTENERS ACTING IN AXIAL TENSION

Axial tension occurring without simultaneous shear exists in fasteners for tension members such as hangers (see Fig. 4.5.1c) or other members

Table 4.9.1 Allowable Tensile Stress on Nominal Cross-Sectional Area of Fasteners (adapted from AISC–Table 1.5.2.1)

	Tension, F_t			
	AISC		AASHTO	
Description of Fastener	*(ksi)*	*(MPa)*	*(ksi)*	*(MPa)*
A502, Grade 1, hot-driven rivets	23.0*	159	Not permitted	
A502, Grade 2, hot-driven rivets	29.0*	200	Not permitted	
A307 bolts	20.0*	138	13.5‡	93
Threaded parts meeting the requirements of AISC–1.4.1 and A449 bolts as provided in AISC–1.4.4	$0.33F_u$*·**			
A325 bolts	44.0†	303	36.0	248
A490 bolts	54.0†	372	48.0	331

For bolts subject to combined tension and shear, the allowable values are obtainable from AISC–1.6.3.
* Static loading only.
** The tensile capacity of the threaded portion of an upset rod shall be larger than the body area of the rod times $0.60F_y$.
† For A325 and A490 bolts subject to tensile fatigue loading see AISC–Appendix B3.
‡ Based on area at the root of threads.

whose line of action is perpendicular to the member to which it is fastened. When such tension members are not perpendicular to their connecting members, the fasteners are subjected to both axial tension and shear. The latter, more typical case, is discussed in Sec. 4.10.

Table 4.9.1 shows values for the allowable tensile stresses according to AISC and AASHTO Specifications. The allowable tensile stress F_t for all fasteners except A307 bolts under AASHTO is based on their nominal cross-sectional area using the *major* or nominal thread diameter. This area is frequently referred to as the *nominal area* based on the specified bolt diameter. In the case of A307 bolts under AASHTO, the allowable tensile stress is based on the area at the root of threads.

Prestress Effect of High-Strength Bolts Under External Tension

In order to understand the effect of an externally applied load on a pretensioned high-strength bolt consider a single bolt and the tributary portion of the connected plates as shown in Fig. 4.9.1a. The pieces being joined are of thickness t and the area of contact between the pieces is A_p. Prior to applying external load the situation is as shown in Fig. 4.9.1b, where the bolt has been installed to have a pretension force T_i (values as in Table 4.4.2). The pieces being joined are compressed an amount C_i. For equilibrium,

$$C_i = T_i \tag{4.9.2}$$

The external load P is then applied and the forces acting are shown in Fig. 4.9.1c. This time equilibrium requires

$$P + C_f = T_f \tag{4.9.3}$$

where the subscript f refers to final conditions after application of the load P.

The force P acting on the system lengthens the bolt an amount δ_b

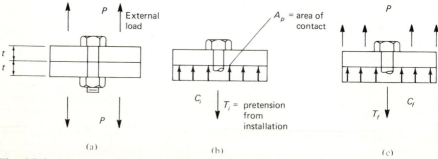

Fig. 4.9.1 Prestress effect on bolted joint.

between the underside of the bolt head and the surface of contact between the two connected plates.

$$\delta_b = \frac{T_f - T_i}{A_b E_b} t \tag{4.9.4}$$

At the same time the compression between the plates decreases and the plate thickness increases an amount δ_p.

$$\delta_p = \frac{C_i - C_f}{A_p E_p} t \tag{4.9.5}$$

If contact is maintained, compatibility of deformation requires $\delta_b = \delta_p$; thus, equating Eqs. 4.9.4 and 4.9.5 gives

$$\frac{T_f - T_i}{A_b E_b} = \frac{C_i - C_f}{A_p E_p} \tag{4.9.6}$$

Next, substitution of Eq. 4.9.2 for C_i and Eq. 4.9.3 for C_f gives

$$\frac{T_f - T_i}{A_b E_b} = \frac{T_i - T_f + P}{A_p E_p} \tag{4.9.7}$$

The modulus of elasticity for the plates and the bolts is essentially the same and thus may be eliminated. Then solving for T_f gives

$$(T_f - T_i)\frac{A_p}{A_b} = T_i - T_f + P \tag{4.9.8}$$

$$T_f\left(1 + \frac{A_p}{A_b}\right) = T_i\left(1 + \frac{A_p}{A_b}\right) + P$$

$$T_f = T_i + \frac{P}{1 + A_p/A_b} \tag{4.9.9}$$

Example 4.9.1

Assume $\frac{7}{8}$-in.-diam A325 bolts are used in a direct tension situation such as in Fig. 4.9.2. With bolts spaced 3 in. apart and having $1\frac{1}{2}$-in. edge distances, the tributary area of contact may reasonably be about 9 sq in. If the maximum external tensile load permitted by AISC is applied, how much does the bolt tension increase?

SOLUTION

(a) Maximum applied tensile force P. From Table 4.9.1,

$$P = F_t A_b = 44.0(0.6013) = 26.46 \text{ kips}$$

(b) Initial tensile force in $\frac{7}{8}$-in.-diam A325 bolt. From Table 4.4.2,

$$T_i = 39 \text{ kips}$$

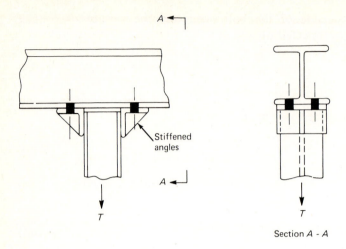

Fig. 4.9.2 Example 4.9.1.

(c) Determine final tensile force in bolt. The ratio of plate contact area to bolt area is

$$\frac{A_p}{A_b} = \frac{9}{0.6013} = 15$$

This neglects subtracting the bolt area from the total tributary area, but little difference results. Using Eq. 4.9.9 then gives

$$T_f = 39 + \frac{26.46}{1+15} = 39 + 1.65 = 40.65 \text{ kips}$$

The increase in tension is 4.2%. The variation in actual pretension from installation may be expected to exceed this amount, so that this increase is not of concern. Furthermore the tributary area used for the example (9 sq in.) is probably the minimum one might encounter in practice, since less than a 3-in. pitch and gage is rarely used.

The important conclusion from this example is that no significant increase in bolt tension arises until the external load equals or exceeds the pretension force, in which case the pieces do not remain in contact and the applied force equals the bolt tension.

If the connection can distort and give rise to "prying forces" these must also be considered. (See AISC Commentary–1.5.2.1 and the treatment in the "Split Beam Tee Connections" part of Sec. 13.6.)

The use of the 44 ksi for nominal stress in direct tension provides a factor of safety compatible with the other factors in design. In the situation of Example 4.9.1 the approximate factor of safety is

$$\text{FS} = \frac{T_i}{P} = \frac{39}{26.46} = 1.5$$

In general, for A325 bolts under AISC–1.5.2 and 1.23.5, the margin against F_t exceeding the proof load is approximately 1.5 for diameters up to 1 in. and approximately 1.3 for diameters over 1 in. Similarly, use of the AASHTO Specification would result in values of 1.8 and 1.5, respectively.

Example 4.9.2

Compute the allowable load T for the connection of Fig. 4.9.2 if $\frac{7}{8}$-in.-diam A325 bolts are used.

SOLUTION

First compute the bolt value R_T (allowable capacity per bolt); then multiply by the number n of bolts.

$$R_T = F_t A_b = 44.0(0.6013) = 26.46 \text{ kips/bolt}$$
$$T = nR_T = 4(26.46) = 106 \text{ kips}$$

Example 4.9.3

Determine the required number of $\frac{3}{4}$-in.-diam A490 bolts required for the connection shown in Fig. 4.9.3. Assume that the pieces making up the connection are adequate and that the nominal tensile stresses on the bolts govern.

SOLUTION

The allowable capacity per bolt is

$$R_T = F_t A_b = 54.0(0.4418) = 23.86 \text{ kips/bolt}$$

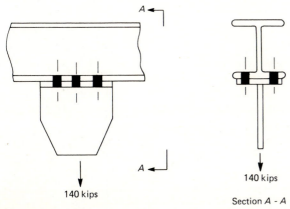

Fig. 4.9.3   Example 4.9.3.

The number of bolts required is

$$n = \frac{T}{R_T} = \frac{140}{23.86} = 5.9$$

Use 6—$\frac{3}{4}$-in.-diam A490 bolts.

4.10 COMBINED SHEAR AND TENSION

In a large number of commonly used connections, both shear and tension occur and must be considered in their design. Figure 4.10.1 shows a few typical connections in which the connectors are simultaneously subjected to both shear and tension. The connection shown in Fig. 4.10.1a is a common one where two angles join the beam web to the column flange. From the moment force indicated in the figure, the upper fasteners are subjected to tension in proportion to the magnitude of the applied moment. However, one may recall from structural analysis that only a small amount of end rotation on a beam is necessary to change from a fixed end to a hinged end condition. In addition the web carries only a small part of the bending moment. Thus one may intuitively sense that

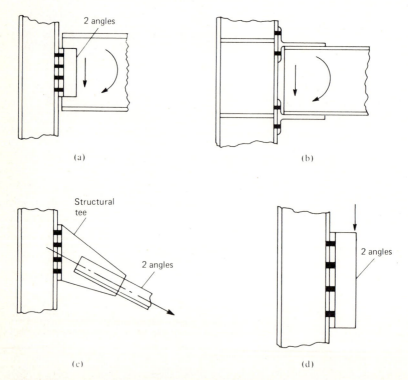

(a)

(b)

(c)

(d)

Fig. 4.10.1 Typical combined shear and tension connections.

the moment shown will be relieved before a significant tension force can be developed in the fasteners. Such connections are used when little end moment is desired to be transmitted. An exception to this occurs in the case of a very deep beam such as a plate girder.

Referring next to Fig. 4.10.1b in which the applied moment is transmitted through the flanges of the beam the situation is different. In this case a large applied moment is intended to be transmitted so the connection is made at the flanges, the elements carrying most of the moment. Chapter 13 deals with this type of connection. Figure 4.10.1c and 4.10.1d typify the two types of fastener loading in combined shear and tension which are developed in the following parts of this section.

Bearing-type Connections

Present practice limits the amount of combined shear and tension in a fastener by *interaction* equations that are based on experimental results [32, 33]. Neglecting initial tension, friction, and bearing, the interaction equation for the strength of a connection with fasteners in combined shear and tension may be approximated by the elliptical relationship

$$\left(\frac{f_{vu}}{F_{vu}}\right)^2 + \left(\frac{f_{tu}}{F_{tu}}\right)^2 \le 1.0 \tag{4.10.1}$$

where f_{vu} = nominal shear stress at failure
f_{tu} = nominal tensile stress at failure
F_{vu} = ultimate shear strength with no tensile stresses present
F_{tu} = ultimate tensile strength with no shear stresses present

By imposing a factor of safety on each of the terms of Eq. 4.10.1, an

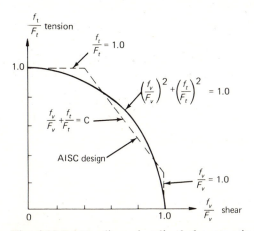

Fig. 4.10.2 Nondimensionalized shear-tension interaction curves: bearing-type connections.

allowable stress equation is obtained (see also Fig. 4.10.2):

$$\left(\frac{f_v}{F_v}\right)^2 + \left(\frac{f_t}{F_t}\right)^2 \leq 1.0 \qquad (4.10.2)$$

where f_v = nominal shear stress due to applied loads

f_t = nominal tensile stress due to applied loads

F_v = allowable shear stress in absence of external tension

F_t = allowable tensile stress in absence of shear

AISC has chosen to use a simpler straight-line interaction which requires a reduction for combined effects only in the most severe loading cases. The straight-line expression

$$\frac{f_v}{F_v} + \frac{f_t}{F_t} \leq C \qquad (4.10.3)$$

where C is a constant. This is compared with the more exact relationship in Fig. 4.10.2. Multiplying Eq. 4.10.3 by F_t and solving for f_t gives

$$f_t \leq F_t C - \frac{F_t}{F_v} f_v \qquad (4.10.4)$$

AISC–1.5.2 gives $F_t/F_v = 44/22 = 2.00$ for A325–N and $54/28 = 1.93$ for A490–N bolts. For A325–X and A490–X, F_t/F_v is 1.47 and 1.35, respectively. In Eq. 4.10.4, AISC has used F_t/F_v equal to 1.8 when threads may be in the shear plane and 1.4 when threads are excluded from the shear plane. The constant C is taken as 1.25. The term F_t' is used instead of f_t to signify an allowable stress in tension which cannot

Table 4.10.1 Allowable Tensile Stress, F_t', when Combined with Shear (for Bearing-type Connections)* (from AISC–Table 1.6.3)

Fastener	F_t'	
	(ksi)	(MPa)
A307 bolts	$26 - 1.8f_v \leq 20$	$179 - 1.8f_v \leq 138$
A325–N bolts (threads *not* excluded)	$55 - 1.8f_v \leq 44$	$379 - 1.8f_v \leq 303$
A325–X bolts (threads excluded)	$55 - 1.4f_v \leq 44$	$379 - 1.4f_v \leq 303$
A490–N bolts (threads *not* excluded)	$68 - 1.8f_v \leq 54$	$469 - 1.8f_v \leq 372$
A490–X bolts (threads excluded)	$68 - 1.4f_v \leq 54$	$469 - 1.4f_v \leq 372$

* Values are applicable for standard holes, oversize holes, and slotted holes when the slots are normal to the direction of loading.

exceed F_t. Thus AISC–1.6.3 prescribes for A325–N,

$$F_t' \leq 55 - 1.8f_v \leq [F_t = 44 \text{ ksi}] \qquad (4.10.5)$$

Equation 4.10.5 illustrates the typical format for use with bearing-type connections. The specific equations are summarized in Table 4.10.1, and the A325 and A490 equations are shown in Fig. 4.10.3.

Under AASHTO–1.7.41, the approach is to use the circle of Eq. 4.10.2 for bearing-type connections, except multiply through by F_v^2, thus

$$f_v^2 + \left(f_t \frac{F_v}{F_t}\right)^2 \leq F_v^2 \qquad (4.10.6)$$

where the ratio F_v/F_t of the allowable values for shear F_v or tension F_t in

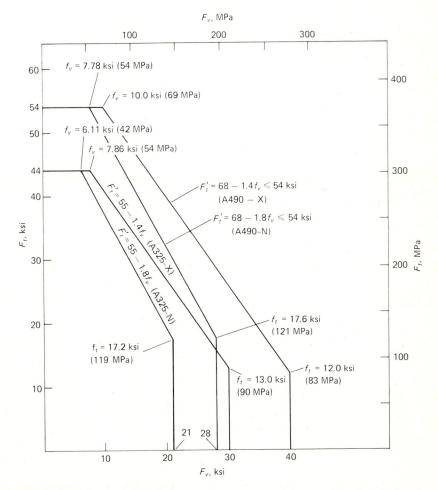

Fig. 4.10.3 Interaction relationship for combined shear and tension in bearing-type connections, AISC–1977.

the absence of the other is $20/36 = 0.555$ for A325 bolts. Under AASHTO, bolts in bearing-type connections *must* have threads excluded from the shear plane. Thus the interaction equation, Eq. 4.10.6, becomes

$$f_v^2 + (0.555F_t)^2 \leq F_v^2 \qquad (4.10.7)$$

Friction-type Connections

Since a higher factor of safety against slip is necessary for friction-type connections than for bearing-type, any reduction in the clamping forces (developed duting the pretensioning of the bolts) due to an externally applied tensile stress f_t will reduce the clamping force, and hence reduce the friction force.

Again, a straight-line interaction relationship is used; but one that is more conservative than the type of Eq. 4.10.5 for bearing-type connections. The constant C is reduced from 1.25 to 1.0 for friction-type connections. In addition, the allowable nominal superimposed tensile stress F_t is replaced in the denominator of the tensile stress ratio by the unit tensile stress F_{tp} based on the pretensioning load. Thus the interaction equation is

$$\frac{f_v}{F_v} + \frac{f_t}{F_{tp}} \leq 1.0 \qquad (4.10.8)$$

where

$$F_{tp} = \frac{T_i}{A_b} \qquad (4.10.9)$$

and T_i = pretension force in the bolt

A_b = nominal cross-sectional area of bolt

Substituting Eq. 4.10.9 into Eq. 4.10.8 gives

$$\frac{f_v}{F_v} + \frac{f_t A_b}{T_i} \leq 1.0 \qquad (4.10.10)$$

or, solving for f_v gives

$$f_v \leq F_v\left(1.0 - \frac{f_t A_b}{T_i}\right) \qquad (4.10.11)$$

Letting f_v be called F_v', the allowable shear stress when external tension is also present, and substituting the values for F_v, the allowable shear stress in the absence of tension for bolts in standard holes, one obtains from Eq. 4.10.11 the limitations of AISC–1.6.3:

$$F_v' \leq 17.5\left(1 - \frac{f_t A_b}{T_i}\right) \qquad \text{A325 bolts in standard holes} \qquad (4.10.12)$$

$$F_v' \leq 22.0\left(1 - \frac{f_t A_b}{T_i}\right) \qquad \text{A490 bolts in standard holes} \qquad (4.10.13)$$

For oversize or slotted holes, the values of F_v from Table 4.6.1 (AISC–1.5.2.1) are to be substituted into Eq. 4.10.11 to obtain the combined stress allowable F'_v equations comparable to Eqs. 4.10.12 and 4.10.13.

The AASHTO–1.7.41 also uses the more conservative straight line interaction relationship for friction-type connections. Substitution of the AASHTO value of 13.5 ksi for F_v with A325 bolts, and letting f_v be called F'_v as above, Eq. 4.10.11 becomes

$$F'_v \leq 13.5 \left(1 - f_t \frac{A_b}{T_i}\right) \qquad \text{(4.10.14)}$$

Examining values of A_b/T_i, for $\frac{3}{4}$-in.-diam A325 bolts,

$$\frac{A_b}{T_i} = \frac{0.44}{28} = 0.017$$

and for $\frac{7}{8}$-in.-diam bolts,

$$\frac{A_b}{T_i} = \frac{0.601}{39} = 0.0154$$

Using the larger value (0.017) in Eq. 4.10.14, one obtains the AASHTO A325 bolt limitation

$$F'_v \leq 13.5 - 0.22 f_t \qquad \text{(4.10.15)}$$

Example 4.10.1

Determine the adequacy of the fasteners in Fig. 4.10.4 if $\frac{7}{8}$-in.-diam A325 bolts and the connection is designed as (a) a friction-type connection and (b) as a bearing-type connection. Neglect all stresses in the pieces making up the connection assuming they do not govern and that threads are excluded from the shear plane.

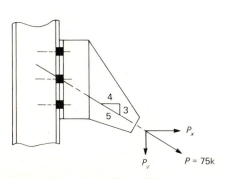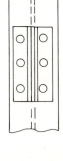

Fig. 4.10.4   Example 4.10.1.

SOLUTION

(a) Check bolts for use in a friction-type connection.

$$P_x = \frac{4}{5}(75) = 60 \text{ kips}$$

$$P_y = \frac{3}{5}(75) = 45 \text{ kips}$$

$$f_v = \frac{P_y}{\sum A} = \frac{45}{6(0.6013)} = 12.5 \text{ ksi} \quad \text{(shear)}$$

$$f_t = \frac{P_x}{\sum A} = \frac{60}{6(0.6013)} = 16.6 \text{ ksi} \quad \text{(tension)}$$

From AISC–1.6.3, Eq. 4.10.12.

$$F'_v = 17.5\left(1 - \frac{f_t A_b}{T_i}\right) = 17.5\left(1 - \frac{16.6(0.6013)}{39}\right) = 13.0 \text{ ksi}$$

Since $f_v < F'_v$, the shear is satisfactory. Also,

$$f_t = 16.6 \text{ ksi} < F_t = 44 \text{ ksi}$$

The connection is satisfactory as a friction-type connection.

(b) Check bolts for use in a bearing-type connection with threads excluded from the shear plane.

$$F'_t = 55.0 - 1.4f_v \leq 44.0$$

$$= 55.0 - 1.4(12.5) = 37.5 \text{ ksi}$$

Since $f_t = 16.64 < 37.5$ and $f_v = 12.5 < 30.0$, the fasteners *are* adequate for a bearing-type connection (A325–X).

Example 4.10.2

Determine the maximum value of P in Example 4.10.1 assuming (a) a friction-type connection and (b) a bearing-type connection (A325–N). Use AISC Spec. and assume standard holes.

SOLUTION

(a) friction-type connection; $\frac{7}{8}$-in.-diam A325 bolts:

$$P_x = \frac{4}{5}P = 0.80P$$

$$P_y = \frac{3}{5}P = 0.60P$$

$$f_v = \frac{0.60P}{6(0.6013)} = 0.166P$$

$$f_t = \frac{0.80P}{6(0.6013)} = 0.222P$$

$$F'_v = 17.5\left(1 - \frac{f_t A_b}{T_i}\right)$$

$$F'_v = 17.5\left(1 - \frac{0.222\,P(0.6013)}{39}\right) = 17.5 - 0.060P$$

Equating f_v to F'_v,

$$0.166P = 17.5 - 0.060P$$

$$P = \frac{17.5}{0.226} = 77.4 \text{ kips}$$

$$f_t = 0.222(77.4) = 17.2 \text{ ksi} < 44.0 \text{ ksi} \qquad \text{OK}$$

The maximum capacity as a friction-type connection is 77.4 kips.
   (b) bearing-type connection (A325–N):

$$F'_t = 55 - 1.8f_v$$

$$= 55 - 1.8(0.166P) = 55.0 - 0.299P$$

Equating f_t to F'_t,

$$0.222P = 55 - 0.299P$$

$$P = \frac{55}{0.521} = 106 \text{ kips}$$

The maximum value of P as limited by $F_v = 21.0$ ksi is

$$f_v = 0.166P = 21.0 \text{ ksi}$$

$$P = 127 \text{ kips} > 106 \text{ kips}$$

Therefore the maximum value of P determined by the fasteners when designed as a bearing-type connection (A325–N) is 106 kips.

Example 4.10.3

Determine the number of $\frac{3}{4}$-in.-diam A325 bolts required to carry a shear of 70 kips combined with a tension of 120 kips. The connection is to be designed such that the resultant force acts through the centroid of the connection. (a) Use a bearing-type connection with thread excluded from the shear plane; (b) use a friction-type connection.

SOLUTION

In this design it is apparent that both the shear and tension forces are of comparable magnitude; thus under such combination neither the full shear allowable F_v nor the full tension allowable F_t may be used, as observed from Fig. 4.10.3. Use of design charts [33] is one possible approach. The following approach [33] may be used when charts are not available.

(a) Bearing-type connection. From AISC–1.6.3, the interaction criterion is

$$F'_t = 55 - 1.4f_v \leq 44 \qquad \text{(A325–X)}$$
$$F_v \leq 30 \text{ ksi}$$

or in general, the F'_t equations may be expressed

$$F'_t = C_1 - C_2 f_v \leq F_t \tag{a}$$

where C_1 and C_2 are constants.

Convert Eq. (a) to a force equation by multiplying by the bolt area A_b; thus

$$F'_t A_b = C_1 A_b - C_2 f_v A_b \leq F_t A_b \tag{b}$$

or, if A_b represents the total area of all bolts, then $F'_t A_b = T$ and $f_v A_b = V$, giving

$$T = C_1 A_b - C_2 V \leq F_t A_b \tag{c}$$

where T and V are the applied tension and shear forces, respectively. Solving Eq. (c) for A_b gives

$$A_b = \frac{T + C_2 V}{C_1} \tag{d}$$

the basic *design equation* for bearing-type connections. The values of C_1 are 55 and 68 for A325 and A490, respectively, and the values of C_2 are 1.4 when threads are excluded from the shear planes and 1.8 when threads are not excluded.

For A325–X bolts, Eq. (d) becomes

$$\text{Required } A_b = \frac{T + 1.4V}{55} \tag{e}$$

For this example, $T = 120$ kips and $V = 70$ kips, and using Eq. (d) which is expected to govern gives

$$\text{Required } A_b = \frac{T + 1.4V}{55} = \frac{120 + 1.4(70)}{55} = 3.96 \text{ sq in.}$$

The use of Eq. (e) satisfies the tension requirement. Shear must also be checked.

$$\text{Max } V = 30(3.96) = 119 \text{ kips} > \text{Actual } V = 70 \text{ kips} \qquad \text{OK}$$

$$\text{Required } n = \frac{3.96}{0.4418} = 9.0 \text{ bolts}$$

Use 10—$\frac{3}{4}$-in.-diam A325–X bolts for a bearing-type connection.

(b) Friction-type connection. From AISC–1.6.3, the interaction expression is

$$F'_v \leq 17.5\left(1 - f_t \frac{A_b}{T_i}\right) \qquad [4.10.12]$$

For $\frac{3}{4}$ in. diam, $A_b/T_i = 0.4418/28 = 0.0158$. Substituting into Eq. 4.10.12 gives

$$F'_v = 17.5 - 17.5(0.0158)f_t$$
$$= 17.5 - 0.276f_t \qquad (f)$$

Convert Eq. (f) into a force equation by multiplying by A_b; thus

$$F'_v A_b = 17.5 A_b - 0.276 A_b f_t \qquad (g)$$

or

$$V = 17.5 A_b - 0.276 T \qquad (h)$$
$$A_b = \frac{V + 0.276 T}{17.5} \qquad (i)$$

Multiplying the numerator and denominator by 4, and calling $4(0.276) \approx 1.1$, gives the interaction design equation, similar to Eq. (d) for bearing-type connections,

$$A_b = \frac{1.1T + 4V}{70} \qquad (j)$$

the basic *design equation* for A325 bolts in friction-type connections using standard holes.

For A490 bolts with standard holes, the denominator becomes $4(22) = 88$,

$$A_b = \frac{1.1T + 4V}{88} \qquad (k)$$

For this example

$$\text{Required } A_b = \frac{1.1T + 4V}{70} = \frac{1.1(120) + 4(70)}{70} = 5.89 \text{ sq in.}$$

The use of Eq. (j) satisfies the shear requirement. The maximum tension must also be checked.

$$\text{Max } T = 44(5.89) > \text{Actual } T = 120 \text{ kips} \qquad \text{OK}$$

$$\text{Required } n = \frac{5.89}{0.4418} = 13.3 \text{ bolts}$$

Use 14—$\frac{3}{4}$-in.-diam A325 bolts for a friction-type connection.

Shear and Tension from Eccentric Loading

In a bracket connection such as in Fig. 4.10.5, the eccentric load produces both shear and tension in the upper fasteners. As in most other connections, the manner in which the pieces behave is complex. However, nominal stresses in the fasteners are usually computed by using one of two approaches: (1) that of neglecting any initial tension in the fasteners or (2) that of considering the initial pretension forces in the fasteners. When fasteners such as A307 bolts are used, the amount of initial tension present is usually small and of an indeterminable amount. Therefore, in this case, the neglecting of any initial tension is reasonable and gives conservative results. On the other hand, when high-strength bolts are used the initial pretension forces exist and should be recognized.

If initial tension does not exist to any appreciable degree, the application of moment Pe (Fig. 4.10.5) will produce a tension that is

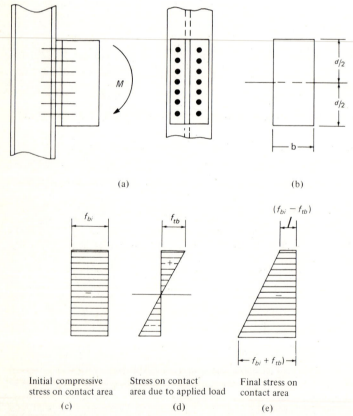

| (a) | (b) |

Initial compressive stress on contact area
(c)

Stress on contact area due to applied load
(d)

Final stress on contact area
(e)

Fig. 4.10.5 Stresses on contact surface of moment-resisting connection, considering initial tension.

maximum at the top bolts. Near the bottom of the connection, compression would exist between the pieces being joined with little effect directly on the bolts. The direct shear would be carried nearly entirely by the bolts since little friction would exist from bolt installation. The use of A307 bolts having little initial tension is rare in important connections having shear in combination with moment-induced tension; thus no further treatment is given to the analysis neglecting initial tension.

Tension from Bending Moment Considering Initial Tension

Consider the moment M applied to the bracket of Fig. 4.10.5 to cause tension on the upper bolts (maximum in the top one of each row of bolts shown as black dots). High-strength bolts used as such fasteners are required to be installed with a prescribed initial tension in them. This tension in each bolt will precompress the plates or sections being joined. For the situation of Fig. 4.10.5, the neutral axis under the action of moment M will occur at the centroid (C.G.) of the contact area; that is, at $d/2$ for the rectangular contact area shown.

The initial bearing pressure f_{bi} as shown in Fig. 4.10.5c is assumed to be uniform over the contact area bd and is equal to

$$f_{bi} = \frac{\sum T_i}{bd} \qquad (4.10.16)$$

where $\sum T_i$ = the pretension load times the number of bolts. The tensile stress f_{tb} at the top due to the applied moment is

$$f_{tb} = \frac{Md/2}{I} = \frac{6M}{bd^2} \qquad (4.10.17)$$

and should not exceed f_{bi} if compression between the pieces is to remain at the top.

The "net" tensile load on a bolt is equal to the product of the tributary area for each fastener times f_{tb}. At the top (most highly stressed) bolt,

$$T_{net} = f_{tb} bp \qquad (4.10.18)$$

Substituting Eq. 4.10.17 into Eq. 4.10.18 gives

$$T_{net} = \frac{6M}{bd^2} bp = \frac{6Mp}{d^2} \qquad (4.10.19)$$

The nominal tensile stress f_t in the top bolt is

$$f_t = \frac{T_{net}}{A_b} = \frac{6Mp}{A_b d^2} \qquad (4.10.20)$$

Assuming the top bolt is approximately $p/2$ from the top, the value of f_t

can be modified to be

$$f_{t\text{(modified)}} = f_t \left(\frac{d-p}{d} \right)$$

$$= \frac{6Mp}{A_b d^2} \left(\frac{d-p}{d} \right) \qquad \text{(4.10.21)}$$

Example 4.10.4

Determine the capacity P for the connection of Fig. 4.10.6 if the fasteners are $\frac{3}{4}$-in.-diam A325–X subject to shear and tension in a bearing-type connection with no threads in the shear plane. AISC Spec.

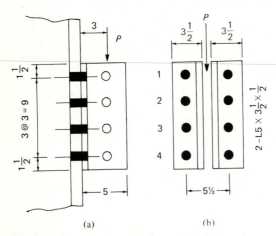

(a)   (b)

Fig. 4.10.6 Shear and moment-induced tension connection for Example 4.10.4.

SOLUTION

Referring to Fig. 4.10.5d, the neutral axis for flexure is at mid-depth of the contact area. Using Eq. 4.10.17,

$$f_{tb} = \frac{6M}{bd^2} = \frac{6P(3)}{7(12)^2} = 0.0179P$$

or if reduced to the stress at the extreme bolt (Bolts 1):

$$f_{tb} = 0.179P \frac{4.5}{6} = 0.0134P$$

Then, according to Eq. 4.10.18, the tensile force on the two most highly stressed bolts is

$$T = f_{tb}bp = 0.0134P(7)(3) = 0.281P$$

and the nominal tensile stress on the two bolts is

$$f_t = \frac{T}{A_b} = \frac{0.281P}{2(0.4418)} = 0.318P$$

The direct-shear component is

$$f_v = \frac{P}{\sum A} = \frac{P}{8(0.4418)} = 0.283P$$

From Table 4.10.1 the AISC allowable tensile stress when shear also is acting is

$$F_t' \le 55 - 1.4f_v \le 44 \text{ ksi}$$

$$0.318P = 55 - 1.4(0.283P)$$

$$P = \frac{55}{0.318 + 0.400} = 76.9 \text{ kips}$$

Check: $f_t = 0.318(76.9) = 24.4 \text{ ksi} < 44 \text{ ksi}$ OK

$\qquad f_v = 0.283(76.9) = 21.8 \text{ ksi} < 22 \text{ ksi}$ OK

Therefore the capacity is 76.9 kips.

Considering Initial Tension—Simplified Procedure

As long as the initial compression between the plates due to bolt installation is not relieved, one may compute tensile stress by the flexure formula

$$f_t = \frac{My}{I} = \frac{My}{\sum Ay^2} \tag{4.10.22}$$

If $d = np$, where $n =$ number of fasteners in one line, Eq. 4.10.21 becomes

$$f_t = \frac{6Mp}{An^2p^2}\left(\frac{np - p}{np}\right) = \frac{12M}{An^3p^2}\left(\frac{p(n-1)}{2}\right) \tag{4.10.23}$$

Note that $p(n-1)/2$ is the distance from mid-depth to the outermost fastener and corresponds to y of Eq. 4.10.22. Further, a single line of fasteners spaced at p apart may be treated as a rectangular resisting section of width A/p and depth np. The moment of inertia of such a section would be

$$I = \frac{1}{12}\left(\frac{A}{p}\right)(np)^3 \tag{4.10.24}$$

which corresponds approximately to the moment of inertia of the bolt areas, $\sum Ay^2$. Thus Eqs. 4.10.22 and 4.10.23 are essentially the same.

Example 4.10.5

Show that the nominal tensile stress for Example 4.10.4 may be obtained using the simplified Eq. 4.10.22.

SOLUTION

$$\sum Ay^2 = 4(0.4418)[(1.5)^2 + (4.5)^2] = 39.8 \text{ in.}^4$$

$$f_t = \frac{My}{\sum Ay^2} = \frac{P(3)(4.5)}{39.8} = 0.340P$$

which is approximately the same as the $0.318P$ using the volume of the stress solid tributary to the extreme fasteners. The simple method is fully as satisfactory as the more complex method.

Example 4.10.6

For the connection of the bracket of Fig. 4.10.7 to the column, determine the number of $\frac{7}{8}$-in.-diam A325 bolts required to transmit the shear and tension forces. Use 3-in. vertical pitch. (a) Use friction-type connection; (b) use bearing-type connection where threads are excluded from the shear plane.

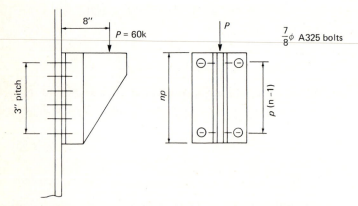

Fig. 4.10.7   Example 4.10.6. Design for shear and tension.

SOLUTION

Since it has been shown that when there is initial tension, the procedure for eccentric shear is also applicable for computing tension; the design equation, Eq. 4.8.26, may also be used here:

$$n = \sqrt{\frac{6M}{Rp}} \qquad\qquad [4.8.26]$$

(a) Friction-type connection. For shear alone,

$$R_{SS} = 17.5(0.6013) = 10.5 \text{ kips/bolt}$$

and for tension alone,

$$R_T = 44(0.6013) = 26.5 \text{ kips/bolt}$$

Noting that $M = 60(8)/2 = 240$ in.-kips per vertical line of fasteners, the number of fasteners required *per line* is

$$n = \sqrt{\frac{6(240)}{26.5(3)}} = 4.3 \text{ required for } M \text{ alone}$$

$$n = \frac{60}{2(10.5)} = 2.9 \text{ required for shear alone}$$

$$n \approx \sqrt{(4.3)^2 + (2.9)^2} = 5.1 \text{ per line}$$

Try 10 bolts (5 per line); check stresses,

$$\sum Ay^2 = 4(0.6013)[(3)^2 + (6)^2] = 108.2 \text{ in.}^4$$

$$f_t = \frac{My}{\sum Ay^2} = \frac{60(8)6}{108.2} = 26.6 \text{ ksi} < 44 \text{ ksi} \qquad \text{OK}$$

$$f_v = \frac{P}{\sum A} = \frac{60}{10(0.6013)} = 10.0 \text{ ksi}$$

For $\frac{7}{8}$-in.-diam A325 bolts, $A_b/T_i = 0.6013/39 = 0.0154$; which when used in the interaction equation, Eq. 4.10.12

$$F_v' = 17.5(1 - f_t A_b/T_i)$$

$$= 17.5(1.0 - 26.6(0.0154)) = 10.3 \text{ ksi}$$

Since $f_v = 10.0$ ksi < 10.3 ksi   OK
Use 10—$\frac{7}{8}$-in.-diam A325 bolts, 5 per row.

(b) Bearing-type connection. For shear alone,

$$R_{ss} = 30(0.6013) = 18.0 \text{ kips/bolt}$$

and for tension alone,

$$R_T = 44(0.6013) = 26.5 \text{ kips/bolt}$$

The estimate using $n = \sqrt{6M/Rp}$ would have been done the same as for part (a), while the number required per line for shear would be 1.7, again indicating less than 5 bolts per line.
Try 4 per row:

$$\sum Ay^2 = 4(0.6013)[(1.5)^2 + (4.5)^2] = 54.1 \text{ in.}^4$$

$$f_t = \frac{My}{\sum Ay^2} = \frac{60(8)4.5}{54.1} = 39.9 \text{ ksi} < 44 \text{ ksi} \qquad \text{OK}$$

$$f_v = \frac{P}{\sum A} = \frac{60}{8(0.6013)} = 12.5 \text{ ksi}$$

$$F_t' = 55 - 1.4(12.5) = 37.5 \text{ ksi}$$

Since $f_t = 39.9$ ksi $> F_t' = 37.5$ ksi   NG
Use 10—$\frac{7}{8}$-in.-diam A325 bolts, 5 per row.

4.11 HIGH-STRENGTH BOLTED CONNECTIONS IN PLASTIC DESIGN

In general, the approach for plastic design is the same as for working stress design. It has already been stated that allowable nominal stresses at service load are based on ultimate strength and provide a factor of safety against connection failure ranging from about 2.2 for long joints to about 3.0 for compact joints.

Under AISC–2.8 when factored loads are used for design, the connection must carry moments and forces that have been factored (1.7 for gravity loads). The allowable stresses are also multiplied by the same factor. The ultimate strength is assumed to be the load factor (say 1.7) times the service load capacity; a safe procedure that still provides a greater safety factor with regard to connectors than for the members joined. Having increased the shear, bearing, and tension capacities of the connectors, the approach to analysis and design is identical with that presented in Secs. 4.7 through 4.10.

SELECTED REFERENCES

1. C. Batho and E. H. Bateman, "Investigations on Bolts and Bolted Joints," Second Report of the Steel Structures Research Committee. London: His Majesty's Stationery Office, 1934.
2 W. M. Wilson and F. P. Thomas, "Fatigue Tests on Riveted Joints," Bulletin 302, Engg. Experiment Station, U. of Illinois, Urbana, Ill., 1938.
3. A. E. R. De Jonge, "Riveted Joints; a Critical Review of the Literature Covering Their Development." New York: American Society of Mechanical Engineers, 1945.
4. "Symposium on High-Strength Bolts," Proc. AISC National Engineering Conference, 1950, pp. 22–43.
5. William H. Munse, "Research on Bolted Connections," *Transactions*, ASCE, 121 (1956), 1255–1266.
6. "Rivets and High-Strength Bolts, A Symposium," *Transactions*, ASCE, 126, Part II (1961), 693–820.
7. *Specifications for Structural Joints Using ASTM A325 or A490 Bolts*, Research Council on Riveted and Bolted Structural Joints of the Engineering Foundation, Feb. 4, 1976.
8. John W. Fisher and John H. A. Struik, *Guide to Design Criteria for Bolted and Riveted Joints*. New York: John Wiley & Sons, Inc., 1974.
9. John L. Rumpf and John W. Fisher, "Calibration of A325 Bolts," *Journal of Structural Division*, ASCE, 89, ST6 (December 1963), 215–234.
10. G. H. Sterling, E. W. J. Troup, E. Chesson and J. W. Fisher, "Calibration Tests of A490 High-Strength Bolts," *Journal of Structural Division*, ASCE, 91, ST5 (October 1965), 279–298.
11. Richard J. Christopher, Geoffrey L. Kulak, and John W. Fisher, "Calibration of Alloy Steel Bolts," *Journal of Structural Division*, ASCE, 92, ST2 (April 1966), 19–40.
12. W. C. Stewart, "What Torque?," *Fasteners*, 1, 4(1944), 8–10.

13. G. A. Maney, "Bolt Measurements by Electrical Strain Gages," *Fasteners*, 2, 1 (1945), 10–13.

14. G. A. Maney, "Predicting Bolt Tension," *Fasteners*, 3, 5 (1946), 16–18.

15. J. W. Fisher, P. O. Ramseier, and L. S. Beedle, "Strength of A440 Steel Joints Fastened with A325 Bolts," *Publications*, IABSE, 23 (1963).

16. John H. A. Struik, Abayomi O. Oyeledun, and John W. Fisher, "Bolt Tension Control with a Direct Tension Indicator," *Engineering Journal*, AISC, 10, 1(First Quarter 1973), 1–5.

17. Desi D. Vasarhelyi, Said Y. Beano, Ronald B. Madison, Zung-An Lu and Umesh C. Vasishth, "Effects of Fabrication Techniques on Bolted Joints," *Transactions*, ASCE, 126, Part II (1961), 764–796.

18. R. A. Hechtman, D. R. Young, A. G. Chin, and E. R. Savikko, "Slip of Joints Under Static Loads," *Transactions*, ASCE, 120 (1955), 1335–1352.

19. Robert T. Foreman and John L. Rumpf, "Static Tension Tests of Compact Bolted Joints," *Transactions*, ASCE, 126, Part II (1961), 228–254.

20. Gordon H. Sterling and John W. Fisher, "A440 Steel Joints Connected by A490 Bolts," *Journal of Structural Division*, ASCE, 92, ST3 (June 1966), 101–118.

21. Desi D. Vasarhelyi and Kah Ching Chiang, "Coefficient of Friction in Joints of Various Steels," *Journal of Structural Division*, ASCE, 93, ST4 (August 1967), 227–243.

22. George C. Brookhart, I. H. Siddiqi, and Desi D. Vasarhelyi, "Surface Treatment of High-Strength Bolted Joints," *Journal of Structural Division*, ASCE, 94, ST3 (March 1968), 671–681.

23. Ronald N. Allen and John W. Fisher, "Bolted Joints With Oversize or Slotted Holes," *Journal of Structural Division*, ASCE, 94, ST9 (September 1968), 2061–2080.

24. Sherwood F. Crawford and Geoffrey L. Kulak, "Eccentrically Loaded Bolted Connections," *Journal of Structural Division*, ASCE, 97, ST3 (March 1971), 765–783.

25. Geoffrey L. Kulak, "Eccentrically Loaded Slip-Resistant Connections," *Engineering Journal*, AISC, 12, 2(2nd Quarter 1975), 52–55.

26. T. R. Higgins, "New Formulas for Fasteners Loaded Off Center," *Engineering News-Record* (May 21, 1964).

27. A. L. Abolitz, "Plastic Design of Eccentrically Loaded Fasteners," *Engineering Journal*, AISC, 3, 2 (July 1966), 122–132.

28. Carl L. Shermer, "Plastic Behavior of Eccentrically-Loaded Connections," *Engineering Journal*, AISC, 8, 2 (April 1971), 48–51.

29. T. R. Higgins, "Treatment of Eccentrically-Loaded Connections in the AISC Manual," *Engineering Journal*, AISC, 8, 2 (April 1971), 52–54.

30. J. W. Fisher, "Behavior of Fasteners and Plates with Holes," *Journal of Structural Division*, ASCE, 91, ST6 (December 1965), 265–286.

31. T. R. Higgins and W. H. Munse, "How Much Combined Stress Can a Rivet Take?," *Engineering News-Record* (Dec. 4, 1952), 40–42.

32. Eugene Chesson, Jr., Norberto L. Faustino, and William H. Munse, "High-Strength Bolts Subjected to Tension and Shear," *Journal of Structural Division*, ASCE, 91, ST5 (October 1965), 155–180.

33. Hans William Hagen and Richard C. Penkul, "Design Charts for Bolts with Combined Shear and Tension," *Engineering Journal*, AISC, 2, 2 (April 1965), 42–45.

PROBLEMS

Note: Use the latest AISC Specification for all problems* except where indicated. All holes are to be assumed standard holes and surface condition is *clean mill scale* (Class A) unless otherwise indicated.

4.1. Determine the tensile capacity of the connection if $\frac{3}{4}$-in.-diam A325 bolts are used and it is designed (a) as a friction-type connection and (b) as a bearing-type connection. Assume the plates to be A36 steel and that the threads are excluded from the shear planes. Specify the minimum values for dimensions A and B for each type of connection. (19-mm bolts)

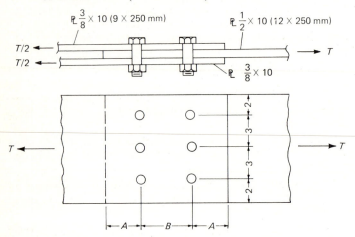

Prob. 4.1

4.2. Rework Prob. 4.1 using A572 Grade 50 steel, and consider threads not excluded from the shear planes (A325–N).

4.3. Rework Prob. 4.1 using A572 Grade 60 steel and $\frac{7}{8}$-in.-diam A325 bolts. (22-mm bolts)

4.4. Rework Prob. 4.1 using A572 Grade 65 steel and $\frac{7}{8}$-in.-diam A490 bolts. (22-mm bolts)

4.5. Determine the factor of safety against slip for the connection of whichever of Probs. 4.1 through 4.3 have been worked. Do you expect slip to occur at service load? If the bolt strength had controlled in each case would slip be expected at service load?

4.6. Determine the factor of safety against slip for the connection of Prob. 4.4. Do you expect slip to occur at service load? If bolt strength had controlled in each case would slip be expected at service load?

* Many problems may be solved either as stated in U.S. customary units or in SI units using the numerical data in parenthesis at the end of the statement. The conversions are only approximate to avoid having the given data imply accuracy greater in SI than for U.S. customary units.

4.7. Determine the capacity T capable of being carried across the butt splice of Prob. 3.19 if $s_1 = s_2 = 2$ in. and A325 bolts are used with no threads in the shear planes, if (a) a friction-type connection is used and (b) a bearing-type connection is used. Specify the end distances required, and evaluate whether or not the 2-in. given stagger distance is sufficient. If the 2 in. is not adequate, specify the value to be used. $(s_1 = s_2 = 50 \text{ mm})$

4.8. Determine the number of $\frac{3}{4}$-in.-diam bolts required to develop the capacity of the double angle tension member when designed (a) as a friction-type (A325-F) and (b) as a bearing-type connection with the threads excluded from the shear planes (A325-X). The angles are A572 Grade 50 steel. Assume a double row of bolts without stagger. Detail the connection. (19-mm bolts; L203×152×25.4; 16-mm gusset plate)

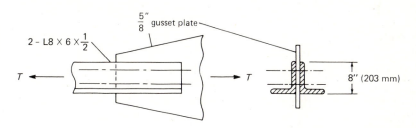

Prob. 4.8

4.9. Rework Prob. 4.8 using $\frac{7}{8}$-in.-diam bolts with threads not excluded from the shear planes (A325-F and A325-N). (22-mm bolts)

4.10. For the single angle tension member shown of A572 Grade 50 steel, how many $\frac{3}{4}$-in.-diam A325 bolts are required for the connection? Assume the connection is to be the friction-type. Use the shortest feasible overlap of pieces for the connection and detail it. (19-mm bolts; L127×89×12.7)

4.11. Rework Prob. 4.10 using $\frac{5}{8}$-in.diam A490 bolts in a bearing-type (A490-X) connection.

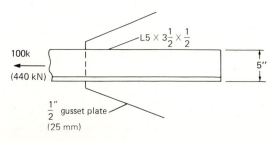

Probs. 4.10 and 4.11

4.12. For the single angle tension member of A36 steel, how many $\frac{7}{8}$-in.-diam A325 bolts are required in a bearing-type connection where threads may exist in the shear plane? Assume there are three $\frac{15}{16}$-in.-diam empty holes in the outstanding leg that are not part of the connection. The bolts carrying the 77 kips are to be located in a single line with the first of the bolts located a distance S ahead of the first empty hole. Detail the connection. (22-mm bolts; L127× 89×12.7)

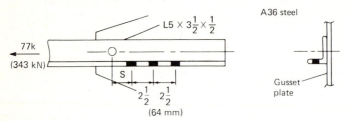

Prob. 4.12

4.13. Design and detail the double lap splice shown to develop maximum tensile capacity. A36 steel is used and $\frac{7}{8}$-in.-diam A325 bolts are to be used in a bearing-type connection with no threads in the shear plane. What is the resulting capacity of the joint? (22-mm bolts)

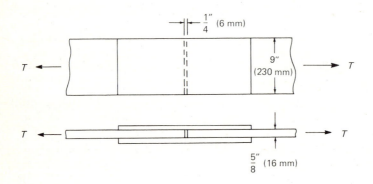

Prob. 4.13

4.14. Determine the safe capacity of this tension chord splice of two channels. The steel is A572 Grade 65, and the $\frac{7}{8}$-in.-diam A325 bolts are in a bearing-type connection with *no* threads in shear planes.

4.15. Compute the maximum stress on the fastener group if $\frac{7}{8}$-in.-diam A325 bolts are used and the threads are excluded from the shear plane. Assume the bracket plate has adequate strength. Use the elastic method with both the actual e and the "effective" e.

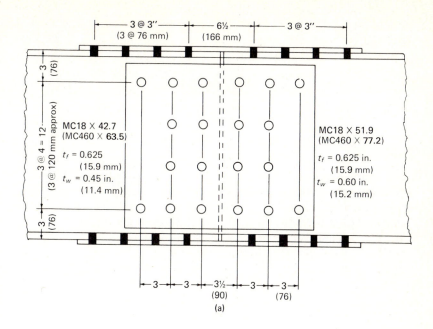

(a)

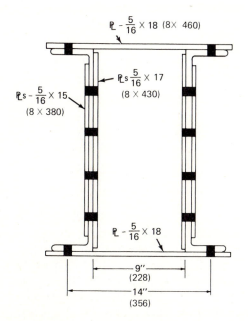

(b) Cross-section

Prob. 4.14

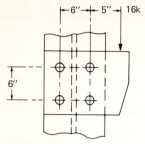

Prob. 4.15

4.16. For the bracket plate (A36 steel) of the accompanying figure, calculate the maximum allowable force P when $\frac{7}{8}$-in.-diam A325 bolts are used. Consider as a bearing-type connection with threads *not* excluded from the shear plane. (22-mm bolts for SI)
  (a) Use elastic (vector) method with the actual e.
  (b) Use elastic (vector) method with "effective e."
  (c) Use the ultimate strength method as illustrated in Example 4.8.4 with an FS = 2.5.
  (d) Use the ultimate strength method coefficient equation (Eq. 4.8.20).
  Tabulate a comparison of the results.

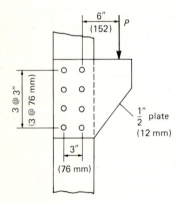

Prob. 4.16

4.17. Repeat Prob. 4.16, except use only 6 bolts instead of 8; that is, 2 @ 3 in. vertically instead of 3 @ 3 in.
4.18. Solve Prob. 4.16 as a friction-type connection instead of bearing-type. Part (c) will be as illustrated in Example 4.8.5 instead of Example 4.8.4. All other requirements are the same.
4.19. Select the proper diameter A490 bolt for a bearing-type connection assuming threads are excluded from the shear plane. (a) Use elastic method with full eccentricity and (b) use ultimate strength analysis with FS = 2.5.

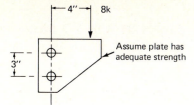

Prob. 4.19

4.20. Assuming the fasteners control the capacity, determine the bolt size required for the connection shown, using A325 bolts in a bearing-type (A325-X) connection. (a) Use elastic method with full eccentricity; and (b) use ultimate strength analysis with FS = 2.5.

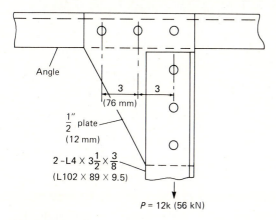

Prob. 4.20

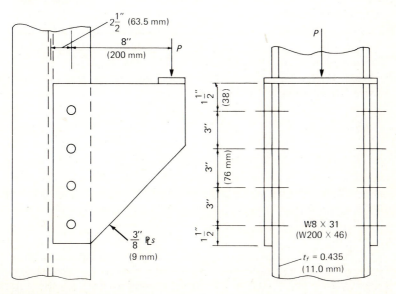

Prob. 4.21

4.21. For the given connection with $\frac{3}{4}$-in.-diam A325 bolts in a friction-type connection, determine allowable capacity P by the following methods and compare results:

(a) Use elastic (vector) method with actual e.

(b) Use elastic (vector) method with "effective e."

(c) Use ultimate strength approach for friction-type connections (see Example 4.8.5).

4.22. Two vertical rows of $\frac{7}{8}$-in.-diam A325 bolts having a 3-in. vertical spacing are acting in single shear. Select the proper number of bolts for a bearing-type (A325-X) connection. Use (a) the elastic (vector) method with the actual eccentricity, and (b) the ultimate strength analysis with an FS = 2.5. (22 mm bolts; 75 mm vertical spacing)

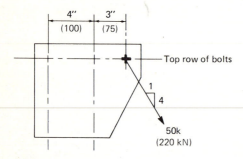

Prob. 4.22

4.23. Repeat Prob. 4.22 using $\frac{3}{4}$-in.-diam A490 bolts. (19-mm bolts)

4.24. For the connection shown using $\frac{7}{8}$-in.-diam A325 bolts in a bearing-type connection with threads excluded from the shear plane:

(a) Determine the force on each fastener and indicate its direction using basic principles of the elastic (vector) method. Is the connection adequate according to the AISC Spec?

(b) What is the capacity of the connection using the ultimate strength analysis with FS = 2.5?

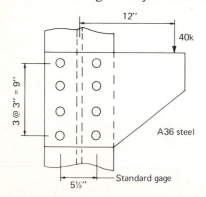

Prob. 4.24

4.25. Determine the number of $\frac{7}{8}$-in.-diam A325 bolts required to resist eccentric shear in a bearing-type connection with threads excluded from the shear plane. Compare the results using (a) the actual moment arm with the elastic method and (b) the ultimate strength analysis with FS = 2.5. What thickness of pieces is required to avoid having bearing control and still use minimum edge distances from AISC–Table 1.16.5.1?

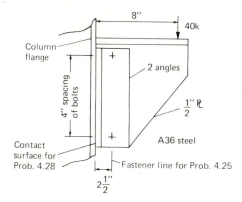

Probs. 4.25 and 4.28

4.26. Assume two angles, $4 \times 3\frac{1}{2} \times \frac{5}{8}$, have been selected to carry their maximum capacity as a tension member of A36 steel. Assume the connection of the angles to the structural tee (WT) web will be along a single gage line as shown. Determine the number and positioning of $\frac{7}{8}$-in.-diam A325 bolts required in a bearing-type connection (A325-X) to attach the WT to the flange of a W section. The flanges of the WT and the W shapes are each $\frac{3}{4}$-in. thick, and the material is A36 steel.

4.27. Repeat Prob. 4.26 using a friction-type connection.

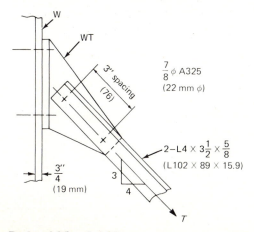

Probs. 4.26 and 4.27

4.28. Determine the number of $\frac{7}{8}$-in.-diam A325 bolts for the combined shear and tension connection of the bracket of Prob. 4.25 to a $\frac{3}{4}$-in.-thick column flange. Give design sketch.

(a) Friction-type connection.

(b) Bearing-type connection, with threads excluded from shear plane.

4.29. Determine the number of $\frac{7}{8}$-in.-diam A325 bolts needed to attach the structural tee (WT) to the column flange. Design (a) a friction-type connection and (b) a bearing-type connection with threads *not* excluded from the shear plane.

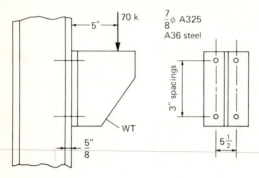

Prob. 4.29

5
Welding

5.1 INTRODUCTION AND HISTORICAL DEVELOPMENT

The process of welding denotes the joining of metal pieces by heating to a plastic or fluid state, with or without pressure. In its simplest form, "welding" has been known and used for several thousand years. Historians have speculated that the early Egyptians may have first used pressure welding about 5500 B.C. in making copper pipes from sheets by overlapping the edges and hammering. Winterton [1] has reported that Egyptian art objects dating about 3000 B.C. have been found on which gold foil has been hammered and fused onto the base copper. This type of welding, called *forge welding*, was man's first process to join pieces of metal together. A well-known early example of forge welding is the Damascus sword which was made by forging layers of iron with different properties. Interestingly, forge welding was sufficiently well developed and important enough to the early Romans that they named one of their gods Vulcan (the god of fire and metalworking) to represent that art. In recent times, the word vulcanizing has been used in reference to treating rubber with sulfur but originally was used to mean "to harden." Today, forge welding is practically a forgotten art in which the village blacksmith was the last major practitioner. Welding as it is known today is much

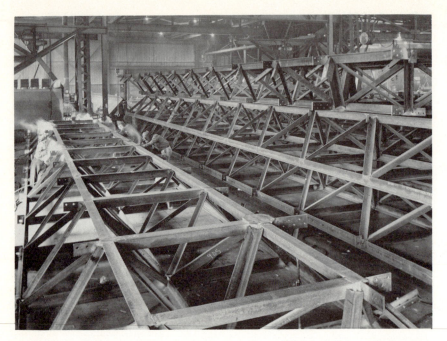

Welding of space trusses. Upjohn Company Office Building, Kalamazoo, Michigan. (Courtesy Whitehead and Kales Company, Detroit)

more complex and highly refined and the remainder of this section will trace some of the important events that have contributed to the art. Specific welding processes are discussed in Sec. 5.2.

Little progress in welding technology had been made until 1877, prior to which time most of the then known processes such as forge welding and brazing had been used for at least 3000 years. The origin of resistance welding began around 1877 when Professor Elihu Thompson began a set of experiments [2, 3] reversing the polarity of transformer coils. He received his first patent [4] in 1885 and the first resistance butt welding machine was demonstrated at the American Institute Fair in 1887. In 1889 Coffin [2] was issued a patent for flash-butt welding and this became one of the important butt welding processes.

Zerner, in 1885, introduced the carbon arc welding process, making use of two carbon electrodes and N. G. Slavinoff [5] in 1888 in Russia was the first to use the metal arc process using uncoated, bare electrodes. Coffin, working independently also investigated the metal arc process and was issued a U.S. Patent in 1892. In 1889, A. P. Strohmeyer [2] introduced the concept of coated metal electrodes to eliminate many of the problems associated with the use of bare electrodes.

Thomas Fletcher [1] in 1887 used a blowpipe burning hydrogen and oxygen and showed that he could successfully cut or melt metal. In 1901–1903 Fouche and Picard developed torches that could be used with acetylene and thus the era of oxyacetylene welding and cutting began.

The period between 1903 and 1918 saw the use of welding primarily as a method of repair, the greatest impetus occurring during World War I (1914–1918). Welding techniques proved to be especially adapted to repairing ships that had been damaged. Winterton [1] reported that in 1917 there were 103 interned enemy ships alone in the United States that were damaged and the number of persons employed in welding operations rose from 8000 to 33,000 during the period 1914–1918.

After 1919, the use of welding as a construction and fabrication technique began to develop with copper-tungsten alloy electrodes being first used for spot welding techniques [1] in 1920. The period 1930–1950 saw many improvements [2, 6] in the development of welding machines. The submerged arc welding process in which the arc is buried under a powdered flux was first used commercially in 1934 and patented in 1935.

Today there are over 50 different welding processes that can be used to join various metals and their alloys. Those of particular interest to the structural engineer are discussed in Sec. 5.2.

5.2 BASIC PROCESSES

As defined by the *Welding Handbook* [7], a welding process is "a materials joining process which produces coalescence of materials by heating them to suitable temperatures, with or without the application of pressure alone, and with or without the use of filler metal." The energy input for heat generation may be categorized according to its source: as electrical, chemical, optical, mechanical, and solid state. Heat is used to melt the base metal and filler material in order that flow of material will occur, i.e., that fusion will take place. In addition, heat is used to increase ductility so that plastic flow can occur even if melting does not take place; further, heating helps to remove contaminating films on the material.

The most common welding processes, particularly for welding structural steel, use electrical energy as the heat source; the most often used is the electric arc. The arc consists of a relatively large current discharge between electrode and base material conducted through a thermally ionized gaseous column, called a plasma [7]. In arc welding, fusion occurs by the flow of material across the arc, without pressure being applied.

Other processes, not ordinarily used for steel structures, involve other energy sources, and some of those processes involve the application of pressure, either in the absence or presence of flow of molten material. Bonding may also occur by diffusion, wherein atomic particles intermix across the interface and melting of the base material does not occur.

There are many welding processes that have special uses for particular metals and for various thicknesses. This section emphasizes those processes that are used in the welding of carbon and low-alloy steel for buildings and bridges. Arc welding is the category of processes that are of particular interest. For some situations involving light-gage steel, resistance welding may also be important.

Shielded Metal Arc Welding (SMAW)

Shielded metal arc welding is one of the oldest, simplest, and perhaps most versatile types for welding structural steel. The SMAW process is often referred to as the *manual stick electrode process*. Heating is accomplished by means of an electric arc between a coated electrode and the materials being joined. The welding circuit is shown in Fig. 5.2.1a.

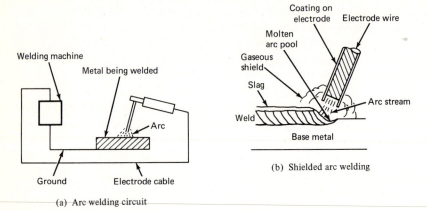

(a) Arc welding circuit

(b) Shielded arc welding

Fig. 5.2.1 Shielded metal arc welding (SMAW).

The coated electrode is consumed as the metal is transferred from the electrode to the base material during the welding process. The electrode wire becomes filler material and the coating is converted partly into a shielding gas, partly into slag, and some part is absorbed by the weld metal. The coating is a clay-like mixture of silicate binders and powdered materials, such as fluorides, carbonates, oxides, metal alloys, and cellulose. The mixture is extruded and baked to produce a dry, hard, concentric coating.

The transfer of metal from electrode to the work being welded is induced by molecular attraction and surface tension, without application of pressure. The shielding of the arc prevents atmospheric contamination of the molten metal in the arc stream and in the arc pool. It prevents nitrogen and oxygen from being picked up and forming nitrides and oxides which may cause embrittlement.

The electrode coating may perform the following functions:

1. Produces a gaseous shield to exclude air and stabilize the arc.
2. Introduces other materials, such as deoxidizers, to refine the grain structure of the weld metal.
3. Produces a blanket of slag over the molten pool and the solidified weld to protect it from oxygen and nitrogen in the air, and also retards cooling.

The electrode material is specified under various American Welding

Society specifications that are listed in AISC–1.4.5. The properties of the filler metal electrodes are summarized in Table 2.3.1. The designations such as E60XX or E70XX indicate 60 ksi and 70 ksi, respectively, for tensile strength. The X's refer to factors such as the welding positions, recommended power supply, type of coating, and type of arc characteristics. Morgan [8] has provided an excellent guide to classification and use of mild steel coated electrodes. Table 5.12.1 indicates which coated electrodes should be used with each particular structural steel.

For welding high-carbon or low-alloy steels, low hydrogen electrodes are required by AWS D1.1 [9] to be used with SMAW for all steels having yield stresses higher than 36 ksi (248 MPa). The low-hydrogen electrode is a rod with a carbonate of soda, or "lime," coating. This electrode requires a different technique from that using the conventional electrode in that a short arc must be made and globular-type, rather than a spray-type, deposition of metal occurs. It is desirable in design because the as-welded mechanical properties have been found to be superior to properties obtained using other types of electrode coatings.

Submerged Arc Welding (SAW)

In the SAW process the arc is not visible because it is covered by a blanket of granular, fusible material, as shown in Fig. 5.2.2. The bare metal electrode is consumable in that it is deposited as filler material. The end of the electrode is kept continuously shielded by the molten flux over which is deposited a layer of unfused flux in its granular condition.

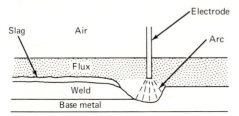

Fig. 5.2.2 Submerged arc welding (SAW).

The flux, which is the special feature of this method, provides a cover that allows the weld to be made without spatter, sparks, or smoke. The granular flux is usually laid automatically along the seam ahead of the advancing electrode. It protects the weld pool against the atmosphere, serves to clean the weld metal, and modifies the chemical composition of the weld metal.

Welds made by the submerged arc process are found to have uniformly high quality; exhibiting good ductility, high impact strength, high density, and good corrosion resistance. Mechanical properties of the weld are consistently as good as the base material.

The combinations of bare-rod electrodes and granular flux are classified under AWS A5.17 or A5.23. They are designated by a prefix F followed by a two-digit number indicating tensile strength and impact strength requirements for the resulting welds, and followed by EXXX classifying the electrode. The designations appear in Table 5.12.1 under the SAW process.

The submerged arc method is commonly used to weld steel in shop fabrication operations using automatic or semiautomatic equipment.

Gas Metal Arc Welding (GMAW)

In the GMAW process the electrode is a continuous wire that is fed from a coil through the electrode holder, a gun-shaped device as shown in Fig. 5.2.3. The shielding is entirely from an externally supplied gas or gas mixture.

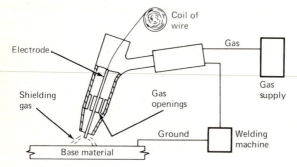

Fig. 5.2.3 Gas metal arc welding (GMAW).

Originally, this method was used only with inert gas shielding, hence, the name MIG (metal inert gas) has been used. Reactive gases alone are generally not practical; the exception is CO_2 (carbon dioxide). The use of CO_2 has become extensively used for welding of steels, either alone or in a mixture with inert gases.

Argon as a shielding gas works for welding virtually all metals; however, it is not recommended for steels because of its expense and the fact that other shielding gases and gas mixtures are acceptable. For welding carbon and some low-alloy steels either (1) 75% argon and 25% CO_2, or (2) 100% CO_2 is recommended [10]. For low-alloy steels where toughness is important, it is recommended [10] to use a mixture of 60–70% helium, 25–30% argon, and 4–5% CO_2.

The shielding gas serves the following functions in addition to protecting the molten metal from the atmosphere.

1. Controls the arc and metal-transfer characteristics.
2. Affects penetration, width of fusion, and shape of the weld region.
3. Affects the speed of welding.
4. Controls undercutting.

By mixing an inert gas with a reactive gas the arc may be made more stable and the spatter during metal transfer may be reduced. The use of CO_2 alone for welding steel is the least expensive procedure because of its lower cost for shielding gas, higher welding speed, better joint penetration, and sound deposits with good mechanical properties. The only disadvantage is that it gives harsh and excessive spatter.

The electrode material for welding carbon steels is an uncoated mild steel, deoxidized carbon manganese steel covered under AWS A5.18 and listed in Table 5.12.1 (see also Table 2.3.1). For welding low-alloy steel a deoxidized low-alloy material is necessary.

The GMAW process using CO_2 shielding is good for the lower carbon and low-alloy steels usually used in buildings and bridges.

Flux Cored Arc Welding (FCAW)

The FCAW process is similar to GMAW, except that the continuously fed filler metal electrode is tubular and contains the flux material within its core. The core material provides the same functions as does the coating in SMAW or the granular flux in SAW. For a continuously fed wire an outside coating would not remain bonded to the wire. Gas shielding is provided by the flux core but additional gas shielding is frequently provided by CO_2 gas.

The electrode-filler metal is a mild steel designated E60T or E70T under AWS A5.20, and for tensile strengths greater than 70 ksi (480 MPa) may be referred to as Grades E80T, E100T, and E110T for tensile strengths of 80, 100, and 110 ksi (550, 690, and 760 MPa), respectively.

Electroslag Welding (ESW)

The ESW process is a machine process used primarily for vertical position welding. It is used to obtain a single pass weld such as for the splice in a heavy column section. Weld metal is deposited into a cavity created by the separated plate edges and water-cooled "shoes." Molten conductive slag protects the weld and melts the filler metal and the plate edges. Since solid slag is not conductive, an arc is needed to start the process by melting the slag and heating the plates.

The arc can be extinguished once the process is well under way so that the welding is done by the heat produced through the resistance of the slag to the flow of current. Since resistance heating is used for all but the initial heat source, the ESW process is not actually an arc welding process.

Stud Welding

The most commonly used process of welding a metal stud to a base material is known as arc stud welding, an essentially automatic process but similar in characteristics to the SMAW process. The stud serves as the electrode and an electric arc is created from the end of the stud to the plate. The stud is contained in a gun which controls the timing during the process. Shielding is accomplished by placing a ceramic ferrule around the end of the stud in the gun. The gun is placed in position and the arc is created, during which time the ceramic ferrule contains the molten metal. After a short instant of time, the gun drives the stud into the molten pool and the weld is completed leaving a small fillet around the stud. Full penetration across the shank of the stud is obtained and the weld is completed usually in less than one second.

5.3 WELDABILITY OF STRUCTURAL STEEL

Most of the ASTM-specification construction steels can be welded without special precautions or special procedures. Section 5.12 discusses the need to select the proper electrode to join a particular grade of steel and a summary of the "matching" electrodes and the base steel is given in Table 5.12.1.

The *weldability* of a steel is a measure of the ease of producing a crack-free and sound structural joint. Some of the readily available structural steels are more suited to welding than others, and are discussed in Chapter 2. Welding procedures should be based on a steel's chemistry instead of the published maximum alloy content since most mill runs are usually below the maximum alloy limits set by its specification. Table 5.3.1 shows the ideal chemical analysis of the carbon steels. Most mild steels fall well within this range, while higher-strength steels may exceed the ideal analysis shown in Table 5.3.1.

Table 5.3.1 Preferred Analysis of Carbon Steel [11] for Good Weldability

Element	Normal Range (%)	Percent Requiring Special Care
Carbon	0.06–0.25	0.35
Manganese	0.35–0.80	1.40
Silicon	0.10 max	0.30
Sulfur	0.035 max	0.050
Phosphorus	0.030 max	0.040

When a mill produces a run of steel it maintains a complete record of its chemical content which follows all shapes made from the particular ingot. If the designer is concerned about the chemistry of a particular grade of steel, he may request a Mill Test Report. Any variation in chemical content above the ideal values may be evaluated and special welding procedures be set up to insure a properly welded joint.

5.4 TYPES OF JOINTS

The type of joint depends on factors such as the size and shape of the members coming into the joint, the type of loading, the amount of joint area available for welding, and the relative costs for various types of welds. There are five basic types of welded joints although many variations and combinations are found in practice. The five basic types are the butt, lap, tee, corner, and edge joints, as shown in Fig. 5.4.1.

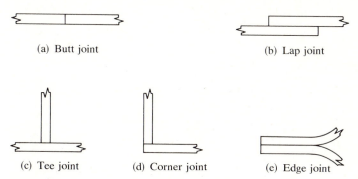

(a) Butt joint (b) Lap joint

(c) Tee joint (d) Corner joint (e) Edge joint

Fig. 5.4.1 Basic types of welded joints.

Butt Joints

The butt joint is used mainly to join the ends of flat plates of the same or nearly the same thicknesses. The principal advantage of this type of joint is to eliminate the eccentricity developed in single lap joints as shown in Fig. 5.4.1b. When used in conjunction with full penetration groove welds, butt joints minimize the size of a connection and are usually more esthetically pleasing than built-up joints. Their principal disadvantage lies in the fact that the edges to be connected must usually be specially prepared (beveled, or ground flat) and very carefully aligned prior to welding. Little adjustment is possible and the pieces must be carefully detailed and fabricated. As a result, most butt joints are made in the shop where the welding process can be accurately controlled.

Lap Joints

The lap joint shown in Fig. 5.4.2, is the most common type. It has two principal advantages:

1. *Ease of fitting.* Pieces being joined do not require the preciseness in fabricating as do the other types of joints. The pieces can be slightly shifted to accommodate minor errors in fabrication or to make adjustments in length.

2. *Ease of joining.* The edges of the pieces being joined do not need special preparation and are usually sheared or flame cut. Lap joints utilize

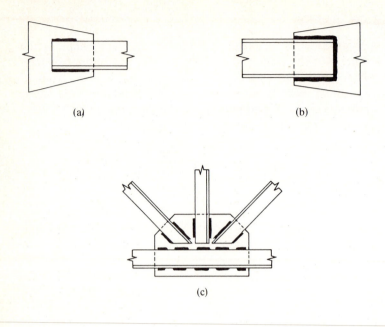

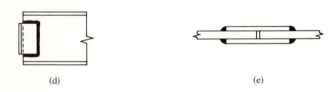

Fig. 5.4.2 Examples of lap joints.

fillet welds and are therefore equally well suited to shop or field welding. The pieces being joined are in most cases simply clamped together without the use of special jigs. Occasionally the pieces are positioned by a small number of erection bolts which may be either left in place or removed after the welding is completed.

A further advantage of the lap joint is the ease in which plates of different thicknesses can be joined, such as in the double lap joint in Fig. 5.4.2e. The reader should especially note the truss connection shown in Fig. 5.4.2c and consider the diffculty in making such a connection by any other type of joint.

Tee Joints

This type of joint is used to fabricate built-up sections such as tees, I-shapes, plate girders, bearing stiffeners, hangers, brackets, and in general, pieces framing in at right angles as shown in Fig. 5.4.1c. This type of joint is especially useful in that it permits sections to be built up of flat plates that can be joined by either fillet or groove welds.

Corner Joints

Corner joints are used principally to form built-up rectangular box sections such as used for columns and for beams required to resist high torsional forces.

Edge Joints

Edge joints are generally not structural but are most frequently used to keep two or more plates in a given plane or to maintain initial alignment.

As the reader can infer from the previous discussions, the variations and combinations of the five basic types of welds are virtually infinite. Since there is usually more than one way to connect one structural member to another, the designer is left with the decision for selecting the best joint (or combination joints) in each given situation.

5.5 TYPES OF WELDS

The four types of welds are the groove, fillet, slot, and plug welds as shown in Fig. 5.5.1. Each type of weld has specific advantages that

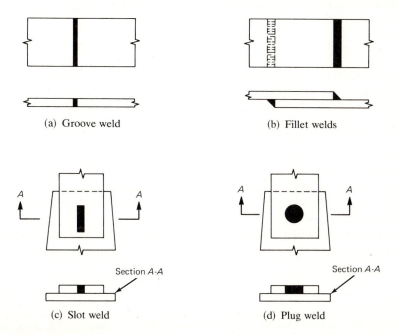

(a) Groove weld (b) Fillet welds

(c) Slot weld (d) Plug weld

Fig. 5.5.1 Types of welds.

determine the extent of its usage. Roughly, the four types represent the following percentages of welded construction: groove welds, 15%; fillet welds, 80%; the remaining 5% are made up of the slot, plug, and other special welds.

Groove Welds

The principal use of groove welds is to connect structural members that are aligned in the same plane. Since groove welds are usually intended to transmit the full load of the members they join, the weld should have the same strength as the pieces joined. Such a groove weld is known as a *complete joint penetration groove weld*. When joints are designed so that groove welds do not extend completely through the thickness of the pieces being joined, such welds are referred to as *partial joint penetration groove welds*. For these, special design requirements apply.

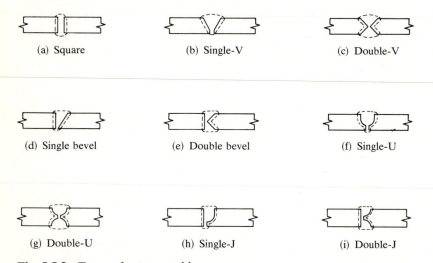

    (a) Square   (b) Single-V   (c) Double-V

    (d) Single bevel   (e) Double bevel   (f) Single-U

    (g) Double-U   (h) Single-J   (i) Double-J

Fig. 5.5.2 Types of groove welds.

There are many variations of groove welds and each is classified according to its particular shape. Most groove welds require a specific edge preparation and are named accordingly. Figure 5.5.2 shows common types of groove welds and indicates the groove preparations required for each. The selection of the proper groove weld is dependent on the welding process used, the cost of edge preparations, and the cost of making the weld. Groove welds may also be used in tee connections as shown in Fig. 5.5.3.

Fillet Welds

Fillet welds owing to their overall economy, ease of fabricating, and adaptability are the most widely used of all the basic welds. A few uses of

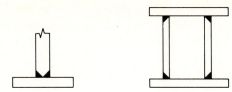

Fig. 5.5.3 Use of groove welds in tee joints.

fillet welds are shown in Fig. 5.5.4. They generally require less precision in the "fitting up" because of the overlapping of pieces, whereas the groove weld requires careful alignment with specified gap (root opening) between pieces. The fillet weld is particularly advantageous to welding in the field or in realigning members or connections that were fabricated

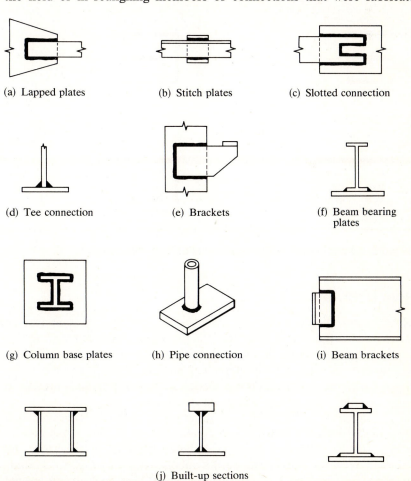

(a) Lapped plates

(b) Stitch plates

(c) Slotted connection

(d) Tee connection

(e) Brackets

(f) Beam bearing plates

(g) Column base plates

(h) Pipe connection

(i) Beam brackets

(j) Built-up sections

Fig. 5.5.4 Typical uses of fillet welds.

within accepted tolerances but which may not fit as accurately as desired. In addition, the edges of pieces being joined seldom need special prepartion such as beveling or squaring since the edge conditions resulting from flame cutting or from shear cutting procedures are generally adequate.

Slot and Plug Welds

Slot and plug welds may be used exclusively in a connection as shown in Figs. 5.5.1c and d, or they may be used in combination with fillet welds as shown in Fig. 5.5.5. A principal use for plug or slot welds is to transmit shear in a lap joint when the size of the connection limits the length available for fillet or other edge welds. Slot and plug welds are also useful in preventing overlapping parts from buckling.

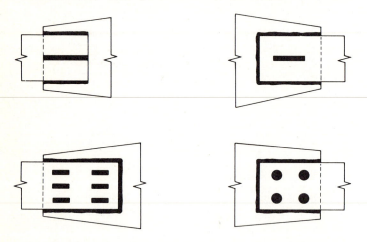

Fig. 5.5.5 Slot and plug welds in combination with fillet welds.

5.6 WELDING SYMBOLS

Before a connection or joint is welded, the designer must in some way be able to instruct the steel detailer and the fabricator as to the type and size of weld required. The basic types of welds and some of their variations are discussed in Sec. 5.5. If individual and detailed instructions were needed each time a connection was made, the task of providing directions for making the joint would indeed be formidable.

The need for a simple and yet accurate method for communicating between the designer and fabricator gave rise to the use of shorthand symbols that characterize the type and size of weld. As a result, the American Welding Society standard symbols, shown in Fig. 5.6.1, indicate the type, size, length, and location of weld, as well as any special instructions.

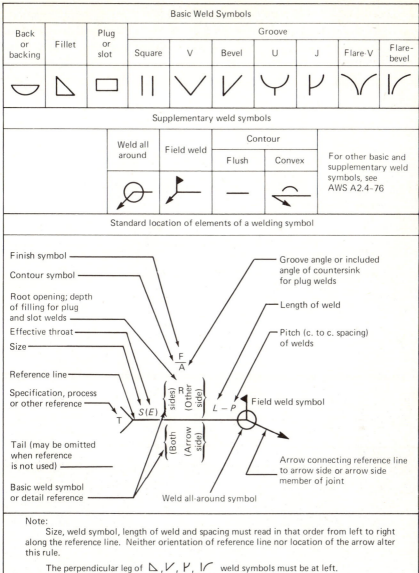

Basic Weld Symbols									
Back or backing	Fillet	Plug or slot	Groove						
			Square	V	Bevel	U	J	Flare-V	Flare-bevel
⌣	△	▭	‖	V	⌐V	Y	ꓕ	V	⌐

Supplementary weld symbols					
		Weld all around	Field weld	Contour	
				Flush	Convex
		⌀	⌐	▬	⌒

For other basic and supplementary weld symbols, see AWS A2.4–76

Standard location of elements of a welding symbol

Finish symbol
Contour symbol
Root opening; depth of filling for plug and slot welds
Effective throat
Size
Reference line
Specification, process or other reference
Tail (may be omitted when reference is not used)
Basic weld symbol or detail reference

Groove angle or included angle of countersink for plug welds
Length of weld
Pitch (c. to c. spacing) of welds
Field weld symbol
Arrow connecting reference line to arrow side or arrow side member of joint
Weld all-around symbol

$\frac{F}{A}$ R   (Other side) (Both sides) (Arrow side) $S(E)$ $L-P$ T

Note:

    Size, weld symbol, length of weld and spacing must read in that order from left to right along the reference line. Neither orientation of reference line nor location of the arrow alter this rule.

    The perpendicular leg of △, V, ꓕ, ⌐ weld symbols must be at left.

Arrow and Other Side welds are of the same size unless otherwise shown.

    Symbols apply between abrupt changes in direction of welding unless governed by the "all around" symbol or otherwise dimensioned.

    These symbols do not explicitly provide for the case that frequently occurs in structural work, where duplicate material (such as stiffeners) occurs on the far side of a web or gusset plate. The fabricating industry has adopted this convention; that when the billing of the detail material discloses the identity of far side with near side, the welding shown for the near side shall also be duplicated on the far side.

Fig. 5.6.1 Standard welding symbols. (Adapted from AISC Manual and AWS A2.4 [12])

Most of the commonly made connections do not require special instructions and are typically specified as shown in Fig. 5.6.2. For a more detailed use of welding symbols the reader is referred to the American Welding Society, AWS A2.4 [12].

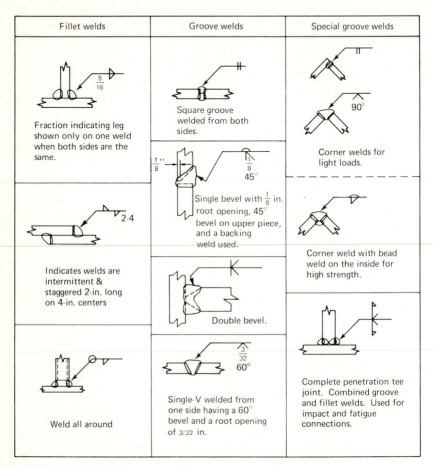

Fillet welds	Groove welds	Special groove welds

Fraction indicating leg shown only on one weld when both sides are the same.

Square groove welded from both sides.

Single bevel with $\frac{1}{8}$ in. root opening, 45° bevel on upper piece, and a backing weld used.

Corner welds for light loads.

Indicates welds are intermittent & staggered 2-in. long on 4-in. centers

Double bevel.

Corner weld with bead weld on the inside for high strength.

Weld all around

Single-V welded from one side having a 60° bevel and a root opening of 3/32 in.

Complete penetration tee joint. Combined groove and fillet welds. Used for impact and fatigue connections.

Fig. 5.6.2 Common uses of welding symbols.

The reader may feel that the number of symbols is burdensome. However, the system of designating welds is broken down into a few basic types that are built up to give a complete set of instructions. Whenever a particular connection is used in many parts of a structure, it may only be necessary to show a typical detail as shown in Fig. 5.6.3a. Whenever special connections are used, they should be detailed sufficiently to leave no doubt as to the designer's intentions, as shown in Fig. 5.6.3b.

In Fig. 5.6.3b the designer specified that the plug weld be made in the shop and ground flush while the double bevel weld connecting the gusset plate to the column be made in the field. Since the designer did not

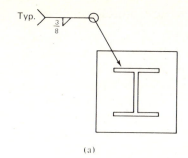

(a)

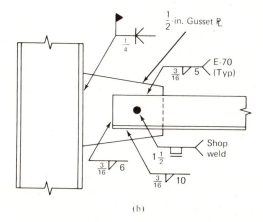

(b)

Fig. 5.6.3   Details showing use of welding symbols.

specify whether the fillet welds attaching the angle to the gusset plate were to be made in the shop or in the field, the steel fabricator would be free to make the decision. However, in this particular detail, it would be better to make the fillet welds in the shop since the plug weld might be over-stressed during the field erection process. In general, the fabricator will make as many welds in the shop as he can, due to economic considerations. Therefore it is important that the designer specify those welds he wants to be *field welded*.

5.7 FACTORS AFFECTING THE QUALITY OF WELDED CONNECTIONS

Obtaining a satisfactory welded connection requires the combination of many individual skills, beginning with the actual design of the weld and ending with the welding operation. The structural engineer needs to be aware of the factors that affect the quality of a weld and design his connections accordingly.

Proper Electrodes, Welding Apparatus, and Procedures

After the proper electrode material is specified to match the strength of the steel in the pieces being joined (see Sec. 5.12), the diameter of the welding electrode must be selected. The particular size of the electrode selected is based on the size of the weld to be made and on the electrical current output of the welding apparatus. Since most welding machines have controls for reducing the current output, electrodes smaller than the maximum capability can easily be accommodated and should be used.

Since the weld metal in arc welding is deposited by the electromagnetic field and not by gravity, the welder is not limited to the flat or horizontal welding positions. The four basic welding positions are shown in Fig. 5.7.1. The designer should avoid whenever possible the overhead position, since it is the most difficult one. Joints welded in the shop are usually positioned in the flat or horizontal positions but field welds may require any welding position depending on the orientation of the connection. The welding position for field welds should be carefully considered by the designer.

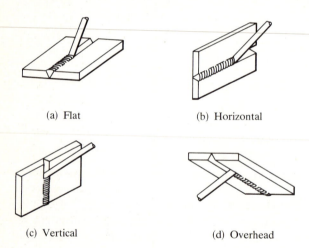

(a) Flat

(b) Horizontal

(c) Vertical

(d) Overhead

Fig. 5.7.1 Welding positions.

Proper Edge Preparation

Typical edge preparations provided for groove welds are shown in Fig. 5.7.2. The root opening R is the separation of the pieces being joined and is provided for electrode accessibility to the base of a joint. The smaller the root opening the greater must be the angle of the bevel. The feathered edge as shown in Fig. 5.7.2a is subject to burn-through unless a backup plate is provided as shown in Fig. 5.7.2b. Backup strips are commonly used when the welding is to be done from one side only. The problem of burn-through is lessened if the bevel is provided a land as

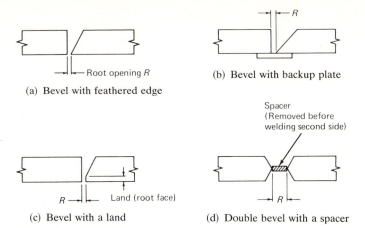

(a) Bevel with feathered edge

(b) Bevel with backup plate

(c) Bevel with a land

(d) Double bevel with a spacer

Fig. 5.7.2   Typical edge preparations for groove welds.

shown in Fig. 5.7.2c. The welder should *not* provide a backup plate when a land is provided since there would be a good possibility that a gas pocket would be formed preventing a full penetration weld. Occasionally a spacer, as shown in Fig. 5.7.2d, is provided to prevent burn-through but is gouged out before the second side is welded.

Control of Distortion

Another factor affecting weld quality is shrinkage. If a single bead is put down in a continuous manner on a plate, it will cause the plate to distort as shown in Fig. 5.7.3. Such distortions will occur unless care is exercised in both the design of the joint and the welding procedure. Figure 5.7.4 shows the result of using unsymmetrical welds as compared to symmetrical welds. Although there are many techniques available for minimizing distortion, the most common one is that of staggering intermittent welds as shown in Fig. 5.7.5a, and then returning to fill in the spaces as shown in Fig. 5.7.5b, a typical sequence being shown. For many structures, such as plate girders, short segments of weld (though not usually regular intermittent welds) may be used at strategic locations to give enough strength to hold all pieces in place; then the continuous lines of weld are placed.

Fig. 5.7.3   Distortion of plate.

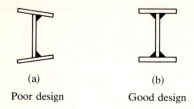

(a)   (b)

Poor design   Good design

Fig. 5.7.4   Effect of weld placement.

To minimize shrinkage and to insure adequate ductility, the American Welding Society (Table 4.2 of Ref. 9) has established minimum preheat and interpass temperatures. For welds requiring more than one progression (pass) of a welding operation along a joint, the interpass temperature is the temperature of the deposited weld when the next pass is begun.

The following summarizes ways of minimizing distortion:

1. Reduce the shrinkage forces by
   (a) Using minimum weld metal; for grooves use no greater root opening than necessary; do not overweld.
   (b) Using as few passes as possible.
   (c) Using proper edge preparation and fit-up.
   (d) Using intermittent weld, at least for preliminary connection.
   (e) Using backstepping; depositing weld segments toward the previously completed weld; i.e., depositing in the direction opposite to the progress of welding the joint.
2. Allow for the shrinkage to occur by
   (a) Tipping the plates so after shrinkage occurs they will be correctly aligned.
   (b) Using prebending of pieces.

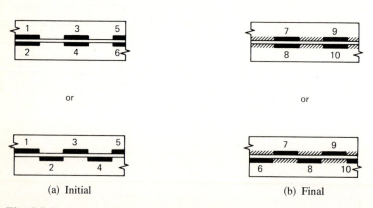

(a) Initial   (b) Final

Fig. 5.7.5   Sequences for intermittent welds.

3. Balance shrinkage forces by
   (a) Using symmetry in welding; fillets on each side of a piece contribute counteracting effects.
   (b) Using scattered weld segments.
   (c) Using peening; stretching the metal by a series of blows.
   (d) Using of clamps, jigs, etc; this forces weld metal to stretch as it cools.

5.8 POSSIBLE DEFECTS IN WELDS

Unless good welding techniques and procedures are used, a number of possible defects may result relating to discontinuities within the weld. Some of the more common defects are: incomplete fusion, inadequate joint penetration, porosity, undercutting, inclusion of slag, and cracks. Examples of these defects are shown in Fig. 5.8.1.

Incomplete Fusion

Incomplete fusion is the failure of the base metal and the adjacent weld metal to fuse together completely. This may occur if the surfaces to be joined have not been properly cleaned and are coated with mill scale, slag, oxides, or other foreign materials. Another cause of this defect is the use of welding equipment of insufficient current, so that base metal does not reach melting point. Too rapid a rate of welding will also have the same effect.

Inadequate Joint Penetration

Inadequate joint penetration means the weld extends a shallower distance through the depth of the groove than specified, as shown in Fig. 5.8.1, where complete penetration was specified. Partial joint penetration is acceptable only when it is so specified.

This defect, relating primarily to groove welds, occurs from use of an unsuitable groove design for the selected welding process, excessively large electrodes, insufficient welding current, or excessive welding rates. Joint designs prequalified by AWS (AWS D1.1–79, Sec. 2) should always be used.

Porosity

Porosity occurs when voids or a number of small gas pockets are trapped during the cooling process. This defect results from using excessively high current or too long an arc length. Porosity may occur uniformly dispersed through the weld, or it may be a large pocket concentrated at the root of a fillet weld or at the root adjacent to a backup plate in a groove weld.

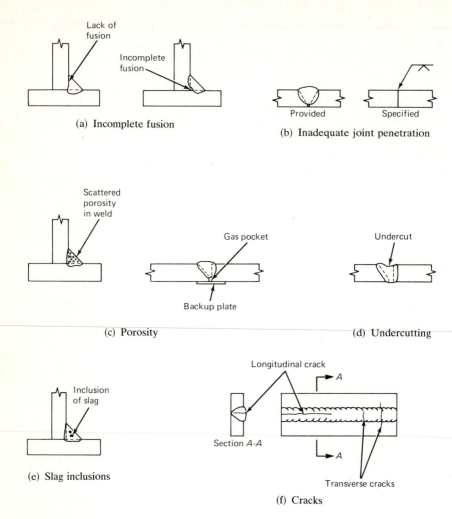

Fig. 5.8.1 Possible weld defects.

The latter is caused by poor welding procedures and careless use of backup plates.

Undercutting

Undercutting means a groove melted into the base material adjacent to the toe of a weld and left unfilled by weld metal. The use of excessive current or an excessively long arc may burn or dig away a portion of the base metal. This defect is easily detected visually and can be corrected by depositing additional weld material.

Slag Inclusion

Slag is formed during the welding process as a result of chemical reactions of the melted electrode coating and consists of metal oxides and other compounds. Having a lower density than the molten weld metal, the slag normally floats to the surface, where upon cooling, it is easily removed by the welder. However, too rapid a cooling of the joint may trap the slag before it can rise to the surface. Overhead welds as shown in Fig. 5.7.1d are especially subject to slag inclusion and must be carefully inspected. When several passes are necessary to obtain the desired weld size, the welder must remove slag between each pass. Failure to properly do so is a common cause of slag inclusion.

Cracks

Cracks are breaks in the weld metal, either longitudinal or transverse to the line of weld, that result from internal stress. Cracks may also extend from the weld metal into the base metal or may be entirely in the base metal in the vicinity of the weld. Cracks are perhaps the most harmful of weld defects; however, tiny cracks called *microfissures* may not have any detrimental effect.

Some cracks form as the weld begins to solidify, generally caused by brittle constituents, either brittle states of iron or alloying elements, forming along the grain boundaries. More uniform heating and slower cooling will prevent the "hot" cracks from forming.

Cracks may also form at room temperature parallel to but under the weld in the base material. These cracks arise in low-alloy steels from the combined effects of hydrogen, a brittle martensite microstructure, and restraint to shrinkage and distortion. Use of low-hydrogen electrodes along with proper preheating and postheating will minimize such "cold" cracking.

5.9 INSPECTION AND CONTROL

The enormous success and growth in recent years in the area of structual welding of buildings and bridges could not have occurred without some means of inspection and control. The welding industry has led in the development of guidelines which, if followed, virtually insure a sound weld. The inspection and control procedure should begin before the first arc is struck, continue throughout the welding procedure, and if necessary, a pretest of the joint should be made to assure its satisfactory performance. Since such close supervision is not possible on every weld made, the following suggestions will serve as a guideline to achieve good structural welds:

    1. Establish good welding procedures.

2. Use only prequalified welders.
3. Use qualified inspectors and have them present.
4. Use special inspection techniques when necessary.

Good welding procedures can be developed from recommendations from the AWS, AISC, and the manufacturers of welding supplies and equipment. The procedure to be followed will depend on the chemical and physical properties of the materials, the types and sizes of weld, and the particular equipment used.

All welders should be required to have passed an American Welding Society Qualification Test before being permitted to make a structural connection. Although this is usually considered adequate, it doesn't prove the ability of the welder to make welds at the actual job site, particularly if the welds are unusual or difficult and were not specified in the Qualification Test. Happily, most welding contractors exercise control over their welders in such situations.

The use of qualified welding inspectors at a jobsite generally has the effect of causing welders to perform their best work, feeling that the inspector is able to recognize the quality of their welds. The welding inspector should be a competent welder himself and be able to recognize possible defects. Any poor or suspicious welds should be cut out and replaced.

The simplest and least expensive method of inspection is *visual* but it is dependent on the competence of the observer. A welding gage such as shown in Fig. 5.9.1a offers a rapid means of checking the size of fillet welds.

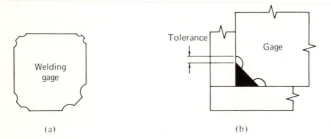

(a)   (b)

Fig. 5.9.1 Checking size of fillet welds.

For more important structures and for welds whose failure could be catastrophic, more rigid inspection techniques should be used. Some of the useful ones are the ultrasonic, radiographic, and magnetic particle methods. The *ultrasonic* method [13] passes ultra-high-frequency sound waves through the weldment. Defects in a particular weld will reflect the sound waves while a weld without defects will not impede passage of the waves. *Radiographic* methods include the use of both X-rays and gamma rays. In this method the radiating source is placed on one side of the weld and a photographic plate on the other. This method is expensive and requires special precautions be taken due to the hazards of radiation. However, the method is reliable and furnishes a permanent record. The *magnetic particle* testing method uses iron powder which is spread around the welded area and polarized by passing an electric current through the weld. Small local poles will be formed at the edges of any defects, and this may be interpreted by an experienced observer.

5.10 ECONOMICS OF WELDED BUILT-UP MEMBERS AND CONNECTIONS

Overall economy in welded connections is difficult to evaluate. Some of the factors to be considered such as the amount of electrode material used can easily be computed while other factors such as the value to be placed on esthetics may be intangible. The actual economy of welded connections must be viewed from a broad aspect and include the overall design of the structural system.

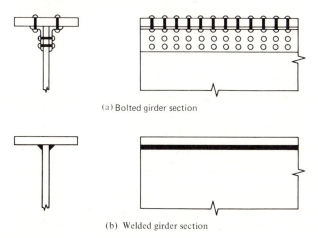

(a) Bolted girder section

(b) Welded girder section

Fig. 5.10.1 Comparison between plate girders.

Welded connections are usually neater in appearance, providing a less cluttered effect in contrast to bolted connections. Figure 5.10.1 shows a comparison between a section of a bolted plate girder and a section of a welded plate girder. Besides the neater appearance of the welded joint, welded connections offer the designer more freedom to be innovative in his entire design concept. The designer is not bound to standard sections but may build up any cross section he feels to be most advantageous. Similarly, he can use the best configuration to transmit the loads from one member to another.

Welded connections generally eliminate the need for holes in members except possibly for erection purposes. Since it is usually the holes at the ends that govern the design of a bolted tension member, a welded connection will generally result in a member with a smaller cross section.

Welded connections can sometime reduce field construction costs by the fact that members may be shifted slightly to accommodate minor errors in fabricating or erection. Also, members may be easily shortened by cutting and rejoining by suitable welding as well as lengthened by splicing a piece of the same cross section.

In addition, several direct factors influence the cost of welding. Generally, welding performed in the shop is less expensive than field welding. Some reasons for this are availability in the shop of automatic welding machines, a more pleasant and less hostile environment (the weather), and the availability of special jigs for holding the pieces to be welded in a more favorable position. Also, work can be scheduled for a continuous operation whereas field welding must often wait for cranes and special erection equipment. Other operations such as the proper preheating of pieces to be welded can be difficult if not impossible to perform in the field. Other factors that influence welding costs are:

1. Cost of preparing the edges to be welded.
2. The amount of weld material required.
3. The ratio of the actual arc time to overall welding time.
4. The amount of handling required.
5. General overhead costs.

The factors listed above are generally unknown to the designer since the fabricator is usually not selected until after the design has been completed. However the designer must still make decisions—should he specify short large fillet welds or long small fillet welds? Should he specify large fillet welds or groove welds? If he decides to use groove welds he must then select the proper and most economical type.

In most instances the designer is not as concerned with the specific costs of each type of weld as he is with *relative* cost of the various types and sizes. As aid to the designer, Donnelly [14] developed relative cost factors relating the cost of fillet and groove welds of common sizes to the cost of a single-pass $\frac{1}{4}$ in. fillet weld. This work is reproduced in Table 5.10.1. The cost of a specific weld can be determined by Eq. 5.10.1 once the designer can obtain the local costs.

Cost of weld = (Length of weld) × (Factor from Table 5.10.1)

$$\times \text{(Local cost)} \qquad \text{(5.10.1)}$$

Currently (1979), welded connections are used for the vast majority of shop connections and a sizable though not a majority of field connections.

Table 5.10.1 Relative Cost Factors for Welded Joints (from Donnelly [14]) (Based on a Cost Factor at 1.00 for a $\frac{1}{4}$ in. Fillet Weld)

Type Size T (in.)	Fillet 45°	Fillet 60°–30° (see sketch)	Single V 60°	Single V 90°	Single Bevel 45°	Single Bevel 60°	Double V 60°	Double V 90°	Double Bevel 45°	Double Bevel 60°	Square Groove	Plug Weld Dia. of Hole (in.)	Plug Weld Cost Factor
1/16	0.55	—	—	—	—	—	—	—	—	—	0.55	7/16	25.20
1/8	0.70	0.80	—	—	—	—	—	—	—	—	0.60	9/16	39.50
3/16	0.85	0.95	—	1.75	—	—	—	—	—	—	0.80	11/16	57.50
1/4	1.00	1.25	1.35	2.70	1.30	1.35	1.20	1.30	—	—	1.00	13/16	78.80
5/16	1.20	1.75	1.95	3.60	1.80	1.95	1.45	1.75	—	—	1.20	15/16	104.00
3/8	1.45	2.55	2.70	4.95	2.50	2.70	1.85	2.50	1.75	2.00	1.50	1 1/16	132.00
7/16	2.00	3.30	3.45	5.10	3.15	3.45	2.40	2.90	2.20	2.50	4.80	1 3/16	165.00
1/2	2.60	4.20	3.80	8.00	3.50	3.90	2.90	3.70	2.70	3.10	5.55		
5/8	3.80	6.30	5.55	11.50	5.05	5.55	4.70	5.40	3.80	4.30	—		
3/4	5.30	8.90	7.85	15.30	7.10	7.90	5.45	7.55	5.05	5.85	—		
7/8	7.10	12.00	10.20	20.15	9.10	10.22	7.05	10.10	6.45	7.55	—		
1	9.15	15.50	12.80	25.60	11.50	12.95	9.15	13.25	8.20	9.65	—		
1 1/8	11.50	19.50	15.70	30.20	14.15	15.95	11.15	16.45	10.00	11.80	—		
1 1/4	14.10	24.00	19.00	35.70	16.90	19.10	13.15	19.85	11.90	14.10	—		
1 3/8	17.00	29.10	22.50	42.85	20.15	22.75	15.70	24.00	14.10	16.70	—		
1 1/2	20.00	34.50	26.40	49.80	23.60	26.75	18.30	28.10	16.40	19.60	—		
1 5/8	—	—	30.90	59.00	27.30	31.20	20.80	32.50	18.70	22.60	—		
1 3/4	—	—	35.40	69.20	31.20	35.60	23.40	36.90	21.00	25.40	—		
1 7/8	—	—	40.00	79.00	35.50	40.40	26.20	41.15	23.30	28.20	—		
2	—	—	45.00	—	40.00	45.50	29.00	46.40	25.60	31.10	—		
2 1/8	—	—	—	—	—	—	—	—	30.40	28.00	—		
2 1/4	—	—	—	—	—	—	—	—	32.90	—	—		
2 3/8	—	—	—	—	—	—	—	—	35.60	—	—		
2 1/2	—	—	—	—	—	—	—	—	38.70	—	—		
2 5/8	—	—	—	—	—	—	—	—	42.00	—	—		
2 3/4	—	—	—	—	—	—	—	—	45.45	—	—		
2 7/8	—	—	—	—	—	—	—	—	49.00	—	—		
3	—	—	—	—	—	—	—	—	—	—	—		

Plug Weld (per 100 holes 1 in. deep)

Notes: Factors apply to all weld processes. A single weld position is assumed for entire weld.

5.11 SIZE AND LENGTH LIMITATIONS FOR FILLET WELDS

Since all welding involves the heating of the metal pieces, prevention of too rapid a rate of cooling is of fundamental importance to achieving a good weld. Consider the two extreme thicknesses of plates in Fig. 5.11.1, each of which has received a bead of fillet weld. Most heat energy given off during the welding process is absorbed by plates being joined. The thicker plate dissipates the heat vertically as well as horizontally whereas the thinner plate is essentially limited to a horizontal dissipation. In other words, the thicker the plate, the faster heat is removed from the welding area, thereby lowering the temperature in the region of the weld. Since a minimum temperature is required to cause the base metal to become molten, it is therefore necessary to provide as a minimum, a weld of sufficient size (and heat content) to prevent the plate from removing the heat at a faster rate than it is being supplied. Unless a proper temperature is maintained in the area being welded a lack of fusion will result.

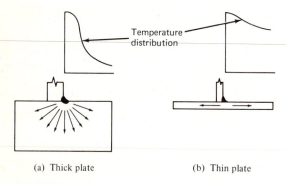

(a) Thick plate   (b) Thin plate

Fig. 5.11.1 Effect of thickness on cooling rate.

Minimum Weld Size

To help insure fusion and minimize distortion the AWS [9] and AISC Specifications provide for a minimum size weld based on the thicker of the pieces being joined. The requirements for fillet weld based on the leg dimension a of the fillet and for partial joint penetration groove weld based on the effective throat (see Sec. 5.13) are given in Table 5.11.1.

Maximum Fillet Weld Size Along Edges

The *maximum* size of fillet weld used *along the edges* of pieces being joined is limited by AISC–1.17.3 in order to prevent the melting of the base material at the location where the fillet would meet the corner of the plate if the fillet were made the full plate thickness. The maximum

Table 5.11.1 Minimum Size Fillet Weld and Minimum Effective Throat for Partial Joint Penetration Groove Welds (from AISC–Table 1.17.2 and AWS D1.1 [9], Tables 2.7 and 2.10.3)

Base Metal Thickness (T) of Thicker Part Joined		Minimum Size* of Fillet Weld		Minimum Effective Throat (t_e) for Partial Joint Penetration Groove Weld	
(in.)	(mm)	(in.)	(mm)	(in.)	(mm)
$T \leq 1/4$	$T \leq 6.4$	1/8†	3‡	1/8	3
$1/4 < T \leq 1/2$	$6.4 < T \leq 12.7$	3/16	5	3/16	5
$1/2 < T \leq 3/4$	$12.7 < T \leq 19.0$	1/4	6	1/4	6
$3/4 < T \leq 1\,1/2$	$19.0 < T \leq 38.1$	5/16	8	5/16	8
$1\,1/2 < T \leq 2\,1/4$	$38.1 < T \leq 57.1$	5/16	8	3/8	10
$2\,1/4 < T \leq 6$	$57.1 < T \leq 152$	5/16	8	1/2	13
$6 < T$	$152 < T$	5/16	8	5/8	16

* Weld size is the leg dimension of fillet weld. The weld size need not exceed the thickness of the thinner part joined. For this exception particular care should be taken to provide sufficient preheat to ensure weld soundness.
† Minimum size for bridge applications is 3/16 in.
‡ For metric, minimum single pass welds must be used.

permitted is (see Fig. 5.11.2):

1. Along edges of material less than $\frac{1}{4}$-in. (6.4-mm) thick, the maximum size may be equal to the thickness of the material.
2. Along edges of material $\frac{1}{4}$ in. (6.4 mm) or more in thickness, the maximum size shall be $\frac{1}{16}$ in. (1.6 mm) less than the thickness of the material, unless the weld is especially designated on the drawings to be built out to obtain full throat thickness.

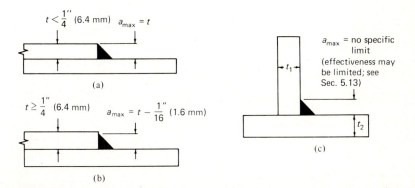

Fig. 5.11.2 Maximum weld size.

Minimum Effective Length of Fillet Welds

When placing a fillet weld, the welder builds up the weld to its full dimension as near the beginning of the weld as he can. However, there is always a slight tapering off in the region where the weld is started and where it ends. AISC–1.17.4 therefore limits the minimum effective length of fillet welds to four times their nominal size. If this criterion is not met, the size of the weld shall be considered to be one-fourth of the effective length.

AISC–1.17.7 recommends the use of end returns, whenever practicable, as shown in Fig. 5.11.3. For other limitations the reader is referred to the AISC Specification.

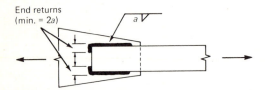

Fig. 5.11.3 Use of end returns.

5.12 ALLOWABLE STRESSES

Since welds must transmit the entire load from one member to another, welds must be sized accordingly and be formed from the correct electrode material.

While distribution of stresses in fillet welds is discussed in Sec. 5.14, it is noted here that fillet welds are assumed for design purposes to transmit loads through *shear stress* on the effective area no matter how the fillets are oriented on the structural connection. Groove welds transmit loads exactly as in the pieces they join.

The electrode material used in welds should have properties of the base material. When properties are comparable the weld metal is referred to as "matching" weld metal. Table 5.12.1 gives "matching" weld metal for many ASTM structural steels used in buildings and bridges.

The allowable stresses on the effective areas are given in Table 5.12.2. For complete joint penetration groove welds the allowable stress is the same as permitted in the base material; however, in tension the "matching" weld metal *must* be used whereas in compression a lesser strength may be used. AWS [9] requires a tensile strength not less than 10 ksi (70 MPa) below the value for "matching" weld.

The allowable stresses for (1) shear on the effective area of all welds and (2) the tensile stress normal to the axis on the effective area of a partial joint penetration groove weld are equal to 0.30 times the electrode tensile strength. However, the stress in the base material may not exceed $0.60F_y$ for tension or $0.40F_y$ for shear.

Table 5.12.1   Matching Filler Metal Requirements (from AWS–Table 4.1.1, Ref. 9)

Group	Base Metal Steel Specification*	Welding Process †‡			
		Shielded Metal Arc Welding (SMAW)	Submerged Arc Welding (SAW)	Gas Metal Arc Welding (GMAW)	Flux Cored Arc Welding (FCAW)
I	ASTM A36, A53 Grade B, A500, A501, A529, A570 Grades D and E, and A709 Grade 36	AWS A5.1 or A5.5 E60XX or E70XX	AWS A5.17 or A5.23 F6X or F7X-EXXX	AWS A5.18 E70S-X or E70U-1	AWS A5.20 E60T-X (except EXXT-2 and EXXT-3)
II	ASTM A242§, A441, A572 Grades 42–55, A588§, and A709 Grades 50 and 50W	AWS A5.1 or A5.5, E70XX‖	AWS A5.17 or A5.23 F7X-EXXX	AWS A5.18 E70S-X or E70U-1	AWS A5.20 E70T-X (except E70T-2 and E70T-3)
III	ASTM A572 Grades 60 and 65	AWS A5.5 E80XX‖	AWS A5.23 F8X-EXXX ¶	Grade E80S¶	Grade E80T¶
IV	ASTM A514 (over 2½ in. thick) and A709 Grades 100 and 100W (2½ to 4 in.)	AWS A5.5 E100XX‖	AWS A5.23 F10X-EXXX¶	Grade E100S¶	Grade E100T¶
V	ASTM A514 (2½ in. and under) and A709 Grades 100 and 100W (2½ in. and under)	AWS A5.5 E110XX‖	AWS A5.23 F11X-EXXX¶	Grade E110S¶	Grade E110T¶

* In joints involving base metals of two different yield points or strengths, filler metal electrodes applicable to the lower strength base metal may be used, except that if the higher strength base metal requires low hydrogen electrodes, they shall be used.

† When welds are to be stress relieved, the deposited weld metal shall not exceed 0.05 percent vanadium.

‡ See AWS D1.1–79, Sec. 4.20 for electroslag and electrogas weld requirements.

§ For bare unpainted applications of ASTM A242 and A588 steel requiring weld metal with atmospheric corrosion resistance and coloring characteristics similar to that of the base metal, the electrode, electrode-flux combination, or grade of weld metal shall be in accordance with AWS D1.1–79, Table 4.1.4.

‖ Low hydrogen classifications only.

¶ Deposited weld metal shall have a minimum impact strength of 20 ft-lb (27.1 J) at 0°F (−18°C) when Charpy V-notch specimens are used. This requirement is applicable only to bridges.

Table 5.12.2 Allowable Stresses on Effective Area of Welds (AISC–Table 1.5.3; also AWS [9]–Table 8.4.1)

Type of Stress on Effective Area*	Allowable Stress	Required Weld Strength Level†‡
Complete Joint Penetration Groove Welds		
Tension normal to effective area	Same as base metal	"Matching" weld metal must be used
Compression normal to effective area	Same as base metal	"Weld metal with a strength level equal to or less than "matching" weld metal may be used
Tension or compression parallel to axis of weld	Same as base metal	
Shear on effective area	0.30×nominal tensile strength of weld metal, except shear stress on base metal shall not exceed 0.40 ×yield stress of base metal	
Partial Joint Penetration Groove Welds§		
Compression normal to effective area	Same as base metal for joints designed to bear	Weld metal with a strength level equal to or less than "matching" weld metal may be used
Tension or compression parallel to axis of weld‖	Same as base metal	
Shear parallel to axis of weld	0.30×nominal tensile strength of weld metal, except shear stress on base metal shall not exceed 0.40×yield stress of base metal	
Tension normal to effective area	0.30×nominal tensile strength of weld metal, except tensile stress on base metal shall not exceed 0.60×yield stress of base metal	
Fillet Welds		
Shear on effective area	0.30×nominal tensile strength of weld metal, except shear stress on base metal shall not exceed 0.40 ×yield stress of vase metal	Weld metal with a strength level equal to or less than "matching" weld metal may be used
Tension or compression parallel to axis of weld‖	Same as base metal	

Table   **5.12.2** (contd.)

Type of Stress on Effective Area*	Allowable Stress	Required Weld Strength Level†‡
	Plug and Slot Welds	
Shear parallel to faying surfaces (on effective area)	0.30 × nominal tensile strength of weld metal, except shear stress on base metal shall not exceed 0.40 × yield stress of base metal	Weld metal with a strength level equal to or less than "matching" weld metal may be used

* For definition of effective area see AISC–1.14.6.
† For "matching" weld metal, see AWS–Table 4.1.1 [9].
‡ Weld metal one strength level stronger than "matching" will be permitted.
§ See AISC–1.10.8 for a limitation on use of partial joint penetration groove welded joints.
‖ Fillet welds and partial joint penetration groove welds joining the component elements of built-up members, such as flange-to-web connections, may be designed without regard to the tensile or compressive stress in these elements parallel to the axis of the welds.

The allowable stresses on fillet welds were significantly increased to the present values in the 1969 AISC Specification. For the E60 and E70 electrodes, the only ones included in 1963 AISC Specification, the increase was about 33 percent. Higgins and Preece [15] have discussed the research on which the present values are based and state that "in recognition of improvements in the art of welding that have taken place in the past 40 years, the more liberal working stress provision—0.3 times the electrode tensile strength—appears fully justified."

The effective areas are discussed in Sec. 5.13 and the limitations on the maximum and minimum size welds were discussed in Sec. 5.11.

5.13   EFFECTIVE AREAS OF WELDS

The allowable stresses on the various types of welds summarized in Sec. 5.12 are nominal stresses acting on the *effective areas*. The effective area of a groove or fillet weld is the product of the *effective throat* dimension t_e, times the length of the weld.

The effective throat dimension depends on the nominal size and the shape of the weld, and may be thought of as the minimum width of the expected failure plane.

Groove Welds

The effective throat dimension of a full joint penetration groove weld is the thickness of the thinner part joined as shown in Figs. 5.13.1a and b.

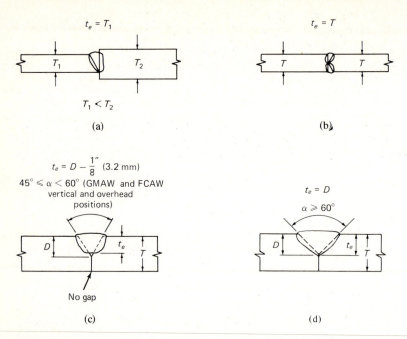

Fig. 5.13.1 Effective throat dimensions for groove welds (AISC–1.14.6) made by SMAW, SAW, GMAW, and FCAW.

The effective throat dimension of partial joint penetration groove welds is the depth of chamfer less $\frac{1}{8}$ in. for grooves having an included angle at the root of the groove less than 60° but not less than 45° when SMAW or SAW processes are used, or when the GMAW or FCAW processes are used in vertical or overhead positions. When the included angle is 60° or more the effective throat is the full depth of chamfer for all four processes mentioned. Effective throat requirements for the groove situations mentioned, as well as others, are given in AISC–1.14.6.

Fillet Welds

The effective throat dimension of a fillet weld is nominally the shortest distance from the root to the face of the weld, as shown in Fig. 5.13.2. Assuming the fillet weld to have equal legs of nominal size a, the effective throat, t_e, is $0.707a$. If the fillet weld is designed to be unsymmetrical (a rare situation) with unequal legs, as shown in Fig. 5.13.2b, the value of t_e must be computed from the diagrammatic shape of the weld. The effective throat dimensions for fillet welds made by the submerged arc (SAW) process are modified by AISC–1.14.6 as follows to account for the inherently superior quality of such welds:

    1. For fillet welds with the leg size equal to or less than $\frac{3}{8}$ in.

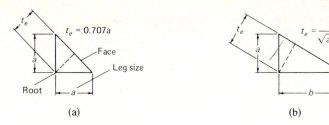

Fig. 5.13.2 Effective throat dimensions for fillet welds (except by submerged arc process).

(9.5 mm), the effective throat dimension shall be taken as equal to the leg size a.

2. For fillet welds larger than $\frac{3}{8}$ in. (9.5 mm), the effective throat dimension shall be taken as the theoretical throat dimension plus 0.11 in. (2.8 mm) (i.e., $0.707a + 0.11$).

Tables 5.13.1 and 5.13.2 summarize the effective throat dimensions and the allowable shear resistance R_w of fillet welds in kips per inch of weld.

Table 5.13.1 Allowable Resistance of Fillet Welds, kips/in.
              (Shielded Metal Arc Welding)

		Allowable Shear in Fillet Welds, R_w, (kips/in. of weld)					
Nominal Size (in.)	Effective Throat (AISC–1.14.6) (in.)	Minimum Tensile Strength of Weld (ksi)					
		60	70	80	90	100	110
1/8	0.088	1.59	1.86	2.12	2.39	2.69	2.92
3/16	0.132	2.38	2.78	3.18	3.58	3.97	4.37
1/4	0.177	3.18	3.71	4.24	4.77	5.30	5.83
5/16	0.221	3.98	4.64	5.30	5.96	6.63	7.30
3/8	0.265	4.77	5.57	6.36	7.16	7.95	8.75
7/16	0.309	5.57	6.49	7.42	8.35	9.28	10.21
1/2	0.353	6.36	7.42	8.48	9.54	10.60	11.66
9/16	0.398	7.16	8.35	9.54	10.74	11.93	13.12
5/8	0.442	7.95	9.28	10.61	11.93	13.26	14.58
11/16	0.486	8.75	10.21	11.67	13.12	14.58	16.04
3/4	0.530	9.54	11.13	12.72	14.31	15.91	17.50

Plug and Slot Welds

The effective shearing area of plug or slot welds is their nominal area (sometimes called *faying surface*) in the shearing plane. The resistance of plug or slot welds is the product of the nominal cross section times the allowable stress.

Table 5.13.2   Allowable Resistance of Fillet Welds, kips/in.
(Submerged Arc Welding)

Nominal Size (in.)	Effective Throat (AISC–1.14.6) (in.)	Allowable Shear in Fillet Welds, R_w, (kips/in. of weld) Minimum Tensile Strength of Weld (ksi)					
		60	70	80	90	100	110
1/8	0.125	2.25	2.62	3.00	3.37	3.75	4.12
3/16	0.187	3.37	3.94	4.50	5.06	5.62	6.19
1/4	0.250	4.50	5.25	6.00	6.75	7.50	8.25
5/16	0.312	5.62	6.56	7.50	8.44	9.37	10.31
3/8	0.375	6.75	7.87	9.00	10.12	11.25	12.37
7/16	0.419	7.55	8.80	10.06	11.32	12.58	13.84
1/2	0.463	8.34	9.73	11.12	12.51	13.90	15.30
9/16	0.508	9.14	10.66	12.18	13.71	15.23	16.75
5/8	0.552	9.93	11.59	13.25	14.90	16.56	18.21
11/16	0.596	10.73	12.52	14.31	16.09	17.88	19.67
3/4	0.640	11.52	13.44	15.36	17.28	19.21	21.13

Example 5.13.1

Determine the effective throat dimension of a $\frac{7}{16}$-in. fillet weld produced by (a) shielded metal arc process and (b) submerged arc process, according to AISC Spec.

SOLUTION

(a) $t_e = 0.707a = 0.707(0.4375) = 0.309$ in.
(b) $t_e = 0.707a + 0.11 = 0.707(0.4375) + 0.11 = 0.419$ in.

Example 5.13.2

Determine the allowable shear resistance of a $\frac{3}{8}$-in. fillet weld produced by (a) shielded metal arc process and (b) submerged arc process. Assume use of E70 electrodes having minimum tensile strength F_u of 70 ksi, and use AISC Spec.

SOLUTION

(a) SMAW process. $t_e = 0.707a = 0.707(0.375) = 0.265$ in.

$R_w = t_e(0.30F_u) = 0.265(0.3)70 = 0.265(21) = 5.57$ kips/in.

(b) SAW process. $t_e = a = 0.375$ in.

$R_w = t_e(0.30F_u) = 0.375(21.0) = 7.87$ kips/in.

The solution to Examples 5.13.1 and 5.13.2 may be checked by referring to Tables 5.13.1 and 5.13.2.

Example 5.13.3

Determine the allowable capacity of a $\frac{3}{4}$-in.-diam plug weld using E70 electrode material. Use AISC Spec.

SOLUTION

Assuming the weld diameter satisfies the limitations of AISC–1.17.9 relating to the dimension of the piece in which the plug weld is made,

$$\text{Capacity} = (\pi D^2/4)(0.30 F_u)$$
$$= 0.4418(21) = 9.28 \text{ kips}$$

Maximum Effective Fillet Weld Size

In Sec. 5.11 the limitations on maximum and minimum fillet weld size and length relating to practical design considerations were given. Those requirements relate to the size of weld that is actually placed. Regarding strength, however, no welds of whatever size may be designed using a stress greater than permitted on the adjacent base material.

Consider the two lines of fillet weld transmitting the shear V across section a–a of Fig. 5.13.3a. Equating the allowable capacity per inch of weld metal to the allowable capacity per inch in shear on the plate (using a shear allowable stress of $0.40 F_y$ as given by Table 5.12.2) gives for the

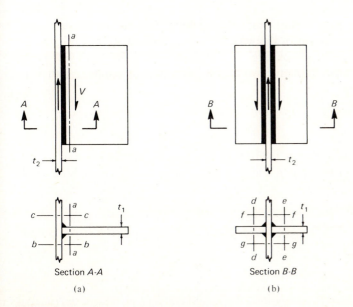

Section *A-A*   Section *B-B*

(a)   (b)

Fig. 5.13.3 Critical sections for possible overstressing of base material.

SMAW process

$$\underbrace{2a(0.707)(0.3F_u)}_{\text{weld}} = \underbrace{0.40F_y t_1}_{\text{plate}} \tag{5.13.1}$$

$$a_{\text{max eff}} = \frac{0.40F_y t_1}{2(0.707)(0.3F_u)} = 0.943 \frac{F_y t_1}{F_u} \tag{5.13.2}$$

where t_1 = thickness of base material
   F_u = tensile strength of electrode material (70 ksi for E70 electrodes)
   F_y = yield stress of base material

Sections $b-b$ and $c-c$ will not be critical since two lines of weld transfer load across two sections; maximum effective weld size for the transfer across $b-b$ and $c-c$ is

$$a(0.707)(0.3F_u) = 0.40F_y t_2 \tag{5.13.3}$$

$$a_{\text{max eff}} = 1.89 \frac{F_y t_2}{F_u} \tag{5.13.4}$$

Considering the four fillets of Fig. 5.13.3b, sections $d-d$ and $e-e$ are the same as section $a-a$ and Eq. 5.13.2 applies. On sections $f-f$ and $g-g$ four fillets are transferring load across two sections:

$$4a(0.707)(0.3F_u) = 2(0.40F_y)t_2$$

and the result is again, Eq. 5.13.2.

Even when fillet welds connect members that are in tension, the transfer of load by means of the weld is a shear transfer to the base pieces when the fillet welds are parallel to the direction of the load. For such cases, the maximum effective weld size concept still applies, with Eq. 5.13.4 applicable.

Example 5.13.4

Determine the capacity per inch of weld, R_w, to be used in design of the flange to web connection in Fig. 5.13.4. The plates are A36 steel and

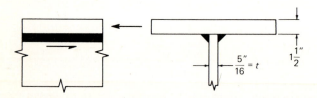

Fig. 5.13.4 Example 5.13.4.

electrodes having $F_u = 70$ ksi are to be used with (a) the shielded metal arc (SMAW) process and (b) the submerged arc (SAW) process.

SOLUTION

Minimum weld size $= a_{min} = \frac{5}{16}$ in. (AISC–1.17.2)

(a) Shielded metal arc process. Equation 5.13.2 applies

$$a_{max\,eff} = 0.943 \frac{F_y t_1}{F_u} = 0.943 \frac{36(t)}{70} = 0.485t$$

$$= 0.485(\tfrac{5}{16}) = 0.152 \text{ in.}$$

$$R_w = 2(0.152)(0.707)21 = 4.50 \text{ kips/in. (790 N/mm) for 2 fillets}$$

Even though a $\frac{5}{16}$-in. fillet must be placed, its strength in design may not exceed the strength assuming $a = 0.152$ in.

(b) Submerged arc process. Here the throat dimension may equal the leg size. Equating weld strength to plate strength gives

$$2(a)(21) = 0.4(36)t$$

$$a_{max\,eff} = 0.343t = 0.343(\tfrac{5}{16}) = 0.107 \text{ in.}$$

$$R_w = 2(0.107)21 = 4.50 \text{ kips/in. (790 N/mm)}$$

Again, the same result is obtained as in (a) since 4.50 kips/in. is the allowable capacity of the web plate, and is well below the actual weld capacity.

5.14 STRESS DISTRIBUTION IN FILLET WELDS

The procedure for designing welded connections uses nominal stresses in the welds as criteria of safety. Just as for bolted and riveted connections, nominal stresses for welds assume the weld is elastic and the plates making up the joint are rigid. When a welded joint is subjected to any one of shear, compression, or tension, the stresses in the welds are assumed to be constant over the length of the welds. When a weld is subjected to moment alone or torsion alone, the stresses are assumed to vary linearly depending on location of the neutral axis. When two or more of the aforementioned types of loading occur simultaneously the resultant stresses are assumed to be the vector sum.

However, the actual stress distribution in a welded connection is complex even in the simplest joint. The welds *and* the pieces being joined must deform together, otherwise a separation will occur. In addition, the true stresses are altered by the existence of residual stresses due to the cooling of the welds, warping stresses arising from poor weld procedures, and stress relieving in the pieces being joined. A detailed and quantitative study of the actual stresses in welded connection is beyond the scope of

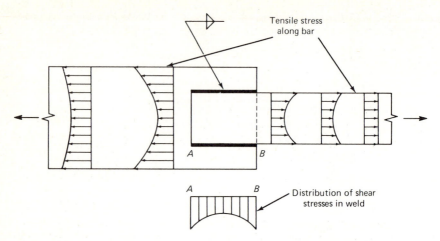

Fig. 5.14.1   Typical stress distribution in a lap joint with longitudinal fillet welds.

this text. However, in order for the reader to obtain an appreciation for the complexity of the problem, a few common types of joints will be shown.

Figure 5.14.1 shows the typical shear distribution in longitudinal fillet welds. The actual magnitude of the variation from points A to B will depend on length of the weld as well as the ratio of widths of the two plates being joined. Figure 5.14.2 shows the typical shear distribution for transverse fillet welds. Again the magnitude of the shear distribution will depend on the relative width of the plates and length of the weld.

The stress distributions in fillet welds used to connect tee-joints is somewhat more complex as indicated in Fig. 5.14.3. Due to the tendency of the fillet to rotate about point C, the maximum tensile stress in the y-direction, f_y, is approximately 4 times the nominal or average stress $f_{y \, avg}$.

If a correct stress analysis were required each time a connection was designed, few welded connections would be made. Instead, once the

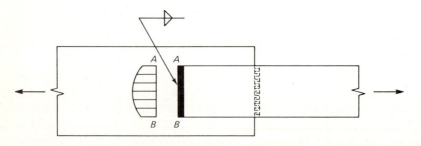

Fig. 5.14.2   Typical stress distribution in a lap joint with transverse fillet welds.

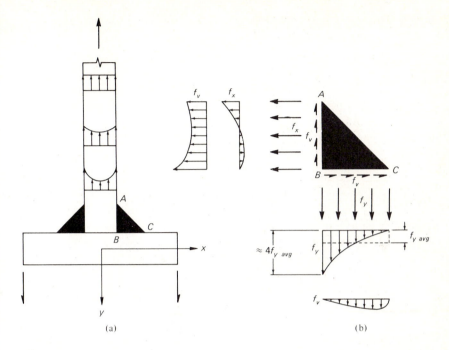

Fig. 5.14.3 Typical stress distribution in a tee joint with fillet welds (shear transverse to fillet).

desired margin of safety is established between service load and ultimate strength, allowable nominal stresses may be set in full recognition of the procedure under which service load stresses will be computed. The designer is then able to make a simple analysis using nominal stresses and select appropriately sized welds with confidence. The fact that the welds will deform plastically after yielding and thus redistribute unusually high stresses is analogous to the assumptions made for rows of bolts as discussed in Sec. 4.5.

It would serve little purpose to develop an "exact" analysis for welded joints, since in most engineering design problems the loading is rarely known to that degree of precision. However, the above discussion should serve to alert the reader to the true distribution of the stresses and should temper his approach to the design of welded connections.

5.15 WELDS CONNECTING MEMBERS SUBJECT TO DIRECT AXIAL STRESS

In the design of welds connecting tension or compression members, the principal task is to insure that the welds are at least as strong as the members they connect and that the connection does not introduce significant eccentricity of loading.

Groove Welds

In the case of full joint penetration groove welds as shown in Fig. 5.5.2, the full strength of the cross section may be developed by selecting the proper electrode material corresponding to the base material as indicated in Table 5.12.1, and specifying an AWS prequalified joint.

Example 5.15.1

Select the required thickness of the plates (A572 Grade 50) and the proper electrode material assuming a single-V-groove weld with submerged arc welding (SAW) for the member in Fig. 5.15.1.

Fig. 5.15.1 Example 5.15.1.

SOLUTION
Since there are no holes to consider, the net section is 6 in. times the thickness, t. The allowable tensile stress F_t in the member is $0.60F_y$, or 30 ksi. The required thickness is therefore

$$\text{Required } t = \frac{T}{6F_t} = \frac{72}{6(30)} = 0.40 \text{ in.}, \qquad \text{use } \tfrac{7}{16} \times 6 \text{ plates}$$

From Table 5.12.1, use F7X-EXXX flux electrode combination and a full joint penetration groove weld.

See AISC Manual, "Welded Joints," or AWS, Table 2.9.1, Ref. 9, for prequalified single-V-groove weld designated B-L2a-S, as indicated ion Fig. 5.15.2.

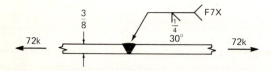

Fig. 5.15.2 Design sketch for Example 5.15.1.

Example 5.15.2

Repeat Example 5.15.1 using A572 Grade 65 plates, a square-groove weld with a zero root opening, and the submerged arc (SAW) process.

SOLUTION

$$\text{Required}\quad t = \frac{72}{6(0.60F_y)} = \frac{72}{6(39)} = 0.31 \text{ in.,}\qquad \text{use } \tfrac{5}{16} \times 6 \text{ plates}$$

From Table 5.12.1, use F8X-EXXX flux electrode combination. See AISC Manual, "Welded Joints," or AWS, Table 2.9.1, Ref. 9, for prequalified square-groove weld designated B-L1-S, as indicated in Fig. 5.15.3.

Fig. 5.15.3 Design sketch for Example 5.15.2.

Note: In Examples 5.15.1 and 5.15.2 it was not necessary to include the weld size or the length of the welds since they are to be made full joint penetration and the full width of the plates unless otherwise specified.

Example 5.15.3

Determine the capacity of the tee connection shown in Fig. 5.15.4 and detail the proper double-bevel-groove weld, assuming the SMAW process. Assume the flange of the tee does not control design.

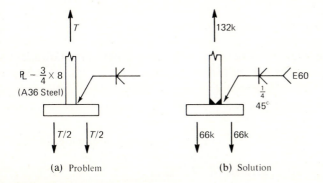

(a) Problem   (b) Solution

Fig. 5.15.4 Example 5.15.3.

SOLUTION

The allowable tensile force is

$$T = 0.60F_yA_g = 22(8)(0.75) = 132 \text{ kips}$$

From Table 5.12.1, use E60 electrodes. See AISC Manual, "Welded Joints," or AWS, Table 2.9.1, Ref. 9, for prequalified double-bevel-groove weld designated TC-U5a or b.

Note: On the basis of strength only, a single $\frac{3}{4}$-in. bevel (TC-U4c) could have been used instead of the double-bevel-groove weld specified. However, welding the stem of the tee from one side only may cause excessive warping and introduces eccentricity into the connection.

Fillet Welds

The design for fillet welds is based on the nominal shear stress on the effective area of the fillet weld as discussed in Sec. 5.13. The selection of the size of the fillet weld is based on the thickness of the pieces being joined and the available length over which the fillet weld can be made. Other factors such as the type of welding equipment used, whether the welds are to be made in the field or in the shop, and the size of other welds being made will also influence the size of fillet specified. Large fillet welds require larger diameter electrodes which in turn require larger and bulkier welding equipment, not necessarily convenient for field use. The most economical size of fillet weld is usually the one that can be made in one pass; about $\frac{5}{16}$ in. for SMAW and $\frac{1}{2}$ in. for SAW. Also, if a certain size of fillet weld is used in adjacent areas to the particular joint in question, it is advisable to use the same size since then the same electrodes and welding equipment could be used and the welder would not have to alter his procedure to accommodate a larger or smaller weld. In addition, inspection of the welds is further simplified.

Example 5.15.4

Determine the size and length of the fillet weld for the lap joint shown in Fig. 5.15.5 using the submerged arc (SAW) process if the plates are A36 steel.

SOLUTION

Referring to Sec. 5.11, AISC–1.17.2 and 1.17.3 give the following limits:

$$\text{Maximum size} = \tfrac{5}{8} - \tfrac{1}{16} = \tfrac{9}{16} \text{ in.}$$
$$\text{Minimum size} = \tfrac{1}{4} \text{ in.}$$

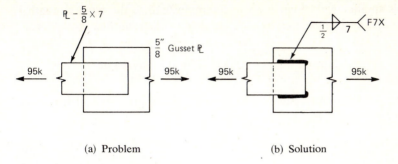

(a) Problem   (b) Solution

Fig. 5.15.5 Example 5.15.4.

Use $\frac{1}{2}$-in. fillet weld, since that is about the maximum size that can be made in one pass by the SAW process.

From AISC–1.14.6, since the nominal size of fillet weld is over $\frac{3}{8}$ in., the effective throat dimension is equal to the theoretical throat plus 0.11 in.

$$\text{Effective throat} = 0.707(0.50) + 0.11 = 0.464 \text{ in,}$$

From Table 5.12.1 use F7X-EXXX flux electrode combination. The capacity of $\frac{1}{2}$-in. fillet weld per inch of length is

$$R_w = (\text{Effective throat})(\text{allowable stress})$$
$$= 0.464(21.0) = 9.73 \text{ kips/in.}$$

The $\frac{5}{8}$-in. plate may not be assumed to carry a shear stress in excess of $0.40F_y$; thus

$$\text{Max } R_w = 0.40F_y t = 14.4(0.625) = 9.0 \text{ kips/in.} \quad \text{(controls)}$$

Length of weld, L_w, required is

$$L_w = \frac{95}{9.0} = 10.6 \text{ in.,} \quad \text{use } \tfrac{1}{2} \text{ in. fillet, 7 in. on each side}$$

Fillet length is extended to satisfy AISC–1.17.4 (par. 2), as in Fig. 5.15.5.

Example 5.15.5

Rework Example 5.15.4 using $\frac{1}{4}$-in. fillet welds.

SOLUTION
Since the nominal size is less than $\frac{3}{8}$ in., the full leg dimension may be used as the effective throat dimension for submerged arc welding.

$$R_w = 0.250(21.0) = 5.25 \text{ kips/in.}$$

For this smaller weld than used in Example 5.15.4, the weld strength is less than plate strength (9.0 kips/in.) so that weld strength controls here.

$$\text{Required } L_w = \frac{95}{5.25} = 18.1 \text{ in.,} \quad \text{say 19 in.}$$

Two possible solutions are shown in Fig. 5.15.6, both of which provide 19 in. of $\frac{1}{4}$-in. fillet welds.

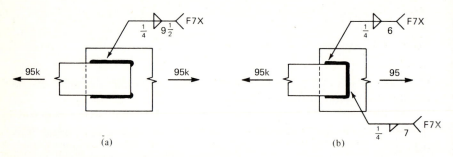

(a)   (b)

Fig. 5.15.6 Solutions to Example 5.15.5.

The solution in Fig. 5.15.6b is preferred since it is more compact and reduces the overall length of the connection.

In a number of cases, members subjected to direct axial stress are themselves unsymmetrical and cause eccentricities in welded connections. Consider the angle tension member shown in Fig. 5.15.7 welded as indicated. The force T applied at some distance from the connection will act along the centroid of the member as shown. The force T will be resisted by the forces F_1, F_2, and F_3 developed by the weld lines. The forces F_1 and F_3 are assumed to act at the edges of the angle rather than more correctly at the center of the effective throat. The force F_2 will act at the centroid of the weld length which is located at $d/2$. Taking moments about point A located on the bottom edge of the member and considering clockwise moments positive,

$$\sum M_A = -F_1 d - F_2 d/2 + Ty = 0 \qquad (5.15.1)$$

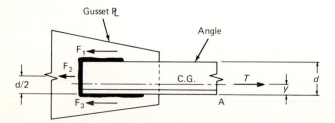

Fig. 5.15.7 Balancing the welds on a tension member connection.

or

$$F_1 = \frac{Ty}{d} - \frac{F_2}{2} \tag{5.15.2}$$

The force F_2 is equal to the resistance of the weld per inch times the length of the weld:

$$F_2 = R_w L_{w2} \tag{(5.15.3)}$$

Horizontal force equilibrium gives

$$\sum F_H = T - F_1 - F_2 - F_3 = 0 \tag{5.15.4}$$

Solving Eqs. 5.15.1 and 5.15.4 simultaneously gives

$$F_3 = T\left(1 - \frac{y}{d}\right) - \frac{F_2}{2} \tag{5.15.5}$$

Designing the connection shown in Fig. 5.15.7 to eliminate eccentricity caused by the unsymmetrical weld is called *balancing the welds*. The procedure for balancing the welds may be summarized as follows:

1. After deciding on the proper weld size and electrode, compute the force resisted by the end weld F_2 (if any) using Eq. 5.15.3.
2. Compute F_1 using Eq. 5.15.2.
3. Compute F_3 using Eq. 5.15.5, or

$$F_3 = T - F_1 - F_2 \tag{5.15.6}$$

4. Compute the lengths, L_{w1} and L_{w3}, on the basis of

$$L_{w1} = \frac{F_1}{R_w} \tag{5.15.7a}$$

and

$$L_{w3} = \frac{F_3}{R_w} \tag{5.15.7b}$$

It is noted that approximately balanced welds are desirable; however, unless the member is subject to fatigue loading, a balanced connection for "single angle, double angle, and similar type members is not required" by AISC–1.15.3.

Example 5.15.6

Design the fillet welds to develop the full strength of the angle shown in Fig. 5.15.8 minimizing the effect of eccentricity. Assume the gusset plate does not govern and the SMAW process is used.

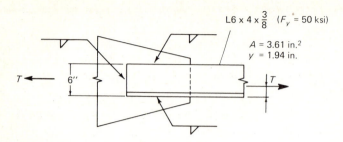

Fig. 5.15.8 Example 5.15.6.

SOLUTION

Using the forces as in Fig. 5.15.8,

$$T = 0.60F_y A_g = 30(3.61) = 108.3 \text{ kips}$$

Min size fillet weld $= \frac{3}{16}$ in. (Table 5.11.1)
Max size fillet weld $= \frac{3}{8} - \frac{1}{16} = \frac{5}{16}$ in. (Fig. 5.11.2)
Use $\frac{3}{16}$-in. fillet weld with E70 electrodes.

$$R_w = \frac{3}{16}(0.707)21 = 2.79 \text{ kips/in.} \quad \text{(controls)}$$

$$\text{Max } R_w = 0.40F_y t = 20(0.375) = 7.5 \text{ kips/in.}$$

Referring to Fig. 5.15.9,

$$F_2 = R_w L_w = 2.79(6) = 16.7 \text{ kips}$$

From moment equilibrium about the back (at F_3) of angle,

$$F_1 = \frac{108.3(1.94) - 16.7(3)}{6} = 26.7 \text{ kips}$$

Summation of forces gives

$$F_3 = T - F_1 - F_2 = 108.3 - 26.7 - 16.7 = 64.9 \text{ kips}$$

$$L_{w1} = \frac{F_1}{R_w} = \frac{26.7}{2.79} = 9.6 \text{ in.}, \quad \text{say 10 in.}$$

$$L_{w3} = \frac{F_3}{R_w} = \frac{64.9}{2.79} = 23.3 \text{ in.}, \quad \text{say 24 in.}$$

Use welds as summarized in Fig. 5.15.10, though for better economy

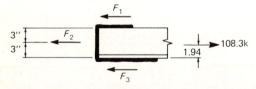

Fig. 5.15.9 Balancing the welds for Example 5.15.6.

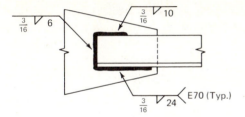

Fig. 5.15.10 Solution for Example 5.15.6.

larger welds would be preferred to reduce connection length. Note that the fillet weld along the back should terminate with about a 1-in. distance to the edge of plate maintained as recommended by the AWS Commentary [Ref. 16, p. 26].

Example 5.15.7

Rework Example 5.15.6 if the weld at the end of the angle is omitted, and the SAW process is used instead of the SMAW process.

SOLUTION

Again, try $\frac{3}{16}$ in. weld on the $F_y = 50$ ksi base material. Using the forces in Fig. 5.15.11,

$$F_1 = \frac{108.3(1.94)}{6} = 35.0 \text{ kips}$$

$$F_3 = T - F_1 = 108.3 - 35.0 = 73.3 \text{ kips}$$

$$R_w = \tfrac{3}{16}(21) = 3.94 \text{ kips/in.} < 7.5 \text{ kips/in. (plate)} \qquad \text{OK}$$

$$L_{w1} = \frac{35.0}{3.94} = 8.9 \text{ in.}, \qquad \text{say 9 in.}$$

$$L_{w3} = \frac{73.3}{3.94} = 18.7 \text{ in.}, \qquad \text{say 19 in.}$$

The length of L_{w3} requires a long joint which probably is uneconomical. Therefore it is advisable to use a larger size fillet weld.

Try a $\frac{5}{16}$ in. weld.

Effective throat dimension $= \frac{5}{16}$ in. (submerged arc process)

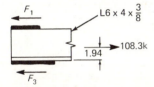

Fig. 5.15.11 Forces acting for Example 5.15.7.

The forces F_1 and F_3 are independent of the weld size.

$$R_w = \tfrac{5}{16}(21) = 6.56 \text{ kips/in.} < 7.5 \text{ kips/in.} \qquad \text{OK}$$

$$L_{w1} = \frac{35.0}{6.56} = 5.3 \text{ in.,} \qquad \text{say 6 in.}$$

$$L_{w3} = \frac{73.3}{6.56} = 11.2 \text{ in.,} \qquad \text{say 12 in.}$$

Use welds as summarized in Fig. 5.15.12.

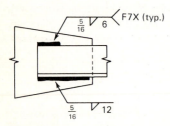

Fig. 5.15.12 Solution for Example 5.15.7.

Slot and Plug Welds

Slot and plug welds have their strength based on the area in the shearing plane between the plates being joined. As indicated in Sec. 5.7, their principal use is in lap joints. Plug welds are also occasionally used to fill up holes in connections, such as beam to column angles where temporary erection bolts had been placed to align the members prior to welding. The strength of such welds may or may not be included in the design strength of a joint. As a rule, plug and slot welds are designed to work together with other welds, usually fillet welds, in lap joints as shown in Fig. 5.5.5.

Example 5.15.8

Determine the value of T permitted by AISC on the connection in Fig. 5.15.13, using A36 steel.

SOLUTION
The resistance R_w per inch supplied by the $\frac{1}{2}$-in. fillet welds is

$$R_w = [0.5(0.707) + 0.11](21) = 9.73 \text{ kips/in.} \quad \text{(controls)}$$

but not to exceed the shear strength of the plate,

$$R_w = 0.40 F_y t = 14.4(0.75) = 10.8 \text{ kips/in.}$$

The resistance T_1 provided by the fillet welds,

$$T_1 = L_w R_w = 10(9.73) = 97.3 \text{ kips}$$

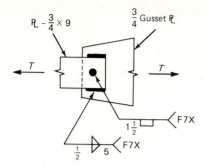

Fig. 5.15.13 Example 5.15.8.

The resistance T_2 provided by the $1\frac{1}{2}$-in.-diam plug weld is

$$T_2 = \frac{\pi(1.5)^2}{4}(21.0) = 37.1 \text{ kips}$$

$$T = T_1 + T_2 = 97.3 + 37.1 = 134.4 \text{ kips}$$

Check tensile capacity of plate:

$$T = 0.60F_yA_g = 22(9)(0.75) = 148.5 \text{ kips} > 134.4$$

The welding controls: $T = 134.4$ kips.

Example 5.15.9

Compute the allowable capacity of the connection shown in Fig. 5.15.14 according to AISC, using A572 Grade 50 steel and the SMAW process.

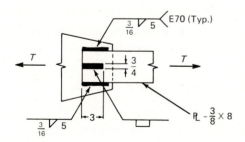

Fig. 5.15.14 Example 5.15.9.

SOLUTION

From Table 5.13.1, a $\frac{3}{16}$-in. E70 fillet weld provides $R_w = 2.78$ kips/in. $[(\frac{3}{16})(0.707)(21)]$. The resistance T_1 provided by the fillet welds is

$$T_1 = L_wR_w = 2(5)2.78 = 27.8 \text{ kips}$$

The resistance T_2 provided by the slot weld is

$$T_2 = \tfrac{3}{4}(3)(21) = 47.3 \text{ kips}$$
$$T = T_1 + T_2 = 27.8 + 47.0 = 74.8 \text{ kips}$$

Check tensile capacity of plate:

$$T = 0.60 F_y A_g = 30(\tfrac{3}{8})(8) = 90 \text{ kips} > 74.8$$

The welding controls: $T = 74.8$ kips.

Example 5.15.10

Design an end connection to develop the full tensile value of a C8 × 13.75 in a lap length of 5 in. The channel of A572 Grade 50 steel is connected to a $\tfrac{3}{8}$-in. plate and fillet welds are limited to $\tfrac{3}{8}$ in. and are to be made by the SMAW process. Use AISC Specification.

SOLUTION
Channel capacity $= T = F_t A = 0.6(50)4.02 = 120.5$ kips
    (a) Fillet weld. Use E70 electrodes.
        Min $a = \tfrac{3}{16}$ in.   (AISC–1.17.5)
        Max $a = 0.303 - \tfrac{1}{16} = 0.24$, say $\tfrac{1}{4}$ in.   (AISC–1.17.3)

While $\tfrac{1}{4}$-in. weld must be used on one end along the channel web, $\tfrac{3}{8}$-in. weld could be used along the flanges. It is better not to mix the fillet sizes, so try $\tfrac{1}{4}$ in. all around.

$$R_W = \tfrac{1}{4}(0.707)(21) = 3.71 \text{ kips/in.}$$

$$\text{Required } L_w = \frac{T}{R_w} = \frac{120.5}{3.71} = 32.5 \text{ in.}$$

Since the length all around is only 26 in., additional capacity from fillet weld in a large slot, slot welds, or plug welds, is necessary.
    (b) Slot weld. Try a slot weld in accordance with AISC–1.17.9.

$$\text{Min width of slot} = (t + \tfrac{5}{16}) \quad \text{(rounded to next odd } \tfrac{1}{16} \text{ in.)}$$

$$= 0.303 + 0.3125 = 0.6155, \quad \text{say } \tfrac{11}{16} \text{ in.}$$

$$\text{Max width of slot} = 2\tfrac{1}{4}(\text{weld thickness}) = 2\tfrac{1}{4}(0.303) = 0.68 \text{ in.}$$

Load to be carried by slot $= 120.5 - (26 - 0.68)3.71 = 26.6$ kips

Try $\tfrac{11}{16}$-in. width of slot:

$$\text{Length required} = \frac{26.6}{(\tfrac{11}{16})(21)} = 1.8 \text{ in.}$$

$$\text{Max length of slot} = 10(\text{weld thickness}) = 10(0.303) = 3.03 \text{ in.}$$

Use a slot weld $\tfrac{11}{16} \times 2$. The final design is shown in Fig. 5.15.15.

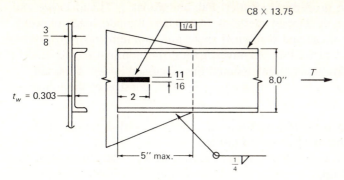

Fig. 5.15.15 Solution for Example 5.15.10.

5.16 ECCENTRIC WELDED CONNECTIONS

The versatility and ease of welding make welded connections desirable for a wide variety of usages such as the types of loading shown in Fig. 5.16.1. As in the case of eccentric loading on bolted connections (Sec. 4.8), the exact elastic analysis of the stresses in eccentric welded connections is impractical. In addition to the complex stress distribution discussed in Sec. 5.14, the variety of possible combinations of loading and grouping of welds makes it practical to evaluate the adequacy of a connection on the basis of nominal stresses. In an effort to obtain

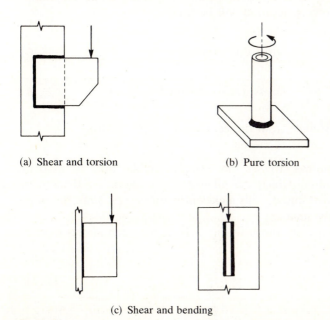

(a) Shear and torsion   (b) Pure torsion

(c) Shear and bending

Fig. 5.16.1 Types of eccentric loading.

improved design procedures, Butler, Pal, and Kulak [17] and Dawe and Kulak [18] have developed methods to predict ultimate loads of welds under eccentric shear and combined shear and moment, respectively.

The general procedure for investigating the nominal stress in weld groups is based on the general assumptions discussed in Sec. 5.14 and the principles of mechanics. The procedure, in brief, consists of the following steps:

1. Establish the *effective* throat dimension t_e and draw the effective cross section of the weld group.
2. Establish a coordinate system and determine the centroid of the weld group.
3. Determine the forces acting on the weld group.
4. Compute the individual stresses in the welds at critical points resulting from direct shear, torsion, and moment.
5. Combine the individual stresses vectorially.

The general procedure outlined above is illustrated in the examples that follow.

Eccentric Shear (Shear and Torsion)

In order to expand the general procedure for combined shear and torsion, consider the connection shown in Fig. 5.16.2a. The effective cross section and the applied force system are shown in Fig. 5.16.2b. In developing the procedure the following notation will be used:

$$f' = \frac{P}{A} = \text{stress due to direct shear} \tag{5.16.1}$$

$$f'' = \frac{Tr}{I_p} = \text{stress due to torsional moment} \tag{5.16.2}$$

where r = radial distance from the centroid to point of stress
I_p = polar moment of inertia

For computing nominal stresses the *locations* of the lines of weld are defined by edges along which the fillets are placed, rather than to the center of the effective throat. This makes little difference, since the throat dimension is usually small.

For the general case shown in Fig. 5.16.2, the components of stress due to direct shear are

$$f'_x = \frac{P_x}{A} \tag{5.16.3a}$$

$$f'_y = \frac{P_y}{A} \tag{5.16.3b}$$

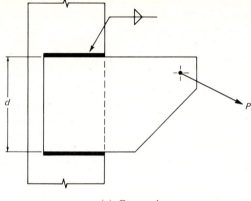

(a) Connection

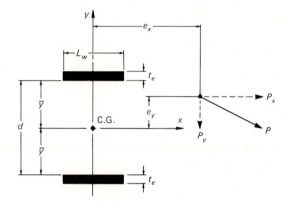

(b) Effective cross section

Fig. 5.16.2 Eccentric bracket connection.

The x- and y-components of f'' resulting from torsion are

$$f''_x = \frac{Ty}{I_p} = \frac{(P_x e_y + P_y e_x)y}{I_p} \tag{5.16.4a}$$

$$f''_y = \frac{Tx}{I_p} = \frac{(P_x e_y + P_y e_x)x}{I_p} \tag{5.16.4b}$$

where

$$I_p = I_x + I_y = \sum I_{xx} + \sum A\bar{y}^2 + \sum I_{yy} + \sum A\bar{x}^2 \tag{5.16.5}$$

In Eq. 5.16.5, $\bar{x}$ and $\bar{y}$ refer to distances from the center of gravity of the weld group to the center of gravity of the individual weld segments. I_{xx} and I_{yy} refer to the moments of inertia of the individual segments with respect to their own centroidal axes.

Thus, for the situation of Fig. 5.16.2, Eq. 5.16.5 becomes

$$I_p = 2\left[\frac{L_w(t_e)^3}{12}\right] + 2[L_w(t_e)(\bar{y})^2] + 2\left[\frac{t_e(L_w)^3}{12}\right]$$

$$= \frac{t_e}{6}[L_w(t_e)^2 + 12L_w(\bar{y})^2 + L_w^3] \qquad (5.16.6)$$

For practical situations, the first term of Eq. 5.16.6 is neglected because, with t_e small, the term is not significant compared to the other terms. Hence

$$I_p \approx \frac{t_e}{6}[12L_w(\bar{y})^2 + L_w^3] \qquad (5.16.7)$$

Then, after evaluating the torsional components in accordance with Eqs. 5.16.4, the x- and y-components of the resultant stress are

$$f_x = f_x' + f_x'' \qquad (5.16.8a)$$

$$f_y = f_y' + f_y'' \qquad (5.16.8b)$$

and the resultant stress f_r is

$$f_r = \sqrt{(f_x)^2 + (f_y)^2}$$

$$= \sqrt{(f_x' + f_x'')^2 + (f_y' + f_y'')^2} \qquad (5.16.9)$$

The resultant stress f_r is considered to be a nominal shear stress on the effective throat area of fillet welds. AISC–1.5.3 gives the allowable values as discussed in Sec. 5.12 (see Table 5.12.2). The safe capacity R_w of a fillet weld per inch of weld is

$$R_w = (t_e) \text{ (allowable unit stress)}$$

or

$$\frac{R_w}{t_e} = \text{allowable unit stress}$$

For satisfactory safety, it is required from analysis that

$$f_r \leq \frac{R_w}{t_e} \qquad (5.16.10)$$

For *investigating* stresses in weld groups, such as in Fig. 5.16.2, I_p may be computed as illustrated in obtaining Eq. 5.16.7. Note that the throat dimension t_e may be factored out of Eq. 5.16.7; this may always be done. Following through the computation of stress components and obtaining f_r, one will find that t_e appears as a denominator multiplier in the resultant stress f_r. By multiplying f_r by t_e, a resultant load per inch is obtained which may be compared with allowable R_w to establish whether or not safety is adequate. In other words, one may frequently desire to use Eq. 5.16.10 in the form

$$f_r t_e \leq R_w \tag{5.16.11}$$

When *designing* welds, the throat dimension t_e is frequently unknown and is to be solved for. In such cases it is usually convenient to compute I_p assuming $t_e = 1$; in other words, one may think of having computed $f_r t_e$, the actual load per inch on the weld group. Thus, for design, it is required that

$$t_e \geq \frac{f_r}{\text{allowable stress}} \tag{5.16.12}$$

where f_r is in *unit stress* (say ksi or MPa). Or Eq. 5.16.12 may be written

$$t_e \geq \frac{f_r(1)}{R_w} \tag{5.16.13}$$

where $f_r(1) = f_r t_e$ with $t_e = 1$, and is in *load per unit length* (say kips/in. or kN/mm).

Treating the welds making up the effective cross section in Fig. 5.16.2 as line welds (i.e., as in deriving Eq. 5.16.7 with $t_e = 1$) and using the general terms b and d, as shown in Fig. 5.16.3, Eq. 5.16.7 becomes

$$I_p \approx \frac{1}{6}\left[12b\left(\frac{d}{2}\right)^2 + b^3\right] = \frac{b}{6}[3d^2 + b^2] \tag{5.16.14}$$

Table 5.16.1 gives I_p values treated as properties of lines for other common weld configurations.

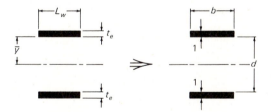

Fig. 5.16.3 Treatment of weld configuration as lines having unit thickness.

Example 5.16.1

Compute the size of fillet weld required using E70 electrodes for the connection shown in Fig. 5.16.4. Assume that the plate does not govern or control weld size.

SOLUTION

The highest stress on the weld group will occur at points A and B. The solution will be illustrated by carrying t_e as a factor throughout the computation.

Table 5.16.1 Properties of Welds Treated as Lines

Section $b = $ width; $d = $ depth	Section Modulus $I_x / \bar{y}$	Polar Moment of Inertia, I_p about Center of Gravity
1.	$S = \dfrac{d^2}{6}$	$I_p = \dfrac{d^3}{12}$
2.	$S = \dfrac{d^2}{3}$	$I_p = \dfrac{d(3b^2 + d^2)}{6}$
3.	$S = bd$	$I_p = \dfrac{b(3d^2 + b^2)}{6}$
4. $\bar{y} = \dfrac{d^2}{2(b+d)}$ $\bar{x} = \dfrac{b^2}{2(b+d)}$	$S = \dfrac{4bd + d^2}{6}$	$I_p = \dfrac{(b+d)^4 - 6b^2d^2}{12(b+d)}$
5. $\bar{x} = \dfrac{b^2}{2b+d}$	$S = bd + \dfrac{d^2}{6}$	$I_p = \dfrac{8b^3 + 6bd^2 + d^3}{12} - \dfrac{b^4}{2b+d}$
6. $\bar{y} = \dfrac{d^2}{b+2d}$	$S = \dfrac{2bd + d^2}{3}$	$I_p = \dfrac{b^3 + 6b^2d + 8d^3}{12} - \dfrac{d^4}{2d+b}$
7.	$S = bd + \dfrac{d^2}{3}$	$I_p = \dfrac{(b+d)^3}{6}$
8. $\bar{y} = \dfrac{d^2}{b+2d}$	$S = \dfrac{2bd + d^2}{3}$	$I_p = \dfrac{b^3 + 8d^3}{12} - \dfrac{d^4}{b+2d}$
9.	$S = bd + \dfrac{d^2}{3}$	$I_p = \dfrac{b^3 + 3b^2 + d^3}{6}$
10.	$S = \pi r^2$	$I_p = 2\pi r^3$

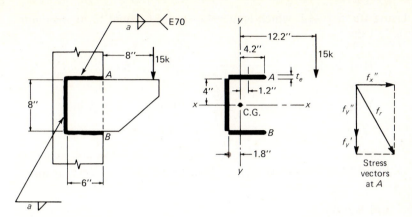

Fig. 5.16.4 Example 5.16.1.

Locate the centroid of the entire configuration by taking moments about the edge of the vertical weld:

$$\bar{x} = \frac{2t_e(6)(3)}{2(6)t_e + 8t_e} = 1.8 \text{ in.}$$

$$I_p = t_e \left\{ \underbrace{\frac{(8)^3}{12} + 2[6(4)^2]}_{I_x} + \underbrace{2\left[\frac{(6)^3}{12}\right] + 2[6(1.2)^2] + 8(1.8)^2}_{I_y} \right\}$$

$$= t_e(313.9) \text{ in.}^4$$

The bracketed multiplier of t_e in the I_p expression is the property of lines as also available in Table 5.16.1.

$$A = t_e[2(6) + 8] = t_e(20.0) \text{ sq in.}$$

$$f'_y = \frac{P_y}{A} = \frac{15}{t_e(20)} = \frac{0.75}{t_e} \text{ ksi } \downarrow$$

$$f''_x = \frac{Ty}{I_p} = \frac{15(12.2)4}{t_e(313.9)} = \frac{2.33}{t_e} \text{ ksi } \rightarrow$$

$$f''_y = \frac{Tx}{I_p} = \frac{15(12.2)4.2}{t_e(313.9)} = \frac{2.45}{t_e} \text{ ksi } \downarrow$$

The vector sum gives the resultant stress f_r,

$$f_r = \sqrt{\frac{(2.33)^2 + (2.45 + 0.75)^2}{t_e^2}} = \frac{3.96}{t_e} \text{ ksi}$$

It will be apparent from the calculation that one may easily obtain the same numerical result by assuming $t_e = 1$.

Using Eq. 5.16.12, which is merely setting f_r equal to its maximum allowable value,

$$\text{Required } t_e = \frac{f_r}{\text{allowable stress}} = \frac{3.97}{21} = 0.189 \text{ in.}$$

$$\text{Required } a = \frac{t_e}{0.707} = \frac{0.189}{0.707} = 0.267 \text{ in.,} \qquad \text{say } \tfrac{5}{16} \text{ in.}$$

Use $\tfrac{5}{16}$-in. E70 fillet welds.

For the general solution of welds under eccentric loading as in Fig. 5.16.4, the AISC Manual has tables, "Eccentric Loads on Weld Groups."

Shear and Bending

Combined shear and bending stresses are computed by vectorially adding the *nominal* shear and bending stresses. The procedure is illustrated by considering the bracket shown in Fig. 5.16.5a and the effective cross section of the weld group shown in Fig. 5.16.5b. Figure 5.16.6 shows the variation of the shear and bending stresses. The reader should note that the actual maximum shear and bending stresses occur at different locations. However, in simplifying the computations, the shear stresses are assumed to be nominally distributed as shown in Fig. 5.16.6c. The nominal shear stress is then added *vectorially* to the maximum bending stress.

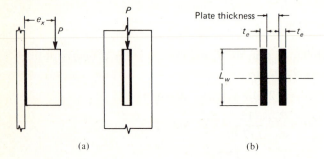

(a)　　　　　　　　　　(b)

Fig. 5.16.5　Welds in shear and bending.

For this particular case the assumed vertical shear stress from Eq. 5.16.3b is

$$f_y' = \frac{P_y}{A} = \frac{P}{2t_e L_w}$$

and the horizontal stress due to bending is

$$f_x'' = \frac{Mc}{I} = \frac{(Pe_x)(L_w/2)}{\left[\dfrac{2t_e(L_w)^3}{12}\right]} = \frac{3Pe_x}{t_e(L_w)^2}$$

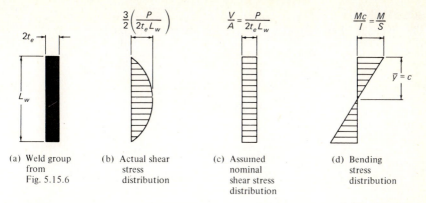

(a) Weld group from Fig. 5.15.6

(b) Actual shear stress distribution

(c) Assumed nominal shear stress distribution

(d) Bending stress distribution

Fig. 5.16.6 *Stresses on vertical lines of weld acting in shear and bending.*

The stress resultant is

$$f_r = \sqrt{(f'_y)^2 + (f''_x)^2}$$

For the flexural stress component, I equals either I_x or I_y, whichever is the axis for bending. The I values may be computed as the properties of line configurations in a manner similar to that used for I_p. For some commonly used configurations, $S = I/\bar{y}$ expressions are given in Table 5.16.1.

Example 5.16.2

Compute the size of E70 fillet weld requred for the connection shown in Fig. 5.16.7a using the SMAW process. Assume that the column and bracket does not control.

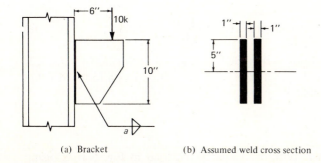

(a) Bracket

(b) Assumed weld cross section

Fig. 5.16.7 Example 5.16.2.

SOLUTION

The average shear stress using a 1-in. effective throat is

$$f_y' = \frac{P}{A} = \frac{10}{2(10)1} = 0.50 \text{ ksi} \downarrow$$

$$I_x = \frac{2(1)(10)^3}{12} = 166.7 \text{ in.}^4$$

$$f_x'' = \frac{Mc}{I} = \frac{10(6)5}{166.7} = 1.80 \text{ ksi} \rightarrow$$

$$f_r = \sqrt{(0.5)^2 + (1.8)^2} = 1.87 \text{ ksi for 1-in. effective throat.}$$

$$\text{Required } t_e = \frac{f_r}{\text{allowable stress}} = \frac{1.87}{21} = 0.089 \text{ in.}$$

$$\text{Required } a = \frac{0.089}{0.707} = 0.126 \text{ in.}$$

Use $\frac{3}{16}$-in. E70 fillet welds.

Design for Lines of Weld Subject to Bending Moment

Even when there are moderate returns at the top of lines of fillet weld, an estimate of the length required may be obtained by using the same approach as used for determining the number of bolts in a line in Sec. 4.8. In Fig. 4.8.8, R/p has units of kips per inch and for welds corresponds to f_r when t_e is unity.

For moment alone on one line of weld,

$$f_r = \frac{M}{S} = \frac{M}{(\frac{1}{6}L_w^2)} \text{ kips/in.} \tag{5.16.15}$$

Since the maximum value of f_r is R_w,

$$R_w = \frac{6M}{L_w^2}$$

or

$$\text{Required } L_w = \sqrt{\frac{6M}{R_w}} \tag{5.16.16}$$

Equation 5.16.16 for welds corresponds to Eq. 4.8.26 for bolts. Since it is correct only for moment alone, R_w should be entered as a reduced value to account for direct shear.

Example 5.16.3

Determine the length L required to carry the load in Fig. 5.16.8, using $\frac{5}{16}$-in. E70 fillet weld.

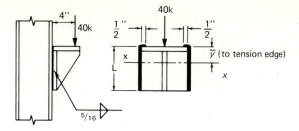

Fig. 5.16.8 Example 5.16.3.

SOLUTION

Estimate L required by using Eq. 5.16.16:

$$R_w = \tfrac{5}{16}(0.707)21 = 4.65 \text{ kips/in.}$$

$$M = 40(4) = 160 \text{ in.-kips per 2 weld lines}$$

$$\text{Required } L \approx \sqrt{\frac{6M}{R_w}} = \sqrt{\frac{6(160/2)}{4.0}} = 11 \text{ in.}$$

where a reduced value of R_w has been used to account for the direct shear effect. The return will add additional strength; try $L = 10$ in.

Investigate configuration using weld properties as lines computed from basic principles,

$$\bar{y} = \frac{2(10)5}{2(10+0.5)} = \frac{100}{21} = 4.76 \text{ in.}$$

Direct shear:

$$f_y' = \frac{40}{20} = 2.00 \text{ kips/in.} \downarrow$$

Since little direct shear is carried by the returns, they are neglected above.

$$I_x = \frac{2L^3}{12} + 2L(5.0 - 4.76)^2 + 2(0.5)(4.76)^2$$

$$= \frac{(10)^3}{6} + 20(0.24)^2 + (4.76)^2 = 190.5 \text{ in.}^3$$

$$S = \frac{I}{\bar{y}} = \frac{190.5}{4.76} = 40.0 \text{ in.}^2$$

Flexure:

$$f_x'' = \frac{M}{S} = \frac{160}{40} = 4.0 \text{ kips/in.} \rightarrow$$

Resultant:

$$f_r = \sqrt{(2.00)^2 + (4.00)^2} = 4.48 \text{ kips/in.} < 4.65 \text{ kips/in.} \qquad \text{OK}$$

Use L = 10 in.

Additional treatment of eccentric loads on welds is to be found in Chapter 13 on connections.

SELECTED REFERENCES

1. K. Winterton, "A Brief History of Welding Technology," *Welding and Metal Fabrication* (November 1962; December 1962).
2. "100 Years of Metalworking-Welding, Brazing and Joining," *The Iron Age* (June 1955).
3. H. Carpmael, *Electric Welding and Welding Appliances.* London: D. Van Nostrand Company, 1920.
4. Preston M. Hall, "77 Years of Resistance Welding." *The Welding Engineer* (February 1954, 54–55); (March 1954, 36–37); (April 1954, 62–63).
5. G. Herden, *Schweiss und Schneid-Technik.* Halle, East Germany: Carl Marhold Verlag, 1960.
6. E. Viall, *Electric Welding.* New York: McGraw-Hill Book Company, Inc., 1921.
7. *Welding Handbook*, 7th ed., Vol. 1, *Fundamentals of Welding.* Miami, Fla.: American Welding Society, 1976.
8. D. W. Morgan, "Classification and Use of Mild Steel Covered Electrodes," *Welding Journal*, December 1976, 1035–1038.
9. *Structural Welding Code*, AWS D1.1-79. Miami, Fla.: American Welding Society, 1979.
10. *Welding Handbook*, 6th ed., Vol. 2, *Welding Processes*; *Gas, Arc, and Resistance.* New York: American Welding Society, 1969.
11. Omer W. Blodgett, *Design of Welded Structures*, James F. Lincoln Arc Welding Foundation, 1966.
12. *Symbols for Welding and Nondestructive Testing*, AWS A2.4-76. Miami, Fla.: American Welding Society, 1976.
13. "Weld Defects Sound Off," *The Iron Age* (March 27, 1969).
14. J. A. Donnelly, "Determining the Cost of Welded Joints," *Engineering Journal*, AISC, 5, 4(October 1968), 146–147.
15. T. R. Higgins and F. R. Preece, "Proposed Working Stresses for Fillet Welds in Building Construction," *Engineering Journal*, AISC, 6, 1 (January 1969), 16–20.
16. *Commentary on Structural Welding Code*, AWS D1.2-77. Miami, Fla.: American Welding Society, 1977.
17. Lorne J. Butler, Shubendu Pal, and Geoffrey L. Kulak, "Eccentrically Loaded Welded Connections," *Journal of Structural Division*, ASCE, 98, ST5 (May 1972), 989–1005.
18. John L. Dawe and Geoffrey L. Kulak, "Welded Connections under Combined Shear and Moment," *Journal of Structural Division*, ASCE, 100, ST4 (April 1974), 727–741.

PROBLEMS

All problems* are to be done according to the latest AISC Specification, unless otherwise indicated. Whenever possible, show all answers on a sketch using appropriate welding symbols.

5.1. Identify each of the following "prequalified" butt joints (under AWS D1.1-79) made by the submerged arc (SAW) process, and specify the proper plate thickness for each. Refer to AISC Manual, "Welded Joints," for diagrams of the AWS prequalified joints. On the weld symbol for each, indicate the groove angle, root opening, and any other pertinent information.
(a) Square-groove weld, both sides with zero root opening.
(b) Single-V-groove weld.
(c) Single-V-groove weld.
(d) Double-V-groove weld.

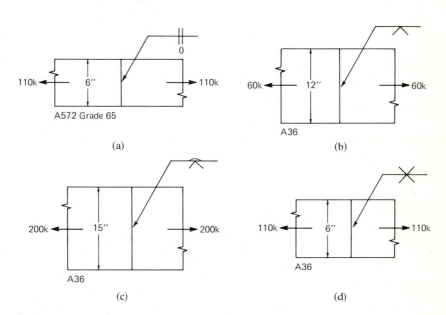

Prob. 5.1

5.2. Identify each of the following "prequalified" butt joints made by the shielded metal arc (SMAW) process, and specify the proper plate thickness, or determine the capacity *T*. (Refer to AWS D1.1-79 or AISC Manual, "Welded Joints.")

* Many problems may be solved either as stated in U.S. customary units or in SI units using the numerical data in parenthesis at the end of the statement. The conversions are approximate to avoid having the given data imply accuracy greater in SI than for U.S. customary units.

(a) Square-groove weld.

(b) Double-bevel-groove weld.

(c) Double-bevel-groove weld, welded from both sides, partial joint penetration.

(d) Single-V-groove weld, welded from one side, partial joint penetration.

For cases (a) and (b), is there a maximum thickness for which the type of weld may be used? Show cross sections for welds in all four cases.

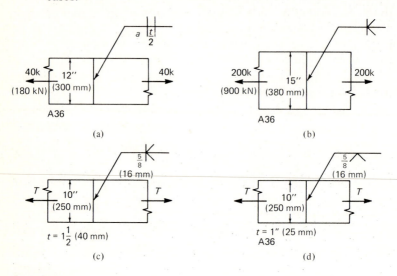

(a)   (b)

(c)   (d)

Prob. 5.2

5.3. Determine the capacity of the connection shown, using the submerged arc (SAW) process: (a) use A36 steel, (b) use A572 Grade 65. Assume appropriate electrode material is to be used.

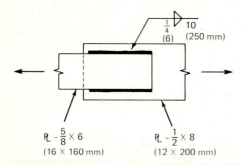

Prob. 5.3

5.4. Determine the fillet welds required to develop the capacity of the connection shown. Specify the proper flux-electrode material for

using the submerged arc (SAW) process: (a) use A572 Grade 42
steel, (b) use A514 steel.

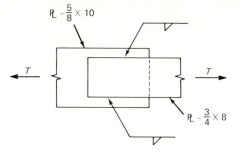

$\text{PL} -\frac{5}{8} \times 10$

T

T

$\text{PL} -\frac{3}{4} \times 8$

Prob. 5.4

5.5. Determine the plate thickness and weld size to be specified for the
joints shown. State weld material to be used for the SMAW
process. Compare A36 and A572 Grade 50 steels for each joint.
Indicate the preferred design for each joint.

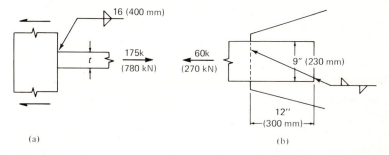

16 (400 mm)

t

175k
(780 kN)

60k
(270 kN)

9" (230 mm)

12"
(300 mm)

(a)

(b)

Prob. 5.5

5.6. Determine the plate thickness and specify the weld for each of the
joints shown. Use the SMAW process and compare results using
A36 and A572 Grade 50 steels. Indicate the preferred design for
each joint.

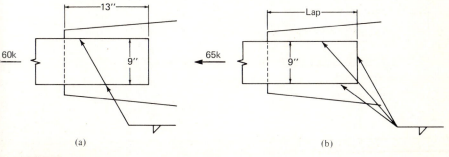

13"

9"

60k

65k

Lap

9"

(a)

(b)

Prob. 5.6

5.7. Design the reinforced lap joint shown, assuming the plates are 7 in. wide. Use A36 steel and the SMAW process. (Refer to AWS D1.1-79 (Fig. 2.10.1) for joint–BTC-P4.)

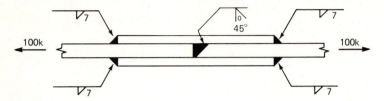

Prob. 5.7

5.8. Select a pair of channels and design the welds assuming the SMAW process: (a) use A36 steel, and (b) use A572 Grade 60 steel.

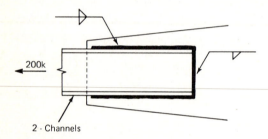

2 - Channels

Prob. 5.8

5.9. Design the plate framing into the W section and the welds, assuming SMAW process is to be used.
(a) Use A572 Grade 42 steel.
(b) Use A572 Grade 65 steel.
(c) Use A572 Grade 42 steel, with fillet welds instead of groove weld.
(d) Use A572 Grade 65 steel, with fillet welds instead of groove weld.

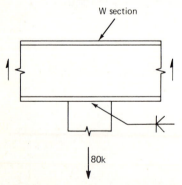

W section

80k

Prob. 5.9

5.10. A $5 \times 3\frac{1}{2} \times \frac{3}{8}$ angle of A572 Grade 50 steel is connected by its long leg to a $\frac{5}{16}$-in. gusset plate. Develop the full tensile capacity of the angle and use a balanced fillet welded connection even though AISC permits neglecting eccentricity in static load cases. The SMAW process is to be used. Use the following arrangements for the design:

(a) $\frac{5}{16}$-in. weld on toe and back, with none on end.

(b) $\frac{1}{4}$-in. weld on toe and $\frac{3}{8}$-in. weld on back, and none on end.

[L127 × 89 × 9.5; 8-mm gusset; 8-mm weld for (a); 6-mm and 10-mm weld for (b)]

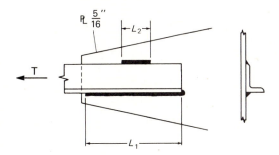

Prob. 5.10

5.11. Design a balanced connection for two $7 \times 4 \times \frac{1}{2}$ angles connected by their long legs to a $\frac{3}{8}$-in. gusset plate. A572 Grade 60 steel is used and welding is by the SMAW process. Detail the joint to balance the loads and still give the shortest possible overlap. $(2 - L178 \times 102 \times 12.7;$ 10-mm gusset plate)

5.12. Design the welds indicated to develop the full strength of the angles and minimize eccentricity. Use the SAW welding process. $(2 - L152 \times 102 \times 6.4)$

(a) Use A36 steel.

(b) Use A572 Grade 50 steel.

(c) Use A572 Grade 65 steel.

(d) Use A36 steel, but omit weld on the end of angles.

(e) Use A572 Grade 65 steel, but omit weld on the ends of angles.

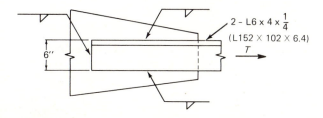

Prob. 5.12

5.13. Assume a 9-in.-wide plate used in a lap joint must carry 125 kips, and a possibility exists of some accidental eccentricity that cannot be computed. To insure a tighter joint, a $1\frac{1}{4}$-in.-diam plug weld is to be used. Determine the thickness of the plate, the amount of lap, and the weld size for the best joint. Assume the gusset plate to which the 9-in. plate is welded does not control any of the design. Use A572 Grade 50 steel and the SMAW process.

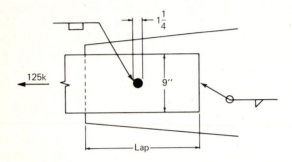

Prob. 5.13

5.14. Determine the minimum length of slot in order to develop the full strength of a C12×20.7 welded to a $\frac{3}{8}$-in. plate. Use the same size fillet weld over the entire length, and assume it is to be placed by the SMAW process.

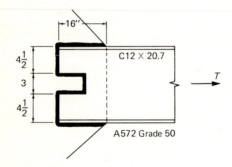

Prob. 5.14

5.15. For the joint shown, what is the shear R_w per inch of weld at the most highly stressed point? (In computing the polar moment of inertia, let the width of the effective section be unity, then $I_p = L^3/12$, as given in Table 5.16.1.) What weld size is indicated if $F_u = 70$ ksi for the weld metal and the SMAW process is used? ($F_u = 480$ MPa)

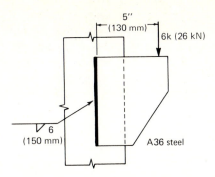

Prob. 5.15

5.16. What is the weld size required for the bracket shown if the shielded metal arc (SMAW) process is used?

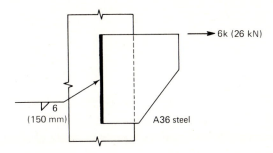

Prob. 5.16

5.17. For the bracket shown, find the safe capacity P based on the weld. Neglect any returns at ends and assume the SMAW process is to be used.

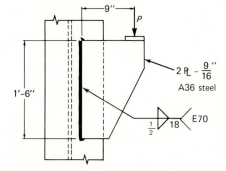

Prob. 5.17

5.18. Ignoring the effect of returns at the lower end of the connection, and allowing a resultant shear $R_w = 3.7$ kips per inch of fillet weld, what is the maximum allowable load P for the given connection?

5.19. Recompute the capacity for the connection of Prob. 5.18, using $\frac{3}{8}$-in. fillet weld on the sides and $\frac{1}{4}$-in. on the end. Neglect the returns. Assume shielded metal arc process is to be used and the plates are of A572 Grade 50 steel.

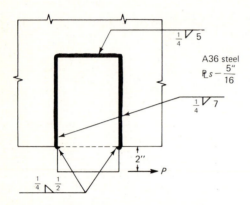

Probs. 5.18 and 5.19

5.20. Determine the capacity P for the bracket shown. The weld size is $\frac{3}{8}$-in. and E70 electrodes are used with shielded metal arc process. Compare answer with AISC Manual tables, "Eccentric Loads on Weld Groups."

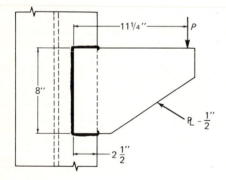

Prob. 5.20

5.21. For the welded bracket shown, use basic principles to find the theoretical weld size required using the SMAW process. Neglect end returns at the upper and lower right corners. Check by AISC Manual tables, "Eccentric Loads on Weld Groups."

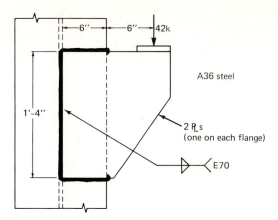

Prob. 5.21

5.22. Derive the general expression for the required weld size on the seat angle (E70 electrodes with shielded metal arc process) in terms of *P*, *L*, and *e*, using the following assumptions:
(a) Ignoring the returns at the top.
(b) Considering an average return of *L*/12.
(c) Using a return equal to twice the weld size.
If $e = 2\frac{3}{8}$ in. and $L = 6$ in., determine using assumption (a) the weld size needed to carry a load $P = 38.3$ kips. For the weld size selected, check the capacity that may be carried using all three assumptions.

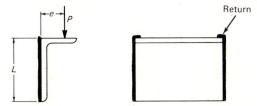

Prob. 5.22

5.23. Determine the length *L* required when using $\frac{5}{16}$ in. fillet weld with SMAW process. Check by using basic principles.

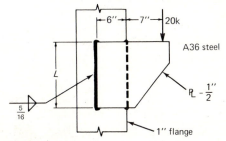

Prob. 5.23

5.24. Determine the length L required to safely carry the 40-kip load using $\frac{5}{16}$-in. weld. Use A572 Grade 50 steel and the SMAW process. Check by using basic principles.

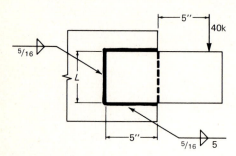

Prob. 5.24

6
Compression Members

PART I: COLUMNS

6.1 GENERAL

In this chapter, members subjected to axial compression stresses are to be treated. Referred to by various terms, such as *column, stanchion, post,* and *strut,* these members are rarely if ever actually carrying axial compression alone. However, whenever the loading is so arranged that either the end rotational restraint is negligible or the loading is symmetrically applied from members framing in at the column ends, and bending may be considered negligible compared to the direct compression, the member can safely be designed as a concentrically loaded column.

It is well known from basic mechanics of materials that only very short columns can be loaded to their yield stress; the usual situation is that buckling, or sudden bending as a result of instability, occurs prior to developing the full material strength of the member. Thus a sound knowledge of compression member stability is necessary for those designing in structural steel.

Exterior tubular columns. U.S. Steel Building, Pittsburgh. (Courtesy United States Steel Corporation)

6.2 EULER ELASTIC BUCKLING AND HISTORICAL BACKGROUND

Column buckling theory originated with Leonhard Euler in 1759 [1]. An initially straight concentrically loaded member, in which all fibers remain elastic until buckling occurs, is slightly bent as shown in Fig. 6.2.1. Though Euler dealt with a member built-in at one end and simply supported at the other, the same logic is applied here to the pin-end column, which having zero end rotational restraint represents the members with least buckling strength.

At any location z, the bending moment M_z (about the x-axis) on the slightly bent member is

$$M_z = Py \qquad (6.2.1)$$

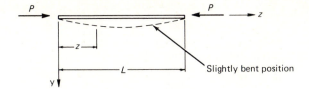

Fig. 6.2.1 Euler column.

and since

$$\frac{d^2y}{dz^2} = -\frac{M_z}{EI} \qquad\qquad (6.2.2)$$

the differential equation becomes

$$\frac{d^2y}{dz^2} + \frac{P}{EI}\, y = 0 \qquad\qquad (6.2.3)$$

After letting $k^2 = P/EI$, the solution of this second-order linear differential equation may be expressed

$$y = A \sin kz + B \cos kz \qquad\qquad (6.2.4)$$

Applying boundary conditions, (a) $y = 0$ at $z = 0$; and (b) $y = 0$ at $z = L$, one obtains for condition (a), $B = 0$; and for condition (b),

$$0 = A \sin kL \qquad\qquad (6.2.5)$$

Satisfaction of Eq. 6.2.5 may be accomplished in three possible ways; (a) constant $A = 0$, i.e., no deflection; (b) $kL = 0$, i.e., no applied load; and (c) $kL = N\pi$, the requirement for buckling to occur. Thus

$$\left(\frac{N\pi}{L}\right)^2 = \frac{P}{EI}$$

$$P = \frac{N^2\pi^2 EI}{L^2} \qquad\qquad (6.2.6)$$

The fundamental buckling mode, a single-curvature deflection ($y = A \sin \pi z/L$ from Eq. 6.2.4.), will occur when $N = 1$; thus the Euler critical load for a column with both ends pinned is

$$P_{cr} = \frac{\pi^2 EI}{L^2} \qquad\qquad (6.2.7)$$

or in terms of average compressive stress, using $I = A_g r^2$

$$F_{cr} = \frac{P_{cr}}{A_g} = \frac{\pi^2 E}{(L/r)^2} \qquad\qquad (6.2.8)$$

Euler's approach was generally ignored for design because test results did not agree with it; columns of ordinary length used in design were not as strong as Eq. 6.2.7 would indicate.

Considère and Engesser [2, 3] in 1889 independently realized that portions of usual length columns become inelastic prior to the occurrence of buckling and that a value of E should be used that could account for some of the compressed fibers being strained beyond the proportional limit. It was thus consciously recognized that in fact *ordinary length* columns fail by inelastic buckling rather than by elastic buckling.

Complete understanding of the behavior of concentrically loaded columns, however, was not achieved until 1946 when Shanley [4, 5] offered the explanation that now seems obvious. He reasoned that it was actually possible for a column to bend and still have increasing axial compression, but that it *begins* to bend upon reaching what is commonly referred to as the *buckling load*, which includes inelastic effects on some or all fibers of the cross section. These inelastic effects are discussed in detail in Sec. 6.4.

An extensive review of the development of column theory from Euler to Shanley is given by N. J. Hoff [6], and a brief summary is given by B. G. Johnston [7].

6.3 BASIC COLUMN STRENGTH

In order to determine a basic column strength certain conditions may be assumed for the ideal column [8]. With regard to material, it may be assumed (1) there are the same compressive stress-strain properties throughout the section; (2) no initial internal stresses exist such as those due to cooling after rolling and those due to welding. Regarding shape and end conditions, it may be assumed (3) the column is perfectly straight and prismatic; (4) the load resultant acts through the centroidal axis of the member until the member begins to bend; (5) the end conditions must be determinate so that a definite equivalent pinned length may be established. Further assumptions regarding buckling may be made, as (6) the small deflection theory of ordinary bending is applicable and shear may be neglected; and (7) twisting or distortion of the cross section does not occur during bending.

Once the foregoing assumptions have been made, it is now agreed [9] that the strength of a column may be expressed by

$$F_{cr} = \frac{P}{A} = \frac{\pi^2 E_t}{(KL/r)^2} \qquad (6.3.1)$$

where P/A = average stress in the member

E_t = tangent modulus at stress P/A

KL/r = effective (equivalent pinned-end) slenderness ratio (K value discussed in Sec. 6.9)

It is well known that long compression members fail by elastic buckling and that short stubby compression members may be loaded until the material yields or perhaps even into the strain-hardening range. However, in the vast majority of usual situations failure occurs by buckling after a portion of the cross section has yielded. This is known as *inelastic buckling.*

Actually, pure buckling under axial load occurs only when the assumptions (1) through (7) aforementioned apply. Columns are usually an integral part of a structure and as such *cannot* behave entirely independently. The practical use of the term *buckling* is that it is the boundary between stable and unstable deflections of a compression member, rather than the instantaneous condition that occurs in the isolated slender elastic rod. Many engineers refer to this "practical buckling load" as the "ultimate load."

As previously mentioned, for many years theoretical determinations of "ultimate load" did not agree with test results. Test results included effects of initial crookedness of the member, accidental eccentricity of load, end restraint, local or lateral buckling, and residual stress. A typical curve of observed ultimate loads was as shown in Fig. 6.3.1. Design formulas, therefore, were based on such empirical results. Various straight-line and parabolic formulas have been used, as well as other more complex expressions, in order to fit the curve of test results in a reasonably accurate, yet practical manner.

As a general approach, Euler elastic buckling governs the strength for large slenderness ratios, yield stress F_y is used for short columns, and a transition curve must be used for inelastic buckling.

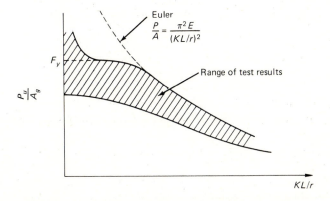

Fig. 6.3.1 Typical range of column strength vs slenderness ratios.

6.4* INELASTIC BUCKLING

Since ordinary length columns buckle when some of their fibers are inelastic, having a modulus of elasticity less than their initial elastic value, the logic of Engesser, Considère, and Shanley is explained in this section.

Basic Tangent Modulus Theory

Euler's theory pertained only to situations where compressive stress below the elastic limit acts uniformly over the cross section when unstable equilibrium occurs. Engesser [2] and Considère [3] were the first to utilize the possibility of a variable modulus of elasticity. In Engesser's tangent modulus theory the column remains straight up to the instant of failure and the modulus of elasticity at failure is the tangent to the stress-strain curve. The relationships are shown in Fig. 6.4.1. The theory prescribed that at a certain stress, $F_{cr} = P_{cr}/A$, the member could acquire an unstable deflected shape and that the deformation at F_{cr} is governed by $E_t = df/d\epsilon$.

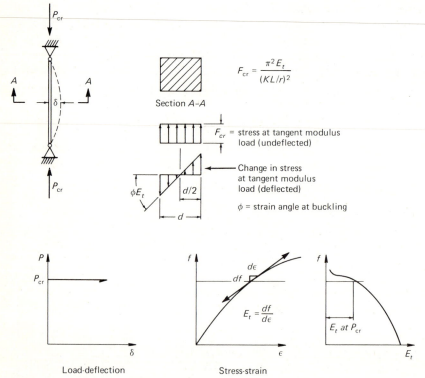

$$F_{cr} = \frac{\pi^2 E_t}{(KL/r)^2}$$

Section A-A

F_{cr} = stress at tangent modulus load (undeflected)

Change in stress at tangent modulus load (deflected)

ϕ = strain angle at buckling

$$E_t = \frac{df}{d\epsilon}$$

E_t at P_{cr}

Load-deflection   Stress-strain

Fig. 6.4.1 Engesser original tangent modulus theory, 1889.

* Sections so marked may be omitted without loss of continuity.

Thus Engesser modified Euler's equation to become

$$F_{cr} = \frac{P_t}{A} = \frac{\pi^2 E_t}{(KL/r)^2}$$

(6.4.1)

where P_t is the tangent modulus load.

This theory, however, still did not agree with test results, giving computed loads lower than measured ultimate capacities. The principal assumption that caused this tangent modulus theory to be considered erroneous is that as the member changes from a straight to bent form, no strain reversal takes place. In 1895 Engesser changed his theory, reasoning that during bending some fibers are undergoing increased strain (lowered tangent modulus) and some fibers are being unloaded (higher modulus at the reduced strain); therefore a combined value should be used for the modulus.

Double Modulus Theory

To examine the process of column bending at stresses beyond the elastic limit, consider the section of Fig. 6.4.2 from which Engesser's double modulus, or "reduced" modulus, is developed. This concept had logic to it which was generally accepted but gave computed loads higher than measured ultimate loads. Not until Shanley's explanation was the inconsistency resolved.

At unstable equilibrium, the stress at the neutral axis (section 1–1 of Fig. 6.4.2) remains as it was prior to the deflection δ occurring. On the loading fibers where strain is increasing, the stress increase is proportional

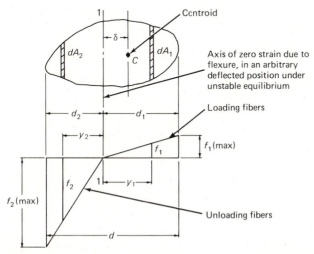

Fig. 6.4.2 Stress distribution in condition of unstable equilibrium (double modulus theory).

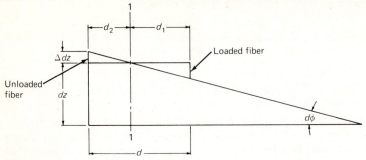

Fig. 6.4.3 An element dz along the axis of the column in the unstable equilibrium position.

to $E_t = df/d\epsilon$, whereas on the unloading fibers the decrease in strain relieves the elastic part of the strain; thus the stress decrease is proportional to the elastic modulus E.

As shown in Fig. 6.4.3, the strain on the cross section will be linear. At the extreme unloaded fiber, applying Hooke's Law the stress becomes

$$f_{2(max)} = (\text{unit strain}) E = \frac{\Delta \, dz}{dz} E \tag{6.4.2}$$

and at the loaded fiber,

$$f_{1(max)} = \frac{\Delta \, dz \, d_1}{d_2} \frac{E_t}{dz} \tag{6.4.3}$$

$$\frac{\Delta \, dz}{d_2} = d\phi \tag{6.4.4}$$

Thus

$$f_{2(max)} = E d_2 \frac{d\phi}{dz} ; \qquad f_{1(max)} = E_t d_1 \frac{d\phi}{dz} \tag{6.4.5}$$

For small curvature,

$$\frac{1}{\text{radius of curvature}} = \frac{M}{E_r I} = \frac{d\phi}{dz} = \frac{d^2 y}{dz^2} \tag{6.4.6}$$

where $E_r = $ Engesser's double modulus.

The internal resisting moment for the stress condition of Fig. 6.4.2 gives

$$M = -Py = \int_0^{d_1} f_1(y_1 - \delta) \, dA_1 + \int_0^{d_2} f_2(y_2 + \delta) \, dA_2 \tag{6.4.7}$$

and from linear stress distribution and Eq. 6.4.5,

$$f_1 = f_{1(max)} \frac{y_1}{d_1} = E_t d_1 \frac{d^2 y}{dz^2} \frac{y_1}{d_1}$$

$$f_2 = f_{2(max)} \frac{y_2}{d_2} = E d_2 \frac{d^2 y}{dz^2} \frac{y_2}{d_2} \tag{6.4.8}$$

Thus Eq. 6.4.7 becomes

$$-Py = E_t \frac{d^2y}{dz^2} \int_0^{d_1} y_1(y_1 - \delta) \, dA_1 + E \frac{d^2y}{dz^2} \int_0^{d_2} y_2(y_2 + \delta) \, dA_2 \quad \text{(6.4.9)}$$

Force equilibrium requires

$$\int_0^{d_1} f_1 \, dA_1 = \int_0^{d_2} f_2 \, dA_2 \quad \text{(6.4.10)}$$

which, using Eq. 6.4.8, gives

$$E_t \frac{d^2y}{dz^2} \int_0^{d_1} y_1 \, dA_1 = E \frac{d^2y}{dz^2} \int_0^{d_2} y_2 \, dA_2 \quad \text{(6.4.11)}$$

Using Eq. 6.4.11, it is seen the terms involving δ cancel each other in Eq. 6.4.9, thus giving

$$-Py = E_t \frac{d^2y}{dz^2} \int_0^{d_1} y_1^2 \, dA_1 + E \frac{d^2y}{dz^2} \int_0^{d_2} y_2^2 \, dA_2$$

$$\therefore \frac{d^2y}{dz^2} \left[E_t \int_0^{d_1} y_1^2 \, dA_1 + E \int_0^{d_2} y_2^2 \, dA_2 \right] + Py = 0 \quad \text{(6.4.12)}$$

Equation 6.4.12 is obviously of the same form as the elastic buckling equation, Eq. 6.2.3. Thus for the double modulus theory,

$$P_{cr} = \frac{\pi^2}{L^2} \left[E_t \int_0^{d_1} y_1^2 \, dA_1 + E \int_0^{d_2} y_2^2 \, dA_2 \right] \quad \text{(6.4.13)}$$

Shanley Concept—True Column Behavior

To understand the actual behavior of a column as explained by Shanley [4] in 1946, consider the rectangular section of Fig. 6.4.4 subjected to axial compression. For loads below the tangent modulus load P_t, the ideal column remains perfectly straight with zero deflection (point A of Fig. 6.4.4a). The load P_t at point A may most correctly be defined as follows [7]: "The tangent modulus load is the smallest value of axial load at which bifurcation of the equilibrium positions can occur regardless of whether or not the transition to the bent position requires an increase of axial load." Consider that at onset of bending (infinitesimal curvature) there will be an infinitesimal increase in axial strain and stress Δf_1. By the time the curvature becomes finite, i.e., the point N moves to N_1, some strain reversal must occur if the column cross section is to develop a resisting moment to maintain equilibrium with the moment due to the external load $P\delta$. For small but finite values of curvature the increment of load represented by stress on the area of increasing strain exceeds the increment of load represented by stress on the area of decreasing strain; thus P is increased by an amount dP (point B of Fig. 6.4.4a). As each

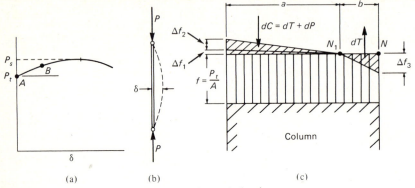

Fig. 6.4.4 Shanley concept—true column behavior.

increment of curvature takes place P will further increase as long as $dC > dT$. The increased compressive force dC is computed using the tangent modulus, E_t, while in the region of strain reversal the elastic modulus, E, is used to compute dT. The double modulus theory, which similarly treated loading and unloading fibers, did not accept $dC > dT$, but rather only considered equilibrium positions near the perfectly straight one.

For practical purposes the increase of capacity from P_t to P_s (Fig. 6.4.4a) can be neglected for design use. Therefore the tangent modulus load may be treated as the critical load, i.e., the load at which bending begins.

6.5 RESIDUAL STRESS

Residual stresses are stresses that remain in a member after it has been formed into a finished product. Such stresses result from plastic deformations, which in structural steel may result from several sources: (1) uneven cooling which occurs after hot rolling of structural shapes; (2) cold bending or cambering during fabrication; (3) punching of holes and cutting operations during fabrication; and (4) welding. Under ordinary conditions those residual stresses resulting from uneven cooling and welding are the most important. Actually the important residual stresses due to welding are really the result of uneven cooling. The mechanism of residual stress due to cooling is discussed in the *Welding Handbook* [10] and the effect of residual stress on compression structural members is summarized by Huber and Beedle [11], as well as by Beedle and Tall [12].

In wide-flange or H-shaped sections, after hot rolling, the flanges, being the thicker parts, cool more slowly than the web region. Furthermore the flange tips having greater exposure to the air cool more rapidly

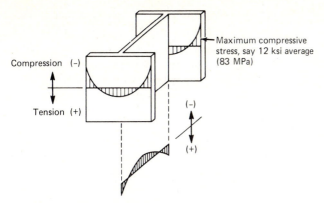

Fig. 6.5.1 Typical residual stress pattern on rolled shapes.

than the region at the junction of flange to web. Consequently, compressive residual stress exists at flange tips and at mid-depth of the web (the regions that cool fastest), while tensile residual stress exists in the flange and the web at the regions where they join. Figure 6.5.1 shows typical residual stress distribution on rolled beams. Considerable variation can be expected as the true pattern will be a function of the dimensions of the section.

   At this point one might wonder whether the general column-strength equation (Eq. 6.3.1) discussed in the preceding section still is applicable. The theory is applicable but all fibers in the cross section cannot be considered as stressed to the same level under the action of the compressive service load. The tangent modulus E_t on one fiber is not the same as that on an adjacent fiber.

   In a rolled steel shape the influence of residual stress on the stress-strain curve is shown in Fig. 6.5.2, using average stress on the gross area

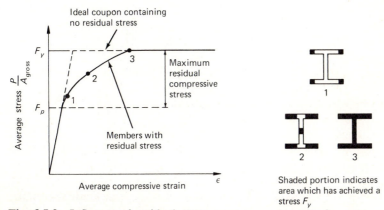

Fig. 6.5.2 Influence of residual stress on average stress-strain curve.

as the ordinate. It is noted that residual stress in an elastic-plastic material such as steel gives the same effect as that obtained for a material such as aluminum, which is not perfectly elastic, when it contains no residual stress. Thus, assuming the tangent modulus concept applies, column strength may be said to be based on inelastic buckling because the average stress-strain curve is nonlinear when maximum column strength is reached.

Whereas it was once believed the nonlinear portion of the average stress-strain curve for axially loaded compression members was due entirely to initial curvature and accidental eccentricity, it has been verified [11] that residual stress is the primary cause and the other factors have a relatively minor effect. Residual stresses at flange tips of rolled shapes have been measured as high as 20 ksi (138 MPa), a high percentage of the minimum specified yield stress for steels such as A36. Residual stresses are essentially independent of yield stress, depending instead on cross-sectional dimensions and configuration since those factors govern cooling rates [13].

Welding of built-up shapes is an even greater contributor to residual stress than cooling of hot-rolled H-shapes [14]. The plates themselves generally have little residual stress initially because of relatively uniform cooling after rolling. However, after the heat is applied to make the welds, the subsequent nonuniform cooling and restraint against distortion cause high residual stresses. Figure 6.5.3 shows typical residual stress patterns on welded H and box built-up shapes.

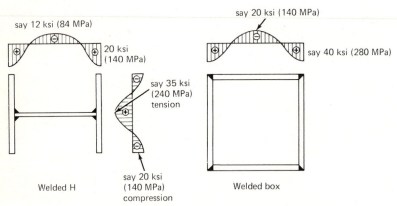

Fig. 6.5.3 Typical residual stress distribution in welded shapes.

One should note that compressive residual stresses typically occurring at flange tips are higher in welded than in rolled H-shaped sections. Thus the column strength of such welded shapes will be lower than corresponding rolled shapes. On the other hand, the welded-box shape, having tensile residual stress in the corner regions that contribute most to the stiffness as a column, will be stronger than a rolled shape having the same slenderness ratio. Sherman [15, 16] has studied residual stresses on rolled tubular members.

Having accepted that residual stresses exist, such information must be used to obtain a column strength curve (average stress vs slenderness ratio) that can form the basis for design. Until the early 1950's, column design was based on many formulas, all of which tried to empirically account for column behavior exhibited by tests. By clearly indicating that the tangent modulus was the proper criterion for strength and by identifying the role of residual stress, the Column Research Council (now Structural Stability Research Council [9]) has made a significant contribution.

6.6* DEVELOPMENT OF COLUMN STRENGTH CURVES INCLUDING RESIDUAL STRESS

The following analytical approach showing the logic behind the SSRC [9] basic column strength equation is essentially as presented by Huber and Beedle [11]. There are two general methods by which column strength may be obtained. Analytically, making use of the residual stress distribution (either measured or assumed) along with the stress-strain diagram for the material (coupon test), the strength may be expressed as a function of the moment of inertia of the unyielded part of the cross section and the slenderness ratio [13]. An alternate method is to experimentally determine an average stress-strain relationship from a short section of rolled shape containing residual stress. Column strength can then be determined using the tangent modulus of this average stress-strain curve along with the slenderness ratio that is proper for strong or weak axis bending. This second approach does not require knowledge about residual stress. Both methods assume symmetrical patterns of residual stress. Yu and Tall [17] have discussed these two approaches in more detail. Johnston [18] and Batterman and Johnston [19] have treated the tangent modulus application to inelastic buckling of columns.

The following development is made with the objective of obtaining a relationship between average externally applied stress and the slenderness ratio. Thus the capacity of a member can be obtained by a simple multiplication of safe stress time gross area, without regard to what the actual stress is at any point in the cross section or what is the true residual stress pattern.

As a starting point consider steel which as a material is perfectly elastic until a certain strain ϵ_y is achieved and then is plastic (i.e., constant stress with increasing strain). A coupon cut from the web of a rolled shape exhibits such behavior, as shown by the dotted lines in Fig. 6.6.1. The solid lines indicate the behavior of an H-shaped rolled section including residual stress.

To account for the effects of early yielding due to residual stress,

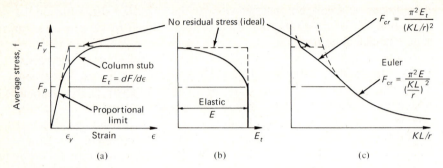

Fig. 6.6.1 Comparison of coupon with H-shaped rolled section containing residual stress.

consider one fiber at a distance x from the axis of zero strain due to bending (Fig. 6.6.2). The bending is taken as an infinitesimal amount consistent with equilibrium at the tangent modulus load. The bending moment contribution from stress on the one fiber is

$$dM = (\text{stress})(\text{area})(\text{moment arm}) = (\phi E_t x)(dA)(x) \tag{6.6.1}$$

which for the entire cross section becomes

$$M = \int_A \phi E_t x^2 \, dA = \phi \int_A E_t x^2 \, dA \tag{6.6.2}$$

From elementary bending theory, the radius of curvature is

$$R = \frac{1}{\phi}$$

$$\phi = \frac{1}{R} = \frac{M}{\text{equivalent } EI} = \frac{M}{E'I} \tag{6.6.3}$$

Thus

$$E'I = \frac{M}{\phi} = \int_A E_t x^2 \, dA$$

$$E' = \frac{1}{I} \int_A E_t x^2 \, dA \tag{6.6.4}$$

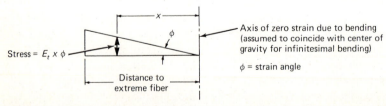

Fig. 6.6.2 Stress on fiber at distance x from axis of zero strain due to bending.

which may be called the *effective modulus* and used in Eq. 6.4.1 as an equivalent value for E_t.

If the idealized elastic-plastic f-ϵ curve of Fig. 6.6.1a (dotted) is used (for $f < F_y$, $E_t = E$ and for $f = F_y$, $E_t = 0$) the bending stiffness of yielded parts becomes zero; however, the buckling strength will be the same as a column whose moment of inertia, I_e, is the moment of inertia of the portion remaining elastic. Equation 6.6.4 then becomes

$$E' = \frac{E}{I} \int_{A \text{ (elastic part only)}} x^2 \, dA = E \frac{I_e}{I} \tag{6.6.5}$$

The stress at which the column may begin to bend, from Eq. 6.3.1, is

$$F_{cr} = \frac{P}{A} = \frac{\pi^2 E(I_e/I)}{(KL/r)^2} \tag{6.6.6}$$

In order for Eq. 6.6.6 to be useful, the relationship between F_{cr} and I_e must be established.

Case A. Buckling about Weak Axis

A reasonable assumption will be that the flanges become fully plastic before the web yields (see Fig. 6.6.3).

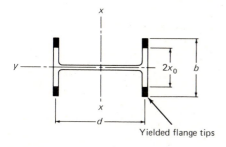

Fig. 6.6.3 Portion of section that has yielded.

Let k = proportion of the flange remaining elastic
$= 2x_0/b = A_e/A_f$

Then Eq. 6.6.5 becomes

$$E \frac{I_e}{I} = E \frac{t_f(2x_o)^3}{12} \left(\frac{12}{t_f b^3} \right) = E k^3 \tag{6.6.7}$$

if the web is neglected in computing I. Applying the tangent modulus definition,

$$E_t = \frac{\text{Nominal incremental stress}}{\text{Incremental elastic strain}}$$

$$= \frac{dP/A}{\dfrac{dP/A_e}{E}} = \frac{A_e E}{A} \tag{6.6.8}$$

$$E_t A = A_e E$$

$$= E(A_w + 2kA_f) \tag{6.6.9}$$

where A_w = web area
A_f = gross area of one flange
A = total gross area of section

Solving Eq. 6.6.9 for k and using Eq. 6.6.7 in Eq. 6.6.6 gives

$$k = \frac{E_t A}{2EA_f} - \frac{A_w}{2A_f} \tag{6.6.10}$$

$$F_{cr} = \frac{\pi^2 E k^3}{(KL/r)^2} = \frac{\pi^2 E}{(KL/r)^2}\left[\frac{AE_t}{2A_f E} - \frac{A_w}{2A_f}\right]^3 \tag{6.6.11}$$

which includes the effect of the elastic web, for buckling with respect to the weak axis $(y - y)$.

Case B. Buckling about Strong Axis

Again, assuming the web is elastic, but neglecting its contribution toward the moment of inertia gives approximately

$$E\frac{I_e}{I} \approx E\frac{2A_e(d/2)^2}{2A_f(d/2)^2} = Ek \tag{6.6.12}$$

If the elastic web is included,

$$E\frac{I_e}{I} = E\left[\frac{2kA_f(d^2/4)+t_w d^3/12}{2A_f(d^2/4)+t_w d^3/12}\right] \tag{6.6.13}$$

$$= E\left[\frac{2kA_f + A_w/3}{2A_f + A_w/3}\right] \tag{6.6.14}$$

Using tangent modulus definition and Eq. 6.6.9,

$$2kA_f = \frac{E_t A}{E} - A_w$$

which, upon eliminating the $2kA_f$ term from Eq. 6.6.14, gives

$$E\frac{I_e}{I} = \left[\frac{E_t A/E - 2A_w/3}{2A_f + A_w/3}\right]E \tag{6.6.15}$$

Thus

$$F_{cr} = \frac{\pi^2 E k}{(KL/r)^2} \tag{6.6.16}$$

is the approximate equation using Eq. 6.6.10 for k, or more exactly using Eq. 6.6.15 in Eq. 6.6.6 gives

$$F_{cr} = \frac{\pi^2 E}{(KL/r)^2}\left[\frac{E_t A/E - 2A_w/3}{2A_f + A_w/3}\right] \tag{6.6.17}$$

for buckling with respect to the strong axis $(x - x)$.

From the foregoing development it is apparent that two equations are necessary to properly determine column strength, one for strong-axis buckling and one for weak-axis buckling. Although the value I_e/I is not itself a function of the residual stress distribution provided that it satisfies the general geometric requirement as shown in Fig. 6.6.3; nevertheless, the critical stress F_{cr}, computed as the buckling load divided by the gross area, has a relationship with KL/r that does depend on residual stress.

Note that if the stress-strain curve for the material is not elastic-plastic, i.e., if E_t is not either E or zero, then the more general equation, Eq. 6.6.4, must be used instead of Eq. 6.6.5.

Example 6.6.1

Establish the column strength curve (F_{cr} vs KL/r) for weak axis buckling of an H-shaped section of steel having a yield stress of 100 ksi (690 MPa) exhibiting perfect elastic-plastic strength in a coupon test (Fig. 6.6.4b), and having the very simplified residual stress pattern shown in Fig. 6.6.4a. Neglect the contribution of the web.

SOLUTION

For any external load the strain on every fiber is the same. Until a fiber reaches the strain ϵ_y at first yield, the applied load is

$$P = \int_A f\,dA = fA$$

After a portion has become plastic, the applied load is

$$P = (A - A_e)F_y + \int_{A_e} f\,dA$$

In this problem for $F_{cr} = P/A \leq 2F_y/3$ the entire section remains elastic, $E_t = E$, in which case E' is EI_e/I and $I_e = I$; thus,

$$F_{cr} = \frac{2F_y}{3} = \frac{\pi^2 E}{(KL/r)^2}$$

$$\frac{KL}{r} = \sqrt{\frac{\pi^2(29,000)}{(2/3)(100)}} = 65.4 \quad \text{(point 1, Fig. 6.6.5)}$$

When $F_{cr} = P/A > 2F_y/3$, the flange tips have yielded, making I_e less than

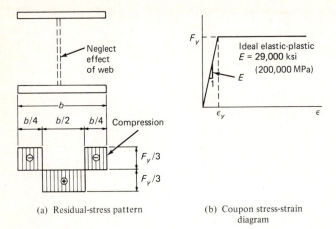

(a) Residual-stress pattern   (b) Coupon stress-strain diagram

Fig. 6.6.4 Data for Example 6.6.1.

I; thus

$$\frac{I_e}{I} = \frac{(b/2)^3}{b^3} = \frac{1}{8}$$

$$F_{cr} = \frac{2F_y}{3} = \frac{\pi^2 E(I_e/I)}{(KL/r)^2} = \frac{\pi^2 E}{8(KL/r)^2}$$

$$\frac{KL}{r} = 23.2 \quad \text{(point 2, Fig. 6.6.5)}$$

for average stress infinitesimally greater than $2F_y/3$. When $F_{cr} = P/A = F_y$,

$$F_{cr} = F_y = \frac{\pi^2 E}{8(KL/r)^2}$$

$$\frac{KL}{r} = 18.9 \quad \text{(point 3, Fig. 6.6.5)}$$

when the total load $P = F_y A$. The results are shown in Fig. 6.6.5. If there

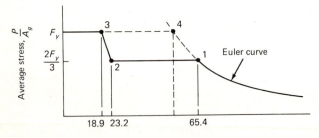

Slenderness ratio, KL/r

Fig. 6.6.5 Column strength curve for Example 6.6.1.

had been no residual stress at $F_{cr} = F_y$,

$$\frac{KL}{r} = 53.5 \quad \text{(point 4, Fig. 6.6.5)}$$

Example 6.6.2

Establish the column-strength curve for the more realistic linear distribution of residual stress shown in Fig. 6.6.6. Consider weak-axis buckling of an H-shaped section of steel for both (a) $F_y = 36$ ksi (250 MPa) and (b) $F_y = 100$ ksi (690 MPa). Neglect the effect of the web.

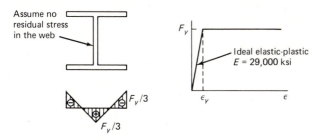

(a) Residual stress pattern   (b) Coupon stress-strain diagram

Fig. 6.6.6   Data for Example 6.6.2.

SOLUTION

For an average superimposed stress $f = P/A \leq 2F_y/3$ the entire section remains elastic (Fig. 6.6.7a); therefore $E_t = E$, and

$$F_{cr} = \frac{2F_y}{3} = \frac{\pi^2 E}{(KL/r)^2}$$

For an average stress due to applied load greater than $2F_y/3$, part of the cross section is plastic and part elastic, as in Fig. 6.6.7b. During this stage, the *change in stress is not the same on all fibers*, because the modulus of elasticity is not the same on all fibers.

$$F_{cr} = \frac{\pi^2 E I_e/I}{(KL/r)^2}$$

$$\frac{I_e}{I} = \frac{2(1/12)(2x_0)^3 t}{2(1/12)b^3 t} = \frac{8x_0^3}{b^3}$$

neglecting the effect of the web,

$$F_{cr} = \frac{8\pi^2 E(x_0/b)^3}{(KL/r)^2} \tag{a}$$

which gives F_{cr} as a function of two variables, x_0/b and KL/r. An

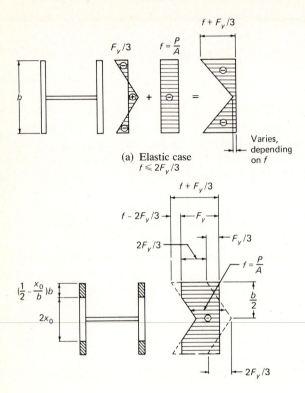

(a) Elastic case
$f \leqslant 2F_y/3$

(b) Elasto-plastic case
$f > 2F_y/3$

f = superimposed stress on elastic fibers

Fig. 6.6.7 Stress distribution with linear residual stress variation.

additional relationship is required. The total load during the elasto-plastic stage can be expressed

$$P_{cr} = 2\left[fbt - 2\left(\frac{1}{2}\right)\left(f - \frac{2F_y}{3}\right)\left(\frac{1}{2} - \frac{x_0}{b}\right)bt \right] \qquad \text{(b)}$$

which is the shaded area of the stress diagram in Fig. 6.6.7b. From similar triangles on the dotted triangle of Fig. 6.6.7b,

$$\frac{f - 2F_y/3}{\left(\frac{1}{2} - \frac{x_0}{b}\right)b} = \frac{2F_y/3}{b/2}$$

Solving for f,

$$f = \left[1 - \frac{x_0}{b}\right]\frac{4F_y}{3} \qquad \text{(c)}$$

Using Eq. (c) to eliminate f from Eq. (b) gives

$$P_{cr} = 2bt\left\{\left(1-\frac{x_0}{b}\right)\frac{4F_y}{3} - \left[\left(1-\frac{x_0}{b}\right)\frac{4F_y}{3} - \frac{2F_y}{3}\right]\left(\frac{1}{2} - \frac{x_0}{b}\right)\right\}$$

$$= A_g F_y\left[1 - \frac{4}{3}\left(\frac{x_0}{b}\right)^2\right] \tag{d}$$

Thus

$$F_{cr} = \frac{P}{A_g} = F_y\left[1 - \frac{4}{3}\left(\frac{x_0}{b}\right)^2\right] \tag{e}$$

which is used in combination with Eq. (a). The results are presented in Fig. 6.6.8.

$\dfrac{x_0}{b}$	F_{cr}	F_{cr} for $F_y = 36$ ksi (ksi)	$\dfrac{KL}{r}$	F_{cr} for $F_y = 100$ ksi (ksi)	$\dfrac{KL}{r}$
0.50	$0.66F_y$	24.0	109.2	66.7	65.4
0.45	0.73	26.3	89.0	73.0	53.4
0.40	0.787	28.3	72.0	78.7	43.1
0.35	0.837	30.2	57.0	83.7	34.2
0.30	0.880	31.7	44.1	88.0	26.5
0.25	0.917	33.0	32.9	91.7	19.7
0.20	0.947	34.1	23.2	94.7	13.9
0.10	0.987	35.5	8.0	98.7	4.8

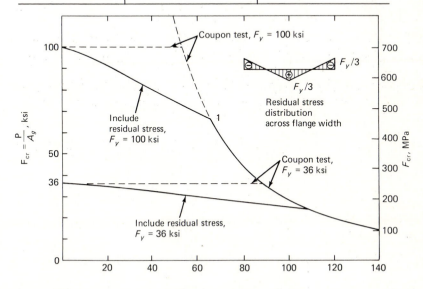

Fig. 6.6.8 Column strength curves showing effect of residual stress ($E = 29{,}000$ ksi). Solution for Example 6.6.2.

If the web of the section were to be included, I_e/I could easily include the web terms. Furthermore, Eq. (b) could also have included the web terms. Such inclusion of the effect of the web brings in the variable A_w/A_f and in most cases the effect is small.

Finally, curves similar to those of Fig. 6.6.8 can be obtained by using an average stress-strain curve for a short segment of rolled shape as referred to earlier in this section. In which case Eqs. 6.6.11 and 6.6.17 can be used with the E_t obtained from the "cross section" stress-strain curve.

6.7 STRUCTURAL STABILITY RESEARCH COUNCIL (SSRC) BASIC STRENGTH CURVE

Based upon the methods discussed in Sec. 6.6, column strength curves can be obtained for weak- or strong-axis buckling with various distributions of residual stress. For most practical situations it has been reported that an assumed linear distribution of residual stress in the flanges results

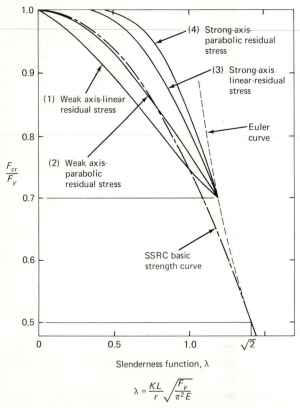

Fig. 6.7.1 Column strength curves for H-shaped sections having compressive residual stress at flange tips. (Adapted from Ref. 9, p. 39)

in a fair average column curve [9]. Furthermore, the development in the previous section (Eqs. 6.6.11 and 6.6.16) shows that for the *same* slenderness ratio, H-shaped column sections allowed to bend in the weak direction can carry less load than columns permitted to bend only in the strong direction. Compressive residual stress which is greatest at the flange tips accounts for this strength difference.

Typical column-strength curves for parabolic and linear distribution of residual stress across the flange are shown in Fig. 6.7.1. For structural carbon steels the average value of the maximum compressive residual stress is 12 to 13 ksi (83 to 90 MPa), corresponding roughly to $0.3F_y$. For the high-strength steels residual stress will generally be a lower fraction of the yield stress.

The basic column-strength curve adopted by the SSRC is based on the parabolic equation proposed by Bleich [20]. Since curves (2), (3), and (4) of Fig. 6.7.1 are essentially parabolas, and since test results have generally given essentially a parabolic variation for buckling in the inelastic range, a parabola representing a compromise among the four curves of Fig. 6.7.1 seems reasonable. The curve proposed by Bleich is

$$F_{cr} = F_y - \frac{F_p}{\pi^2 E} (F_y - F_p) \left(\frac{KL}{r}\right)^2 \tag{6.7.1}$$

where F_p = stress at the proportional limit. Since the deviation from elastic behavior on the average stress-strain curve is accounted for by residual stress, the F_p was replaced by

$$F_p = F_y - F_r = \text{yield stress} - \text{residual stress}$$

and Eq. 6.7.1 becomes

$$F_{cr} = F_y \left[1 - \frac{F_r}{\pi^2 E} \left(\frac{F_y - F_r}{F_y}\right) \left(\frac{KL}{r}\right)^2\right] \tag{6.7.2}$$

In order for Eq. 6.7.2 to properly represent a compromise between the strengths for the weak and strong axes of bending, the residual stress must be assumed higher than $0.3F_y$. The SSRC chose $F_p = 0.5F_y$ so as to give a smooth transition from the Euler curve for elastic buckling to the parabolic curve representing inelastic buckling. The two curves are tangent at $F_{cr}/F_y = 0.5$. The SSRC curve then becomes

$$F_{cr} = F_y \left[1 - \frac{F_y}{4\pi^2 E} \left(\frac{KL}{r}\right)^2\right] \tag{6.7.3}$$

In Fig. 6.7.1 where the SSRC curve is compared with other curves that distinguish between residual stress patterns and axes of bending, it may appear that the SSRC compromise curve lies too far above curve (1) over much of its range. When it is realized that typically the residual stress pattern is somewhere between linear and parabolic the true weak axis strength curve is actually closer to the SSRC curve.

From Fig. 6.7.1 one may note that to provide the same degree of safety for all columns, different strength curves would be required depending on the expected residual stress distribution, the shape of the section, and the axis of bending when the column buckles. As discussed in the next section, AISC uses only the SSRC column-strength curve rather than multiple curves.

6.8 AISC DESIGN EQUATIONS

The basic strength curve of the Structural Stability Research Council, Eq. 6.7.3, divided by a factor of safety is used as the allowable stress equation of the AISC Specification for slenderness ratios where *inelastic buckling* controls. Thus, using a critical stress of $0.5F_y$ (assumed as the proportional limit) in Eq. 6.7.3 gives the maximum value of KL/r, defined as C_c,

$$0.5F_y = F_y\left[1 - \frac{F_y}{4\pi^2 E}\left(\frac{KL}{r}\right)^2\right]$$

$$C_c = \frac{KL}{r} = \sqrt{\frac{2\pi^2 E}{F_y}} = \frac{755}{\sqrt{F_y}} \qquad (6.8.1)^*$$

when $E = 29,000$ ksi and F_y is in ksi.

Therefore Eq. 6.7.3 upon replacing $2\pi^2 E/F_y$ by C_c^2 becomes AISC Formula (1.5–1) for slenderness ratios, $KL/r \leq C_c$,

$$F_a = \frac{F_y}{\text{FS}}\left[1 - \frac{(KL/r)^2}{2C_c^2}\right] \qquad (6.8.2)$$

where F_a = allowable stress on gross area under service, or working, load; KL/r is the slenderness ratio of the *equivalent* pinned-end column; and FS = factor of safety.

While Eq. 6.8.2 is applicable for ordinary design of rolled H-shaped sections, AISC also provides for a reduced efficiency for thin-walled sections that may exhibit instability (local buckling) in the plate elements comprising the cross section when the width-thickness ratio limits of AISC–1.9 are not satisfied.

AISC Appendix C introduces a reduction factor Q to Eqs. 6.8.1 and 6.8.2 when AISC–1.9 is not satisfied. Thus when local buckling of one or more plate components of the cross section may occur prior to overall column buckling, the allowable stress is

$$F_a = \frac{QF_y}{\text{FS}}\left[1.0 - \frac{(KL/r)^2}{2C_c^2}\right] \qquad (6.8.3)$$

---

* For SI, $C_c = \dfrac{1987}{\sqrt{F_y}}$, with F_y in MPa $\qquad\qquad$ (6.8.1)

which is AISC Formula (C5–1). Also,

$$C_c = \sqrt{\frac{2\pi^2 E}{QF_y}} \qquad (6.8.4)$$

Since $Q = 1$ for all rolled H-shaped sections (standard W, S, and M shapes), the development of the logic behind the use of the Q factor is reserved for Sec. 6.18 in Part II on plate buckling.

Since all columns contain some initial curvature and accidental eccentricity the factor of safety should reflect such conditions. For short members that are negligibly affected by small eccentricity of load or by residual stress, the factor of safety need not be greater than is deemed desirable for tension members, i.e., 1.67 under the AISC Specification. The larger the slenderness ratio the greater the effects of accidental eccentricity, initial crookedness, and the effective length factor K (see Sec. 6.9). Thus AISC prescribes an increasing FS with slenderness ratio to a maximum of a 15% increase over the basic value, i.e., a maximum factor of 1.92. Arbitrarily, to obtain a smooth transition from FS = 1.67 for $KL/r = 0$ to FS = 1.92 for $KL/r = C_c$, a cubic equation approximation to a quarter sine wave was used:

$$FS = \frac{5}{3} + \frac{3}{8}\frac{KL/r}{C_c} - \frac{1}{8}\left(\frac{KL/r}{C_c}\right)^3 \qquad (6.8.5)$$

which is used with Eq. 6.8.3. Equation 6.8.5 is illustrated graphically in Fig. 6.8.1.

For columns with slenderness ratios exceeding $KL/r = C_c$, elastic buckling controls strength according to the Euler equation (Eq. 6.3.1 with $E_t = E$),

$$F_a = \frac{\pi^2 E}{FS(KL/r)^2} \qquad (6.8.6)$$

which gives allowable stress on gross area for long columns. For $KL/r = C_c$, Eq. 6.8.5 gives FS = 23/12 (approximately 1.92), and Eq. 6.8.6 becomes

$$F_a = \frac{12\pi^2 E}{23(KL/r)^2} \qquad (6.8.7)$$

which is AISC Formula (1.5–2), and applies for long columns used as main load carrying members.

For secondary and bracing members where a lower FS is logical, the AISC increases the allowable stress for such members having KL/r exceeding 120,

$$F_a = \frac{\text{Eq. 6.8.3 or 6.8.7}}{\left(1.6 - \dfrac{L/r}{200}\right)} \qquad (6.8.8)$$

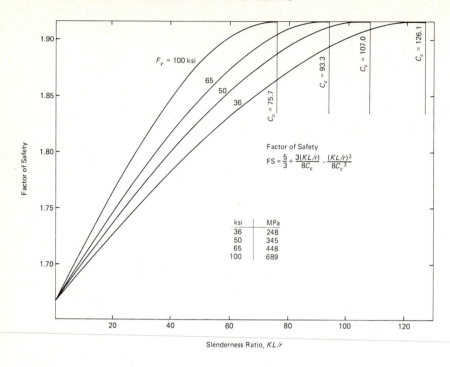

Fig. 6.8.1 Factor of safety for members under axial compression according to the AISC Specification.

which is AISC Formula (1.5–3). Note that K is taken as unity for Eq. 6.8.8.

After the basis for the allowable stress formulas is understood, routine designs should utilize values from AISC Appendix Tables 3–36, 3–50, 4, and 5 which are an integral part of the Specification. The allowable stresses are plotted in Fig. 6.8.2 to provide visual presentation.

The reader is reminded that in the AISC Specification the strength of columns based on inelastic buckling is assumed to be representable by the single SSRC column strength curve (Fig. 6.7.1), whereas the actual strength depends on the residual stress pattern, the cross-sectional configuration, and the axis of bending. Multiple design curves may well be used in future design specifications. For a discussion of this, see the *SSRC Guide* [9] (pp. 65–72).

For the most part, the AISC procedure using the SSRC curve has been based on studies using rolled wide-flange sections. A study of the implications of designing other shaped sections as columns using the AISC formulas is outside the scope of this book; however, for studies of tubular sections refer to Sherman [15, 16], Snyder and Lee [21, 22], and

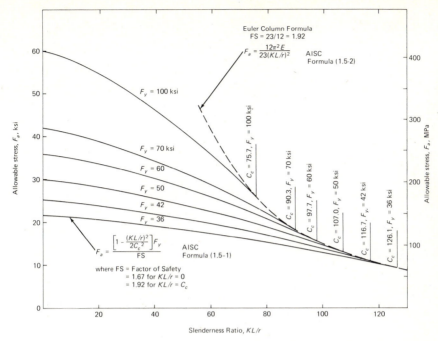

Fig. 6.8.2 Allowable stresses on gross area due to axial compression. (AISC–1.5.1.3)

Chen and Ross [23]; for round columns see Galambos [24]; and for angles and tees see Kennedy and Murty [25].

6.9 EFFECTIVE LENGTH

Discussion of column strength to this point has assumed hinged ends or ends with no moment restraint. Zero moment restraint at ends constitutes the weakest situation for compression members when translation of one end relative to the other is prevented. For such pinned-end columns the equivalent pinned-end length, known as the *effective length*, equals the actual length, i.e., $K = 1.0$.

For most real situations moment restraint at the ends does exist and the inflection points in the buckled shape curve may occur at locations other than the ends of the member. The distance between the points of inflection, either real or imaginary, is the effective or equivalent pinned length for the column.

It is difficult in most situations to evaluate the degree of moment restraint offered by members that frame into the given column, by the footing and soil under it, and indeed the full interaction of all the members of a steel frame.

Whether or not the designer is able to determine the degree of end restraint accurately, he needs to understand the difference between the *braced frame* and the *unbraced frame*.

A *braced frame*, according to AISC–1.8.2, is one in which "lateral stability is provided by adequate attachment to diagonal bracing, to shear walls, to an adjacent structure having adequate lateral stability, or to floor slabs or roof decks secured horizontally by walls or bracing systems parallel to the plane of the frame." For example, a column in a braced frame would have no sideways movement of its top relative to its bottom. The buckling of a braced frame would result in a buckled column shape having at least one point of contraflexure between the ends of the member, such as cases (a), (b), and (c) of Fig. 6.9.1. Taking the effective length *KL* as the actual length *L* for such cases is conservative and reasonable, since the *K* factor will always be less than 1.0 and greater than 0.5 for actual braced frames.

An *unbraced frame*, according to AISC–1.8.3, is one in which "lateral stability is dependent upon the bending stiffness of rigidly connected beams and columns." The buckling of an unbraced frame is one of sidesway where, for example, the top of the column moves to the side

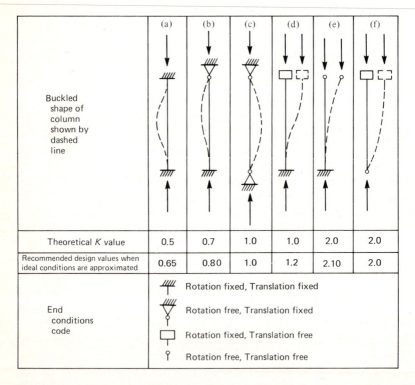

Fig. 6.9.1 Effective-length factors for centrally loaded columns with various idealized end conditions. (Adapted from Ref. 9, p. 74)

relative to the bottom. Cases (d), (e), and (f) of Fig. 6.9.1 are sidesway buckling cases. For the unbraced frame, an analysis is necessary to determine the effective length factor K, *which is always greater than* 1.0. Except perhaps for the flagpole-type situation of Fig. 6.9.1, case (e), an arbitrary selection of K is not satisfactory for design. One acceptable "rational" approach is to use the Alignment Chart which is presented as Fig. 14.3.1, and is discussed in detail in Chapter 14.

Although presented here for members under axial compression without bending, the effective length concept, particularly for the unbraced frame, has greater practical applicability in the design of beam-columns (see Chapter 12). Except for the design of members in braced frames where K is conservatively taken as 1.0, proper design of frames requires an understanding of frame buckling and behavior, an introduction to which is given in Chapters 14 and 15.

For the reader generally familiar with the effective length concept, in addition to the Alignment Charts of Fig. 14.3.1, sources are available for evaluating effective length for stepped columns [26], for columns having an axial load applied at an intermediate point along the length [27], and for gabled frames [28]. Adjustments of the Alignment Chart values to account for inelastic behavior have been proposed by Yura [29], and discussed by Disque [30] and Smith [31]. Details of this adjustment appear in Sec. 14.3 where the Alignment Chart development appears.

For truss compression members, end restraint may be present and joint translation is prevented so that K might logically be less than 1.0. Under static loading, stresses in all the members remain in the same proportion to one another for various loads. If all members are designed for minimum weight they will achieve ultimate capacity simultaneously under live load. Thus restraint offered by members framing at a joint disappears or at least is greatly reduced. The SSRC, therefore, recommends using $K = 1.0$ for members of a truss designed for fixed-position loading. When designing for moving load systems on trusses, K can be reduced to 0.85 because conditions causing maximum stress in the member under consideration will not cause maximum stress in the members framing in to provide restraint [9].

6.10 DESIGN OF ROLLED SHAPES (W, S, AND M) SUBJECT TO AXIAL COMPRESSION

In this section reference will be made to the AISC Steel Manual, Part 1 which gives properties of the rolled shapes and Part 3 which contains column load tables. Whether or not the final choice of a section must satisfy the AISC Specification, use of this general procedure and use of the Part 3 column load tables will enable the designer to get at the least a good preliminary estimate of member size.

General Procedure

The working stress method for design of compression members, whether rolled shapes or built-up sections, whether using the AISC Specification or some other, is based on an allowable stress on gross area. The allowable stress in all cases is a function of slenderness ratio and yield stress of the material. Since the allowable stress depends on slenderness ratio KL/r, where r depends on the section selected, the design of compression members is an indirect process unless load tables are available. The general procedure is:

1. Assume an allowable stress.
2. Based on computed area requirement, select a section. (Note that local buckling requirements of AISC–1.9 must be satisfied. This is discussed in Part 2 of this chapter, particularly Sec. 16.6.)
3. Based on KL/r for the section selected, compute allowable stress.
4. Compute P/A and compare with allowable stress.
5. If P/A is less than allowable or not more than 2 to 3% greater, generally the design would be considered acceptable, otherwise repeat Steps 1–5.

Example 6.10.1

Select the lightest W section of A36 steel to serve as a pin-ended main member column 16 ft long to carry an axial compression load of 195 kips in a braced structure (see Fig. 6.10.1). Use the AISC Specification, and indicate first three choices.

SOLUTION

Since the assumption of hinged ends is made, the effective length equals the actual length, i.e., $K = 1.0$. Considering $KL = 16$ ft as a moderately

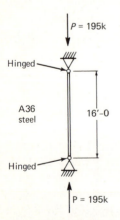

Fig. 6.10.1 Example 6.10.1.

long length, the slenderness ratio may be estimated at about 70 to 80. For rolled W shapes, $Q = 1.0$, so that the special provisions of AISC Appendix C are not involved.

Estimate allowable stress from Fig. 6.8.2, or use AISC Formula (1.5–1) with an FS estimated from Fig. 6.8.1. Using the latter, estimate $FS = 1.85$. Thus

$$F_a = \frac{36}{1.85}\left[1 - \frac{80^2}{2(126.1)^2}\right] = \frac{36}{1.85}[1 - 0.201] = 15.6 \text{ ksi}$$

$$\text{Required } A = \frac{195}{15.55} = 12.5 \text{ sq in.}$$

Since buckling in the weak direction (based on $K_y L_y/r_y$) will control the strength for W shapes when KL is the same with respect to both x- and y-axes, the lightest sections for this problem will be those having the least r_x/r_y. A high r_x indicates excessive strength with respect to the strong axis, which may be utilized only by providing additional bracing in the weak direction.

Using the column load tables for W shapes in Part 3 of the AISC Manual, one might tentatively select a W8×48, the lightest W8 that has at least the required area. Furthermore these W8 sections have the lowest r_x/r_y for a given area.

Try W8×48: $A = 14.1$ sq in.

$$\frac{KL}{r_y} = 92.3; \qquad F_a = 13.9 \text{ ksi}$$

$$f_a = \frac{P}{A} = \frac{195}{14.1} = 13.8 \text{ ksi} < F_a \qquad \text{OK}$$

Note AISC Table 3–36 gives a tabulation of allowable stress F_a for values of KL/r.

Since the computed stress P/A does not exceed the allowable value, and since no other section having this area has a lower r_x/r_y, the W8×48 is the lightest available section. For deeper sections, heavier sections will be required, as follows:

Section	Area (sq in.)	KL/r_y	F_a (ksi)	P/A (ksi)	
W8×48	14.1	92.3	13.9	13.8	1st choice
W10×49	14.4	75.6	15.8	13.5	2nd choice
W12×50	14.7	98.0	13.2	13.3	3rd choice
W14×53	15.6	100.0	13.0	12.5	

Some designers might justifiably argue that the W10×49 is better than W8×48 since for only one pound per foot extra a considerably greater margin of safety is achieved.

Example 6.10.2

Select the lightest W section of A36 steel to serve as a main member 30 ft long and carry an axial compression of 160 kips. The member may be assumed pinned at top and bottom and in addition has weak direction support at mid-height (see Fig. 6.10.2). The column is part of a braced structure and the AISC Specification is to be used.

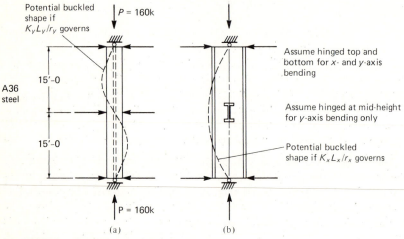

Fig. 6.10.2. Example 6.10.2.

SOLUTION

The effective length factors K for buckling in either the strong or the weak direction equal unity; i.e., $K_x = K_y = 1.0$. Since the AISC column load tables are computed assuming $K_y L_y / r_y$ controls, enter such tables with the effective length $K_y L_y$. Thus enter with

$$P = 160 \text{ kips}; \qquad K_y L_y = 15 \text{ ft}$$

Starting with the shallow W8 sections and working toward the deeper sections, find

$$\begin{array}{lll} \text{W8} \times 40 & P = 169 \text{ kips} & r_x/r_y = 1.73 \\ \text{W10} \times 39 & P = 162 & r_x/r_y = 2.16 \\ \text{W12} \times 40 & P = 163 & r_x/r_y = 2.66 \end{array}$$

Since the actual support conditions are such that $K_x L_x = 2 K_y L_y$, if $r_x/r_y \geq 2$, weak-axis controls and tabular loads give the correct answer. Thus W10×39 and W12×40 are obviously acceptable.

Since r_x/r_y for W8×40 is less than 2, strong-axis bending controls. The capacity can be checked by determining the maximum unbraced length for weak-axis bending that corresponds to $K_x L_x = 30$ ft. For equal strength about x- and y-axes and omitting K since it equals unity,

$$\frac{L_x}{r_x} = \frac{L_y}{r_y}$$

$$\text{Equivalent } L_y = \frac{L_x}{r_x/r_y} = \frac{30}{1.73} = 17.3 \text{ ft}$$

For $K_y L_y$ equal to 17.3, the W8×40 can carry only 149 kips; therefore it is not acceptable.

When using load tables it is wise to verify any solution thus found:

$$\text{W10}\times 39, \quad KL/r_y = 15(12)/1.98 = 90.9, \qquad F_a = 14.1 \text{ ksi}$$
$$P/A = 160/11.5 = 13.9 \text{ ksi} < 14.1 \text{ ksi} \qquad \text{OK}$$

Use W10×39.

Example 6.10.3

Select the lightest W section of A572 Grade 60 steel to serve as a main member 22 ft long carrying an axial compression of 300 kips. Assume the member hinged at the top and fixed at the bottom for bending about either principal axis (see Fig. 6.10.3). The member is part of a braced structure and the AISC Specification is to be used.

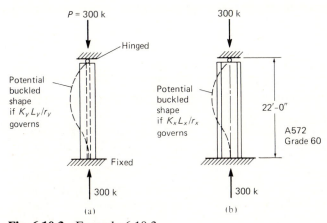

Fig. 6.10.3 Example 6.10.3.

SOLUTION

For this problem there are no AISC column load tables available for direct selection of sections. To get an estimate of the required section, use AISC Manual column load tables for $F_y = 50$. Since the member is fixed at one end, in accordance with SSRC recommendation (Fig. 6.9.1), assume $K = 0.8$. Select for $K_y L_y = 0.8(22) = 17.6$ ft and

$$P \approx 300\left(\frac{50}{60}\right) = 250 \text{ kips}$$

Try

$$W10 \times 49: \quad P = 265 \text{ kips} \quad \text{for} \quad F_y = 50 \text{ ksi}$$
$$W12 \times 53: \quad P = 280 \text{ kips} \quad \text{for} \quad F_y = 50 \text{ ksi}$$

Check $W10 \times 49$ for $P = 300$ kips with $F_y = 60$ ksi,

$$C_c = 97.7 \text{ (from AISC Appendix A, Table 5)}$$
$$\frac{KL/r}{C_c} = \frac{17.6(12)/2.54}{97.7} = \frac{83.1}{97.7} = 0.851$$
$$C_a = 0.335 \quad \text{(from AISC Appendix A, Table 4)}$$
$$F_a = C_a F_y = 0.335(60) = 20.1 \text{ ksi}$$

The above procedure using AISC Appendix A tables is equivalent to directly using AISC Formula (1.5–1).

$$f_a = \frac{P}{A} = \frac{300}{14.4} = 20.8 \text{ ksi} > F_a \qquad \text{NG}$$

Check $W12 \times 53$,

$$\frac{KL}{r} = \frac{17.6(12)}{2.48} = 85.2; \qquad C_a = 0.329$$
$$F_a = C_a F_y = 0.329(60) = 19.7 \text{ ksi}$$
$$f_a = \frac{P}{A} = \frac{300}{15.6} = 19.2 \text{ ksi} < 19.5 \text{ ksi} \qquad \text{OK}$$

Use $W12 \times 53$, A572 Grade 60.

Example 6.10.4

Design column A of the unbraced frame shown in Fig. 6.10.4 as an axially loaded compression member carrying 275 kips using A36 steel. In the plane perpendicular to the frame the system is braced, with supports at top and bottom of a 21-ft height.

SOLUTION

While it is rare that a frame member would be designed as axially loaded only, it may occasionally be proper for some interior members with symmetrical loading.

(a) The column size must be estimated in order to determine the K_x

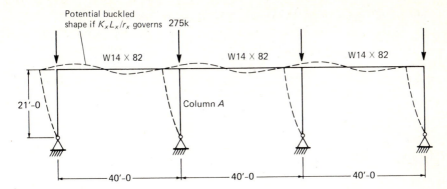

Fig. 6.10.4 Unbraced frame for Example 6.10.4.

factor for the frame. It is given that $K_yL_y = 21$ ft for the plane perpendicular to the frame, so that a preliminary size might be determined from AISC Manual column load tables.

$$K_yL_y = 21 \text{ ft} \qquad \text{Find W12}\times 65, \qquad P = 286 \text{ kips}$$

(b) Using $I = 533$ in.[4] for W12×65, compute G_{top} in accordance with AISC Commentary Sec. 1.8 (or text Fig. 14.3.1b). This illustrates the traditional use of the Alignment Chart and does not include the modification for inelastic behavior recommended by Yura [29]. See Chapter 15 examples for use of the modification.

$$G_{top} = \frac{\Sigma \, I/L \text{ for columns}}{\Sigma \, I/L \text{ for beams}} = \frac{533/21}{2(882)/40} = \frac{25.4}{44.1} = 0.58$$

$$G_{bottom} = 10 (\text{according to AISC–C1.8})$$

$$K_x = 1.8$$

$$K_xL_x = 1.8(21) = 37.8 \text{ ft}$$

For equal strength about each axis,

$$\frac{r_x}{r_y} = \frac{37.8}{21} = 1.8$$

(c) Select sections from column load tables,

W12×65: $r_x/r_y < 1.8$ Equivalent $K_yL_y = 37.8/1.75 = 21.6$ ft

$$P = 282 \text{ kips} \qquad \text{OK}$$

W14×61: $r_x/r_y > 1.8$ $K_yL_y = 21$ ft $P = 226 \text{ kips} \qquad \text{NG}$

(d) Check W12×65, $I = 533$ in.[4]

Since G_{top} and G_{bottom} are not changed from the preliminary calculation, no recheck is needed; thus $K_x = 1.8$. If weak axis had not been close to controlling so as to give the correct section, as it was in this case, a new computation of G_{top} would have been necessary to obtain a revised K_x.

$$\frac{K_x L_x}{r_x} = \frac{1.8(21)(12)}{5.28} = 85.9; \qquad \frac{K_y L_y}{r_y} = \frac{1.0(21)(12)}{3.02} = 83.4$$

$F_a = 14.7$ ksi from AISC Appendix A, Table 3-36.

$$f_a = \frac{275}{19.1} = 14.4 \text{ ksi} < F_a \qquad \text{OK}$$

Use W12×65, A36 steel.

6.11 DESIGN FORMULAS FROM SPECIFICATIONS OTHER THAN AISC

A multitude of column formulas have been used by various building codes through the years dating back to the widely used straight-line formula proposed by T. H. Johnson [32] in 1886. Generally such formulas were attempts to empirically match the behavior of columns as demonstrated by tests (see the reverse curve pattern of Fig. 6.3.1) and were made simple and conservative.

AASHTO–1977

The equations of AASHTO [33] for working stress design of concentrically loaded columns follow the logic of the Structural Stability Research Council but use a higher safety factor than AISC to account for the dynamic nature of loads. The equations are, in terms of C_c, as previously defined, for $KL/r \leq C_c$ (AASHTO–1.7.1),

$$F_a = \frac{F_y}{2.12}\left[1 - \frac{K(L/r)^2}{2C_c^2}\right] \qquad (6.11.1)$$

For $KL/r > C_c$ (AASHTO–1.7.1),

$$F_a = \frac{\pi^2 E}{2.12(KL/r)^2} \qquad (6.11.2)$$

The factor of safety for columns (2.12) is taken as 16 percent greater than the basic value of 1.82 used for tension members.

Secant Formula

This long-time rational formula could be applied to the entire range of slenderness ratios. In essence such a formula gives the correct relationship if one assumes the deviation from elastic buckling is entirely due to initial curvature and accidental eccentricity. As discussed previously, residual stress is now agreed to be the main cause of inelastic buckling, so that continued use of the unwieldy secant formula has lost much of its appeal.

The equation for maximum stress on a structural member subject to axial compression and uniform bending along its length, as shown in Fig. 6.11.1, is

$$f = \frac{P}{A} + \frac{M}{S} \sec \frac{kL}{2}$$

(6.11.3)

where $k = \sqrt{P/EI}$. The derivation of the second term of Eq. 6.11.3 occurs in Chapter 12, Sec. 12.2, on beam-columns; however, it is obtained from the differential equation solution which includes the constant primary bending moment, Pe, due to equal end eccentricity, as well as the secondary moment due to axial load times beam deflection.

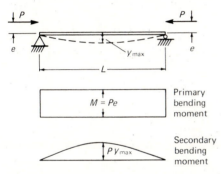

Fig. 6.11.1 Loading and bending moment for secant formula.

Placing the upper limit for stress f at the yield point F_y gives, from Eq. 6.11.3,

$$F_y = \frac{P}{A}\left(1 + \frac{eA}{S} \sec \frac{L}{2}\sqrt{\frac{P}{EI}}\right)$$

(6.11.4)

Assuming an initial value of 0.25 for the eccentricity ratio eA/S even when the loading is presumably concentrically applied and recognizing $I = Ar^2$ gives

$$F_y = \frac{P}{A}\left(1 + 0.25 \sec \frac{L}{2r}\sqrt{\frac{P}{EA}}\right)$$

(6.11.5)

which for adequate safety should not exceed F_y even when P is increased by a factor of safety (FS). Thus the allowable average stress on the gross section is

$$F_a = \frac{P}{A} = \frac{F_y/\text{FS}}{1 + 0.25 \sec \frac{L}{2r}\sqrt{\frac{F_a(\text{FS})}{E}}}$$

(6.11.6)

which is similar to the long column equation used until recent years by the highway and railroad bridge specifications. In Eq. 6.11.6, L was the effective length, KL, with K taken as 0.75 for riveted ends and 0.875 for pinned ends, and the factor of safety was taken about 1.80.

Other column formulas may be found adequately discussed elsewhere [34, 35].

Example 6.11.1

For the W8×48 selected for the hinged end member in Example 6.10.1, determine allowable stress according to the secant formula, Eq. 6.11.6 with a factor of safety of 1.77 for A36 steel.

SOLUTION

$$L = 0.875(16) = 14 \, \text{ft}$$

$$FS = 1.77 \, \text{for A36}$$

$$\frac{L}{2r} = \frac{14(12)}{2(2.08)} = 40.3$$

$$F_a = \frac{36/1.77}{1 + 0.25 \sec (40.3)\sqrt{\dfrac{F_a(1.77)}{29,000}}} = \frac{20.3}{1 + 0.25 \sec 0.314\sqrt{F_a}}$$

Solve by trial.

$$\begin{array}{lll}
\text{Assume } F_a = 15 \, \text{ksi}, & \sec 1.217 = 2.90, & F_a = 11.8 \, \text{ksi} \\
F_a = 14 \, \text{ksi}, & \sec 1.172 = 2.58, & F_a = 12.3 \\
F_a = 12.9 \, \text{ksi}, & \sec 1.128 = 2.33, & F_a = 12.8 \, \text{ksi}
\end{array}$$

This shows the allowable stress to be lower than that allowed by the AISC. The example is intended to demonstrate the procedure and difficulty of using the secant formula.

6.12* SHEAR EFFECT

When built-up members are connected together by means of lacing bars, the objective is to make all of the components act as a unit. As a compression member bends, a shearing component of the axial load arises. The magnitude of the shear effect in reducing column strength is proportional to the amount of deformation that can be attributed to shear. Solid-webbed sections, such as W shapes, have less shear deformation than do latticed columns (Fig. 6.12.1) using lacing bars and/or batten plates.

Furthermore, as shown later, shear has an insignificant effect on reducing column strength for solid-webbed shapes and may safely be

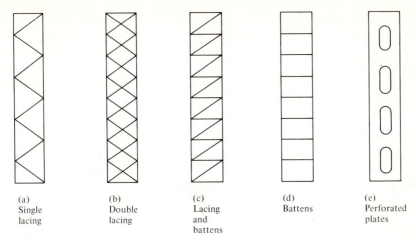

(a)
Single
lacing

(b)
Double
lacing

(c)
Lacing
and
battens

(d)
Battens

(e)
Perforated
plates

Fig. 6.12.1 Types of latticed columns.

neglected. The shear effect should not, however, be neglected for latticed columns.

To include the effect of shear, the curvature due to shear should be added to the buckling curvature to obtain the total curvature. It is well known that

$$V = \frac{dM_z}{dz} = P\frac{dy}{dz} \tag{6.12.1}$$

after recognizing that $M_z = Py$ from Eq. 6.2.1 (see Fig. 6.2.1).

The slope θ, due to shear deformation is

$$\theta = \frac{\text{shear stress}}{\text{shear modulus}} = \frac{\beta V}{AG} \tag{6.12.2}$$

where β is a factor to correct for nonuniform stress across various cross-sectional shapes. The shear contribution to curvature becomes

$$\frac{d\theta}{dz} = \frac{\beta}{AG}\frac{dV}{dz} = \frac{P\beta}{AG}\frac{d^2y}{dz^2} \tag{6.12.3}$$

The total curvature is the sum of Eqs. 6.2.3 and 6.12.3,

$$\frac{d^2y}{dz^2} = -\frac{Py}{EI} + \frac{P\beta}{AG}\frac{d^2y}{dz^2}$$

which gives

$$\frac{d^2y}{dz^2} + \frac{P}{EI}\left[\frac{1}{1 - P\beta/AG}\right]y = 0 \tag{6.12.4}$$

which is of the same form as Eq. 6.2.3; therefore the modified form of the

Euler critical load is

$$P_{cr} = \frac{\pi^2 EI}{L^2} \frac{1}{\left[1 + \underbrace{\frac{\beta}{AG} \frac{\pi^2 EI}{L^2}}\right]}$$ (6.12.5)

modification for
shear effect

In accordance with the previous discussion on basic column strength, G and E can be replaced by the tangent modulus values, G_t and E_t, and $E_t/G_t = 2(1+\mu)$, and L can be replaced by the effective length KL. Further, combining the shear effect with KL gives

$$F_{cr} = \frac{P_{cr}}{A} = \frac{\pi^2 E_t}{(\alpha KL/r)^2}$$ (6.12.6)

where $\alpha = \sqrt{1 + 2(1+\mu)\pi^2 \beta/(KL/r)^2}$. Thus the shear effect may be accounted for by an adjustment to the effective length. For W shapes when bending about the weak axis, β averages about 2. Using $\mu = 0.3$ for steel, typical values for α are

$$
\begin{aligned}
KL/r = \ 50 \qquad & \alpha = 1.01 \\
= \ 70 \qquad & = 1.005 \\
= 100 \qquad & = 1.003
\end{aligned}
$$

For slenderness ratios less than about 50, yielding controls, so that the shear effect on solid H-shaped columns is equivalent to an increase in effective length of less than 1%, which can be safely neglected.

Latticed Columns

The lacing or batten plates used to tie together the main longitudinal compression elements are themselves subject to axial deformation. For instance, from Fig. 6.12.2a, the lengthening of the diagonal gives a slope γ_1 over the panel length a and from Fig. 6.12.2b the shortening of the

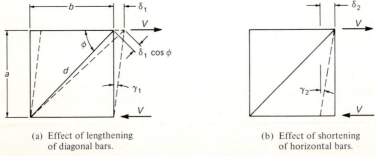

(a) Effect of lengthening   (b) Effect of shortening
of diagonal bars.   of horizontal bars.

Fig. 6.12.2 Shear deformation in laced column.

horizontal bars gives a slope γ_2 over panel length a. Because the lacing elements are relatively small in cross section, the stiffness of such members to resist the transverse shear components is considerably less than for solid-webbed members. Detailed treatment of the columns with lacing, battens, or perforated plates is available in the *SSRC Guide* [9] and elsewhere [36–38].

The Structural Stability Research Council [9] reports the suggestion of Bleich [20] that "a conservative estimate of the influence of 60° or 45° lacing as generally specified in bridge design practice can be made by modifying the effective length factor" K to a new factor αK, as follows:

$$\text{for } \frac{KL}{r} > 40, \qquad \alpha = \sqrt{1 + 300/(KL/r)^2} \qquad (6.12.7)$$

$$\text{for } \frac{KL}{r} \le 40, \qquad \alpha = 1.1 \qquad (6.12.8)$$

Such effective length modification will rarely effect the design of short columns in braced systems.

The SSRC [9] suggests an even simpler empirical procedure where the percent reduction in allowable stress due to shear in laced columns is $0.04F_y$, with F_y in ksi. For example, when F_y is 36 ksi, the percent reduction in allowable stress would be $0.04(36) = 1.4\%$.

6.13* DESIGN OF LATTICED MEMBERS

Under most specifications, latticed members are designed according to detailed empirical rules most of which are related to local buckling requirements. Two examples follow that illustrate some of the provisions of AISC–1.18 as well as general procedures for built-up sections. The reader is referred to Blodgett [39] who has summarized the AISC provisions along with other information concerning built-up section design.

Example 6.13.1

Design a laced column as shown in Fig. 6.13.1, consisting of four angles to carry 600 kips axial compression, with an effective length KL of 30 ft. Assume all connections will be welded and use A572 Grade 50 steel.

SOLUTION

(a) Establish the depth of section h. The radius of gyration of a four angle column depends only on h and essentially is independent of thickness. Thus selecting h establishes the slenderness ratio, and vice versa. Appendix Table A1 shows the relationships between the radius of gyration and the geometry of the cross section. Thus from text Appendix

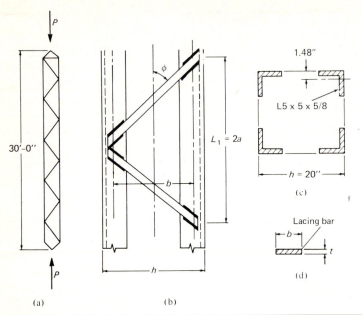

Fig. 6.13.1 Details of Example 6.13.1.

Table A1,

$$r \approx 0.42h$$

$$\frac{KL}{r} = \frac{360}{0.42h} = \frac{857}{h}$$

h	KL/r	F_a (AISC Table 3–50)	Required Area (sq in.)	Angles	
24	35.7	26.4 ksi	22.7	$L6 \times 6 \times \frac{1}{2}$ $L5 \times 5 \times \frac{5}{8}$	$A = 23.0$ 23.44
22	39	25.97	23.1	Same as above	
21	40.7	25.79	23.25	$L5 \times 5 \times \frac{5}{8}$	
20	42.8	25.43	23.55	$L5 \times 5 \times \frac{5}{8}$	

It would appear that 20 in. is preferred, since dimension is probably more important than 0.44 sq in. of cross section. Investigate $4 - L5 \times 5 \times \frac{5}{8}$, assuming $Q = 1.0$,

$$I_x = I_y = 4[13.6 + 5.86(10.0 - 1.48)^2] = 1756 \text{ in.}^4$$

$$r_x = r_y = \sqrt{\frac{I}{A}} = \sqrt{\frac{1756}{23.44}} = 8.66 \text{ in.} \quad (\text{Approx. } 0.42(20) = 8.40 \text{ in.})$$

$$\frac{KL}{r} = \frac{360}{8.66} = 41.6, \qquad F_a = 25.6 \text{ ksi} \qquad \text{(Table 3–50)}$$

$$f_a = \frac{P}{A} = \frac{600}{23.44} = 25.6 \text{ ksi} = F_a \qquad \text{OK}$$

Use $4 - L5 \times 5 \times \frac{5}{8}$ with $h = 20$ in.

(b) Local buckling. For $Q = 1.0$ to apply, angles must satisfy local buckling requirements of AISC–1.9. In this case,

$$\left[\frac{b}{t} = \frac{5}{0.625} = 8\right] < \left[\frac{76}{\sqrt{50}} = 10.7\right] \qquad \text{OK}$$

Development of the provisions of AISC–1.9 appears in Chapter 6, Part II on plate strength.

(c) Design single lacing. According to AISC–1.18.2.6, "The inclination of lacing bars to the axis of the member shall preferably be not less than 60 degrees for single lacing . . ."

For $\phi = 60°$ (Fig. 6.13.1b), $b = 20 - 2(3) = 14$ in., assuming use of distance between standard gage lines for bolts. This would be approximately center-to-center of welded connection. Thus

$$L_1 = 2b \tan 30° = 2(14)0.577 = 16.2 \text{ in.}; \qquad \text{use 16 in.}$$

For a single angle,

$$\frac{L_1}{r_z} = \frac{16}{0.97} = 16.5 < 41.7 \text{ (slenderness ratio for entire member)} \qquad \text{OK}$$

According to AISC–1.18.2.6, "Lacing shall be proportioned to resist a shearing stress normal to the axis of the member equal to 2 percent of the total compressive stress in the member."

$$V = 0.02(600) = 12 \text{ kips} \quad \text{(6 kips per side)}$$

The force in one bar is

$$P = V/\cos 30° = 6/0.866 = 6.93 \text{ kips}$$

$$\frac{L}{r} \leq 140 \text{ for single lacing}$$

$$r = \sqrt{\frac{I}{A}} = \sqrt{\frac{\frac{1}{12}bt^3}{bt}} = 0.288t \qquad \text{(Fig. 6.13.1d)}$$

$$t_{min} = \frac{16}{0.288(140)} = 0.397 \text{ in.} \qquad \text{Use } \tfrac{7}{16} \text{ in.}$$

$$\text{For } t = \tfrac{7}{16} \text{ in.}, \qquad \frac{L}{r} = \frac{16}{0.288(0.4375)} = 127$$

From Table 3–50, $F_a = 9.59$ ksi for secondary members,

$$\text{Required } A = \frac{6.93}{9.59} = 0.72 \text{ sq in.}$$

$$\text{Width } b = \frac{0.72}{0.4375} = 1.65 \text{ in.}$$

Since no holes are required for connectors, tension on net section need not be investigated for this design.

Use bars $\frac{7}{16} \times 1\frac{3}{4}$.

(d) Design the plates at ends (AISC–1.18.2.5). The tie plates should extend along the length of the member a distance equal to the distance b from the end of the member. Use a length of 14 in.

$$t \geq \tfrac{1}{50}(b) = \tfrac{1}{50}(14) = 0.28 \text{ in.}$$

Use tie plates $\frac{5}{16} \times 14 \times 1'\text{-}8''$.

(e) Examine the effect of shear on the effective length, Eqs. 6.12.7 or 6.12.8.

$$\frac{KL}{r} = 41.6 > 40, \qquad \text{Use Eq. 6.12.7}$$

$$\alpha = \sqrt{1 + 300/(KL/r)^2}$$

$$= \sqrt{1 + 300/(41.6)^2} = 1.08$$

Thus the effective length should have been increased 8% due to shear. The neglect of end restraint probably is, in most cases, equal to about the same increase in effective length.

Example 6.13.2

Redesign the column of Example 6.13.1 using a welded perforated box shape (Fig. 6.13.2).

SOLUTION

One important advantage of a welded shape is the number of individual components in the shape is minimized. Four plates can be used for this welded shape, whereas a bolted or riveted section requires four angles in addition to plates.

Using text Appendix Table A1 for a box shape,

$$r \approx 0.40h$$

$$\frac{KL}{r} = \frac{360}{0.4h} = \frac{900}{h}$$

$$\text{Area} \approx 4ht$$

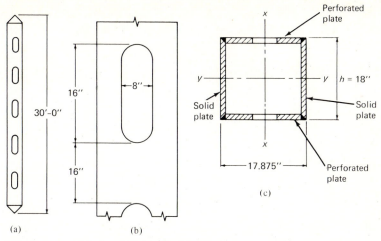

Fig. 6.13.2 Details of Example 6.13.2.

h (in.)	KL/r	F_a (AISC Table 3–50) (ksi)	Area Required (sq. in.)	Plate Thickness (in.)
21	42.8	25.43	23.55	0.28
	$\dfrac{b}{t}$ (solid plate) $= 21/(\tfrac{5}{16}) = 67 > 33.7$ NG			
18	50	24.35	24.6	0.342
16	56.2	23.36	25.7	0.402
15	60	22.72	26.4	0.44

Without perforations, $2\text{PLs} - \tfrac{7}{16} \times 16$ and $2\text{PLs} - \tfrac{7}{16} \times 15$ are probably acceptable, satisfying b/t ratios in accordance with AISC–1.9.2.

Usually, however, such shapes are used on bridges where access is required for maintenance so that perforations may be desirable.

Assume perforations to be 8 in. wide (frequently one-half or less of total width) and $h \approx 18$ in.

Net area available $= 2(18)(\tfrac{7}{16}) + 2(17 - 8)(\tfrac{1}{2}) = 24.75$ sq in.

$$I_x = 2(18)(\tfrac{7}{16})(8.719)^2 + \tfrac{2}{12}[(17)^3 - (8)^3]\tfrac{11}{2} = 1564 \text{ in.}^4$$

$$I_y = \tfrac{2}{12}(18)^3(\tfrac{7}{16}) + 2(9)(\tfrac{1}{2})(8.75)^2 = 1119 \text{ in.}^4$$

$$r_y = \sqrt{\frac{1119}{24.75}} = 6.72 \text{ in.}$$

$$\frac{KL}{r_y} = \frac{360}{6.72} = 53.6, \qquad F_a = 23.8 \text{ ksi} \qquad \text{(AISC Table 3–50)}$$

$$f_a = \frac{P}{A_n} = \frac{600}{24.75} = 24.2 \text{ ksi} \qquad \text{(1.8\% overstress)}$$

Accept this overstress.

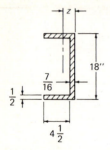

Fig. 6.13.3 Half-section through perforated column of Example 6.13.2.

Check b/t ratio for perforated plate:

$$\frac{b}{t}=\frac{17}{0.5}=34<45 \qquad \text{OK}$$

Access hole size (AISC–1.18.2.7):

$$\text{Length of hole}=2(8)=16 \text{ in. max}$$

$$\text{Clear spacing between holes}=17 \text{ in. min}$$

It is to be noted that among the rules summarized by the *SSRC Guide* (Ref. 9 p. 376) is one which insures that the portions of the column section separated by the perforations can act individually without buckling. The SSRC suggestion is that

$$\frac{\text{Length of perforation}}{\text{Radius of gyration of flange}}\leq\frac{1}{3}\left(\frac{KL}{r}\begin{array}{l}\text{of entire}\\\text{section}\end{array}\right)\leq 20 \text{ max}$$

Referring to Fig. 6.13.3,

$$z=\frac{4.5(2.6875)+7.87(7/32)}{4.5+7.87}=1.12 \text{ in.}$$

$$I=\tfrac{1}{3}(4.5)^3-12.37(1.12)^2=14.90 \text{ in.}^4$$

$$r=\sqrt{\frac{14.90}{12.37}}=1.10 \text{ in.}$$

$$\left[\frac{L}{r}=\frac{16}{1.10}=14.5\right]<\left[\frac{53.6}{3}=17.9\right]<20 \qquad \text{OK}$$

When this slenderness ratio restriction is satisfied the SSRC suggests that only then may the full net section be considered effective.
Use 2 ℞s $\frac{7}{16}\times18$ (solid) and 2 ℞s $\frac{1}{2}\times17$ (perforated) with holes 16″ by 8″ spaced 2′-8″ center-to-center.

PART II: PLATES

6.14* INTRODUCTION TO STABILITY OF PLATES

All column sections whether rolled shapes or built-up sections are composed of plate elements. Up to this point in the chapter consideration has been given only to the possibility of buckling of the member based on the slenderness ratio for the entire cross section. It may be, however, that a local buckling will occur first in one of the plate elements that make up the cross section. Such local buckling means that the buckled element will no longer take its proportionate share of any *additional* load the column is to carry; in other words, the efficiency of the cross section is reduced.

The theory of bending of plates and elastic stability of plates are subjects that should be studied in depth by the advanced student in structural engineering. The brief treatment that follows is intended to give the reader the general idea of plate buckling necessary to properly use and understand current steel specifications. The general approach and terminology follow that of Timoshenko [40, 41].

Before one can treat the stability problem, the differential equation for bending of plates is required, just as the differential equation for the bending of a beam, Eq. 6.2.2, was used in the slender column stability treatment in Sec. 6.2.

Differential Equation for Bending of Homogeneous Plates

First, the strains will be obtained in terms of displacements. Let h = plate thickness and u, v, and w equal the displacements in the x, y, and z directions, respectively. Referring to Fig. 6.14.1, consider an element of a plate $dx\,dy$, and assume no stretching of the neutral plane at $z = 0$. Examining a slice $dx\,dy\,dz$ of the plate element located at a distance z from the neutral plane shows, in Fig. 6.14.2, the coodinate unit strains ϵ_x,

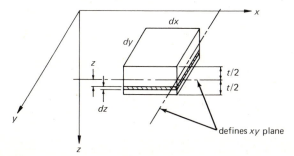

Fig. 6.14.1 Plate element, coordinate system definition.

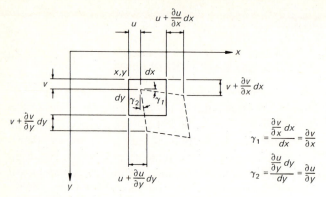

Fig. 6.14.2 Deformations of plate element in xy plane.

ϵ_y, and the shearing strain γ_{xy}. Thus

$$\epsilon_x = \frac{u + \dfrac{\partial u}{\partial x}\,dx - u}{dx} = \frac{\partial u}{\partial x} \tag{6.14.1a}$$

$$\epsilon_y = \frac{\partial v}{\partial y} \tag{6.14.1b}$$

$$\gamma_{xy} = \gamma_1 + \gamma_2 = \frac{\partial v}{\partial x} + \frac{\partial u}{\partial y} \tag{6.14.1c}$$

Expressing the displacements in the plane of the plate in terms of the lateral deflection w, as shown in Fig. 6.14.3, and recognizing that positive slope gives a negative displacement u or v, one establishes

$$-u = z\frac{\partial w}{\partial x}; \qquad -v = z\frac{\partial w}{\partial y} \tag{6.14.2}$$

Substitution of Eqs. 6.14.2 into Eqs. 6.14.1 gives strains in terms of

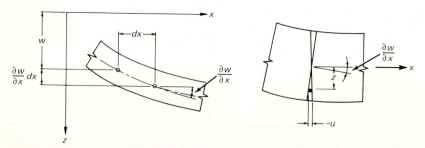

Fig. 6.14.3 Deformation of plate element perpendicular to the xy plane.

curvatures for x-direction bending, y-direction bending, and twisting

$$\epsilon_x = \frac{\partial u}{\partial x} = -z\frac{\partial^2 w}{\partial x^2} \tag{6.14.3a}$$

$$\epsilon_y = \frac{\partial v}{\partial y} = -z\frac{\partial^2 w}{\partial y^2} \tag{6.14.3b}$$

$$\gamma_{xy} = \frac{\partial v}{\partial x} + \frac{\partial u}{\partial y} = -z\left(\frac{\partial^2 w}{\partial x\,\partial y} + \frac{\partial^2 w}{\partial x\,\partial y}\right) = -2z\frac{\partial^2 w}{\partial x\,\partial y} \tag{6.14.3c}$$

Next, making use of Hooke's Law expressing strains in terms of the stresses σ_x, σ_y, normal stresses in the x- and y-directions, and τ_{xy}, the shear stress,

$$\epsilon_x = \frac{1}{E}\left[\sigma_x - \mu\sigma_y\right] \tag{6.14.4a}$$

$$\epsilon_y = \frac{1}{E}\left[-\mu\sigma_x + \sigma_y\right] \tag{6.14.4b}$$

$$\gamma_{xy} = \tau_{xy}/G \tag{6.14.4c}$$

where $\mu =$ Poisson's ratio (see Sec. 2.6) and $G =$ shear modulus of elasticity.

For any stress condition, such as $\sigma_y = -\sigma_x$ that gives pure shear on an element rotated 45° to the x-axis, the work done by the equivalent systems of Fig. 6.14.4 must be a constant:

$$\tfrac{1}{2}\sigma_x\epsilon_x - \tfrac{1}{2}\sigma_x\epsilon_y = \tfrac{1}{2}\tau\gamma \tag{6.14.5}$$

Substituting Eqs. 6.14.4 into Eq. 6.14.5 gives

$$\frac{\sigma_x}{E}(\sigma_x - \mu\sigma_y + \mu\sigma_x - \sigma_y) = \frac{\tau^2}{G}$$

If $\sigma_y = -\sigma_x$, maximum $\tau = \sigma_x$; thus

$$\frac{1}{E}(1 + \mu + \mu + 1) = \frac{1}{G}$$

$$G = \frac{E}{2(1+\mu)} \tag{6.14.6}$$

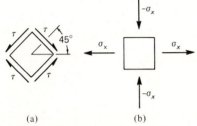

(a)　　　　　(b)

Fig. 6.14.4 Equivalent systems.

Solving Eqs. 6.14.4 for stresses in terms of strains and substituting Eqs. 6.14.3 for the strains give stresses in terms of curvatures,

$$\sigma_x = \frac{-zE}{1-\mu^2}\left(\frac{\partial^2 w}{\partial x^2} + \mu\frac{\partial^2 w}{\partial y^2}\right) \tag{6.14.7a}$$

$$\sigma_y = \frac{-zE}{1-\mu^2}\left(\mu\frac{\partial^2 w}{\partial x^2} + \frac{\partial^2 w}{\partial y^2}\right) \tag{6.14.7b}$$

$$\tau_{xy} = -2zG\frac{\partial^2 w}{\partial x\,\partial y} \tag{6.14.7c}$$

Next it is necessary to relate curvatures and bending moments. Referring to Fig. 6.14.5, and using the right-hand rule for positive twist, it is seen that the moments per unit width are

$$M_x = \int_{-t/2}^{t/2} z\sigma_x\,dz = \frac{-Et^3}{12(1-\mu^2)}\left(\frac{\partial^2 w}{\partial x^2} + \mu\frac{\partial^2 w}{\partial y^2}\right) \tag{6.14.8a}$$

$$M_y = \int_{-t/2}^{t/2} z\sigma_y\,dz = \frac{-Et^3}{12(1-\mu^2)}\left(\mu\frac{\partial^2 w}{\partial x^2} + \frac{\partial^2 w}{\partial y^2}\right) \tag{6.14.8b}$$

$$M_{xy} = -\int_{-t/2}^{t/2} \tau_{xy}z\,dz = +2G\left(\frac{t^3}{12}\right)\frac{\partial^2 w}{\partial x\,\partial y} \tag{6.14.8c}$$

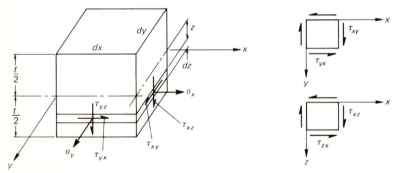

Fig. 6.14.5 A plate element in bending. (Note that forces on faces at $x=0$ and $y=0$ not shown.)

Note that plate bending involves double curvature (a dish-shaped deflection surface for the plate). The narrower and longer in plan a plate is, the more the bending causes curvature to be one directional. A beam is a special case of a plate since it has a narrow width and long span. For beams the Poisson's ratio (μ) effect is neglected. For instance, if the member is narrow in the y-direction and long in the x-direction, Eq. 6.14.8a for the plate would become

$$M_x = \frac{-Et^3}{12}\frac{d^2 w}{dx^2} \tag{6.14.9}$$

where the partial derivatives disappear because w is no longer a function of y. If Eq. 6.14.9 is multiplied by the width b to change from moment per unit width in the y-direction to total moment, Eq. 6.14.9 would be the differential equation for beams,

$$M_x = -EI\frac{d^2w}{dx^2}$$

(6.14.10)

where $I = t^3b/12$.

In theory of plates, the sign convention is that M_x is the bending moment caused by *stresses* σ_x acting in the x-direction. In beam theory, the usual practice is to have the subscript on M refer to the axis of zero stress in bending; that is, the neutral axis. Thus, for the span in the x-direction and the width in the y-direction as assumed for Eq. 6.14.10, M_y (instead of M_x) would have been used in beam theory, since the y-axis is the neutral axis about which bending occurs.

To continue the derivation of the plate bending differential equation, consider the equilibrium of all forces and moments acting on the plate element. Moment summation about the x- and y-axes and force summation in the z-direction gives three equations. Figure 6.14.6 is a free body of the plate element showing only the forces involved with moments about the y-axis.

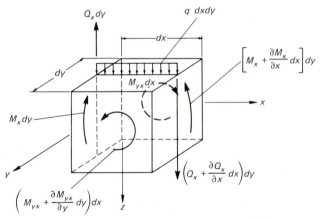

Fig. 6.14.6 Free-body forces involved in rotation about the y-axis. (Forces involved in rotation about the x-axis not shown.)

Taking moments about the y-axis gives

$$\cancel{M_x\,dy} + \frac{\partial M_x}{\partial x}\,dx\,dy - \cancel{M_x\,dy} + \cancel{M_{yx}\,dx} + \frac{\partial M_{yx}}{\partial y}\,dy\,dx - \cancel{M_{yx}\,dx}$$

$$-\left(Q_x\,dy + \frac{\cancel{\partial Q_x}}{\partial x}\cancel{dx\,dy}\right)dx - q\,\cancel{dx\,dy}\,\frac{dx}{2} = 0$$

neglect   mom. arm   neglect

Neglecting infinitesimals of higher order and dividing by $dx\,dy$ gives

$$\frac{\partial M_x}{\partial x}+\frac{\partial M_{yx}}{\partial y}-Q_x=0 \qquad (6.14.11)$$

Similarly for moments about the x-axis,

$$\frac{\partial M_y}{\partial y}-\frac{\partial M_{xy}}{\partial x}-Q_y=0 \qquad (6.14.12)$$

Force equilibrium in the z-direction gives

$$\frac{\partial Q_x}{\partial x}+\frac{\partial Q_y}{\partial y}+q=0 \qquad (6.14.13)$$

Using Eqs. 6.14.11 and 6.14.12 for Q_x and Q_y, and substitution into Eq. 6.14.13 gives

$$\frac{\partial^2 M_x}{\partial x^2}+\frac{\partial^2 M_y}{\partial y^2}-2\frac{\partial^2 M_{xy}}{\partial x\,\partial y}=-q \qquad (6.14.14)$$

Defining $D=Et^3/[12(1-\mu^2)]$ and substituting Eq. 6.14.8 into Eq. 6.14.14 gives the differential equation for bending of homogeneous plates,

$$D\left(\frac{\partial^4 w}{\partial x^4}+2\frac{\partial^4 w}{\partial x^2\,\partial y^2}+\frac{\partial^4 w}{\partial y^4}\right)=q \qquad (6.14.15)$$

Equation 6.14.15, if written for a beam of width b, is the differential equation for load,

$$EI\frac{d^4 w}{dx^4}=qb \qquad (6.14.16)$$

where qb is the load per unit length along the span of the beam.

Buckling of Uniformly Compressed Plate

The following approach is essentially that of Timoshenko [41] as modified by Gerstle [42]. Realizing that q is a general term representing the transverse load component causing plate bending, it is desired to find the transverse component of compressive force N_x when the plate is deflected into a slightly buckled position. Taking summation of forces in the z-direction on the plate element of Fig. 6.14.7b gives

$$N_x\,dy\,\frac{\partial w}{\partial x}-\left(N_x+\frac{\partial N_x}{\partial x}\,dx\right)dy\left(\frac{\partial w}{\partial x}+\frac{\partial^2 w}{\partial x^2}\,dx\right)=q\,dx\,dy$$

$$-\left(N_x\frac{\partial^2 w}{\partial x^2}+\frac{\partial N_x}{\partial x}\frac{\partial w}{\partial x}+\frac{\partial N_x}{\partial x}\,dx\,\frac{\partial^2 w}{\partial x^2}\right)dy\,dx=q\,dx\,dy \qquad (6.14.17)$$

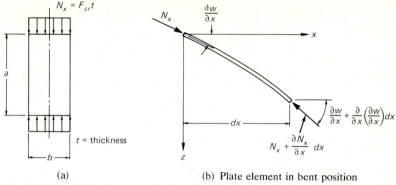

(a)

(b) Plate element in bent position

Fig. 6.14.7 Uniformly compressed plate.

which upon neglecting the higher-order infinitesimal terms gives

$$q = -N_x \frac{\partial^2 w}{\partial x^2} \qquad (6.14.18)$$

The differential equation, Eq. 6.14.15, then becomes

$$\frac{\partial^4 w}{\partial x^4} + 2 \frac{\partial^4 w}{\partial x^2 \partial y^2} + \frac{\partial^4 w}{\partial y^4} = -\frac{N_x}{D} \frac{\partial^2 w}{\partial x^2} \qquad (6.14.19)$$

which is a partial differential equation where w is a function of both x and y. The deflection w can be expressed as the product of an x function (X) and a y function (Y). Further, buckling may be assumed to give sinusoidal variation in the x-direction. Thus

$$w = X(x)Y(y) \qquad (6.14.20)$$

Letting

$$X(x) = \sin \frac{m\pi x}{a}$$

where the X function satisfies the zero deflection and zero moment conditions of simple support at $x = 0$ and $x = a$. Substitution of Eq. 6.14.20 into Eq. 6.14.19 gives, after cancelling the common term $\sin m\pi x/a$,

$$\left(\frac{m\pi}{a}\right)^4 Y - 2\left(\frac{m\pi}{a}\right)^2 \frac{d^2 Y}{dy^2} + \frac{d^4 Y}{dy^4} = +\frac{N_x}{D}\left(\frac{m\pi}{a}\right)^2 Y$$

$$\frac{d^4 Y}{dy^4} - 2\left(\frac{m\pi}{a}\right)^2 \frac{d^2 Y}{dy^2} + \left[\left(\frac{m\pi}{a}\right)^4 - \frac{N_x}{D}\left(\frac{m\pi}{a}\right)^2\right] Y = 0 \qquad (6.14.21)$$

an ordinary fourth-order homogeneous differential equation.

The solution may be expressed in the form

$$Y = C_1 \sinh \alpha y + C_2 \cosh \alpha y + C_3 \sin \beta y + C_4 \cos \beta y \qquad (6.14.22)$$

where

$$\alpha = \sqrt{\left(\frac{m\pi}{a}\right)^2 + \sqrt{\frac{N_x}{D}\left(\frac{m\pi}{a}\right)^2}} \quad \text{and} \quad \beta = \sqrt{-\left(\frac{m\pi}{a}\right)^2 + \sqrt{\frac{N_x}{D}\left(\frac{m\pi}{a}\right)^2}}$$

Thus the entire plate deflection equation is

$$w = \left(\sin \frac{m\pi x}{a}\right)(C_1 \sinh \alpha y + C_2 \cosh \alpha y + C_3 \sin \beta y + C_4 \cos \beta y)$$

$$(6.14.23)$$

which must satisfy boundary conditions. Assuming the x-axis as an axis of symmetry through the plate, i.e., identical support conditions along the two edges parallel to the direction of loading, the odd function coefficients C_1 and C_3 must be zero. Thus

$$w = (C_2 \cosh \alpha y + C_4 \cos \beta y) \sin \frac{m\pi x}{a} \qquad (6.14.24)$$

Using simple support conditions at $y = b/2$ and $y = -b/2$, requires that at $y = \pm b/2$,

$$w = 0 = \left(C_2 \cosh \alpha \frac{b}{2} + C_4 \cos \beta \frac{b}{2}\right) \sin \frac{m\pi x}{a}$$

$$(6.14.25)$$

$$\frac{\partial^2 w}{\partial y^2} = 0 = \left(C_2 \alpha^2 \cosh \alpha \frac{b}{2} - C_4 \beta^2 \cos \beta \frac{b}{2}\right) \sin \frac{m\pi x}{a}$$

For a solution other than $C_2 = C_4 = 0$ it is necessary for the determinant of the coefficients to be zero. Thus

$$(\alpha^2 + \beta^2) \cosh \alpha \frac{b}{2} \cos \beta \frac{b}{2} = 0 \qquad (6.14.26)$$

Since $\alpha^2 \neq -\beta^2$ unless $N_x = 0$ (a trivial solution), and since $\cosh \alpha (b/2) > 1$, the only way Eq. 6.14.26 can be satisfied in the real problem is for

$$\cos \beta \frac{b}{2} = 0$$

Therefore

$$\beta \frac{b}{2} = \frac{\pi}{2}, \ \frac{3\pi}{2}, \ \frac{5\pi}{2}, \ \text{etc.}$$

Using the lowest value of $\beta(b/2)$ and substituting into β as defined below Eq. 6.14.22 gives

$$\frac{b}{2}\sqrt{-\left(\frac{m\pi}{a}\right)^2 + \sqrt{\frac{N_x}{D}\left(\frac{m\pi}{a}\right)^2}} = \frac{\pi}{2}$$

$$\frac{N_x}{D}\left(\frac{m\pi}{a}\right)^2 = \left[\frac{\pi^2}{b^2} + \left(\frac{m\pi}{a}\right)^2\right]^2$$

$$N_x = D\left[\frac{\pi^2 a}{b^2 m\pi} + \frac{m\pi}{a}\right]^2$$

$$N_x = \frac{D\pi^2}{b^2}\left[\frac{1}{m}\frac{a}{b} + m\frac{b}{a}\right]^2 \tag{6.14.27}$$

Since $N_x = F_{cr}t$ and $D = Et^3/[12(1-\mu^2)]$, the elastic buckling unit stress may be expressed as

$$F_{cr} = k\frac{\pi^2 E}{12(1-\mu^2)(b/t)^2} \tag{6.14.28}$$

where for the specific case treated here

$$k = \left[\frac{1}{m}\frac{a}{b} + m\frac{b}{a}\right]^2 \tag{6.14.29}$$

The buckling coefficient k is a function of the type of stress (in this case uniform compression on two opposite edges) and the edge support conditions (in this case simple support on four edges), in addition to the aspect ratio a/b which appears directly in the equation.

The equation for plate buckling, Eq. 6.14.28, is entirely general in terms of k and the development leading up to it for this one case may be considered illustrative of the procedure. The integer m indicates the number of half-waves that occur in the x direction at buckling. Figure 6.14.8 shows that there is a minimum value of k for any given number of half-waves, i.e., the weakest condition. It is noted that this weakest

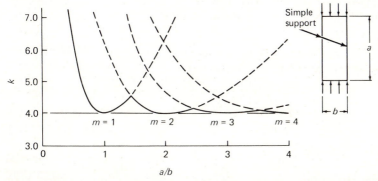

Fig. 6.14.8 Buckling coefficient for uniformly compressed plate—simple support longitudinal edges (Eq. 6.14.29).

situation occurs when the length is an even multiple of width, and that multiple equals the number of half-waves.

Thus, setting $a/b = m$ gives $k = 4$. Further, as m becomes larger the k equation becomes flatter and approaches a constant value of 4 for large a/b ratio. This gives for the elastic buckling stress equation of plate elements under uniform compression along two edges and simply supported along the two edges parallel to the load,

$$F_{cr} = \frac{4\pi^2 E}{12(1 - \mu^2)(b/t)^2} \qquad (6.14.30)$$

6.15* STRENGTH OF PLATES UNDER UNIFORM EDGE COMPRESSION

Since rolled shapes, as well as built-up shapes, are composed of plate elements, the column strength of the section based on its overall slenderness ratio can only be achieved if the plate elements do not buckle locally. Local buckling of plate elements can cause premature failure of the entire section, or at the least it will cause stresses to become nonuniform and reduce the overall strength.

In Sec. 6.14 the basic approach to elastic stability of plates was developed. The theoretical elastic buckling stress for a plate was shown to be expressible as

$$F_{cr} = k \frac{\pi^2 E}{12(1 - \mu^2)(b/t)^2} \qquad [6.14.28]$$

where k is a constant depending on type of stress, edge support conditions, and length to width ratio (aspect ratio) of the plate, E the modulus of elasticity, μ Poisson's ratio, and b/t the width-to-thickness ratio.

In general, plate compression elements can be separated into two categories: (1) stiffened elements; those supported along two edges parallel to the direction of compressive stress; and (2) unstiffened elements; those supported along one edge and free on the other edge parallel to direction of compressive stress. Refer to Fig. 6.15.1 for typical examples of these two situations.

For the elements shown in Fig. 6.15.1 various degrees of edge rotation restraint are present. Figure 6.15.2 shows the variation in k with aspect ratio a/b for most of the idealized edge conditions, i.e., clamped (fixed), simply supported, and free.

Actual plate ultimate strength in compression is dependent on many of the same factors that affect overall column strength, particularly residual stress. Figure 6.15.3 shows typical behavior of a compressed plate loaded to its ultimate load. Assuming ideal elastic-plastic material containing no residual stress the stress distribution remains uniform until the elastic buckling stress F_{cr} is reached. Further increase in load can be

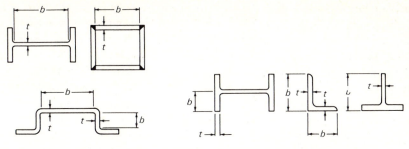

(a) Elements supported along two edges (stiffened elements)

(b) Elements supported along one edge (unstiffened elements)

Fig. 6.15.1 Stiffened and unstiffened compression elements.

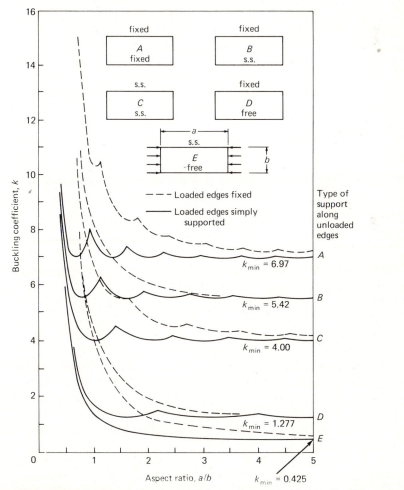

Fig. 6.15.2 Elastic buckling coefficients for compression in flat rectangular plates. (Adapted from Gerard and Becker [43])

achieved but the portion of the plate farthest from its side supports will deflect out of its original plane. This out-of-plane deflection causes the stress distribution to be nonuniform even though the load is applied through ends which are rigid and perfectly straight.

Figure 6.15.3 shows that plate strength under edge compression consists of the sum of two components; (1) elastic or inelastic buckling stress represented by Eq. 6.14.28, and (2) post-buckling strength. Also one should note the higher post-buckling strength as the width-to-thickness ratio b/t becomes larger. For low values of b/t, not only will post-buckling strength vanish, but the entire plate may have yielded and reached the strain-hardening condition, so that F_{cr}/F_y may become greater than unity. For plates *without* residual stress (referring to Fig. 6.15.4) three regions must be considered for establishing strength; elastic buckling (Euler hyperbola), yielding (segments AB, $A'B$, and $A''B$), and strain hardening.

If F_{cr}/F_y is defined as $1/\lambda^2$, which for plates then becomes

$$\lambda = \frac{b}{t} \sqrt{\frac{F_y(12)(1-\mu^2)}{\pi^2 Ek}} \qquad (6.15.1)$$

it is observed from Fig. 6.15.4 that, when compared with columns (curve a), plates (curves b and c) achieve a strain hardening condition at relatively higher values of λ. In the earlier discussion on columns the value of λ at which strain hardening commences (λ_0) was assumed to be zero because of its relatively small value. The values of λ_0 for columns and plates under uniform edge compression for $F_y = 36$ ksi (248 MPa) are given as follows from Haaijer and Thürlimann [44]:

Columns	$\lambda_0 = 0.173 \, (KL/r = 15.7)$
Long hinged flanges	$\lambda_0 = 0.455 \, (b/t = 8.15)$
Fixed flanges	$\lambda_0 = 0.461 \, (b/t = 14.3)$
Hinged webs	$\lambda_0 = 0.588 \, (b/t = 32.3)$
Fixed webs	$\lambda_0 = 0.579 \, (b/t = 42.0)$

From the above, the important factor determining λ_0 is whether the plate element is supported along one or both edges parallel to loading, while the degree of rotational restraint along the loaded edge (simply supported or fixed) has essentially no effect. Thus curves b and c of Fig. 6.15.4 each can represent two cases, where point A' has been taken at $\lambda = 0.46$ and point A'' at $\lambda = 0.58$.

Since plates as well as rolled shapes contain residual stress the true strength should be represented by a transition curve, Fig. 6.15.4, between the Euler curve and the point at which strain hardening commences. It has been suggested [44] that the transition curve for compression

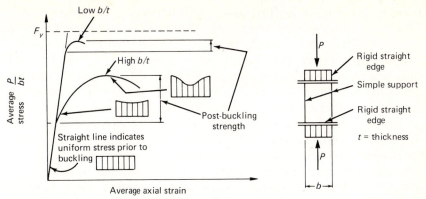

Fig. 6.15.3   Behavior of plate under edge compression.

elements may by taken as

$$\frac{F_{cr}}{F_y} = 1 - \left(1 - \frac{F_p}{F_y}\right)\left(\frac{\lambda - \lambda_0}{\lambda_p - \lambda_0}\right)^n \qquad (6.15.2)$$

where λ_p = slenderness function at F_p = proportional limit.

The SSRC parabola for overall column strength, Eq. 6.7.3, is also obtainable from Eq. 6.15.2 when $n = 2$, $\lambda_0 = 0$, and $\lambda_p = \sqrt{2}$ at $F_{cr}/F_y = 0.5$ on the Euler hyperbola with

$$\lambda = \frac{KL}{r}\sqrt{\frac{F_y}{\pi^2 E}}$$

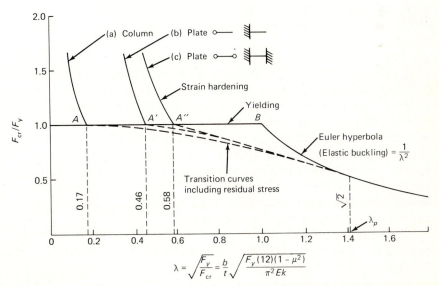

Fig. 6.15.4   Plate buckling compared with column buckling. (Adapted from Ref. 44)

Since the influence of residual stress on plate buckling is less severe than for columns, a value of n greater than 2 would seem reasonable.

When considering inelastic behavior, the modulus of elasticity used for calculating strain in the direction of maximum stress, σ_x, should be the tangent modulus E_t. Examination of Eq. 6.14.4a shows that for inelastic strains in the x-direction but elastic strain in the y-direction, E cannot be factored out. Bleich [20] has shown the solution for this case of using different E values, and suggests arbitrarily using $\sqrt{E_t/E}$ as a multiplier for Eq. 6.14.28. This alternative to Eq. 6.15.2 is essentially the same thing, but without specifically defining the variation of E_t.

Additional treatment of inelastic buckling of plates is available in the work of von Karman, Sechler, and Donnell [45], Winter [46], Jombock and Clark [47], Clark and Rolf [48], and Bulson [59].

In summary, the strength of plates under edge compression may be governed by (1) strain hardening, low values of λ; (2) yielding, at $\lambda =$ say 0.5 to 0.6; (3) inelastic buckling, represented by the transition curve (some fibers elastic and some yielded); (4) elastic buckling represented by the Euler hyperbola, at λ about 1.4; and (5) post-buckling strength with stress redistribution and large deformation, say for λ greater than 1.5.

For design purposes, performance criteria must be established to decide what range of λ values may be acceptable in design and how conservative (and simple) or liberal (and relatively complicated) should be the specification expressions for plate strength.

6.16 AISC BASIC PROVISIONS TO PREVENT PLATE BUCKLING FOR WORKING STRESS DESIGN

For a better understanding of the background for these requirements the reader is invited to delve into the subject of plate stability and strength as introduced in Secs. 6.14 and 6.15. However, it may be sufficient for many purposes merely to understand that components such as flanges, webs, angles, and cover plates, which are combined to form a column section may themselves buckle locally prior to the entire section achieving its maximum capacity. Typical elements are shown in Fig. 6.15.1. The buckled deflection of uniformly compressed plates is shown in Fig. 6.16.1 where two categories are apparent: (1) "unstiffened" plate elements having one free edge parallel to loading; and (2) "stiffened" plate elements supported along both edges parallel to loading.

Plates in compression behave essentially the same as columns and the basic elastic buckling expression corresponding to the Euler equation for columns has been derived as Eq. 6.14.28,

$$F_{cr} = k\frac{\pi^2 E}{12(1-\mu^2)(b/t)^2}$$ [6.14.28]

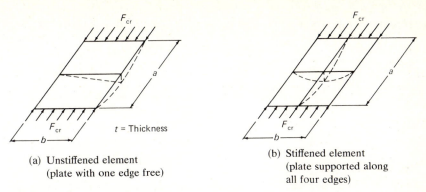

Fig. 6.16.1 Buckled deflection of uniformly compressed plates.

where k is a constant depending on type of stress, edge conditions, and length to width ratio; μ is Poisson's ratio, and b/t is the width-to-thickness ratio (see Fig. 6.16.1). Typical k values are given in Fig. 6.15.2.

It is known that for low b/t values, strain hardening is achieved without buckling occurring, for medium values of b/t residual stress and imperfections give rise to inelastic buckling represented by a transition curve, and for large b/t buckling occurs in accordance with Eq. 6.14.28. Actual strength for plates with large b/t ratio exceeds buckling strength, i.e., they exhibit post-buckling strength. Thus strength for plates may be shown in a dimensionless fashion as in Fig. 6.16.2.

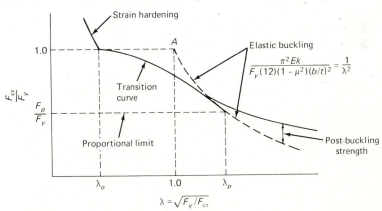

Fig. 6.16.2 Dimensionless representation of plate strength in edge compression.

To establish design requirements, it must be ascertained what performance is desired. Prior to 1969, AISC design procedures provided for designing columns composed of elements capable of reaching yield stress in compression without local buckling occurring. The local buckling of a column component may more logically be restricted so as not to occur prior to achieving full strength of the column based on its overall slenderness ratio. The performance requirement would then be

$$F_{cr} \geq F_{cr} \tag{6.16.1}$$

$$\begin{array}{cc} \text{Component} & \text{Overall} \\ \text{element,} & \text{column} \\ \text{i.e., plate} & \end{array}$$

meaning that acceptable b/t ratios would vary depending on the overall slenderness ratio of the column. If post-buckling strength were considered, the relationship would be additionally complicated. Generally, neglect of post-buckling strength for hot-rolled plates $\frac{3}{16}$ in. or more in thickness involves little error.

For many years the AISI Specification [51] for cold-formed steel has used the approach of Eq. 6.16.1. The AISC Specification in Appendix C has, since 1969, included provisions to consider post-buckling strength, and permits general treatment of plate elements with higher width-to-thickness ratios (slenderness ratios) than the ordinary limitations of AISC–1.9.

To simplify design procedures, the AISC–1.9 basic provisions essentially require the critical buckling stress to be no less than yield stress in plate elements. Buckling, in other words, is prevented prior to achieving an average stress of F_y by application of the basic limits; buckling at lower stress may be considered by using AISC Appendix C. The basic limitation to prevent buckling is

$$F_{cr} = \frac{k\pi^2 E}{12(1-\mu^2)(b/t)^2} \geq F_y \tag{6.16.2}$$

Using $\mu = 0.3$ for steel, and $E = 29,000,000$ psi and F_y in psi,

$$\frac{b}{t} \leq 5120 \sqrt{\frac{k}{F_y, \text{ psi}}} \tag{6.16.3}*$$

which is represented by point A at $\lambda = 1.0$ on Fig. 6.16.2, a point lying above the transition curve. Thus a reduced value of λ should be used to minimize the deviation between F_y and the transition curve which accounts for residual stress and imperfections. Thus $\lambda = 0.7$ is taken as a rational value, which gives for b/t

$$\frac{b}{t} \leq 5120\lambda \sqrt{\frac{k}{F_y}} = 3580 \sqrt{\frac{k}{F_y}} \tag{6.16.4}$$

where F_y is in psi. Table 6.16.1 shows width-to-thickness ratios for various situations of uniform compression. The coefficients used by AISC since 1969 tend to imply greater accuracy in criteria than is justified. The

---

* For SI, $\dfrac{b}{t} \leq 425 \sqrt{\dfrac{k}{F_y}}$ with F_y in MPa, $\tag{6.16.3}$

Table 6.16.1 Width-to-Thickness Ratio Requirements for Plate Elements Subject to Uniform Compression

Structural Elements (1)	Buckling Coefficients (Fig. 6.15.2) (2)	b/t Eq. 6.16.4 (3)	AISC–1.9, 1978 F_y(psi) (4)	F_y(ksi) (5)	F_y(MPa) (6)
Unstiffened:					
(a) Single angles	0.425	$2340/\sqrt{F_y}$	$2400/\sqrt{F_y}$	$76/\sqrt{F_y}$	$200/\sqrt{F_y}$
(b) Flanges	0.70*	$3000/\sqrt{F_y}$	$3000/\sqrt{F_y}$	$95/\sqrt{F_y}$	$250/\sqrt{F_y}$
(c) Stems of tees	1.277	$4050/\sqrt{F_y}$	$4000/\sqrt{F_y}$	$127/\sqrt{F_y}$	$333/\sqrt{F_y}$
Stiffened:					
(a) Uniform thickness flanges, such as tubular sections			$7500/\sqrt{F_y}$‡	$238/\sqrt{F_y}$	$625/\sqrt{F_y}$
(b) Perforated cover plates	6.97†	$9460/\sqrt{F_y}$	$10,000/\sqrt{F_y}$	$317/\sqrt{F_y}$	$832/\sqrt{F_y}$
(c) Others	5.0**	$8010/\sqrt{F_y}$	$8000/\sqrt{F_y}$	$253/\sqrt{F_y}$	$664/\sqrt{F_y}$

* Arbitrarily selected to fall about mid-way between simply supported and fixed along the supported edge.
** Edge restraint estimated at about $\frac{1}{3}$ fixed ($k = 4.0$ for simple support, and $k = 6.97$ for fixed—see Fig. 6.15.2).
† Consider full fixity—use of net plate width will provide adequate reserve.
‡ Hollow sections generally receive negligible torsional restraint by the thin supporting edges; thus a coefficient somewhat less than 8000 is used.

Table 6.16.2 Basic Width-to-thickness Ratios for Plate Elements, AISC—1978

Structural Elements	AISC–1978–1.9 F_y (ksi)					
	36	42	50	60	65	100
Unstiffened:						
(a) Single angles	12.7	11.7	10.7	9.8	9.4	7.6
(b) Flanges	15.8	14.7	13.4	12.3	11.8	9.5
(c) Stems of tees	21.2	19.6	18.0	16.4	15.8	12.7
Stiffened:						
(a) Uniform thickness flanges, as for tubular sections	39.7	36.7	33.7	30.7	29.5	23.8
(b) Perforated plates	52.8	48.9	44.8	40.9	39.3	31.7
(c) Others	42.2	39.0	35.8	32.7	31.4	25.3

original coefficients were established using F_y in psi; after some rounding they formed the basis of AISC–1963 specifications. The present values are obtained by dividing by $\sqrt{1000}$ so that F_y can be used in ksi.

Table 6.16.2 gives AISC–1.9 values for width-to-thickness ratios.

6.17 AISC PLASTIC STRENGTH PLATE BUCKLING PROVISIONS

Since plastic design is not treated until Chapter 7, it is only necessary to state here that the performance requirement of plastic design for plate elements under edge compression is that they be able to undergo strain in excess of that at first yield, ϵ_y. Referring to Fig. 6.17.1, local buckling must not occur prior to achieving a compressive strain well into the plastic range and approaching strain hardening. Referring to Fig. 6.16.2, λ must be less than λ_0, i.e., reduced width-to-thickness ratios must be used compared with those for working stress method wherein it was necessary only to achieve a compressive strain, ϵ_y.

Development of theoretical expressions for local buckling under large plastic strains is outside the scope of this text. It is logical, however, from the brief discussion in this chapter that λ (Fig. 6.15.4) should be reduced to λ_0, about 0.46 for unstiffened compression elements and 0.58 for stiffened compression elements.

For *unstiffened elements*, Eq. 6.14.4 with $\lambda = 0.46$ gives

$$\frac{b}{t} \leq 2350\sqrt{\frac{k}{F_y, \text{psi}}} \quad \text{or} \quad 74.3\sqrt{\frac{k}{F_y, \text{ksi}}} \tag{6.17.1}$$

When $k = 0.425$ (its least value), Eq. 6.17.1 gives

$$\frac{b}{t} \leq \frac{48.5}{\sqrt{F_y, \text{ksi}}} \tag{6.17.2}*$$

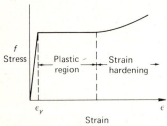

Fig. 6.17.1 Plastic and strain-hardening regions of stress-strain relationship for steel.

* For SI, with F_y in MPa,

$$\frac{b}{t} \leq \frac{127}{\sqrt{F_y}} \tag{6.17.2}$$

Since residual stress effects disappear in the plastic range and material imperfections have less effect, the Eq. 6.17.2 limitation is overly severe. For plastic design, AISC–2.7 specifies limits in a tabular form (see Table 6.17.1) rather than by an equation such as Eq. 6.17.2. The AISC limits are more liberal than Eq. 6.17.2 for the lower yield stresses and agree with that equation for the higher yield stresses.

For *stiffened elements*, Eq. 6.16.4 with $\lambda = 0.58$ gives

$$\frac{b}{t} \leq 2965 \sqrt{\frac{k}{F_y, \text{psi}}} \quad \text{or} \quad 93.7 \sqrt{\frac{k}{F_y, \text{ksi}}} \tag{6.17.3}$$

when $k = 4$, the minimum value assuming edge rotational restraint as the hinged condition (actually the k lies somewhere between values for Cases A and C of Fig. 6.15.2), Eq. 6.17.3 gives

$$\frac{b}{t} \leq \frac{187}{\sqrt{F_y, \text{ksi}}} \tag{6.17.4}$$

AISC–2.7 prescribes the limit for the stiffened compression element as

$$\frac{b}{t} \leq \frac{190}{\sqrt{F_y, \text{ksi}}} \tag{6.17.5}*$$

A tabulation of these limits appears in Table 6.17.1. Additional discussion relative to local buckling limitations to develop plastic strength is given by Lay [49] and McDermott [50].

Table 6.17.1 Width-to-Thickness Ratio Requirements for Plate Elements to Accommodate Plastic Strain in Axial Compression, AISC–2.7

		Unstiffened Elements		Stiffened Elements
F_y (ksi)	F_y (MPa)	Eq. 6.17.2	AISC–2.7	Eq. 6.17.5 and AISC–2.7
36	250	8.1	8.5	31.7
42	290	7.5	8.0	29.3
45	310	7.2	7.4	28.3
50	340	6.9	7.0	26.9
55	380	6.5	6.6	25.6
60	410	6.3	6.3	24.5
65	450	6.0	6.0	23.6

* For SI,

$$\frac{b}{t} \leq \frac{500}{\sqrt{F_y, \text{MPa}}} \tag{6.17.5}$$

6.18* AISC PROVISIONS TO ACCOUNT FOR THE BUCKLING AND POST-BUCKLING STRENGTHS OF PLATE ELEMENTS

Prior to the 1969 AISC, whenever width-to-thickness ratios exceeded the limits of AISC–1.9, the excess width could be neglected in computing the area and radius of gyration of the cross section. Strength computed on this reduced section is usually lower than actual strength, but is not uniformly so. Actual behavior dictates an alternative procedure, such as that used since 1969.

As discussed in Sec. 6.15 and 6.16, plate elements in compression, either "stiffened" or "unstiffened" (see Fig. 6.16.1), have strength after buckling has occurred, i.e., post-buckling strength. Stiffened elements have a large post-buckling strength while unstiffened elements have only a little. However, since the ultimate strength of such elements can be evaluated, there is good reason to provide for its use as has been done in the *Specification for the Design of Cold-Formed Steel Structural Members* [51] since it was first introduced in 1946.

From Fig. 6.18.1a it is apparent that the strength of a stiffened element might be expressed as

$$P_{\text{ult}} = t \int_0^b f(x) \, dx \tag{6.18.1}$$

involving an integration of a nonuniform stress distribution; or alternatively, an "effective width" concept (Fig. 6.18.1b) may be used:

$$P_{\text{ult}} = t b_E f_{\text{max}} \quad \text{(stiffened element)} \tag{6.18.2}$$

where b_E = effective width over which the maximum stress may be considered uniform and give the correct total capacity.

Figure 6.18.1c shows that Eq. 6.18.1 is equally valid for the unstiffened element except that the stress distribution is not symmetrical

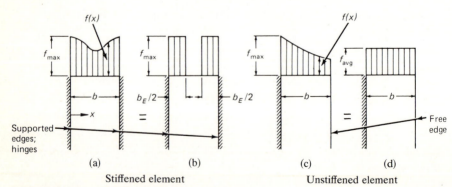

Fig. 6.18.1 Plate elements under axial compression, showing actual stress distribution and an equivalent system.

about the center of the element. If a reduced stress $f_{avg} < f_{max}$ is used, the unstiffened element capacity could be written (Fig. 6.18.1d),

$$P_{ult} = tbf_{avg} \quad \text{(unstiffened element)} \tag{6.18.3}$$

The AISC and the AISI [51] have chosen to treat thin elements according to Eqs. 6.18.2 and 6.18.3, although actually either equation could be used for either type of element. Because of the large post-buckling strength of the stiffened element, one can imagine that it *has* buckled and part of the element is no longer active. On the other hand, the unstiffened element, with relatively little post-buckling strength may be thought of as not buckling because of the use of a reduced stress.

Effect on Overall Column Strength

For design stresses, it is desired to use gross section properties; thus for stiffened elements

$$\frac{P_{ult}}{A_{gross}} = \frac{A_{eff}}{A_{gross}} f_{max} = Q_a f_{max} \tag{6.18.4}$$

and for unstiffened elements,

$$\frac{P_{ult}}{A_{gross}} = \frac{f_{avg}}{f_{max}} (f_{max}) \frac{A_{gross}}{A_{gross}} = Q_s f_{max} \tag{6.18.5}$$

where Q_a and Q_s may be thought of as shape, or form, factors.

A compression system composed of both stiffened and unstiffened elements would be treated as unstiffened for establishing the stress f_{avg}; then the effective width for the stiffened elements is determined using $f_{max} = f_{avg}$. Thus the total capacity would be

$$P_{ult} = f_{avg} A_{eff} \tag{6.18.6}$$

which upon dividing by A_{gross} gives

$$\frac{P_{ult}}{A_{gross}} = \frac{f_{avg}}{f_{max}} (f_{max}) \frac{A_{eff}}{A_{gross}} = Q_s Q_a f_{max} \tag{6.18.7}$$

From Eqs. 6.18.4, 6.18.5, and 6.18.7, it is clear that the effect of premature local buckling before the strength of the overall column has been achieved is to multiply the maximum achievable stress by the form factors Q. Neglecting the possibility of strain hardening, the maximum stress is the yield stress, which is therefore to be multiplied by Q. For the basic SSRC parabola, Eq. 6.7.3, this will give

$$F_{cr} = QF_y \left[1 - \frac{QF_y}{4\pi^2 E} \left(\frac{KL}{r} \right)^2 \right] \tag{6.18.8}$$

where $Q = Q_s Q_a$. From Eq. 6.18.8, when $F_{cr} = QF_y/2$ the slenderness ratio is

$$C_c = \sqrt{\frac{2\pi^2 E}{QF_y}}$$

(6.18.9)

Applying the factor of safety to Eq. 6.18.8 gives AISC Formula (C5–1), which is also Eq. 6.8.3. Whenever KL/r exceeds C_c, the effect of local buckling on overall column stability is insignificant, so that the Euler equation divided by the factor of safety 23/12 is used without any form factor adjustment (see Fig. 6.8.8).

Since the form factor Q is used in the allowable stress equations, the radius of gyration r is to be that of the *gross* section.

Form Factor Q_s for Unstiffened Elements

Referring back to Fig. 6.16.2, a Q_s value less than 1.0 means that $\lambda > \lambda_0$. A transition parabola, such as Eq. 6.15.2, could have been used to compute the reduced stress. For simplification, however, a straight line has been used as shown by curve A in Fig. 6.18.2. It assumes as did Fig. 6.16.4 that $\lambda = 0.7$ is the maximum value for which $F_{cr} = F_y$, and that the proportional limit occurs at $\lambda_p = \sqrt{2}$, the same as for overall buckling. However, because of some post-buckling strength the Euler-type curve has been raised above the theoretical (curve C) so that on the AISC design curve (curve B) when $\lambda_p = \sqrt{2}$, $Q_s = F_{cr}/F_y = 0.65$. Many different expressions could have been used with the same logical results.

AISC Appendix C gives the stress reduction equations for the case of flanges and for the stems of tees. The logic for these reduction equations is the same as described for single angles and shown in Fig. 6.18.2. These other equations will be found to be approximately proportional to $\sqrt{k}$, as seen by reference to Eq. 6.16.4 and Table 6.16.1. The table contains the k values used for the other unstiffened cases.

The limiting proportions for channels and tees given in AISC Appendix Table C1 are to preclude torsional buckling as a failure mode. This concept is discussed in Sec. 8.11.

Although the derivation and discussion has defined $Q_s = F_{cr}/F_y$, when overall buckling of a column occurs (based on its KL/r) the average stress $P/A = F_a(\text{FS})$ is always less than F_y. This means local buckling of an unstiffened element will reduce the efficiency of the cross section only when F_{cr} for the plate element is less than $F_a(\text{FS})$.

Thus, in general, for columns

$$Q_s = \frac{F_{cr}}{F_a(\text{FS})} \geq \frac{F_{cr}}{F_y}$$

(6.18.10)

and for beam compression flanges,

$$Q_s = \frac{F_{cr}}{F_b(\text{FS})} \geq \frac{F_{cr}}{F_y}$$

(6.18.11)

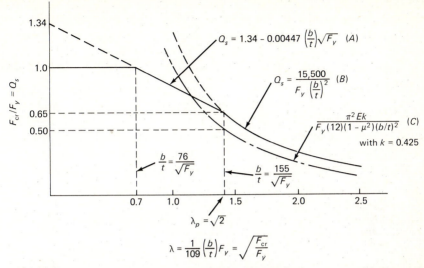

Fig. 6.18.2 Plate strength for unstiffened compression element (single angle) with one edge hinged and the other free (AISC–1978).

Form Factor Q_a for Stiffened Elements

The concept of using an effective width over which stress may be considered uniform, even though it is actually nonuniform, was developed by von Kármán [45] and later modified by Winter [46]. Winter's equation, which has been used by the AISI specification [51] since 1946, is

$$\frac{b_E}{t} = 1.9\sqrt{\frac{E}{f}}\left[1.0 - \frac{0.475}{(b/t)}\sqrt{\frac{E}{f}}\right] \qquad (6.18.12)$$

where f = stress acting on the element (f_{max} of Fig. 6.18.1)
b/t = actual width-to-thickness ratio
The form and constants of the equation were essentially determined to agree with experimental results.
Substituting $E = 29{,}000$ ksi gives

$$\frac{b_E}{t} = \frac{324}{\sqrt{f}}\left[1.0 - \frac{81}{(b/t)\sqrt{f}}\right] \qquad (6.18.13)$$

which would give the correct effective width under the action of a stress f in ksi; i.e., no factor of safety would be involved.
Reexamine Eq. 6.18.2:

$$P_{ult} = b_E t f_{max} \qquad [6.18.2]$$

which if divided by a factor of safety, FS, would give the safe service load,

P_w; thus

$$P_w = b_E t(f_{max}/FS) \tag{6.18.14}$$

Now let $F_a = f_{max}/FS$, which in Eq. 6.18.14 gives

$$P_w = b_E t F_a = A_{eff} F_a \tag{6.18.15}$$

The important point is that b_E of Eq. 6.18.15 is the same effective width as in Eq. 6.18.2; i.e., the effective width under a stress f_{max}. In other words, proper use of Eq. 6.18.13 to include the effect of a factor of safety requires that b_E/t actually be computed for a stress equal to FS times f.

Substitution of $1.65f(FS = 1.65)$ into Eq. 6.18.13 gives

$$\frac{b_E}{t} = \frac{253}{\sqrt{f}}\left[1 - \frac{63}{(b/t)\sqrt{f}}\right] \tag{6.18.16}$$

Experience with light-gage steel has shown that the coefficient 63 can be reduced. AISC uses a coefficient of 50.3,

$$\frac{b_E}{t} = \frac{253}{\sqrt{f}}\left[1 - \frac{50.3}{(b/t)\sqrt{f}}\right] \tag{6.18.17*}$$

which is AISC Formula (C3–1) with f in ksi, which applies for "flanges of square and rectangular sections of uniform thickness."

The difference between Eqs. 6.18.16 and 6.18.17 can be partly explained by the difference in the assumed rotational edge restraint (moment along the supported edges, Fig. 6.18.1b). In sections comprising hot-rolled elements as covered by AISC, the degree of restraint provided by adjacent elements is probably greater than on typical cold-formed shapes (where t is usually $<\frac{3}{16}$ in.).

Additional discussion of the effective width for stiffened elements in compression is available in the work of Korol and Sherbourne [52, 53], Dawson and Walker [54], and Abdel-Sayed [53]. Sharp [56] has considered stiffened elements with one edge stiffened by a lip. Kalyanaraman, Pekoz, and Winter [57] have proposed an effective width expression for unstiffened elements.

Since a column cross section may include unstiffened elements which under present design procedures utilize reduced average stress rather than an effective width, the controlling stress permitted on unstiffened elements is used as the applicable maximum stress acting on the stiffened elements. Thus the service load stress is

$$f = \frac{F_{cr}}{FS} \quad \text{based on unstiffened elements} \tag{6.18.18}$$

---

* For SI with f in MPa,

$$\frac{b_E}{t} = \frac{21.0}{\sqrt{f}}\left[1 - \frac{4.18}{(b/t)\sqrt{f}}\right] \tag{6.18.17}$$

Then, utilizing the higher rational value (see Eq. 6.18.10) of Q_s,

$$Q_s = \frac{F_{cr}}{F_a(\text{FS})}$$

$$f = Q_s F_a \tag{6.18.19}$$

as prescribed by AISC–C3.

Finally, Q_a as defined by Eq. 6.18.4 is

$$Q_a = \frac{(\text{effective width})t}{\text{actual area}} = \frac{A_{\text{eff}}}{A_{\text{gross}}} \tag{6.18.4}$$

where $A_{\text{eff}} = A_{\text{gross}} - \Sigma\,(b - b_E)t$.

Design Properties

In computing nominal stress $f_a = P/A$ and $f_b = M/S$ under service loads the following rules apply, in accordance with AISC Appendix–C4.

For axial compression:

1. Use gross area for P/A.
2. Use radius of gyration of gross area for KL/r.

This agrees with the philosophy stated earlier in this section when developing Eqs. 6.18.4 through 6.18.7.

For flexure:

1. Use reduced section properties for beam with flanges containing stiffened elements.
2. For sections having gross area symmetrical about the axis of bending; in lieu of computing a neutral axis shift due to reduced effective area on the compression flange, the same reduction may be assumed for the tension flange.

Since the allowable stresses for flexure do not contain the form factors Q, it is entirely appropriate to use section properties based on effective area.

For beam-columns:

1. Use gross area for P/A.
2. Use reduced section properties for flexure involving stiffened compression elements for M_x/S_x and M_y/S_y.
3. Use Q_a and Q_s for determining F_a.
4. For F_b, use formulas based on lateral-torsional buckling as discussed in Chapter 9, but the maximum value is $Q_s F_b$ when unstiffened compression elements are involved, obtained by the same reasoning as Eq. 6.18.19.

6.19 DESIGN OF COMPRESSION MEMBERS AS AFFECTED BY LOCAL BUCKLING PROVISIONS

Design of single and double angle struts, structural tees, welded built-up I-shapes, and most other built-up sections, including box-type built-up sections, involves close consideration of local buckling limitations. Rolled W, S, or M shapes as treated previously are proportioned such that for working stress design the width-to-thickness limitations of AISC–1.9 are satisfied ($Q = 1.0$), and even in plastic design, many sections satisfy local buckling limitations of AISC–2.7.

The following examples illustrate the treatment of situations where local buckling may occur prior to development of column strength based on the overall slenderness ratio, KL/r.

Example 6.19.1

A double angle compression chord member for the truss of Fig. 6.19.1 is composed of 2—L9×4×$\frac{1}{2}$, with short legs back-to-back, having a length of 28 ft. Bracing is provided in the plane of the truss every 7 ft, but bracing between trusses is such that lateral support occurs only at the ends of the 28-ft length. Neglect contribution of roofing to lateral support. What is the maximum axial compression service load which this member can be permitted to carry safely? Use A36 steel and AISC Spec.

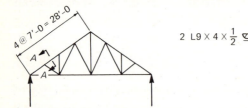

Section A–A

2 L9 × 4 × $\frac{1}{2}$ $\bar{y} = 0.81$ in. for gross section

$\frac{3}{8}$-in. gusset plate

Fig. 6.19.1   Example 6.19.1.

SOLUTION

The 9-in. legs of this compression member are unstiffened elements according to AISC–1.9.1.

Check width/thickness ratio:

$$\left[\frac{b}{t} = \frac{9}{0.5} = 18\right] > \left[\frac{76}{\sqrt{F_y}} = 12.7\right]$$

Thus local buckling will control and the section efficiency is reduced. Using AISC–Appendix–C2, the reduced stress factor Q_s is

$$Q_s = 1.340 - 0.00447\left(\frac{b}{t}\right)\sqrt{F_y}$$

$$= 1.340 - 0.00447(18)\sqrt{36} = 0.857$$

For axial compression, properties of the gross section are used. The AISC Manual properties for double-angle struts give

$$A = 12.5 \text{ sq in.}$$

$$r_x = 1.05 \text{ in.}, \qquad r_y = 4.55 \text{ in. for } \tfrac{3}{8}\text{-in. gusset plate}$$

Assuming $K = 1.0$ for truss members as per Sec. 6.9,

$$\frac{KL_x}{r_x} = \frac{1.0(7)(12)}{1.05} = 80, \qquad \frac{KL_y}{r_y} = \frac{1.0(28)(12)}{4.55} = 74$$

Using AISC Formula (C5–1),

$$C_c = \sqrt{\frac{2\pi^2 E}{Q_s F_y}} = \frac{757}{\sqrt{Q_s F_y}} = \frac{757}{\sqrt{0.857(36)}} = 136.2$$

$$FS = \frac{5}{3} + \frac{3}{8}\left(\frac{80}{136.2}\right) - \frac{1}{8}\left(\frac{80}{136.2}\right)^3 = 1.862$$

$$F_a = Q_s F_y\left[1 - \frac{(KL/r)^2}{2C_c^2}\right]\frac{1}{FS}$$

$$= 0.857(36)\left[1 - \frac{1}{2}\left(\frac{80}{136.2}\right)^2\right]\frac{1}{1.862} = 13.7 \text{ ksi}$$

Safe permissible load:

$$P = A_g F_a = 12.5(13.7) = 171 \text{ kips}$$

Example 6.19.2

Select an economical double angle compression member for use as a spreader strut used for hoisting large loads, as shown in Fig. 6.19.2a. The total lifted load is to be 60 tons. Use $F_y = 60$ ksi with AISC Spec.

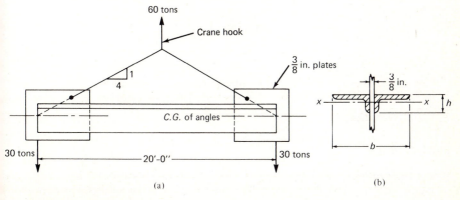

Fig. 6.19.2   Example 6.19.2.

SOLUTION

Assume $K = 1.0$, and referring to Fig. 6.19.2b, use text Appendix Table A1 to estimate r,

$$r_x \approx 0.29h$$
$$r_y \approx 0.24b$$

If 8-in. legs are used, $r_x \approx 0.29(8) = 2.32$ in.

Estimate $\quad \dfrac{KL}{r} = \dfrac{240}{2} = 120 > C_c$

$$F_a = 10.37 \text{ ksi} \quad \text{(from AISC Appendix A, Table 3-50)}$$

Required $A = \dfrac{240}{10.37} = 23$ sq in.

Try L8×8×$\frac{5}{8}$, $r_{min} = 2.49$ in., $\quad KL/r = 96.4$, $\quad F_a = 16.1$ ksi

Revised required $A = \dfrac{240}{16.1} = 14.9$ sq in.

Try L8×8×$\frac{1}{2}$, $r_{min} = 2.50$ in., $A = 15.5$ sq. in., $\quad KL/r = 96$

$$\left[\frac{b}{t} = \frac{8}{0.5} = 16 \right] > \left[\frac{76}{\sqrt{F_y}} = \frac{76}{\sqrt{60}} = 9.8 \right]$$

$$Q_s = 1.340 - 0.00447(16)\sqrt{60} = 0.786$$

$$C_c = \frac{757}{\sqrt{0.786(60)}} = 110$$

Using Fig. 6.8.1, estimate $FS \approx 1.92$:

$$F_a = 0.786(60)\left[1 - \frac{1}{2}\left(\frac{96}{110}\right)^2 \right]\frac{1}{1.92} = 15.2 \text{ ksi}$$

$$f_a = \frac{240}{15.5} = 15.5 > 15.2 \quad \text{but within 3\%, say OK}$$

Use 2—L8×8×$\frac{1}{2}$.

Example 6.19.3

Select the thinnest 10×10 structural tube to carry an axial compression of 200 kips over a length $KL = 18$ ft. The member is part of a braced system. Use $F_y = 65$ ksi and AISC Spec.

SOLUTION

Use text Appendix Table A1 to estimate radius of gyration,

$$r \approx 0.4h \approx 0.4(10) = 4.0 \text{ in.}$$

$$\frac{KL}{r} = \frac{18(10)}{4.0} = 45; \qquad \frac{KL/r}{C_c} = \frac{45}{93.8} = 0.480$$

$$C_a = 0.483 \quad \text{(AISC, Appendix A, Table 4)}$$

$$F_a = C_a F_y = 0.483(65) = 31.4 \text{ ksi}$$

$$\text{Required } A = \frac{200}{31.4} = 6.37 \text{ sq in.}$$

Try $10 \times 10 \times \frac{1}{4}$ structural tube (Fig. 6.19.3): $A = 9.48$, $r = 3.95$ in.

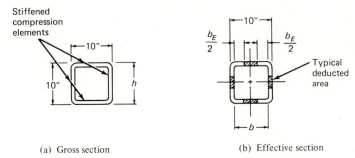

(a) Gross section   (b) Effective section

Fig. 6.19.3 Example 6.19.3.

Check b/t on stiffened elements:

$$\left[\frac{b}{t} = \frac{10.0 - 2(0.25)}{0.25} = 38\right] > \left[\frac{238}{\sqrt{F_y}} = 29.5\right]$$

Local buckling will reduce the efficiency of this section. Compute the form factor Q in accordance with AISC Appendix C. The effective width/thickness ratio is

$$\frac{b_E}{t} = \frac{253}{\sqrt{f}}\left[1 - \frac{50.3}{(b/t)\sqrt{f}}\right] \qquad \text{[6.18.15]}$$

where $f = F_a Q_s$. In this case, a section with no unstiffened elements has $Q_s = 1.0$; Thus $f = F_a$. Try 31.4 ksi as obtained from the first estimate of KL/r.

$$\frac{b_E}{t} = \frac{253}{\sqrt{31.4}}\left[1 - \frac{50.3}{(38)\sqrt{31.4}}\right] = 34.5$$

for all four sides of the tube. The effective area is

$$A_{\text{eff}} = 9.48 - 4\left[\frac{b}{t} - \frac{b_E}{t}\right]t^2$$

$$= 9.48 - 4(38 - 34.5)(0.25)^2 = 8.60 \text{ sq in.}$$

and the form factor is

$$Q_a = \frac{A_{\text{eff}}}{A_{\text{gross}}} = \frac{8.60}{9.48} = 0.907$$

The allowable stress on gross section is then determined, using

$$C_c' = \sqrt{\frac{2\pi^2 E}{Q_a Q_s F_y}} = \frac{757}{\sqrt{0.907(65)}} = 98.6$$

$$\frac{KL/r}{C_c'} = \frac{45}{98.6} = 0.456$$

$$C_a = 0.491 \quad \text{(AISC, Appendix A, Table 4)}$$

$$F_a = C_a Q F_y = 0.491(0.907)(65) = 28.9 \text{ ksi}$$

For a more correct result, b_E/t could have been recomputed using $f = 28.9$ ksi.

$$f_a = \frac{P}{A_{\text{gross}}} = \frac{200}{9.48} = 21.1 \text{ ksi} < 28.9 \text{ ksi} \qquad \text{OK}$$

The stress is low but the next lightest 10×10 will be overstressed.

Use $10 \times 10 \times \frac{1}{4}$ structural tube, $F_y = 65$ ksi.

For design, in addition to the properties of rolled structural tubing available in the AISC Manual, there is available the *Manual of Cold Formed Welded Structural Steel Tubing*,* containing column and beam load tables.

Example 6.19.4

Determine the axial compression capacity for the nonstandard shape of Fig. 6.19.4 for an effective length, $KL = 8$ ft. Use $F_y = 100$ ksi and AISC Spec.

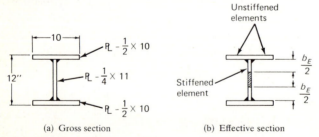

(a) Gross section (b) Effective section

Fig. 6.19.4 Example 6.19.4.

* First Edition, 1974. Published by the Welded Steel Tube Institute, Inc., Structural Tube Division, 522 Westgate Tower, Cleveland, Ohio 44116.

SOLUTION

In axial compression this section contains unstiffened elements (flanges) and a stiffened element (the web). Unstiffened elements must be considered first in order to determine the effective stress level.

(a) Properties of the gross section.

$$I_y = \text{negligible (web)} + 2(\tfrac{1}{12})(0.5(10)^3(\text{flanges}) = 83.4 \text{ in.}^4$$

$$A = 11(0.25) + 2(5.0) = 12.75 \text{ in.}^2$$

$$r_y = \sqrt{I/A} = \sqrt{83.4/12.75} = 2.56 \text{ in.}$$

(b) Unstiffened elements, AISC–1.9.1.2:

$$\left[\frac{b}{t} = \frac{5.0}{0.5} = 10.0\right] > \left[\frac{95}{\sqrt{100}} = 9.5\right]$$

Thus local buckling causes reduced efficiency; $Q_s < 1.0$:

$$Q_s = 1.415 - 0.00437\left(\frac{b}{t}\right)\sqrt{F_y} \qquad \text{AISC Formula (C2–3)}$$

$$= 1.415 - 0.00437(10)\sqrt{100} = 0.978$$

(c) Stiffened element, AISC–1.9.2.2:

$$\left[\frac{b}{t} = \frac{11.0}{0.25} = 44\right] > \left[\frac{253}{\sqrt{100}} = 25.3\right]$$

Thus $Q_a < 1.0$. The stress presumed to act on the stiffened element is that controlled by the unstiffened elements and the overall buckling strength of the member as a column. Try the upper limit of $0.60F_y$,

$$f = 0.6F_yQ_s = 0.6(100)(0.978) = 58.7 \text{ ksi}$$

$$\frac{b_E}{t} = \frac{253}{\sqrt{f}}\left[1 - \frac{44.3}{(b/t)\sqrt{f}}\right] \qquad \text{AISC Formula (C3–2)}$$

$$= \frac{253}{\sqrt{58.7}}\left[1 - \frac{44.3}{44\sqrt{58.7}}\right] = 28.7$$

$$A_{\text{eff}} = A_{\text{gross}} - \left(\frac{b}{t} - \frac{b_E}{t}\right)t^2$$

$$= 12.75 - (44 - 28.7)(0.25)^2 = 11.79 \text{ sq in.}$$

$$Q_a = \frac{A_{\text{eff}}}{A_{\text{gross}}} = \frac{11.79}{12.75} = 0.925$$

$$Q = Q_sQ_a = 0.978(0.925) = 0.904$$

$$\frac{KL}{r} = \frac{8(12)}{2.56} = 37.5$$

$$C'_c = \sqrt{\frac{2\pi^2 E}{Q_a Q_s F_y}} = \frac{757}{\sqrt{0.904(100)}} = 79.6$$

$$\frac{KL/r}{C'_c} = \frac{37.5}{79.6} = 0.471; \qquad C_a = 0.486$$

$$F_a = C_a Q F_y = 0.486(0.904)100 = 43.9 \text{ ksi}$$

Reevaluate b_E/t for $f = 43.9$ ksi; $b_E/t = 32.4$

$$A_{eff} = 12.03 \text{ sq in.}; \qquad Q_a = 0.943; \qquad Q = 0.922$$

$$F_a = 44.6 \text{ ksi}$$

Allowable $P = A_g F_a = 12.75(44.6) = 569$ kips

SELECTED REFERENCES

1. L. Euler, *De Curvis Elasticis, Additamentum I, Methodus Inveniendi Lineas Curvas Maximi Minimive Proprietate Gaudentes.* Lausanne and Geneva, 1744 (pp. 267–268); and "Sur le Forces des Colonnes," *Memoires de l'Academie Royale des Sciences et Belles Lettres,* Vol. 13, Berlin, 1759; English translation of the letter by J. A. Van den Broek, "Euler's Classic Paper 'On the Strength of Columns'," *American Journal of Physics,* 15 (January–February 1947), 309–318.

2. F. Engesser, "Ueber die Knickfestigkeit gerader Stabe," *Zeitschrift fur Architektur und Ingenieurwesen,* 35 (1889), 455; also "Die Knickfestigkeit gerader Stabe," *Zentralblatt der Bauverwaltung,* Berlin (December 5, 1891), 483.

3. A Considère, "Resistance des pièces comprimèes," Congrès International des Procèdés de Construction, Paris, 1891, Vol. 3, p. 371.

4. F. R. Shanley, "The Column Paradox," *Journal of the Aeronautical Sciences,* 13, 5 (December 1946), 678–679.

5. F. R. Shanley, "Inelastic Column Theory," *Journal of the Aeronautical Sciences,* 14, 5 (May 1947), 261–264.

6. N. J. Hoff, "Buckling and Stability," *J. Royal Aeronaut. Soc.,* Vol. 58, Aero Reprint No. 123 (January 1954).

7. Bruce G. Johnston, "Buckling Behavior Above the Tangent Modulus Load," *Journal of Engineering Mechanics Division,* ASCE, 87, EM6 (December 1961), 79–98.

8. Bruce G. Johnston, "A Survey of Progress, 1944–51," Bulletin No. 1, Column Research Council, January 1952.

9. Bruce G. Johnston, ed., *Structural Stability Research Council, Guide to Stability Design Criteria for Metal Structures,* 3rd ed. New York: John Wiley & Sons, Inc., 1976.

10. *Welding Handbook,* Vol. 1, American Welding Society, 7th ed., 1976, Chap. 6.

11. A. W. Huber and L. S. Beedle, "Residual Stress and the Compressive Strength of Steel," *Welding Journal*, (December 1954), 589s–614s.
12. Lynn S. Beedle and Lambert Tall, "Basic Column Strength," *Journal of Structural Division*, ASCE, 86, ST7 (July 1960), 139–173.
13. C. H. Yang, L. S. Beedle, and B. G. Johnston, "Residual Stress and the Yield Strength of Steel Beams," *Welding Journal*, (April 1952), 205s–229s.
14. N. R. Nagaraja Rao, F. R. Estuar, and L. Tall, "Residual Stresses in Welded Shapes," *Welding Journal*, (July 1964), 295s–306s.
15. Donald R. Sherman, "Residual Stress Measurement in Tubular Members," *Journal of Structural Division*, ASCE, 95, ST4 (April 1969), 635–647.
16. Donald R. Sherman, "Residual Stresses and Tubular Compression Members," *Journal of Structural Division*, ASCE, 97, ST3 (March 1971), 891–904.
17. Ching K. Yu and Lambert Tall, "Significance and Application of Stub Column Test Results," *Journal of Structural Division*, ASCE, 97, ST7 (July 1971), 1841–1861.
18. Bruce G. Johnston, "Inelastic Buckling Gradient," *Journal of Engineering Mechanics Division*, ASCE, 90, EM6 (December 1964), 31–47.
19. Richard H. Batterman and Bruce G. Johnston, "Behavior and Maximum Strength of Metal Columns," *Journal of Structural Division*, ASCE, 93, ST2 (April 1967), 205–230.
20. Friedrich Bleich, *Buckling Strength of Metal Structures.* New York: McGraw-Hill Book Company, Inc., 1952.
21. Julian Snyder and Seng-Lip Lee, "Buckling of Elastic-Plastic Tubular Columns," *Journal of Structural Division*, ASCE, 94, ST1 (January 1968), 153–173.
22. Seng-Lip Lee and Julian Snyder, "Stability of Strain-Hardening Tubular Columns," *Journal of Structural Division*, ASCE, 94, ST3 (March 1968), 683–707.
23. Wai F. Chen and David A. Ross, "Tests of Fabricated Tubular Columns," *Journal of Structural Division*, ASCE, 103, ST3 (March 1977), 619–634.
24. Theodore V. Galambos, "Strength of Round Steel Columns," *Journal of Structural Division*, ASCE, 91, ST1 (February 1965), 121–140.
25. John B. Kennedy and Madugula K. S. Murty, "Buckling of Steel Angle and Tee Struts," *Journal of Structural Division*, ASCE, 98, ST11 (November 1972), 2507–2522.
26. John P. Anderson and James H. Woodward, "Calculation of Effective Lengths and Effective Slenderness Ratios of Stepped Columns," *Engineering Journal*, AISC, 9, 3 (October 1972), 157–166.
27. Balbir S. Sandhu, "Effective Length of Columns with Intermediate Axial Load," *Engineering Journal*, AISC, 9, 3 (October 1972), 154–156.
28. Le-Wu Lu, "Effective Length of Columns in Gable Frames," *Engineering Journal*, AISC, 2, 1 (January 1965), 6–7.
29. Joseph A. Yura, "The Effective Length of Columns in Unbraced Frames," *Engineering Journal*, AISC, 8, 2 (April 1971), 37–42. Disc. 9, 3 (October 1972), 167–168.
30. Robert O. Disque, "Inelastic K-factor for Column Design," *Engineering Journal*, AISC, 10, 2 (2nd Quarter 1973), 33–35.
31. C. V. Smith, Jr., "On Inelastic Column Buckling," *Engineering Journal*, AISC, 13, 3 (3rd Quarter 1976), 86–88.

32. T. H. Johnson, "On the Strength of Columns," *Transactions*, ASCE, 15 (July 1886), 517–536. Also Appendix.

33. American Association of State Highway and Transportation Officials (AASHTO), *Standard Specifications for Highway Bridges*, 12th ed. 1977. Also Interim Specifications, 1978, and 1979.

34. Jack C. McCormac, *Structural Steel Design*, 2nd ed. New York: Intext Educational Publishers (Harper and Row), 1971.

35. D. H. Young, "Rational Design of Steel Columns," *Transactions*, ASCE, 101 (1936), 422–500.

36. Cyrus Omid'varan, "Discrete Analysis of Latticed Columns," *Journal of Structural Division*, ASCE, 94, ST1 (January 1968), 119–132.

37. Fung J. Lin, Ernst C. Glauser, and Bruce G. Johnston, "Behavior of Laced and Battened Structural Members," *Journal of Structural Division*, ASCE, 96, ST7 (July 1970), 1377–1401.

38. Bruce G. Johnston, "Spaced Steel Columns," *Journal of Structural Division*, ASCE, 97, ST5 (May 1971), 1465–1479.

39. Omer W. Blodgett, *Design of Welded Structures*. Cleveland, Ohio: James F. Lincoln Arc Welding Foundation, 1966.

40. S. Timoshenko and S. Woinowsky-Krieger, *Theory of Plates and Shells*, 2nd ed. New York: McGraw-Hill Book Company, Inc., 1959 (pp. 79–82).

41. Stephen P. Timoshenko and James M. Gere, *Theory of Elastic Stability*, 2nd ed. New York: McGraw-Hill Book Company, Inc., 1961 (pp. 319–328, 351–356).

42. Kurt H. Gerstle, *Basic Structural Design*. New York: McGraw-Hill Book Company, Inc., 1967 (pp. 88–90).

43. George Gerard and Herbert Becker, *Handbook of Structural Stability*, Part I—*Buckling of Flat Plates*, Tech. Note 3871, National Advisory Committee for Aeronautics, Washington, D.C., July 1957.

44. Geerhard Haaijer and Bruno Thürlimann, "On Inelastic Buckling in Steel," *Transactions*, ASCE, 125 (1960), 308–344.

45. Theodore von Kármán, E. E. Sechler, and L. H. Donnell, "The Strength of Thin Plates in Compression," *Transactions*, ASME, 54, APM-54-5 (1932), 53.

46. G. Winter, "Strength of Thin Compression Flanges," *Transactions*, ASCE, 112 (1947), 527–576.

47. J. R. Jombock and J. W. Clark, "Postbuckling Behavior of Flat Plates," *Journal of Structural Division*, ASCE, 87, ST5 (June 1961), 17–33.

48. John W. Clark and Richard L. Rolf, "Buckling of Aluminum Columns, Plates, and Beams," *Journal of Structural Division*, ASCE, 92, ST3 (June 1966), 17–38.

49. Maxwell G. Lay, "Flange Local Buckling in Wide-Flange Shapes," *Journal of Structural Division*, ASCE, 91, ST6 (December 1965), 95–116.

50. John F. McDermott, "Local Plastic Buckling of A514 Steel Members," *Journal of Structural Division*, ASCE, 95, ST9 (September 1969), 1837–1850.

51. *Specification for the Design of Cold-Formed Steel Structural Members*, American Iron and Steel Institute, New York, 1968 (with Addendum No. 1, November 19, 1970 and Addendum No. 2, February 4, 1977).

52. Robert M. Korol and Archibald N. Sherbourne, "Strength Predictions of Plates in Uniaxial Compression," *Journal of Structural Division*, ASCE, 98, ST9 (September 1972), 1965–1986.

53. Archibald N. Sherbourne and Robert M. Korol, "Post-Buckling of Axially

Compressed Plates," *Journal of Structural Division*, ASCE, 98, ST10 (October 1972), 2223–2234.

54. Ralph G. Dawson and Alastair C. Walker, "Post-Buckling of Geometrically Imperfect Plates," *Journal of Structural Division*, ASCE, 98, ST1 (January 1972), 75–94.

55. George Abdel-Sayed, "Effective Width of Thin Plates in Compression," *Journal of Structural Division*, ASCE, 95, ST10 (October 1969), 2183–2203.

56. Maurice L. Sharp, "Longitudinal Stiffeners for Compression Members," *Journal of Structural Division*, ASCE, 92, ST5 (October 1966), 187–211.

57. V. Kalyanaraman, Teoman Pekoz, and George Winter, "Unstiffened Compression Elements," *Journal of Structural Division*, ASCE, 103, ST9 (September 1977), 1833–1848.

58. Reidar Bjorhovde, "The Safety of Steel Columns," *Journal of Structural Division*, ASCE, 104, ST3 (March 1978), 463–477.

59. P. S. Bulson, *The Stability of Flat Plates*. New York: American Elsevier Publishing Company, 1969.

PROBLEMS

Note: All problems* are to be done according to the latest AISC Specification unless otherwise indicated. For each problem, draw the potential buckled shape on a figure showing the column and its restraints for both x and y principal directions.

6.1. Select the lightest W section to carry an axial compression load P of 100 kips. The member is part of a braced frame and has an effective length KL of 22 ft. (a) Use A 36 steel; (b) Use A572 Grade 50 steel. ($P = 450$ kN; $KL = 6.5$ m)

6.2. Select the lightest W section to carry a compressive load of 825 kips on an effective length of 25 ft. Use A572 Grade 50 steel. If there is a 7% differential in price per pound between this steel and A36, is the section economical when designed using the A572? ($P = 3700$ kN; $KL = 7.6$ m)

6.3. Determine the allowable column load for the built-up section shown, if the steel is A572 Grade 50, and the effective length $K_yL_y = 14$ ft and $K_xL_x = 42$ ft. ($K_yL_y = 4.2$ m; $K_xL_x = 14.5$ m)

$\frac{5}{8} \times 16$ (8 × 406 mm)

$\frac{3}{4} \times 26$ (19 × 660 mm)

$\frac{5}{8} \times 16$ (8 × 406 mm)

Prob. 6.3

---

* Many problems may be solved either as stated in U.S. customary units or in SI units using the numerical data in parenthesis at the end of the statement. The conversions are approximate to avoid having the given data imply accuracy greater in SI than for U.S. customary units.

6.4. Re-solve Prob. 6.1 considering that the member is fixed (built-in) at the bottom of the 22-ft length but remains pinned at the top. ($L = 6.5$ m)

6.5. Select the lightest W section to serve as a main member column 28 ft. long, in a braced frame, with additional lateral support in the weak direction at mid-height, to carry an axial load of 215 kips. Assume bottom and top of column are pinned. Use A36 steel and indicate first and second choices for the sections. ($L = 8.5$ m; $P = 930$ kN)

6.6. Select the most economical W section to carry an axial compression of 150 kips for a member having $K_xL_x = K_yL_y = 18$ ft in a braced frame, assuming that relative costs of various steels are as follows: A36, 1.0; A572 Grade 50, 1.14; A572 Grade 60, 1.20. ($P = 670$ kN; $KL = 5.5$ m).

6.7. Redesign the column of Prob. 6.6 assuming additional weak direction support at mid-height.

6.8. Select the lighest W section of A36 steel to carry an axial compression load of 310 kips. The member is part of a braced frame and may be assumed pinned at the top and bottom of a 30-ft length, and in addition has support in the weak direction 14 ft from the bottom. ($P = 1370$ kN; $L = 9.1$ m; support 4.2 m from bottom)

6.9. Select the lightest W section of A36 steel to carry an axial compression load of 410 kips. Assume the member to be part of a braced frame. The idealized support conditions are that the member is hinged in both principal directions at the top of a 30-ft height; supported in the weak direction at 14 and 22 ft from the bottom; and fixed in both directions at the bottom. ($P = 1820$ kN; $L = 9.1$ m; support at 4.2 m and 6.7 m from bottom)

6.10. Select the lightest W section for the column shown in the accompanying figure. The member is built into a wall so that it may be considered as continuously braced in the weak direction. Use A36 steel. *Note:* Not all of the available W sections are included in the AISC Manual column load tables.

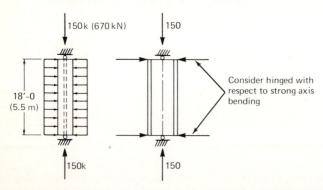

Prob. 6.10

6.11. Redesign the column of Prob. 6.2 assuming there is no residual stress and that column buckling strength may be defined by Euler's equation, $F_{cr} = \pi^2 E/(KL/r)^2$, but still use the AISC factor of safety. Use A572 Grade 50 steel. (Remember F_{cr} cannot exceed F_y.)

6.12. Redesign the column of Prob. 6.2 using the secant formula (Eq. 6.11.6) with FS = 1.89.

6.13. Using the tangent modulus theory, draw to scale the column strength curve (average unit stress on gross area vs slenderness ratio) for a steel with $F_y = 50$ ksi but having a stress-strain diagram for a column stub as shown in the accompanying figure. Assume no residual stress. Using the strength curve values reduced by the AISC factor of safety (text Eq. 6.8.5), select the lightest W section for the loading and support conditions of Prob. 6.2. ($F_y = 350$ MPa)

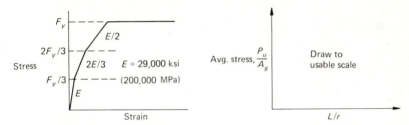

Prob. 6.13

6.14. Using the tangent modulus theory: (a) construct a column strength curve (P_{cr}/A_g versus KL/r) for an H-shaped section. Assume weak-axis bending ($K_y L_y/r_y$) controls and neglect the effect of the web. Assume the idealized stress-strain relationship shown in the accompanying figure is to be used for each fiber of the cross section; (b) select the lightest W section to carry 300 kips with an effective length of 30 ft using your constructed strength curve divided by the AISC factor of safety for columns. In other words, the allowable stress F_a equals P_{cr}/A_g divided by Eq. 6.8.5, where P_{cr}/A_g is the constructed curve of (a); (c) compare with a solution using completely the AISC Specification. Use $F_y = 50$ ksi. ($P = 1300$ kN; $L = 9.1$ m; $F_y = 350$ MPa)

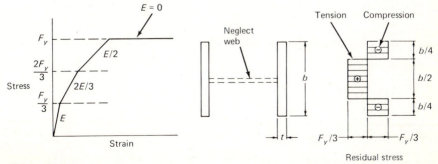

Prob. 6.14

6.15. Calculate critical ordinates and draw to scale a column strength curve using tangent modulus theory. Plot average stress on gross area vs slenderness ratio for weak-axis buckling of wide-flange shapes. Assume each fiber has an idealized elastic-plastic stress-strain diagram with $F_y = 50$ ksi. Neglect the effect of the web and consider each flange to have a residual stress pattern as shown. Using the developed curve along with the AISC factor of safety, select the lightest W section for the loading and support conditions of Prob. 6.2. (Draw conclusions regarding the results of Probs 6.2, and 6.13, if worked, and discuss briefly the merits of making a comparison.)

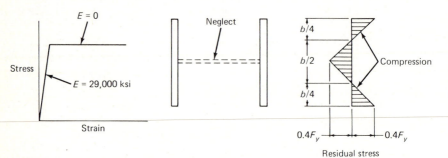

Prob. 6.15

6.16. Design an interior column (use W shape) for a multistory rigid frame. No bracing is provided in the plane of the frame. In the plane perpendicular to the frame, bracing is provided at top, bottom, and mid-height of columns and simple flexible beam to column connections are used. The axial compressive load is 1500 kips and bending moments are neglected. Use A572 Grade 50 steel. ($P = 6700$ kN)

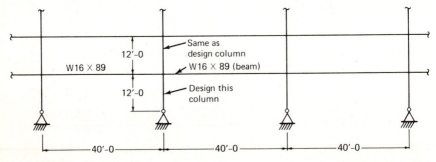

Prob. 6.16

6.17. Design a welded laced column (use single lacing) composed of four angles to carry an axial compression of 500 kips on an effective length, $KL = 28$ ft. Use either A36 or A572 Grade 50 steel, whichever is most economical, if the A572 steel costs 9% more per pound of fabricated steel than A36. Satisfy local buckling limits of AISC–1.9. ($P = 2200$ kN; $KL = 8.5$ m)

6.18. Repeat Prob. 6.17 if the axial load is 400 kips and the effective length is 32 ft. ($P = 1800$ kN; $KL = 9.7$ m)

6.19. Design a solid welded box section for the conditions of Prob. 6.17.

6.20. Design a solid welded box section for the conditions of Prob. 6.18.

6.21. Design a box section with two perforated plates for the conditions of Prob. 6.17.

6.22. Design a box section with two perforated plates for the conditions of Prob. 6.18.

6.23. Determine the safe compressive load on a WT12×38 (structural tee) of A36 steel when used in a truss location where it is braced in the plane of the truss at 20-ft intervals and braced transverse to the plane of the truss at 10-ft intervals. Apply provisions of AISC Appendix C if necessary.

6.24. Repeat Prob. 6.23 using A572 Grade 65 steel. If this material costs 15% more than A36, is its use justified?

6.25. Select the lightest double angle compression member of A36 steel to carry 240 kips on an equivalent pinned length of 20 ft. Assume the angles are attached to $\frac{3}{8}$-in. gusset plates.

6.26. Select the lightest two-angle main compression member connected to a $\frac{3}{8}$-in. gusset plate and to carry a concentric load of 180 kips on an effective length of 16 ft for both principal axes. Use A572 Grade 50 steel.

6.27. Select the most economical double angle member to carry a compressive load of 80 kips on an effective length of 12 ft for both principal axes. Consider A36, A572 Grade 50, and A572 Grade 60 steels. Assume relative costs are A36(1.0), Grade 50(1.07), and Grade 60(1.10). Indicate orientation of the legs.

6.28. Design with A36 steel a top chord of a roof truss consisting of a double angle section connected to $\frac{1}{2}$-in. gusset plates, and to carry a load of 160 kips (including wind loading). The member is braced in the plane of the truss by adjoining web members connecting in at 5-ft intervals. The chord is braced transverse to the plane of the truss at 10-ft intervals. Neglect bending due to roof loads. (*Note:* Refer to AISC–1.5.6.)

6.29. Select the lightest structural tee (WT) for use as a top-chord compression member. Neglect bending. The member has a 9-ft effective length for buckling in either the x–x or y–y plane. The axial compressive load is 135 kips. Use A572 Grade 50 steel.

6.30. Select the lightest structural tee (WT) to serve as the compression chord of a truss to carry 85 kips. In the plane of the truss the chord is braced by adjoining web members that frame in at 5-ft intervals. Perpendicular to the plane of the truss, the chord is braced at 10 ft by a system of lateral purlin supports. Use the most economical of A36 or A572 Grade 65 steels if Grade 65 costs 12% more than A36.

6.31. Compute the allowable axial compressive load on a $10 \times 10 \times \frac{1}{4}$ structural tube having an effective length $KL = 8$ ft. Use (a) A36 steel; (b) steel with $F_y = 90$ ksi.

6.32. Compute the allowable axial compressive load on an $8 \times 6 \times \frac{1}{4}$ structural tube having an effective length $K_y L_y = 7$ ft with respect to weak-axis bending, and $KL_x = 10$ ft with respect to strong-axis bending. (a) Use A36 steel; (b) A514 Grade 90 steel.

6.33. Redesign the column of Prob. 6.1, selecting a structural tube instead of a W section.

6.34. Redesign the column of Prob. 6.6, selecting a structural tube instead of a W section.

6.35. Redesign the column of Prob. 6.7, selecting a structural tube instead of a W section.

6.36. Compute the allowable axial compression load for the nonstandard sections shown in the accompanying figure if they are of A572 Grade 60 steel, and the effective length is 12 ft for $K_y L_y$ and 6 ft for $K_x L_x$.

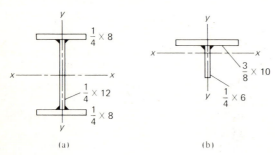

(a)　　　　　　(b)

Prob. 6.36

7
Beams: Laterally Supported

7.1 INTRODUCTION

A beam is generally considered to be any member subjected principally to transverse gravity loading. The term transverse loading is taken to include end moments. Thus beams in a structure may also be referred to as *girders* (usually the most important beams which are frequently at wide spacing); *joists* (usually less important beams which are closely spaced, frequently with truss type webs); *purlins* (roof beams spanning between trusses); *stringers* (longitudinal bridge beams spanning between floor beams); *girts* (horizontal wall beams serving principally to resist bending due to wind on the side of an industrial building; frequently supporting corrugated siding); and *lintels* (members supporting a wall over window or door openings). Other terms, such as header, trimmer, and rafter, are sometimes used, but beam identification by these terms is not generally applied.

A beam is a combination of a tension element and a compression element. The concepts of tension members and compression members are now combined in the treatment as a beam. In this chapter, the compression element (one flange) that is integrally braced perpendicular to its plane through its attachment to the stable tension flange by means

Beams, including open-web joists, channels, and W (wide-flange) shapes, along with tubular columns. (Photo by C. G. Salmon)

of the web, is assumed also to be braced laterally in the direction perpendicular to the plane of the web. Thus overall buckling of the compression flange as a column cannot occur prior to its full participation to develop the ultimate moment capacity of the section. While it is likely true that most beams used in practical situations are adequately braced laterally so that such stability need not be considered, the percentage of stable situations is probably not as high as assumed. The important treatment of lateral stability is found in Chapter 9.

7.2 SIMPLE BENDING OF SYMMETRICAL SHAPES

The most common design situations involve selection of rolled wide-flange shapes from the AISC tables, which often becomes routine and may lead the designer into overconfidence in treatment of beams. It is well known that the flexure formula $(f = Mc/I)$ is applicable to ordinary situations. The stresses on the common sections of Fig. 7.2.1 may be

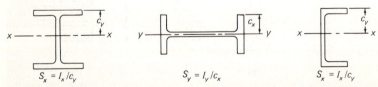

Fig. 7.2.1 Elastic section modulus expressions for symmetrical shapes.

computed by the simple flexure formula when loads are acting in one of the principal directions. When any section with at least one axis of symmetry and loaded through the shear center is subjected to a bending moment in an arbitrary direction, the components M_{xx} and M_{yy}, in the principal directions, can be obtained and the stress computed as

$$f = \frac{M_{xx}}{S_x} + \frac{M_{yy}}{S_y} \tag{7.2.1}$$

where S is the *section modulus*, defined as the moment of inertia I divided by the distance c from the center of gravity to the extreme fiber. The subscripts x and y indicate the axis about which the moment of inertia is computed and from which the distance c is measured (see Fig. 7.2.1). For members without at least one axis of symmetry the reader is referred to Sec. 7.8.

7.3 DESIGN FOR STRENGTH CRITERION

Assuming adequate lateral stability of the compression flange, beam design is based on development of the maximum bending strength of the section. The stress distribution on a typical wide-flange shape subjected to increasing bending moments is shown in Fig. 7.3.1. This behavior is based on the material remaining elastic until the yield point is reached; thereafter additional strain induces no increase in stress. Such stress-strain behavior is shown in Fig. 7.3.2 and is the accepted idealization for structural steels with yield stresses of about $F_y = 65$ ksi (448 MPa) and less. These steels are referred to as exhibiting elastic-plastic behavior. When the yield stress is achieved at the extreme fiber (Fig. 7.3.1b), the moment capacity is referred to as the *yield moment* and may be computed as

$$M_y = F_y S_x \tag{7.3.1}$$

When the condition of Fig. 7.3.1d is reached, every fiber has a strain equal to or greater than $\epsilon_y = F_y/E_s$, i.e., it is in the plastic range. The moment capacity is therefore referred to as the *plastic moment*, and is

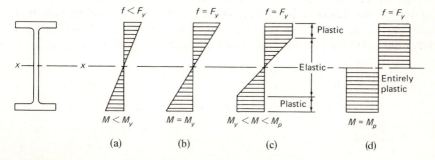

Fig. 7.3.1 Stress distribution at different stages of loading.

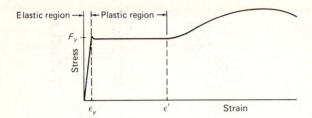

Fig. 7.3.2 Stress-strain diagram for most structural steels.

computed

$$M_p = F_y \int_A y \, dA = F_y Z \tag{7.3.2}$$

where $Z = \int y \, dA$ may be called the *plastic modulus*.

It will be observed that the ratio M_p/M_y is a property of the cross-sectional shape and is independent of the material properties. This ratio is referred to as the *shape factor* f,

$$f = \frac{M_p}{M_y} = \frac{Z}{S} \tag{7.3.3}$$

For wide-flange (W) shapes in flexure about the strong axis $(x - x)$ the shape factor ranges from about 1.09 to about 1.18 with the usual value being about 1.12. One may conservatively say the ultimate bending capacity (plastic moment) of W sections is at least 10 percent greater than the capacity at first yield (M_y).

Design procedures beginning with the 1963 AISC Specification have recognized that beams behave in the manner discussed above. During the period since 1946 extensive testing has adequately verified that plastification of the entire cross section does occur [1].

Example 7.3.1

Determine the shape factor for a rectangular beam of width b and depth d.

SOLUTION
Referring to Fig. 7.3.3a, the moment at first yield, M_y, is

$$M_y = \int_A f y \, dA$$

and

$$f = F_y(y)/(d/2) = F_y \frac{2y}{d}$$

$$M_y = 2 \int_0^{d/2} \frac{2F_y}{d} y^2 b \, dy = F_y \frac{bd^2}{6}$$

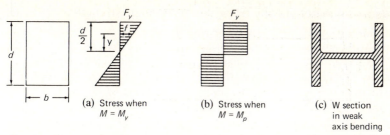

Fig. 7.3.3 A rectangular section and a W section in weak-axis bending.

From Fig. 7.3.3b,

$$M_p = \int_A f y \, dA$$

$$= 2 \int_0^{d/2} F_y b y \, dy = F_y \frac{bd^2}{4}$$

The shape factor is then

$$f = \frac{M_p}{M_y} = 1.5$$

which illustrates that there is a greater reserve beyond first yield in the bending of a rectangular section than in an I-shaped section bending about its strong axis. The reader is alerted to the fact that the W shape bent about its *weak* axis (y–y) is essentially a rectangular section (two rectangles separated by a distance) (Fig. 7.3.3c).

Working Stress (or Elastic Design) Method

Under the working stress method the service load bending moments at various sections are computed using traditional elastic concepts, including statically indeterminate structural analysis whenever statics alone is not sufficient. Designs are made such that computed stresses under the action of the service load moments do not exceed specified allowable stresses. The allowable bending stress, F_b, is taken as some fraction of the yield stress. Thus a correctly designed beam would have a service load capacity computed as

$$M = F_b S$$

Typical values of basic allowable bending stresses are

$$F_b = 0.60 F_y \quad \text{(AISC)}$$
$$F_b = 0.55 F_y \quad \text{(AASHTO)}$$

Such values provide a factor of safety against achieving first yield of 1.67 and 1.82 for AISC and AASHTO, respectively. If, however, first yielding at extreme fiber is not directly related to the ultimate strength of the cross section, then the true safety factor against failure will vary depending on how much the ultimate capacity exceeds the first yield. In other words, the ratio of ultimate moment capacity, M_u, to service load moment, M, may more correctly be called the safety factor. When the ultimate moment can reach M_p, the plastic moment, the ratio M_p/M is referred to as the load factor (LF).

The AISC Specification recognizes the ability of most rolled beams to achieve the plastic moment; such beams are referred to as "compact sections" and the allowable stress is increased 10 percent and 25 percent for bending about the strong and weak axes, respectively. This is because M_p usually exceeds M_y by at least these percentages.

Why are there some beams that cannot achieve the plastic moment capacity? Local buckling of the compression flanges, as discussed in Sec. 6.16, may occur prior to achieving the high compressive strain necessary to develop M_p. Even though the local buckling provisions of AISC–1.9 are satisfied, achievement of only M_y is assured (i.e., buckling below $f = F_y$ is prevented). The extreme fiber strain is assured only of reaching $\epsilon_y = F_y/E_s$. To achieve greater strain, the values of b/t must be further restricted. To undergo large plastic strain the more severe restrictions discussed in Sec. 6.17 prescribed for "compact sections" must be satisfied, as given in Table 7.3.1.

Table 7.3.1 Width-Thickness Limits for "Compact Sections" (AISC–1.5.1.4.1)

F_y (ksi)	Unstiffened Elements (Uniform Compression) $\dfrac{b_f}{2t_f} \leq \dfrac{65}{\sqrt{F_y}}$	Stiffened Elements (Uniform Compression) $\dfrac{b}{t_f} \leq \dfrac{190}{\sqrt{F_y}}$	Stiffened Elements (Bending) $\dfrac{d}{t_w} \leq \dfrac{640}{\sqrt{F_y}}$
36	10.8	31.7	107
42	10.0	29.3	98.8
45	9.7	28.3	95.4
50	9.2	26.9	90.5
55	8.8	25.6	86.3
60	8.4	24.5	82.6
65	8.1	23.6	79.4
70*	7.8	22.7	76.5

* Steel having F_y greater than 70 ksi (483 MPa) is *not* applicable under AISC for consideration as a "compact section."

The basic working stress (or elastic design) method of AISC for beams may be summarized (AISC–1.5.1.4) as follows:

1. For laterally supported "compact sections" that are symmetrical about, and loaded in, the plane of their minor axis (i.e., sections that have a shape factor of at least 1.10 and are capable of achieving a plastic moment capacity M_p),

$$F_b = 0.66F_y \qquad (7.3.4)$$

2. For sections in category (1) whose flanges (unstiffened elements) are stiffer than required by AISC–1.9.1 (i.e., $b_f/2t_f < 95/\sqrt{F_y}$) but not stiff enough to accommodate the plastic strain required of a "compact section" (i.e., $b_f/2t_f > 65/\sqrt{F_y}$, a linear transition between $0.60F_y$ and $0.66F_y$ is provided in AISC–1.5.1.4.2. For such cases,

$$F_b = F_y\left[0.79 - 0.002\left(\frac{b_f}{2t_f}\right)\sqrt{F_y}\right] \qquad (7.3.5)$$

the variation is shown in Fig. 7.3.4.

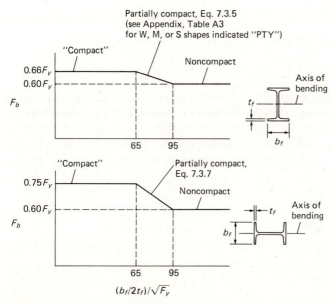

Fig. 7.3.4 Allowable stress on H- or I-shaped sections qualifying as "compact sections" except for an excessive width-to-thickness ratio at the unstiffened compression flange element.

3. For doubly symmetrical I- and H-shaped members, satisfying the unstiffened element local buckling requirement of AISC–1.5.1.4.1(2.), and bent about the minor axis,

$$F_b = 0.75F_y \qquad (7.3.6)$$

in recognition of the higher shape factor for a rectangle in bending, as shown in Example 7.3.1. Lateral bracing is not required for this case.

4. For sections in category (3) whose flanges (unstiffened elements) are stiffer than required by AISC–1.9 (i.e., $b_f/2t_f < 95/\sqrt{F_y}$) but not stiff enough to accommodate the plastic strain of a "compact section" (i.e., $b_f/2t_f > 65/\sqrt{F_y}$), a linear transition between $0.60F_y$ and $0.75F_y$ is provided for in AISC–1.5.1.4.3. For such cases,

$$F_b = F_y\left[1.075 - 0.005\left(\frac{b_f}{2t_f}\right)\sqrt{F_y}\right] \qquad (7.3.7)$$

as shown in Fig. 7.3.4.

5. For all sections having adequate lateral bracing but where the maximum moment strength is considered to be reached when the extreme fiber strain equals yield strain, $\epsilon_y = F_y/E_s$. This means all sections not qualifying as "compact" or "partially compact." This basic value is

$$F_b = 0.60F_y \qquad (7.3.8)$$

Since plastic behavior is not utilized for material with F_y greater than 70 ksi[*], the higher yield stress steels are governed by a maximum allowable stress of $0.60F_y$.

It is to be noted that the AASHTO working stress method does not recognize the ability of any section to develop a moment capacity exceeding M_y; therefore the maximum allowable stress is $F_b = 0.55F_y$. Further, because of fatigue considerations involved in highway bridge design, allowable stresses may be significantly less than $0.55F_y$ depending on the maximum and minimum stresses and the number of cycles of loading for which the structure is to be designed.

Plastic Design Method

In the plastic design method, the service loads are multiplied by a *load factor* LF to obtain the ultimate loads that the structure must be capable of carrying at plastic collapse. The ultimate moments are then obtained for the collapse condition. In the statically determinate structure, achieving a plastic moment at one location is sufficient to create a collapse

---

[*] AISC–1.5.1.4.1 recognizes all steels in AISC–1.4.1.1, except A514 steel, as eligible for "compact section" criteria; the highest F_y for recognized steels is 70 ksi for A607 Grade 70.

mechanism. The section at which the plastic moment occurs will continue to deform without inducing additional resistance after it reaches M_p. This condition of increasing deformation with a constant resisting moment is called a *plastic hinge*.

Referring to Fig. 7.3.5, it is seen that as the bending moment is increased above the service load value, the strain angle ϕ (rotation in radians/inch) is entirely an elastic (up to M_y) or partially elastic (M_y to M_p) (see Fig. 7.3.1) relationship until M_p is reached. Thereafter an unstable situation or mechanism occurs such that the deflection increases without restraint. At the ultimate condition the elastic deformation due to bending on the segments between the ends and midspan is negligible compared to the rotation ϕ_u occurring at the plastic hinge. Thus the analysis may treat the collapse situation as two rigid bodies with an angular discontinuity ϕ_u at midspan. As will be shown later in Sec. 10.2, it is only for the statically determinate situations that one can expect every point on the ultimate moment diagram to be directly proportional to the elastic moments. Redistribution of the moments occurs during loading beyond the elastic range in usual statically indeterminate situations; that is, the moment diagram after a plastic hinge has occurred will no longer be proportional to the elastic moment diagram.

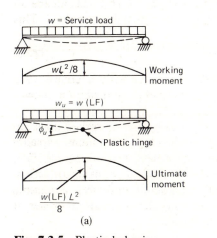

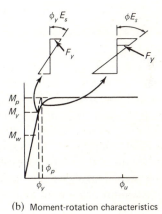

(b) Moment-rotation characteristics

Fig. 7.3.5 Plastic behavior.

Under AISC, plastic design requires provision of adequate lateral stability since it is paramount in the method that the plastic moment be achieved under ultimate loading. The load factor prescribed for gravity loading is 1.70.

Just as in the working stress method, where a "compact section" is necessary to use $0.66F_y$, so in plastic design similar local buckling provisions are given in AISC–2.7.

At the present time (1979), plastic behavior of steels having strengths as high as A607 Grade 70 is recognized by the AISC "compact section" provisions of AISC–1.5.1.4.1, while A572 Grade 65 is the highest strength steel recognized by the plastic design provisions of AISC–Part 2. Steels having yield stresses above 65 to 70 ksi (448 to 483 MPa) do not exhibit the sharp yield point of the lower strength steels.

Example 7.3.2

Using the working stress method, select the lightest W or M section to carry a uniformly distributed superimposed load of 1 kip/ft (14.6 kN/m) on a 20-ft (4.3 m) simply supported span (Fig. 7.3.6). The compression flange of the beam is fully supported against lateral movement. Use the following steels; (a) A36; (b) A572 Grade 50; (c) A572 Grade 65.

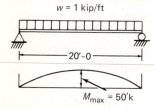

Fig. 7.3.6 Example 7.3.2.

SOLUTION

(a) A36 steel. Assume "compact section" since the vast majority of sections satisfy the necessary conditions; thus the allowable stress is

$$F_b = 0.66F_y = 24 \text{ ksi} \quad (165 \text{ MPa})$$

Note that rounded (i.e., 0.66 times 36 ksi = 23.8 ksi; use 24 ksi) values are accepted values in accordance with the Appendix A, Table 1 of the AISC Specification. Due to the superimposed load,

$$M = \frac{wL^2}{8} = \frac{1(20)^2}{8} = 50 \text{ ft-kips} \quad (67.9 \text{ kN·m})$$

$$\text{Required } S_x = \frac{M}{F_b} = \frac{50(12)}{24} = 25 \text{ in.}^3 \quad (410\,000 \text{ mm}^3)$$

Select from the AISC Manual "Allowable Stress Design Selection Table," (or from text, Appendix, Table A3), the lightest section having at least $S_x = 25$ in.^{3}:

$$\text{Try W12} \times 22: \qquad S_x = 25.4 \text{ in.}^3$$

Check "compact section" requirements:

$$\frac{b_f}{2t_f} = \frac{4.03}{2(0.425)} = 4.74 < 10.8 \quad \text{(Table 7.3.1)} \qquad \text{OK}$$

$$\frac{d}{t_w} = \frac{12.31}{0.26} = 47.3 < 107 \qquad \text{OK}$$

Check the flexural stress:

$$M = \frac{1.022(20)^2}{8} = 51.1 \text{ ft-kips} \quad (69.3 \text{ kN·m})$$

$$f_b = \frac{M}{S_x} = \frac{51.5(12)}{25.4} = 24.1 \approx F_b = 24 \text{ ksi} \quad (165 \text{ MPa}) \qquad \text{OK}$$

Note that in the check of stress the moment is increased to include the beam weight.

Use W12×22, $F_y = 36$ ksi (W310×33,* $F_y = 248$ MPa).

(b) A572 Grade 50 steel.

$$F_b = 0.66F_y = 33 \text{ ksi} \quad (228 \text{ MPa})$$

$$\text{Required } S_x = \frac{50(12)}{33} = 18.2 \text{ in.}^3 \quad (298\,000 \text{ mm}^3)$$

Try M14×17.2 with $S_x = 21.0$ as lightest section satisfying the section modulus requirement. Check "compact section."

$$\frac{b_f}{2t_f} = \frac{4.0}{2(0.272)} = 7.35 < 9.2 \qquad \text{OK}$$

$$\frac{d}{t_w} = \frac{14.00}{0.21} = 66.7 < 90.5 \qquad \text{OK}$$

$$f_b = \frac{50(12)}{21.0} \left(\frac{1.017}{1.0}\right) = 29.1 \text{ ksi} < 33 \text{ ksi} \qquad \text{OK}$$

Use M14×17.2, $F_y = 50$ ksi (M360×25.6, $F_y = 345$ MPa). Though this section is understressed, no lighter section will work.

(c) A572 Grade 65 steel.

$$F_b = 0.66F_y = 43 \text{ ksi} \quad (296 \text{ MPa})$$

$$\text{Required } S_x = \frac{50(12)}{43} = 14.0 \text{ in.}^3 \quad (229\,000 \text{ mm}^3)$$

* In SI designation, 310 mm nominal depth and 33 kg/m mass.

Try W12×14: $S_x = 14.9$ in.[3] Check "compact section" limits.

$$\frac{d}{t_w} = \frac{11.91}{0.200} = 59.6 < 79.4 \qquad \text{OK}$$

$$\left[\frac{b_f}{2t_f} = \frac{3.97}{2(0.225)} = 8.8 \right] > \left[\frac{65}{\sqrt{65}} = 8.1 \right] \qquad \text{NG}$$

Thus $F_b \neq 0.66F_y$. However,

$$8.8 < \left[\frac{95}{\sqrt{65}} = 11.8 \right]$$

Therefore, using Eq. 7.3.5,

$$F_b = F_y[0.79 - 0.002(b_f/2t_f)\sqrt{F_y}]$$
$$= F_y[0.79 - 0.002(8.82)\sqrt{65}] = 0.648F_y = 42.1 \text{ ksi}$$
$$f_b = \frac{50(12)}{14.8} \left(\frac{1.014}{1.0} \right) = 40.8 \text{ ksi} < 42.1 \text{ ksi} \qquad \text{OK}$$

Use W12×14, $F_y = 65$ ksi (W310×21, $F_y = 448$ MPa).

Example 7.3.3

Repeat Example 7.3.2 using the plastic design method of AISC–Part 2.

SOLUTION
The load factor 1.7 is applied to the service load first and the required plastic moment is then computed.

(a) A36 Steel.

$$w_u = 1.0(1.70) = 1.70 \text{ kips/ft}$$
$$\text{Required } M_p = 1.70(20)^2/8 = 85 \text{ ft-kips} \quad (115 \text{ kN·m})$$

From Eq. 7.3.2,

$$\text{Required } Z = \frac{M_p}{F_y} = \frac{85(12)}{36} = 28.4 \text{ in.}^3 \quad (465\,000 \text{ mm}^3)$$

Select from AISC Manual "Plastic Design Selection Table," W12×22, $Z_x = 29.3$ in.[3] The sections tabulated are those that also satisfy the local buckling requirements of AISC-2.7 for A36 steel, i.e., $b/2t_f \leq 8\frac{1}{2}$ and $d/t_w \leq 68.7$.
Use W12×22, $F_y = 36$ ksi (W310×33, $F_y = 248$ MPa).

(b) A572 Grade 50 steel.

$$\text{Required } Z = \frac{M_p}{F_y} = \frac{85(12)}{50} = 20.4 \text{ in.}^3 \quad (334\,000 \text{ mm}^3)$$

Try W10×19, $Z = 21.6$ in.^{3} (the M14×17.2 is the next lightest possibility but does not satisfy local buckling requirements.)

$$\frac{b_f}{2t_f} = \frac{4.02}{2(0.395)} = 5.1 < 7.0 \quad \text{OK}$$

$$\left[\frac{d}{t_w} = \frac{10.24}{0.25} = 41 \right] < \left[\frac{412}{\sqrt{50}} = 58.3 \right] \quad \text{OK}$$

Use W10×19, $F_y = 50$ ksi (W250×28, $F_y = 345$ MPa).

(c) A572 Grade 65.

$$\text{Required } Z = \frac{M_p}{F_y} = \frac{85(12)}{65} = 15.7 \text{ in.}^3 \quad (257\,000 \text{ mm}^3)$$

Since W10×15, W12×16, and W8×18 do not satisfy one or both of the $b_f/2t_f$ and d/t_w local buckling requirements; try W10×17, $Z = 16.2$ in.^{3}

$$\frac{b_f}{2t_f} = \frac{4.01}{2(0.330)} = 6.1 \approx 6.0 \text{ limit,} \quad \text{say} \quad \text{OK}$$

$$\left[\frac{d}{t_w} = \frac{10.11}{0.24} = 42.1 \right] < \left[\frac{412}{\sqrt{65}} = 51.1 \right] \quad \text{OK}$$

Use W10×17, $F_y = 65$ ksi (W250×25, $F_y = 448$ MPa).

A comparison of results is given below:

	WSD	Plastic Design
$F_y = 36$ ksi	W12×22	W12×22
$F_y = 50$ ksi	M14×17.2	W10×19
$F_y = 65$ ksi	W12×14	W10×17

The requirement of plastic design that the section selected must have the capability of developing a plastic hinge resulted in heavier sections for the last two cases under plastic design than for working stress design.

7.4 DEFLECTION

As beams are used on long spans (i.e., as the ratio of the span length to the depth of the section becomes large) or shallower sections of high-strength steels are used, deflection restrictions may control the design.

Excessive deflection of a floor or roof system may cause damage to attached nonstructural elements such as partitions, may impair the usefulness of the structure by, for instance, distorting door jambs so that doors will not open and close, or may cause "bouncy" floors. These are all *serviceability* requirements, in many cases having no correlation with the *strength* of the floor system. Excessive deflection also often indicates a flexibility in the floor system that is directly related to vibration and noise transmission problems. Wright and Walker [2] and Murray [3] have discussed vibration of floor systems and the related human response.

On roofs a major deflection-related problem is the ponding of water; this is specifically treated later in this section.

Numerous structural analysis methods are available for computing deflections on uniform and variable moment of inertia sections in statically determinate and indeterminate structures. In general, the maximum deflection in an elastic member may be expressed as

$$\Delta_{max} = \beta_1 \frac{WL^3}{EI} \tag{7.4.1}$$

where W = total load on the span
  L = span length
  E = modulus of elasticity (29,000 ksi or 200 000 MPa for steel)
  I = moment of inertia
  β_1 = coefficient which depends upon the degree of fixity at supports, the variation in moment of inertia along the span, and the distribution of loading. (For a simply supported beam, $\beta_1 = 5/384$; other values are available in the AISC Manual section, "Beam Diagrams and Formulas.")

For continuous beams, the midspan deflection in the common situation of a uniform loading on a prismatic beam with unequal end moments (see Fig. 7.4.1) may be expressed* as

$$\Delta_{midspan} = \frac{5L^2}{48EI}[M_s - 0.1(M_a + M_b)] \tag{7.4.2}$$

Equation 7.4.2 will give satisfactory results when considered to be the maximum deflection for nearly all practical loadings for beams having uniform moment of inertia. Equation 7.4.2 may be verified by the use of a method such as conjugate beam.

---

* See Chu-Kia Wang and Charles G. Salmon, *Reinforced Concrete Design*, 3rd Ed. (Harper & Row, Publishers, New York, 1979), p. 482.

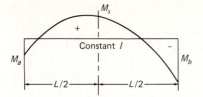

Fig. 7.4.1 Typical bending moment diagram for uniformly loaded beam.

For uniformly loaded simply supported beams, Eq. 7.4.1 becomes

$$\Delta_{max} = \frac{5wL^4}{384EI} \qquad (7.4.3)$$

which upon substitution of $M = wL^2/8$, $f = Mc/I$, and $c = d/2$, gives

$$\Delta_{max} = \frac{10fL^2}{48Ed} \qquad (7.4.4)$$

Equation 7.4.4 can be used as a good approximation for any simply supported beam as long as the maximum stress occurs near midspan. Refer to Table 7.4.1 for typical values.

Table 7.4.1 Deflection Relationships According to Eq. 7.4.4

Δ_{max}	L/d	$L/d(f = 22$ ksi$)$	$L/d(f = 30$ ksi$)$
$L/360$	$387/f$	17.6	12.9
$L/300$	$464/f$	21.1	15.5
$L/240$	$580/f$	26.3	19.3
$L/200$	$695/f$	31.6	23.2

AISC–1.13.1 states, "Beams and girders supporting floors and roofs shall be proportioned with due regard to the deflection produced by design loads." In addition, live-load deflection where plastered ceilings are supported is limited to $L/360$.

For the $L/360$ limitation, Eq. 7.4.4, using $E = 29{,}000$ ksi becomes

$$\frac{L}{d} \leq \frac{48(29{,}000)}{(10)360f} = \frac{387}{f} \qquad (7.4.5)^*$$

where f is in ksi.

When considering deflections, it should be remembered that the

---

* For SI, $\dfrac{L}{d} = \dfrac{2667}{f}$ for f in MPa. $\qquad$ (7.4.5)

dead-load deflections usually can be accounted for during construction by either cambering (negative bending) or thickening the slab or floor topping. It is only the deflection that occurs due to loads applied after construction is completed that may crack ceilings, partitions, or walls.

Specification requirements for limiting deflections are meager because there is no single or standard value for the tolerable deflection. The acceptable amount must of necessity depend on the type and arrangement of materials being supported.

As a guide only, the AISC Commentary suggests the following limitations:

Floor beams and girders, fully stressed, not subject to shock or vibration:

$$\frac{L}{d} \leq \frac{800}{F_y, \text{ksi}} \tag{7.4.6}$$

Floor beams and girders, subject to shock or vibratory loads, supporting large open areas free of partitions or other sources of damping:

$$\frac{L}{d} \leq 20 \tag{7.4.7}$$

Roof purlins, fully stressed, except sloped roofs steeper than slope of 3 in 12:

$$\frac{L}{d} \leq \frac{1000}{F_y, \text{ksi}} \tag{7.4.8}$$

Based on service load deflections where $F_b = 0.66F_y$ or assuming $f = F_y/1.515$, the AISC Commentary suggestions of 800 and 1000 correspond to L/d values of $528/f$ and $660/f$, respectively. Using Eq. 7.4.4 the simple beam deflection limitations would be approximately $L/260$ and $L/210$.

On continuous spans it is, of course, the actual deflection that is of importance, not the L/d ratio. For continuous beam deflection, a comparison of Eqs. 7.4.2 and 7.4.4 shows that Eq. 7.4.4 can also be used for continuous beams if the stress f is computed using the equivalent bending moment

$$M_e = M_s - 0.1(M_a + M_b) \tag{7.4.9}$$

Ponding of Water on Flat Roofs

When members of a flat roof system deflect, a bowl-shaped volume is created which is capable of retaining water. As water begins to accumulate, deflection increases to provide an increased volumetric capacity. This cyclical process continues until either (1) the succeeding deflection increments become smaller and equilibrium is reached; or (2) succeeding deflection increments are increasing, the system is divergent, and collapse occurs. This retention of water which results solely from the deflection of flat roof framing is what is referred to as *ponding.*

To prevent ponding of excessive water accumulation on flat roofs, AISC–1963 required supporting members to satisfy the limitation

$$\frac{L}{d} \leq \frac{600}{f_b} \qquad (7.4.10)$$

where f_b is the computed bending stress in ksi. Using Eq. 7.4.4, this would correspond roughly to a simple span deflection limitation of $L/240$.

The problem of ponding is much more complex than indicated by the above limitation. Marino [4] has provided an extensive treatment that forms the basis for the AISC provisions. The flat roof is treated as a two-way system of secondary members (say, purlins) elastically supported by primary members (say, girders) which are rigidly supported by walls or columns, as shown in Fig. 7.4.2. Figure 7.4.3a shows the primary member with midspan deflection equal to Δ. Figure 7.4.3b represents a secondary member intersecting the primary member at midspan; its supports have deflected an amount Δ, with δ representing the secondary member deflection relative to its ends.

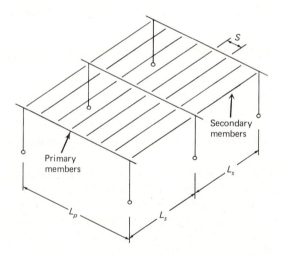

Fig. 7.4.2 Flat roof arrangement for ponding analysis. (From Ref. 4)

The loads assumed for the analysis are shown in Fig. 7.4.4. In each case the deflection curve is assumed as a half sine wave. The loading on the critical midspan secondary member assumes uniform distribution of water of depth equal to $\Delta(\Delta_0$, initial deflection not including that due to water, plus Δ_w, the additional deflection due to water) due to the deflection of the primary member. There is additional loading which

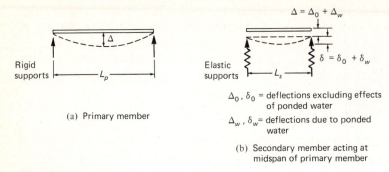

(a) Primary member

$\Delta = \Delta_0 + \Delta_w$

$\delta = \delta_0 + \delta_w$

Δ_0, δ_0 = deflections excluding effects of ponded water

Δ_w, δ_w = deflections due to ponded water

(b) Secondary member acting at midspan of primary member

Fig. 7.4.3 Deflection of flat roof elements.

varies as a half sine wave due to the relative deflection of the secondary member.

The primary member is loaded with a half sine curve due to its own deflection plus the reaction from the half sine wave loading caused by the deflection of the secondary member. This reaction from the midspan secondary member is

$$\frac{2}{\pi}\gamma L_s(\delta_0 + \delta_w)$$

per foot of width along the primary member. The secondary member framing in near the support of the primary member will have a reduced self-deflection δ_{w1} due to the water; thus the reaction to the primary member at its end is

$$\frac{2}{\pi}\gamma L_s(\delta_0 + \delta_{w1})$$

per foot of width L_p.

Referring to Fig. 7.4.5, compute the bending moment on the primary member. First determine the beam reaction assuming the curves of load

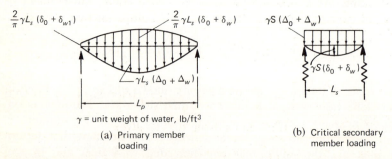

$\frac{2}{\pi}\gamma L_s(\delta_0 + \delta_{w1})$ $\frac{2}{\pi}\gamma L_s(\delta_0 + \delta_w)$ $\gamma S(\Delta_0 + \Delta_w)$

$\gamma L_s(\Delta_0 + \Delta_w)$

$\gamma S(\delta_0 + \delta_w)$

γ = unit weight of water, lb/ft^3

(a) Primary member loading

(b) Critical secondary member loading

Fig. 7.4.4 Assumed ponding loads.

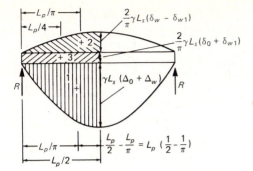

Fig. 7.4.5 Loading for computing bending moment on primary member.

distribution are sine curves,

$$R = \frac{\gamma L_s L_p}{\pi}(\Delta_0 + \Delta_w) + \frac{2\gamma L_s L_p}{\pi^2}(\delta_w - \delta_{w1}) + \frac{\gamma L_s L_p}{\pi}(\delta_0 + \delta_{w1})$$

$$= \frac{\gamma L_s L_p}{\pi}\left[(\Delta_0 + \Delta_w) + \frac{2}{\pi}(\delta_w - \delta_{w1}) + (\delta_0 + \delta_{w1})\right] \qquad (7.4.11)$$

The midspan bending moment is

$$M = \frac{\gamma L_s L_p}{\pi}\left[(\Delta_0 + \Delta_w)\left(\frac{L_p}{2} - \frac{L_p}{2} + \frac{L_p}{\pi}\right)\right.$$

$$\left. + \frac{2}{\pi}(\delta_w - \delta_{w1})\frac{L_p}{\pi} + (\delta_0 + \delta_{w1})\left(\frac{L_p}{2} - \frac{L_p}{4}\right)\right]$$

$$= \frac{\gamma L_s L_p^2}{\pi}\left[\frac{1}{\pi}(\Delta_0 + \Delta_w) + \frac{2}{\pi^2}(\delta_w - \delta_{w1}) + \frac{1}{4}(\delta_0 + \delta_{w1})\right] \qquad (7.4.12)$$

The deflection is found by finding the bending moment on the conjugate beam. Since the original loading was a half sine curve for load terms 1 and 2, the M/EI diagram for those terms is likewise a sine curve. The third term, representing uniform loading gives a parabola for its M/EI loading. Assuming the parabola as a sine curve, the moment at midspan of the conjugate beam is obtained by multiplying by L_p^2/π^2; thus

$$\Delta_w = \frac{\gamma L_s L_p^4}{\pi^3 EI_p}\left[\frac{1}{\pi}(\Delta_0 + \Delta_w) + \frac{2}{\pi^2}(\delta_w - \delta_{w1}) + \frac{1}{4}(\delta_0 + \delta_{w1})\right]$$

$$= \frac{\gamma L_s L_p^4}{\pi^4 EI_p}\left[(\Delta_0 + \Delta_w) + \frac{2}{\pi}(\delta_w - \delta_{w1}) + \frac{\pi}{4}(\delta_0 + \delta_{w1})\right] \qquad (7.4.13)$$

Letting $C_p = \gamma L_s L_p^4/(\pi^4 EI_p)$, and solving for Δ_w gives

$$\Delta_w = \frac{C_p}{1 - C_p}\left(\Delta_0 + \frac{\pi}{4}\delta_0 + \frac{2}{\pi}\delta_w + \frac{\pi}{4}\delta_{w1} - \frac{2}{\pi}\delta_{w1}\right) \qquad (7.4.14)$$

Similarly, the midspan deflection on the secondary member may be expressed as

$$\delta_w = \frac{\gamma S L_s^4}{\pi^4 E I_s} \left[\frac{\pi^2}{8} (\Delta_0 + \Delta_w) + \delta_0 + \delta_w \right] \tag{7.4.15}$$

Letting $C_s = \gamma S L_s^4 / (\pi^4 E I_s)$ and solving for δ_w gives

$$\delta_w = \frac{C_s}{1 - C_s} \left(\frac{\pi^2}{8} \Delta_0 + \delta_0 + \frac{\pi^2}{8} \Delta_w \right) \tag{7.4.16}$$

At the support of the primary member, δ_w is δ_{w1} and $\Delta_0 = \Delta_w = 0$. Thus, from Eq. 7.4.16,

$$\delta_{w1} = \frac{C_s}{1 - C_s} \delta_0 \tag{7.4.17}$$

Furthermore the initial deflections, δ_0 / Δ_0, are in the same ratio as $S L_s^4 / E I_s$ is to $L_s L_p^4 / E I_p$; thus

$$\frac{\delta_0}{\Delta_0} = \frac{C_s}{C_p} \tag{7.4.18}$$

Thus substitution of Eq. 7.4.17 into 7.4.14 and then solving Eqs. 7.4.14 and 7.4.16 for Δ_w and δ_w gives

$$\Delta_w = \frac{\alpha_p \Delta_0 \left[1 + \frac{\pi}{4} \alpha_s + \frac{\pi}{4} \rho (1 + \alpha_s) \right]}{1 - \frac{\pi}{4} \alpha_p \alpha_s} \tag{7.4.19}$$

$$\delta_w = \frac{\alpha_s \delta_0 \left[1 + \frac{\pi^3}{32} \alpha_p + \frac{\pi^2}{8\rho} (1 + \alpha_p) + 0.185 \alpha_s \alpha_p \right]}{1 - \frac{\pi}{4} \alpha_p \alpha_s} \tag{7.4.20}$$

where $\rho = \delta_0 / \Delta_0 = C_s / C_p$
$\alpha_p = C_p / (1 - C_p)$
$\alpha_s = C_s / (1 - C_s)$

Note that even if the supporting system is assumed to be elastic no matter how large the deflection, collapse occurs as $\alpha_p \alpha_s \to 4/\pi$. Restrictions are necessary, however, to keep the system safely within the actual elastic range.

Since deflection is proportional to stress,

$$\frac{f_w}{f_0} = \frac{\text{stress under a deflection, } \Delta_w}{\text{stress initially under deflection, } \Delta_0} = \frac{\Delta_w}{\Delta_0} \tag{7.4.21}$$

Using a factor of safety of 1.25, the stress due to ponding may be limited,

$$f_w \leq 0.8F_y - f_0 \qquad (7.4.22)$$

which gives in terms of deflections, using Eq. 7.4.21,

$$\Delta_w \leq \left(\frac{0.8F_y - f_0}{f_0}\right)\Delta_0 \qquad (7.4.23)$$

or, substituting Eq. 7.4.19 for Δ_w, gives the final criterion for safety of the primary member,

$$\left[\frac{0.8F_y - f_0}{f_0}\right]_p \geq \frac{\alpha_p\left[1 + \dfrac{\pi}{4}\alpha_s + \dfrac{\pi}{4}\rho(1 + \alpha_s)\right]}{1 - \dfrac{\pi}{4}\alpha_p\alpha_s} \qquad (7.4.24)$$

For the secondary member, the criterion becomes

$$\left[\frac{0.8F_y - f_0}{f_0}\right]_s \geq \frac{\alpha_s\left[1 + \dfrac{\pi^3}{32}\alpha_p + \dfrac{\pi^2}{8\rho}(1 + \alpha_p) + 0.185\alpha_s\alpha_p\right]}{1 - \dfrac{\pi}{4}\alpha_p\alpha_s} \qquad (7.4.25)$$

Equations 7.4.24 and 7.4.25 are presented with some of the foregoing discussion in the AISC Commentary, along with charts for evaluating the criteria. Knowing the stress index $U = (0.8F_y - f_0)/f_0$ and C_s, the value of C_p satisfying Eq. 7.4.24 can be determined by a chart (AISC–Fig. C1.13.3.1). Knowing U for the secondary member and C_p, the value of C_s satisfying Eq. 7.4.25 can be determined by chart (AISC–Fig. C1.13.3.2). f_0 is the initial stress in the member under consideration.

AISC–1.13.3 gives a simple, but very conservative criterion

$$C_p + 0.9C_s \leq 0.25 \qquad (7.4.26)$$

and

$$I_d \geq 25S^4/10^6 \qquad (7.4.27)$$

where $C_p = \dfrac{\gamma L_s L_p^4}{\pi^4 E I_p} = \dfrac{62.4 L_s L_p^4(144)}{\pi^4(29,000,000)I_p} = \dfrac{32 L_s L_p^4}{10^7 I_p}$

$C_s = \dfrac{32 S L_s^4}{10^7 I_s}$

L_p = length of primary member, ft
L_s = length of secondary member, ft
S = spacing of secondary member, ft
I_p = moment of inertia of primary member, in.[4]
I_s = moment of inertia of secondary member, in.[4]
I_d = moment of inertia of steel deck supported on secondary members, in.[4] per ft

The criterion $C_p + 0.9C_s \leq 0.25$ assumes the members essentially fully stressed before onset of ponding. For instance,

$$\frac{0.8F_y - f_0}{f_0} = \frac{0.8F_y - 0.66F_y}{0.66F_y} = 0.212$$

$$= \frac{0.8F_y - 0.6F_y}{0.6F_y} = 0.33$$

A stress index 0.25 is a reasonable lower limit for that quantity. Examination of AISC Commentary Figs. C1.13.3.1 and C1.13.3.2 will show the above criteria to be conservative, the more so when stresses at onset of ponding are low. Burgett [5] has provided some graphs for a fast check using the AISC criteria.

Equation 7.4.27 pertains to roof decking that is supported on secondary members. Since this contributes little to ponding it may be treated as a one-way system in the manner presented by Chinn [6].

Consider the decking to span a distance S, be supported on rigid supports, and be loaded with a half sine curve whose maximum ordinate is $\gamma(\Delta_0 + \Delta_w)$ per ft of width. The beam reaction is

$$R = \gamma(\Delta_0 + \Delta_w)\frac{S}{\pi}$$

Computing the midspan bending moment and then using conjugate beam to get the midspan deflection gives

$$\Delta_w = \frac{\gamma S^4}{\pi^4 EI}(\Delta_0 + \Delta_w) \tag{7.4.28}$$

$$\Delta_w = \Delta_0 \frac{\gamma S^4/(\pi^4 EI)}{[1 - \gamma S^4/(\pi^4 EI)]} \tag{7.4.29}$$

which becomes infinite when

$$I = \frac{\gamma S^4}{\pi^4 E} \tag{7.4.30}$$

Applying a factor of 1.25, the member will be safe from ponding failure when

$$I \geq \frac{\gamma S^4}{1.25\pi^4 E} = \frac{62.4(144)S^4}{1.25\pi^4(29,000,000)} = \frac{24.8S^4}{10^6} \tag{7.4.31}$$

Thus AISC requires for roof deck supported on secondary members,

$$I_d \geq \frac{25S^4}{10^6} \tag{7.4.32}$$

where S is in ft and I_d is in in.^{4}

When roof decking *is* the secondary member then it should be treated according to Eq. 7.4.26.

More elaborate mathematical treatment of ponding has been given by Salama and Moody [7], Sawyer [8, 9], Chinn, Mansouri, and Adams [10], Avent and Stewart [11], and Avent [12].

Example 7.4.1

Select the lightest W section to support a superimposed load of 1.5 kips per ft, of which 1.0 kips per ft is live load, on a simply supported span of 42 ft. Limit live load deflection to $L/360$. Adequate lateral support is provided. Use A572 Grade 50 steel, and working stress method of AISC Spec.

SOLUTION

Estimating beam weight at 70 lb per ft,

$$M = \tfrac{1}{8}(1.57)(42)^2 = 346 \text{ ft-kips}$$

$$F_b = 0.66F_y = 33 \text{ ksi}$$

$$\text{Required } S_x = \frac{346(12)}{33} = 126 \text{ in.}^3$$

The stress due to live load would be $33(1.0)/1.57 = 21$ ksi. From Table 7.4.1 for $L/360$,

$$\frac{L}{d} \leq \frac{387}{21} = 18.4$$

$$\text{Min } d = 42(12)/18.4 = 27.4 \text{ in.}$$

From AISC "Allowable Stress Design Selection Table," try W24×68, $S_x = 154$ in.^{3}, a compromise choice between the 27.4-in. minimum depth requirement and the $S = 126$ in.^{3} requirement.

$$\text{Live load stress} = \frac{346(12)}{154}\left(\frac{1.0}{1.57}\right) = 27.0\left(\frac{1.0}{1.57}\right) = 17.2 \text{ ksi}$$

$$\frac{L}{d} \leq \frac{387}{17.2} = 22.5, \qquad d_{min} = 22.4 \text{ in.} < 24 \text{ in.} \qquad \text{OK}$$

The W24×68 is acceptable. The section is "compact" which makes $F_b = 0.66F_y = 33$ ksi as assumed; however, the stress acting is only 27.0 ksi since deflection controlled the choice rather than strength.

An alternative more direct procedure is to solve for the required moment of inertia. For this example simply supported beam, the maximum deflection is given by Eq. 7.4.3. Using $w = 1.0$ kip/ft (the given live load),

$$\Delta = \frac{5wL^4}{384EI} = \frac{5(1.0)(42)^4 1728}{384(29,000)I} = \frac{2414}{I}$$

When the limit on Δ is $L/360$ due to live load,

$$\text{Required } I = \frac{2414}{\Delta} = \frac{2414(360)}{42(12)} = 1724 \text{ in.}^4$$

When a section having $S \geq 126$ in.^{3} is selected from the "Allowable Stress Design Selection Table," the I value must be checked against the 1724 in.^{4} requirement.

W24×68, $\quad I = 1830$ in.$^4 > 1724$ in.^{4} required $\quad$ OK

Use W24×68, $F_y = 50$ ksi.

Example 7.4.2

A flat roof for an industrial building has bays 45 ft by 35 ft. Without regard to ponding, the girders have been selected as W24×84 (A36 steel) and the joists spanning 35 ft are 24 J 6 ($F_b = 22$ ksi).*
    Given:

W24×84, $\quad I_p = 2370$ in.^{4}
$\qquad\qquad f_b = 23.5$ ksi
24 J 6 $\qquad I_s = 367(12)/22 = 200$ in.^{4}
(spacing 5′ = 0″) $\quad$ (computed from load table data)
$\qquad\qquad f_b = 20$ ksi

Assume that one-fourth of live load is acting at onset of ponding (i.e., assume that this is 60 percent of the total service load) and check adequacy with regard to ponding.

SOLUTION

(a) Check the girder:

$$f_0 = 0.6(23.5) = 14.1 \text{ ksi}$$

$$U = \text{stress index} = \frac{0.8F_y - f_0}{f_0} = \frac{0.8(36) - 14.1}{14.1} = 1.04$$

$$C_p = \frac{32(35)(45)^4}{10^7(2370)} = 0.194$$

---

* Refer to *Standard Specifications and Load Tables for Open Web Steel Joists*, J-Series and H-Series, adopted by the Steel Joist Institute, 1978. (Suite 204-B, 1703 Parham Road, Richmond, Va. 23229).

$$C_s = \frac{32(5)(35)^4}{10^7(200)} = 0.12$$

$$C_p + 0.9C_s = 0.194 + 0.9(0.12) = 3.0 > 0.25$$

This appears inadequate by this conservative check. Using chart in AISC Commentary, Fig. C1.13.3.1 gives

For $U_p = 1.04$ and $C_s = 0.12$
Find allowable $C_p \approx 0.40 > 0.194$ OK

Thus girder is OK

    (b) Check the joist:

$$f_0 = 0.6(20) = 12.0 \text{ ksi}$$

$$U = \text{stress index} = \frac{0.8F_y - f_0}{f_0} = \frac{0.8(36) - 12.0}{12.0} = 1.40$$

By accurately checking using Fig. C1.13.3.2 of the AISC Commentary,

For $U_s = 1.40$ and $C_p = 0.194$

Find allowable $C_s \approx 0.38 > 0.12$ OK

Thus both members satisfy ponding provisions. When the stress index is above about 0.5, the simple expression of AISC-1.13.3 will be grossly conservative.

7.5 SHEAR ON ROLLED BEAMS

Whereas long beams may be governed by deflection and medium length beams are usually controlled by flexural stress, short-span beams may be governed by shear.

    To review the development of the shear stress equation for symmetrical sections, consider the slice dz of the beam of Fig. 7.5.1a, shown as a free body in Fig. 7.5.1b. If the unit shear stress v at a section y_1 from the neutral axis is desired, it is observed from Fig. 7.5.1c that

$$dC' = vt\, dz \qquad (7.5.1)$$

The horizontal forces arising from bending moment are

$$C' = \int_{y_1}^{y_2} f\, dA$$

$$C' + dC' = \int_{y_1}^{y_2} (f + df)\, dA$$

Subtracting,

$$dC' = \int_{y_1}^{y_2} df\, dA \qquad (7.5.2)$$

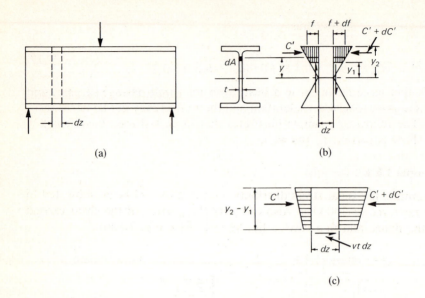

Fig. 7.5.1 Flexural stresses involved in derivation of shear stress equation.

$$df = \frac{dMy}{I} \tag{7.5.3}$$

$$dC' = \int_{y_1}^{y_2} \frac{dMy}{I}\, dA = \frac{dM}{I} \int_{y_1}^{y_2} y\, dA \tag{7.5.4}$$

Substituting Eq. 7.5.4 into Eq. 7.5.1 and solving for the shear stress v gives

$$v = \frac{dM}{dz}\left(\frac{1}{tI}\right) \int_{y_1}^{y_2} y\, dA \tag{7.5.5}$$

and upon recognizing that $V = dM/dz$, and letting

$$Q = \int_{y_1}^{y_2} y\, dA$$

the familiar equation

$$v = \frac{VQ}{It} \tag{7.5.6}$$

is obtained where Q is the first moment of area about the x-axis of the area between the extreme fiber and the particular location at which the shearing stress is to be determined, i.e., the shaded area lying between the limits y_1 and y_2 in Fig. 7.5.1b.

Under usual procedures of steel design, the shear stress is computed as the average value over the gross area of the web neglecting the effect of any fastener holes; thus

$$f_v = \frac{V}{A_w} = \frac{V}{dt_w} \tag{7.5.7}$$

Note that large holes cut in a beam web to permit passage of pipes and ducts require special consideration and their effect may *not* be neglected.

The following example illustrates that in an I-shaped beam most of the shear is carried by the web.

Example 7.5.1

Determine the shear stress distribution on a W24×94 beam subjected to a shear force of 200 kips. Also compute the portion of the shear carried by the flange and that carried by the web. (See Fig. 7.5.2)

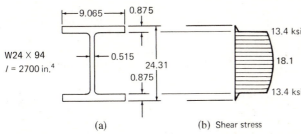

Fig. 7.5.2 Example 7.5.1.

SOLUTION

(a) Stress at junction of flange and web.

$V = 200$ kips

$Q = 9.065(0.875)(12.155 - 0.4375) = 92.9$ in.^{3}

$$v = \frac{200(92.9)}{2700(0.515)} = 13.4 \text{ ksi (web)}, \qquad v = 0.76 \text{ ksi (flange)}$$

(b) Stress at neutral axis.

$Q = 92.9 + (12.155 - 0.875)^2(0.515)(0.5) = 92.9 + 32.8$

$\quad = 125.7$ in.^{3}

$$v = \frac{200(125.7)}{2700(0.515)} = 18.1 \text{ ksi}$$

(c) Shear carried by flanges and web. Using an approximate linear variation,

V(flanges) $= 2(\frac{1}{2})(0.76)(0.875)(9.065) = 6$ kips

V(web) $\quad = 200 - 6 = 194$ kips

In this case, 97 percent of the shear is carried by the web.

(d) Average shear stress f_v on web.

$$f_v = \frac{V}{dt_w} = \frac{200}{24.31(0.515)} = 16.0 \text{ ksi}$$

which is 11.6% below the maximum value.

Because of this relatively small difference, specifications usually permit and prescribe calculation of average shear stress and correspondingly adjust the allowable values. In computing the average shear stress for rolled beams the overall depth is usually used (see AISC–1.5.1.2), while on plate girders the web plate depth is used. Logic would suggest use of the web area lying between the flanges for all cases but the difference is not significant.

AISC—Working Stress Requirement

Rolled beams whose webs do not exhibit instability due to shear stress, or a combination of shear and bending stress, are permitted (AISC–1.5.1.2) to carry an average shear stress of

$$f_v = \frac{V}{A_w} \leq F_v = 0.40F_y \qquad (7.5.8)$$

It is the low depth to web thickness ratio that insures achievement of shear yield, thus eliminating the web buckling problem and the need for stiffeners. The use of $F_v = 0.40F_y$ implies h/t ratios that satisfy the AISC Formula (1.10–1) for omission of intermediate stiffeners. The development of Formula (1.10–1) appears in Chapter 11 on plate girders, but from that formula, h/t limits that insure use of $F_v = 0.40F_y$ can be obtained and these are given in Table 7.5.1.

Table 7.5.1 Maximum h/t Values Without Stiffeners*
(Based on AISC–1.10.5)

F_y (ksi)	F_y (MPa)	h/t
36	248	63.3
42	290	58.6
45	310	56.6
50	345	53.7
55	379	51.2
60	414	49.0
65	448	47.1
100	689	38.0

$h = T$

h, for plate girders

For rolled sections

* From Eq. 11.9.7.

The use of $0.40F_y$ as the maximum allowable shear stress arises by considering theories of failure when shear stresses are simultaneously acting with bending normal stresses. Using the generally accepted Huber-von Mises-Hencky "energy of distortion" theory (see Sec. 2.6), the shear yield stress τ_y equals the compression-tension yield stress F_y divided by $\sqrt{3}$, i.e.,

$$\tau_y = \frac{F_y}{\sqrt{3}}$$ (7.5.9)

A nominal 1.67 factor of safety would give an allowable shear stress F_v of about $0.35F_y$. There are two reasons why the allowable stress is permitted to exceed this value. First, for low h/t ratios, the shear stresses can actually go into the strain hardening range without the beam exhibiting any distress. Secondly, for about sixty years steel specifications have permitted shear stresses to be two-thirds of the values permitted for bending. Considering the many years of experience with the lower factor of safety and the relatively minor consequences of shear yielding, the value $0.40F_y$ seems reasonable.

Example 7.5.2

Select the lightest W section of A36 steel to carry 20 kips/ft on a simply supported span of 5 ft. Lateral bracing is adequate for stability.

SOLUTION
Since the loading is heavy and the span is short, the designer should investigate the shear as well as flexure.

$$M = \tfrac{1}{8}(20)(5)^2 = 62.5 \text{ ft-kips}$$
$$V = \tfrac{1}{2}(20)(5) = 50 \text{ kips}$$

Assuming "compact section," $F_b = 0.66F_y = 24$ ksi

$$\text{Required } S_x = \frac{62.5(12)}{24} = 31.2 \text{ in.}^3$$

Try W12×26 from AISC "Allowable Stress Design Selection Table" as the lightest beam having $S_x \geq 31.2$ in.^{3}
Check shear.

$$f_v = \frac{V}{dt_w} = \frac{50}{12.22(0.23)} = 17.8 \text{ ksi}$$

$$F_v = 0.40F_y = 0.4(36) = 14.4 \text{ ksi} < 17.8 \text{ ksi} \qquad \text{NG}$$

The required web area A_w for shear is

$$\text{Required } A_w = \frac{V}{F_v} = \frac{50}{14.4} = 3.47 \text{ sq in.}$$

For a web thickness about 0.25 in., the required depth of section would be 13.9 in. Try W14×26, $S_x = 35.3$ in.[3] For shear,

$$f_v = \frac{V}{A_w} = \frac{50}{13.91(0.255)} = 14.1 \text{ ksi} < 14.4 \text{ ksi} \qquad \text{OK}$$

Note that for $F_v = 0.40F_y$, the h/t ratio cannot exceed the value in Table 7.5.1.

$$\frac{h}{t} = \frac{\text{unsupported height, } T}{\text{web thickness, } t_w} = \frac{12.125}{0.255} = 47.5$$

This is less than the limit value of 63.3 and confirms the $0.40F_y$ allowable value. The detailed discussion of the use of intermediate stiffeners when h/t exceeds the value from Table 7.5.1 appears in Chapter 11 on plate girders.
Use W14×26, $F_y = 36$ ksi (W360×39, $F_y = 248$ MPa).

AISC—Plastic Design Requirement

The ultimate shear stress f_{vu}, computed as an average on gross web area, when the web is not reinforced by diagonal stiffeners or a doubler plate, is permitted according to AISC–2.5 to be

$$f_{vu} = \frac{V_u}{dt_w} \leq 0.55F_y \qquad (7.5.10)$$

which is consistent with the "energy of distortion" theory which says $\tau_y = F_y/\sqrt{3} = 0.58F_y$.

7.6 WEB CRIPPLING AND BEARING PLATES

Web crippling is a localized yielding that arises from high compressive stresses occurring in the vicinity of concentrated loads. Such situations arise when concentrated loads are applied to beams, beam bearing supports, and reactions of beam flanges at connections to columns.

Referring to Fig. 7.6.1, according to AISC–1.10.10.1, a conservative assumption is used that the load is distributed at 45 degrees into the critical section at the toe of the fillet at a distance k from the face of the beam. To establish the reliability of such a procedure two factors should be considered. First, before a crippling failure occurs the distribution of load spreads out over a distance of $N+5k$ to $N+7k$, particularly when the bearing length N is small, according to tests [13]; and secondly, the yield stress in the region between the critical section and the interior face of the flange is likely to be less than the yield stress in the web, primarily because the thinner web material has been hot worked more during forming. The important requirement is not that yielding be prohibited

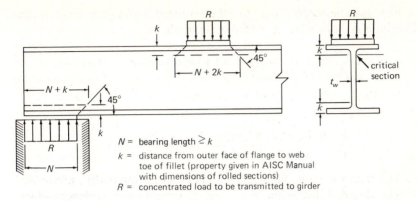

N = bearing length $\geq k$

k = distance from outer face of flange to web
toe of fillet (property given in AISC Manual
with dimensions of rolled sections)

R = concentrated load to be transmitted to girder

Fig. 7.6.1 Web crippling considerations for establishing bearing length.

everywhere but that it not occur over a significant length of critical section such that a crippling failure can occur.

AISC—Working Stress Requirement

AISC–1.10.10.1 is based empirically on the 1935 work of Lyse and Godfrey [14], who realized that this crippling phenomenon is not the same as the overall buckling of the web. Investigators since that time have shown the AISC-procedure to be conservative [13, 15].

The compressive stress f_c at the web toes of the fillets resulting from concentrated loads not supported by stiffeners (Fig. 7.6.1) shall not exceed the following values.

For interior loads,

$$f_c = \frac{R}{t_w(N+2k)} \leq 0.75 F_y \qquad (7.6.1)$$

For end reactions,

$$f_c = \frac{R}{t_w(N+k)} \leq 0.75 F_y \qquad (7.6.2)$$

The 1.33 nominal safety factor against yielding in addition to the conservatively low values of $2k$ and k make these specification requirements reasonably consistent with other procedures.

AISC—Plastic Design Requirement

AISC–2.6 prescribes that at every point of load concentration where a plastic moment is expected to develop under collapse conditions a web stiffener must be used.

No reference is made to other points of load concentration, and it seems proper for those locations to be treated in accordance with the working stress procedure of AISC–1.10.10.1.

Web crippling problems in beam to column connections are treated in Chapter 13, while the design of bearing stiffeners for cases where stresses exceed the allowables to prevent web crippling is discussed in Sec. 11.11.

Example 7.6.1

Determine the size of bearing plate required for an end reaction of 30 kips on a W10×26 beam of A36 steel according to AISC Specification. The beam rests on a wall of concrete with a 28-day compressive strength of 3000 psi.

SOLUTION

Solving Eq. 7.6.2 for the bearing length N gives

$$N = \frac{R}{0.75F_y t} - k = \frac{30}{27(0.260)} - \frac{7}{8} = 3.4 \text{ in.}$$

Try a 4-in. bearing plate.

As a practical matter, probably 3 in. should be considered minimum bearing length. In this example, if a 4-in. length is used, and the beam is resting on a masonry wall, the plate width B (Fig. 7.6.2) required is

$$B = \frac{R}{NF_p}$$

where F_p = allowable unit stress on the support (see AISC–1.5.5). Assuming a wall with concrete having a 28-day compressive strength f'_c, of 3000 psi,

$$F_p = 0.35 f'_c = 1050 \text{ psi}$$

$$\text{Required } B = \frac{30}{4(1.05)} = 7.1 \text{ in., say } 7\tfrac{1}{2}\text{in.}$$

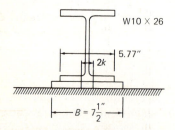

Fig. 7.6.2 Example 7.6.1.

To determine the plate thickness, bending must be considered. A rectangular section has a shape factor, f, equal to 1.5 (see Sec. 7.3) which is 30 to 40 percent greater than that of rolled W shapes; therefore the allowable bending stress on plates (AISC–1.5.1.4.3) is

$$F_b = 0.75F_y$$

which conservatively increases the basic $0.60F_y$ by 25 percent for rectangular plates. A reason for conservatism is that uniform bearing pressure assumes a relatively rigid plate; thus it is really deflection limitations that justify holding the allowable bending stress to $0.75F_y$.

The critical section for bending is taken at the toe of flange-to-web fillet, a distance of k from center of web, and the beam flange is assumed not to participate. The bending moment and required thickness are then computed:

Uniform bearing pressure $p = \dfrac{30{,}000}{4(7.5)} = 1000 \text{ psi}$

$$M = \frac{p(B/2 - k)^2 N}{2} = \frac{1.07(3.75 - 0.875)^2 N}{2} = 4.13 \, N$$

Section modulus $S = \frac{1}{6}Nt^2$

$$\text{Required } S = \frac{M}{F_b} = \frac{4.13 \, N}{0.75(36)} = 0.153 \, N$$

$$\frac{Nt^2}{6} = 0.153 \, N, \qquad t = \sqrt{6(0.153)} = 0.96 \text{ in.}$$

Use bearing plate, $1 \times 4 \times 0' - 7\frac{1}{2}''$.

Solving for the plate thickness in general,

$$\frac{Nt^2}{6} = \frac{p(B/2 - k)^2 N}{2F_b}$$

$$t = \sqrt{\frac{3p(B/2 - k)^2}{F_b}} \tag{7.6.3}$$

7.7 HOLES IN BEAMS

Flange Holes

For tension members the effect of fastener holes has been discussed in Chapter 3, where holes are deducted and net section is used. For compression members, since the fasteners occupy most of the space in the hole, the fasteners are assumed in design to completely fill the holes and a deduction for holes has never been the practice in building construction.

It seems clear that the effect of fastener holes must surely be to weaken a beam and not strengthen it. However, unless the section area is significantly reduced by holes the bending capacity at *service loads* seems largely unaffected, primarily because of stress concentrations adjacent to holes.

Within the accuracy of the design methods employed, it does not seem justifiable to calculate a shift in neutral axis as a result of holes in the tension flange. The total *service load* capacity of the tension flange without holes is essentially the same as that with holes; the stresses are merely distributed differently. Thus the neutral axis does not shift.

Present AISC Specifications (AISC–1.10.1) consider no shift in neutral axis and deduct area for fastener holes only when the area of holes exceeds 15 percent of the gross flange, and then only the excess area over 15 percent of gross flange is deducted.

A reduction in this case seems justifiable because the presence of holes in the tension flange reduces the plastic moment capacity of the section. The plastic moment is computed using yield stress acting over the net area. For cases where ultimate capacity in bending is M_y or less (see Fig. 7.3.1) neglect of the holes gives no major change in safety factor; stress concentrations adjacent to holes provide carrying capacity to replace that lost by making holes. If the ultimate capacity is M_p, the holes will reduce the factor of safety because M_p will reduce almost in proportion to the area of the holes. Thus AISC–1.10.1 seems to consider that a slight reduction in safety factor is not sufficiently important to warrant deduction for holes, but when the area lost by holes is large some correction should be made to prevent too great a decrease in the safety factor.

Other specifications follow the more conservative procedures that have been used for many years. The American Association of State Highway and Transportation Officials (AASHTO) and the American Railway Engineering Association (AREA) use the net section (holes deducted from both flanges) for computing tension flange stress, and use gross section (no holes from either flange) for computing compression stress.

Examples of procedure for considering fastener holes appear in the Chapter 13 section on beam and girder splices.

Web Holes

The AISC Specification permits neglect of fastener holes located in the web, largely for the same reasons fastener holes in the flange may be neglected. Large holes cut into beam webs are entirely another matter. These holes require special analysis and usually will have to be reinforced by attaching extra plate material, often including stiffeners, around the sides of the hole. Treatment of this is outside the scope of this text;

however, the reader may refer to the work of Bower [16, 17, 20], Frost and Leffler [18], Mandel, Brennan, Wasil, and Antoni [19], Cooper and Snell [21], Chan and Redwood [22, 23], Wang, Snell, and Cooper [24], Larson and Shah [25], and Cooper, Snell, and Knostman [26]. Design tables for rectangular holes have been given by Redwood [27], and a design example is presented by Kussman and Cooper [28].

7.8 GENERAL FLEXURAL THEORY

Thus far consideration has been given only to symmetrical shapes loaded symmetrically for which $f = Mc/I$ is correct. The following development treats the general bending of arbitrary prismatic beams, i.e., beams having any cross-sectional shape with no variation along the length. They are also assumed to be free from twisting.

Consider the straight uniform cross section beam of Fig. 7.8.1 acted upon by pure moment applied in the plane ABCD which makes the angle ϕ with the xz plane. The moments are represented by vectors normal to the plane of action (positive moment defined by using right-hand rule for rotation).

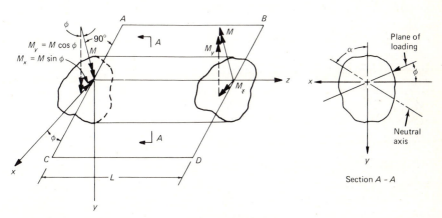

Fig. 7.8.1 Prismatic beam under pure bending.

Examine next a portion of the beam of length z as shown in Fig. 7.8.2a. To satisfy equilibrium on the free body of Fig. 7.8.2a requires

$$\sum F_z = 0, \qquad \int_A \sigma \, dA = 0 \qquad (7.8.1)$$

$$\sum M_x = 0, \qquad M_x = \int_A y\sigma \, dA \qquad (7.8.2)$$

$$\sum M_y = 0, \qquad M_y = \int_A x\sigma \, dA \qquad (7.8.3)$$

It is to be noted from Fig. 7.8.2b that the moments M_x and M_y are both positive in accordance with the customary convention of calling positive

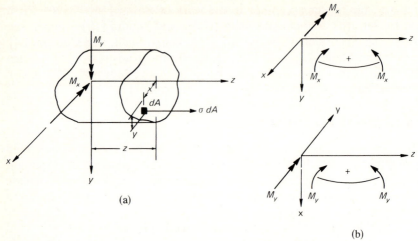

Fig. 7.8.2   Free body of a portion of a beam having length z.

bending that which causes compression in the top portion of the beam. Also, the subscripts for M designate the axis *about which* bending occurs; i.e., the direction of the moment vector.

Bending in the yz Plane Only

If bending occurs in the yz plane, the stress is then proportional to y. Thus

$$\sigma = k_1 y \tag{7.8.4}$$

Using Eqs. 7.8.1 through 7.8.3 gives

$$k_1 \int_A y\, dA = 0$$

$$M_x = k_1 \int_A y^2\, dA = k_1 I_x$$

$$M_y = k_1 \int_A xy\, dA = k_1 I_{xy}$$

From the first of the above expressions, $\int_A y\, dA = 0$, meaning x must be a centroidal axis. The stress may then be computed as

$$\sigma = \frac{M_x y}{I_x} \quad \text{or} \quad \frac{M_y y}{I_{xy}}$$

and the angle ϕ must be such that

$$\tan \phi = \frac{M_x}{M_y} = \frac{I_x}{I_{xy}} \tag{7.8.5}$$

As a practical matter it would be unlikely that an unsymmetrical beam section would be located in a plane making the angle ϕ with the xz plane and have bending occur in the yz plane. Equation 7.8.5 also shows that if a section is used with at least one axis of symmetry (for which $\int_A xy\, dA = I_{xy} = 0$), $\tan \phi = \infty$, $\phi = 90$ degrees, meaning the loading and the bending both occur in the yz plane.

An important conclusion here is that only if $I_{xy} = 0$ does the bending occur in the plane of loading. For example, on unsymmetrical shapes such as angles or zees loaded in the xz or yz plane, the plane of loading and the plane of bending will be different.

Bending in the *xz* Plane Only

If bending occurs in the xz plane the stress is then proportional to x. Thus

$$\sigma = k_2 x \tag{7.8.6}$$

and from Eqs. 7.8.1 through 7.8.3 are obtained:

$$k_2 \int_A x\, dA = 0$$

which means y must be a centroidal axis. Also,

$$M_x = k_2 \int_A xy\, dA = k_2 I_{xy}$$

$$M_y = k_2 \int_A x^2\, dA = k_2 I_y$$

and

$$\tan \phi = \frac{M_x}{M_y} = \frac{I_{xy}}{I_y} \tag{7.8.7}$$

In the case where $I_{xy} = 0$, $\tan \phi = 0$, i.e., the loading and bending both occur in the xz plane.

Bending in neither *xz* nor *yz* Planes

This is the realistic case when considering unsymmetrical sections in flexure. Since it is assumed all stresses are within the elastic limit, the total stress is the sum of the stresses due to bending in each of the xz and yz planes. Thus

$$\sigma = k_1 y + k_2 x \tag{7.8.8}$$

and

$$M_x = k_1 I_x + k_2 I_{xy} \tag{7.8.9}$$

$$M_y = k_1 I_{xy} + k_2 I_y \tag{7.8.10}$$

Solving Eqs. 7.8.9 and 7.8.10 for k_1 and k_2 and substituting into Eq. 7.8.8 gives

$$\sigma = \frac{M_x I_y - M_y I_{xy}}{I_x I_y - I_{xy}^2} y + \frac{M_y I_x - M_x I_{xy}}{I_x I_y - I_{xy}^2} x \tag{7.8.11}$$

which is the general flexure equation. The assumptions inherent in Eq. 7.8.11 are (a) a straight beam; (b) constant cross section; (c)x- and y-axes are mutually perpendicular centroidal axes; and (d) that stress is proportional to strain and the maximum value is within the proportional limit.

Principal Axes

The principal axes are mutually perpendicular centroidal axes for which the moment of inertia is either a maximum or minimum. Furthermore, these axes are the only mutually perpendicular axes for which the product of inertia I_{xy} is zero. When a section has an axis of symmetry, that axis is a principal axis and Eq. 7.8.11 becomes

$$\sigma = \frac{M_x}{I_x} y + \frac{M_y}{I_y} x \tag{7.8.12}$$

When there is no axis of symmetry, Eq. 7.8.12 can still be used if the principal axes are located and the quantities M_x, M_y, I_x, I_y, x, and y are all corrected so as to refer to the principal axes. Usually such transformations offer no advantage over direct use of Eq. 7.8.11.

Inclination of the Neutral Axis

When the loads acting on a flexural member pass through the centroid of the cross section but are inclined with respect to either of the principal axes, the stresses may be determined by using Eq. 7.8.11 or Eq. 7.8.12. However, note should be made that the neutral axis is not necessarily perpendicular to the plane of loading. As shown from Eqs. 7.8.5 and 7.8.7 and Fig. 7.8.1,

$$\tan \phi = \frac{M_x}{M_y} \tag{7.8.13}$$

Since the stress along the neutral axis is equal to zero, σ may be set equal to zero in Eq. 7.8.11. With $\sigma = 0$, solving for $-x/y$ gives

$$-\frac{x}{y} = \left(\frac{M_x I_y - M_y I_{xy}}{I_x I_y - I_{xy}^2} \right) \left(\frac{I_x I_y - I_{xy}^2}{M_y I_x - M_x I_{xy}} \right) \tag{7.8.14}$$

From Fig. 7.8.1, Section $A - A$, it is seen that at any point on the neutral axis, $\tan \alpha = -x/y$. Dividing both numerator and denominator of the right side of Eq. 7.8.14 by M_y gives

$$\tan \alpha = \frac{\dfrac{M_x}{M_y} I_y - I_{xy}}{I_x - \dfrac{M_x}{M_y} I_{xy}} \tag{7.8.15}$$

Substitution of Eq. 7.8.13 into Eq. 7.8.15 gives

$$\tan \alpha = \frac{I_y \tan \phi - I_{xy}}{I_x - I_{xy} \tan \phi} \tag{7.8.16}$$

When investigating a section with one axis of symmetry, $I_{xy} = 0$; Eq. 7.8.16 then becomes

$$\tan \alpha = \frac{I_y}{I_x} \tan \phi \tag{7.8.17}$$

Example 7.8.1

A W18×50 used as a beam is subjected to loads inclined at 5° from the vertical axis as shown in Fig. 7.8.3. Locate the inclination of the neutral axis.

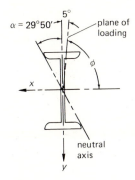

Fig. 7.8.3 Biaxially loaded beam of Example 7.8.1.

SOLUTION

$$I_x = 800 \text{ in.}^4; \qquad I_y = 40.1 \text{ in.}^4$$

$$\tan 85° = \tan \phi$$

Using Eq. 7.8.17,

$$\tan \alpha = \frac{I_y}{I_x} \tan \phi = \frac{40.1}{800} \tan 85° = 0.573$$

$$\alpha = 29.8° \ (29° \ 50')$$

Example 7.8.2

Compute the maximum flexural stress in a $6 \times 4 \times \frac{1}{2}$ angle with the long leg vertically downward when loaded with 0.5 kip/ft on a simply supported span of 10 ft (see Fig. 7.8.4). Compare the value assuming the angle is

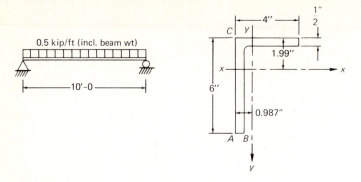

Fig. 7.8.4 Data for Example 7.8.2.

completely free to bend in any direction with that obtained assuming bending in only the vertical plane.

SOLUTION

(a) Angle free to bend in any direction. Using Eq. 7.8.11,

$$I_x = 17.4 \text{ in.}^4, \qquad I_y = 6.27 \text{ in.}^4$$

$$I_{xy} = [6(3.00 - 1.99)(-0.987 + 0.25) + 3.5(2.25 - 0.987)(-1.99 + 0.25)]$$
$$= [(-4.47) + (-7.69)] = -6.08 \text{ in.}^4$$

$$M_x = \tfrac{1}{8}(0.5)(10)^2 = 6.25 \text{ ft-kips} = 75 \text{ in.-kips}$$
$$M_y = 0$$

Stress at point A:

$$f_A = \frac{M_x(I_y y - I_{xy} x)}{I_x I_y - I_{xy}^2} = \frac{6.25(12)[6.27(+4.01) - (-6.08)(-0.987)]}{17.4(6.27) - (6.08)^2}$$

$$= \frac{75(19.14)}{72.1} = 75(0.265) = +19.9 \text{ ksi} \quad \text{(tension)}$$

Stress at point B:

$$f_B = \frac{75[6.27(+4.01) - (-6.08)(-0.487)]}{72.1}$$

$$= 75(0.308) = +23.1 \text{ ksi} \quad \text{(tension)}$$

Stress at point C:

$$f_C = \frac{75[6.27(-1.99) - (-6.08)(-0.987)]}{72.1}$$

$$= 75(-0.256) = -19.2 \text{ ksi} \quad \text{(compression)}$$

(b) Angle free to bend in any direction. Use alternate method suggested by Gaylord and Gaylord [29] (pp. 143–147). Compute the

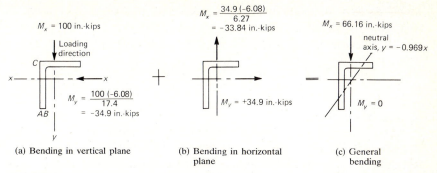

Fig. 7.8.5 Solution by superposition of bending in the vertical and horizontal planes.

stresses assuming first that the beam bends only in the yz plane. Conveniently let $M_x = 100$ in.-kips. Then, according to Eq. 7.8.5, $M_y = M_x I_{xy}/I_x$ must also be acting (Fig. 7.8.5a) if bending is to occur only in the yz plane.

Since the real loading has only M_x, M_y must be removed by application of an equal and opposite moment, further considering that bending occurs only in the xz plane. This means, according to Eq. 7.8.7, the simultaneous application of $M_x = M_y I_{xy}/I_y$ (see Fig. 7.8.5b).

Bending in vertical plane:

$$M_x = 75.0\left(\frac{100}{66.16}\right) = 113.4 \text{ in.-kips}$$

$$f_{A_1} = f_{B_1} = \frac{113.4(4.01)}{17.4} = 26.1 \text{ ksi} \quad \text{(tension)}$$

$$f_{C_1} = \frac{113.4(1.99)}{17.4} = 13.0 \text{ ksi} \quad \text{(compression)}$$

Bending in horizontal plane:

$$M_y = -113.4(-6.08)/17.4 = 39.6 \text{ in.-kips}$$

$$f_{A_2} = f_{C_2} = \frac{39.6(0.987)}{6.27} = 6.2 \text{ ksi} \quad \text{(compression)}$$

$$f_{B_2} = \frac{39.6(0.487)}{6.27} = 3.1 \text{ ksi} \quad \text{(compression)}$$

Total stresses in general bending:

$$f_A = +26.1 - 6.2 = +19.9 \text{ ksi} \quad \text{(tension)}$$

$$f_B = +26.1 - 3.1 = +23.0 \text{ ksi} \quad \text{(tension)}$$

$$f_C = -13.0 - 6.2 = -19.2 \text{ ksi} \quad \text{(compression)}$$

which agree with the values, 19.9, 23.1 and 19.2 ksi, respectively, as computed by the general formula.

The general equation for stress at any point is seen to be

$$f = \frac{113.4y}{17.4} + \frac{39.6x}{6.27}$$

where if $f = 0$, the neutral axis is $y = -0.969x$.

This superimposition method permits the designer to visualize what is taking place. If attached construction constrains an unsymmetrical section to bend in the vertical plane, the restraining moment capacity can be computed by using Eq. 7.8.5.

(c) Angle restrained to bend in the vertical plane:

$$f_A = f_B = \frac{75(4.01)}{17.4} = +17.3 \text{ ksi} \quad \text{(tension)}$$

$$f_C = \frac{75(-1.99)}{17.4} = -8.6 \text{ ksi} \quad \text{(compression)}$$

Unless the horizontal constraints actually act, the tensile stress at point B is underestimated by 25 percent, and the compressive stress at C is underestimated by 55 percent.

Frequently designers assume $f = My/I$ is applicable without considering whether or not adequate horizontal restraints are present. Although usually some degree of restraint is present, care should be exercised when investigating unsymmetrical beams. Neglect of the lateral (horizontal) component is always on the unsafe side.

7.9 BIAXIAL BENDING OF SYMMETRICAL SECTIONS

Flexural stresses on sections with at least one axis of symmetry and loaded through the centroid may be computed using Eq. 7.8.12, which when modified to give maximum stress becomes

$$\sigma = \frac{M_x}{S_x} + \frac{M_y}{S_y} \tag{7.9.1}$$

where $S_x = I_x/(d/2)$ and $S_y = I_y/(b/2)$ are the section modulus values.

The investigation of stresses by Eq. 7.9.1 is a simple task; however, the selection of a beam of minimum weight to satisfy a maximum stress limitation involves an indirect process.

Assume the maximum combined stress σ may not exceed a prescribed allowable value, F_b; thus Eq. 7.9.1 becomes

$$\frac{M_x}{S_x} + \frac{M_y}{S_y} \le F_b \tag{7.9.2}$$

and multiplying by S_x and dividing by F_b gives

$$S_x \geq \frac{M_x}{F_b} + \frac{M_y}{F_b}\left(\frac{S_x}{S_y}\right)$$ (7.9.3)

For selecting standard rolled shapes one finds that for a given depth of section the ratio S_x/S_y is relatively constant. A typical range of such values appears in Table 7.9.1. The lightest sections in any depth will be the narrowest ones and thus have the higher values of S_x/S_y.

Table 7.9.1 Typical S_x/S_y Values

Shape	Depth d (in.)	S_x/S_y
M	6, 8	5–7
M	10, 12	8–11
M	14, 16	11–12
Light W and M	4–8	3
W	8, 10	3–4
W	12	3–6
W	14 (up to 84 lb/ft)	4–8
W	14 (over 84 lb/ft)	$2\frac{1}{2}$–3
W	16, 18, 21	5–9
W	24, 27	6–10
W	30, 33, 36	7–12
S	6–8	d
S	10–18	$0.75d$
C	up to 7	$1.5d$
C	8–10	$1.25d$
C	12, 15	d

An alternate method is suggested by Gaylord and Gaylord [29] (p. 168), for use with I-shaped sections (W, M, and S). The quantity S_x/S_y is computed approximately using the beam properties neglecting the web effect. This gives, using $b =$ width and $d =$ depth,

$$\frac{S_x}{S_y} = \frac{I_x(b/2)}{(d/2)I_y} = \frac{2bt(d/2)^2(b/2)}{(d/2)(2)(1/12)tb^3} = \frac{3d}{b}$$ (7.9.4)

which is suggested to be increased to $3.5d/b$ to account for a greater error in the numerator than in the denominator resulting from the neglect of web.

Example 7.9.1

Select the lightest W or M section to carry moments $M_x = 60$ ft-kips and $M_y = 25$ ft-kips. Use the working stress method and consider that adequate bracing is provided so that the allowable combined stress is $0.60F_y$. Use steel with $F_y = 50$ ksi.

SOLUTION

In this case $F_b = 0.60F_y = 30$ ksi. Note that under AISC–1.5.1.4.5 the maximum allowable stress in unsymmetrical bending is $0.60F_y$, which assumes the ultimate capacity is achieved when the extreme fiber reaches F_y. When the section is "compact" it is probably more logical to use an allowable stress between $0.66F_y$ (F_{bx} for strong-axis bending) and $0.75F_y$ (F_{by} for weak-axis bending). A straight line interaction would be obtained if F_{bx} and F_{by} were used in the denominators of Eq. 7.9.3 instead of the single value F_b. While AISC–1.5.1.4 seems explicit in allowing only $0.60F_y$ for this case, AISC–1.6.1 implies the straight line interaction for the limiting case of zero axial compression when using the beam-column formulas. Using Eq. 7.9.3,

$$\text{Required } S_x \geq \frac{60(12)}{30} + \frac{25(12)}{30}\left(\frac{S_x}{S_y}\right)$$

$$\geq 24 + 10 S_x/S_y$$

From Table 7.9.1 the ratio S_x/S_y can be expected to be on the order of 3 to 4; thus $S_x \approx 55 - 65$ in.^{3}

Using the AISC "Allowable Stress Design Selection Table," try W16×40, $S_x = 64.7$. For this beam, $S_x/S_y = 64.7/8.25$, and it is apparent by inspection the beam is inadequate. Continuing with the "Allowable Stress Design Selection Table," it is found by inspection that the lightest W shapes (boldface type) are inadequate up to about W21×68, which has about $S_x/S_y \approx 9$ (Table 7.9.1). Thus 60 lb/ft seems to be the weight range.

Try W10×60:

$$S_x/S_y = 66.7/23.0 = 2.9$$

$$\text{Required } S_x = 24 + 10(2.9) = 53 \text{ in.}^3 < 66.7 \text{ in.}^3 \qquad \text{OK}$$

For W10×49,

$$S_x/S_y = 54.6/18.7 = 2.92$$

$$\text{Required } S_x = 24 + 10(2.92) = 53.2 \text{ in.}^3 < 54.6 \text{ in.}^3 \qquad \text{OK}$$

Use W10×49, $F_y = 50$ ksi.

If one had assumed a 10-in. depth as desirable and used Eq. 7.9.4 with a coefficient of 3.5 instead of 3,

$$\text{Required } S_x = 24 + 10(3.5 \, d/b)$$

For W10 sections, d/b is either 1 or 1.25.

$$\text{Required } S_x \approx 24 + 35 = 59 \text{ in.}^3$$

which would also have required a check of two sections. The more general approach using typical S_x/S_y values seems most useful unless a specific depth is desired.

7.10 UNSYMMETRICAL SECTIONS

With the increasing use of multitypes of steel, and composite construction combining the compressive strength of concrete with the tensile strength of steel (see Chapter 16), the need to consider unsymmetrical sections increases. While very irregular shapes must be dealt with individually on a trial-and-error basis, sections of the general I-shape that merely have one flange larger than the other can be treated in a systematic manner.

The following development of a useful design equation is similar to that presented by Gaylord and Gaylord [29] (pp. 182–183). The primary shape is assumed to have its centroid at mid-depth, though it need not be I-shaped. Referring to Fig. 7.10.1, the shift in centroid $\bar{y}$ which results from adding a plate of area A_f to one flange is

$$\bar{y} = \frac{A_f y_1}{A_i + A_f} \tag{7.10.1}$$

which if $\gamma = A_f/A_i$ becomes

$$\bar{y} = \frac{\gamma y_1}{1 + \gamma} \tag{7.10.2}$$

The moment of inertia of the compound section is

$$I = I_i + A_i \bar{y}^2 + A_f (y_1 - \bar{y})^2 \tag{7.10.3}$$

which upon substituting Eq. 7.10.2 becomes

$$I = I_i + A_i \left(\frac{\gamma y_1^2}{1 + \gamma} \right) \tag{7.10.4}$$

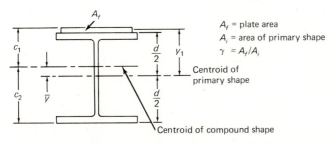

A_f = plate area
A_i = area of primary shape
γ = A_f/A_i
Centroid of primary shape
Centroid of compound shape

Fig. 7.10.1 Compound shape having one axis of symmetry.

For the most common situation of an I-shaped section (W, S, or M), the radius of gyration is approximately known (see text, Appendix, Table A1) to range from $0.38d$ to $0.42d$; thus, using an average value,

$$I_i = A_i r_i^2 = A_i (0.4d)^2 \tag{7.10.5}$$

Next, convert the moment of inertia I into section modulus form, letting $S_1 = I/c_1 \approx I/(y_1 - \bar{y})$; thus, using Eq. 7.10.2 gives

$$S_1 = I \left(\frac{1 + \gamma}{y_1} \right)$$

and putting this into Eq. 7.10.4 gives

$$S_1 = \frac{I_i}{y_1} + \frac{I_i \gamma}{y_1} + A_i \gamma y_1 \qquad (7.10.6)$$

and when the second term of Eq. 7.10.6 is modified by substituting Eq. 7.10.5 into it; and further, since $S_i \approx I_i / y_1$,

$$S_1 = S_i + A_i \left[\frac{(0.4d)^2 \gamma}{y_1} + \gamma y_1 \right] \qquad (7.10.7)$$

and finally, saying $y_1 \approx d/2$,

$$S_1 = S_i + A_f \left(\frac{0.16d^2}{d/2} + \frac{d}{2} \right)$$
$$= S_i + A_f d \ (0.32 + 0.5) \qquad (7.10.8)$$

Solving for A_f gives the equation for the cover plate area,

$$A_f = 1.22 \left(\frac{S_1 - S_i}{d} \right) \qquad (7.10.9)$$

For design with a primary section of I-shape,

$$A_f \approx 1.2 \left(\frac{S_1 - S_i}{d} \right) \qquad (7.10.10)$$

If the primary section is other than I-shaped, the radius of gyration, r_i, can be estimated and used in Eq. 7.10.7 in place of the $0.4d$ which was used.

Example 7.10.1

Determine the cover plate required on a W12×65 beam if a total moment of 160 ft-kips is to be carried. Assume the allowable stress in tension is 22 ksi while the allowable in compression is 16 ksi. (Such a condition could commonly arise because of lateral stability problems as will be discussed in Chapter 9, or it might occur when designing using a hybrid plate girder having flanges of two different steels.)

SOLUTION
For the W12×65, $S_i = 87.9$ in.³ and $d = 12.12$ in.
    The required S_1 to the compression face where the flange plate will be is

$$\text{Required } S_1 = \frac{M}{F_b} = \frac{160(12)}{16} = 120 \text{ in.}^3$$

$$\text{Required } A_f \approx 1.2 \left(\frac{120 - 87.9}{12} \right) = 3.2 \text{ sq. in.}$$

One could select a plate and check the stresses. Or, preferably before doing so, examine the section modulus S_2 required to satisfy the tension stress requirement. That section modulus may be stated as

$$S_2 = I/(d/2 + \bar{y}) \approx I/(y_1 + \bar{y})$$

or

$$S_2 = S_1/(1 + 2\gamma) \tag{7.10.11}$$

For tension stress,

$$\text{Required } S_2 = \frac{160(12)}{22} = 87.3 \text{ in.}^3$$

Based on required S_2 and $\gamma = A_f/A_i = 3.2/19.1 = 0.168$,

$$\text{Required } S_1 = 87.3[1 + 2(0.168)] = 117 \text{ in.}^3$$

Since this is less than 120 in.^{3} required based on compression stress, select plate for $A_f = 3.2$ sq in. and check stress.
Try ℞—$\frac{5}{16} \times 10$: $A_f = 3.125$ sq in.

$$\bar{y} = \frac{3.125(6.06 + \frac{5}{32})}{19.1 + 3.125} = 0.874 \text{ in.}$$

$$I = 533 + 19.1(0.874)^2 + 3.125(6.216 - 0.874)^2 = 637 \text{ in.}^4$$

$$f(\text{compression}) = \frac{160(12)5.50}{637} = 16.6 \text{ ksi} > 16 \text{ ksi} \qquad 3.8\% \text{ high}$$

$$f(\text{tension} = \frac{160(12)6.92}{637} = 20.9 \text{ ksi} < 22 \text{ ksi} \qquad \text{OK}$$

Use ℞—$\frac{3}{8} \times 10$ in preference to $\frac{5}{16} \times 10$ which is overstressed by more than three percent.

SELECTED REFERENCES

1. Joint Committee of Welding Research Council and the American Society of Civil Engineers, *Commentary on Plastic Design in Steel*, 2nd ed., ASCE Manual and Reports on Practice No. 41, New York. 1971.
2. Richard N. Wright and William H. Walker, "Vibration and Deflection of Steel Bridges," *Engineering Journal*, AISC, 9, 1 (January, 1972), 20–31.
3. Thomas M. Murray, "Design to Prevent Floor Vibrations," *Engineering Journal*, AISC, 12, 3 (Third Quarter, 1975), 82–87.
4. Frank J. Marino, "Ponding of Two-Way Roof Systems," *Engineering Journal*, AISC, 3, 3 (July 1966), 93–100.
5. Lewis B. Burgett, "Fast Check for Ponding," *Engineering Journal*, AISC, 10, 1 (First Quarter, 1973), 26–28.
6. James Chinn, "Failure of Simply-Supported Flat Roofs by Ponding of Rain," *Engineering Journal*, AISC, 2, 2 (April 1965), 38–41.

7. A. E. Salama and M. L. Moody, "Analysis of Beams and Plates for Ponding Loads," *Journal of Structural Division*, ASCE, 93, ST1 (February 1967), 109–126.

8. D. A. Sawyer, "Ponding of Rainwater on Flexible Roof Systems," *Journal of Structural Division*, ASCE, 93, ST1 (February 1967), 122–147.

9. D. A. Sawyer, "Roof-Structure Roof-Drainage Interaction," *Journal of Structural Division*, ASCE, 94, ST1 (January 1968), 175–198.

10. James Chinn, Abdulwahab H. Mansouri, and Staley F. Adams, "Ponding of Liquids on Flat Roofs," *Journal of Structural Division*, ASCE, 95, ST5 (May 1969), 797–807.

11. R. Richard Avent and William G. Stewart, "Rainwater Ponding on Beam-Girder Roof Systems," *Journal of Structural Division*, ASCE, 101, ST9 (September 1975), 1913–1927.

12. R. Richard Avent, "Deflection and Ponding of Steel Joists," *Journal of Structural Division*, ASCE, 102, ST7 (July 1976), 1399–1410.

13. J. D. Graham, A. N. Sherbourne, R. N. Khabbaz, and C. D. Jensen, *Welded Interior Beam-to-Column Connections*. New York: American Institute of Steel Construction, Inc., 1959.

14. I. Lyse and H. J. Godfrey, "Investigation of Web Buckling in Steel Beams," *Transactions*, ASCE, 100 (1935), 675–706.

15. B. G. Johnston and G. G. Kubo, "Web Crippling at Seat Angle Supports," Fritz Laboratory Report No. 192A2, Lehigh University, Bethlehem, Pa., 1941.

16. John E. Bower, "Elastic Stresses Around Holes in Wide-Flange Beams," *Journal of Structural Division*, ASCE, 92, ST2 (April 1966), 85–101.

17. John E. Bower, "Experimental Stresses in Wide-Flange Beams with Holes," *Journal of Structural Division*, ASCE, 92, ST5 (October 1966), 167–186.

18. Ronald W. Frost and Robert E. Leffler, "Fatigue Tests of Beams with Rectangular Web Holes," *Journal of Structural Division*, ASCE, 97, ST2 (February 1971), 509–527.

19. James A. Mandel, Paul J. Brennan, Benjamin A. Wasil, and Charles M. Antoni, "Stress Distribution in Castellated Beams," *Journal of Structural Division*, ASCE, 97, ST7 (July 1971), 1947–1967.

20. John E. Bower, Chairman, "Suggested Design Guides for Beams with Web Holes," *Journal of Structural Division*, ASCE, 97, ST11 (November 1971), 2707–2728.

21. Peter B. Cooper and Robert R. Snell, "Tests on Beams with Reinforced Web Openings," *Journal of Structural Division*, ASCE, 98, ST3 (March 1972), 611–632.

22. Peter W. Chan and Richard G. Redwood, "Stresses in Beams with Circular Eccentric Web Holes," *Journal of Structural Division*, ASCE, 100, ST1 (January 1974), 231–248.

23. Richard G. Redwood and Peter W. Chan, "Design Aides for Beams with Circular Eccentric Web Holes," *Journal of Structural Division*, ASCE, 100, ST2 (February 1974), 297–303.

24. Tsong-Miin Wang, Robert R. Snell, and Peter B. Cooper, "Strength of Beams with Eccentric Reinforced Holes," *Journal of Structural Division*, ASCE, 101, ST9 (September 1975), 1783–1800.

25. Marvin A. Larson and Kirit N. Shah, "Plastic Design of Web Openings in

Steel Beams," *Journal of Structural Division*, ASCE, 102, ST5 (May 1976), 1031–1041.

26. Peter B. Cooper, Robert R. Snell, and Harry D. Knostman, "Failure Tests on Beams with Eccentric Web Holes," *Journal of Structural Division*, ASCE, 103, ST9 (September 1977), 1731–1738.

27. R. G. Redwood, "Tables for Plastic Design of Beams with Rectangular Holes," *Engineering Journal*, AISC, 9, 1 (January 1972), 2–19.

28. Richard L. Kussman and Peter B. Cooper, "Design Example for Beams with Web Openings," *Engineering Journal*, AISC, 13, 2 (Second Quarter 1976), 48–56.

29. E. H. Gaylord, Jr., and C. N. Gaylord, *Design of Steel Structures*. New York: McGraw-Hill Book Company, Inc, 1957, Chap. 5.

PROBLEMS

All problems* are to be done in accordance with the latest AISC Specification unless otherwise indicated, and the term "beam section" is intended to include standard W, S, and M shapes.

7.1. Select the lightest beam section to carry a uniformly distributed superimposed load of 1 kip per ft on a simply supported span of 35 ft. Assume full lateral support of compression flange and that deflection need not be considered. Use the working stress method with the following steels: (a) A36; (b) A572 Grade 50; (c) A572 Grade 65; (d) A514 Grade 100. Assume all standard sections are available (which they are not) in each of these steels. (Loading = 15 kN/m; Span = 11 m)

7.2. Repeat Prob. 7.1, increasing the span to 55 ft. (Span = 16 m)

7.3. Repeat Prob. 7.1, using plastic design for all applicable steels. Does plastic design offer any saving over using working stress method? By using only the cross-section dimensions, compute the shape factor for the A36 section selected.

7.4. Repeat Prob. 7.2, using plastic design for all applicable steels. Does plastic design offer any saving over using working stress method? By using only the cross-section dimensions, compute the shape factor for the A36 section selected.

7.5. Determine the allowable superimposed concentrated load at mid-span for a special beam of A36 steel whose web is $\frac{3}{8} \times 24$ and flange plates are $\frac{1}{2} \times 12$. Assume the beam has full lateral support of the compression flange. (a) Use the working stress method. (b) What is the safe P if computed according to plastic design method? Can plastic design be used?

---

* Many problems may be solved either as stated in U.S. customary units or in SI units using the numerical data in parenthesis at the end of the statement. The conversions are approximate to avoid having the given data imply accuracy greater in SI than for U.S. customary units.

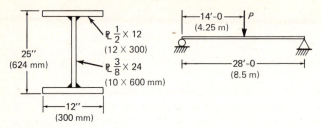

Prob. 7.5

7.6. Repeat Prob. 7.5 if the flange plates are $\frac{3}{4} \times 12$ instead of $\frac{1}{2} \times 12$. In addition to parts (a) and (b) use working stress with A607 Grade 70 steel. (Plate, 16×300 mm)

7.7. Repeat Prob. 7.1, assuming that 80 percent of the superimposed loading is live load and that live load deflection may not exceed $L/360$.

7.8. Repeat Prob. 7.2, assuming that 60 percent of the superimposed loading is live load and that live load deflection may not exceed $L/300$.

7.9. Select the lightest beam section to carry 20 kips/ft on a simple span of 7 ft. Compression flange is adequately supported laterally. Use the working stress method with the following steels: (a) A36; (b) A572 Grade 50; (c) A572 Grade 60; and (d) A514 Grade 100. ($w = 300$ kN/m; $L = 2.1$ m)

7.10. Repeat Prob. 7.9, using plastic design and only A36 steel.

7.11. A W24×94 beam 6-ft-long center-to-center of supports (see accompanying figure) underpins a column that brings 280 kips to its top flange 2 ft 6 in. from the left support. The length of the base plate under the column is 12 in. measured along the beam. The length of each of the plates under the ends of the beam is 8 in. There are no end connections. Review this A36 steel beam according

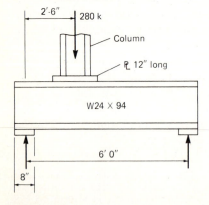

Prob. 7.11

to the working stress method for (a) flexure, (b) shear, and (c) web crippling. Specify changes (if any) necessary to satisfy AISC Spec.

7.12. A W16×77 section of A36 steel is to be used as a simply supported beam on a span of 10 ft with the compression flange laterally supported. Wall bearing length is 10 in. (a) What slowly *moving* concentrated load may be safely carried by this beam? (b) If two 1-in. diam holes exist in each flange of the beam at midspan, what would be the safe load for this beam? Use working stress method.

7.13. Determine the size bearing plate required for an end reaction of 55 kips on a W14×53 beam of A572 Grade 60 steel. The beam rests on a 4-in.-thick concrete wall ($f'_c = 3500$ psi). Specify thickness in multiples of $\frac{1}{4}$ in. and the length and width to whole inches.

7.14. For a W12×87 of A36 steel, calculate the shear capacity V, the reaction R_i that can be carried per inch of bearing length, and the bearing length N_e that would be needed if the end reaction equals the shear capacity V. Use working stress method.

7.15. For the Zee section shown in the accompanying figure, assume the loading in the plane of the web (yz plane). (a) Determine the safe uniform loading (total) assuming bending occurs in the yz plane and the maximum allowable stress is 30 ksi. (b) Using the loading determined in (a) compute the bending stress at points designated by letters if the beam is free to bend and not restrained to bend in the yz plane. Use the general flexure equation, Eq. 7.8.11. (c) Repeat (b) but use the method described in Example 7.8.2b. (d) If your instructor specifically assigns and discusses this part, locate principal axes, transform the moment into components M'_x and M'_y about the principal axes, compute moments of inertia I'_x and I'_y with respect to these axes, and use $f = M'_x/S'_x + M'_y/S'_y$. (e) State conclusions.

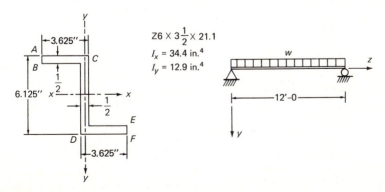

Z6 X $3\frac{1}{2}$ X 21.1

$I_x = 34.4$ in.^{4}

$I_y = 12.9$ in.^{4}

Prob. 7.15

7.16. Repeat Prob. 7.15, using a $7 \times 4 \times \frac{1}{2}$ angle with long leg pointed downward as shown in the accompanying figure. Assume loading in the yz plane and neglect any torsional effects.

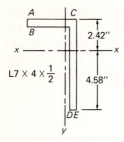

Prob. 7.16

7.17. Repeat Prob. 7.15, using a combination section, consisting of a W16×50 and a C8×11.5, as shown in the accompanying figure. Assume loading in the yz plane and neglect any torsional effects.

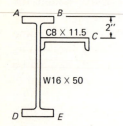

Prob. 7.17

7.18. Repeat Prob. 7.15, using a combination section, consisting of a C10×15.3 and a $4 \times 3 \times \frac{1}{4}$ angle, as shown in the accompanying figure. Assume loading in the yz plane and neglect any torsional effects.

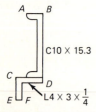

Prob. 7.18

7.19. Suppose the given $8 \times 6 \times \frac{1}{2}$ angle, positioned with its long leg pointing downward, is used as a beam to span 12 ft. The uniform gravity loading (including angle weight) is 0.6 kips/ft, and the horizontal leg is to be restrained laterally so the angle will bend vertically. Consider only the effect of unsymmetrical bending and neglect any torsional effect. Assuming the attachment to the horizontal leg is simply supported, what moment capacity against lateral load must it be designed to carry?

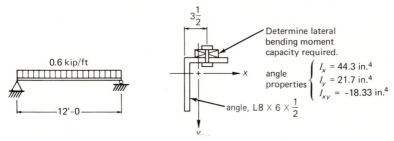

Prob. 7.19

7.20. Select the lightest W8 section of A36 steel to use as a purlin on a roof sloped 30° to the horizontal. The span is 21 ft, the load is uniform 0.37 kip/ft (superimposed), and lateral stability is assured by attachment of the roofing to the compression flange. Assume load acts through the beam centroid and that any torsional effect can be resisted by the roofing and therefore neglected in selecting the beam. Use the working stress method.

7.21. Select the lightest W section to carry moments, $M_x = 145$ ft-kips and $M_f = 30$ ft-kips (lateral moment resisted by top flange). Assume $M_y \approx 2M_f$ and that torsion is neglected. Use A572 Grade 50 steel and assume lateral stability does not govern and the allowable stress under biaxial bending is $0.60F_y$.

7.22. Select the lightest W section to carry $M_x = 275$ ft-kips and $M_f = 100$ ft-kips (lateral moment resisted by top flange). Use A36 steel and assume lateral stability does not govern and $F_b = 0.60F_y$ for biaxial bending.

7.23. Repeat Prob. 7.21, selecting a combination wide-flange and channel.

7.24. Design the lightest combination of beam section and one cover plate to carry a moment, $M_x = 500$ ft-kips if the allowable extreme fiber stresses are $F_{b\ tension} = 30$ ksi, and $F_{b\ compression} = 23$ ksi. How much saving in weight is there over using a larger beam section without plate?

7.25. Design a built-up I-shaped welded girder with different size flanges. The moment to be carried is 450 ft-kips and the allowable extreme fiber stresses are $F_{b\ tension} = 30$ ksi and $F_{b\ compression} = 15$ ksi. (As will be discussed in Chapter 9, this could happen when lateral stability of compression flange controls.)

8
Torsion

8.1 INTRODUCTION

In structural design, torsional stress may on occasion be a significant stress for which provision must be made. The most efficient shape for carrying a torque is a hollow circular shaft; extensive treatment of torsion and torsion combined with bending and axial force is to be found in most texts on mechanics of materials [1].

Frequently torsion is a secondary, though not necessarily a minor effect that must be considered in combination with other types of behavior. The shapes that make good columns and beams, i.e., those that have their material distributed as far from their centroids as practicable, are not equally efficient in resisting torsion. It is found that thin-walled circular and box sections are stronger torsionally than sections with the same area arranged as channel, I, tee, angle, or zee shapes.

When a simple circular solid shaft is twisted, the shearing stress at any point on a transverse cross section varies directly as the distance from the center of the shaft. Thus, during twisting, the cross section which is initially planar remains a plane and rotates only about the axis of the shaft.

In 1853 the French engineer Adhémar Jean Barré de Saint-Venant

presented to the French Academy of Sciences the classical torsion theory
that forms the basis for present-day analysis.* Saint-Venant showed that
when a noncircular bar is twisted, a transverse section that was planar
prior to twisting does *not* remain plane after twisting. The original
cross-section plane surface becomes a warped surface. In torsion problems it is necessary to recognize the out-of-plane, or warping effect, in
addition to the rotation, or pure twisting, effect.

In this chapter primary emphasis is given to the recognition of
torsion on the usual structural members, such as I-shaped, channel, angle,
and zee sections; how the torsional stresses may be approximated and
how such members may be selected to resist torsional effects.

Also included is a brief treatment of torsional stiffness and the
computation of torsional stresses on closed thin-walled sections as well as
torsional buckling.

8.2 PURE TORSION OF HOMOGENEOUS SECTIONS

A review of shear stress under pure torsion and of torsional stiffness
seems a desirable beginning prior to considering structural shapes in
locations where the warping of the cross section is restrained.

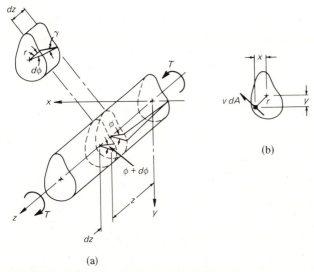

(a)

(b)

Fig. 8.2.1 Torsion of a prismatic shaft.

Consider a torsional moment T acting on a solid shaft of homogeneous material and uniform cross section, as shown in Fig. 8.2.1. Assume no

* For a summary of Saint-Venant's work, see Isaac Todhunter and Karl Pearson, *A History
of the Theory of Elasticity and of the Strength of Materials*, Vol. II, 1893 (reprinted by Dover
Publications, Inc., New York, 1960, pp. 17–51).

out-of-plane warping, or at least that out-of-plane warping has negligible effect on the angle of twist ϕ. This assumption will be nearly correct so long as the cross section is small compared to the length of the shaft and also that no significant reentrant corners exist. It is further assumed that no distortion of the cross section occurs during twisting. The rate of twist (twist per unit length) may therefore be expressed as

$$\theta = \text{rate of twist} = \frac{d\phi}{dz} \tag{8.2.1}$$

which can be thought of as torsional curvature (rate of change of angle). Since it is the relative rotation of the cross sections at z and $z + dz$ that causes strain, the magnitude of displacement at a given point is proportional to the distance r from the center of twist. The strain angle γ, or unit shear strain, at any element r from the center is

$$\gamma \, dz = r \, d\theta$$
$$\gamma = r(d\phi/dz) = r\theta \tag{8.2.2}$$

Using the shear modulus G, Hooke's law gives the unit shear stress v as

$$v = \gamma G \tag{8.2.3}$$

Thus as shown in Fig. 8.2.1b, the elemental torque is

$$dT = rv \, dA = r\gamma G \, dA = r^2(d\phi/dz)G \, dA \tag{8.2.4}$$

The total resisting moment for equilibrium is

$$T = \int_A r^2 \frac{d\phi}{dz} G \, dA$$

and since $d\phi/dz$ and G are constants at any section,

$$T = \frac{d\phi}{dz} G \int_A r^2 \, dA = GJ \frac{d\phi}{dz} \tag{8.2.5}$$

where $J = \int_A r^2 \, dA$. Equation 8.2.5 may be thought of as analogous to flexure, i.e., bending moment M equals rigidity EI times curvature, d^2y/dz^2. Here torsional moment T equals torsional rigidity GJ times torsional curvature (rate of change of angle).

Shear stress may then be computed using Eqs. 8.2.2 and 8.2.3,

$$v = \gamma G = r \frac{d\phi}{dz} G \tag{8.2.6}$$

and

$$\frac{d\phi}{dz} = \frac{T}{GJ}$$

which gives

$$v = \frac{Tr}{J} \tag{8.2.7}$$

Thus as long as the assumptions of this development reasonably apply, torsional shear stress is proportional to the radial distance from the center of twist.

Circular Sections

For the specific case of the circular section of diameter t, no warping of the sections occurs (i.e., no assumption is required) and $J =$ polar moment of inertia $= \pi t^4/32$. Thus, for maximum shear stress at $r = t/2$,

$$v_{max} = \frac{16T}{\pi t^3} \qquad (8.2.8)$$

Rectangular Sections

The analysis as applied to rectangles becomes complex since the shear stress is affected by warping, though essentially the angle of twist is unaffected.

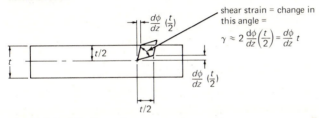

Fig. 8.2.2 Torsion of a rectangular section.

As an approximation, consider the element of Fig. 8.2.2 subjected to shear, in which

$$\gamma = t \frac{d\phi}{dz} \qquad (8.2.9)$$

For a thin rectangle, neglecting end effects, the shear stress may be expressed as

$$v = \gamma G = tG \frac{d\phi}{dz} \qquad (8.2.10)$$

or using Eq. 8.2.5,

$$v = \frac{Tt}{J} \qquad (8.2.11)$$

From the theory of elasticity [1–3], the maximum shear stress, v_{max}, occurs at the midpoint of the long side of a rectangle and acts parallel to it. The magnitude is a function of the ratio b/t (length/width) and may be expressed as

$$v_{max} = \frac{k_1 T}{bt^2} \qquad (8.2.12)$$

and the torsional constant J may be expressed as

$$J = k_2 bt^3 \tag{8.2.13}$$

where the values of k_1 and k_2 may be found in Table 8.2.1.

I-shaped, Channel, and Tee Sections.

As will be observed from a study of Table 8.2.1 the values of k_1 and k_2 become nearly constant for large ratios b/t. Thus the torsional constants for sections composed of thin rectangles may be computed as the sum of the values for the individual components. Such an approach will give an approximation which neglects the contribution in the fillet region where the components are joined. For most common structural shapes this approximation causes little error, thus

$$J \approx \sum \tfrac{1}{3} bt^3 \tag{8.2.14}$$

where b is the long dimension and t the thin dimension of the rectangular elements.

More accurate expressions for various structural shapes have been developed by Lyse and Johnston [4], Chang and Johnston [5], Kubo, Johnston, and Eney [6], and El-Darwish and Johnston [7].

In addition to the torsional properties in the AISC Manual, torsional design aid publications are available by Heins and Seaburg [8], Hotchkiss [9], and Heins and Kuo [10].

Table 8.2.1 Values of k_1 and k_2 for Eqs. 8.2.12 and 8.2.13

b/t	1.0	1.2	1.5	2.0	2.5	3.0	4.0	5.0	∞
k_1	4.81	4.57	4.33	4.07	3.88	3.75	3.55	3.44	3.00
k_2	0.141	0.166	0.196	0.229	0.249	0.263	0.281	0.291	0.333

8.3 SHEAR STRESSES DUE TO BENDING OF THIN-WALLED OPEN CROSS SECTIONS

Before treating the computation of stresses due to torsion of thin-walled open sections restrained from warping, a review of shear stress resulting from general flexure will be developed. Recognition of a torsion situation precedes concern about calculation of resulting stresses. Extensive treatment of thin-walled members of open cross section is given by Timoshenko [11].

Referring to the general thin-walled section of Fig. 8.3.1, where x and y are centroidal axes, consider equilibrium of the element $t\, ds\, dz$ acted upon by flexural stress σ_z and shear stress τ, both of which result

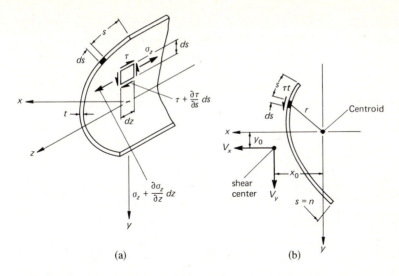

Fig. 8.3.1 Stresses on thin-walled open sections in bending.

from bending moment. The shear stress τ multiplied by the thickness t may be termed the *shear flow* τt. Force equilibrium in the z direction requires

$$\frac{\partial(\tau t)}{\partial s}\, ds\, dz + t\frac{\partial \sigma_z}{\partial z}\, dz\, ds = 0 \tag{8.3.1}$$

or

$$\frac{\partial(\tau t)}{\partial s} = -t\frac{\partial \sigma_z}{\partial z} \tag{8.3.2}$$

1. Assume moment is applied in the yz plane only, i.e., $M_y = 0$. The flexural stress due to bending, as given by Eq. 7.8.11, is

$$\sigma_z = \frac{M_x}{I_x I_y - I_{xy}^2}(I_y y - I_{xy} x) \tag{7.8.11}$$

$$\frac{\partial \sigma_z}{\partial z} = \frac{\partial M_x / \partial z}{I_x I_y - I_{xy}^2}(I_y y - I_{xy} x) \tag{8.3.3}$$

Recognizing that $V_y = \partial M_x / \partial z$, and substituting Eq. 8.3.3 into Eq. 8.3.2 gives

$$\frac{\partial \tau t}{\partial s} = \frac{-t V_y}{I_x I_y - I_{xy}^2}(I_y y - I_{xy} x) \tag{8.3.4}$$

Integrating to find τt at a distance s from a free edge gives the shear flow τt as

$$\tau t = \frac{-V_y}{I_x I_y - I_{xy}^2}\left[I_y \int_0^s yt\, ds - I_{xy}\int_0^s xt\, ds\right] \tag{8.3.5}$$

2. Assume moment is applied in the xz plane only, i.e., $M_x = 0$. The flexural stress due to bending as given by Eq. 7.8.11 is

$$\sigma_z = \frac{M_y}{I_x I_y - I_{xy}^2}(-I_{xy}y + I_x x) \qquad [7.8.11]$$

Taking $\partial \sigma_z / \partial z$, recognizing that $V_x = \partial M_y / \partial z$, and integrating to get the shear flow τt, gives in a manner similar to Eq. 8.3.5,

$$\tau t = \frac{+V_x}{I_x I_y - I_{xy}^2}\left[I_{xy}\int_0^s yt\, ds - I_x \int_0^s xt\, ds\right] \qquad (8.3.6)$$

3. Moments applied in both yz and xz planes. If shear stresses are desired they can be computed by superimposing the results from Eqs. 8.3.5 and 8.3.6.

It is to be observed from Fig. 8.3.1b that equilibrium requires that the shear in the y direction, V_y, equals the components of τt in the y direction summed over the entire section. Similarly V_x equals the summation of τt components in the x direction. Rotational equilibrium must also be satisfied; the moment about the centroid of the section is (see Fig. 8.3.1b)

$$\int_0^n (\tau t)r\, ds$$

which will be zero in some cases (such as I-shaped and Z-shaped sections). If such rotational equilibrium is automatically satisfied when the flexural shears act through the centroid, then no torsion will occur simultaneously with bending.

8.4 SHEAR CENTER

The shear center is the location in a cross section where no torsion occurs when flexural shears act in planes passing through that location. In other words, loads applied through the shear center will cause no torsional stresses to develop, i.e.,

$$\int_0^n (\tau t)r\, ds = 0 \qquad (8.4.1)$$

Since the shear center does not necessarily coincide with the centroid of the section, it is necessary to be able to locate the shear center in order to evaluate the torsional stress. For I-shaped and Z-shaped sections, the shear center coincides with the centroid, but for channels and angles it does not.

Referring to Fig. 8.3.1b, consider the shears V_x and V_y acting at distances from the centroid y_0 and x_0, respectively, such that the torsional moment with respect to the centroid is the same as $\int_0^n (\tau t)r\, ds$; thus

$$V_y x_0 - V_x y_0 = \int_0^n (\tau t)r\, ds \qquad (8.4.2)$$

In other words, the torsional moment is $(V_y x_0 - V_x y_0)$ when the loads are applied in planes passing through the centroid but is zero if the loads are in planes passing through the shear center, i.e., the point whose coordinates are $x_0 y_0$.

It is observed that the location of shear center is independent of the magnitude or type of loading, but is dependent only on the cross-sectional configuration.

To determine the shear center location, first let one of the shears be zero, say $V_y = 0$; then from Eq. 8.4.2,

$$y_0 = -\frac{1}{V_x} \int_0^n (\tau t) r \, ds \tag{8.4.3}$$

where according to Eq. 8.3.6,

$$\tau t = \frac{V_x}{I_x I_y - I_{xy}^2} \left[I_{xy} \int_0^s yt \, ds - I_x \int_0^s xt \, ds \right]$$

Alternately, letting $V_x = 0$ gives from Eq. 8.4.2,

$$x_0 = \frac{1}{V_y} \int_0^n (\tau t) r \, ds \tag{8.4.4}$$

where according to Eq. 8.3.5,

$$\tau t = \frac{-V_y}{I_x I_y - I_{xy}^2} \left[I_y \int_0^s yt \, ds - I_{xy} \int_0^s xt \, ds \right]$$

Example 8.4.1

Locate the shear center for the channel section of Fig. 8.4.1.

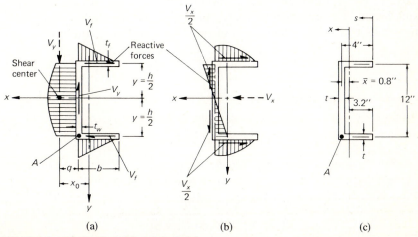

| (a) | (b) | (c) |

Fig. 8.4.1 Channel of Example 8.4.1.

SOLUTION

It is to be noted than many practical cases can be solved without resort to the general formulas, Eqs. 8.4.3 and 8.4.4. Since the shear center location is a problem in equilibrium, moments may most conveniently be taken through a point that eliminates the greatest number of forces. Thus, letting $V_x = 0$ and taking moments about point A of Fig. 8.4.1a, changes the equilibrium equation, Eq. 8.4.2, to

$$V_y q = V_f h = \int_0^b (\tau t) h \, ds \tag{a}$$

where according to Eq. 8.3.5,

$$\tau t = \frac{-V_y}{I_x} \int_0^s yt \, ds = \frac{-V_y}{I_x} yts \tag{b}$$

For these thin-walled sections, the length s along which integration is performed is measured at mid-thickness.

Substituting Eq. (b) into Eq. (a), and using $y = -h/2$, give

$$V_y q = \int_0^b \frac{-V_y}{I_x} \left(\frac{-h}{2}\right) ths \, ds \tag{c}$$

$$= \frac{V_y th^2}{2I_x} \int_0^b s \, ds = \frac{V_y th^2 b^2}{4I_x}$$

Thus the shear center location along the x-axis is

$$q = \frac{th^2 b^2}{4I_x} \tag{d}$$

measured in the positive x direction to the left of the channel web.

For the shear-center coordinate measured along the y-axis, apply V_x and let $V_y = 0$, and because of symmetry V_x must act at $y = 0$ for equilibrium. To demonstrate, let V_x be applied at the distance y_0 below the x-axis and take moments about point A. Satisfying equilibrium,

$$V_x \left(\frac{y}{2} - y_0\right) = \int_0^b (\tau t) y \, ds \tag{e}$$

where according to Eq. 8.3.6,

$$\tau t = \frac{-V_x}{I_y} \int_0^s xt \, ds \tag{f}$$

To illustrate numerically, use $b = 4$ in. and $h = 12$ in., in which case the centroid of the channel (refer to Fig. 8.4.1c) is located

$$\bar{x} = \frac{\sum Ax}{\sum A} = \frac{2b(b/2)t}{(h+2b)t} = \frac{2(4)(2)}{12+2(4)} = 0.8 \text{ in.}$$

Then, $s = x + 3.2$ in.

$$I_y = [\tfrac{1}{3}(4)^3 2 - 20(0.8)^2]t = 29.87t$$

$$\tau t = \frac{-V_x t}{29.87t} \int_0^s (s - 3.2) \, ds = \frac{-V_x}{29.87} \left(\frac{s^2}{2} - 3.2s\right)$$

Substitution of τt into Eq. (e) gives

$$V_x \left(\frac{h}{2} - y_0\right) = \int_0^4 \frac{-V_x}{29.87} \left(\frac{s^2}{2} - 3.2s\right) h \, ds$$

$$= \frac{-V_x h}{29.87} \left(\frac{s^3}{6} - 3.2 \frac{s^2}{2}\right) \Big|_0^4 = \frac{+V_x h}{2}$$

Thus it is shown that y_0 is zero. The shear center may also be located as follows. First compute, by integrating over each stress distribution of Fig. 8.4.1, the shear forces acting in each of the component elements of the section. Then the shear center is located such that V_x or V_y counteract all of the shear forces acting on the components to produce equilibrium. In solving for the shear center location, the solution may be made as illustrated, and then checked by verifying that the forces are in equilibrium.

8.5 TORSIONAL STRESSES IN I-SHAPED STEEL SECTIONS

The structural engineer must recognize a torsion situation and be able to apply approximate design methods and perform a stress analysis when necessary, even though only occasionally will torsion be severe enough to control the design of a section. Rolled steel sections under uniform and nonuniform torsion have been studied analytically and experimentally by many investigators. The development in this section is similar to that of Timoshenko [11], Lyse and Johnston [4], Kubo, Johnston and Eney [6], Goldberg [12], and Chu and Johnson [13]. Discussion of some of the practical aspects, along with summary of solutions for various loading and support cases, is given by Hotchkiss [9]; and charts are given in the handbook by Heins and Seaburg [8].

Application of load in a plane other than the one through the shear center (see Fig. 8.5.1) will cause the member to twist unless external restraints prevent such twisting. The torsional stress due to twisting consists of both shear and flexural stresses. These stresses must be superimposed on the shear and flexural stresses that exist in the absence of torsion.

Torsion may be categorized into two types: pure torsion, or as it is often called, *Saint-Venant's torsion*, and warping torsion. Pure torsion assumes that a cross-sectional plane prior to application of torsion remains a plane and only element rotation occurs during torsion. A circular

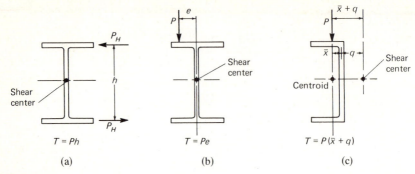

Fig. 8.5.1 Common torsional loadings.

shaft subjected to torsion is a situation where pure torsion exists as the only type. Warping torsion is the out-of-plane effect that arises when the flanges are laterally displaced during twisting, analogous to bending from laterally applied loads.

1. *Pure torsion (Saint-Venant's torsion).* Just as flexural curvature (change in slope per unit length) can be expressed as $M/EI = d^2y/dz^2$, i.e., moment divided by flexural rigidity equals flexural curvature, in pure torsion the torsional moment divided by the torsional rigidity GJ equals the torsional curvature (change in angle of twist per unit length). Recalling previously derived Eq. 8.2.5 for T, which now becomes the component, M_s, due to pure torsion,

$$M_s = GJ\frac{d\phi}{dz} \tag{8.5.1}$$

where M_s = pure torsional moment (Saint-Venant torsion)
G = shear modulus of elasticity = $E/[2(1+\mu)]$, in terms of the tension-compression modulus of elasticity, E, and Poisson's ratio, μ
J = torsional constant (see Sec. 8.2)

In accordance with Eq. 8.2.7, stress due to M_s is proportional to the distance from the center of twist.

2. *Warping torsion.* A beam subjected to torsion M_z, as in Fig. 8.5.2, will have its compression flange bent in one direction laterally while its tension flange is bent in the other. Whenever the cross section is such that it would warp (become a nonplanar section) if not restrained, the re-strained system has stresses induced. The torsional situation of Fig. 8.5.2 illustrates a beam that is prevented from twisting at each end but the top flange deflects laterally by an amount u_f. This lateral flange bending causes flexural normal stresses (tension and compression) as well as shear stresses across the flange width.

Thus torsion may be thought of as being composed of two parts: (1) rotation of elements, the pure torsion part, and (2) translation producing lateral bending, the warping part.

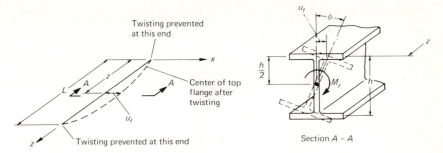

Fig. 8.5.2   Torsion of an I-shaped section.

3. *Differential equation for torsion on* I- *and channel-shaped sections.*
Consider the deflected position of a flange centerline, as in Fig. 8.5.2,
where u_f is the lateral deflection of one of the flanges at a section a
distance z from the end of the member; ϕ is the twist angle at the same
section, and V_f (Fig. 8.5.3) is the horizontal shear force developed in the
flange at that section due to lateral bending. It is noted that an important
assumption is that the web remains a plane during rotation, so that the
flanges deflect laterally an equal amount. Thus it is assumed the web is
thick enough compared to the flanges so that it does not bend during
twisting as a result of high torsional resistance of the flanges. Except for
thin-web plate girders, it has been shown [6, 14] that assuming no lateral
bending in the web, i.e., no effect on the warping torsion component, is
sufficiently correct for practical purposes. Since rarely are thin-web plate
girders used without stiffeners, and certainly not when torsional stress
exists, such cases are not of practical importance.

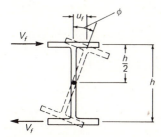

Fig. 8.5.3   Warping shear force on I-shaped section.

From geometry,

$$u_f = \phi \frac{h}{2} \qquad (8.5.2)$$

for small values of ϕ. For understanding of torsion on I- and channel-
shaped sections, Eq. 8.5.2 is the single most important relationship. The
twist angle is directly proportional to the lateral deflection. Torsion

boundary conditions are analogous to lateral bending boundary conditions.

Differentiating three times with respect to z in Eq. 8.5.2 gives

$$\frac{d^3 u_f}{dz^3} = \left(\frac{h}{2}\right)\frac{d^3 \phi}{dz^3}$$

(8.5.3)

For one flange the curvature relationship is

$$\frac{d^2 u_f}{dz^2} = \frac{-M_f}{EI_f}$$

(8.5.4)

where M_f is the lateral bending moment on one flange, I_f is the moment of inertia for one flange about the y-axis of the beam, and the minus sign arises from positive bending as shown in Fig. 8.5.2. Also, since $V = dM/dz$,

$$\frac{d^3 u_f}{dz^3} = \frac{-V_f}{EI_f}$$

(8.5.6)

Using Eqs. 8.5.3 and 8.5.6 gives

$$V_f = -EI_f\left(\frac{h}{2}\right)\frac{d^3 \phi}{dz^3}$$

(8.5.7)

Referring to Fig. 8.5.3, the torsional moment component M_w, causing lateral bending of the flanges, equals the flange shear force times the moment arm h. This assumes no shear resistance to warping is contributed by the web,

$$M_w = V_f h = -EI_f \frac{h^2}{2}\frac{d^3 \phi}{dz^3}$$

(8.5.8)

$$= -EC_w \frac{d^3 \phi}{dz^3}$$

(8.5.9)

where $C_w = I_f h^2 / 2$, often referred to as the *warping torsional constant.*

The total torsional moment is composed of the sum of the rotational part, M_s, and the lateral bending part, M_w, which from Eqs. 8.5.1 and 8.5.9 give

$$M_z = M_s + M_w = GJ\frac{d\phi}{dz} - EC_w \frac{d^3 \phi}{dz^3}$$

(8.5.10)

the differential equation for torsion. The torsional moment M_z depends on the loading and in usual situations will be a polynomial in z. Values of the torsion constant J and warping constant C_w are to be found in text, Appendix Table A2.

Rewrite Eq. 8.5.10, dividing by EC_w,

$$\frac{d^3 \phi}{dz^3} - \frac{GJ}{EC_w}\frac{d\phi}{dz} = \frac{-M_z}{EC_w}$$

(8.5.11)

Letting $\lambda^2 = GJ/EC_w$ ($\lambda = 1/a$ of *Torsion Analysis* [8]), and for the homogeneous solution of Eq. 8.5.11 let $\phi_h = Ae^{mz}$,

$$\frac{d^3\phi}{dz^3} - \lambda^2 \frac{d\phi}{dz} = 0 \tag{8.5.12}$$

which upon substitution of the homogeneous solution gives

$$Ae^{mz}(m^3 - \lambda^2 m) = 0 \tag{8.5.13}$$

which requires

$$m(m^2 - \lambda^2) = 0; \qquad \therefore m = 0, \; m = \pm\lambda$$

Thus

$$\phi_h = A_1 e^{\lambda z} + A_2 e^{-\lambda z} + A_3 \tag{8.5.14}$$

which upon using the hyperbolic function identities and regrouping the constants may be expressed as

$$\phi_h = A \sinh \lambda z + B \cosh \lambda z + C \tag{8.5.15}$$

where

$$\lambda = \frac{1}{a} = \sqrt{\frac{GJ}{EC_w}}$$

For the particular solution, since M_z is in general some function of z,

$$M_z = f(z)$$

Let $\phi_p = f_1(z)$, and substitute into Eq. 8.5.11, giving

$$\frac{d^3 f_1(z)}{dz^3} - \lambda^2 \frac{df_1(z)}{dz} = -\frac{1}{EI_w} f(z) \tag{8.5.16}$$

where terms on the left-hand side must be paired with terms on the right side. Rarely will $f_1(z)$ be required to contain higher than second-degree terms.

Example 8.5.1

Develop the expressions for the twist angle ϕ, as well as the first, second, and third derivatives, for the case of concentrated torsional moment applied at midspan when the ends are torsionally simply supported, using the differential equation.

SOLUTION

Referring to Fig. 8.5.4, it is apparent that M_z is constant and equal to $T/2$. Thus let

$$\phi_p = C_1 + C_2 z \quad \text{(any polynomial)} \tag{a}$$

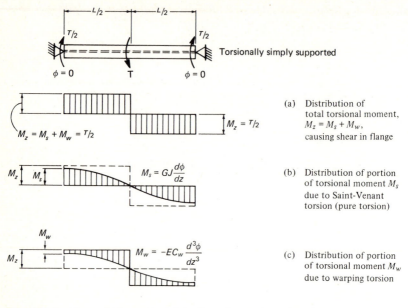

(a) Distribution of total torsional moment, $M_z = M_s + M_w$, causing shear in flange

(b) Distribution of portion of torsional moment M_s due to Saint-Venant torsion (pure torsion)

(c) Distribution of portion of torsional moment M_w due to warping torsion

Fig. 8.5.4 Case of Example 8.5.1. Concentrated torsional moment at midspan; torsionally simply supported. (Adapted from Ref. 9)

Using Eq. 8.5.11 gives

$$-\lambda^2 C_2 = -\frac{1}{EC_w}\left(\frac{T}{2}\right); \qquad \therefore C_2 = \frac{T}{2GJ}$$

The other constant C_1 may be combined with C of Eq. 8.5.15. The complete solution for this loading is therefore

$$\phi = A \sinh \lambda z + B \cosh \lambda z + C + \frac{T}{2GJ} z \qquad \textbf{(b)}$$

Consider the boundary conditions for torsional simple support. Thinking of the lateral bending of the flange (since ϕ is proportional to u_f), simple support conditions mean zero moment and deflection at each end, or for torsion

$$\phi = 0 \quad \text{at} \quad z = 0 \quad \text{and} \quad z = L$$

$$\frac{d^2\phi}{dz^2} = \phi'' = 0 \quad \text{at} \quad z = 0 \quad \text{and} \quad z = L$$

In this case the differential equation is discontinuous at $L/2$; thus, using zero slope of the flange at $L/2$, i.e., $\phi' = 0$, along with $\phi = 0$ and $\phi'' = 0$ at $z = 0$ will permit solution for the three constants of Eq. (b).

From $\phi = 0$ at $z = 0$,

$$0 = B + C \qquad \textbf{(c)}$$

Using $\phi'' = 0$ at $z = 0$,

$$\phi'' = A\lambda^2 \sinh \lambda z + B\lambda^2 \cosh \lambda z$$

$$0 = B \tag{d}$$

Thus from Eq. (c),

$$C = 0$$

Using $\phi' = 0$ at $z = L/2$,

$$0 = A\lambda \cosh \lambda L/2 + \frac{T}{2GJ} \tag{e}$$

$$A = -\frac{T}{2GJ\lambda} \left(\frac{1}{\cosh \lambda L/2}\right)$$

Finally, Eq. (b) becomes

$$\phi = \frac{T}{2GJ\lambda} \left[\lambda z - \frac{\sinh \lambda z}{\cosh \lambda L/2}\right] \tag{f}$$

also

$$\phi' = \frac{T}{2GJ} \left[1 - \frac{\cosh \lambda z}{\cosh \lambda L/2}\right] \tag{g}$$

$$\phi'' = \frac{T\lambda}{2GJ} \left[\frac{-\sinh \lambda z}{\cosh \lambda L/2}\right] \tag{h}$$

$$\phi''' = \frac{T\lambda^2}{2GJ} \left[\frac{-\cosh \lambda z}{\cosh \lambda L/2}\right] \tag{i}$$

Thus the solution of the differential equation is illustrated. The stress equations making use of the derivatives are developed in the next section.

4. *Torsional stresses.* The shear stress v_s resulting from the Saint-Venant torsion M_s is computed in accordance with the form of Eq. 8.2.11,

$$v_s = \frac{M_s t}{J} \tag{8.2.11}$$

and using Eq. 8.5.1 gives

$$v_s = Gt \frac{d\phi}{dz} \tag{8.5.17}$$

whose distribution is shown in Fig. 8.5.5a.

The shear stress v_w that results from warping varies parabolically across the width of the rectangular flange as shown in Fig. 8.5.5b and may be computed as

$$v_w = \frac{V_f Q_f}{I_f t_f} \tag{8.5.18}$$

where $Q_f =$ statical moment of area about the y-axis.

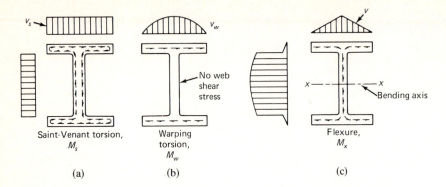

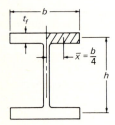

Fig. 8.5.5 Direction and distribution of shear stress in I-shaped sections.

The negligible shear carried by the web is not considered. For maximum shear stress v_w, which actually acts at the face of web but may be approximated as acting at the mid-width of the flange, take Q_f (see Fig. 8.5.6) as

$$Q_f = A\bar{x} = \frac{bt_f}{2}\left(\frac{b}{4}\right)$$

Substituting Q_f and V_f from Eq. 8.5.7 into Eq. 8.5.18 gives

$$v_w = E\frac{b^2 h}{16}\frac{d^3\phi}{dz^3} \qquad (8.5.19)$$

taking the absolute value.

Fig. 8.5.6 Dimensions for computation of statical moment of area, Q_f.

The normal stress (tension or compression) due to lateral bending of the flanges (i.e., warping of the cross section as shown in Fig. 8.5.7) may be expressed as

$$f_{bw} = \frac{M_f x}{I_f} \qquad (8.5.20)$$

which is distributed linearly across the flange width as shown in Fig. 8.5.7. The bending moment M_f, the lateral moment acting on one flange, may be obtained by substituting Eq. 8.5.2 into Eq. 8.5.4 and noting that $I_f h^2 / 2$

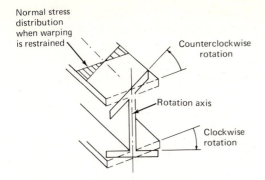

Normal stress distribution when warping is restrained

Counterclockwise rotation

Rotation axis

Clockwise rotation

Fig. 8.5.7   Warping of cross section.

is warping torsional constant C_w,

$$M_f = EI_f\left(\frac{h}{2}\right)\frac{d^2\phi}{dz^2} = \frac{EC_w}{h}\frac{d^2\phi}{dz^2} \qquad (8.5.21)$$

The minus sign is dropped since tension occurs on one side while compression occurs on the other.

The maximum stress occurs at $x = b/2$, which when used with Eq. 8.5.21 gives for Eq. 8.5.20,

$$f_{bw} = EI_f\left(\frac{h}{2}\right)\frac{d^2\phi}{dz^2}\left(\frac{b}{2I_f}\right)$$

$$f_{bw} = \frac{Ebh}{4}\frac{d^2\phi}{dz^2} \qquad (8.5.22)$$

In summary, three kinds of stresses arise in any I-shaped or channel section due to torsional loading: (a) shear stresses v_s in web and flanges due to rotation of the elements of the cross section (Saint-Venant torsional moment, M_s); (b) shear stresses v_w in the flanges due to lateral bending (warping torsional moment, M_w); and (c) normal stresses (tension and compression) f_{bw} due to lateral bending of the flanges (lateral bending moment on flange, M_f).

Example 8.5.2

A W18×71 beam on a 24-ft simply supported span is loaded with a concentrated load of 20 kips at midspan. The ends of the member are simply supported with respect to torsional restraint (i.e., $\phi = 0$) and the concentrated load acts with a 2-in. eccentricity from the plane of the web (see Fig. 8.5.8). Compute combined bending and torsional stresses.

SOLUTION

The differential equation solution for this type of loading and end

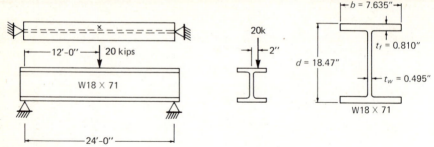

Fig. 8.5.8 Data for Example 8.5.2.

restraint was obtained in Example 8.5.1. The solution as obtained is

$$\phi = \frac{T}{2GJ\lambda}\left[\lambda z - \frac{\sinh \lambda z}{\cosh \lambda L/2}\right]$$

In accordance with the derivation (see Fig. 8.5.4), T is the applied torsional moment,

$$T = 20(2) = 40 \text{ in.-kips}$$

Recalling from Eq. 8.5.11,

$$\lambda = \sqrt{\frac{GJ}{EC_w}} = \sqrt{\frac{3.39}{2.6(4685)}} = \frac{1}{59.9} = 0.01668$$

where

$$\frac{E}{G} = \frac{2E(1+\mu)}{E} = 2.6 \qquad \text{for } \mu = 0.3$$

$$J \approx \sum \frac{bt^3}{3}, \qquad \text{Eq. 8.2.14}$$

$$= \tfrac{1}{3}[2(7.635)(0.810)^3 + (18.47 - 1.620)(0.495)^3] = 3.39 \text{ in.}^4$$

$$C_w = \frac{I_f h^2}{2} = \frac{(7.635)^3(0.810)}{12}\frac{(18.47 - 0.810)^2}{2}$$

$$= 4685 \text{ in.}^6$$

The above values of J and C_w compare with $J = 3.48$ and $C_w = 4700$ given in the AISC Manual computed for rectangular flanges using a more exact expression for J including the effect of the fillets at the junction of flange to web. Though different sources give slightly different values for these torsional constants, any of the values are satisfactory for design purposes.

The function values required are

$$\lambda L = 24(12)/59.9 = 4.80$$

z	λz	$\sinh \lambda z$	$\cosh \lambda z$
0.1L	0.480	0.499	1.118
0.2L	0.960	1.116	1.498
0.3L	1.441	1.994	2.231
0.4L	1.922	3.343	3.489
0.5L	2.402	5.477	5.567

(a) Pure torsion (Saint-Venant torsion). Using Eq. 8.5.17,

$$v_s = Gt \, d\phi/dz$$

$$\frac{d\phi}{dz} = \frac{T}{2GJ} \left(1 - \frac{\cosh \lambda z}{\cosh \lambda L/2} \right)$$

$$v_s = \frac{Tt}{2J} \left(1 - \frac{\cosh \lambda z}{5.567} \right) = \frac{40t}{2(3.39)} \left(1 - \frac{\cosh \lambda z}{5.567} \right)$$

The shear stress v_s is a maximum at $z = 0$ and zero at $z = L/2$:

$$v_s \text{ (flange at } z = 0) = \frac{40(0.810)}{2(3.39)} \left(1 - \frac{1}{5.567} \right) = 3.92 \text{ ksi}$$

$$v_s \text{ (web at } z = 0) = 3.92 \frac{0.495}{0.810} = 2.40 \text{ ksi}$$

(b) Lateral bending of flanges (warping torsion). Use Eq. 8.5.19 for shear stress in flanges,

$$v_w = E \frac{b^2 h}{16} \frac{d^3 \phi}{dz^3}$$

$$\frac{d^3 \phi}{dz^3} = \frac{T \lambda^2}{2GJ} \left(\frac{-\cosh \lambda z}{\cosh \lambda L/2} \right)$$

$$v_w = \frac{T}{2C_w} \frac{b^2 h}{16} \left(\frac{-\cosh \lambda z}{\cosh \lambda L/2} \right)$$

This shear stress acts at mid-width of the flange, and the maximum value occurs at $z = L/2$ while the minimum value is at $z = 0$,

$$v_w \text{ (flange at } z = L/2) = \frac{40}{2(4700)} \left(\frac{(7.635)^2 17.660}{16} \right) = 0.27 \text{ ksi}$$

$$v_w \text{ (flange at } z = 0) = 0.27 \frac{1.0}{5.567} = 0.05 \text{ ksi}$$

For normal stress in flanges due to warping, use Eq. 8.5.22:

$$f_{bw} = \frac{Ebh}{4} \frac{d^2\phi}{dz^2}$$

$$\frac{d^2\phi}{dz^2} = \frac{M\lambda}{2GJ}\left[\frac{-\sinh \lambda z}{\cosh \lambda L/2}\right]$$

$$f_{bw} = \frac{M(2.6)\lambda bh}{8J}\left[\frac{\sinh \lambda z}{\cosh \lambda L/2}\right]$$

which is a maximum at $z = L/2$ and zero at $z = 0$. Thus

$$f_{bw} \text{ (flanges at } z = L/2) = \frac{40(2.6)(7.635)(17.660)}{8(3.39)(59.9)}\left[\frac{5.477}{5.567}\right] = 8.49 \text{ ksi}$$

(c) Ordinary flexure. Maximum normal stress is

$$f_b \text{ (at } z = L/2) = \frac{PL}{4S_x} = \frac{20(24)(12)}{4(127)} = 11.34 \text{ ksi}$$

The shear stresses due to flexure are constant from $z = 0$ to $L/2$ and are computed by

$$v = \frac{VQ}{It} = \frac{10Q}{1170t} = \frac{Q}{117t}$$

For maximum flange shear stress, taking the more correct value at the face of the web rather than the value at mid-width of the flange,

$$Q = \left(\frac{7.635 - 0.495}{2}\right)(0.810)\left(\frac{17.660}{2}\right) = 25.53 \text{ in.}^3$$

$$v \text{ (flange at } z = 0) = \frac{25.53}{117(0.810)} = 0.27 \text{ ksi}$$

For maximum web shear stress,

$$Q = 7.635(0.810)\left(\frac{17.660}{2}\right) + \frac{16.850}{2}(0.495)\left(\frac{16.850}{4}\right) = 72.18 \text{ in.}^3$$

$$v \text{ (web at } z = 0) = \frac{72.18}{117(0.495)} = 1.25 \text{ ksi}$$

A summary of stresses showing combinations is given in Table 8.5.1

8.6 ANALOGY BETWEEN TORSION AND PLANE BENDING

Because the differential equation solution is time consuming, and really suited only for analysis, design of a beam to include torsion is most conveniently done by making the analogy between torsion and ordinary bending.

Table 8.5.1 Summary of Stresses for Example 8.5.2

Type of Stress	Support $(z = 0)$	Midspan $(z = L/2)$
Normal Stress:		
Vertical bending, f_b	0	11.34
Torsional bending, f_{bw}	0	8.49
		19.83 ksi
Shear stress, web:		
Saint-Venant torsion, v_s	2.40	0
Vertical bending, v	1.25	1.25
	3.65 ksi	
Shear stress, flange:		
Saint-Venant torsion, v_s	3.92	0
Warping torsion, v_w	0.05	0.27
Vertical bending, v	0.27	0.27
	4.24 ksi	

Consider that the applied torsional moment T of Fig. 8.6.1 can be converted into a couple P_H times h. The force P_H can then be treated as a lateral load acting on the flange of a beam.

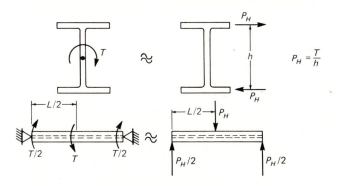

Fig. 8.6.1 Analogy between flexure and torsion.

The substitute system will have constant shear over one-half the span, a diagram as given in Fig. 8.5.4a. The true distribution of lateral shear which contributes to lateral deflection is only that part due to warping as shown in Fig. 8.5.4c. Thus the substitute system overestimates the lateral shear force and consequently overestimates the lateral bending moment M_f which causes normal stresses (tension and compression).

In most practical design situations when it is desirable to include the effect of torsion, the compressive normal stress due to the warping component is the quantity of most importance. The shear stress contributions are normally not of significance.

Example 8.6.1

Compute the stresses on the W18×71 beam of Example 8.5.2 and Fig. 8.5.8 using the flexural analogy rather than the differential equation solution.

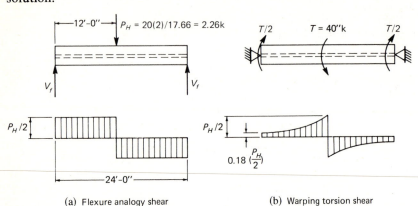

(a) Flexure analogy shear   (b) Warping torsion shear

Fig. 8.6.2 Comparison of lateral shear on flange due to warping torsion with that from simple lateral flexure analogy.

SOLUTION

The substitute system is as shown in Fig 8.6.2a. The lateral bending moment is then

$$M_f = V_f(L/2) = 1.13(12) = 13.6 \text{ ft-kips}$$

acting on one flange. Twice the moment acting on the entire section gives

$$f_{bw} = \frac{2M_f}{S_y} = \frac{2(13.6)(12)}{15.8} = 20.6 \text{ ksi}$$

For torsional shear stress, since $M_z = T/2 = 20$ in.-kips,

$$v_s = \frac{M_z t}{J} = \frac{20(0.810)}{3.39} = 4.78 \text{ ksi} \quad (\text{flange})$$

$$v_s = 4.78\left(\frac{0.495}{0.810}\right) = 2.92 \text{ ksi} \quad (\text{web})$$

For lateral bending flange shear stress,

$$v_w = \frac{V_f Q_f}{I_f t_f} = \frac{1.13(5.90)}{(30.0)0.810} = 0.27 \text{ ksi}$$

where $Q_f = (7.635/2)(0.810)(7.635/4) = 5.90 \text{ in.}^3$

The results of the two methods are compared as follows:

Type of Stress	Flexural Analogy	Diff. Equation
Normal stress $= f_b + f_{bw} = 11.3 + 20.6 =$	31.9 ksi	19.83 ksi
Web shear stress $= v + v_s = 1.25 + 2.92 =$	4.17 ksi	3.65 ksi
Flange shear stress $= v + v_s + v_w = 0.27 + 4.78 + 0.27 =$	5.32 ksi	4.24 ksi

It is apparent that use of the flexure analogy without modification is a very conservative approach. In some situations it is so excessively conservative as to be practically useless. Furthermore, the most important design item, the lateral bending normal stress f_{bw} is overestimated by the greatest amount.

The relationship between the flexural analogy and the true torsion problem is best illustrated by referring to Fig. 8.5.4a. Note that the full torsional shear resulting from M_s and M_w is analogous to the lateral flexure problem. Figure 8.5.4b shows the portion of the shear that goes into rotation of elements, while Fig. 8.5.4c shows the portion contributing to lateral flange bending. If one could correctly assess how the shear due to warping torsion compares with the lateral flexure situation, design for torsion could be greatly simplified without being grossly conservative.

Figure 8.6.2b shows the accurate variation of V_f for the problem of Example 8.6.1, computed according to Eq. 8.5.7, whereupon

$$V_f = \frac{T}{2h} \left(\frac{\cosh \lambda z}{\cosh \lambda L/2} \right) \tag{8.6.1}$$

in which the shear from the lateral bending analogy, $T/2h$, is modified by the hyperbolic function.

The lateral bending moment can thus be expressed for this problem as

$$M_f = \beta \frac{T}{2h} \left(\frac{L}{2} \right) \tag{8.6.2}$$

or, in general, the change in lateral moment between the support and location of zero shear is

$$\Delta M_f = \beta \times (\text{area under flexure analogy shear diagram}) \tag{8.6.3}$$

where β is a reduction factor that depends on λL.

It is to be noted that if Eq. 8.6.2 is multiplied by h, and the concentrated moment T is thought of as a concentrated load, the analogous moment $M_f h$ (sometimes referred to as *bimoment*) equals β times the simple beam moment. Thus the modified flexure analogy gives

$$M_f h = \beta \left(\frac{TL}{4} \right) \tag{8.6.4}$$

for the case of Fig. 8.6.1.

Tables 8.6.1 through 8.6.5 give "exact" values for β for several common loading and restraint conditions. For other cases Table I of Ref. 9 (M_w of Ref. 9 equals $M_f h$ above) or the curves of *Torsion Analysis* [8] may be used. In Tables 8.6.3 and 8.6.4, m is the applied torsional loading per unit length (say, in.-kips/ft).

Example 8.6.2

Recompute the stresses due to torsion on the beam of Example 8.6.1, using the modified flexural analogy method utilizing the β values from Table 8.6.1.

Table 8.6.1 β Values, Concentrated Load, Torsional Simple Support

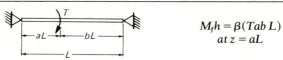

$$M_f h = \beta(Tab\,L)$$
$$\text{at } z = aL$$

λL	β values				
	$a = 0.5$	$a = 0.4$	$a = 0.3$	$a = 0.2$	$a = 0.1$
0.5	0.98	0.98	0.98	0.99	0.99
1.0	0.92	0.93	0.94	0.95	0.97
2.0	0.76	0.77	0.80	0.84	0.91
3.0	0.60	0.62	0.65	0.72	0.83
4.0	0.48	0.50	0.54	0.62	0.76
5.0	0.39	0.41	0.45	0.54	0.70
6.0	0.33	0.34	0.39	0.47	0.65
8.0	0.25	0.26	0.30	0.37	0.55
10.0	0.20	0.21	0.24	0.31	0.48

Table 8.6.2 β Values, Concentrated Load, Torsionally Fixed Supports

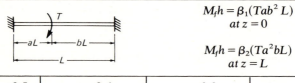

$$M_f h = \beta_1(Tab^2\,L)$$
$$\text{at } z = 0$$

$$M_f h = \beta_2(Ta^2 bL)$$
$$\text{at } z = L$$

λL	$a = 0.5$	$a = 0.4$		$a = 0.3$		$a = 0.2$	
	$\beta_1 = \beta_2$	β_1	β_2	β_1	β_2	β_1	β_2
0.5	0.99	1.00	0.99	1.00	0.99	1.00	0.99
1.0	0.98	0.98	0.98	0.98	0.98	0.99	0.98
2.0	0.92	0.93	0.92	0.94	0.92	0.96	0.92
3.0	0.85	0.86	0.84	0.88	0.84	0.91	0.85
4.0	0.76	0.78	0.75	0.81	0.75	0.86	0.77
5.0	0.68	0.70	0.67	0.74	0.67	0.80	0.69
6.0	0.60	0.63	0.59	0.67	0.60	0.75	0.62
8.0	0.48	0.51	0.47	0.56	0.49	0.65	0.52
10.0	0.39	0.42	0.39	0.47	0.41	0.56	0.44

Table 8.6.3 β Values, Uniform Load, Torsional Simple Support

$$M_f h = \beta\left(\frac{m}{2}\, abL^2\right)$$

$$at\ z = aL$$

λL	β values				
	$a = 0.5$	$a = 0.4$	$a = 0.3$	$a = 0.2$	$a = 0.1$
0.5	0.97	0.97	0.98	0.98	0.98
1.0	0.91	0.91	0.91	0.91	0.92
2.0	0.70	0.71	0.71	0.72	0.74
3.0	0.51	0.51	0.52	0.54	0.57
4.0	0.37	0.37	0.38	0.41	0.44
5.0	0.27	0.27	0.29	0.31	0.34
6.0	0.20	0.20	0.22	0.24	0.28
8.0	0.12	0.12	0.13	0.16	0.19
10.0	0.08	0.08	0.09	0.11	0.14

Table 8.6.4 β Values, Uniform Load, Torsionally Fixed Supports

$$M_f h = \beta\left(\frac{m}{12}\, L^2\right)$$

$$at\ z = 0\ and\ z = L$$

λL	0.5	1.0	2.0	3.0	4.0	5.0	6.0	8.0
β	0.99	0.98	0.94	0.88	0.81	0.74	0.67	0.56

Table 8.6.5 β Values, Concentrated Load, Torsionally Fixed Supports

$$M_f h = \beta\ (positive\ moment\ by\ flexure\ theory)$$
$$= \beta[2Ta^2b^2L]$$
$$at\ z = aL$$

λL	$a = 0.5$	$a = 0.3$	$a = 0.1$
0.5	0.99	1.00	1.00
1.0	0.98	0.99	1.01
2.0	0.92	0.95	1.05
3.0	0.85	0.91	1.10
4.0	0.76	0.85	1.16
5.0	0.68	0.79	1.21
6.0	0.60	0.73	1.25

SOLUTION

The flexure analogy gives

$$M_f = 13.6 \text{ ft-kips}$$

as previously computed.

$$\lambda L = 4.80 \quad \text{(as computed in Example 8.5.2)}$$

From Table 8.6.1 at $a = 0.5$, $\beta \approx 0.41$, i.e., use about 41 percent of the flexure analogy value. Thus the modified flexure analogy gives

$$M_f = 13.6(0.41) = 5.58 \text{ ft-kips}$$

$$f_{bw} = \frac{2M_f}{S_y} = \frac{2(5.58)12}{15.8} = 8.48 \text{ ksi}$$

which compares favorably with $f_{bw} = 8.49$ ksi as computed by the differential equation solution using $\lambda L = 4.80$. For this case that exactly fits a table case, the β modified flexure analogy is the "exact" value obtained from the differential equation solution value.

8.7 PRACTICAL SITUATIONS OF TORSIONAL LOADING

There are relatively few occasions in actual practice where the torsional load can cause significant twisting, and frequently these situations arise during construction. In most building construction the members are laterally restrained by attachments along the length of the member and therefore they are not free to twist. Even though torsional loading exists, it may be self-limiting because the rotation cannot exceed the end slope of the transverse attached members.

Torsional loading exists on spandrel beams, where the torsional loading may be uniformly distributed; it exists where a beam frames into a girder on one side only, or where unequal reactions come to opposite sides of a girder. The design of crane runway girders involves the combination of biaxial bending and torsion, and is illustrated in Sec. 8.8 for laterally stable beams. Any situation where the loading or reaction acts eccentrically to the shear center gives rise to torsional stresses.

Analysis for Torsional Moment

The determination of the torsional moment in a framing system involves an elastic analysis where the joints may be rigid or semirigid. While the details of such an analysis are outside the scope of this text, some discussion is necessary so that at least the problem is understood. Goldberg [12] has discussed this subject and presented an approximate method suitable for design. Spandrel girders have been treated by Lothers [15].

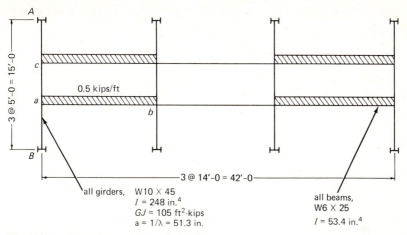

Fig. 8.7.1   Plan view of floor framing.

Consider an example of a floor framing system (similar to Goldberg's [12]) as shown in Fig. 8.7.1. Spandrel beam *AB* is subjected to torsion because of the floor beams framing on only one side. Contrary to some common belief, however, the torsional moment is *not* equal to the beam reaction times its eccentricity from the centerline of the girder web. Moment is transmitted across the joint, and the end moment on the beam must equal the torsional moment on the girder. To attack such a problem one must first determine the relationship between the angle of twist ϕ and the applied torsional moment *T*.

For example, in Fig. 8.7.1 the loading system causes equal torsional moments at the $\frac{1}{3}$ points on member *AB*. Assuming the girder torsionally simply supported at ends *A* and *B*, using either the differential equation solution formulas or the curves of *Torsion Analysis* [8] (Case 3), for $\lambda L = 15(12)/51.3 = 3.51$, one finds the angle of twist ϕ at point *a*,

$$\phi \frac{GJ}{TL} \approx 0.09$$

or

$$\phi_{aa} = 16.2 \frac{T_a}{GJ}$$

for *T* applied at *a*. In addition the value of ϕ at *a* for *T* applied at *c* is

$$\phi_{ac} = 0.07(180) \frac{T_c}{GJ} = 12.6 \frac{T_c}{GJ}$$

Finally, for $T_a = T_c = T$,

$$\phi_a = (16.2 + 12.6) \frac{T}{GJ} = 28.8 \frac{T}{GJ}$$

The twist angle ϕ_a must be compatible with the end slope of the beam; using slope deflection,

$$M_{ab} = M_{Fab} + \frac{2EI}{L_{ab}}\left[-2\phi_a - \phi_b + \frac{3\Delta}{L_{ab}}\right]$$

where M_{Fab} = the fixed-end moment for beam ab at a
ϕ_a = beam slope at a
ϕ_b = beam slope at b
Δ = relative deflection between a and b

After having established the necessary slope deflection equations for moments, joint equilibrium and shear conditions are necessary, after which simultaneous equations must be solved. One such joint equation is that at joint a,

$$T + M_{ab} = 0$$

After solving for the slopes, then the torsional moments can be found; for the torsional moment at a,

$$T = \phi_a GJ/28.8$$

In the Goldberg [12] example using members having properties similar to those of Fig. 8.7.1, the value of T obtained was 1.55 in.-kips using an approximate method of satisfying deformation compatibility.

Suppose one had taken the simple beam reaction for member ab, $0.5(7) = 3.5$ kips, and assume use of an AISC "Framed Beam Connection" (see Fig. 13.2.1). If the eccentricity had been taken to the bolt line on outstanding leg ($2\frac{1}{4}$ in.), the torsional moment would have been far too great, while if the eccentricity had been taken as one-half the W10×45 web thickness (0.350/2), the torsional moment would have been far too small.

The proper torsional moment can only be obtained (even approximately) by considering deformation compatibility.

Torsional End Restraint

If a torsional situation is deemed to require design or analysis, the torsional end restraint must be evaluated. Under AISC–1.2 three types of construction are permitted: "rigid-frame," "simple" or "conventional," and "semi-rigid framing." For practical purposes, most construction comes under Type I—rigid frame, i.e., full flexural moment restraint at joints, or Type II—simple connections, i.e., negligible flexural restraint at joints.

The torsional restraint conditions for Type I and Type II construction are given in Fig. 8.7.2. Again, the lateral bending analogy will help in visualizing the torsional restraint conditions. Figure 8.7.2a shows the analogy situation of zero deflection and zero moment which correspond torsionally to $\phi = 0$ and $d^2\phi/dz^2 = 0$. It is to be noted that $\phi = 0$ only if

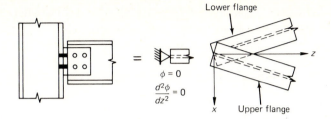

(a) AISC simple framing connection (Type II)
$$M_z = M_s + M_w \ ; M_f = 0$$

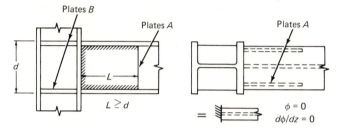

(b) AISC rigid framing connection (Type I) with
    additional stiffening plates.
$$M_z = M_w \ ; M_s = 0$$

Fig. 8.7.2 Torsional restraint conditions. (Adapted from Fig. 8 of Ref. 9)

the simple connection extends over a significant portion of the beam depth.

   Figure 8.7.2b shows the analogy situation of zero deflection and zero slope which correspond torsionally to $\phi = 0$ and $d\phi/dz = 0$. Hotchkiss [9] states the ends of the beam must be boxed in so as to assure $d\phi/dz = 0$. Stiffener plates (plates A, Fig. 8.7.2b) may be welded between the toes of the flanges and extended along the beam for a length equal at least to the beam depth. Furthermore, if the column has torsionally flexible flanges, column stiffeners (plates B of Fig. 8.7.1b) should be provided opposite the beam flanges. Vacharajittiphan and Trahair [16] also discuss torsional restraint at I-section joints.

   The structural engineer should remember that in practical situations where no special precautions are taken at the ends, the torsional restraint is neither simple ($d^2\phi/dz^2 = 0$) nor fixed ($d\phi/dz = 0$) but is, however, usually such that the end twist is nearly zero ($\phi = 0$).

8.8 DESIGN FOR COMBINED BENDING AND TORSION —LATERALLY STABLE BEAMS

In order to illustrate the use of some of the concepts developed in the previous sections, several design examples are presented. In all these examples it will be assumed that the combined maximum compression stress is limited to $0.60F_y$ in accordance with AISC–1.5.1.4.6b wherein the laterally unbraced length does not exceed $76b_f/\sqrt{F_y}$ nor $20{,}000/(F_y\, d/A_f)$. Treatment of cases where this length is exceeded is in Chapter 9.

In essence, present design procedures are based on elastic behavior; therefore ultimate strength including any effect of residual stress on torsional strength is not considered.

Example 8.8.1

Select the lightest W section of A36 steel to carry a superimposed uniform loading of 1.9 kips per ft on a simple supported span of 28 ft, as shown in Fig. 8.8.1. The loading is applied eccentrically 7 in. from the web, and assume that the ends have torsional simple support.

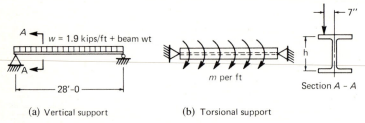

$w = 1.9$ kips/ft + beam wt

28'-0

(a) Vertical support

m per ft

(b) Torsional support

7''

h

Section A - A

Fig. 8.8.1 Conditions for Example 8.8.1.

SOLUTION

Estimating the beam weight as 0.13 kips/ft, the flexural bending moment is

$$M_x = \tfrac{1}{8}wL^2 = \tfrac{1}{8}(2.03)(28)^2 = 199 \text{ ft-kips}$$

The torsional uniformly distributed moment is

$$m = 1.9(7) = 13.3 \text{ in.-kips/ft}$$

Consider m/h as the uniformly distributed lateral load acting on *one flange* of the beam, and using the flexure analogy, gives

$$M_f = \frac{1}{8}\frac{m}{h}L^2 = \frac{1}{8}\frac{13.3}{h}(28)^2 = \frac{1303}{h} \text{ ft-kips}$$

without regard to the modification factor β.

As a first approximation, assume $h = 14$ in. and $\beta = 0.5$ (approximation from Table 8.6.3, for $\lambda L \approx 3$); Thus the modified flexure analogy gives

$$M_f = \beta\frac{1303}{h} = 0.5\frac{1303}{14} = 46.5 \text{ ft-kips}$$

The design acceptability criterion is

$$\frac{M_x}{S_x} + \frac{M_y}{S_y} \le 0.60F_y$$

and using the procedure discussed in Sec. 7.9 gives

$$\text{Required } S_x = \frac{M_x}{0.60F_y} + \frac{M_y}{0.60F_y}\left(\frac{S_x}{S_y}\right)$$

$$= \frac{199(12)}{22} + \frac{2(46.5)(12)}{22}(3) = 109 + 152 = 261 \text{ in.}^3$$

in which the ratio S_x/S_y is estimated at 3 (Table 7.9.1) for medium weight W14 sections, and M_f is doubled to give the moment acting on two flanges.

This would indicate a W14×176 with an $S_x = 281$ in.^{3} It will be observed that actual S_x/S_y for sections in this range is about 2.6. The required S_x is then reduced to about 241, which would indicate W14× 159.

Using torsional properties in the AISC Manual for the W14×159, $\lambda = \sqrt{GJ/EC_w} = 1/68.3$. Thus

$$\lambda L = 28(12)/68.3 = 4.9$$

Using Fig. 8.6.3, β is reduced to about 0.3, which further reduces required S_x to about 190 in.^{3} Try W14×132.

$$\lambda L = 28(12)/73.2 = 4.59$$

$$\beta \approx 0.31 \quad (\text{Table 8.6.3})$$

$$M_f = \beta \frac{mL^2}{8h} = 0.31 \frac{13.3(28)^2}{8(14.66 - 1.030)} = 29.6 \text{ ft-kips}$$

$$\text{Max } f_b = \frac{M_x}{S_x} + \frac{M_f}{S_y/2} = \frac{199(12)}{209} + \frac{29.6(12)}{74.5/2}$$

$$= 11.4 + 9.5 = 20.9 \text{ ksi}$$

Use W14×132. Where high torsional strength is required, the wide W14 sections are most suitable. For the *same weight per foot*, deeper sections give a reduced stress from ordinary flexure but an increased stress from torsional warping. The W21×132 ($f_b = 19.5$ ksi) and the W24×131 ($f_b = 20.3$ ksi) give about the same maximum normal stress as the above selected beam.

The differential equation solution gives for the normal stress due to warping torsion 9.4 ksi as compared with 9.5 ksi computed above. The maximum flange shear stress is 10.3 ksi, while that in the web is 8.7 ksi, both computed from the differential equation solution. These are well within allowable values.

Example 8.8.2

Design a beam with torsionally fixed ends to carry two concentrated loads eccentric to the plane of the web by 6 in. as shown in Fig. 8.8.2. Assume for conservatism that for ordinary flexure the beam is simple supported. Use A36 steel and assume the maximum combined stress is limited to $0.60F_y$.

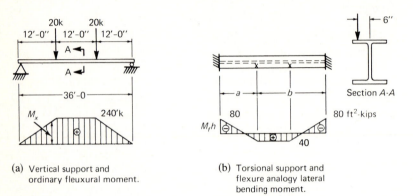

(a) Vertical support and ordinary fleuxural moment.

(b) Torsional support and flexure analogy lateral bending moment.

Fig. 8.8.2 Conditions for Example 8.8.2.

SOLUTION

Estimating beam weight as 0.15 kips/ft, the flexural bending moment M_x is

$$M_x = 20(12) + \tfrac{1}{8}(0.15)(36)^2 = 264 \text{ ft-kips}$$

The concentrated torsional moment is

$$T = 20(0.5) = 10 \text{ ft-kips}$$

Considering T/h as the analogous concentrated loads, the fixed end moments are computed; thus

$$M_f h \text{ (at ends)} = \frac{Tab^2}{L^2} + \frac{Ta^2b}{L^2} = \frac{10(12)(24)}{(36)^2}(24+12)$$

$$= 53.3 + 26.7 = 80 \text{ ft}^2\text{-kips}$$

and in the positive moment zone,

$$M_f h \text{ (at concentrated loads)} = \frac{TL}{3} - 80.0 = 10(12) - 80.0$$

$$= 40 \text{ ft}^2\text{-kips}$$

without regard to the β reduction factor; the flexure analogy gives $M_f h$ values as shown in Fig. 8.8.2b. The calculation just illustrated is more appropriate than using the expression in Table 8.6.5 because that expression is for *one* concentrated load. The β values from Table 8.6.5 are reasonable, however, since the effect of one load on the torsional stress at the other load is small.

Estimating average λL at about 3, and using $a = 0.3$ in Table 8.6.2 for end moments, the modified analogous fixed-end moments become

$$M_f h \text{ (at ends)} = 0.88(53.3) + 0.84(26.7) = 47 + 22$$
$$= 69 \text{ ft}^2\text{-kips}$$

For the positive moment at 12 ft from the support, refer to Table 8.6.5 and estimate β as 0.9, though the exact case being considered is not convered in any of the tables of β values. Thus

$$M_f h \text{ (at } z = 0.3L) \approx 0.9(40) = 36 \text{ ft}^2\text{-kips}$$

which is known to be conservatively high (see Fig. 8.8.3) because the 40 value includes the effects of both concentrated torsional moments. (*Torsion Analysis* [8], Case 6, indicates that T applied at $z = 0.3L$ has negligible effect at $0.7L$)

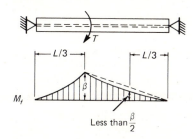

Fig. 8.8.3 M_f variation for concentrated M.

Assuming $h = 14$ in.,

$$M_f = 36(12)/14 = 30.9 \text{ ft-kips}$$

$$\text{Required } S_x = \frac{M_x}{22} + \frac{2M_f}{22}\left(\frac{S_x}{S_y}\right)$$

$$= \frac{264(12)}{22} + \frac{2(30.9)(12)}{22}(2.5) = 144 + 84 = 228 \text{ in.}^3$$

W14×145: $S_x = 232$ in.^{3}

$$\lambda L = 36(12)/73.6 = 5.87$$

Using Table 8.6.5, find $\beta = 0.74$ which gives

$$M_f(\text{at } z = 0.3L) = \beta\frac{M_f}{h} = 0.74\frac{40(12)}{(14.78 - 1.09)} = 25.9 \text{ ft-kips}$$

$$\text{Required } S_x = 144 + \frac{2(25.9)(12)}{22}(2.7) = 220 \text{ in.}^3$$

Use W14×145, since there is no lighter satisfactory W14 section.

For a more accurate check using *Torsion Analysis* [8], Case 6 for $\lambda L = 5.87$ at $z = 0.3L$,

$$M_f h = (\text{Ref. 8 coeff.}) \frac{T}{\lambda} = 0.37(10)(73.6)/12 = 22.7 \text{ ft}^2\text{-kips}$$

$$M_f = 22.7(12)/13.69 = 19.9 \text{ ft-kips}$$

$$f_b = \frac{264(12)}{216} + \frac{2(19.9)(12)}{77.0} = 19.2 \text{ ksi} < 22 \text{ ksi} \qquad \text{OK}$$

This more exact check indicates a W14×132 would be satisfactory ($f_b = 21.6$ ksi).

Also the stress at the supports should be checked. Using Table 8.6.2, find $\beta_1 \approx 0.68$ and $\beta_2 = 0.61$ for $\lambda L = 5.87$ and $a = 0.3$. Then

$$M_f h = 0.68(53.3) + 0.61(26.7) = 36.2 + 15.3 = 52.5 \text{ ft}^2\text{-kips}$$

$$M_f = 52.5(12)/13.69 = 46.0 \text{ ft-kips}$$

$$f_b = \frac{2M_f}{S_y} = \frac{2(46.0)12}{87.3} = 12.6 \text{ ksi} < 22 \text{ ksi} \qquad \text{OK}$$

These two examples illustrate that using approximate β values, along with the flexure analogy for lateral bending due to warping torsion, gives sufficiently quick and accurate results for ordinary design. Furthermore, the designer can better visualize what is happening using the flexure analogy rather than working with the hyperbolic functions for ϕ.

For additional treatment of combined torsion and flexure, particularly on channel and zee sections, the reader is referred to the work of Lansing [17]. For nonprismatic open section members, Evick and Heins [18] present solution techniques and give some design information.

Another topic, outside the scope of this text, is the secondary lateral bending moment that arises from the torsional deflection of the compression flange laterally. In the deflected position the compressive force resulting from ordinary flexural moment, M_x, times the lateral flange deflection gives rise to the secondary lateral moment which in turn causes greater lateral deflection. Discussion of this topic appears elsewhere [19] and is similar to the secondary bending moment that occurs in beam-columns, a subject treated in Chapter 12.

8.9 TORSION IN CLOSED THIN-WALLED SECTIONS

In general, it will be found that for situations where a high torsional stiffness is required, a closed section is preferred over the ordinary open section, such as the I-shape or channel. An excellent general discussion of the torsion phenomena with comparative behavior of open and closed sections is given by Tamberg and Mikluchin [20]. Some practical com-

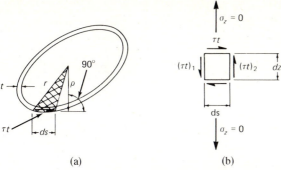

Fig. 8.9.1 Shear flow in a closed thin-wall section.

ments relating to closed sections in torsion are given by Siev [21]. It is the high torsional stiffness exhibited by closed sections that makes them ideal for aircraft structural components and curved girders in bridges and buildings. This subject is treated in a number of textbooks [1–3] so that only a brief treatment follows.

In the closed section of Fig. 8.9.1, it is assumed that the walls are thin so that the shearing stress may be assumed uniformly distributed across the thickness t. If the shear stress is τ, then τt is the shear force per unit distance along the wall, usually referred to as *shear flow*. Since only torsional stress is presently being considered, the normal stresses (σ_z of Fig. 8.9.1b) are zero. Since $\sigma_z = 0$, the shear flow τt cannot vary along the wall; i.e., τt is constant.

Referring to Fig. 8.9.1a, the increment of torsional moment contributed by each element is

$$dT = \tau t \rho \, ds \qquad (8.9.1)$$

Integrating gives the full torsional moment, which is in effect the same as Eq. 8.2.5,

$$T = \tau t \int_s \rho \, ds \qquad (8.9.2)$$

Note that $\frac{1}{2}\rho \, ds$ is the cross-hatched area of the triangular segment in Fig. 8.9.1. Thus the integral

$$\int_s \rho \, ds = 2A \qquad (8.9.3)$$

where $A =$ area enclosed by the walls. Finally,

$$T = 2\tau t A \qquad (8.9.4)$$

If a cut is made in the wall of a closed thin-wall section (Fig. 8.9.2), a relative movement (as in Fig. 8.9.2b) will be produced between the two

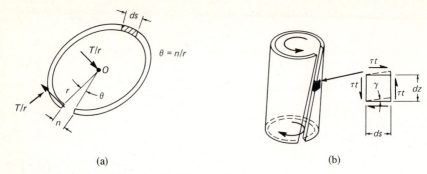

Fig. 8.9.2 Forces on a cut thin-wall section.

sides in the axial direction of the member. The unit shear strain along the perimeter is

$$\gamma = \tau/G \tag{8.9.5}$$

The internal strain energy for any elemental length ds along the perimeter is

$$dW_i = \tfrac{1}{2}\tau t \gamma \, ds \tag{8.9.6}$$

$$= \frac{1}{2}\left(\frac{T}{2A}\right)\frac{\tau}{G}\, ds \tag{8.9.7}$$

The twisting moment, T, about point 0 can now be replaced by a couple, T/r. The external work done by the couple is

$$dW_e = \frac{1}{2}\left(\frac{T}{r}\right)n = \frac{T\theta}{2} \tag{8.9.8}$$

Equating internal and external work per unit length gives

$$\frac{T\theta}{2} = \frac{T}{4AG}\int_s \tau \, ds \tag{8.9.9}$$

$$\theta = \frac{\int_s \tau \, ds}{2AG} = \frac{\tau t \int_s ds/t}{2AG} \tag{8.9.10}$$

since τt is a constant.

In order to obtain more useful forms of the equations, recall from Sec. 8.2 that

$$T = GJ\theta \tag{8.2.5}$$

and using Eq. 8.9.10 gives

$$T = GJ\frac{\tau t \int ds/t}{2AG} \tag{8.9.11}$$

and eliminating T between Eqs. 8.9.4 and 8.9.11 gives, when solving for the torsional constant J,

$$J = \frac{4A^2}{\int_s ds/t} \tag{8.9.12}$$

Multicell Sections

In computing the shear stresses in multicell sections under torsion, equilibrium must be satisfied; thus the shear flow into a junction (such as the one shown shaded in Fig. 8.9.3a) must equal the flow out; the angle of twist θ must be identical for each cell; the sum of the individual torsional resistances must be equal to the applied torsional moment.

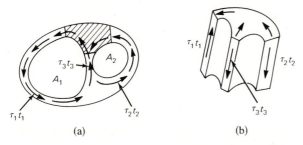

(a)   (b)

Fig. 8.9.3 Shear flow in multicell thin-wall section.

In Fig. 8.9.3 the force equilibrium at joints is satisfied by

$$\tau_3 t_3 = \tau_1 t_1 - \tau_2 t_2 \tag{8.9.13}$$

Secondly, deformation compatibility is obtained when

$$\left. \begin{array}{c} \theta = \dfrac{\displaystyle\int_1 \tau\, ds}{2A_1 G} \\[4mm] \theta = \dfrac{\displaystyle\int_2 \tau\, ds}{2A_2 G} \end{array} \right\} \tag{8.9.14}$$

Finally, moment equilibrium requires

$$T = 2A_1 \tau_1 t_1 + 2A_2 \tau_2 t_2 + \cdots + 2A_n \tau_n t_n \tag{8.9.15}$$

The development of the equations for torsion stiffness and stress in closed thin-walled sections may be alternatively developed using the "membrane analogy" developed by Prandtl [1–3].

The principal aim in this text is to develop the basic expressions and illustrate the high degree of torsional stiffness that closed sections exhibit as compared to open ones. The following two examples are intended for this purpose.

Example 8.9.1

Compare the torsional resisting moment T and the torsional constant J for the sections of Fig. 8.9.4 all having about the same cross-sectional area. The maximum shear stress τ is 14 ksi.

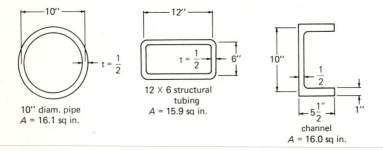

10″ diam. pipe
$A = 16.1$ sq in.

12 X 6 structural
tubing
$A = 15.9$ sq in.

channel
$A = 16.0$ sq in.

Fig. 8.9.4 Sections for Example 8.9.1.

SOLUTION

(a) Circular thin-wall section. Using Eq. 8.9.4,

$$T = 2\tau t A = 2(14)(0.5)[\pi(10)^2/4]\tfrac{1}{12} = 91.6 \text{ ft-kips}$$

$$J = \frac{4A^2}{\displaystyle\int ds/t} = \frac{4(25\pi)^2}{20\pi} = 393 \text{ in.}^4$$

where $\int ds/t = 2\pi(5)/0.5 = 20\pi$.

(b) Rectangular box section.

$$T = 2\tau t A = 2(14)(0.5)(72)\tfrac{1}{12} = 84.0 \text{ ft-kips}$$

$$J = \frac{4A^2}{\displaystyle\int ds/t} = \frac{4(72)^2}{(36/0.5)} = 288 \text{ in.}^4$$

(c) Channel section. Since for this open section,

$$\tau = \frac{Tt}{J}$$

the maximum shear stress will be in the flange. Also,

$$J = \sum \frac{bt^3}{3}$$

$$J = \tfrac{1}{3}[10(0.5)^3 + 2(5.5)(1)^3] = 4.08 \text{ in.}^4$$

$$T = \frac{J\tau}{t_f} = \frac{4.08(14)}{(1)(12)} = 4.8 \text{ ft-kips}$$

The circular section is best for torsional capacity, the rectangular box is next; these closed sections having torsional constant, J, 96 and 71 times that of the channel, respectively. The resisting moments are 19 and 18, respectively, times that of the channel.

Example 8.9.2

Compute the torsional moment capacity and the torsional constant for the multicell girder cross section of Fig. 8.9.5 if the maximum shear stress τ cannot exceed 14 ksi.

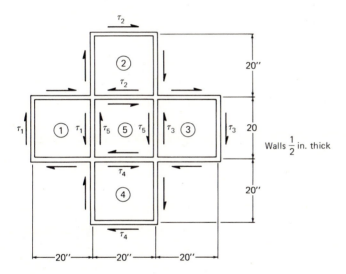

Fig. 8.9.5 Section for Example 8.9.2.

SOLUTION

From symmetry,

$$\tau_1 = \tau_2 = \tau_3 = \tau_4$$

Also, the net shear stress in the interior walls is $\tau_1 - \tau_5$.
From the compatibility of deformation, Eqs. 8.9.14 give:
For cell ①,

$$\int \tau \, ds = 3(20)\tau_1 + 20(\tau_1 - \tau_5) = 2G\theta(400)$$

For cell (5),

$$\int \tau \, ds = 4[20(\tau_5 - \tau_1)] = 2G\theta(400)$$

The above two equations are then solved for τ_1 and τ_5:

$$
\begin{aligned}
80\tau_1 - 20\tau_5 &= 800G\theta \\
-80\tau_1 + 80\tau_5 &= 800G\theta \\
\hline
60\tau_5 &= 1600G\theta
\end{aligned}
$$

$$\tau_5 = 26.67G\theta$$
$$\tau_1 = 16.67G\theta$$

The exterior wall shear stresses equal $16.67G\theta$, while the interior wall stress is $(26.67 - 16.67)G\theta$. For $\tau_{max} = 14$ ksi,

$$G\theta = \frac{14}{16.67} = 0.84$$

Using Eq. 8.9.15,

$$
\begin{aligned}
T &= \sum 2A\tau t \\
&= 4[2(400)(16.67)G\theta](0.5) + 2(400)(26.67)G\theta(0.5) \\
&= 26,700G\theta + 10,700G\theta = 37,400G\theta \\
&= 37,400(0.84)\tfrac{1}{12} = 2620 \text{ ft-kips}
\end{aligned}
$$

The torsional constant J is

$$J = \frac{T}{G\theta} = 37,400 \text{ in.}^4$$

The design of closed sections for bending and torsion is outside the scope of this text, and the reader is referred to the work of Felton and Dobbs [22].

8.10 TORSION IN SECTIONS WITH OPEN AND CLOSED PARTS

Generally this problem is treated by combining the principles discussed separately for open and closed parts. The procedure to be used for determining resisting moment, stiffness, and shear center location for such sections is presented with examples by Chu and Longinow [23]. The following is a summary of pertinent equations:

Total resisting moment is

$$T = \sum_{i=1}^{n} 2\tau_i t_i A_i + GJ\theta \qquad (8.10.1)$$

where $J = \sum \tfrac{1}{3} bt^3$ for open parts only.

In addition, each of the closed cells must satisfy Eq. 8.9.14:

$$\int_s \tau_i t_i \frac{ds}{t_i} = 2GA_i\theta \qquad (8.10.2)$$

8.11 TORSIONAL BUCKLING

Since the differential equation for torsion was developed earlier in this chapter, and buckling of axially loaded columns has previously been treated, torsional buckling may now be treated. The strength of most centrally loaded columns is reached at the tangent-modulus Euler load with a reduced efficiency if local buckling occurs before overall column buckling occurs, as discussed in Chapter 6. However, some thin-walled sections such as angles, tees, zees, and channels, with relatively low torsional stiffness may, under axial compression, buckle torsionally while the longitudinal axis remains straight.

The subject of torsional buckling is treated extensively by Timoshenko and Gere [24] and Bleich [25].

Using concepts previously developed, it is the objective here merely to illustrate how such buckling can occur and the types of situations for which the designer should be cautious.

Consider the doubly symmetrical section in the shape of a cross given in Fig. 8.11.1, whose shear center and centroid coincide. Recalling the Euler equation,

$$EI\frac{d^2y}{dz^2} + Py = 0 \qquad [6.2.3]$$

which differentiated twice becomes

$$EI\frac{d^4y}{dz^4} = -P\frac{d^2y}{dz^2} \qquad (8.11.1)$$

Since $EI\,d^4y/dz^4$ is the loading, the Euler column can be thought of as a beam laterally loaded with the fictitious loading $-P(d^2y/dz^2)$. Thus with the section put in the slightly buckled position the compressive force $\sigma_z t\,dr$ on the element $dr\,dz$ is statically equivalent to a lateral load whose intensity per unit length is

$$-(\sigma_z t\,dr)\frac{d^2(r\phi)}{dz^2}$$

The increment of torsional moment about the z-axis tributary to the length dz equals the load times the moment arm r; thus

$$dm_z = -(\sigma_z tr\,dr)\frac{r\,d^2\phi}{dz^2}\,dz \qquad (8.11.2)$$

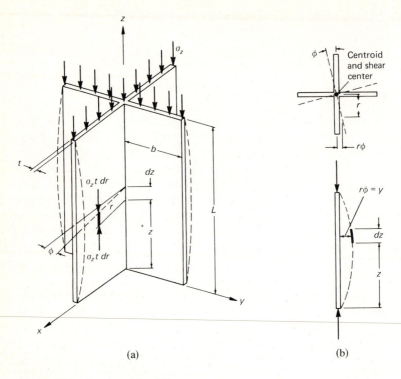

(a)   (b)

Fig. 8.11.1 Torsional buckling.

The total torsional moment for the slice dz of the column is

$$m_z = -\sigma_z \frac{d^2\phi}{dz^2} dz \int_A r^2 t \, dr \qquad (8.11.3)$$

Equation 8.11.3 represents the contribution to the torsional moment M_z tributary to the element dz:

$$dM_z = m_z \, dz \qquad (8.11.4)$$

The differential equation for torsion on I-shaped sections, Eq. 8.5.10, is

$$M_z = GJ\frac{d\phi}{dz} - EC_w\frac{d^3\phi}{dz^3} \qquad [8.5.10]$$

which when differentiated once becomes

$$\frac{dM_z}{dz} = GJ\frac{d^2\phi}{dz^2} - EC_w\frac{d^4\phi}{dz^4} \qquad (8.11.5)$$

Referring to Fig. 8.5.2, positive M_z at the section z gives a clockwise rotation; whereas in Fig. 8.11.1 at the section z there is counterclockwise rotation. Thus the Eq. 8.11.4 relationship requires a minus sign for use in Eq. 8.11.5. The differential equation for torsional buckling is then

$$\sigma_z \frac{d^2\phi}{dz^2} \int_A r^2 t \, dr = GJ \frac{d^2\phi}{dz^2} - EC_w \frac{d^4\phi}{dz^4}$$

or

$$EC_w \frac{d^4\phi}{dz^4} - \left(GJ - \sigma_z \int_A r^2 t \, dr\right) \frac{d^2\phi}{dz^2} = 0 \qquad (8.11.6)$$

in which $\int_A r^2 t \, dr = I_p$, the polar moment of inertia *about the shear center*. When the centroid coincides with the shear center, Eq. 8.11.6 alone determines the buckling condition.

It will be found (text Appendix Table A2) that the warping rigidity EC_w is zero for shapes consisting of thin rectangular elements intersecting at a common point.

For other cases, Eq. 8.11.6 may be written as

$$\frac{d^4\phi}{dz^4} + p^2 \frac{d^2\phi}{dz^2} = 0 \qquad (8.11.7)$$

when

$$p^2 = \frac{\sigma_z I_p - GJ}{EC_w}$$

for which the general solution is

$$\phi = A_1 \sin pz + A_2 \cos pz + A_3 z + A_4 \qquad (8.11.8)$$

Considering the pin-end column, with rotation about z prevented at each end, but with warping not restricted at the ends, gives in a manner similar to the Euler column derivation in Chapter 6, that

$$A_2 = A_3 = A_4 = 0$$

and since A_1 cannot also be zero,

$$\sin pL = 0, \qquad pL = n\pi$$

The elastic buckling stress at which torsional buckling occurs is

$$\frac{\pi^2}{L^2} = \frac{\sigma_z I_p - GJ}{EC_w}$$

$$\sigma_{z\,\text{critical}} = \frac{EC_w \pi^2}{I_p L^2} + \frac{GJ}{I_p} \qquad (8.11.9)$$

which is accurate for doubly symmetrical sections whose shear center and centroid coincide, such as I- and Z-shaped sections. For the common single-angle strut, since the distance from centroid to shear center is small, Eq. 8.11.9 will provide a reasonable approximation for the torsional buckling stress. Expressions for the warping constant C_w and the torsion constant J for various shapes are to be found in text Appendix Table A2.

The reader should not lose sight of the fact that the most probable buckling mode is still that occurring at the tangent-modulus Euler load because of lateral bending about the x- or y-axis. Thus the problem involves three critical values of axial load; bending about either principal axis and twisting about the longitudinal axis. On wide-flange sections, torsional buckling may be important on sections with extra wide flanges and short lengths [24].

In the general case where the shear center does not coincide with the centroid, the buckling failure is actually a combination of torsion and flexure. For this case, the three differential equations, (1) buckling by lateral bending about the x-axis; (2) buckling by lateral bending about the y-axis; and (3) twisting about the shear center, are interdependent. Thus three simultaneous differential equations must be solved to get the buckling loads. The development and solution of these equations is outside the scope of this text and is adequately treated elsewhere [24, 25].

For design purposes, since torsional buckling is frequently not considered by standard design specifications and usually may be assumed not to control, a check may be made to determine whether an equivalent radius of gyration is below r_x and r_y. Setting Eq. 8.11.9 equal to Euler's equation, defining r_E as the equivalent radius of gyration gives

$$\frac{\pi^2 E}{(L/r_E)^2} = \frac{EC_w \pi^2}{I_p L^2} + \frac{GJ}{I_p}$$

$$r_E = \sqrt{\frac{C_w}{I_p} + \frac{GJL^2}{EI_p \pi^2}}$$

which for steel with $E/G = 2.6$ gives

$$r_E = \sqrt{\frac{C_w}{I_p} + 0.04 \frac{JL^2}{I_p}} \qquad (8.11.10)$$

for doubly symmetrical sections. It has been demonstrated that only for short lengths will r_E be lower than r_x and r_y for W shapes [25].

The equivalent radius of gyration for a cross section which is symmetrical only about the y-axis is given by

$$\frac{1}{r_e^2} = \frac{1}{2r_E^2} + \frac{1}{2r_y^2} + \sqrt{\left(\frac{1}{2r_E^2} - \frac{1}{2r_y^2}\right)^2 + \left(\frac{y_0}{r_E r_y r_p}\right)^2} \qquad (8.11.11)$$

where r_e = equivalent radius of gyration for torsional-flexural buckling
r_E = equivalent radius of gyration for torsional buckling in accordance with Eq. 8.11.10
r_y = radius of gyration for axis of symmetry

r_p = polar radius of gyration = $\sqrt{I_p/A}$, where I_p = polar moment of inertia about shear center

y_0 = distance from center of gravity of cross section to its shear center

Example 8.11.1

For the sections given in Fig. 8.11.2, determine under what conditions, if any, torsional or flexural-torsional buckling is likely to occur under axial compression loading. Assume the members pinned at the ends of the unbraced lengths, and free to warp at the ends, fully recognizing that these two assumptions minimize buckling strength.

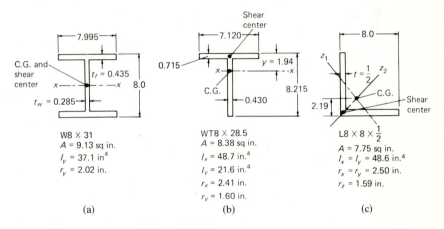

Fig. 8.11.2 Sections for Example 8.11.1.

SOLUTION

(a) W8×31. Since the centroid and shear center coincide, use Eq. 8.11.10 to get the equivalent r_E:

$$J = \tfrac{1}{3}[2(7.995)(0.435)^3 + 7.13(0.285)^3] = 0.494 \text{ in.}^4$$

(AISC Manual, $J = 0.538 \text{ in.}^4$)

$$C_w = h^2 I_y/4 = (7.565)^2(37.1)/4 = 531 \text{ in.}^6$$

(AISC Manual $C_w = 530 \text{ in.}^6$)

$$I_p = I_x + I_y = 110 + 37.1 = 147.1 \text{ in.}^4$$

$$r_E = \sqrt{\frac{C_w}{I_p} + 0.04\frac{JL^2}{I_p}} = \sqrt{\frac{531}{147.1} + 0.04\frac{0.494L^2(144)}{147.1}}$$

$$= \sqrt{3.61 + 0.0193L^2}$$

where L is the unsupported length, in feet. Only when L is less than 4.9 ft does r_y exceed r_E; only for a very short column is torsional buckling a possibility, and even when r_E approaches its minimum value (when $L=0$), which is $r_E(\text{min}) = 1.90$ in., it is only 6% less than r_y.

The result using this section is typical of standard W and S shapes and indicates torsional buckling may properly be neglected for them.

(b) WT8$\times$28.5. The centroid and shear center do not coincide, but the section has one axis of symmetry; using Eq. 8.11.11:

$$J = \tfrac{1}{3}[(7.12)(0.715)^3 + 7.50(0.30)^3] = 1.066 \text{ in.}$$

Referring to text Appendix Table A2,

$$C_w = \frac{1}{36}\left[\frac{(7.120)^3(0.715)^3}{4} + (7.50)^3(0.430)^3\right] = 1.85 \text{ in.}^6$$

which is actually small enough to be neglected.

$$I_p = I_x + I_y + Ay_0^2 = 48.7 + 21.6 + 8.38(1.37)^2 = 86.0 \text{ in.}^4$$

For use in Eq. 8.11.11, $y_0 = 1.37$ in., $r_y = 1.60$ in.

$$r_p = \sqrt{I_p/A} = \sqrt{86.0/8.38} = 3.20 \text{ in.}$$

$$r_E = \sqrt{\frac{1.85}{86.0} + \frac{0.04(1.066)L^2(144)}{86.0}} = \sqrt{0.0215 + 0.0714L^2}$$

with L in feet.

Suppose $L = 10$ ft, $r_E = 2.68$ in. Equation 8.11.11 becomes

$$\frac{1}{r_e^2} = \frac{1}{2r_E^2} + \frac{1}{2r_y^2} + \sqrt{\left(\frac{1}{2r_E^2} - \frac{1}{2r_y^2}\right)^2 + \left(\frac{y_0}{r_E r_y r_p}\right)^2}$$

$$= \frac{1}{14.32} + \frac{1}{5.12} + \sqrt{\left(\frac{1}{14.32} - \frac{1}{5.12}\right)^2 + \left(\frac{1.37}{2.68(1.60)(3.20)}\right)^2}$$

$$r_e = 1.53 \text{ in.}$$

In this case r_e is less than r_y and flexural-torsional buckling is critical for ordinary lengths.

(c) L8$\times$8$\times\tfrac{1}{2}$. For this section, Eq. 8.11.11 may be used with the z_2 axis of symmetry.

$$J = \tfrac{1}{3}(2)(7.75)(0.5)^3 = 0.646 \text{ in.}^4$$

$$C_w = \text{neglect}$$

$$I_p = 2I_x + 2Ay_0^2 = 2(48.6) + 2(7.75)(2.74)^2 = 213.6 \text{ in.}^4$$

$$r_p = \sqrt{213.6/7.75} = 5.25 \text{ in.}$$

$$x = y = 2.19 \text{ in.}, \quad y_0 = (2.19 - 0.25)/0.707 = 2.74 \text{ in.}$$

$$r_{z_2} = \sqrt{2(2.50)^2 - (1.58)^2} = 3.16 \text{ in.}$$

$$r_E = \sqrt{0.04 \frac{0.646}{213.6} L^2(144)} = \sqrt{0.0174L^2} = 0.132L$$

with L in feet.

Let $L = 10$ ft, $r_E = 1.32$ in. Equation 8.11.11 then becomes

$$\frac{1}{r_e^2} = \frac{1}{2(1.32)^2} + \frac{1}{2(3.16)^2} + \sqrt{\left(\frac{1}{2(1.32)^2} - \frac{1}{2(3.16)^2}\right)^2 + \left(\frac{2.74}{1.32(3.16)(5.25)}\right)^2}$$

$$r_e = 1.29 \text{ in.}$$

Thus, for this angle r_e is less than $r_{z_1} = 1.59$ in., so that flexural-torsional buckling controls for the 10-ft length. In general, the single-angle strut is controlled by flexural-torsional buckling; it is for this reason that caution concerning its use is given in the AISC Manual. Usami and Galambos [26] have given a detailed treatment of the related problem of an eccentrically loaded single angle strut.

As a conclusion to this treatment, the designer is cautioned about using open sections in compression having less than two axes of symmetry, particularly when high width-to-thickness ratios for the elements exist. AISC–1.9.1 on local buckling provides some control over the more critical cases, since local buckling of sections such as angles, flanges, and tees is closely related to torsional buckling. AISC Specification Appendix Table C1 gives limiting proportions for channels and tees to preclude torsional buckling.

SELECTED REFERENCES

1. Fred B. Seely and James O. Smith, *Advanced Mechanics of Materials*, 2nd ed., John Wiley and Sons, Inc., New York, 1952, Chap. 9.
2. S. Timoshenko, *Strength of Materials*, Part II, 2nd ed., D. Van Nostrand Company, Inc., New York, 1941, Chap. 6.
3. William McGuire, *Steel Structures*, Prentice-Hall, Inc., Englewood Cliffs, N.J., 1968, pp. 346–400.
4. I. Lyse and B. G. Johnston, "Structural Beams in Torsion," *Transactions*, ASCE, 101 (1936), 857–926.
5. F. K. Chang and Bruce G. Johnston, "Torsion of Plate Girders," *Transactions*, ASCE, 118 (1953), 337–396.
6. Gerald G. Kubo, Bruce G. Johnston, and William J. Erney, "Nonuniform Torsion of Plate Girders," *Transactions*, ASCE, 121 (1956), 759–785. (Good summary of torsion theory.)
7. I. A. El Darwish and Bruce G. Johnston, "Torsion of Structural Shapes," *Journal of Structural Division*, ASCE, 91, ST 1 (February 1965), 203–227. (Errata: ST 1 (February 1966), 471.) See also *Transactions*, ASCE, 131 (1966), 428–429 for summary of equations.

8. C. P. Heins, Jr. and P. A. Seaburg, *Torsion Analysis of Rolled Steel Sections*, Handbook 1963, Bethlehem Steel Corp., 1963.

9. John G. Hotchkiss, "Torsion of Rolled Steel Sections in Building Structures," *Engineering Journal*, AISC, 3, 1 (January 1966), 19–45.

10. Conrad P. Heins, Jr. and John T. C. Kuo, "Torsional Properties of Composite Girders," *Engineering Journal*, AISC, 9, 2 (April 1972), 79–85.

11. S. Timoshenko, "Theory of Bending, Torsion , and Buckling of Thin-Walled Members of Open Cross Section," *J. Franklin Inst.*, 239, 3, 4, and 5 (1945), 201–219; 249–268, and 343–361.

12. John E. Goldberg, "Torsion of I-Type and H-Type Beams," *Transactions*, ASCE, 118 (1953), 771–793.

13. Kuang-Han Chu and Robert B. Johnson, "Torsion in Beams with Open Sections," *Journal of Structural Division*, ASCE, 100, ST7 (July 1974), 1397–1419.

14. J. N. Goodier and M. V. Barton, "The Effects of Web Deformation on the Torsion of I-Beams," *J. Appl. Mech.*, (March 1944), p. A–35.

15. J. E. Lothers, "Torsion in Steel Spandrel Girders," *Transactions*, ASCE, 112 (1947), 345–376.

16. Porpan Vacharajittiphan and Nicholas S. Trahair, "Warping and Distortion at I-Section Joints," *Journal of Structural Division*, ASCE, 100, ST3 (March 1974), 547–564.

17. Warner Lansing, "Thin-Walled Members in Combined Torsion and Flexure," *Transactions*, ASCE, 118 (1953), 128–146. (Particular emphasis on channel and zee sections.)

18. Donald R. Evick and Conrad P. Heins, Jr., "Torsion of Nonprismatic Beams of Open Section," *Journal of Structural Division*, ASCE, 98, ST12 (December 1972), 2769–2784.

19. Basil Sourochnikoff, "Strength of I-Beams in Combined Bending and Torsion," *Transactions*, ASCE, 116 (1951), 1319–1342.

20. K. G. Tamberg and P. T. Mikluchin, "Torsional Phenomena Analysis and Concrete Structure Design," *Analysis of Structural Systems for Torsion*, SP-35, American Concrete Institute, 1973, 1–102.

21. Avinadav Siev, "Torsion in Closed Sections," *Engineering Journal*, AISC, 3, 1 (January 1966), 46–54.

22. Lewis P. Felton and M. W. Dobbs, "Optimum Design of Tubes for Bending and Torsion," *Journal of Structural Division*, ASCE, 93, ST4 (August 1967), 185–200.

23. Kuang-Han Chu and Anatole Longinow, "Torsion in Sections with Open and Closed Parts," *Journal of Structual Division*, ASCE, 93, ST6 (December 1967), 213–227.

24. S. P. Timoshenko and J. M. Gere, *Theory of Elastic Stability*, 2nd ed. New York: McGraw-Hill Book Company Inc., 1961, pp. 225–250.

25. Friedrich Bleich, *Buckling Strength of Metal Structures*. New York: McGraw-Hill Book Company, Inc., 1952.

26. Tsutomu Usami and Theodore V. Galambos, "Eccentrically Loaded Single Angle Columns," *Publications*, International Association for Bridge and Structural Engineering, 31-II (1971), 153–184.

PROBLEMS

8.1. For the channel shown in the accompanying figure, separately apply V_x and V_y through the centroid of the section. For each shear compute and draw to scale the shear flow τt distribution along each of the elements of the cross section. On the two separate diagrams of shear flow distribution compute the total shear force in each element of the cross section in terms of the applied shear V_x or V_y. Using these computed shear forces calculate the two coordinates of the shear center.

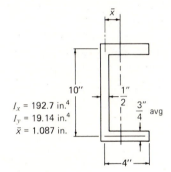

$I_x = 192.7 \text{ in.}^4$
$I_y = 19.14 \text{ in.}^4$
$\bar{x} = 1.087 \text{ in.}$

Prob. 8.1

8.2. Repeat the requirements of Prob. 8.1 for the channel with sloping flanges. Comment on the effect of using average thickness instead of the actual sloping flanges for determining shear center on standard rolled channels.

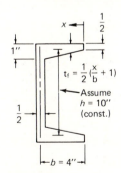

$t_f = \frac{1}{2}\left(\frac{x}{b} + 1\right)$

Assume $h = 10''$ (const.)

$b = 4''$

Prob. 8.2

8.3. Repeat the requirements of Prob. 8.1 for the channel with unequal flanges as shown.

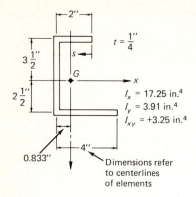

$l_x = 17.25$ in.^{4}
$l_y = 3.91$ in.^{4}
$l_{xy} = +3.25$ in.^{4}

Dimensions refer
to centerlines
of elements

Prob. 8.3

8.4. Repeat the requirements of Prob. 8.1 for the angle section of the accompanying figure.

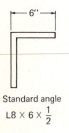

Standard angle
L8 × 6 × $\frac{1}{2}$

Prob. 8.4

8.5. Repeat the requirements of Prob. 8.1 for the zee section of the accompanying figure.

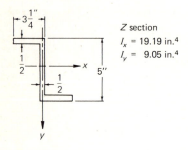

Z section
$l_x = 19.19$ in.^{4}
$l_y = 9.05$ in.^{4}

Prob. 8.5

8.6. Locate the shear center for the combined W and channel crane girder section. Is there significant error in assuming the shear center lies at the centroid? Use average thickness and constant depth for the channel.

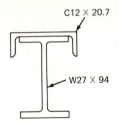

Prob. 8.6

8.7. An MC18×58 is to be used on a 24-ft simply supported span to carry a total load of 0.8 kips/ft, with the load applied in the plane of the web. Suppose the flanges are to have attachments so that the channel will bend vertically about the x-axis. What lateral moment capacity M_y should the attachments be capable of resisting? What percent of M_x does this represent?

8.8. (a) Develop the torsion differential equation solution for the W section with torsionally fixed ends, having an eccentrically applied concentrated load at midspan.

    (b) Compute the torsion constant J, the warping constant C_w, and λ.

    (c) Compute the combined bending stress, including warping torsion and ordinary flexure components.

    (d) Compute the maximum shear stress in the web, including the Saint-Venant torsion, and flexural shear.

    (e) Compute the maximum shear stress in the flange, including Saint-Venant torsion, warping shear, and vertical flexure flange shear. Give a tabular summary of all stresses.

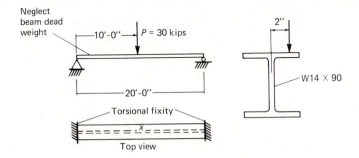

Prob. 8.8

8.9. Develop the torsion differential equation solution for the cantilever beam with an eccentric concentrated load at its end, and compute constants and stresses as given in items (b) through (e) of Prob. 8.8. Is there any relationship between this problem and Prob. 8.8?

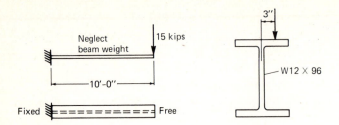

Prob. 8.9

8.10. Develop the torsion differential equation solution for the uniformly loaded beam with loading applied eccentrically to the web. Consider the ends torsionally simply supported. Compute constants and stresses as given in items (b) through (e) of Prob. 8.8.

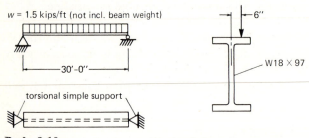

Prob. 8.10

8.11. Repeat Prob. 8.10, considering the ends torsionally fixed.

8.12. Repeat Prob. 8.8, with the load applied in the plane of the y-axis of a channel, C15×50.

8.13. Select a W14 section, using the β modified flexure analogy approach, to carry a 25-kip concentrated load at midspan of a 20-ft simply supported span. The 25-kip load is eccentric to the web by 4 in. and the allowable combined compressive stress is 22 ksi. Allowable shear stress is 14.4 ksi. The ends of the beam are assumed to have torsional simple support. Check the stresses in the selected section using the "exact" solution for assumed conditions as developed in Example 8.5.2.

8.14. Select the lightest W section to carry uniform superimposed loading of 2.0 kips/ft, acting 7-in. eccentric to the plane of the web, on a fixed end (both for vertical and torsional loading) span of 28 ft. Use the β modified flexure analogy method with A36 steel where the maximum compressive stress is limited to 22 ksi. If "exact" solution was obtained in Prob. 8.11, check compressive stress also by that equation.

8.15. Select the lightest W section to carry a superimposed uniform loading of 1.75 kips/ft, acting 5 in. eccentric to the plane of the web, on a simply supported (both for vertical and torsional loading) span of 26 ft. Use the β modified flexure analogy method with A572 Grade 50 steel and assume the maximum compressive stress may not exceed 30 ksi. If "exact" solution was obtained in Prob. 8.10, check compressive stress also by that equation.

8.16. Repeat Prob. 8.15, assuming the ends fixed with respect to torsional restraint only. If "exact" solution was obtained in Prob. 8.11, check compressive stress also by that equation.

8.17. Repeat Prob. 8.15 using A36 steel with allowable compressive stress of 22 ksi. If A572 Grade 50 steel costs 7 percent more per pound that A36 steel, is it economical to use the higher strength material?

8.18. Given the 40-ft simply supported span carrying two symmetrically placed concentrated loads as shown in the accompanying figure. If the loads are eccentric to the web by 5 in., and the member is torsionally simply supported, select the lightest W14 section suitable using the β modified flexure analogy method, assuming the maximum combined stress is limited to 22 ksi.

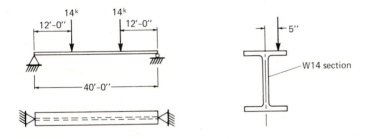

Prob. 8.18

8.19. The 30-ft simply supported (for M_x) span is to carry two symmetrically placed concentrated loads of 22 kips located 10 ft from the supports. The loads are 6-in. eccentric to the web, and full fixity is assumed for torsional restraint. Select, using the β modified flexure analogy method, the lightest W section such that the maximum combined stress does not exceed 22 ksi.

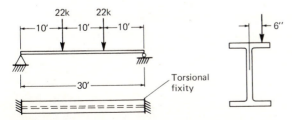

Prob. 8.19

8.20. Estimate, using the modified flexure analogy design procedure used for W shapes and crane girders, the uniformly distributed load capacity for a C15×50 of A36 steel. Assume that the allowable compressive stress is $0.60F_y$. The ends of the beam are restrained against rotation ($\phi = 0$), but the ends are free to warp. Check stresses using the differential equation solution.

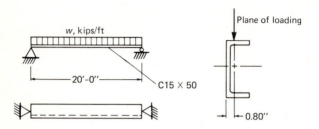

Prob. 8.20

8.21. A simply supported beam is to carry 0.8 kip/ft on a span of 24 ft. A channel section is to be used, and since it is not laterally restrained, torsion is to be considered in the design. Assume the member is torsionally simply supported. After computing the torsional moment, use the generally accepted approximate flexure analogy to make a selection of an American Standard channel (C-section) from the AISC Manual. Assume the allowable combined stress to be $0.60F_y$. Investigate the stress on the selected section using the exact solution based on the stated loading and support condition. Use A36 steel.

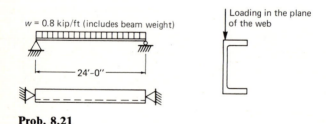

Prob. 8.21

8.22. Assume a single W section is to serve as a crane runway girder which carries a vertical loading, as shown. In addition, design must include an axial compressive force of 10 kips and a horizontal force of 3 kips on each wheel applied at 3 in. above the top of the compression flange. Assume torsional simple support at the ends of the beam. Select the lightest W14 section of A36 steel using the β modified flexure analogy approach, limiting the maximum combined compressive stress to 22 ksi.

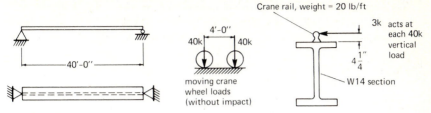

Prob. 8.22

8.23. Compare the torsional constants J for each of the following sections. If the maximum shear stress is 14 ksi, compute the torsional moment capacity of each section. Can this be done for the W30× 99? Explain.

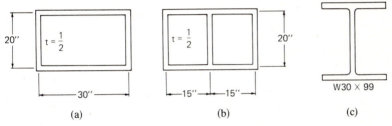

Prob. 8.23

8.24. If the maximum shear stress is limited to 14 ksi, compute the torsional moment capacity for the section given. What is the percentage change in capacity if the interior walls are omitted (assuming local buckling does not control)?

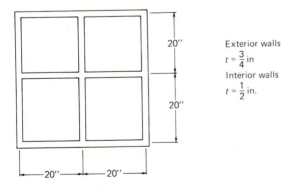

Exterior walls
$t = \dfrac{3}{4}$ in

Interior walls
$t = \dfrac{1}{2}$ in.

Prob. 8.24

8.25. Investigate the possibility of torsional buckling occurring on the following beams: W16×31. W6×16, M8×6.5. For each, plot equivalent slenderness ratio, r_E, against length of column. Give conclusions.

8.26. Estimate the buckling load, assuming zero residual stress and elastic buckling so that Euler's equation applies, for the following sections on a pinned-end length of 10 ft. At what length would torsional buckling be likely to control?
(a) MC10×6.5
(b) L4×4×$\frac{1}{4}$
(c) WT7×15
(d) Zee section shown

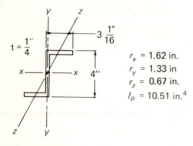

r_x = 1.62 in.
r_y = 1.33 in
r_z = 0.67 in.
I_p = 10.51 in.^{4}

Prob. 8.26

9
Lateral-Torsional Buckling of Beams

9.1 RATIONAL ANALOGY TO PURE COLUMNS

Emphasis in this chapter is on the lateral stability considerations associated with bending about the strong axis of the section. In beams, as in axially loaded columns, it is not possible to achieve perfect loading, i.e., beams are never perfectly straight, not perfectly homogeneous, and are usually not loaded in exactly the plane that is assumed for design and analysis.

Consider the compression zone of the laterally unsupported beam of Fig. 9.1.1. With the loading in the plane of the web, according to ordinary beam theory, points A and B are equally stressed. Imperfections in the beam and accidental eccentricity in loading actually result in different stresses at A and B. Furthermore, residual stresses as discussed in Chapter 6 contribute to unequal stresses across the flange width at any distance from the neutral axis.

In a qualitative way one may look upon the compression flange of a beam as a column, with all the considerations treated in Chapter 6. The rectangular flange as a column would ordinarily buckle in its weak direction, by bending about an axis such as 1–1 of Fig. 9.1.1b, but the web provides continuous support to prevent such buckling. At higher

Large girder with shear stud connectors on top flange to be embedded in concrete to provide continuous lateral support (see Fig. 9.2.1b). (Photo by C. G. Salmon)

compressive loads the rectangular flange will tend to buckle by bending about axis 2–2 of Fig. 9.1.1b. It is this sudden buckling of the flange about its strong axis in a lateral direction that is commonly referred to as lateral buckling. The analogy between the compression flange of a beam and a column is intended to present only the general behavior for lateral buckling.

In order to evaluate this behavior more precisely, one must realize that the compression flange is not only braced in its weak direction by its

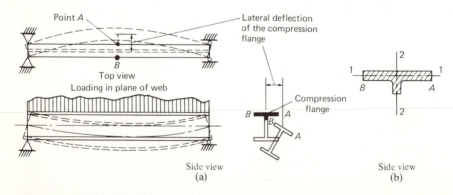

Fig. 9.1.1 Beam laterally supported only at its ends.

attachment via the web to the stable tension flange, but the web also provides continuous moment and shear restraint along the junction of the flange and web. Thus the bending stiffness of the web brings the entire section into action when lateral motion commences.

9.2 LATERAL SUPPORT

Rarely does a beam exist with its compression flange entirely free of all restraint. Even when it does not have a positive connection to a floor or roof system, there is still friction between the beam flange and whatever it supports. There are two categories of lateral support that are definite and adequate; these are:

1. Continuous lateral support by embedment of the compression flange in a concrete floor slab (Fig. 9.2.1a and b).
2. Lateral support at intervals (Fig. 9.2.1c through g) provided by cross beams, cross frames, ties, or struts, framing in laterally, where the lateral system is itself adequately stiff and braced.

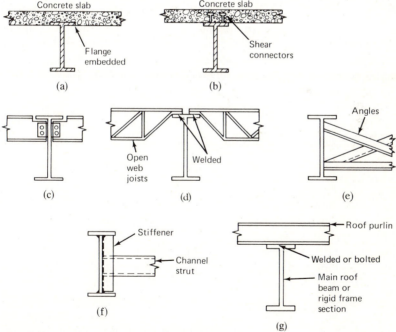

Fig. 9.2.1 Types of definite lateral support.

It is necessary to examine not only the individual beam for adequate bracing, but also the entire system. Figure 9.2.2a shows beam *AB* with a cross beam framing in at midlength, but buckling of the entire system is still possible unless the system is braced, such as shown in Fig. 9.2.2b.

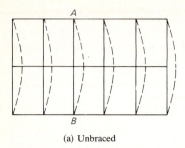

(a) Unbraced

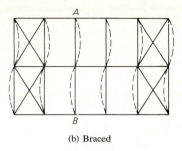

(b) Braced

Fig. 9.2.2 Lateral buckling of a roof or floor system.

All too frequently in design, the engineer encounters situations that are none of these well-defined cases. A common unknown situation occurs when heavy beams have light-gage steel decking spot welded to them; certainly providing a degree of restraint all along the member. However, the stiffness and lateral strength may be questioned. Other questionable cases are (a) where bracing frames into the beam in question, but it is at or near the tension flange; (b) timber or light-gage decking floor systems that rest on but are not solidly attached to the beams; (c) rigid frames that are enclosed in light gage metal sheathing.

However, it is better to assume no lateral support in doubtful situations. Alternatively, it may be possible in some cases to evaluate it as an elastic restraint. Such an analytical approach is discussed in Sec. 9.10.

Lateral support considerations must not be ignored; probably most failures in steel structures are the result of inadequate bracing against lateral instability of some type. The engineer is also reminded to consider carefully the construction stage when all of the restraints which may eventually act are not yet in place.

9.3* I-SHAPED BEAMS UNDER UNIFORM MOMENT

Specifications generally use the situation of uniform moment on a laterally unbraced segment as the basic lateral-buckling situation. Using the analogy to the column, the case of uniform moment creates constant compression in one flange over the entire unbraced length, while in other loading situations the compression stress varies along the unbraced length, giving a lower average compression force and less likelihood of buckling.

Because of the importance of stability in structural design, the authors believe it desirable to show the development of the basic equation for elastic lateral-torsional buckling of I-shaped beams under pure moment. More detailed treatment of this case as well as other common loading cases is to be found in the work of Timoshenko [1], Bleich [2], deVries [3], Hechtman et al. [4], Austin et al. [5], Clark and Jombock [6], and Salvadori [7].

Differential Equation for Elastic Lateral Buckling

Referring to Fig. 9.3.1, which shows the beam in a buckled position, it is observed that the applied moment M_0 in the yz plane will give rise to moment components $M_{x'}$, $M_{y'}$, and $M_{z'}$, about the x', y', and z'-axes, respectively. This means there will be bending curvature in both the $x'z'$ and $y'z'$ planes as well as torsional curvature about the z'-axis. Assuming small deformation, the bending in the $y'z'$ plane (considering the direction cosine is 1 between y'- and y-, and z'- and z-axes) may be written

$$EI_x \frac{d^2v}{dz^2} = M_{x'} = M_0 \qquad (9.3.1)$$

where v is the displacement of the centroid in the y direction (see Fig. 9.3.1b).

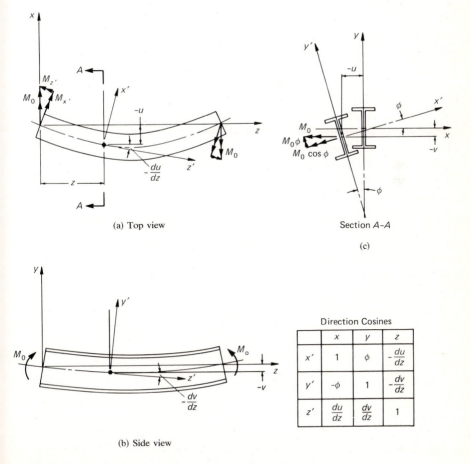

(a) Top view

Section A–A

(c)

(b) Side view

Direction Cosines

	x	y	z
x'	1	ϕ	$-\dfrac{du}{dz}$
y'	$-\phi$	1	$\dfrac{dv}{dz}$
z'	$\dfrac{du}{dz}$	$\dfrac{dv}{dz}$	1

Fig. 9.3.1 I-shaped beam in slightly buckled position.

Also, the curvature in the $x'z'$ plane is

$$EI_y \frac{d^2u}{dz^2} = M_{y'} = M_0 \phi \tag{9.3.2}$$

as is seen from Fig. 9.3.1c, where u is the displacement of the centroid in the x direction.

The differential equation for torsion of I-shaped beams was developed in Chapter 8 as Eq. 8.5.10, as follows:

$$M_{z'} = GJ \frac{d\phi}{dz} - EC_w \frac{d^3\phi}{dz^3} \tag{8.5.10}$$

From Fig. 9.3.1 and the direction cosines, the torsional component of M_0 when the beam is slightly buckled is proportional to the slope of the beam in the xz plane:

$$M_{z'} = -\frac{du}{dz} M_0 \tag{9.3.3}$$

which gives for the torsional differential equation

$$-\frac{du}{dz} M_0 = GJ \frac{d\phi}{dz} - EC_w \frac{d^3\phi}{dz^3} \tag{9.3.4}$$

Two assumptions are inherent in Eqs. 9.3.1 and 9.3.2, both of which relate to the assumption of small deformation. It is assumed that properties $I_{x'}$, and $I_{y'}$, equal I_x and I_y, respectively; and also that I_x is large compared to I_y, so that Eq. 9.3.1 is not linked to Eqs. 9.3.2 and 9.3.4. Thus displacement v in the plane of bending does not affect the torsional function ϕ.

Differentiating Eq. 9.3.4 with respect to z gives

$$-\frac{d^2u}{dz^2} M_0 = GK \frac{d^2\phi}{dz^2} - EC_w \frac{d^4\phi}{dz^4} \tag{9.3.5}$$

From Eq. 9.3.2,

$$\frac{d^2u}{dz^2} = \frac{M_0 \phi}{EI_y}$$

which when substituted into Eq. 9.3.5 gives

$$EC_w \frac{d^4\phi}{dz^4} - GJ \frac{d^2\phi}{dz^2} - \frac{M_0^2}{EI_y} \phi = 0 \tag{9.3.6}$$

which is the differential equation for the angle of twist.

To obtain a solution for Eq. 9.3.6, divide by EC_w and let

$$2\alpha = \frac{GJ}{EC_w} \quad \text{and} \quad \beta = \frac{M_0^2}{E^2 C_w I_y} \tag{9.3.7}$$

Equation 9.3.6 then becomes

$$\frac{d^4\phi}{dz^4} - 2\alpha \frac{d^2\phi}{dz^2} - \beta\phi = 0 \qquad (9.3.8)$$

Let

$$\left.\begin{array}{l} \phi = Ae^{mz} \\[2mm] \dfrac{d^2\phi}{dz^2} = Am^2 e^{mz} \\[2mm] \dfrac{d^4\phi}{dz^4} = Am^4 e^{mz} \end{array}\right\} \qquad (9.3.9)$$

Substitution of Eq. 9.3.9 into Eq. 9.3.8 gives

$$Ae^{mz}(m^4 - 2\alpha m^2 - \beta) = 0 \qquad (9.3.10)$$

Since e^{mz} cannot be zero and A can be zero only if no buckling has occurred, the bracket expression of Eq. 9.3.10 must be zero:

$$m^4 - 2\alpha m^2 - \beta = 0$$

which gives for the solution

$$m^2 = \alpha \pm \sqrt{\beta + \alpha^2}$$

or

$$m = \pm\sqrt{\alpha \pm \sqrt{\beta + \alpha^2}} \qquad (9.3.11)$$

It is apparent from Eq. 9.3.11 that m will consist of two real and two complex roots because

$$\sqrt{\beta + \alpha^2} > \alpha$$

Let

$$n^2 = \alpha + \sqrt{\beta + \alpha^2} \quad \text{(both real roots)} \qquad (9.3.12)$$
$$q^2 = -\alpha + \sqrt{\beta + \alpha^2} \quad \text{(real part of complex roots)} \qquad (9.3.13)$$

Using the four values for m, the expression for ϕ from Eq. 9.3.9 becomes

$$\phi = A_1 e^{nz} + A_2 e^{-nz} + A_3 e^{iqz} + A_4 e^{-iqz} \qquad (9.3.14)$$

The complex exponential functions may be expressed in terms of circular functions,

$$\left.\begin{array}{l} e^{iqz} = \cos qz + i \sin qz \\[2mm] e^{-iqz} = \cos qz - i \sin qz \end{array}\right\} \qquad (9.3.15)$$

By using Eqs. 9.3.15 and defining new constants A_3 and A_4 which equal $(A_3 + A_4)$ and $(A_3 i - A_4 i)$, respectively, one obtains

$$\phi = A_1 e^{nz} + A_2 e^{-nz} + A_3 \cos qz + A_4 \sin qz \qquad (9.3.16)$$

The constants A_1 through A_4 are determined by the end support conditions. For the case of torsional simple support, i.e., beam ends may not twist, but are free to warp, the conditions are

$$\phi = 0, \qquad \frac{d^2\phi}{dz^2} = 0 \qquad \text{at } z = 0 \quad \text{and} \quad z = L$$

For $\phi = 0$ at $z = 0$, Eq. 9.3.16 gives

$$0 = A_1 + A_2 + A_3 \tag{9.3.17}$$

For $d^2\phi/dz^2 = 0$ at $z = 0$,

$$0 = A_1 n^2 + A_2 n^2 - A_3 q^2 \tag{9.3.18}$$

Multiplying Eq. 9.3.17 by n^2 and subtracting Eq. 9.3.18 gives

$$0 = A_3(q^2 + n^2), \qquad \therefore A_3 = 0$$

Then, from Eq. 9.3.17,

$$A_1 = -A_2 \tag{9.3.19}$$

Thus Eq. 9.3.16 becomes

$$\phi = A_1(e^{nz} - e^{-nz}) + A_4 \sin qz \tag{9.3.20}$$

which may be written

$$\phi = 2A_1 \sinh nz + A_4 \sin qz \tag{9.3.21}$$

At $z = L$, $\phi = 0$; therefore, from Eq. 9.3.21,

$$0 = 2A_1 \sinh nL + A_4 \sin qL \tag{9.3.22}$$

Also, at $z = L$, $d^2\phi/dz^2 = 0$, which gives

$$0 = 2A_1 n^2 \sinh nL - A_4 q^2 \sin qL \tag{9.3.23}$$

Multiplying Eq. 9.3.22 by q^2 and adding to Eq. 9.3.23 gives

$$2A_1(n^2 + q^2) \sinh nL = 0 \tag{9.3.24}$$

Since $(n^2 + q^2)$ cannot be zero, and $\sinh nL$ can be zero only if $n = 0$, therefore A_1 must be zero:

$$A_1 = -A_2 = 0$$

Finally, from Eqs. 9.3.21 and 9.3.22,

$$\phi = A_2 \sin qL = 0 \tag{9.3.25}$$

If lateral-torsional buckling occurs, A_4 cannot be zero, so that

$$\sin qL = 0 \tag{9.3.26}$$

$$qL = N\pi$$

where N is any integer.

The elastic-buckling condition is defined by

$$q = \frac{N\pi}{L} \tag{9.3.27}$$

which for the fundamental buckling mode $N = 1$.

The value of M_0 which satisfies Eq. 9.3.27 is said to be the critical moment:

$$q = \sqrt{-\alpha + \sqrt{\beta + \alpha^2}} = \frac{\pi}{L} \tag{9.3.28}$$

Squaring both sides, and substituting the definitions of α and β, from Eqs. 9.3.7,

$$-\frac{GJ}{2EC_w} + \sqrt{\frac{M_0^2}{E^2 C_w I_y} + \left(\frac{GJ}{2EC_w}\right)^2} = \frac{\pi^2}{L^2} \tag{9.3.29}$$

Solving for $M_0 = M_{cr}$ gives

$$M_{cr}^2 = E^2 C_w I_y \left[\left(\frac{\pi^2}{L^2} + \frac{GJ}{2EC_w} \right)^2 - \left(\frac{GJ}{2EC_w} \right)^2 \right] \tag{9.3.30}$$

$$M_{cr} = \sqrt{\frac{\pi^4 E^2 C_w I_y}{L^4} + \frac{\pi^2 EI_y GJ}{L^2}} \tag{9.3.31}$$

Equation 9.3.31 may be converted to unit stress by dividing by the section modulus, S_x,

$$F_{cr} = \sqrt{\frac{\pi^4 E^2 C_w I_y}{S_x^2 L^4} + \frac{\pi^2 EI_y GJ}{S_x^2 L^2}} \tag{9.3.32}$$

Factoring out the second term, Eq. 9.3.32 becomes

$$F_{cr} = \frac{\pi \sqrt{EI_y GJ}}{S_x L} \sqrt{\frac{\pi^2 EC_w}{L^2 GJ} + 1} \tag{9.3.33}$$

which may also be expressed using the symbol C_4 of the *SSRC Guide* [8],

$$F_{cr} = \frac{C_4 \sqrt{EI_y GJ}}{S_x L} \tag{9.3.34}$$

where after substituting $\lambda^2 = 1/a^2 = GJ/EC_w$,

$$C_4 = \pi \sqrt{1 + \left(\frac{\pi}{\lambda L} \right)^2} \tag{9.3.35}$$

Equation 9.3.35 is the C_4 expression for the case of uniform moment; the general expression for other cases is given in Sec. 9.4.

Lateral Buckling in the Plastic Range

When design procedures consider moments greater than $M_y = SF_y$ and up to M_p, the plastic moment, lateral buckling is a distinct possibility. Both working stress and plastic design procedures inherently require lateral bracing at locations where plastic hinges are expected to occur in the failure mechanism. Upon reaching a plastic hinge at any section, the extreme fibers will be strained near or into the strain hardening region.

Studies of inelastic lateral buckling have been made by Galambos [9], Lay and Galambos [10, 11], Massey and Pittman [12], Bansal [14], and Hartmann [15]. The current (1978) status is summarized by Yura et al. [56].

If the rigidities EI_y and GJ are taken to include the values in the inelastic range as well as the elastic range, the lateral-torsional buckling equilibrium equation for pure moment, Eq. 9.3.31, may also be used for the plastic range. Since for beams where plastic moments are assumed to develop, the distances between lateral support points will be relatively short, it has been determined [10] that the term involving torsional rigidity GJ may be neglected. Thus Eq. 9.3.31, neglecting the second term, becomes

$$M_{cr} = \frac{\pi^2 E}{L^2} \sqrt{C_w I_y} \tag{9.3.36}$$

Since M_{cr} must reach M_p, substitute $M_p = ZF_y$ for M_{cr}. Also $C_w = I_y h^2/4$ and $I_y = Ar_y^2$. Solving Eq. 9.3.36 then gives the maximum slenderness ratio,

$$\frac{L}{r_y} = \sqrt{\frac{\pi^2 E}{2F_y}\left(\frac{hA}{Z}\right)} \tag{9.3.37}$$

for uniform plastic moment, as in Fig. 9.3.2. The exteme fiber strain will be approaching or already into the strain hardening range; thus the strain hardening modulus should be used instead of the modulus of elasticity E. Equation 9.3.37 then becomes

$$\frac{L}{r_y} = \sqrt{\frac{\pi^2 E_{st}}{2F_y}} \sqrt{\frac{hA}{Z}} \tag{9.3.38}$$

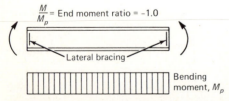

Fig. 9.3.2   Uniform plastic moment over unbraced length.

Examination of the cross-sectional dimension term $\sqrt{hA/Z}$ indicates this term varies from about 1.5 to 1.6.

For situations where there is variation of moment along the unbraced length, including double curvature, Eq. 9.3.38 may be multiplied by factors to correct for other variables.

$$\frac{L}{r_y} = \sqrt{\frac{\pi^2 E_{st}}{2F_y}} \sqrt{\frac{hA}{Z}} \, \xi_\rho \xi_\alpha \xi_s \xi_\gamma \qquad (9.3.39)$$

where ξ_ρ = moment gradient correction, which from the *Commentary on Plastic Design* [16] may be approximated as $1.34 + 0.24\, M/M_p$
ξ_α = partial yielding correction; varies from about 1.8 for $M/M_p = 0$ to about 1.3 for $M/M_p = -0.7$
ξ_s = correction for torsional stiffness (pure torsion); approximately $1.08 + 0.04\, M/M_p$
ξ_γ = correction for end fixity, say average is 1.25 for continuous beams with constant EI and equal unbraced segment lengths

Using the approximations indicated, and taking $\sqrt{hA/Z} = 1.60$, Eq. 9.3.39 becomes

$$\frac{L}{r_y} = \sqrt{\frac{\pi^2 E}{2F_y}} \sqrt{\frac{E_{st}}{E}} (1.60)\left(3.3 + 2.2\frac{M}{M_p}\right) \qquad (9.3.40)$$

Simplification of Eq. 9.3.40 requires a value for the strain hardening modulus of elasticity E_{st}. Unlike the elastic modulus E, E_{st} is not constant for all yield strengths. Lay and Galambos [10] used $E_{st} = E/F_y$ in ksi, giving 805 ksi for A36 steel and 446 ksi for $F_y = 65$ ksi. Lukey and Adams [13] established $E_{st} = 450$ ksi from a statistical study where the mean value was 600 ksi and the standard deviation was 150 ksi. Using the mean value less one standard deviation indicates that approximately 85 percent of the time E_{st} could be expected to exceed 450 ksi.

Using $E_{st} = (E/F_y)$ ksi and $E = 29,000$ ksi, Eq. 9.3.40 becomes

$$\frac{L}{r_y} = \frac{2000 + 1330M/M_p}{F_y, \text{ksi}} \qquad (9.3.41)^*$$

Alternatively, if E_{st} is taken as 900 ksi as in the Lehigh Lecture Notes [17], Eq. 9.3.40 would become

$$\frac{L}{r_y} = \frac{352}{\sqrt{F_y, \text{ksi}}}(1.0 + 0.67M/M_p) \qquad (9.3.42)^*$$

* For SI units, with F_y in MPa,

$$\frac{L}{r_y} = \frac{290 + 190M/M_p}{F_y} \qquad (9.3.41)$$

$$\frac{L}{r_y} = \frac{924}{\sqrt{F_y}}(1.0 + 0.67M/M_p) \qquad (9.3.42)$$

Both Eqs. 9.3.41 and 9.3.42 would be generally valid for $+1.0 > M/M_p > -0.6$ (i.e., large moment gradient including double curvature).

The major assumption in Eqs. 9.3.41 and 9.3.42 was that the extreme fiber strain would be near the strain hardening range. Referring to Fig. 9.3.3, elastic lateral buckling is the major concern when rotation capacity θ needs only to reach θ_p at the formation of a plastic hinge. Equation 9.3.37 with $\sqrt{hA/Z} = 1.6$ would give $L/r_y = 605/\sqrt{F_y}$, which would correspond to reaching θ_p. On the other hand, Eqs. 9.3.41 or 9.3.42 assume conservatively that θ must reach θ_{st} (Fig. 9.3.3).

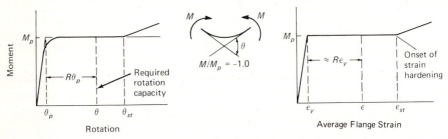

Fig. 9.3.3 Deformation requirements for developing plastic strength.

The present AISC requirement for "compact section" is based on Eq. 9.3.42 using $M/M_p = 0$ (see Sec. 9.5). Recent tests at the University of Texas [14] indicate the following equation is satisfactory to provide adequate rotation,

$$\frac{L}{r_y} = \frac{3600 + 2200M/M_p}{F_y, \text{ksi}} \qquad (9.3.43)*$$

Equation 9.3.43 is of the form of Eq. 9.3.41 but has coefficients about 75 percent higher. A rotation capacity θ (see Fig. 9.3.3) less than θ_{st} is likely to be satisfactory for design.

9.4* DESIGN PROCEDURES—WORKING STRESS METHOD

The elastic lateral buckling formulas apply to those cases where the ultimate moment capacity is achieved under an applied moment which is less than that causing a stress F_y at the extreme fiber. Elastic or inelastic buckling controls the maximum safe stress which may be permitted to act.

The Structural Stability Research Council (SSRC) *Guide* [8] discusses methods of considering lateral buckling strength for doubly symmetric I-shaped beams and girders. The two most important of these are (1) the rational "Basic" design procedure utilizing the theoretical elastic stability

---

*For SI units, with F_y in MPa,

$$\frac{L}{r_y} = \frac{522 + 319M/M_p}{F_y} \qquad (9.3.43)$$

solutions for critical stress and (2) the double formula simplified procedure used by the AISC Specification.

For working stress method the critical stress must be divided by the
desired factor of safety (1.67 for AISC) to obtain the allowable stress.

"Basic" Design Procedure

This rational method involves using the theoretical elastic stability solution for the actual loading and restraint condition, which as a practical
matter can only be done by the use of tables and charts. For the beam
under pure moment, torsionally simply supported, this would mean using
Eq. 9.3.32 or one of its alternate forms.

For any loading and restraint condition, the critical stress may be
expressed [8]

$$F_{cr} = \frac{C_4 \sqrt{EI_y GJ}}{S_x L} \tag{9.4.1}$$

where for the pure moment case (Eq. 9.3.35),

$$C_4 = \pi \sqrt{1 + \left(\frac{\pi}{\lambda L}\right)^2} \tag{9.4.2}$$

but which in general may be written [8]

$$C_4 = C_1 \pi \left[\sqrt{1 + \left(\frac{\pi}{\lambda L}\right)^2 (C_2^2 + 1)} \pm C_2 \frac{\pi}{\lambda L} \right] \tag{9.4.3}$$

where C_1 is a coefficient dependent on the type of load and support
condition, and C_2 is a coefficient to account for the position of the load
vertically with respect to the shear center. C_2 is zero for end-moment
loading and is zero when the transverse gravity loading is applied at the
beam shear center. The plus sign for C_2 is used when the gravity load is
applied at the bottom flange and the minus sign is used when the gravity
load is applied at the top flange. Values of C_1 and C_2 are available for
common cases in Fig. 9.4.1.

An alternative to using Eq. 9.4.3 for various loading cases is to use
the case of uniform bending moment (moment without shear) (C_4 in
accordance with Eq. 9.4.2) as the standard of reference. Figure 9.4.2
adapted from the *SSRC Guide* [8] permits application of a correction
multiplier to Eq. 9.4.2.

Special note should be made that for cases 6 through 10 of Fig. 9.4.1,
torsional simple support with loading on the top flange, the critical stress
is lower than that for the beam under pure moment (see Fig. 9.4.2).

Nethercot and Rockey [18] have given a procedure for computing
M_{cr} (and $F_{cr} = M_{cr}/S_x$) using a modified version of Eq. 9.4.1 along with a
method to account for a wide variety of loadings.

Case No.	Loading	Bending Moment Diagram	Condition of Restraint Against Rotation about Vertical Axis at Ends	Equiv. Length Factor, K	Value of Coefficients C_1[*]	C_2
		Beams Restrained Against Lateral Displacement at Both Ends of Span				
1.	$M \quad L \quad M$	M	Simple support	1.0	1.0	0
			Fixed	0.5	1.0	0
2.	$M \quad M/2$	$M \quad M/2$	Simple support	1.0	1.3	0
			Fixed	0.5	1.3	0
3.	M	M	Simple support	1.0	1.8	0
			Fixed	0.5	1.8	0
4.	$M \quad M/2$	$M \quad M/2$	Simple support	1.0	2.4	0
			Fixed	0.5	2.3	0
5.	$M \quad M$	$M \quad M$	Simple support	1.0	2.6	0
			Fixed	0.5	2.3	0
6.	W	$\dfrac{WL}{8}$	Simple support	1.0	1.1	0.45
			Fixed	0.5	1.0	0.29
7.	W	$\dfrac{WL}{12}$ $M_{cr} = WL/24$	Simple support	1.0	1.3	1.55
			Fixed	0.5	0.9	0.82
8.	P	$\dfrac{PL}{4}$	Simple support	1.0	1.4	0.55
			Fixed	0.5	1.1	0.42
9.	P	$\dfrac{PL}{8}$ $\dfrac{PL}{8}$	Simple support	1.0	1.7	1.42
			Fixed	0.5	1.0	0.84
10.	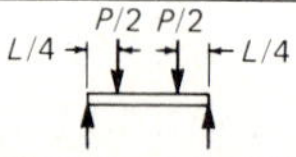 $L/4$ $P/2$ $P/2$ $L/4$	 $\dfrac{PL}{8}$	Simple support	1.0	1.0	0.42
		Cantilever Beams				
11.	 P	 PL	Warping restrained at supported end	1.0	1.3	0.64
12.	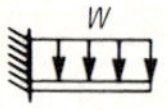 W	 $\dfrac{WL}{2}$	Warping restrained at supported end	1.0	2.1	

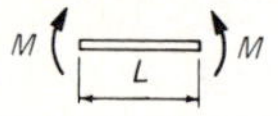

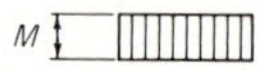

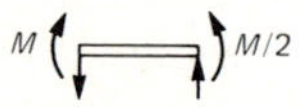

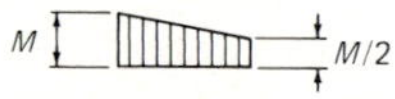

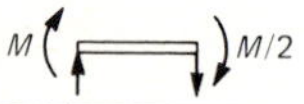

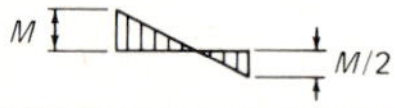

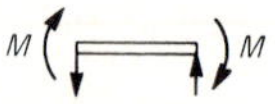

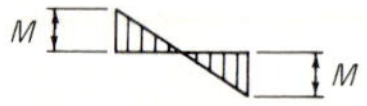

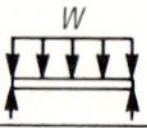

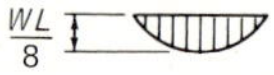

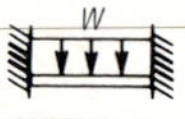

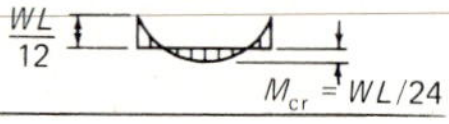

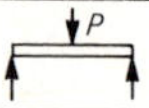

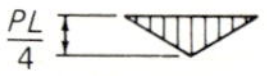

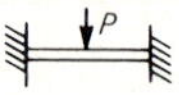

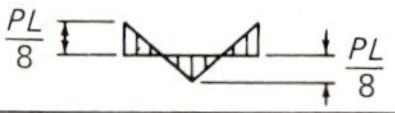

[*] Approximate minimum values for C_1 are given. Range of values and sources of data are to be found in Ref. 19.

Fig. 9.4.1 Usable values of coefficients in Eq. 9.4.1 for elastic buckling strength of beams. (Adapted from Clark and Hill [19])

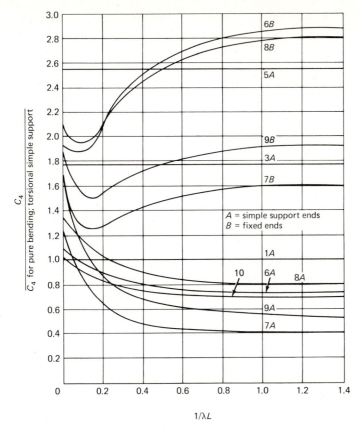

Fig. 9.4.2 Curves of $C_4/(C_4$ for pure bending with torsional simple support at ends) vs $1/\lambda L$. The curves are numbered in accordance with cases of Fig. 9.4.1. (Adapted from Ref. 8) (Lateral loading cases assume loads are applied at the top flange.)

It is to be noted that in the above equations the length L is to be taken as the effective length KL between points of torsional simple support (i.e., rotation about beam axis z is prevented, but rotation about y-axis (warping) is not prevented). As a practical matter the full length equals the effective length when at the ends of the beam there is zero rotational restraint about the y-axis, and the effective length equals one-half the actual length when full rotational restraint (torsional fixity) about the y-axis exists at the beam ends.

For cases of unequal end moment without transverse loads, Salvadori [20] has shown that the following expression for C_1 is valid:

$$C_1 = 1.75 + 1.05\left(\frac{M_1}{M_2}\right) + 0.3\left(\frac{M_1}{M_2}\right)^2 \leq 2.3 \qquad (9.4.4)$$

where M_1 is the smaller of the rotational end moments on the member. A positive ratio indicates double curvature while a negative ratio indicates single curvature.

Because of residual stress, accidental eccentricity of loading, and initial crookedness, truly elastic buckling occurs only when the critical stress computed by Eq. 9.4.1 is less than about $F_y/2$, just as discussed for compression members in Chapter 6. Thus computed buckling stresses above $F_y/2$ should be adjusted to account for inelastic buckling. The *SSRC Guide* [8] suggests the use of the same parabolic variation between $F_y/2$ and F_y that is used for columns. Use the Euler equation (Eq. 6.3.1 with $E_t = E$)

$$F_{cr} = \frac{\pi^2 E}{\left(\dfrac{KL}{r}\right)^2} \tag{9.4.5}$$

with F_{cr} computed from Eq. 9.4.1 for lateral-torsional buckling to determine an equivalent slenderness ratio. Since KL is already known, the equivalency really refers to the radius of gyration r_e. Thus, solving Eq. 9.4.5 gives

$$\frac{KL}{r_e} = \pi\sqrt{\frac{E}{F_{cr} \text{ by Eq. 9.4.1}}} \tag{9.4.6}$$

This equivalent slenderness ratio can be used with the basic SSRC column formula (Eq. 6.7.3), dividing the result by the desired safety factor.

For the case of uniform moment, F_{cr} from Eq. 9.3.32 is set equal to the Euler buckling stress, wherein KL is conservatively taken as L, and r is the equivalent radius of gyration r_e. Thus

$$\frac{\pi^2 E}{\left(\dfrac{L}{r_e}\right)^2} = \sqrt{\frac{\pi^2 E^2 C_w I_y}{S_x^2 L^4} + \frac{\pi^2 E I_y G J}{S_x^2 L^2}} \tag{9.4.7}$$

Noting that $C_w = I_y h^2/4$ and $G = E/2.6$ and solving for r_e^2 gives

$$r_e^2 = \frac{L^2}{\pi^2 E}\frac{\pi E}{S_x L^2}\sqrt{\frac{I_y^2 h^2}{4} + \frac{I_y L^2 J}{2.6}} \tag{9.4.8}$$

$$r_e^2 = \frac{I_y}{2 S_x}\sqrt{h^2 + \frac{0.156 L^2 J}{I_y}} \tag{9.4.9}$$

This equivalent radius of gyration may be used as r in the Euler hyperbola (Eq. 9.4.5) unless F_{cr} exceeds $F_y/2$ in which case r_e is used in the SSRC parabola (Eq. 6.7.3). For lateral-torsional buckling under loading other than uniform moment, r_e^2 may be multiplied by the correction factor from the curves of Fig. 9.4.2.

Double Formula Procedure

In this approach it is recognized that many beams having a high torsional stiffness (shallow, thick-walled sections), Eq. 9.3.32 can be simplified because the first term under the radical may be neglected compared to the second. For thin-walled sections having low torsional stiffness, the lateral bending stiffness of the section (based on L/r_y) predominates and the second term under the radical of Eq. 9.3.32 may be neglected.

While Eq. 9.3.32 may be used directly when the torsional properties C_w and J are available, good approximations may be obtained with a simplified version of Eq. 9.3.32.

As indicated by Eq. 8.2.14, the torsional constant J may be expressed for an I-shaped section,

$$J = 2\left(\frac{bt_f^3}{3}\right) + \frac{1}{3}\,dt_w^3 = \frac{t_f^2}{3}\left[2bt_f + dt_w\left(\frac{t_w}{t_f}\right)^2\right] \tag{9.4.10}$$

where t_f = flange thickness, b = flange width, t_w = web thickness, and d = web depth. If $t_w/t_f \approx 0.5$, and if the web area is about 20% of the area of the beam ($A_w = 0.2A$), Eq. 9.4.10 becomes

$$J = \frac{t_f^2}{3}[2bt_f + dt_w - 0.75\,dt_w]$$

$$A = 2bt_f + dt_w$$

$$A_w = dt_w$$

$$0.75A_w = 0.75(0.2A) = 0.15A$$

then

$$J \approx \frac{t_f^2}{3}[A - 0.15A] \approx 0.28At_f^2 \tag{9.4.11}$$

For other variables,

$$C_w = I_y h^2/4$$

$$G = \text{shear modulus} = \frac{E}{2(1+\mu)} = \frac{E}{2.6}$$

$$I_y = Ar_y^2$$

$$S_x = 2Ar_x^2/d$$

$r_x \approx 0.41d$ (see text Appendix Table A1)

$$h = 0.95d$$

where μ = Poisson's ratio (0.3 for steel)
r_y = radius of gyration with respect to the y-axis
r_x = radius of gyration with respect to the x-axis
d = overall depth of beam
h = distance between centroids of flanges

Substitution into Eq. 9.3.32 gives

$$F_{cr} = \sqrt{\frac{\pi^4 E^2 (I_y h^2/4) I_y}{[2A(0.41d)^2/d]^2 L^4} + \frac{\pi^2 E I_y (E/2.6)(0.28 A t_f^2)}{[2A(0.41d)^2/d]^2 L^2}} \qquad (9.4.12)$$

or

$$F_{cr} = \sqrt{\left(\frac{\pi^2 E I_y h}{4A(0.41)^2 dL^2}\right)^2 + \left(\frac{\pi E \sqrt{I_y}\sqrt{0.28}\sqrt{A}\,t_f}{2A(0.41)^2 dL\sqrt{2.6}}\right)^2} \qquad (9.4.13)$$

Then, substituting $I_y = A r_y^2$ and $h = 0.95d$ gives

$$F_{cr} = \sqrt{\left(\frac{\pi^2 (0.95) E}{4(0.41)^2 (L/r_y)^2}\right)^2 + \left(\frac{\pi\sqrt{0.28}\,E}{2(0.41)^2\sqrt{2.6}\,(Ld/r_y t_f)}\right)^2} \qquad (9.4.14)$$

or

$$F_{cr} = \sqrt{\left(\frac{13.95E}{(L/r_y)^2}\right)^2 + \left(\frac{3.07E}{Ld/r_y t_f}\right)^2} \qquad (9.4.15)$$

The two formula approach uses whichever of two squared terms predominates within the square root of Eq. 9.4.15. The reader is reminded that Eq. 9.4.15 is actually an alternate form of Eq. 9.4.1 where C_4 is given by Eq. 9.4.2.

Torsionally strong sections: For shallow thick flanged sections, torsional strength predominates and the first term of Eq. 9.4.15 may be neglected. Thus

$$F_{cr} = \frac{3E}{Ld/r_y t_f} \qquad (9.4.16)$$

Torsionally weak sections: For deep sections having relatively thin flanges and web (such as plate girder sections), column strength of the compression flange predominates and the second term of Eq. 9.4.15 may be neglected. Thus

$$F_{cr} = \frac{14E}{(L/r_y)^2} \qquad (9.4.17)$$

A conservative approach would then be to compute the buckling stress based on Eqs. 9.4.16 and 9.4.17, accepting the larger of the two values for design purposes.

Alternatively to neglecting either term, the entire equation, Eq. 9.4.15, may be used as given here with rounded coefficients,

$$F_{cr} = \sqrt{\left[\frac{3E}{Ld/r_y t_f}\right]^2 + \left[\frac{14E}{(L/r_y)^2}\right]^2} \qquad (9.4.18)$$

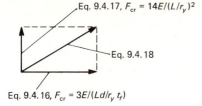

Fig. 9.4.3 in upper area shows:

Eq. 9.4.17, $F_{cr} = 14E/(L/r_y)^2$

Eq. 9.4.18

Eq. 9.4.16, $F_{cr} = 3E/(Ld/r_y\, t_f)$

Fig. 9.4.3 Lateral-torsional buckling stress as the resultant of two vectors.

Referring to Fig. 9.4.3, Eq. 9.4.18 may be visualized as the resultant of two vectors at 90° to each other. If the resultant is not used, the next best approach is to use the larger of the two legs of the right triangle.

Again, as in the SSRC "Basic" method, critical stress F_{cr} values exceeding $F_y/2$ should be adjusted for inelastic buckling, using Eqs. 9.4.5 and 6.7.3. The AISC Specification is based on the procedure using the larger of two formulas, as described in the next section.

Example 9.4.1

Consider the 50-ft simple span W27×84 beam carrying a uniformly distributed load of 1.1 kip/ft, as shown in Fig. 9.4.4. Lateral support is provided at the ends and at 20 ft from each end. Use A36 steel.

(a) Determine the allowable stress and adequacy of the section using the SSRC "Basic" Design Procedure, utilizing the differential equation solution as corrected in accordance with Fig. 9.4.1. Use a 1.67 factor of safety.

(b) Repeat (a) but use the SSRC Double Formula Procedure.

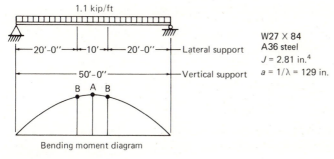

W27 X 84
A36 steel
$J = 2.81$ in.^{4}
$a = 1/\lambda = 129$ in.

Fig. 9.4.4 Example 9.4.1.

SOLUTION

(a) SSRC "Basic" Procedure. The allowable stress is based on the critical stress F_{cr} given by Eq. 9.4.1 for *constant moment* along the unbraced segment. For this problem, the moment is not constant; for the

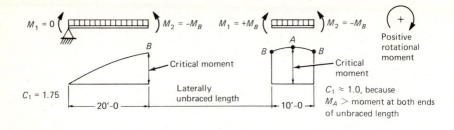

Segment 1   Segment 2

Fig. 9.4.5 C_1(AISC, C_b) values for Example 9.4.1.

highest moment at point A, the 10-ft laterally unbraced length is used as shown by Segment 2 in Fig. 9.4.5; for the moment at B the 20 ft laterally unbraced length is used as shown by Segment 1 in Fig. 9.4.5. The moment gradient on any such segment may be accounted for by a correction factor C_1 (called C_b by AISC) given by Eq. 9.4.4 and applied as a multiplier to Eq. 9.4.1.

Begin by computing F_{cr} for the basic case of constant moment using Eq. 9.4.1 for the longer 20-ft segment,

$$F_{cr} = \frac{C_4\sqrt{EI_y GJ}}{S_x L} \qquad [9.4.1]$$

where

$$C_4 = \pi\sqrt{1 + \left(\frac{\pi}{\lambda L}\right)^2} \qquad [9.4.2]$$

$$= \pi\sqrt{1 + \left(\frac{129\pi}{20(12)}\right)^2} = 6.16$$

where $G = E/2.6$ in Eq. 9.4.1 above, $I_y = 106$ in.⁴, $S_x = 213$ in.³

$$F_{cr} = \frac{6.16(29{,}000)\sqrt{106(2.81)/2.6}}{(213)(240)} = 37.4 \text{ ksi}$$

Next, since the moment varies from zero to a maximum over the 20-ft segment, apply the C_1 correction factor, Eq. 9.4.4

$$C_1 = 1.75 + 1.05\left(\frac{M_1}{M_2}\right) + 0.3\left(\frac{M_1}{M_2}\right)^2 \le 2.3 \qquad [9.4.4]$$

$$C_1 = 1.75 \text{ for this 20-ft segment}$$

$$F_{cr} = C_1 \text{ (Basic } F_{cr}) = 1.75(37.4) = 65.5 \text{ ksi}$$

Since this value exceeds $0.5F_y = 18$ ksi, inelastic buckling controls.

Use Eq. 9.4.5 to determine an equivalent slenderness ratio,

$$F_{cr} = \frac{\pi^2 E}{(KL/r_e)^2} = 65.5 \text{ ksi}$$

$$\frac{KL}{r_e} = \sqrt{\frac{\pi^2 (29,000)}{65.5}} = 66.1$$

As for most situations, torsional simple support is assumed at the ends of the unbraced lengths so that $K = 1$. $L/r_e = 66.1$ is then substituted into the SSRC basic column formula, Eq. 6.7.3, giving

$$F_{cr} = F_y \left[1 - \frac{F_y}{4\pi^2 E} \left(\frac{KL}{r} \right)^2 \right]$$

$$= 36 \left[1 - \frac{36(66.1)^2}{4\pi^2 (29,000)} \right] = 36(0.863) = 31.1 \text{ ksi}$$

Using FS $= 1.67$, the allowable stress may be obtained,

$$F_b = \frac{F_{cr}}{1.67} = \frac{31.1}{1.67} = 18.6 \text{ ksi} \quad \text{(for Segment 1)}$$

The bending moment at point B is

$$M_B = 0.5(1.1)(20)30 = 330 \text{ ft-kips}$$

$$f_b = \frac{330(12)}{213} = 18.6 \text{ ksi} = [F_b = 18.6 \text{ ksi}] \qquad \text{OK}$$

Note is made that instead of using Eq. 9.4.4 for C_1, which is not applicable to all cases, Eq. 9.4.3 or Fig. 9.4.2 may be used. In this example, Case 3 of Fig. 9.4.2 approximates the loading on Segment 1; thus

$$\frac{\text{Actual } C_4}{\text{Pure bending } C_4} \approx 1.78$$

which for this loading case is essentially the same as the 1.75 from the C_1 equation.

Segment 2 should also be checked. For that segment the moment is essentially constant so that $C_1 = 1$ is quite accurate. Because $L = 10$ ft, C_4 is different for this segment,

$$C_4 = \pi \sqrt{1 + \left(\frac{\pi}{\lambda L} \right)^2} = \pi \sqrt{1 + \left(\frac{129\pi}{10(12)} \right)^2} = 11.07$$

Using Eq. 9.4.1,

$$F_{cr} = \frac{11.07(29,000)\sqrt{106(2.81)/2.6}}{213(120)} = 134.4 \text{ ksi}$$

Since this exceeds $F_y/2$, use Eq. 9.4.5,

$$\frac{KL}{r_e} = \sqrt{\frac{\pi^2(29,000)}{134.4}} = 46.1$$

Using the SSRC parabola, Eq. 6.7.3,

$$F_{cr} = 36\left[1 - \frac{36}{4\pi^2(29,000)}(46.1)^2\right] = 33.6 \text{ ksi}$$

$$F_b = \frac{F_{cr}}{1.67} = \frac{33.6}{1.67} = 20.1 \text{ ksi} \quad \text{(for Segment 2)}$$

The bending moment at point A is

$$M_A = \tfrac{1}{8}(1.1)(50)^2 = 343.8 \text{ ft-kips}$$

$$F_b = \frac{343.8(12)}{213} = 19.4 \text{ ksi} < 20.1 \text{ ksi} \qquad \text{OK}$$

(b) SSRC Double Formula Procedure. In this method, the larger of Eqs. 9.4.16 and 9.4.17 is used as F_{cr}. Examine the 20-ft segment,

$$F_{cr} = \frac{3E}{Ld/r_y t_f} = \frac{3(29,000)}{20(12)26.71/[2.07(0.640)]} = 18.0 \text{ ksi} \quad \text{[9.4.16]}$$

$$F_{cr} = \frac{14E}{(L/r_y)^2} = \frac{14(29,000)}{(240/2.07)^2} = 30.2 \text{ ksi} \quad \textit{controls} \qquad \text{[9.4.17]}$$

In order to compare with the "Basic" method of part (a), the multiplier $C_1 = 1.75$ should be included to account for moment gradient. Thus

$$F_{cr} = C_1 \text{ (Basic } F_{cr})$$

$$= 1.75(30.2) = 52.9 \text{ ksi}$$

Since this exceeds $F_y/2$, use Eq. 9.4.5,

$$\frac{KL}{r_e} = \sqrt{\frac{\pi^2 29,000}{529}} = 73.6$$

Using the SSRC parabola, Eq. 6.7.3,

$$F_{cr} = 36\left[1 - \frac{36}{4\pi^2(29,000)}(73.6)^2\right] = 29.9 \text{ ksi}$$

$$F_b = \frac{F_{cr}}{1.67} = \frac{29.9}{1.67} = 17.9 \text{ ksi}$$

This compares with 18.6 ksi obtained using the "Basic" procedure.

9.5 AISC WORKING STRESS CRITERIA FOR I-SHAPED BEAMS SUBJECTED TO STRONG-AXIS BENDING

This section considers the full range of strength from laterally stable beams to situations where lateral buckling causes considerable strength reduction. Remember that service load allowable stresses are based on the strength capable of being achieved. Loading in the plane of the web of an I-shaped section is assumed.

Case 1; Plastic moment is reached $(M_u = M_p)$ *along with large plastic rotation capacity.* The requirements to achieve this condition are that local buckling of compression elements (flange and web) is prevented (see Table 7.3.1) and lateral-torsional buckling does not occur. This latter will be accomplished when L/r_y is kept low enough to permit the necessary plastic strain to occur. As discussed in Chapter 7, a section capable of achieving the plastic moment condition is designated in the working stress method as a "compact section." Since less plastic strain (rotation capacity, Fig. 9.3.4) is utilized in working stress method than in plastic design, the slenderness ratio L/r_y limit for "compact section" can be more liberal than indicated by Eq. 9.3.42.

The slenderness ratio limit based on the high rotation capacity requirement was developed in Sec. 9.3 as

$$\left(\frac{L}{r_y}\right)_{\max} = \frac{352}{\sqrt{F_y, \text{ksi}}}\left(1.0 + 0.67\frac{M}{M_p}\right) \qquad [9.3.42]$$

Letting the end moment ratio M/M_p equal zero (a liberalizing assumption for bending moment of one sign over the laterally unbraced length), letting $L = L_c$ (that is, the maximum laterally unbraced length for the member to be treated as a "compact section,") and recognizing that $r_y \approx 0.22b_f$, where b_f is the flange width, gives for Eq. 9.3.42 approximately

$$L_c = \frac{76b_f}{\sqrt{F_y, \text{ksi}}} \quad \text{or} \quad \frac{200b_f}{\sqrt{F_y, \text{MPa}}} \qquad (9.5.1)$$

which is given by AISC–1.5.1.4.1(5.).

The development of Eq. 9.5.1 involved considering the lateral buckling strength to be primarily the column strength of the compression flange. However, it is possible that lateral stability for a section is primarily controlled by torsional strength and the section may not be capable of developing its plastic moment even though Eq. 9.3.42 is satisfied. Hence for these sections where torsional strength predominates, the slenderness ratio limit is based on the elastic buckling equation, Eq. 9.4.16,

$$F_{cr} = \frac{3E}{Ld/r_y t_f} \qquad [9.4.16]$$

Since $r_y \approx 0.22b_f$ and $A_f = b_f t_f$, Eq. 9.4.16 becomes

$$F_{cr} = \frac{0.66E}{Ld/A_f} \tag{9.5.2}$$

When $F_{cr} = F_y$, $L = L_c$, and $E = 29,000$ ksi, Eq. 9.5.2 becomes approximately

$$L_c = \frac{20,000}{(d/A_f)F_y} \tag{9.5.3}$$

which is given by AISC–1.5.1.4.1(5.). The laterally unsupported length of the compression flange may not exceed the *smaller* of Eqs. 9.5.1 or 9.5.3 if the section is to be laterally stable enough to reach M_p (i.e., to be treated as a "compact section").

A summary of the "compact section" lateral support requirements is given in Table 9.5.1. The length L_c is a property of the section and specified F_y. Values of L_c for rolled W, S, and M shapes are given in text Appendix Table A3.

Table 9.5.1 Lateral Bracing Requirements for "Compact Sections," AISC–1.5.1.4.1(5.)

F_y, ksi	Both limits may not be exceeded		F_y, MPa
	$\dfrac{L_c}{b_f}$	$L_c\left(\dfrac{d}{A_f}\right)$	
36	12.7	556	250
42	11.7	476	290
45	11.3	444	310
50	10.7	400	350
55	10.2	364	380
60	9.8	333	410
65	9.4	308	450
70*	9.1	286	480

*I-shaped sections loaded in the plane of the web may be assumed to develop M_p only when F_y does not exceed 70 ksi (480 MPa). A607 Grade 70 is presently the highest strength steel applicable under AISC–1.5.1.4.1.

Case 2: $M_p > M_u \geq M_y$. Beams in this category are those capable of achieving yield strain $\varepsilon_y = F_y/E_s$ at the extreme fiber prior to lateral-torsional buckling, but lateral-torsional buckling occurs before M_p can be reached. For Case 2 situations, AISC does not permit an interpolated strength between M_y and M_p. Instead the strength is chosen to be M_y and the corresponding allowable stress is $0.60F_y$.

In order to reach a strength $M_u = M_y$ prior to lateral-torsional buckling, the maximum unbraced length L_u that a section can have is controlled by

$$F_{cr} = \sqrt{\left(\frac{3E}{Ld/r_y t_f}\right)^2 + \left(\frac{14E}{(L/r_y)^2}\right)^2} \qquad \text{[9.4.18]}$$

If Eq. 9.4.18 is considered too complex to be used in design, the next best approximation is to use the larger of the two terms.

The first term, representing torsional strength, was used previously in this section as Eq. 9.5.2. In order to modify Eq. 9.5.2 to account for moment gradient the term C_b is inserted, giving

$$F_{cr} = \frac{0.66EC_b}{Ld/A_f} \qquad (9.5.4)$$

Examination of Eq. 9.4.1, the original source of Eq. 9.4.18, shows C_4 as a multiplier, and Eq. 9.4.3 shows C_4 directly proportional to C_1, which is now called C_b. Dividing Eq. 9.5.4 by FS $= 1.67$ and taking $E = 29,000$ ksi gives the allowable stress F_b in ksi:

$$F_b = \frac{12,000C_b}{Ld/A_f} \text{ ksi} \qquad (9.5.5)^*$$

which is AISC Formula (1.5–7).

The second term of Eq. 9.4.18, representing the column strength of the compression portion of the section, is

$$F_{cr} = \frac{14E}{(L/r_y)^2} \qquad (9.5.6)$$

Since only the compression flange is the column, logically the radius of gyration should be that of the compression flange and some portion of the adjacent web. AISC–1.5.1.4.5(2.) introduces the symbol r_T for this compression portion, defined as "radius of gyration of a section comprising the compression flange plus one-third of the compression web area, taken about an axis in the plane of the web." For doubly symmetrical I-shaped sections, r_T may be expressed

$$r_T = \sqrt{\frac{\frac{1}{12}t_f b_f^3}{t_f b_f + \frac{1}{6}t_w(d - 2t_f)}} \qquad (9.5.7)$$

The values of r_T are given in the AISC Manual along with other dimensions and properties for rolled shapes. As an approximation for this development of AISC formulas r_T may be taken as $1.2r_y$.

---

* For SI units,

$$F_b = \frac{83,000C_b}{Ld/A_f} \text{ MPa} \qquad (9.5.5)$$

Replacing r_y by $r_T/1.2$ in Eq. 9.5.6 would make that equation very closely the same as the Euler column equation,

$$F_{cr} = \frac{\pi^2 E}{(L/r_T)^2} \tag{9.5.8}$$

This makes it more visible that Eq. 9.5.6. is representative of column strength of the compression flange.

Treatment of column strength as it relates to beams requires application of the principles of column design as discussed in Chapter 6. Recall the use of two design equations, one based on inelastic buckling strength represented by the SSRC parabola (Eq. 6.7.3) for buckling stress F_{cr} above $F_y/2$, and the other based on the Euler elastic buckling strength for buckling stress F_{cr} equal to or less than $F_y/2$.

The SSRC parabola with $r = r_T$ and the effective length factor taken as 1.0 is

$$F_{cr} = F_y \left[1 - \frac{F_y}{4\pi^2 E} \left(\frac{L}{r_T}\right)^2 \right] \tag{9.5.9}$$

Unlike the axial compression loading of Chapter 6 where the magnitude of compression was normally constant over the unbraced length, the compression flange of a beam will have its magnitude varying as the bending moment varies. Thus, for beams the term C_b to account for moment gradient is included in Eq. 9.5.9,

$$F_{cr} = F_y \left[1 - \frac{F_y}{4\pi^2 E C_b} \left(\frac{L}{r_T}\right)^2 \right] \tag{9.5.10}$$

where $C_b = C_1$, Eq. 9.4.4.

To obtain the allowable stress, the strength equation is divided by a factor to get the calculation into the service load range. For small L/r_T the strength should be $M_p = F_y Z$, whereas for large L/r_T the strength will be $M_u = F_{cr} S$. As discussed in Chapter 7 when M_p is achievable, the allowable value for elastic service load stresses is $0.66F_y$. By the same reasoning, F_b based on Eq. 9.5.10 should be $\frac{2}{3}$ of F_{cr}, giving

$$F_b = \left[\frac{2}{3} - \frac{(\frac{2}{3})F_y}{4\pi^2 E C_b} \left(\frac{L}{r_T}\right)^2 \right] F_y \tag{9.5.11}$$

For $E = 29{,}000$ ksi and F_y in ksi, Eq. 9.5.11 becomes

$$F_b = \left[\frac{2}{3} - \frac{F_y (L/r_T)^2}{1720(10^3) C_b} \right] F_y \tag{9.5.12}$$

The coefficient in the second term has been adjusted from 1720 to 1530 (i.e., $\frac{3}{4}$ instead of $\frac{2}{3}$ is used on that term) to allow for the fact that as L/r_T gets larger the strength drops below M_y and the extra benefit of the shape factor disappears. When M_u is less than M_y the F_b should be $0.60F_{cr}$

rather than $0.66F_{cr}$. Thus Eq. 9.5.12 becomes

$$F_b = \left[\frac{2}{3} - \frac{F_y(L/r_T)^2}{1530(10^3)C_b}\right]F_y \qquad (9.5.13)^*$$

which is AISC Formula (1.5–6a).

For large L/r_T when elastic buckling controls, Eq. 9.5.7 would logically be the basis for design. Dividing Eq. 9.5.7 by 1.67 (the basic FS for beams) and inserting C_b for moment gradient gives

$$F_b = \frac{\pi^2 E C_b}{1.67(L/r_T)^2} \qquad (9.5.14)$$

which gives closely the following:

$$F_b = \frac{170,000C_b}{(L/r_T)^2}\text{ ksi} \qquad (9.5.15)^*$$

which is AISC Formula (1.5–6b).

The use of 1.67 here for beams rather than the 1.92 used for long axially loaded columns is explainable (a) by realizing that torsional strength is neglected in Eq. 9.5.15, and (b) Eq. 9.5.6 is inherently conservative when lateral-torsional buckling strength is primarily column strength, in that for sections consisting of thin flanges and deep thin web $r_x \approx 0.38d$ would have been more correct than the value $0.41d$ used in the development of Eq. 9.5.6.

Thus lateral-torsional buckling controls the design when the laterally unbraced length L exceeds the *larger* value of L defined by Eqs. 9.5.5 or 9.5.13 equaling an allowable stress of $0.60F_y$. When the *larger* value of F_b from these two equations exceeds $0.60F_y$ it implies that M_y can be reached prior to lateral-torsional buckling.

The maximum unbraced length L for which $F_b = 0.60F_y$ may be obtained by setting Eqs. 9.5.5 and 9.5.13 equal to $0.60F_y$. Thus

$$0.60F_y = \frac{12,000C_b}{Ld/A_f} \qquad [9.5.5]$$

or

$$0.60F_y = \left[\frac{2}{3} - \frac{F_y(L/r_T)^2}{1530(10^3)C_b}\right]F_y \qquad [9.5.13]$$

* In SI units,

$$F_b = \left[\frac{2}{3} - \frac{F_y(L/r_T)^2}{10,550(10^3)C_b}\right]F_y \quad \text{MPa} \qquad (9.5.13)$$

$$F_b = \frac{1170(10^3)C_b}{(L/r_T)^2} \quad \text{MPa} \qquad (9.5.15)$$

From the first,

$$L = \frac{20,000C_b}{(d/A_f)F_y} \tag{9.5.16}$$

and from the second,

$$L = r_T \sqrt{\frac{102,000C_b}{F_y}} \tag{9.5.17}$$

The *larger* value of L from Eqs. 9.5.16 and 9.5.17 is the maximum unbraced length for a Case 2 design situation; i.e., where the maximum strength M_u achievable lies between M_y and M_p.

When $C_b = 1$ (the conservative case of constant bending moment), the larger value of L from Eqs. 9.5.16 and 9.5.17 is termed L_u, a property of the section for any specified yield stress F_y,

but not less than
$$\left.\begin{array}{c} L_u = \dfrac{20,000}{(d/A_f)(F_y, \text{ksi})} \\[2em] L_u = r_T\sqrt{\dfrac{102,000}{F_y, \text{ksi}}} \end{array}\right\} \tag{9.5.18}*$$

The AISC Manual gives values for L_u for $F_y = 36$ ksi and 50 ksi. Values of L_u for $F_y = 36$, 42, 45, 50, 55, 60, and 65 ksi are given in this text, Appendix Table A3.

Case 3: $M_u < M_y$. For this case, lateral-torsional buckling occurs before M_y can be reached and the allowable stress is less than $0.60F_y$.

Whenever the laterally unsupported length exceeds the *larger* L value of Eqs. 9.5.16 and 9.5.17, the allowable stress F_b is the *larger* stress value of Eqs. 9.5.5 and 9.5.13 for strength controlled by inelastic lateral-torsional buckling, or is the *larger* stress value of Eqs. 9.5.5 and 9.5.15 for strength controlled by elastic lateral-torsional buckling.

The slenderness ratio L/r_T at which the criteria for inelastic and elastic lateral-torsional buckling prevention give the same allowable stress

---

* For SI units, with F_y in MPa,

but not less than
$$\left.\begin{array}{c} L_u = \dfrac{2900}{(d/A_f)F_y} \\[2em] L_u = \sqrt{\dfrac{14,800}{F_y}} \end{array}\right\} \tag{9.5 18}$$

is obtained by equating Eqs. 9.5.13 and 9.5.15; thus,

$$\frac{2F_y}{3} - \frac{F_y^2(L/r_T)^2}{1530(10^3)C_b} = \frac{170,000C_b}{(L/r_T)^2}$$

from which is obtained

$$\frac{L}{r_T} = \sqrt{\frac{510,000C_b}{F_y}} \qquad (9.5.19)$$

Table 9.5.2 Limits on L/r_T and Coefficients for AISC Formulas (1.5–6)

F_y	$\sqrt{\dfrac{102,000C_b}{F_y}}$ Eq. 9.5.17			$\dfrac{1530(10^3)C_b}{F_y^2}$, for F_b(1.5–6a) from Eq. 9.5.13			$\sqrt{\dfrac{510,000C_b}{F_y}}$ Eq. 9.5.19		
	C_b			C_b			C_b		
ksi	1.0	1.75	2.3	1.0	1.75	2.3	1.0	1.75	2.3
36	53	70	80	1181	2070	2720	119	158	180*
42	49	65	74	867	1517	1995	110	146	167
50	45	60	68	612	1070	1408	101	134	153
60	41	54	62	425	744	977	92	122	139
65	40	53	61	362	634	832	89	118	135
100	32	42	49	153	268	352	71	94	107

* AISC–1.8.4 limits slenderness ratios for compression members to 200. $L/r_y = 200$ when $L/r_T \approx 200/1.2 = 167$.

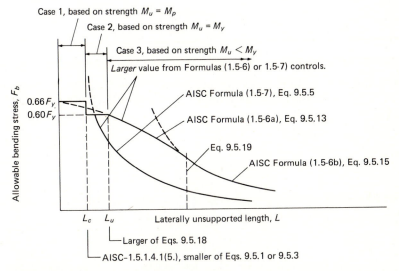

Fig. 9.5.1 Summary of AISC criteria for allowable flexural stress for working stress design of members having an axis of symmetry in, and loaded in, the plane of the web.

Elastic behavior occurs for slenderness ratios greater than Eq. 9.5.19 and Eq. 9.5.15 controls over Eq. 9.5.13. Table 9.5.2 gives L/r_T upper and lower limits and denominator coefficients for the second term in Eq. 9.5.13 (AISC Formula 1.5–6a) for various values of F_y.

A summary of the AISC allowable flexural stress vs laterally unbraced length relationships is given in Fig. 9.5.1.

Example 9.5.1

Calculate and plot the allowable stress relationship for a W14×30 of A36 steel taking $C_b = 1$.

SOLUTION

(a) Calculate L_c. Using the *smaller* value from Eqs. 9.5.1 and 9.5.3 as prescribed in AISC–1.5.1.4.1(5.),

$$\frac{76b_f}{\sqrt{F_y}} = \frac{76(6.73)}{\sqrt{36}(12)} = 7.1 \text{ ft}$$

or

$$\frac{20,000}{(d/A_f)F_y} = \frac{20,000}{(5.34)36(12)} = 8.7 \text{ ft}$$

Thus $L_c = 7.1$ ft.

(b) Calculate L_u. Using the *larger* of Eqs. 9.5.18,

$$\frac{20,000}{(d/A_f)F_y} = 8.7 \text{ ft} \quad \text{(from above)}$$

or

$$r_T\sqrt{\frac{102,000}{F_y}} = \frac{1.74}{12}\sqrt{\frac{102,000}{36}} = 7.7 \text{ ft}$$

Thus $L_u = 8.7$ ft.

(c) Unbraced lengths larger than L_u. Use the *larger* value from Eq. 9.5.5 or 9.5.13, as prescribed in AISC–1.5.1.4.5(2.), Formulas (1.5–6a) and (1.5–7),

$$F_b(1.5\text{–}7) = \frac{12,000C_b}{Ld/A_f}$$

$$= \frac{12,000(1.0)}{L(5.34)12} = \frac{187}{L} \text{ ksi}$$

$$F_b(1.5\text{–}6a) = \left[\frac{2}{3} - \frac{F_y(L/r_T)^2}{1530(10^3)C_b}\right]F_y$$

$$= \left[\frac{2}{3} - \frac{36(L/1.74)^2 144}{1530(10^3)(1.0)}\right]36$$

$$= 24.0 - \frac{L^2}{24.8} \text{ ksi}$$

According to Eq. 9.5.19, AISC Formula (1.5–6a) and (1.5–6b) give the same allowable stress at

$$L = r_T\sqrt{\frac{510,000C_b}{F_y}} = \frac{1.74}{12}\sqrt{\frac{510,000(1.0)}{36}} = 17.3\text{ ft}$$

For L larger than 17.3 ft, the allowable stress is given by the larger of

$$F_b(1.5\text{–}7) = \frac{187}{L}\text{ ksi}\quad\text{(computed above)}$$

or

$$F_b(1.5\text{–}6b) = \frac{170,000C_b}{(L/r_T)^2}$$

$$= \frac{170,000(1.0)}{(L/1.74)^2 144} = \frac{3570}{L^2}\text{ ksi}$$

The complete allowable stress relationship for this W14×30 with $C_b = 1$ is shown in Fig. 9.5.2.

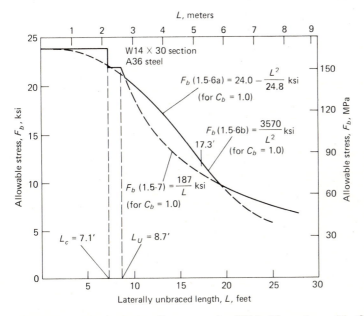

Fig. 9.5.2 Allowable bending stress for W14×30 section with $C_b = 1$, as computed in Example 9.5.1.

More Precise Analysis

AISC–1.5.1.4.5(2.) permits the use of "a more precise analysis," which according to the AISC Commentary includes the SSRC "Basic" Design Procedure. As a slight modification of the procedure described in Sec.

9.4, the AISC Commentary suggests using an equivalent radius of gyration r_e given by Eq. 9.4.9. Then, instead of using the SSRC parabola, Eq. 6.7.3, or the Euler curve, whichever is applicable, and dividing by a factor of safety to obtain allowable stress, the r_e may be used directly in AISC Formulas (1.5–6a) or (1.5–6b), whichever is applicable. There is no appreciable difference between these approaches.

9.6 AISC WORKING STRESS CRITERIA FOR WEAK-AXIS BENDING OF I-SHAPED SECTIONS

The treatment thus far in this chapter has dealt with lateral-torsional buckling where instability might occur in a direction perpendicular to the plane of strong-axis bending (that is, the buckling occurs in the weaker direction). When an I-shaped beam is bent about its weak axis (y-axis), that is, bending in a plane *perpendicular* to the plane of the web, making the y-axis the neutral axis, lateral-torsional instability is no longer of concern. The beam will tend to deflect only in the direction of the loading since that is the principal axis orientation offering least resistance. Since lateral instability will not occur on doubly symmetrical I-shaped sections bent about their weak axis, the only factor that might prevent them from achieving the plastic moment condition would be local buckling of the compression portion (unstiffened element) of the flanges.

The AISC requirements for weak-axis bending of these shapes are treated in Sec. 7.3.

9.7 AISC—PLASTIC DESIGN CRITERIA

Since the basic approach of plastic design as discussed in Secs. 1.8 and 7.3 is to develop the plastic strength of the structure, lateral-torsional buckling must not be allowed to occur prior to the plastic strength being reached. In other words, a reduced strength of section based on laterally unbraced length is not acceptable. Lateral bracing must be provided. AISC–2.9 states "Members shall be adequately braced to resist lateral and torsional displacements at the plastic hinge locations associated with the failure mechanism."

In accordance with AISC–2.9 unbraced length slenderness ratios must not exceed

$$\frac{L_{cr}}{r_y} \le \frac{1375}{F_y, \text{ksi}} + 25 \qquad \text{(9.7.1)}^*$$

when $+1.0 > M/M_p > -0.5$ (i.e., large moment gradient, including double

---

* In SI units, with F_y in MPa,

$$\frac{L_{cr}}{r_y} \le \frac{9480}{F_y} + 25 \qquad \text{(9.7.1)}$$

curvature), and

$$\frac{L_{cr}}{r_y} \le \frac{1375}{F_y, \text{ksi}}$$ (9.7.2)

when $-0.5 \ge M/M_p > -1.0$ (i.e., small moment gradient approaching uniform moment).

In Eqs. 9.7.1 and 9.7.2, M/M_p is the end moment ratio over a laterally unbraced length where a plastic hinge associated with the failure mechanism occurs at one end. The ratio is negative for single-curvature bending and positive for double curvature.

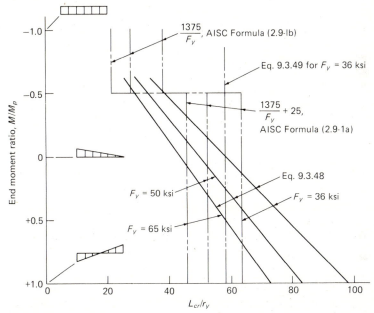

Fig. 9.7.1 Comparison of AISC critical slenderness ratio for plastic design with theoretical developed equation.

A comparison of the above provisions with the equations developed theoretically in Sec. 9.3 appears in Fig. 9.7.1. It may be noted that Eq. 9.3.42, which, converted into Eq. 9.5.1, is used for satisfying the lateral stability requirement for a "compact section" in the working stress method, compares favorably (as it should) with Eq. 9.7.1 for plastic design. Equations 9.7.1 and 9.7.2 were developed for situations where lateral support (and restraint of twisting) is provided *only at the ends* of the unbraced segment. In practical situations, attachments to the top flange of continuous beams do in fact help to prevent twisting even where the top flange is not the one carrying compression. Thus in the negative moment region of continuous beams having the top flange laterally braced, it seems reasonable to treat the point of inflection as a braced point.

For unbraced segments where a plastic hinge is not expected to occur in the failure mechanism, the working stress equations, Formulas (1.5–6) and (1.5–7) are to be applied. To get the unit stresses into the working stress range, the stresses P/A or M/S, computed using factored loads (i.e., design ultimate loads), are to be divided by the applicable load factor. Furthermore, if the last plastic hinge of the collapse mechanism is the only plastic hinge occurring over the unbraced length, the working stress provisions apply just as when no plastic hinge is expected to form.

For weak-axis bending, lateral stability is not a factor and the slenderness ratio limitations are not applicable.

AISC–2.4, regarding beam-columns, provides an equation, Formula (2.4–4), for computing the reduced ultimate bending strength of a member subject to moment alone and having a slenderness ratio exceeding the limits of Eqs. 9.7.1 and 9.7.2. The maximum ultimate moment is given by

$$M_m = \left[1.07 - \frac{(L/r_y)\sqrt{F_y}}{3160}\right]M_p \le M_p \tag{9.7.3}$$

It was felt that for the beam-column interaction one might desire to exceed the slenderness limitations ordinarily used for continuous beam design. However, since a beam-column in a plastically designed structure will be an integral part of a frame, continuity tends to provide restraint against lateral-torsional buckling not accounted for in AISC Formulas (1.5–6) and (1.5–7).

The use of the larger of AISC Formulas (1.5–6) or (1.5–7) has already been shown to be conservative, and the SSRC "Basic" design method of computing an equivalent KL/r is overly complex for ordinary use and does not accurately consider inelastic behavior. The more accurate analysis of Galambos [9], shown in Fig. 9.7.2, shows Eq. 9.7.3 to be a realistic expression, and it is recommended in Ref. 17 for plastic design.

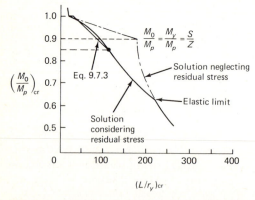

Fig. 9.7.2 Effects of residual stress and lateral-torsional buckling on the ability to develop M_p under pure bending. (From Galambos [9])

Illustration of these provisions for plastic design appears in Chapter 10 for continuous beams and in Chapter 12 for beam-columns.

9.8 EFFECTIVE LATERALLY UNBRACED LENGTH

Design equations, such as those of AISC–1.5.1.4.5(2.) are based on the most critical situation of uniform bending moment over the unbraced segment with assumed torsional simple support at its ends. This would mean that when considering torsional behavior only, the effective unbraced length KL equals the actual unbraced length L. As is seen from Fig. 9.4.1, the effective length factor K varies from 0.5 for torsional fixity to 1.0 for torsional simple support.

Nearly any type of lateral brace or end connection will prevent rotation about the longitudinal axis of the beam ($\phi = 0$ in Fig. 9.3.1c). However, only rarely will there be much restraint against rotation of the flanges about a vertical axis as discussed in regard to Fig. 8.5.7 (i.e., warping restraint), so that $K = 1.0$ is reasonable for most cases.

Nonuniform Moment

To account for loading other than uniform moment, the SSRC suggests [8] (permitted under AISC–1.5.1.4.5(2.)) computing an equivalent slenderness ratio, as discussed in Sec. 9.4, rather than directly determining an effective length. Figures 9.4.1 and 9.4.2 may be used for determining the buckling stress according to Eq. 9.4.1 and then use Eq. 9.4.5 to get the equivalent slenderness ratio.

For linearly varying moment gradient, the C_b factor already discussed provides the appropriate correction.

Continuous Beams

A continuous beam will have lateral end restraint moments develop as a result of the beam being continuous over adjacent spans. If the adjacent spans are shorter than the span in question, or at least braced laterally at closer intervals; or the adjacent spans are less severely loaded, some lateral moment end restraint may develop.

Typically, however, such end restraint about the y–y axis should not be assumed to be present since alternate unbraced spans could buckle in opposite directions. For determination of lateral buckling loads on continuous beams, the reader is referred to the work of Salvadori [21] and Hartmann [22].

In past practice, the authors have considered it acceptable to treat an inflection point as a braced point when a lateral buckling formula had no provision for moment gradient. The AISC Formulas (1.5–6) and (1.5–7) do include a C_b term to account for moment gradient, including an unbraced segment containing an inflection point. When the C_b term is used the inflection point should not be treated as braced.

For establishing ability to qualify as a "compact section" the equations for L_c, Eqs. 9.5.1 and 9.5.3, do not provide for moment gradient. Equation 9.5.1 was derived on the basis of plastic strength and for a moment gradient varying from M_p at one end of the unbraced length to zero at the other. Equation 9.5.3 is based on torsional strength. There is no doubt the AISC L_c equations are conservative if an inflection point is *not* considered a braced point; however, they are likely to be unconservative if the inflection point is treated as braced. While there are differing opinions regarding whether the inflection point should be considered as a braced point, an important factor in making this decision is whether or not there is torsional restraint at or in the vicinity of the inflection point. Good torsional restraint provided, for example, by floor system attachments to the top flange are likely to be adequate to consider the inflection point as a braced point for use with the AISC "compact section" L_c requirements. Some years ago Milek [23] stated the point of inflection should *not* be considered as braced. Now, after considerable study of the lateral stability of continuous beams Milek's view appears overly conservative. The authors have concluded that in routine situations treatment of the point of inflection as a braced point is reasonable when moment gradient is not explicitly provided for by the design formulas (such as in application of the AISC expression for L_c). The combination of torsional restraint provided by the floor system attachments and the continuity at the support point of maximum negative moment will normally be adequate to justify considering the inflection point to be a braced point.

When AISC Formulas (1.5–6) and (1.5–7) are considered inadequate to evaluate lateral-torsional buckling strength for an unbraced length spanning across an inflection point when using the working stress method, the SSRC "Basic" method discussed in Sec. 9.3 may be used. For plastic design or for the working stress "compact section" requirement, one might use Eq. 9.3.43, taking M as the moment at the first laterally braced point beyond the inflection point.

Regarding endorsement of these suggestions, AISC–1.5.1.4.5(2) authorizes "a more precise analysis" for allowable stresses not exceeding $0.60F_y$. However, neither AISC–1.5.1.4.1 for "compact section" under working stress design nor AISC–2.9 for plastic design gives permission for using a more exact approach. Thus, for inelastic lateral-torsional buckling strength the designer seemingly has no choice but to use the given equations. For such equations, the value of L to be used should be established rationally in recognition that moment gradient is not a variable. The use of the inflection point as a braced point when torsional restraint is present seems realistic for equations that do not involve the moment gradient.

Cantilever Beams

Unlike the flagpole type of column where the effective pinned-end length is twice the actual length, the lateral buckling of a cantilever beam is not even as severe as the unbraced segment under uniform moment. If one considers the analogy to a column as discussed in Sec. 9.1, such a result will be logical. Since the moment at the end of the cantilever is zero, the compression force in the flange decreases from a maximum at one end to zero at the free end.

Clark and Hill [19] and the *SSRC Guide* [8] report that it is conservative to use the full length as the effective laterally unsupported length for lateral buckling of cantilevers. The authors suggest that the actual cantilever length be used for the AISC formulas; but the *SSRC Guide* [8] states that for cantilevers having complete *fixity about both axes* at the support end, C_b is 1.3 for a concentrated load on the free end, and 2.05 for uniform loading. Such fixity rarely exists so that taking $KL = L$ is recommended. For nonuniform cantilevers, refer to the work of Massey and McGuire [24].

9.9 DESIGN OF LATERALLY UNSUPPORTED BEAMS —AISC WORKING STRESS METHOD

Several examples are presented to illustrate working stress design procedure to include lateral-torsional buckling as a factor determining the allowable stress. Other considerations, such as deflection, shear, and web crippling were treated and illustrated in Chapter 7. For some additional practical treatment of AISC beam design, see Stockwell [25] and Brandt [26].

Example 9.9.1

A simply supported beam is loaded as shown in Fig. 9.9.1. The beam has transverse lateral support at the ends and every 7'-6" along the span. Select the lightest W section of A36 steel, using AISC Specification.

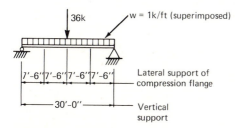

Fig. 9.9.1 Example 9.9.1.

SOLUTION

(a) Estimate whether or not lateral supports are close enough to qualify beam as "compact section" under AISC–1.5.1.4.1. For "compact section," Eqs. 9.5.1 and 9.5.3,

$$L_c = \text{max unbraced length} = \text{smaller of } \frac{76b_f}{\sqrt{F_y}} \text{ and } \frac{20,000}{(d/A_f)F_y}$$

$$= 12.7b_f \quad (\text{for } F_y = 36 \text{ ksi})$$

$$b_{f\min} = \frac{L_c}{12.7} = \frac{7.5(12)}{12.7} = 7.1 \text{ in.} \quad \text{if } L_c = 7.5 \text{ ft}$$

(b) Assume compact section; the allowable bending stress

$$F_b = 0.66F_y = 0.66(36) = 24 \text{ksi}$$

$$M = \tfrac{1}{8}(1.0)(30)^2 + \tfrac{1}{4}(36)(30) = 382.5 \text{ ft-kips without beam weight}$$

$$\text{Required } S_x = \frac{M}{F_b} = \frac{382.5(12)}{24} = 191 \text{ in.}^3$$

(c) Selection from "Allowable Stress Design Selection Table," AISC Manual. This procedure is efficient if designer is certain "compact section" requirements can be satisfied. Select W24×84, $S_x = 196$ in.^{3} (One may also use text Appendix Table A3.)

$$M_{DL} = (0.084)(30)^2/8 = 9.54 \text{ ft-kips}$$

$$f_b = \frac{M}{S_x} = \frac{(382.5 + 9.45)(12)}{196} = 24.0 = F_b \qquad \text{OK}$$

$$b_f = 9.02 > 7.1 \text{ in.} \qquad \text{OK}$$

Other requirements for compact section are satisfied.
Use W24×84.

Example 9.9.2

Select the lightest W section for the simply supported beam of Fig. 9.9.2. Lateral support is provided at the ends and at midspan. Assume deflection limitations need not be considered. Use A36 steel and the AISC Specification.

SOLUTION

(a) Use AISC Manual beam curves, "Allowable Moments in Beams." These curves are plots of the allowable moment $F_b S_x$ vs laterally unbraced length L for rolled W and M shapes for $F_y = 36$ ksi and $F_y = 50$ ksi. The allowable stress F_b is the larger of the values obtained from AISC Formulas (1.5–6) and (1.5–7), Eqs. 9.5.13, 9.5.15, and 9.5.5, with the value of C_b taken as 1.0 for the curves.

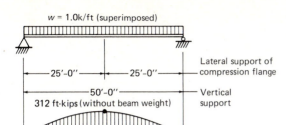

Fig. 9.9.2 Data for Examples 9.9.2 and 9.9.4.

Assuming the beam weight about 90 lb/ft, the midspan bending moment is

$$M = \tfrac{1}{8}(1.09)(50)^2 = 341 \text{ ft-kips}$$

$$L = \text{Laterally unbraced length} = 25 \text{ ft}$$

$$C_b = 1.75 \text{ (AISC–1.5.1.4.5(2.), Eq. 9.4.4 for } C_1)$$

An examination of AISC Formula (1.5–6a) will show that C_b can be combined with L if $L/\sqrt{C_b}$ replaces L in that equation, Eq. 9.5.13. In AISC Formula (1.5–7) C_b can be combined with L by replacing L by L/C_b. Since the curves are for $C_b = 1$, when C_b is not 1.0 the curves may be entered as follows:

1. Use L/C_b if Formula (1.5–7) controls F_b.
2. Use $L/\sqrt{C_b}$ if Formula (1.5–6a) controls F_b.

Since for W and M rolled shapes torsional strength is usually the predominant factor, most such beams have F_b controlled by AISC Formula (1.5–7), it is recommended that the first entry to the curves be with L/C_b. As the values are read from the curves, the user can tell from the shape of the curve whether it is the hyberbola of Formula (1.5–7) or the parabola of Formula (1.5–6a), and adjust the entry appropriately.

For this example, in place of L enter with

$$\frac{L}{C_b} = 14.3 \text{ ft} \quad \text{or} \quad \frac{L}{\sqrt{C_b}} = 18.9 \text{ ft}$$

depending on the formula suspected as controlling the allowable stress. In using the curves, the solid line *above* the intersection of M and L indicates the lightest section satisfying these requirements.

Entering with $M = 341$ ft-kips and $L/C_b = 14.3$ ft, find

$$\text{W27} \times 84, \qquad M > 341 \text{ ft-kips}$$

$$\text{W24} \times 84, \qquad \text{Might work}$$

Note that the W27×84 is controlled by the parabolic curve, F_b(1.5–6a), and reentering the curves with $L/\sqrt{C_b} = 18.9$ ft indicates W27×84 to be inadequate.

(b) Check W24×84, $S_x = 196$ in.³

Using AISC Formula (1.5–7), Eq. 9.5.5,

$$F_b = \frac{12,000C_b}{Ld/A_f} = \frac{12,000(1.75)}{25(12)3.47} = 20.2 \text{ ksi} < 0.60F_y \text{ maximum}$$

Though from the AISC Manual curves it is not expected to govern, examine AISC Formula (1.5–6a), Eq. 9.5.13,

$$\frac{L}{r_T} = \frac{25(12)}{2.31} = 129.9; \qquad C_b = 1.75$$

$$\left(\frac{L}{r_T}\right)_{\substack{\text{lower} \\ \text{limit}}} = \sqrt{\frac{102,000C_b}{F_y}} = 70$$

$$\left(\frac{L}{r_T}\right)_{\substack{\text{upper} \\ \text{limit}}} = \sqrt{\frac{510,000C_b}{F_y}} = 157$$

Since $70 < 129.9 < 157$, parabolic Formula (1.5–6a) applies,

$$F_b = \left[\frac{2}{3} - \frac{F_y(L/r_T)^2}{1530(10^3)C_b}\right]F_y$$

which for A36 steel becomes

$$F_b = 24.0 - \frac{(L/r_T)^2}{2070} = 24.0 - \frac{(129.9)^2}{2070} = 15.8 \text{ ksi}$$

which is less than given by Formula (1.5–7) and therefore does not control.

Bending stress under applied load, corrected for beam weight, is

$$f_b = \frac{M}{S_x} = \frac{339(12)}{196} = 20.7 \text{ ksi} > F_b = 20.2 \text{ ksi}$$

Use W24×84, since there is only about 2.5% overstress.

Example 9.9.3

Select an economical W section for the beam of Fig. 9.9.3. Lateral support is provided at the vertical supports, concentrated load points, and at the end of the cantilever. Use A36 steel and the AISC Specification.

SOLUTION

Three cases must be considered since each of the three laterally unbraced lengths is different and is subject to a different maximum bending moment. Assume the segment containing the largest moment governs.

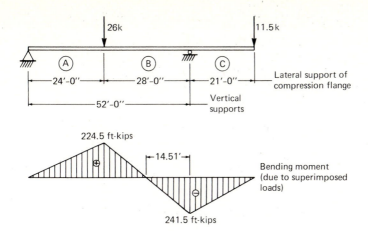

Fig. 9.9.3   Data for Example 9.9.3.

(a) Segment ©:

$$M = 241.5 \text{ ft-kips}, \qquad C_b = 1.75, \qquad L = 21 \text{ ft}$$

Use the AISC Manual curves, "Allowable Moments in Beams," a portion of which is shown in Fig. 9.9.4. As discussed previously, the C_b factor may be combined with L, as follows:

$$\frac{L}{C_b} = \frac{21}{1.75} = 12 \text{ ft} \qquad \text{if } F_b \text{ (1.5–7) governs}$$

or

$$\frac{L}{\sqrt{C_b}} = \frac{21}{\sqrt{1.75}} = 15.9 \text{ ft} \quad \text{if } F_b(1.5\text{–}6a) \text{ governs}$$

Try entering curves with $M = 241.5$ ft-kips with $L/C_b = 12.0$ ft, select

$$\text{W21} \times 68, \qquad M = 247 \text{ ft-kips}, \qquad L_u = 12.4 \text{ ft}$$

Since $L = 21$ ft exceeds L_c (8.7 ft) but $L/C_b = 12.0$ ft does not exceed L_u (12.4 ft),

$$F_b = 0.60 F_y$$

Note that in the curves the solid black dot (•) represents L_c and the open circle (○) represents L_u. Also, the values of L_c and L_u for $F_y = 36$ and 50 ksi are given in the AISC Manual. The L_c and L_u values for $F_y = 36$, 42, 45, 50, 55, 60, and 65 ksi are given in text Appendix Table A3.

The additional moment due to a beam weight about 76 lb/ft should be included at this stage.

At the $(-M)$ region,

$$M_{DL} = \tfrac{1}{2}(0.07)(21)^2 = 15.4 \text{ ft-kips}$$

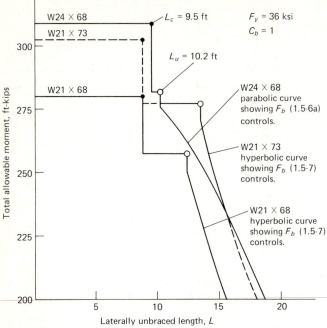

Fig. 9.9.4 Portion of AISC curves, "Allowable Moments in Beams" used in Example 9.9.3.

At the 26 kips load,

$$M_{DL} = \tfrac{1}{2}(0.07)(24)(28) - 15.4\left(\frac{24}{52}\right) = 16.4 \text{ ft-kips}$$

Reexamine curves with

$$M = 241.5 + 15.4 = 257 \text{ ft-kips}$$

Segment Ⓒ looks OK.

(b) Segment Ⓑ: From AISC–1.5.1.4.5(2.)

$$C_b = 1.75 + 1.05\left(\frac{M_1}{M_2}\right) + 0.3\left(\frac{M_1}{M_2}\right)^2 = 2.99$$

Use maximum $C_b = 2.3$.

$$\frac{L}{C_b} = \frac{28}{2.3} = 12.2 \text{ ft}; \qquad \frac{L}{\sqrt{C_b}} = \frac{28}{\sqrt{2.3}} = 18.5 \text{ ft}$$

Since $L/C_b < (L_u = 12.4 \text{ ft})$, this still looks acceptable for $M = 257$ ft-kips.

(c) Segment Ⓐ: $C_b = 1.75$

$$\frac{L}{C_b} = \frac{24}{1.75} = 13.7 \text{ ft}; \qquad \frac{L}{\sqrt{C_b}} = \frac{24}{\sqrt{1.75}} = 18.1 \text{ ft}$$

Here L/C_b exceeds $L_u = 12.4$ ft for W21×68, meaning F_b will be less than $0.60F_y$.

$$M = 224.5 + 16.4 = 241 \text{ ft-kips}$$

From the curves, the point defined by $M = 241$ and $L/C_b = 13.7$ ft lies above the hyperbolic curve for W21×68, indicating that W21×68 is not adequate. A reentry to the curves shows the next solid line curve above the W21×68 is that for W24×68; however, that curve is a parabolic one and for that section

$$M = 241 \text{ ft-kips} \quad \text{at} \quad L/\sqrt{C_b} = 18.1 \text{ ft}$$

The W24×68 is not adequate for Segment Ⓐ. The next section is W21×73 with the hyperbolic curve governing.

$$M > 241 \text{ ft-kips} \quad \text{at} \quad L/C_b = 13.7 \text{ ft}$$

(d) Check stresses for W21×73, $S_x = 151$ in.^{3}
For Segment Ⓐ:

$$F_b(1.5\text{-}7) = \frac{12{,}000C_b}{Ld/A_f} = \frac{12{,}000(1.75)}{24(12)3.46} = 21.1 \text{ ksi}$$

or for $L/r_T = 24(12)/2.13 = 135.2 < 157$. Thus parabolic Formula (1.5–6a) is to be used.

$$F_b(1.5\text{-}6a) = 24.0 - \frac{(L/r_T)^2}{1181C_b}$$

$$= 24.0 - \frac{(135.2)^2}{1181(1.75)} = 15.2 \text{ ksi}$$

Thus the larger, $F_b = 21.1$ ksi, governs.

$$f_b = \frac{M}{S_x} = \frac{241(12)}{151} = 19.1 \text{ ksi} < F_b \qquad \text{OK}$$

For Segment Ⓒ:

$$F_b(1.5\text{-}7) = \frac{12{,}000(1.75)}{21(12)3.46} = 24.1 \text{ ksi} > 0.60F_y$$

$$F_b = 0.60F_y = 22 \text{ ksi} \quad \text{governs}$$

(Note that $F_b = 0.60F_y$ may also be determined from $L/C_b = 12$ ft $< L_u$.)

$$f_b = \frac{M}{S_x} = \frac{257(12)}{151} = 20.4 \text{ ksi} < F_b = 22 \text{ ksi} \qquad \text{OK}$$

For Segment Ⓑ:

$$M = 257 \text{ ft-kips} \qquad \text{same as Segment Ⓒ}$$

$$L/C_b = 12.2 \text{ ft} < L_u; \qquad F_b = 22 \text{ ksi}$$

Stress check identical with Segment ©.
Use W21×73.

Example 9.9.4

Repeat Example 9.9.2 (See Fig. 9.9.2) except use A572 Grade 60 steel.

SOLUTION

Since there are no AISC Manual tables or curves available for direct use when $F_y = 60$ ksi, a more general approach must be used.

Note the following in regard to the allowable stresses F_b:

1. If AISC Formula (1.5–7) controls, the curves, "Allowable Moments in Beams" give the correct capacity, since that formula does not involve the yield stress F_y.
2. If AISC Formula (1.5–6) controls, with L/r_T not exceeding about 85, the capacity will be roughly proportional to the yield stress F_y.
3. If the lateral support conditions are such that $L < L_u$, or that $L < L_c$, the capacity is directly proportional to the yield stress F_y.

(a) Use the AISC Manual, "Allowable Moments in Beams." For approximate entry to the $F_y = 50$ ksi curves, convert the applied moment into equivalent moment on a beam having $F_y = 50$ ksi.

Actual $M = 340$ ft-kips (for $F_y = 60$ ksi) (from Example 9.9.2, including estimated beam weight)

$$M\text{(for } F_y = 50 \text{ ksi)} \approx 340\left(\frac{50}{60}\right) = 283 \text{ ft-kips}$$

$$\frac{L}{C_b} = \frac{25}{1.75} = 14.3 \text{ ft} \qquad \text{if } F_b(1.5\text{–}7) \quad \text{governs}$$

$$\frac{L}{\sqrt{C_b}} = \frac{25}{\sqrt{1.75}} = 18.9 \text{ ft} \qquad \text{if } F_b(1.5\text{–}6a) \quad \text{governs}$$

At the point for $M = 283$ ft-kips and $L = 14.3$ ft, it appears the following sections might prove satisfactory:

W16×71   [hyperbolic curve, $F_b(1.5\text{–}7)$]

Also additional sections,

W21×73   [hyperbolic curve, $F_b(1.5\text{–}7)$]

W24×68   [hyperbolic curve, $F_b(1.5\text{–}7)$]

The solid line for W24×68 indicates that section to be the lightest one satisfying the entry requirements.

At this stage, with AISC Formula (1.5–7) governing, the entry must be revised to use $M = 340$ ft-kips, since F_b would not be a function of F_y.
For $M = 340$ and $L = 14.3$ ft,

$$\text{W24} \times 76 \quad [\text{parabolic curve, } F_b(1.5\text{–}6a)]$$

This section, however, requires entering with $L = 18.9$ ft. Revise to

$$\text{W27} \times 84 \quad [\text{parabolic curve, } F_b(1.5\text{–}6a)]$$

Of course, with the parabolic curve controlling, M would be reduced closer to 283 ft-kips for the $F_y = 50$ ksi curves. From all of this, satisfactory sections are expected in the weight range of 70 to 80 lb/ft. The shallower sections have $F_b(1.5\text{–}7)$ controlling while the deeper sections have $F_b(1.5\text{–}6a)$ controlling.

With this preliminary examination of the curves accomplished, check stresses on the W27×84, the section most likely to be acceptable.

(b) Check stresses on W27×84, $S_x = 213$ in.[3]. From text Appendix Table A3, $L_c = 6.6$ ft and $L_u = 8.6$ ft for $F_y = 60$ ksi.

$$\frac{L}{r_T} = \frac{25(12)}{2.49} = 120.5$$

From Table 9.5.2 with $C_b = 1.75$,

$$54 < 120.5 < 122 \qquad \therefore F_b(1.5\text{–}6a) \text{ applies.}$$

$$F_b(1.5\text{–}6a) = \left[\frac{2}{3} - \frac{F_y(L/r_T)^2}{1530(10^3)C_b}\right] F_y$$

$$= 40.0 - \frac{(L/r_T)^2}{425C_b} \qquad \text{for } F_y = 60 \text{ ksi}$$

$$= 40.0 - \frac{(120.5)^2}{425(1.75)} = 20.5 \text{ ksi}$$

Though it is not expected to control, check

$$F_b(1.5\text{–}7) = \frac{12{,}000C_b}{Ld/A_f}$$

$$= \frac{12{,}000(1.75)}{25(12)4.19} = 16.7 \text{ ksi}$$

The larger value, $F_b = 20.5$ ksi, controls.

$$M = \tfrac{1}{8}(1.084)(50)^2 = 339 \text{ ft-kips}$$

$$f_b = \frac{M}{S_x} = \frac{339(12)}{213} = 19.1 \text{ ksi} < F_b = 20.5 \text{ ksi} \qquad \text{OK}$$

Based on the preliminary work with the AISC beam curves, there may be shallower and lighter weight sections that will work. Those shallower sections have higher torsional strength which will raise F_b to be closer to $0.60F_y$ (i.e., 36 ksi). A summary of the check on other sections is shown in Table 9.9.1.

Table 9.9.1 Summary of Stress Check for Example 9.9.4

Beam	$(Ld/A_f)/C_b$	$F_b(1.5\text{–}7)$	L/r_T	$F_b(1.5\text{–}6)$	$f_b = M/S_x$	Comment
W14×74	307	36.0	110	—	36.0	OK. Deflection?
W16×77	362	33.2	108	24.2	30.1	OK
W18×76	416	28.8	102	26.1	27.6	OK
W21×83	526	22.8	140	15.3	23.8	4% high NG
W24×84	595	20.2	130	17.6	20.7	2.7% high OK
W27×84	718	16.7	120.5	20.5	19.1	OK

For lightest section; use W14×74 if deflection can be tolerated; or W27×84 if minimum deflection is desired.

Example 9.9.5

What W section can be used for the beam of Example 9.9.4 if lateral support is provided every 5 ft?

SOLUTION
In this case, since the stability has been improved by reducing the unbraced length, the deeper sections can safely carry greater load. The deeper sections will also be the lightest ones.

Assume $L < L_c$, in which case the allowable stress will be $0.66F_y$ if the section can satisfy local buckling prevention requirements for a "compact section."

$$M = \tfrac{1}{8}(1.06)(50)^2 = 331 \text{ ft-kips}$$

$$\text{Required } S_x \geq \left[\frac{M}{F_b} = \frac{331(12)}{39.6} = 100 \text{ in.}^3 \right]$$

Try W24×55: $S_x = 114$ in.^{3}, $d/A_f = 6.66$, $b_f = 7.005$ in.

Check whether "compact section" requirements, AISC–1.5.1.4.1, are satisfied.

Lateral support:

$$\frac{20,000}{F_y(d/A_f)} = \frac{20,000}{60(6.66)12} = 4.2 \text{ ft}$$

or

$$\frac{76b_f}{\sqrt{F_y}} = \frac{76(7.005)}{\sqrt{60}(12)} = 5.7 \text{ ft}$$

Thus $L_c = 4.2$ ft, since the *smaller* value governs.

The lateral support is not adequate for $F_b = 0.66F_y$, since $(L = 5 \text{ ft}) > L_c$.

Check whether $L = 5$ exceeds L_u, the maximum unbraced length for $F_b = 0.60F_y$. The W24×55 may still work since the $S_x = 114$ was 14% greater than required for $F_b = 0.66F_y$. Using Eqs. 9.5.18,

$$\frac{20,000}{(d/A_f)F_y} = 4.2 \text{ ft} \quad \text{(computed above)}$$

or

$$r_T\sqrt{\frac{102,000}{F_y}} = \frac{1.68}{12}\sqrt{\frac{102,000}{60}} = 5.8 \text{ ft}$$

Thus $L_u = 5.8$ ft, since the *larger* value governs.

The lateral support is adequate for $F_b = 0.60F_y$, since $L < 5.8$ ft. Note that the 5-ft segment adjacent to midspan is governing and C_b will be close to 1.0. If L_u had been only slightly less than 5.0 ft, computation of C_b may have been prudent, but in this case the refinement of C_b is unnecessary.

$$M = \tfrac{1}{8}(1.055)(50)^2 = 330 \text{ ft-kips}$$

$$f_b = \frac{M}{S_x} = \frac{330(12)}{114} = 34.7 \text{ ksi} < F_b = 36 \text{ ksi} \qquad \text{OK}$$

<u>*Use* W24×55</u>, the lightest section satisfying the loading and support conditions.

Example 9.9.6

Using the "more precise analysis" permitted by AISC–1.5.1.4.5(2.)—i.e., the SSRC "Basic" design procedure as discussed in Sec. 9.4, determine whether a W24×76 would be acceptable for the beam of Example 9.9.4 (Fig. 9.9.2). $F_y = 60$ ksi.

SOLUTION

$$M = \tfrac{1}{8}(1.076)(50)^2 = 336 \text{ ft-kips}$$

Assume the loading acts at the centroid of the beam, and use Eqs. 9.4.1 and 9.4.3,

$$F_{cr} = \frac{C_4\sqrt{EI_yGJ}}{S_xL} \qquad \qquad \text{[9.4.1]}$$

where C_4 is given by the following ($C_2 = 0$):

$$C_4 = C_1\pi\sqrt{1 + \left(\frac{\pi}{\lambda L}\right)^2} \qquad \qquad \text{[9.4.3]}$$

Torsional properties of beams may be obtained from the AISC Manual. For W24×76,

$$J = 2.84 \text{ in.}^4 \qquad C_w = 11,100 \text{ in.}^6$$
$$\lambda = \sqrt{J/(2.6C_w)} = 1/103.8$$

The actual loading and bracing most closely corresponds to Case 3 of Fig. 9.4.1; for which case $C_1 = C_b = 1.75$, according to AISC Spec.

$$C_4 = 1.75\pi\sqrt{1+(103.8\pi/300)^2} = 8.12$$

$$F_{cr} = \frac{8.12E\sqrt{I_yJ/2.6}}{S_xL}$$

$$= \frac{8.12(29,000)\sqrt{82.5(2.68)/2.6}}{176(300)} = 41.1 \text{ ksi}$$

Since $41.1 \text{ ksi} > F_y/2$, use Eq. 9.4.6 to find an equivalent slenderness ratio,

$$\frac{KL}{r_e} = \pi\sqrt{\frac{29,000}{41.1}} = 83$$

The basic SSRC parabola is then used for the inelastic range of buckling,

$$F_{cr} = F_y\left[1-\frac{F_y}{4\pi^2E}\left(\frac{KL}{r_e}\right)^2\right]$$

$$= 60\left[1-\frac{60}{4\pi^2(29,000)}(83)^2\right] = 38.1 \text{ ksi}$$

$$F_b = \frac{F_{cr}}{1.67} = \frac{38.1}{1.67} = 22.8 \text{ ksi}$$

$$f_b = \frac{336(12)}{176} = 23.0 \text{ ksi}$$

A W24×76 would be overstressed slightly according to the "rational method" indicating conservatism of AISC formulas. Whenever both terms under the square root sign in Eq. 9.3.32 are significant, the AISC Formulas are conservative.

The AISC Commentary alternative (though essentially the same philosophically) to the procedure just illustrated is to compute r_e according to Eq. 9.4.9,

$$r_e^2 = \frac{I_y}{2S_x}\sqrt{h^2+\frac{0.156L^2J}{I_y}}$$

$$= \frac{82.5}{2(176)}\sqrt{(23.92-0.680)^2+\frac{0.156(300)^2 2.68}{82.5}} = 7.40 \text{ in.}$$

$$r_e = \sqrt{7.40} = 2.72 \text{ in.}$$

Then r_e replaces r_T in AISC Formulas (1.5–6),

$$\frac{L}{r_e} = \frac{300}{2.72} = 110.2 < 122 \qquad \therefore \text{Use } F_b\,(1.5\text{–}6a)$$

$$F_b = 40.0 - \frac{(L/r_T)^2}{425C_b}$$

$$= 40.0 - \frac{(110.2)^2}{425(1.75)} = 23.6 \text{ ksi}$$

This compares with $F_b = 22.8$ ksi using the SSRC "Basic" procedure.

9.10/*LATERAL BUCKLING OF CHANNELS, ZEES, AND UNSYMMETRICAL I-SHAPED SECTIONS

The basic development of lateral buckling strength-related criteria has assumed that loads are applied vertically through the shear center. Furthermore, the resistance to lateral buckling considered that the shear forces which developed in the flanges were equal and the center of twist was located at mid-height.

Channels

Unless loaded through the shear center, a channel is subjected to combined bending and torsion. Since the shear center is not in the plane of the web (see Fig. 8.5.1), usual loadings through the centroid or in the plane of the web give rise to such combined stress. For loads in a plane parallel to the web, lateral buckling must be considered, even if the torsional moment may properly be neglected. The *SSRC Guide* [8] states "if an otherwise laterally unsupported channel has concentrated loads brought in by other members that frame into it, such loads can be considered as being applied at the shear center, *provided that the span of the framing member is measured from the channel shear center and the framing connections are designed for the moment and shear at the connection.*"

For design purposes, Hill [27] indicates that the lateral buckling equations for symmetrical I-shaped sections may be applied for channels (AISC–1.5.1.4.5(2.) accepts use of Formula (1.5–7) only). Such a procedure is stated to err on the unsafe side by about 6 percent in extreme cases. For more exact determination of lateral buckling strength, Eq. 9.3.34 may be used for theoretical elastic buckling stress along with the Eq. 9.4.5 approach of using an equivalent slenderness ratio. The torsion-bending constant, C_w, is different for channels than for I-shaped sections and can be obtained from the AISC Manual or computed from formulas in text Appendix Table A2.

Zees

The zee-section lateral buckling strength is complicated by the fact that loading in the plane of the web causes unsymmetrical bending, resulting because a principal axis does not lie in that plane. The general treatment of buckling under biaxial bending is found in Sec. 9.12. The effect of biaxial bending on zee sections was found [27] to reduce the critical moment, M_x, to 90–95 percent of the value given by Eq. 9.3.34. In addition, the torsion-bending constant C_w, is different than for channels or I-shaped sections.

For design purposes, in view of the fact that unbraced zees are relatively rare, AISC does not provide for them. The authors suggest use of the "rational method" using Eq. 9.3.34, or as an alternative it is suggested to use one-half the values obtained from AISC Formula (1.5–6).

Unsymmetrical I-Shapes

I-shaped sections symmetrical about the y-axis, but unsymmetrical about the x-axis, are summarized in the *SSRC Guide* [8] and by Clark and Hill [19]. The additional variable involved is y_0, the distance from the centroid of the girder cross section to the shear center (positive if the shear center lies between the centroid and compression flange, otherwise negative).

An approximate expression, for uniform moment, as derived by Hill [28] is

$$F_{cr} = \frac{\pi^2 E I_y}{S_c L^2} \left[y_0 + \sqrt{y_0^2 + \frac{C_w}{I_y} \left[1 + \left(\frac{\lambda L}{\pi}\right)^2 \right]} \right]$$ (9.10.1)

where L = effective length for torsional support and where $\lambda = \sqrt{GJ/EC_w}$. Equation 9.10.1 corresponds to Eq. 9.3.34 for symmetrical sections.

According to AISC–1.5.1.4.5(2.), such sections with an axis of symmetry in the plane of loading may be dealt with using Formulas (1.5–6) and (1.5–7). Such an approach will be conservative provided the compression flange area is larger than the tension flange area. Equation 9.10.1 is also conservative under the same conditions; but is considered unsafe when the compression flange area is smaller than the tension flange area.

When the tension flange has the greater area, an equation derived by Winter [29] may be used in the "rational method,"

$$F_{cr} = \frac{\pi^2 E d}{2 S_c L^2} \left[I_c - I_t + I_y \sqrt{1 + \frac{4GJ}{Ed^2 I_y} \left(\frac{L}{\pi}\right)^2} \right]$$ (9.10.2)

where I_c and I_t are the moments of inertia of the compression and tension flanges, respectively, computed about the y-axis, and S_c is the section modulus S_x referred to the compression flange.

Example 9.10.1

Determine the allowable bending stress for a channel, C12×20.7, on a span of 24 ft with concentrated loads at the $\frac{1}{3}$-span points as shown in Fig. 9.10.1. Use steel with $F_y = 50$ ksi.
(a) Use AISC procedure.
(b) Use "Basic" design procedure (i.e., equivalent slenderness ratio method) suggested by the Structural Stability Research Council (SSRC).

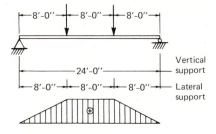

Fig. 9.10.1 Example 9.10.1.

SOLUTION

(a) AISC Method. Assume loading to be through the shear center. For C12×20.7, $r_y = 0.799$ in., $d/A_f = 8.13$, $S_x = 21.5$ in.^{3} Since concentrated load points in this case provide lateral bracing, the center unbraced segment controls; thus $C_b = 1.0$.

Since only AISC Formula (1.5–7) may be applied,

$$F_b(1.5\text{–}7) = \frac{12,000}{Ld/A_f} = \frac{12,000}{8(12)8.13} = 15.4 \text{ ksi}$$

(b) SSRC "Basic" design procedure. For this procedure the torsional properties are required. From AISC Manual, $J = 0.371$ in.^{4} and $C_w = 112$ in.^{6} (These may also be computed using formulas in text Appendix Table A2.)

$$\lambda = \frac{1}{a} = \sqrt{\frac{GJ}{EC_w}} = \sqrt{\frac{0.371}{1.2(112)}} = \frac{1}{28.0}$$

Using Eq. 9.3.34,

$$F_{cr} = \frac{C_4\sqrt{EI_yGJ}}{S_xL} \qquad [9.3.34]$$

where

$$C_4 = \pi\sqrt{1 + \left(\frac{\pi}{\lambda L}\right)^2} = \pi\sqrt{1 + \left(\frac{28.0\pi}{96}\right)^2} = 4.26$$

$$F_{cr} = \frac{4.26(29,000)\sqrt{3.88(0.371)/2.6}}{21.5(96)} = 44.5 \text{ ksi}$$

Since $F_{cr} > F_y/2$, use Eq. 9.4.6 to obtain an equivalent slenderness ratio,

$$\frac{KL}{r_e} = \pi\sqrt{\frac{E}{F_{cr}}} = \pi\sqrt{\frac{29,000}{44.5}} = 80$$

Applying this with the basic SSRC Column Formula, Eq. 6.7.3, which gives

$$F_{cr} = 50\left[1 - \frac{50}{4\pi^2(29,000)}(80)^2\right] = 36.0 \text{ ksi}$$

$$F_b = 36.0/1.67 = 21.5 \text{ ksi}$$

which indicates an allowable stress about 40 percent higher than AISC permits.

Thus the AISC Method is conservative for this channel, probably justifiably to offset the neglect of torsional moment when the load does not pass through shear center.

Example 9.10.2

Repeat Example 9.10.1 using a girder (Fig. 9.10.2) consisting of flange plates, $\frac{3}{4} \times 10$ for compression, and $\frac{3}{4} \times 5$ in tension, with a $\frac{3}{8} \times 13\frac{1}{2}$ web. The span is 42 ft, braced at the one-third points. $F_y = 50$ ksi.

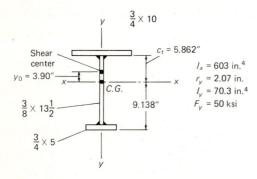

Fig. 9.10.2 Example 9.10.2.

SOLUTION

(a) AISC Method, $C_b = 1.0$

$$r_T = \sqrt{\frac{t_f b_f^3/12}{t_f b_f + (c_t - t_f)t_w/3}}$$

$$r_T = \sqrt{\frac{0.75(10)^3/12}{10(0.75) + 5.112(0.375)/3}} = 2.77 \text{ in.}$$

$$\frac{L}{r_T} = \frac{14(12)}{2.77} = 60.6$$

which exceeds $L/r_T = 45$ (Table 9.5.2, $F_y = 50$ ksi), indicating the parabolic equation, Formula (1.5–6a) is to be used.

$$F_b(1.5\text{–}6a) = 33.3 - \frac{(60.6)^2}{612} = 27.3 \text{ ksi}$$

$$F_b(1.5\text{–}7) = \frac{12{,}000 C_b}{Ld/A_f} = \frac{12{,}000(1.0)}{14(12)(15.0)/7.5} = 35.7 \text{ ksi} > 0.60 F_y$$

Thus $F_b = 0.60 F_y = 30$ ksi and stability is not controlling.

(b) SSRC "Basic" method using equivalent slenderness ratio. Use Eq. 9.10.1:

$$F_{cr} = \frac{\pi^2 E I_y}{S_c L^2}\left[y_0 + \sqrt{y_0^2 + \frac{C_w}{I_y}\left[1 + \left(\frac{\lambda L}{\pi}\right)^2\right]} \right] \qquad [9.10.1]$$

Obtain torsional constants using text Appendix Table A2,

$$J = \sum \tfrac{1}{3}bt^3 = \tfrac{1}{3}[(10+5)(0.75)^3 + 13.5(0.375)^3] = 2.35 \text{ in.}^4$$

$$C_w = \frac{t_f h^2}{12}\left[\frac{b_1^3 b_2^3}{b_1^3 + b_2^3}\right] = \frac{0.75(14.25)^2}{12}\left[\frac{(10)^3(5)^3}{1000+125}\right] = 1410 \text{ in.}^6$$

Using $E/G = 2.6$,

$$\lambda = \frac{1}{a} = \sqrt{\frac{GJ}{EC_w}} = \sqrt{\frac{2.35}{2.6(1410)}} = \frac{1}{39.5}$$

y_0 = distance from centroid to shear center = $+3.90$ in.

$S_c = I_x/$distance to compression flange = $603.2/5.862 = 103$ in.^{3}

$$F_{cr} = \frac{\pi^2(29{,}000)(70.3)}{103(168)^2}\left[3.90 + \sqrt{(3.90)^2 + \frac{1410}{70.3}\left[1 + \left(\frac{168}{\pi(39.5)}\right)^2\right]}\right]$$

$$= 85.7 \text{ ksi} > F_y/2$$

Even though F_{cr} exceeds F_y, inelastic buckling may still control.

The equivalent slenderness ratio

$$\frac{KL}{r_e} = \pi\sqrt{\frac{E}{F_{cr}}} = \pi\sqrt{\frac{29{,}000}{85.7}} = 58$$

which in the SSRC basic formula, Eq. 6.7.3, with a 1.67 factor of safety gives

$$F_b = 50\left[1 - \frac{50}{4\pi^2(29{,}000)}(58)^2\right]\frac{1}{1.67} = 25.6 \text{ ksi}$$

which is less than obtained by AISC Formula (1.5–7), indicating that the AISC Formula is probably not conservative.

An alternate application of AISC Formula (1.5–7) was proposed by de Vries [3] where

$$A_f = b_f t_f = \frac{5I_y}{b_f^2}$$

where I_y is the y-axis moment of inertia for the full section if the flanges are equal; or twice the moment of inertia of the compression flange about the y-axis if the flanges are different.

In this case,

$$I_y = 2\left[\frac{(0.75)(10)^3}{12}\right] = 125 \text{ in.}^4$$

$$\text{Equivalent } A_f = \frac{5(125)}{(10)^2} = 6.25 \text{ sq in.}$$

which if used in AISC Formula (1.5–7) gives

$$F_b(1.5\text{–}7) = \frac{12,000}{168(15)/(6.25)} = 29.8 \text{ ksi} < 0.60F_y$$

The use of de Vries suggestion gives the more correct result. The SSRC parabola to account for inelastic buckling is conservative for situations like this one where torsional stiffness predominates in the lateral buckling resistance.

9.11* LATERAL BRACING DESIGN

The questions of what constitutes bracing and how to design bracing continue to be major concerns of practicing engineers. The subject is included in this chapter because a major item of concern in lateral bracing design is the restraint required to prevent lateral-torsional buckling in beams. The development of this section, however, is applicable to the bracing of columns as well as beams. In Sec. 6.9 the concept of *braced* and *unbraced* systems was briefly discussed in regard to the effective length factor *K*. In the following discussion, the emphasis in on *braced* systems; that is, the overall structural system is braced by cross bracing or attachment to an adjoining system that is braced. The bracing requirements for frames are treated in Chapter 14. Bracing for individual beams or columns may consist of cross bracing where the axial stiffness of the bracing elements is utilized; it may be provided at discrete locations by flexural members framing in transverse to the member being braced, wherein both axial and flexural stiffnesses of the bracing member are utilized; or such bracing may be provided continuously by material such as light gage roof decking or wall panels.

Little is available in specifications but some information on point bracing is in the works of Zuk [30], Winter [31], Massey [32], Pincus [33], Galambos [34], Driscoll et al. [17], Higgins [35], Schmidt [36], Lay and

Galambos [37], Taylor and Ojalvo [38], Hartmann [36], and Mutton and Trahair [40]. In addition, Yura [41] has provided simplified practical treatment of the subject. What follows is largely a combination of the work of Winter, Galambos, and Yura.

Point Bracing for Elastic Columns and Beams

Consider the axially loaded column of Fig. 9.11.1a where the top and bottom of the member are assumed to be supported in such a way that no side movement occurs at one end relative to the other. Such restraint would constitute a *braced* system. The bracing to create such restraint may be considered as a spring at the top that is capable of developing a horizontal reaction equal to the spring constant k times the deflection Δ. When the brace has a large spring constant (that is, the brace is very stiff) the deflection Δ could be close to zero and yet the spring may provide a large enough horizontal force to prevent any side motion (sidesway) at the top. This would be the situation in Fig. 9.11.1b. The equilibrium requirement is shown in Fig. 9.11.1c, wherein a sidesway is shown. If one imagines this as a slightly deflected position, then in order to have equilibrium, it is required that

$$P\Delta = QL = (k\Delta)L \qquad (9.11.1)$$

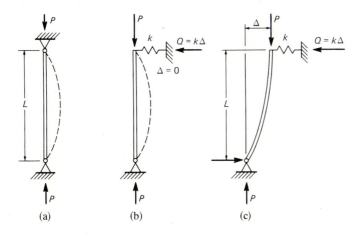

Fig. 9.11.1 Bracing for a single-story column.

If $(k\Delta)L$ is less than $P\Delta$, sidesway occurs. If $(k\Delta)L$ is greater than $P\Delta$, no sidesway occurs and the column would be considered braced. The ideal brace, then, would be one that has just enough stiffness k to prevent movement (at the top in this example); that is,

$$k = \frac{P}{L} \qquad (9.11.2)$$

The maximum load for which bracing would be required is the elastic buckling load P_{cr}, or the load causing yielding or inelastic buckling if that is lower than the elastic P_{cr}. Thus the largest required stiffness k_{ideal} is

$$k_{ideal} = \frac{P_{cr}}{L} \tag{9.11.3}$$

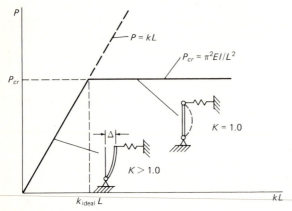

Fig. 9.11.2 Brace stiffness relative to concept of "braced" ($K = 1.0$) and "unbraced" ($K > 1.0$) systems for column hinged at top and bottom.

The concept is shown by the plot of P vs kL in Fig. 9.11.2, wherein when k exceeds k_{ideal}, P_{cr} is reached and the column buckles without end translation (sidesway); in other words, it is a *braced* system. When k is less than k_{ideal}, a sidesway deflection will occur such that $P = kL$; in other words, a so-called *unbraced* system. The major treatment of unbraced systems is in Chapter 14, devoted to rigid frames.

Next, extend the concept to a two-story column within a braced system, as shown in Fig. 9.11.3. When no displacement occurs at midheight, i.e., full bracing is provided, the column will buckle at a load nearly equal to

$$P_{cr} = \frac{\pi^2 EA}{(L/r)^2} \tag{9.11.4}$$

In other words, one may imagine that a hinge exists at mid-height. Buckling occurs when the column snaps into the two half-wave mode of Fig. 9.11.3c.

Taking moments about the imaginary hinge location with the column deflected by an amount Δ, as in Fig. 9.11.3b, gives

$$P_{cr}\Delta = \frac{Q}{2}L \tag{9.11.5}$$

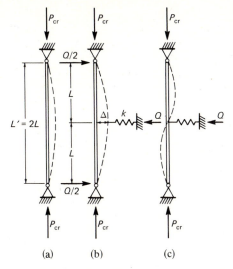

Fig. 9.11.3 Mid-height brace for a two-story column.

Since $Q = k\Delta$,

$$P_{cr}\Delta = \frac{(k\Delta)L}{2} \tag{9.11.6}$$

As in the one-story column, if k_{ideal} is the necessary stiffness to create a nodal point (zero deflection) at mid-height of the two-story column, then

$$k_{ideal} = \frac{2P_{cr}}{L} \tag{9.11.7}$$

For situations with more than two equal spans, the same procedure may be used to obtain k_{ideal}. Examination of Fig. 9.11.4 for three equal spans will show that the spring forces Q can act either in the same or in opposite directions. Assuming they act in the *same* direction (Fig. 9.11.4a), using imaginary hinges at one-third span points, and taking moments when slightly deflected at the brace points, gives

$$QL = (k\Delta)L = P_{cr}\Delta; \qquad k_{ideal} = \frac{P_{cr}}{L} \tag{9.11.8}$$

Assuming the forces Q acting in *opposite* directions (Fig. 9.11.4b) gives

$$QL/3 = (k\Delta)L = P_{cr}\Delta; \qquad k_{ideal} = \frac{3P_{cr}}{L} \tag{9.11.9}$$

The configuration requiring the highest spring constant is the correct one, that which will permit the highest critical load. If a lesser stiffness is used, an alternate buckling mode will occur at a lower load, accompanied by displacement at the springs.

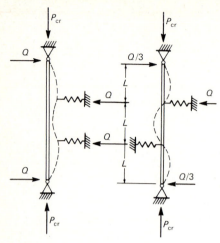

Fig. 9.11.4 Column braced to make three equal spans.

By the same process k_{ideal} may be determined for any number of equal spans. In general,

$$k_{ideal} = \frac{\beta P_{cr}}{L} \qquad (9.11.10)$$

where β varies from 1 for one span to 4 for infinite equal spans. The variation is given in Fig. 9.11.5.

Thus Eqs. 9.11.3, 9.11.7, 9.11.9, and 9.11.10 give the ideal brace *stiffness* to prevent translation at the points where the braces act.

In addition to stiffness, a brace must provide adequate *strength*. The strength Q required of an ideal brace is

$$Q = k_{ideal} \Delta \qquad (9.11.11)$$

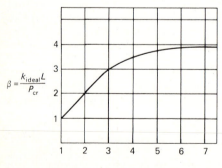

$$\beta = \frac{k_{ideal} L}{P_{cr}}$$

Number of Equal Spans

Fig. 9.11.5 Variation of required spring constant for column with number of equal unbraced spans.

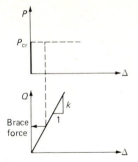

Fig. 9.11.6 Brace force relative to column load for ideal system.

but until buckling occurs, Δ is zero (see Fig. 9.11.6); therefore there will be no brace force in the ideal system until buckling occurs.

Compression members in real structures are not perfectly straight, perfectly aligned vertically, nor perfectly loaded as assumed in calculations; there is always an initial crookedness. In other words, Δ is not zero even when there is no compressive load P acting. Reexamine the single-story column of Fig. 9.11.1 assuming there is an initial deflection Δ_0 that exists even when P is zero. Then, as shown in Fig. 9.11.7, equilibrium requires

$$(k\Delta)L = P(\Delta + \Delta_0) \tag{9.11.12}$$

for $P = P_{cr}$,

$$k_{reqd} = \frac{P_{cr}}{L}\left(1 + \frac{\Delta_0}{\Delta}\right) \tag{9.11.13}$$

Since $k_{ideal} = P_{cr}/L$, Eq. 9.11.13 then becomes

$$k_{reqd} = k_{ideal}\left(1 + \frac{\Delta_0}{\Delta}\right) \tag{9.11.14}$$

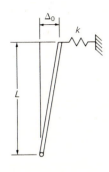

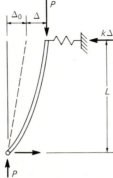

(a) No load applied   (b) When load is applied

Fig. 9.11.7 Column with initial crookedness Δ_0.

which is the *stiffness* requirement for compression members having initial crookedness Δ_0.

The *strength* requirement is then

$$Q = k_{\text{reqd}}\Delta = k_{\text{ideal}}\left(1+\frac{\Delta_0}{\Delta}\right)\Delta$$

$$Q = k_{\text{ideal}}(\Delta+\Delta_0) \tag{9.11.15}$$

Normal tolerances on crookedness of compression members would vary from 1/500 to 1/1000 of the length [31]. The AISC *Code of Standard Practice** indicates acceptable out-of-plumbness to be $L/500$. Considering accidental eccentricity of loading, Winter suggests taking Δ_0 from 1/250 to 1/500 of the length. If $\Delta = \Delta_0$ (as suggested by Winter) and $\Delta_0 = L/500$, the design requirements would be as follows:

For *stiffness*,

$$k_{\text{reqd}} = 2k_{\text{ideal}} \tag{9.11.16}$$

where $k_{\text{ideal}} = \beta P_{\text{cr}}/L$.

For *strength*,

$$Q_u = k_{\text{ideal}}(2\Delta_0)$$

$$Q_u = k_{\text{ideal}}(0.004L) \tag{9.11.17}$$

Note that Eq. 9.11.17 is for ultimate strength and k_{ideal} is based on buckling load P_{cr}. In strength design (for example, using plastic design with factored loads) Eqs. 9.11.16 and 9.11.17 could be used directly. When service loads are used for P in the working stress method, a factor of safety FS must be applied to the stiffness requirement; thus in working stress design where service load P replaces P_{cr} in all equations:

For *stiffness*, with FS = 2,

$$k_{\text{reqd}} = 4k_{\text{ideal}} \tag{9.11.18}$$

where $k_{\text{ideal}} = \beta P/L$.

For *strength*,

$$Q = \frac{Q_u}{\text{FS}} = \frac{\beta P_u}{\text{FS }L}\,(0.004L)$$

$$= k_{\text{ideal}}(0.004L) \tag{9.11.19}$$

where $k_{\text{ideal}} = \beta P/L$.

Where the strength of the compression member being braced is controlled by $F_{\text{cr}} \leq F_y$, the foregoing method is applicable. Under the working stress method, that would correspond to all columns and could apply to the compression flanges of beams where the allowable stress does not exceed $0.60F_y$. If large plastic strain must be accommodated at

---

* *Code of Standard Practice for Steel Buildings and Bridges*, American Institute of Steel Construction, Adopted September 1, 1976 (Section 7.11.3.1).

the bracing points, the design suggestions of Lay and Galambos [37] for inelastic steel beams, as discussed later, should be applied.

To summarize the procedure:

1. Establish the bracing locations and compute $P_{cr} = \pi^2 EI/L^2$ for the compression element (either entire column, or compression flange of beam) being braced.
2. Estimate β from Fig. 9.11.5 based on number of equal unbraced lengths.
3. Compute $k_{ideal} = \beta P_{cr}/L$.
4. Select bracing area A_b so that a stiffness $2k_{ideal}$ will be obtained. For axial stiffness,

$$k = 2k_{ideal}$$

$$\frac{A_b E_b}{L_b} = \frac{2\beta \pi^2 E_c I_c}{L_c^2}$$

$$\text{Required } A_b = 2\beta \pi^2 \left(\frac{E_c}{E_b}\right)\left(\frac{L_b}{L_c}\right)\frac{A_c}{(L_c/r)^2} \qquad (9.11.20)$$

where the subscripts b and c refer to the brace and compression element, respectively. If $E_c = E_b$ and $L_b \approx L_c$, then Eq. 9.11.20 becomes

$$\text{Required } A_b \approx 2\beta \pi^2 \frac{A_c}{(L_c/r)^2} \qquad (9.11.21)$$

5. Verify that the required force can be carried by the brace. Using Eq. 9.11.17,

$$\text{Required } Q_u = k_{ideal}(0.004L)$$

$$= \frac{\beta P_{cr}}{L}(0.004L)$$

$$= 0.004\beta P_{cr} \qquad (9.11.22)$$

The ultimate strength of the brace, including consideration of buckling strength, must equal or exceed the value given by Eq. 9.11.22.

Example 9.11.1

Design a brace (Brace A) to provide lateral support for a W27×84 beam (Beam A) positioned as shown in Fig. 9.11.8. Assume the braces are 7.5 ft long, are attached near the compression flange, and are located along the supported beam at the one-third points of a 48-ft span. Use the AISC working stress method and A36 steel.

SOLUTION

Since the bracing locations are given, the first step here is to compute P_{cr} for the compression zone of the beam or beams to be braced. In this

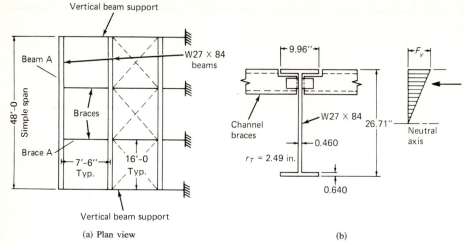

Fig. 9.11.8 Data for Example 9.11.1.

example, only one beam is to be braced. Assume maximum flange stress equals yield stress; that is, allowable stress is $0.60F_y$. When the extreme fiber stress reaches F_y, the total ultimate compressive load P_u is

$$P_u = 36(9.96)(0.640) + \tfrac{1}{2}(36)(13.355 - 0.64)(0.46) = 333 \text{ kips}$$

To obtain the critical load, the slenderness ratio for lateral movement of the compression flange is

$$\frac{L}{r_T} = \frac{16(12)}{2.49} = 77$$

An expression is required to obtain the stress F_{cr} for this beam when lateral buckling is imminent. The SSRC basic parabola (Eq. 6.7.3) would be appropriate. Alternatively, F_{cr} may be obtained by using AISC Formula (1.5–6a) with the factor of safety removed. It is suggested, therefore, to multiply Formula (1.5–6a) by $\tfrac{3}{2}$ so that it gives $F_{cr} = F_y$ for L/r_T approaching zero. Thus

$$F_{cr} = F_y\left[1.0 - \frac{3}{2}\frac{F_y(L/r_T)^2}{1530(10^3)C_b}\right]$$

$$= 36.0\left[1.0 - \frac{(L/r_T)^2}{28,300}\right] = 36.0\left[1.0 - \frac{(77)^2}{28,300}\right] = 28.5 \text{ ksi}$$

$$P_{cr} \text{ for compression zone} \approx \frac{F_{cr}A}{2} = \frac{28.5(24.8)}{2} = 353 \text{ kips}$$

There is no need to design bracing for a compressive force exceeding P_u; use P_u since it is less than P_{cr}.

For three equal unbraced lengths, estimate $\beta = 3$ from Fig. 9.11.5. Using Eq. 9.11.21,

$$\text{Required } A_b = 2\beta\pi^2\frac{A_c}{(L_c/r)^2}$$

$$= 2(3)\pi^2\frac{(24.8/2)}{(77)^2} = 0.124 \text{ sq in.}$$

While any nominal size will satisfy this area requirement, it must be remembered that as a compression element it must satisfy AISC–1.8.4 When several beams are to be braced, the required A_b for the most heavily loaded brace would be based on the total compression area of all beams to be braced.

$$\text{Min } r_y = \frac{L}{200} = \frac{7.5(12)}{200} = 0.45 \text{ in.}$$

Select C4×5.4 as the lightest section with $r_y \geq 0.45$ in. ($r_y = 0.449$ in.)
    Check stiffness:

$$k_{ideal} = \frac{\beta P_{cr}}{L} = \frac{3(333)}{16(12)} = 5.2 \text{ kips/in.}$$

$$k_{act} = \frac{AE}{L_b} = \frac{1.59(29,000)}{7.5(12)} = 503 \text{ kips/in.}$$

which far exceeds the minimum of 2 times k_{ideal} to account for initial crookedness.
    Check strength:

$$\frac{L_b}{r_y} = \frac{7.5(12)}{0.45} = 200$$

$$F_{cr} = \frac{\pi^2 E}{(L/r_y)^2} = \frac{\pi^2(29,000)}{(200)^2} = 7.15 \text{ ksi}$$

$$\text{Actual } Q_u = F_{cr}A_b = 7.15(1.59) = 11.2 \text{ kips}$$

$$\text{Required } Q_u = 0.004\beta P_{cr}$$

$$= 0.004(3)333 = 4.0 \text{ kips} < 11.2 \text{ kips} \qquad \text{OK}$$

Use C4×5.4 for brace *A.*
It is noted that for this case the AISC–1.8.4 maximum slenderness ratio of 200 for compression members assured proper bracing size.

Example 9.11.2

For the beam laterally braced by joists as shown in Fig. 9.11.9, determine the amount of weld required so that the joist will adequately brace the beam. The service load bending moment is 125 ft-kips, and the steel is A36.

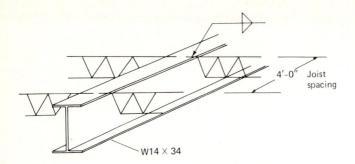

Fig. 9.11.9 Beam laterally braced by joists.

SOLUTION

(a) Determine stiffness required. Whereas in Example 9.11.1 the compression capacity P_{cr} of the flange was used, in this example the desired P_{cr} to be accommodated is taken as the service load compression force in the flange times a desired FS = 2; thus

$$P = \frac{M}{d} = \frac{125(12)}{14} = 107 \text{ kips}$$

$$P_{cr} = P(\text{FS}) = 107(2) = 214 \text{ kips}$$

For 4 or more point braces along the compression flange, $\beta = 4$; thus

$$k_{\text{ideal}} = \frac{\beta P_{cr}}{L} = \frac{4(214)}{4} = 214 \text{ kips/ft}$$

$$k_{\text{reqd}} = 2k_{\text{ideal}} = 428 \text{ kips/ft}$$

Assuming the joists have metal deck adequately attached to them, there will develop in the decking an in-plane shear, known as *diaphragm action*, that will restrain the relative axial movement of two adjacent joists, thereby restraining relative lateral movement between two adjacent laterally braced points on the beam being braced, such as points A and B of Fig. 9.11.10. The adequacy of the attachment of the decking to the joists will determine the degree to which diaphragm action prevents the relative motion of points, such as A and B. If there is zero relative motion, then the requirement for lateral support may be based on $\beta = 1$, i.e., the same as the bracing of a single-story column. In which case, the required stiffness would be

$$k_{\text{reqd}} = 2k_{\text{ideal}} = \frac{2\beta P_{cr}}{L} = \frac{2(1)(214)}{4} = 107 \text{ kips/ft}$$

Using diaphragm action can greatly reduce the stiffness requirement for point bracing. In this case, if diaphragm action can be developed, the stiffness requirement is one-fourth as much as required without any

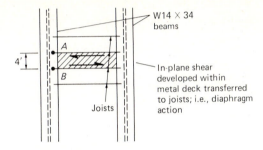

Fig. 9.11.10   Diaphragm action of metal deck attached to joists.

diaphragm action. Metal deck with $2\frac{1}{2}$-in. concrete fill would provide many times the required stiffness. Even the diaphragm action of the metal deck without the concrete slab is likely to provide more than the 107 kips/ft required. Treatment of diaphragm bracing is generally outside the scope of this text; however, a number of references [42–53] on diaphragm action are included at the end of this chapter.

(b) Determine weld required to attach joists to top flange of W14× 34. The force required is, according to Eq. 9.11.22,

$$\text{Required } Q_u = 0.004\beta P_{cr} \qquad \text{[9.11.22]}$$

Note that dividing both sides by FS gives the working stress equation,

$$\text{Required } Q = 0.004\beta P$$

Conservatively take $\beta = 4$ as for a series of point braces,

$$\text{Required } Q = 0.004(4)107 = 1.7 \text{ kips}$$

Using a $\frac{3}{16}$-in. fillet weld with E70 electrodes and the SMAW process, the allowable capacity per inch is

$$R_w = \tfrac{3}{16}(0.707)21 = 2.78 \text{ kips/in.}$$

$$\text{Required } L_w = \frac{Q}{R_w} = \frac{1.7}{2.78} = 0.6 \text{ in.}$$

Use $\frac{3}{16}$-in. weld, E70, $L = 0' - 0\frac{3}{4}''$.

Bracing Requirements for Inelastic Steel Beams

When ability to accommodate large plastic strain is desired at bracing points, such as when plastic design or the "compact section" provision in working stress design is used, the procedure in the previous section may not be adequate. Lay and Galambos [37] have developed a set of rules for design in cases where such high plastic strain (rotation capacity) is required to be accommodated.

In effect, bracing requirements are based on a rotation capacity R

consistent with the beam unbraced length slenderness ratio given by Eq. 9.3.42. It has been found that within the laterally unbraced length "local buckling causes a curtailment of the load capacity of the member and therefore defines the rotation capacity of the beam" [37].

The derivation of Lay and Galambos [37] has determined the maximum lateral moment that can develop in the compression flange under a *uniform* moment, $M_x = M_p$, by using the strain distributions on the compression flange due to (a) compression due to $M_x = M_p$ and (b) the lateral bending strains when local buckling occurs on the "compression" side.

The design recommendations are:

1. For axial strength, the required cross-sectional area is

$$\text{Required } A_b = \left[\frac{\alpha_{st} - 1}{\alpha_e - \sqrt{\alpha_e}}\right]\left[\frac{2}{3}\right]\frac{A_c}{(L_{av}/b)} \tag{9.11.23}$$

where $L_{av} = \dfrac{2L_L L_R}{L_L + L_R}$

L_L = unbraced length to left of braced point
L_R = unbraced length to right of braced point
b = width of compression flange
α_{st} = strain at strain hardening divided by yield strain, ϵ_{st}/ϵ_y (A value of 12 may be reasonable for steels up to $F_y = 60$ ksi [37].)
α_e = elastic modulus divided by strain hardening modulus of elasticity, E_s/E_{st}

2. The axial stiffness requirement is satisfied when

$$\frac{L_b}{L_{av}} \leq 0.57\left[\frac{\alpha_{st} - 1}{\alpha_e - \sqrt{\alpha_e}}\right]\left[\frac{\text{actual } A_b}{\text{required } A_b}\right]\left[\frac{L_a}{b}\right] \tag{9.11.24}$$

where L_a = longer of the two adjacent unbraced lengths.

In addition to the axial strength and stiffness requirements, Lay and Galambos indicated [37] that when only the compression flange is braced there are additional flexural strength and stiffness requirements that must be satisfied. These flexural requirements (not given here) give overly large and deep bracing members. Recent studies [39] indicate that flexural requirements are unnecessary for lateral bracing locations away from beam vertical reactions. When the compression flange is braced, point restraint giving the necessary axial strength and stiffness is sufficient. Lateral bracing at vertical supports undoubtedly does need some flexural strength and stiffness to prevent a beam from tipping, but ordinary framing at such locations generally provides adequate flexural strength and stiffness.

Example 9.11.3

For the beam of Example 9.11.1 (Fig. 9.11.8) determine the brace size required if plastic hinge rotation is required at the bracing points.

SOLUTION

(a) Determine the section required for axial *strength*. Use Eq. 9.11.23,

$$\text{Required } A_b = \left(\frac{\alpha_{st}-1}{\alpha_e-\sqrt{\alpha_e}}\right)\left(\frac{2}{3}\right)\frac{A_b}{(L_{av}/b)}$$

where $\alpha_{st} = \epsilon_{st}/\epsilon_y$ which may be taken as 12 for $F_y = 36$ ksi and can probably also be used for steels to about $F_y = 60$ ksi. For $\alpha_e = E/E_{st}$, use $E_{st} = 450$ ksi, giving $\alpha_e = 29,000/450 = 64$. Certainly, the use of such values is accurate enough for design purposes. For A36 steel, $\alpha_{st} = 12$, $\alpha_e = 64$.

$$\frac{\alpha_{st}-1}{\alpha_2-\sqrt{\alpha_e}} = \frac{12-1}{64-\sqrt{64}} = 0.2$$

$$L_{av} = \frac{2L_LL_R}{L_L+L_R} = \frac{2(16)(16)(12)}{16+16} = 192 \text{ in.}$$

$$\text{Required } A_b = 0.2\left(\frac{2}{3}\right)\frac{A_c}{192/9.96} = 0.007A_c$$

$$= 0.007(24.8/2) = 0.09 \text{ sq in.}$$

Practically any rolled shape will satisfy this small requirement.

(b) Examine the requirement for axial *stiffness*. Use Eq. 9.11.24, assuming conservatively the area A_b provided equals exactly that required,

$$\frac{L_b}{L_{av}} \leq 0.57\left[\frac{\alpha_{st}-1}{\alpha_e-\sqrt{\alpha_e}}\right]\left[\frac{\text{actual } A_b}{\text{required } A_b}\right]\frac{L_a}{b}$$

$$\leq 0.57(0.2)(1.0)(192/9.96) = 4.06$$

$$L_b \leq 4.06L_{av} = 4.06(16) = 65 > 7.5 \text{ ft for brace} \qquad \text{OK}$$

As in Example 9.11.1, the slenderness ratio limit of 200 for compression members will give a member adequate for bracing purposes.
Use C4×5.4.

A conclusion of the section on lateral bracing for beams and columns is that the requirements for such bracing are easily met. It is more important to provide a brace *of some size* than to be overly concerned about what the size should be.

Empirical procedures have long been used in lieu of a rational investigation of the strength and stiffness requirements for braces. A summary of some of these rules is given by Lay and Galambos [37]. The typical rule-of-thumb has been to use a brace having a strength equal to or greater than 2 percent of the compressive strength of the compression element being braced. This seems to be a conservative alternative to an analytical study.

9.12* BIAXIAL BENDING OF DOUBLY SYMMETRICAL SECTIONS

A designer frequently encounters the situation where a lateral bending moment M_y is applied in combination with the vertical moment M_x. The design of roof purlins and crane girders are two of the most common situations. How does the moment M_y affect lateral stability? When all moment is applied in the weak direction, stability is not a problem and if "compact section" limitations on local buckling are satisfied, an allowable stress of $0.75F_y$ is permitted by the AISC Specification.

The subject of lateral buckling in biaxial bending has been discussed little in the literature. The following design suggestion is that presented by Gaylord and Gaylord [54].

From Eq. 9.3.31 for pure bending M_x with respect to the strong axis,

$$M_{cr}^2 = M_x^2 = \frac{\pi^2}{L^2}\left[\frac{\pi^2 E^2 C_w I_y}{L^2} + EI_y GJ\right] \qquad (9.12.1)$$

or

$$\frac{M_x^2}{EI_y GJ} = \frac{\pi^2}{L^2}\left[1+\left(\frac{\pi}{\lambda L}\right)^2\right] \qquad (9.12.2)$$

where $\lambda = 1/a = \sqrt{GJ/EC_w}$.

When simultaneously a moment M_y is applied, Eq. 9.12.2 becomes [55]

$$\frac{M_x^2}{EI_y GJ} + \frac{M_y^2}{EI_x GJ} = \frac{\pi^2}{L^2}\left[1+\left(\frac{\pi}{\lambda L}\right)^2\right] \qquad (9.12.3)$$

which is applicable to I-shaped sections with two axes of symmetry. Furthermore, it is applicable for sections with point symmetry (such as the zee), and is approximately valid for channels when M_y does not exceed $0.25M_x$ [54].

Combinations of M_x and M_y which satisfy Eq. 9.12.3 will plot as an ellipse, as shown in Fig. 9.12.1 for a W14×74. In biaxial bending it is assumed for design that ultimate capacity is reached when the extreme fiber stress reaches F_y. Therefore, no matter how stable the beam, the combination of moments must satisfy

$$\frac{M_x}{S_x} + \frac{M_y}{S_y} = F_y \qquad (9.12.4)$$

assuming ideal elastic-plastic stress-strain conditions.

Since the relationships shown in Fig. 9.12.1 for W14×74 are typical,

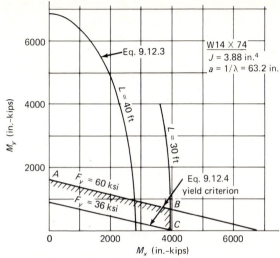

Fig. 9.12.1 Lateral buckling strength for biaxial bending of doubly symmetrical I-shaped sections.

certain conclusions may be drawn. Consider a laterally unbraced length of 30 ft and steel with $F_y = 60$ ksi. Ideally, the ultimate condition is defined by the lines AB (yielding controls) and BC (buckling controls).

The most important observation is that for ordinary laterally unbraced lengths (say 25 ft or less) the line BC is nearly vertical; therefore *simultaneous application of M_y does not appreciably affect the critical moment M_x.*

Based on the aforementioned conclusion, the recommended design procedure [53] is as follows:

1. For yielding controlling (line AB of Fig. 9.12.1):

$$\frac{M_x}{S_x} + \frac{M_y}{S_y} \le 0.60 F_y \tag{9.12.5}$$

2. For stability controlling (line BC of Fig. 9.12.1):

$$\frac{M_x}{S_x} \le \frac{\text{Allowable stresses based on}}{\text{AISC Formulas (1.5–6) or (1.5–7)}} \tag{9.12.6}$$

The AISC Specification has no specific provision for the above situation; it will always be conservative to require that the combined stress be within the allowable based on lateral buckling strength.

Example 9.12.1

Design a W section to serve as a crane girder to carry a moment $M_x = 301$ ft-kips (without impact) and a top flange moment $M_f = 30$ ft-kips. The moment M_f is based on a lateral force acting on the top flange equal to 10 percent of the lifted load and crane trolley weight in accordance with AISC–1.3.4. The moment M_f about the y-axis is assumed to be resisted by one flange; in effect, this accounts for the torsional effect by using the flexure analogy (see Sec. 8.6). The approximation of equivalent systems is shown in Fig. 9.12.2. Assume the simple span of 24 ft is laterally braced only at the ends. Use $F_y = 50$ ksi.

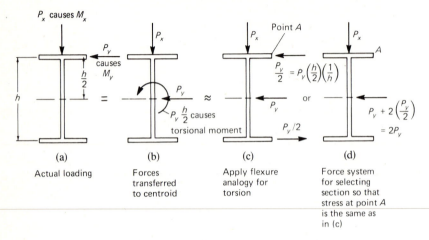

Fig. 9.12.2 Approximate equivalent system for biaxial bending and torsion.

SOLUTION

(a) AISC—using Eqs. 9.12.5. and 9.12.6 as criteria.

$$M_x = 301(1.25) = 376 \text{ ft-kips (using 25 percent impact)}$$
$$M_f = 30 \text{ ft-kips}$$

Obtain approximate section modulus S_x assuming that yielding controls; thus $F_b = 0.60F_y = 30$ ksi. Solving Eq. 9.12.5 for S_x,

$$\text{Required } S_x = \frac{M_x}{30} + \frac{M_y}{30}\left(\frac{S_x}{S_y}\right)$$

$$= \frac{(376+10)(12)}{30} + \frac{2(30)12}{30}(\approx 6) = 298 \text{ in.}^3$$

where a dead-load moment of 10 ft-kips is assumed using a 24-ft simple span; further, two times the moment carried by one flange is taken to be resisted by the entire section, in accordance with Fig. 9.12.2d. Try W24×131: $S_x = 329$ in.^{3}

$$\text{Required } S_x = \frac{385.3(12)}{30} + \frac{60(12)(6.2)}{30} = 303 \text{ in.}^3$$

Try W24×117: $S_x = 291$ in.^{3}

$$\text{Required } S_x = \frac{384(12)}{30} + \frac{60(12)(6.26)}{30} = 304 \text{ in.}^3$$

Check W24×131 yield controlling criterion, Eq. 9.12.5:

$$f_{bx} = \frac{M_x}{S_x} = \frac{385(12)}{329} = 14.0 \text{ ksi}$$

$$f_{by} = \frac{2M_y}{S_y} = \frac{60(12)}{53.0} = \underline{13.6 \text{ ksi}}$$

$$f_b = 27.6 \text{ ksi} < 0.60F_y \qquad \text{OK}$$

Check stability controlling, Eq. 9.12.6, conservatively assuming $C_b = 1.0$:

$$\frac{Ld}{A_f} = 24(12)(1.98) = 570$$

$$F_b(1.5\text{-}7) = \frac{12,000}{570} = 21.1 \text{ ksi}$$

$$\frac{L}{r_T} = \frac{24(12)}{3.40} = 84.7$$

Using Table 9.5.2, $C_b = 1.0$ and

$$F_b(1.5\text{-}6a) = 33.3 - \frac{(L/r_T)^2}{612} = 33.3 - \frac{(84.7)^2}{612} = 21.6 \text{ ksi} \qquad \text{(controls)}$$

$$\text{Since } f_{bx} = 14.0 \text{ ksi} < 21.6 \text{ ksi} \qquad \text{OK}$$

Use W24×131.

(b) AISC—using conservative approach of requiring combined stress to satisfy the allowable stress based on lateral buckling strength.

Try W24×146:

$$f_{bx} + f_{by} = 12.5 + 11.9 = 24.4 \text{ ksi}$$

$$F_b(1.5\text{-}7) = \frac{12,000}{24(12)(1.76)} = 23.7 \text{ ksi} \quad \text{(controls)}$$

$$F_b(1.5\text{-}6a) = 33.3 - \frac{(84.0)^2}{612} = 21.8 \text{ ksi}$$

$$f_b = 24.4 \text{ ksi} \approx F_b = 23.7 \text{ ksi} \text{ (i.e., within 3\%)} \qquad \text{OK}$$

Use W24×146. (The authors believe this to be an excessively conservative approach.)

(c) AISC—but use the corrected flexure analogy for the torsional component of the loading, in accordance with Sec. 8.6.

Try W24×117: $J = 6.72$ in.^{4}, $C_w = 40,800$ in.^{6}, $\lambda = 1/125$.

Of the 60 ft-kips used for M_y in the preceding parts, one-half is the actual lateral moment, M_y (Fig. 9.12.2b), while the other half is to approximate the torsional effect so that identical stress acts at point A in Figs. 9.12.2c and d.

Using Table 8.6.1, assuming torsional simple support, the true torsional effective M_y is obtained as follows:

$$\lambda L = 24(12)/125 = 2.30$$

Find $\beta \approx 0.71$ for $a = 0.5$.

$$M_y \text{ (for torsion)} \approx 0.71(30) = 21.3 \text{ ft-kips}$$

Check yield condition:

$$f_{bx} + f_{by} = \frac{384(12)}{291} + \frac{(30+21.3)(12)}{46.5}$$
$$= 15.8 + 13.2 = 29.0 \text{ ksi} < 0.60F_y \qquad \text{OK}$$

Check stability condition:

$$f_{bx} = 15.8 \text{ ksi} < F_b(1.5\text{–}7) = \frac{12,000}{24(12)(2.23)} = 18.7 \text{ ksi} \qquad \text{OK}$$

No need to check F_b (1.5–6), though in this case it gives the higher and therefore the correct allowable value; F_b (1.5–6) = 21.4 ksi.

Once again it is demonstrated that a better understanding or behavior may permit a saving in weight.

Use W24 × 117.

SELECTED REFERENCES

1. S. P. Timoshenko and James M. Gere, *Theory of Elastic Stability*, 2nd ed. New York: McGraw-Hill Book Company, Inc., 1961, Chap. 6.
2. Friederich Bleich, *Buckling Strength of Metal Structures*. New York: McGraw-Hill Book Company, Inc., 1952, Chap. 4.
3. Karl de Vries, "Strength of Beams as Determined by Lateral Buckling," *Transactions*, ASCE, 112 (1947), 1245–1320.
4. R. A. Hechtman, J. S. Hattrup, E. F. Styer, and J. L. Tiedemann, "Lateral Buckling of Rolled Steel Beams," *Transactions*, ASCE, 122 (1957), 823–843.
5. W. J. Austin, S. Yegian, and T. P. Tung, "Lateral Buckling of Elastically End-Restrained I-Beams," *Transactions*, ASCE, 122 (1957), 374–388.
6. J. W. Clark and J. R. Jombock, "Lateral Buckling of I-Beams Subjected to Unequal End Moments," *Journal of Engineering Mechanics Division*, ASCE, 83, EM3 (July 1957).
7. Mario G. Salvadori, "Lateral Buckling of I-Beams," *Transactions*, ASCE, 120 (1955), 1165–1182.
8. Bruce G. Johnston, ed. *Structural Stability Research Council, Guide to Stability Design Criteria for Metal Structures*, 3rd ed. New York: John Wiley & Sons, Inc., 1976, Chap. 6.

9. T. V. Galambos, "Inelastic Lateral Buckling of Beams," *Journal of Structural Division*, ASCE, 89, ST5 (October 1963), 217–242.

10. M. G. Lay and T. V. Galambos, "Inelastic Steel Beams Under Uniform Moment," *Journal of Structural Division*, ASCE, 91, (December 1965), 67–93.

11. M. G. Lay and T. V. Galambos, "Inelastic Beams Under Moment Gradient," *Journal of Structural Division*, ASCE, 93, ST1 (February 1967), 381–399.

12. Campbell Massey and F. S. Pitman, "Inelastic Lateral Stability Under a Moment Gradient," *Journal of Engineering Mechanics Division*, ASCE, 92, EM2 (April 1966), 101–111.

13. A. F. Lukey and P. F. Adams, "Rotation Capacity of Beams Under Moment Gradient," *Journal of Structural Division*, ASCE, 95, ST6 (June 1969), 1173–1188.

14. J. Bansal, "The Lateral Instability of Continuous Beams," *AISI Report No. 3*, American Iron and Steel Institute, New York, August 1971.

15. A. J. Hartmann, "Inelastic Flexural-Torsional Buckling," *Journal of Engineering Mechanics Division*, ASCE, 97, EM4 (August 1971), 1103–1119.

16. Joint Committee of Welding Research Council and the American Society of Civil Engineers, *Commentary on Plastic Design in Steel*, 2nd ed., ASCE Manual and Reports on Practice No. 41, New York, 1971, Chap. 6, pp. 90–92.

17. G. C. Driscoll et al., "Plastic Design of Multi-Story Frames, Lecture Notes and Design Aids," Fritz Engineering Laboratory Reports, Nos. 273.20 and 273.24, Lehigh University, Bethelem, Pa., 1965.

18. D. A. Nethercot and K. C. Rockey, "A Unified Approach to the Elastic Lateral Buckling of Beams," *Engineering Journal*, AISC, 9, 3 (July 1972), 96–107.

19. J. W. Clark and H. N. Hill, "Lateral Buckling of Beams," *Transactions*, ASCE, 127, Part II (1962), 180–201.

20. Mario G. Salvadori, "Lateral Buckling of Eccentrically Loaded I-Columns," *Transactions*, ASCE, 121 (1956), 1163–1178.

21. M. G. Salvadori, "Lateral Buckling of Beams of Rectangular Cross Section Under Bending and Shear," Proc. 1st U.S. Congress of Applied Mechanics, 1951, pp. 403–406.

22. A. J. Hartmann, "Elastic Lateral Buckling of Continuous Beams," *Journal of Structural Division*, ASCE, 93, ST4 (August 1967), 11–26.

23. William A. Milek, "One Engineer's Opinion," *Engineering Journal*, AISC, 3, 2 (April 1966), 88–90.

24. Campbell Massey and Peter J. McGuire, "Lateral Stability of Nonuniform Cantilevers," *Journal of Engineering Mechanics Division*, ASCE, 97, EM3 (June 1971), 673–686.

25. Frank W. Stockwell, Jr., "Simplified Approach to AISC Bending Formulas," *Engineering Journal*, AISC, 11, 3 (Third Quarter, 1974), 65–66.

26. G. Donald Brandt, "Direct Feasible and Optimal Design of Laterally Unsupported Beams," *Engineering Journal*, AISC, 14, 2 (Second Quarter, 1977), 78–84.

27. H. N. Hill, "Lateral Buckling of Channels and Z-Beams," *Transactions*, ASCE, 119 (1954), 829–841.

28. H. N. Hill, "Lateral Stability of Unsymmetrical I-Beams," *Journal of the Aeronautical Sciences*, 9 (1942), 175–180.

29. G. Winter, "Lateral Stability of Unsymmetrical I-Beams and Trusses in Bending," *Transactions*, ASCE, 108 (1943), 247–268.

30. William Zuk, "Lateral Bracing Forces on Beams and Columns," *Journal of Engineering Mechanics Division*, ASCE, 82, EM3 (July 1956), Proc. Paper No. 1032, 16 pp.

31. George Winter, "Lateral Bracing of Columns and Beams," *Transactions*, ASCE, 125 (1960), 807–845.

32. Campbell Massey, "Lateral Bracing Force of Steel I-Beams," *Journal of Engineering Mechanics Division*, ASCE, 88, EM6 (December 1962), 89–113.

33. George Pincus, "On the Lateral Support of Inelastic Columns," *Engineering Journal*, AISC, 1, 4 (October 1964), 113–115.

34. Theodore V. Galambos, "Lateral Support for Tier Building Frames," *Engineering Journal*, AISC, 1, 1 (January 1964), 16–19; Disc, 1, 4 (October 1964), 141.

35. T. R. Higgins, "Column Stability Under Elastic Support," *Engineering Journal*, AISC, 2, 2 (April 1965), 46–49.

36. Tor B. Urdal, "Bracing of Continuous Columns," *Engineering Journal*, AISC, 6, 3 (July 1969), 80–83.

37. Maxwell G. Lay and T. V. Galambos, "Bracing Requirements for Inelastic Steel Beams," *Journal of Structural Division*, ASCE, 92, ST2 (April 1966), 207–228.

38. Arthur C. Taylor, Jr. and Morris Ojalvo, "Torsional Restraint of Lateral Buckling" *Journal of Structural Division*, ASCE, 92, ST2 (April 1966), 115–129.

39. A. J. Hartmann, "Experimental Study of Flexural-Torsional Buckling," *Journal of Structural Division*, ASCE, 96, ST7 (July 1970), 1481–1493.

40. Bruce R. Mutton and Nicholas S. Trahair, "Stiffness Requirements for Lateral Bracing," *Journal of Structural Division*, ASCE, 99, ST10 (October 1973), 2167–2182.

41. Joseph A. Yura, Notes presented at University of Wisconsin Extension Institute, "Design of Bracing," September 17–18, 1970; December 11–12, 1974, Milwaukee, Wisconsin.

42. George Pincus and Gordon P. Fisher, "Behavior of Diaphragm-Braced Columns and Beams," *Journal of Structural Division*, ASCE, 92, ST2 (April 1966), 323–350.

43. Samuel J. Errera, George Pincus, and Gordon P. Fisher, "Columns and Beams Braced by Diaphragms," *Journal of Structural Division*, ASCE, 93, ST1 (February 1967), 295–318.

44. Larry D. Luttrell, "Strength and Behavior of Light-Gage Steel Shear Diaphragms," *Cornell Engineering Research Bulletin* No. 67-1, July 1967.

45. T. V. S. R. Apparao, Samuel J. Errera, and Gordon P. Fisher, "Columns Braced by Girts and a Diaphragm," *Journal of Structural Division*, ASCE, 95, ST5 (May 1969), 965–990.

46. Arthur H. Nilson and Albert R. Ammar, "Finite Element Analysis of Metal Deck Shear Diaphragms," *Journal of Structural Division*, ASCE, 100, ST4 (April 1974), 711–726.

47. David A. Nethercot and Nicholas S. Trahair, "Design of Diaphragm-Braced I-Beams," *Journal of Structural Division*, ASCE, 101, ST10 (October 1975), 2045–2061.

48. Amir Simaan and Teoman B. Pekoz, "Diaphragm Braced Members and Design of Wall Studs," *Journal of Structural Division*, ASCE, 102, ST1 (January 1976), 77–92.

49. Samuel J. Errera and Tamirisa V. S. R. Apparao, "Design of I-Shaped Beams with Diaphragm Bracing," *Journal of Structural Division*, ASCE, 102, ST4 (April 1976), 769–781.

50. J. Michael Davies, "Calculation of Steel Diaphragm Behavior," *Journal of Structural Division*, ASCE, 102, ST7 (July 1976), 1411–1430.

51. Samuel J. Errera and Tamirisa V. S. R. Apparao, "Design of I-shaped Columns with Diaphragm Bracing," *Journal of Structural Division*, ASCE, 102, ST9 (September 1976), 1685–1701.

52. John T. Easley, "Strength and Stiffness of Corrugated Metal Shear Diaphragms," *Journal of Structural Division*, ASCE, 103, ST1 (January 1977), 169–180.

53. *Tentative Recommendations for the Design of Steel Deck Diaphragms*, Steel Deck Institute, Westchester, Illinois, October 1972, 26 pp.

54. E. H. Gaylord, Jr. and C. N. Gaylord, *Design of Steel Structures*. New York: McGraw-Hill Book Company, Inc., 1957 (pp. 169–170).

55. C. O. Dohrenwend, "Action of Deep Beams Under Combined Vertical, Lateral and Torsional Loads," *Journal of Applied Mechanics*, 8 (1941), A–130.

56. Joseph A. Yura, Theodore V. Galambos, and Mayasandra K. Ravindra, "The Bending Resistance of Steel Beams," *Journal of Structural Division*, ASCE, 104, ST9 (September 1978), 1355–1370.

PROBLEMS

All problems* are to be done in accordance with the latest AISC Specification working stress provisions, unless otherwise indicated. Assume lateral support consists of translational restraint but not moment (rotational) restraint, unless otherwise indicated. After selecting a section in a design problem, compute the stress f_b due to the loading and compare with allowable stress F_b.

9.1. Plot the allowable bending stress F_b vs laterally unsupported length, using C_b of 1.0 and 2.3, for the following beams: (a) W16×26; (b) W14×145; (c) welded girder with flange plates $\frac{5}{8}$×20 and web plate $\frac{5}{16}$×60. Use A36 steel and show both AISC Formulas (1.5–6) and (1.5–7) for comparison, but show the controlling part with solid lines and the noncontrolling part with dashed lines. (For SI, (a) W410×39; (b) W360×216; (c) flanges 16×500 and web 8×1500)

---

* Many problems may be solved either as stated in U.S. customary units or in SI units using the numerical data in parenthesis at the end of the statement. The conversions are only approximate to avoid having the given data imply accuracy greater in SI than for U.S. customary units.

9.2. On each of the plots for the three beams of Prob. 9.1, superimpose the allowable stress using the SSRC "Basic Design Procedure" of Sec. 9.4. (Use Eq. 9.4.1 for pure bending, Eq. 9.4.6, and the SSRC column formula with a factor of safety of 1.67.) Discuss the comparison of AISC with "Basic Design Procedure."

9.3. Determine the allowable concentrated load P acting at midspan of a W21×62 (W530×92) section on a simply supported 20-ft span. Neglect the weight of the beam and assume lateral supports exist only at the ends of the span. Use A36 steel. Compare results using (a) AISC Spec.; (b) SSRC "Basic" design procedure with FS = 1.67; and (c) AISC Commentary method using Eq. 9.4.9 for r_e.

9.4. Select the lightest W sections of A36 steel for the following conditions:

Span	Live Load	Lateral Support
(a) 20 ft	2 kips/ft	Continuous
(b) 20 ft	2 kips/ft	Ends and midspan
(c) 20 ft	2 kips/ft	Ends only

(For SI, span = 6 m; load = 30 kN/m)

9.5. Select the lightest W section to carry a uniform live load of 0.8 kip/ft on a simply supported span of 35 ft. Select beams for two cases of lateral support, (1) continuously braced, and (2) braced at ends and midspan only. Assume no deflection restrictions. Compare for (a) A36 steel; (b) A572 Grade 65; and (c) A514 Grade 100. (For SI, load = 12 kN/m; span = 10.6 m)

9.6. Select the lightest W sections of A36 steel for a 35-ft simply supported beam carrying a live load of 0.65 kips/ft, and having lateral support (a) every 5 ft; (b) ends and midspan; (c) ends only. (For SI, span = 11 m, load = 9.5 kN/m, for (a) lateral support at 1.5 m maximum spacing at midspan.)

9.7. Select the lightest W sections for the situation shown in the accompanying figure, under the following conditions:
(a) A36 steel; continuous lateral support
(b) A36 steel; lateral support at ends only
(c) A36 steel; lateral support at ends and at point A
(d) A572 Grade 60 steel; lateral support at ends and point A

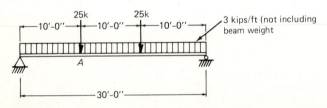

Prob. 9.7

What is the maximum price differential between the Grade 60 steel and A36 steel that would justify using the higher-strength steel? (For SI, span = 9 m; concentrated loads = 110 kN; uniform load = 40 kN/m)

9.8. Select the lightest W section for a simply supported beam on a span of 48 ft, carrying 2.5 kips/ft (not including beam weight), with lateral support at 16-ft intervals. Assume there is no deflection limit. Compare results for A36 steel and A572 Grade 60 steel. (Span = 14.4 m; uniform load = 36 kN/m; lateral support at 4.8 m intervals.)

9.9. Select the lightest W section for the situation shown in the accompanying figure using (a) A572 Grade 50 steel and (b) A572 Grade 65 steel, assuming lateral support at the ends and at point A only. (For SI, span = 8 m; concentrated loads = 115 kN; uniform load = 15 kN/m)

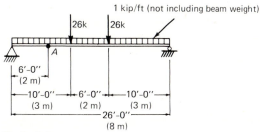

Prob. 9.9

9.10. Select the lightest W section for the conditions shown in the accompanying figure. Assume there is no deflection limitation. Use (a) A36 steel and (b) A572 Grade 60 steel.

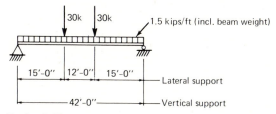

Prob. 9.10

9.11. A floor beam, laterally supported at the ends only and supporting vibration inducing heavy machinery, is subject to the loads shown in the accompanying figure. Select the lightest W section of A36 steel (Consider AISC–1.3.3). Compare the result when there is no deflection limit with that when L/d is limited to a maximum of 20 under full stress (see AISC Commentary–C.1.13.2).

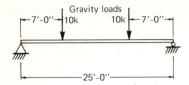

Prob. 9.11

9.12. Select the lightest W section to serve as a uniformly loaded library floor beam carrying a live load moment of 240 ft-kips on a simply supported span of 48 ft. Lateral support is at 12-ft intervals. Assume live load deflection may not exceed $L/300$. Assume $C_b = 1.0$ Use (a) A36 steel and (b) A572 Grade 65 steel. (Live load moment = 325 kN·m; span = 14.4 m; lateral support at 3.6-m intervals.)

9.13. A beam is to serve as a floor beam on a simple span of 20 ft. The live load consists of a movable concentrated load (no impact) of 50 kips. Live load deflection may not exceed $L/360$.
  (a) Select the lightest W section of A36 steel when continuous lateral support is provided.
  (b) Repeat (a) if lateral support is provided only at the ends.

9.14. A W10×33 is to be used as a simply supported beam on a span of 25 ft with lateral support at the ends only. The beam is required to support a plastered ceiling. If the dead load is 0.15 kip/ft (including beam weight), what is the allowable uniform live load on the beam, using A36 steel? What percentage increase in live load can be gained if the beam is A572 Grade 60 steel? Comment.

9.15. Select the lightest W section to carry a library floor on a simply supported span of 28 ft. The superimposed loading is 1.5 kips/ft and lateral support is provided at the ends and at midspan. A plastered ceiling is to be under this beam and the architect requires that nothing deeper than a nominal W12 be used. Use A36 steel and A572 Grade 60 steel.

9.16. Investigate the beam of the accompanying figure for bending and shear if the section is A572 Grade 50 steel. Compare with allowable AISC values. External lateral support for the beam is provided only at the vertical supports and at the tip of the cantilever. If one additional lateral support were provided at the 12-kip load, how much lighter, if any, could the W14 section be made?

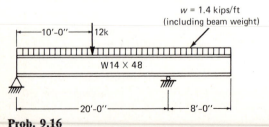

Prob. 9.16

9.17. Select the lightest W section for each of the situations shown in the accompanying figure. Assume lateral support is provided at the reactions and at the concentrated loads. Use A36 steel.

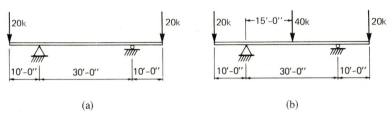

(a)　　　　　　　　　　　　　　　(b)

Prob. 9.17

9.18. Select the lightest W section for the beam of the accompanying figure. Lateral support is provided at concentrated load points, reactions, and at end of cantilever. Use (a) A36 steel and (b) A572 Grade 65 steel.

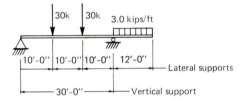

Prob. 9.18

9.19. Design the beam of Prob. 9.11, using the SSRC "Basic" design procedure, utilizing Figs. 9.4.1 and 9.4.2. Use FS = 1.67. Compare with AISC method (Prob. 9.11) and explain the difference.

9.20. Design the beam of Prob. 9.12 (A36 steel only) using the SSRC "Basic" design procedure with FS = 1.67. Compare with AISC method (Prob. 9.12) and explain any difference.

9.21. Investigate the beam of Prob. 9.16 using the SSRC "Basic" design procedure instead of AISC Formulas (1.5–6) and (1.5–7).

9.22. Determine the allowable bending stress for the channel section and loading in Prob. 8.20 if lateral support exists only at the ends. *Neglecting torsion*, how much larger if any, section would be required? Use A36 steel.

9.23. Assuming lateral support only at the ends, and further assuming the combined stress due to bending and torsion should not exceed the allowable based on lateral buckling, select a channel for the conditions of Prob. 8.21. Consider torsion in accordance with Chapter 8.

9.24. Design a built-up I-shaped beam with different sized flanges for the conditions of Prob. 9.7, part (b). What percent weight can be saved, if any, by using different sized flanges? Use beam depth and web thickness approximately the same as for the lightest rolled W shape that satisfies loading conditions. A36 steel.

9.25. For the beam selected for Prob. 9.4b estimate the size of bracing (i.e., select a section) required. The bracing frames into both sides and is attached to the compression flange. Length of bracing is 6 ft.

9.26. For the beams selected in Prob. 9.12, estimate the size bracing required. Assume bracing is 12 ft long, frames into both sides of beam, and is attached to the compression flange. Preferably select channels.

9.27. Determine the adequacy of a W24×84 (with rail, 20 lb/ft) serving as a crane support girder of A36 steel. The simple span is 20 ft with lateral support at the ends only. Use accepted good practice in accounting for the torsional effect of lateral loading. Maximum moments occurring near midspan are

$M_x = 100$ ft-kips (includes live load plus impact)

$M_f = 10$ ft-kips (assumed resisted by one flange using top flange lateral loading in accordance with AISC–1.3.4).

9.28. Select the lightest W8 section to be used in an inclined position such that the plane of the web makes an angle of 30 degrees with the plane of loading. The beam is to be of A572 Grade 50 steel, and it has lateral support only at the ends of the 22-ft simple span. The uniform gravity load is 0.4 kips/ft, in addition to the beam weight.

9.29. Design the lightest W section to serve as a crane support girder with the following requirements: crane capacity = 10 tons; maximum gravity reaction per end-truck wheel = 18 kips; weight of trolley = 6 kips. Assume lateral support at the ends only and that deflection need not be restricted. Note AISC–1.3.3 and 1.3.4. Use A36 steel.

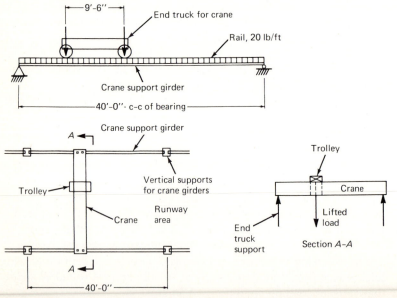

Prob. 9.29

9.30. Redesign the crane support girder of Prob. 9.29, using a combination channel and W section.

9.31. Design the crane runway girder indicated on the accompanying figure. The two cranes are each 30-ton capacity, with the end-truck wheel spacing as given in the figure. The rails are ASCE 60 lb with clamps (see AISC Manual). The dead load of each crane is 15 kips, equally distributed to its four wheels. Each crane trolley weighs 3 kips. Use A36 steel and (a) select a single W section and (b) select a combination W section and channel, where the channel would have its web flat against the top flange of the W section.

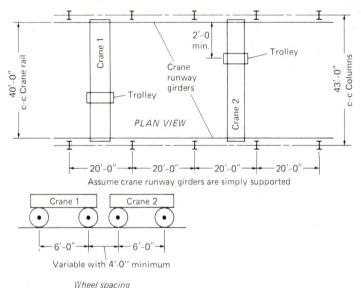

Assume crane runway girders are simply supported

Wheel spacing

Prob. 9.31

10
Continuous Beams

10.1 INTRODUCTION

This chapter brings together theoretical concepts relating to both the working stress and the plastic design methods that have been presented in Chapters 7 and 9. In addition, the plastic design section presents the analysis procedure for the method as applied to continuous structures. It is assumed the reader is familiar with elastic methods of statically indeterminate analysis as required for the working stress method.

Before proceeding into the chapter, the reader should review Secs. 7.3 and 7.5 where the strength of the cross section in flexure and shear is treated, and where the working stress and plastic design approaches are contrasted. Also in Sec. 7.3 are examples of the two methods as applied to simple span beams.

The reader is also expected to be familiar with the lateral stability criteria considered in Chapter 9, particularly Secs. 9.1, 9.2, and 9.5–9.9.

For the reader who wishes a package study of plastic design, Secs. 7.3 and 7.5 give plastic moment and shear strength, including the concepts of plastic moment, plastic hinge, and shape factor. Section 7.3 contains the basic approach of using factored loads to obtain the required ultimate strength, as well as a design example for a simple beam. The plastic analysis and design sections of this chapter can then form the body of the study. When lateral stability considerations are applicable the plastic strength information in Secs. 9.3, 9.7, and 9.11 may be studied.

10.2 PLASTIC STRENGTH OF A STATICALLY INDETERMINATE SYSTEM

In Sec. 7.3 it was demonstrated that on a simply supported beam the maximum moment at a single location increased under increasing load until a collapse condition was reached, at which point the maximum moment was the plastic moment M_p. The plastic condition was assumed to exist at one point only with the remainder of the beam elastic.

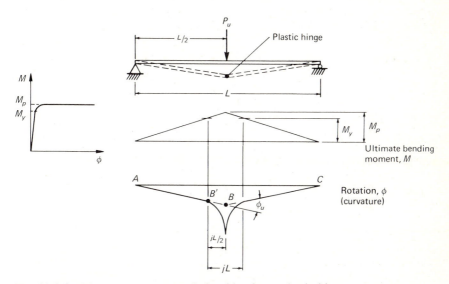

Fig. 10.2.1   Moment-curvature relationships for a plastic hinge.

Consider Fig. 10.2.1, where a simple beam is loaded with its ultimate concentrated load. As the load P is increased from zero such that the bending moment exceeds M_y, a portion of the length is inelastic. When the moment reaches M_p the inelastic length is jL, and the curvature at midspan is very large. The variation is as shown in Fig. 10.2.1; however, for practical purposes the beam may be treated as two rigid portions (AB and BC) connected by a hinge at B, known as a *plastic hinge*, and having a concentrated angle change, or hinge rotation, ϕ_u.

The actual length of a plastic hinge is dependent on the shape of the cross section and can vary from about one-tenth to as much as one-third of the span. As an example, consider the relationship between the ultimate moments and the beam segments AB and AB' in Fig. 10.2.1:

$$\frac{AB}{M_p} = \frac{AB'}{M_y} \quad \text{or} \quad \frac{L/2}{M_p} = \frac{L/2 - jL/2}{M_y} \tag{10.2.1}$$

Solving for j,

$$j = 1 - \frac{M_y}{M_p} = 1 - \frac{1}{f} \tag{10.2.2}$$

where f is the shape factor as discussed in Sec. 7.3.

Example 10.2.1

Compute the length of the plastic hinge for the beam shown in Fig. 10.2.1 for (a) a W16×40 and (b) a rectangular beam having a width b and a depth d.

SOLUTION

(a) For W16×40,

$$f = \frac{M_p}{M_y} = \frac{Z}{S} = \frac{72.9}{64.7} = 1.13$$

From Eq. 10.2.2,

$$jL = L\left(1 - \frac{1}{f}\right) = L\left(1 - \frac{1}{1.13}\right) = 0.115\,L$$

(b) For rectangular section,

$$f = \frac{Z}{S} = \frac{\dfrac{bd^2}{4}}{\dfrac{bd^2}{6}} = 1.5$$

$$jL = L\left(1 - \frac{1}{1.5}\right) = 0.333L$$

Even though the distance jL may be as much as one-third of the span, as shown in Example 10.2.1, the simple assumption of a plastic hinge at a point has been amply demonstrated by tests. Beedle [1] and Massonnet and Save [2] have extensive discussions of theoretical and experimental verification of plastic design procedures so individual research references are not included here.

Plastic Limit Load—Equilibrium Method

At the ultimate condition with the plastic limit load P_u acting, the requirements of equilibrium are still applicable. Consider first a statically determinate simply supported beam, as shown in Fig. 10.2.2. A collapse condition is achieved when the load P_u is large enough to cause the plastic moment M_p to occur at one location (in this case under the load). When the sufficient number of plastic hinges have been developed to allow instantaneous hinge rotations without developing increased resistance, a mechanism is said to have occurred.

Example 10.2.2

Determine the ultimate load P_u for the W21×62 beam of Fig. 10.2.2. Assume $F_y = 36$ ksi.

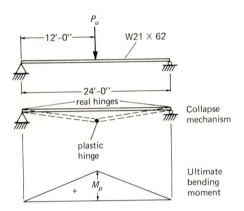

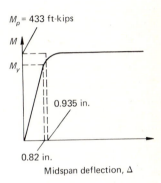

Fig. 10.2.2 Example 10.2.2.

SOLUTION

Using equilibrium,

$$M_p = \frac{P_u L}{4}$$

$$P_u = \frac{4M_p}{L} = \frac{4(F_y Z)}{L} = \frac{4(36)(144)}{24(12)} = 72 \text{ kips}$$

Using the simplified procedure of considering the behavior as ideally elastic-plastic, the deflection occurring when M_p is reached is based on the elastic equation, which is strictly valid only until M_y is reached. At a maximum moment of M_y,

$$\Delta_y = \frac{PL^3}{48EI} = \frac{M_y L^2}{12EI} = \frac{F_y IL^2}{c(12)EI} = \frac{F_y L^2}{12cE}$$

$$= \frac{36(24)^2(144)}{12(10.5)(29,000)} = 0.82 \text{ in.}$$

Assuming a linear extension until $M = M_p$,

$$\Delta_p = 0.82\frac{Z}{S} = 0.82\frac{144}{127} = 0.82(1.14) = 0.93 \text{ in.}$$

The ultimate deflection when M_p is achieved will be higher than this.

However, it is the service load deflection that is normally of concern, and this is correctly computed by the usual elastic procedures.

Note that the ultimate moment diagram for this statically determinate problem is the same shape as that which occurs if the maximum moment is down in the working stress range. At whatever load level, from an infinitesimal load to ultimate load, the bending moment at every point remains in a constant proportion to the load.

Example 10.2.3

Determine the ultimate capacity of the fixed end W16×40 beam of Fig. 10.2.3. Assume $F_y = 36$ ksi.

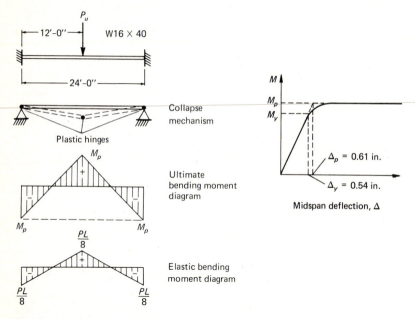

Fig. 10.2.3   Example 10.2.3.

SOLUTION

Since the beam is statically indeterminate, three plastic hinges are required to form a mechanism. Using equilibrium,

$$2M_p = \frac{P_u L}{4}$$

$$M_p = F_y Z = 36(72.9)\tfrac{1}{12} = 219 \text{ ft-kips}$$

$$P_u = 8M_p/L = 8(219)/24 = 73 \text{ kips}$$

In order to determine the load-deflection diagram, it is necessary to know the loading history up to the collapse condition. In this case the elastic bending moments, as shown in Fig. 10.2.3. give equal positive and negative bending moment; therefore the three plastic hinges form simultaneously. Moments will increase simultaneously in direct proportion until the collapse condition is reached. Again, as for the statically determinate case, as load is applied the bending moment at every point remains in a constant proportion to the load. This special statically indeterminate case offers no difference from a statically determinate one.

$$\Delta_y = \frac{PL^3}{192EI} = \frac{M_yL^2}{24EI} = \frac{F_yL^2}{24Ec}$$

$$= \frac{36(24)^2(144)}{24(29,000)8} = 0.54 \text{ in.}$$

Assuming a linear extension until $M = M_p$,

$$\Delta_p = 0.54\frac{Z}{S} = 0.54\frac{72.9}{64.7} = 0.61 \text{ in.}$$

Example 10.2.4

Determine the load-deflection diagram for the W16×40 beam of Fig. 10.2.4 for loading up to the collapse condition. Assume $F_y = 36$ ksi.

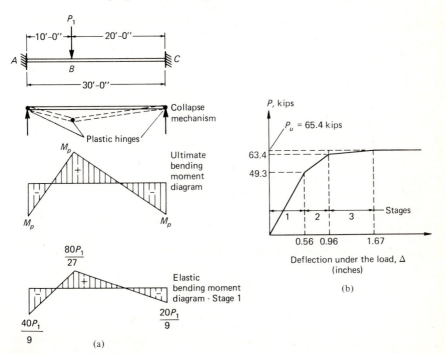

Fig. 10.2.4. Example 10.2.4.

SOLUTION

The ultimate capacity P_u can be found directly without knowing the sequence of plastic hinge formation. In this case, since the elastic moments are different at the three locations where plastic hinges will form, the plastic hinges will not form simultaneously.

At ultimate, from the equilibrium requirement,

$$M_p = \frac{P_u(10)(20)}{30} - M_p$$

$$P_u = 2M_p\frac{3}{20} = 0.3M_p$$

$$M_p = 219 \text{ ft kips} \quad \text{(from Example 10.2.3)}$$

$$P_u = 0.3(219) = 65.7 \text{ kips}$$

The load-deflection diagram, however, requires examination of the loading stages.

Stage 1. From the elastic bending moments, the first plastic hinge will form at point A.

$$\frac{40P_1}{9} = M_p = 219 \text{ ft-kips}$$

$$P_1 = 9(218)/40 = 49.3 \text{ kips}$$

$$\Delta_1 = \frac{P(10)^3(20)^3}{3(30)^3EI} = \frac{49.3(8000)(1728)}{3(27)(29,000)(518)} = 0.56 \text{ in.}$$

Stage 2. With $P = 49.3$ kips applied one can consider that part of the available moment capacity at points B and C has been used up. The available capacity remaining is

$$M_B = M_p - 80(49.3)/27 = 219 - 146.1 = 72.9 \text{ ft-kips}$$

$$M_C = M_p - 20(49.3)/9 = 219 - 109.6 = 109.4 \text{ ft-kips}$$

As the load is increased above $P = 49.3$ kips, the added load acts on a different elastic system. The moments caused by the additional load are not distributed over the span in the same manner as moments caused by the first 49.3 kips; thus the term "redistribution of moments" is applied. Figure 10.2.5a shows the elastic system and moments for stage 2. It is apparent that the next plastic hinge will form under the load.

$$\frac{140P_2}{27} = 72.9$$

$$P_2 = 27(72.9)/140 = 14.1 \text{ kips}$$

$$\Delta_2 = \frac{Pa^2b^3(3L+a)}{12L^3EI} = \frac{14.1(10)^2(20)^3(90+10)(1728)}{12(30)^3(29,000)(518)} = 0.40 \text{ in.}$$

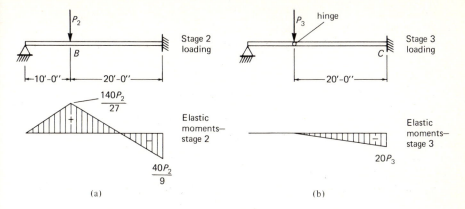

Fig. 10.2.5 Stages 2 and 3 for Example 10.2.4.

Stage 3. With a total of $P_1 + P_2 = 49.3 + 14.1 = 63.4$ kips applied, the moment capacity available at C is

$$M_C = M_p - 109.4 \text{ (Stage 1)} - 40(14.1)/9 \text{(Stage 2)} = 46.9 \text{ ft-kips}$$

$$20P_3 = 46.9; \qquad P_3 = 2.3 \text{ kips}$$

$$\Delta_3 = \frac{PL^3}{3EI} = \frac{2.3(20)^3(1728)}{3(29,000)(518)} = 0.71 \text{ in.}$$

The complete load-deflection diagram appears in Fig. 10.2.4b. The total ultimate load equals

$$P_u = P_1 + P_2 + P_3 = 49.3 + 14.1 + 2.3 = 65.7 \text{ kips}$$

Note that this statically indeterminate case, unlike the previous special case, exhibits the multistage load-deflection diagram typical of statically indeterminate systems. As load is increased moments vary in a different ratio for each stage of loading. Thus the ultimate moment diagram is *not* equal to a constant times the elastic moments. Only for statically determinate cases, and a few special statically indeterminate ones can the ultimate diagram be obtained by multiplying the elastic moments by a constant.

Example 10.2.5

Using the equilibrium method for the continuous beam of Fig. 10.2.6, determine the ultimate capacity P_u. Assume $F_y = 36$ ksi and that plastic strength is the sole governing factor.

SOLUTION
Whereas in the previous examples only one mechanism was possible, and therefore obvious, there are many situations where the collapse mechanism is not obvious. Each of the three possible mechanisms will be investigated for this example.

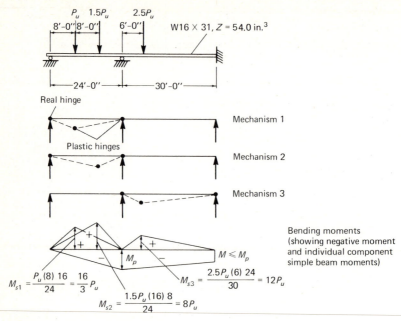

Fig. 10.2.6 Example 10.2.5.

Mechanism 1: Positive moment under the load P_u,

$$M_p = M_{s1} + \tfrac{1}{2}M_{s2} - \tfrac{1}{3}M_p$$
$$\tfrac{4}{3}M_p = \tfrac{16}{3}P_u + \tfrac{1}{2}(8P_u)$$
$$M_p = 7P_u$$

Mechanism 2: Positive moment plastic hinge at the load $1.5P_u$,

$$M_p = \tfrac{1}{2}M_{s1} + M_{s2} - \tfrac{2}{3}M_p$$
$$\tfrac{5}{3}M_p = \tfrac{8}{3}P_u + 8P_u$$
$$M_p = \tfrac{32}{5}P_u = 6.4P_u$$

Mechanism 3: Positive moment plastic hinge at the load $2.5P_u$ with negative moment plastic hinges at the ends of the 30-ft span,

$$M_p = M_{s3} - M_p$$
$$2M_p = 12P_u$$
$$M_p = 6P_u$$

Since the plastic moment capacity required is largest for Mechanism 1, it controls:

$$P_u = \frac{M_p}{7} = \frac{36(54.0)}{7(12)} = 23 \text{ kips}$$

In this case the collapse mechanism occurs in the 24-ft span while the 30-ft span remains stable and elastic. The use of cover plates as discussed in Example 10.3.3 may be a more economical solution.

Plastic Limit Load—Energy Method

The principle of virtual work may also be applied to obtain the limit load P_u in an analysis of a given structure, or to find the required plastic moment M_p in a design problem.

Consider that as the ultimate load is reached the beam moves through a virtual displacement δ. For equilibrium, the external work done by the load moving through the virtual displacement must equal the internal strain energy due to the plastic moment rotating through small angles (hinge rotations).

Example 10.2.6

Determine the ultimate load in Example 10.2.2 by the virtual-work principle. Referring to Fig. 10.2.7,

$$W_e \text{ (external work)} = W_i \text{ (internal work)}$$

$$P_u \delta = M_p 2\theta$$

$$P_u \left(\frac{\theta L}{2} \right) = 2M_p \theta$$

$$P_u = 4M_p/L$$

the same as previously obtained.

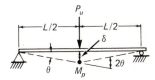

Fig. 10.2.7 Example 10.2.6.

Example 10.2.7

Determine the ultimate load in Example 10.2.3 by the virtual-work principle. Referring to Fig. 10.2.8,

$$W_e = W_i$$

$$P_u \frac{\theta L}{2} = 2M_p \theta + M_p (2\theta)$$

$$P_u = 8M_p/L$$

which agrees with the previous solution.

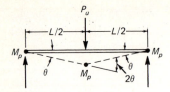

Fig. 10.2.8   Example 10.2.7.

Example 10.2.8

Determine the ultimate load for Example 10.2.4 by the virtual-work method. Referring to Fig. 10.2.9,

$$W_e = W_i$$

$$P_u \frac{2\theta L}{3} = M_p(2\theta + \theta + 3\theta)$$

$$P_u = \frac{3}{2L}(6)M_p = 9M_p/L$$

For $L = 30$ ft, $P_u = 0.3M_p$ as before.

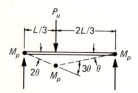

Fig. 10.2.9   Example 10.2.8.

Example 10.2.9

Determine the ultimate load P_u for the continuous beam of Example 10.2.5. Referring to Fig. 10.2.10,

Mechanism 1:

$$P_u(2\theta)(8) + 1.5P_u\theta(8) = M_p(3\theta + \theta)$$

$$28P_u = 4M_p$$

$$M_p = 7P_u$$

Mechanism 2:

$$P_u(\theta)(8) + 1.5P_u 2\theta(8) = M_p(3\theta + 2\theta)$$

$$32P_u = 5M_p$$

$$M_p = 6.4P_u$$

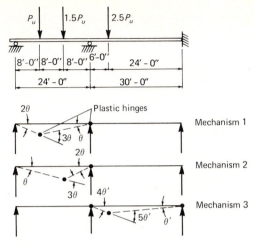

Fig. 10.2.10   Example 10.2.9.

Mechanism 3:

$$2.5P_u(4\theta')(6) = M_p(4\theta' + 5\theta' + \theta')$$
$$60P_u = 10M_p$$
$$M_p = 6P_u$$

The equilibrium and energy procedures are equally applicable for frames where degrees of freedom in sidesway must be considered. For single-story frames, the reader is referred to Chapter 15. In multistory braced frames, the design procedure is to cause the plastic hinges to form in the girders rather than in the columns; thus the girders are treated as continuous beams. Where bracing contributes an axial force to the girder, the beam-column principles of Chapter 12 apply.

10.3 DESIGN EXAMPLES—PLASTIC DESIGN

An elementary example of the design of a simple beam by plastic design was presented in Sec. 7.3. For additional elementary material on the philosophy, procedure, and experimental verification of plastic design, References 3 through 10 are suggested. Plastic design examples are presented first because the ultimate strength approach is the most realistic approach and is the authors' recommended procedure for most continuous beams. The working stress method might be the method of choice for the situations in which long distances between lateral supports exist. Furthermore the working stress method involves procedures that attempt to bring the result of that method close to the plastic design result.

In plastic design, plastic strength is the basis for design rather than a limiting stress. Factors such as shear and deflection, as discussed in

Chapter 7, must also be considered. Local buckling and lateral-torsional buckling must be prevented until the collapse mechanism has been achieved.

Example 10.3.1

Select the lightest W section for the 2-span continuous beam of Fig. 10.3.1, using plastic design. Use A36 steel. External lateral support is provided at the ends and midpoint of each span.

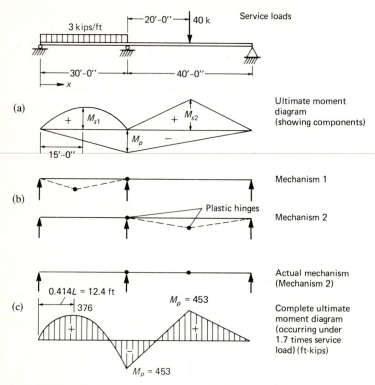

Fig. 10.3.1 Example 10.3.1.

SOLUTION

Temporarily neglecting the weight of the beam, apply the load factor (LF) to the service loads:

$$w_u = 3(1.7) = 5.1 \text{ kips/ft}; \qquad P_u = 40(1.7) = 68 \text{ kips}$$

Next, consider the various possible mechanisms for failure. In this case there are two, as shown in Fig. 10.3.1b.

(a) *Mechanism 1:* For a span with uniform load, simply supported at one end, and expected to develop a plastic hinge at the other, the location of maximum positive moment is not self-evident. At an unknown distance x from the discontinuous end, the net positive moment is

$$M_x = \frac{w}{2} x(L-x) - \frac{x}{L} M_p \qquad \text{(a)}$$

Taking $dM_x/dx = 0$ and solving for x, the location of maximum moment is

$$x = \frac{L}{2} - \frac{M_p}{wL} \qquad \text{(b)}$$

Substituting Eq. (b) into Eq. (a) gives

$$M_x\,(\text{max}) = \frac{w}{2}\left(\frac{L}{2} - \frac{M_p}{wL}\right)\left(L - \frac{L}{2} + \frac{M_p}{wL}\right) - \frac{M_p}{L}\left(\frac{L}{2} - \frac{M_p}{wL}\right) \qquad \text{(c)}$$

When a plastic hinge develops, $M_x\,(\text{max}) = M_p$; thus Eq. (c) becomes

$$M_p = \frac{wL^2}{8} - \frac{M_p^2}{2wL^2} - \frac{M_p}{2} + \frac{M_p^2}{wL^2} \qquad \text{(d)}$$

Substitute $M_s = wL^2/8$ into Eq. (d), giving

$$M_p^2 - 24M_p M_s + 16M_s^2 = 0 \qquad \text{(e)}$$

Solving the quadratic for M_p gives

$$M_p = 0.686 M_s \qquad \text{(10.3.1)}$$

and the location of the plastic hinge is at

$$x = \frac{L}{2} - \frac{M_p}{wL}\left(\frac{L}{L}\right)\frac{8}{8} = \frac{L}{2} - \frac{M_p}{M_s}\left(\frac{L}{8}\right)$$

$$= L\left(\frac{1}{2} - \frac{0.686}{8}\right) = 0.414L$$

The results of this development are shown in Fig. 10.3.2.

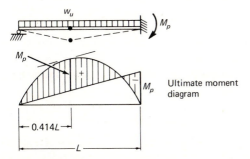

Fig. 10.3.2 Collapse condition for uniformly loaded span continuous at one end and simply supported at the other.

For the specific example, Mechanism 1 (Fig. 10.3.1b) requires

$$M_p = 0.686M_{s1}; \qquad M_{s1} = \frac{w_u L^2}{8} = \frac{5.1(30)^2}{8} = 574 \text{ ft-kips}$$

$$M_p = 0.686(574) = 394 \text{ ft-kips}$$

(b) *Mechanism 2:* For the 40-ft span,

$$M_{s2} = \frac{P_u L}{4} = \frac{68(40)}{4} = 680 \text{ ft-kips}$$

Equilibrium requires

$$M_p = M_{s2} - \frac{M_p}{2} = 680 - \frac{M_p}{2}$$

$$M_p = \tfrac{2}{3}(680) = 453 \text{ ft-kips}$$

The largest M_p requirement obtained from the possible mechanisms determines the controlling condition. Thus Mechanism 2 governs. A check by determining the complete ultimate moment diagram should be made to insure that M_p is the maximum moment. If a location is found where the computed ultimate moment exceeds M_p, the mechanism used is not correct. In this case, Fig. 10.3.1c shows $M_p = 453$ ft-kips is the maximum, so that Mechanism 2 is the correct one.

The maximum positive moment on the 30-ft span using Eq. (a) is

$$M = \frac{5.1}{2}(12.4)(17.6) - 0.414M_p$$

$$= 557 - 0.4(453) = 376 \text{ ft-kips} < 453 \qquad \text{OK}$$

(c) Select section:

$$\text{Required } Z = \frac{M_p}{F_y} = \frac{453(12)}{36} = 151 \text{ in.}^3$$

Though a W24×62 is the section indicated by the AISC "Plastic Design Selection Table," a section with a large value of r_y will be needed because of the large laterally unbraced length. Try W14×109, $Z = 192$ in.^{3}, $r_y = 3.74$ in.

(d) Check lateral support. According to AISC–2.9, the lateral support requirements of that section need not be applied in the region of the last hinge to form. However, even though the last hinge forms under the 40-kip load, the authors believe it to be appropriate to apply AISC Formula (2.9–1a) to the 20-ft segment between the concentrated load and the interior support where significant rotation must occur to develop the last hinge:

$$L_{cr} = \left[\frac{1375}{F_y} + 25\right]r_y = \left[\frac{1375}{36} + 25\right]r_y = 63.2r_y$$

for $M/M_p = +1.0$ (double curvature). For the W14×109, $r_y = 3.74$ in.

$$L_{cr} = \frac{63.2(3.74)}{12} = 19.7 \text{ ft.} \approx 20 \text{ ft} \qquad \text{OK}$$

For the exterior 20-ft segment, where at one end of the segment the final plastic hinge occurs under the 40-kip load, AISC–1.5.1.4.5(2.) must be satisfied:

$$F_b (1.5\text{-}7) = \frac{12,000C_b}{Ld/A_f} = \frac{12,000(1.75)}{240(1.14)} > 0.60F_y$$

$$f_b = \frac{M_p}{(LF)S_x} = \frac{453(12)}{1.7(173)} = 18.5 \text{ ksi} < 22 \text{ ksi} \qquad \text{OK}$$

For the shorter unbraced lengths on the 30-ft span, the flexural stresses f_b are equal to or less than the above value and therefore acceptable. Note that moments should be adjusted to include factored beam weight; in this case there is clearly adequate reserve capacity without actually making the calculation.

(e) Check shear. At interior end of 40-ft span,

$$V_u = 20 + \frac{453}{40} = 20 + 11.3 = 31.3 \text{ kips}$$

$$v_u = \frac{V_u}{t_w d} = \frac{31.3}{0.525(14.32)} = 4.2 \text{ ksi} < 0.55F_y = 19.8 \text{ ksi} \qquad \text{OK}$$

As expected, except on short spans with heavy loading, shear does not govern.

(f) Deflection. If deflection appears to be a problem, service load moments must be determined and the deflection computed. Deflection at the collapse condition is not easily determined, nor is it generally of interest. In this case,

$$\frac{L}{d} = \frac{40(12)}{14} = 34$$

which is somewhat high, so that if there is a deflection limitation, service load deflection should be computed.

Use W14×109 – A36 Steel.

Example 10.3.2

Select the lightest uniform section required for the continuous beam of Fig. 10.3.3 using the AISC plastic design method. Use $F_y = 60$ ksi, and assume lateral support is provided at the vertical supports and at each concentrated load.

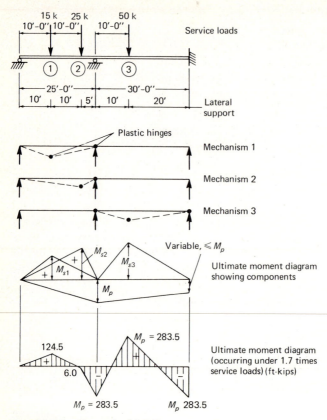

Fig. 10.3.3   Example 10.3.2.

SOLUTION

The equilibrium method is illustrated, although the energy method is equally applicable. Apply the 1.7 load factor and compute the simple beam moments under the factored loads.

$$M_{s1} = \frac{15(1.7)(10)15}{25} = 153 \text{ ft-kips}$$

$$M_{s2} = \frac{25(1.7)(20)5}{25} = 170 \text{ ft-kips}$$

$$M_{s3} = \frac{50(1.7)10(20)}{30} = 567 \text{ ft-kips}$$

(a) *Mechanism 1:* Assume moment under 15-kip load equals M_p,

$$M_p = M_{s1} + \tfrac{1}{2}M_{s2} - \tfrac{10}{25}M_p$$

$$\tfrac{35}{25}M_p = 153 + \tfrac{1}{2}(170) = 238$$

$$M_p = \tfrac{25}{35}(238) = 170 \text{ ft-kips}$$

(b) *Mechanism 2:* Assume moment under 25-kip load equals M_p,

$$M_p = \tfrac{1}{3}M_{s1} + M_{s2} - \tfrac{20}{25}M_p$$
$$\tfrac{45}{25}M_p = \tfrac{1}{3}(153) + 170 = 221$$
$$M_p = \tfrac{25}{45}(221) = 123 \text{ ft-kips}$$

(c) *Mechanism 3:* Assume moment under 50-kip load equals M_p,

$$M_p = M_{s3} - M_p$$
$$2M_p = 567; \qquad M_p = 283.5 \text{ ft-kips} \quad \text{(Controls)}$$

If a section is selected that has $M_p = 283.5$ ft-kips, the resulting moments under the 15- and 25-kip loads are

$$M_1 = M_{s1} + \tfrac{1}{2}M_{s2} - \tfrac{10}{25}M_p$$
$$= 238 - \tfrac{10}{25}(283.5) = 124.5 \text{ ft-kips} < 283.5 \qquad \text{OK}$$
$$M_2 = \tfrac{1}{3}M_{s1} + M_{s2} - \tfrac{20}{25}M_p$$
$$= 221 - \tfrac{20}{25}(283.5) = -6.0 \text{ ft-kips} < 283.5 \qquad \text{OK}$$

(d) Select beam (refer to text Appendix Table A4).

$$\text{Required } Z = \frac{283.5(12)}{60} = 56.7 \text{ in.}^3$$

Try $S15 \times 42.9$ ($Z = 69.3$ in.^{3}) or $W16 \times 45$ ($Z = 82.3$ in.^{3}) as sections that satisfy the local buckling requirements of AISC–2.7.
 Try $W16 \times 45$: $Z = 82.3$ in.^{3}, $r_y = 1.57$ in.

$$\frac{b_f}{2t_f} = \frac{7.035}{2(0.565)} = 6.2 < 6.3 \text{ limit} \qquad \text{OK}$$

$$\frac{d}{t_w} = \frac{16.13}{0.345} = 46.7 < 53.2 \qquad \text{OK}$$

Although it is likely the plastic hinge under the 50-kip load is the last to form, the designer may not be sure. The authors suggest that it is prudent to apply the critical length criteria of AISC–2.9 to all segments having an M_p at either end, eliminating the need for knowing the sequence of plastic hinge formation. The 20-ft segment will obviously control,

$$L_{cr} = \left[\frac{1375}{F_y} + 25\right]r_y = \left[\frac{1375}{60} + 25\right]r_y = 48\, r_y$$
$$= 48(1.57)\tfrac{1}{12} = 6.3 \text{ ft} < 20 \text{ ft} \qquad \text{NG}$$

$$\text{Required } r_y = \frac{240}{48} = 5.0 \text{ in.}$$

Lateral stability controls. Since an $r_y = 5.0$ in. cannot be achieved by any realistic design, additional lateral bracing must be provided.

Assume lateral bracing at 5-ft intervals in the 30-ft span. Then, at the most critical location,

$$\frac{M}{M_p} = \frac{-141.75}{283.5} = -0.5$$

which is the dividing condition between using AISC Formulas (2.9–1a) or (2.9–1b).

$$L_{cr}(2.9-1a) = 48r_y = 6.3 \text{ ft} > 5 \text{ ft} \qquad \text{OK}$$

as previously computed.

$$L_{cr}(2.9-1b) = \frac{1375}{F_y} r_y = \frac{1375}{60} r_y = 23r_y$$

$$= 23(1.57)\tfrac{1}{12} = 3.0 \text{ ft} < 5 \text{ ft} \qquad \text{NG}$$

Based on the development in Sec. 9.3, Formula (2.9–1a) is more nearly correct.

Use 5-ft spacing of lateral bracing on the 30-ft span.

On the 25-ft span, check the 10-ft unbraced segment; for W16×45,

$$L_c = 9.8b_f = 9.8(7.035)\tfrac{1}{12} = 5.7 \text{ ft} < 10 \text{ ft} \qquad \text{NG}$$

Not "compact" for working stress method.

$$F_b(1.5-7) = \frac{12{,}000C_b}{Ld/A_f} = \frac{12{,}000(1.75)}{120(4.06)} > 0.60F_y = 36 \text{ ksi}$$

$$f_b = \frac{M_u}{(LF)S_x} = \frac{124.5(12)}{1.7(72.7)} = 12.1 \text{ ksi} < 36 \text{ ksi} \qquad \text{OK}$$

Shear and deflection will not be critical for this problem.

Use W16×45, $F_y = 60$ ksi.

Example 10.3.3

Redesign the beam of Example 10.3.2, using either cover plates or two different sections butt-spliced together. $F_y = 60$ ksi.

SOLUTION

Assume that the change of section occurs near the inflection point to the right of the central support, giving rise to different plastic moments M_{p1} and M_{p2} at the two restrained supports (see Fig. 10.3.4).

(a) Consider first the 25-ft section supporting the 15- and 25-kip loads. Assume the positive movement under the 15-kip load reaches M_{p1},

$$M_{p1} = M_{s1} + \tfrac{1}{2}M_{s2} - \tfrac{10}{25}M_{p1}$$

$$= 153 + 85 - 0.4M_{p1}$$

$$M_{p1} = 170 \text{ ft-kips}$$

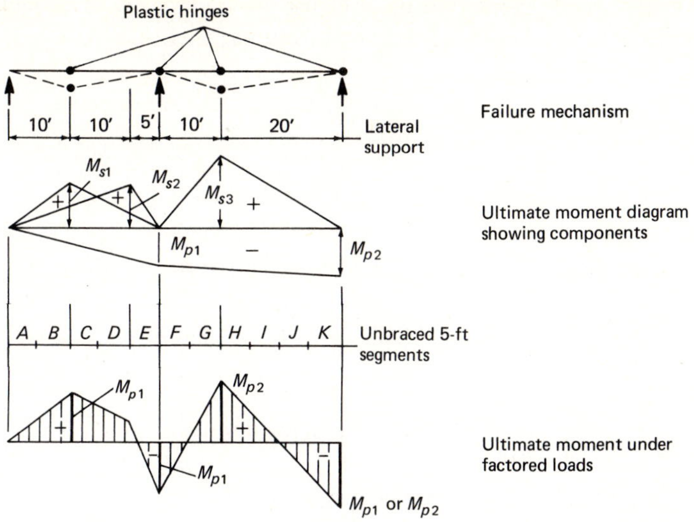

Fig. 10.3.4 Example 10.3.3.

Under the 25-kip load the moment is

$$M = \tfrac{1}{3}M_{s1} + M_{s2} - \tfrac{20}{25}M_{p1}$$
$$= 51 + 170 - 0.8(170) = 85 \text{ ft kips} < 170 \qquad \text{OK}$$

(b) Consider lateral support requirements. For the 10-ft unbraced length between the 15- and 25-kip loads, the end-moment ratio under factored loads is

$$\frac{M}{M_p} = \frac{-85}{170} = -0.5$$

which is the boundary between use of AISC Formulas (2.9–1a) and (2.9–1b).

$$L_{cr}(2.9\text{–}1b) = \frac{1375}{F_y}\, r_y = 23 r_y$$

$$L_{cr}(2.9\text{–}1a) = \left[\frac{1375}{F_y} + 25\right] r_y = 48 r_y$$

For $L = 10$ ft, the required r_y is $120/23 = 5.2$ in., which cannot be achieved from any realistic design. Increase the lateral support to every 5 ft. This will be necessary in the 30-ft span also, as shown in Example 10.3.2.

Assuming the plastic hinge $(+M)$ between segments B and C (see

Fig. 10.3.4) is the last hinge to form, only the working stress compact section requirements of AISC–1.5.1.4.1 need to be satisfied; i.e., $L < L_c$.

$$\frac{L_c}{b_f} = \frac{76}{\sqrt{F_y}} = \frac{76}{\sqrt{60}} = 9.8$$

Required $b_f = \frac{60}{9.8} = 6.12$ in.   for 5-ft unbraced length

This restriction applies for segments B and C adjacent to the last plastic hinge.

For segments E and F, rotation capacity must be assured by satisfying AISC Formulas (2.9–1).

$$\frac{M}{M_p} = \frac{+85}{170} = +0.5 \quad \text{(reverse curvature)}$$

$$L_{cr}(2.9\text{–}1a) = 48r_y \quad \text{(previously computed)}$$

Required $r_y = \frac{60}{48} = 1.25$ in.

(c) Select section for 25-ft span:

$$\text{Required } Z = \frac{170(12)}{60} = 34 \text{ in.}^3$$

Required $b_f = 6.12$ in. (for AISC–1.5.1.4.1)

Required $r_y = 1.25$

From text Appendix Table A4,

W10×30,	$Z = 36.6$ in.³,	$r_y = 1.37$ in.
W8×48,	$Z = 49.0$ in.³,	$r_y = 2.08$ in.
W12×35,	$Z = 51.2$ in.³,	$r_y = 1.54$ in.

(While this section is indicated as not satisfying the local buckling requirements for plastic design, it is satisfactory when $F_y = 55$ ksi and may be close enough for this case.)

W16×45,	$Z = 82.3$ in.³,	$r_y = 1.57$ in.

Check flange width: [to satisfy AISC–1.5.1.4.1(2.)]

W10×30,	$b_f = 5.81$ in. < required 6.12 in.	NG
W12×35,	$b_f = 6.56$ in. > required 6.12 in.	OK
W16×45,	$b_f = 7.04$ in. > required 6.12 in.	OK

Though text Appendix Table A4 includes only sections satisfying local buckling requirements of AISC–2.7, illustrate check anyway:

For W12×35,

$$\frac{b_f}{2t_f} = \frac{6.560}{2(0.520)} = 6.31 \approx \text{maximum } 6.3 \qquad \text{OK}$$

$$\frac{d}{t_w} = \frac{12.50}{0.300} = 41.7 < \frac{412}{\sqrt{F_y}} = \frac{412}{\sqrt{60}} = 53.1 \qquad \text{OK}$$

Use W12×35, $F_y = 60$ ksi, for the 25-ft span. For the 30-ft span, top and bottom cover plates will be added, since the 35-lb W12 is the heaviest within its group having the same proportions. Sections within a group having the same proportions can be economically butt joined by welding together, but not for sections having different proportions.

Since the section used has a greater plastic moment capacity than necessary, the ultimate moment diagram under factored loads is as shown in Fig. 10.3.5b. The diagram assumes the first plastic hinge is at the support, and that it does form under the application of the factored loads. The $M_{p1} = 256$ ft-kips as provided is then used in the analysis of the 30-ft span.

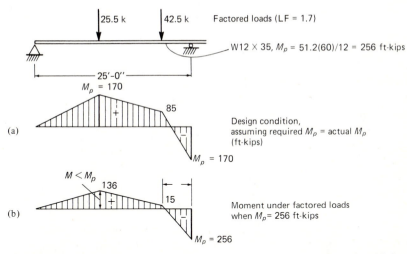

Fig. 10.3.5 Comparison of design condition and actual condition for 25-ft span of Example 10.3.3.

In the actual condition, unbraced segment E is more severely loaded than in the design condition. Recheck $M/M_p = +16/256$, which is still >-0.5, so that AISC Formula (2.9–1a) still applies and is satisfied.

(d) Determine plastic moment requirement on the 30-ft span. When M_{p2} forms under the 50-kip load, equilibrium requires

$$M_{p2} = M_{s3} - \tfrac{1}{3}(M_{p2} - M_{p1}) - M_{p1} \qquad \text{(long cover plates)}$$

if M_{p2} is developed at the fixed end. This would require cover plates to extend to the support. If the cover plates are not extended to the support, equilibrium requires

$$M_{p2} = M_{s3} - M_{p1} \quad \text{(short cover plates)}$$

If long cover plates are used,

$$M_{p2} = 567 - \tfrac{1}{3}M_{p2} + \tfrac{1}{3}(256) - 256; \qquad \tfrac{4}{3}M_{p2} = 396.3, \qquad M_{p2} = 297 \text{ ft-kips}$$

If shorter cover plates are used,

$$M_{p2} = 567 - 256 = 311 \text{ ft-kips}$$

Since there is little difference in requirements, the cover plates will be made as short as possible. The cover plates, top and bottom, must provide

$$M_p \text{ (cover)} = 311 - 256 = 55 \text{ ft-kips}$$

(e) Select plate size. For an area A_p representing the cover plate at one flange, and a distance d center-to-center of cover plates, the plastic modulus Z becomes

$$Z = 2A_p\left(\frac{d}{2}\right) = A_p d$$

$$\text{Required } A_p = \frac{Z}{d} = \frac{55(12)}{60d} = \frac{11.0}{d} = \frac{11.0}{12.75} = 0.86 \text{ sq in.}$$

assuming $d \approx 12.5 + 0.25$ ($\frac{1}{4}$-in. plates). If the plate width is somewhat less than the flange width of W12×35, say 6 in., the area A_p provided will be more than needed. One might reduce the width of plate to 5 in.

Check local buckling on plate as a stiffened element welded along two sides, according to AISC–2.7:

$$\frac{b}{t} = \frac{5}{0.25} = 20 < \frac{190}{\sqrt{F_y}} = \frac{190}{\sqrt{60}} = 24.5 \qquad \text{OK}$$

Use plates $\frac{1}{4}$×5, top and bottom. These would probably be welded continuously along the sides. Discussion of weld sizes and other requirements is contained in Chapter 5.

(f) Determine plate length. Referring to Fig. 10.3.6 showing the design condition for the 30-ft span, the reader may observe that cover plates will be required for only a short distance, L_1, which might easily be scaled from the diagram. For this straight line moment diagram, it is easily computed:

$$L_1 = \left(1 - \frac{256 + 256}{256 + 311}\right)(10 + 20) = 0.097(30) = 2.9 \text{ ft}$$

Extension of the plates in both directions must satisfy AISC–1.10.4 regarding flange development; thus the welded connection must develop

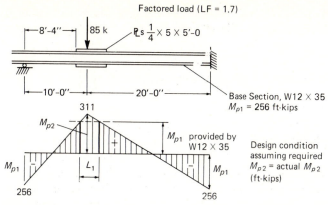

Fig. 10.3.6 Design condition for 30-ft span of Example 10.3.3.

the cover plate's proportion of the capacity. The tension or compression capacity required from a plate is computed:

$$\text{Total required } Z = \frac{310(12)}{60} = 62 \text{ in.}^3$$

$$\% \ Z \text{ from plates} = \frac{A_p}{A_p + A_f}(100) = \frac{1.25}{1.25 + 3.41}(100) = 27\%$$

neglecting the web effect.

$$\left(\begin{array}{c}\text{Force in plate}\\\text{at max moment}\end{array}\right) \approx \frac{0.27 Z F_y}{d} = \frac{0.27(62)(60)}{12.75} = 78.8 \text{ kips}$$

AISC–1.10.4 requires that the necessary force in the plate be developed in the distance a' from the end of plate. Assuming no weld across the end of the plate and continuous weld along the sides, the distance $a' = 2(\text{width}) = 2(5) = 10$ in. At 10 in. from termination of the plate, the welds must have developed the proportionate part of the total flange force needed. For $\frac{3}{16}$ in. fillet welds placed by the submerged arc process (AISC–1.14.6), and using E 70 electrodes, the capacity per inch is

$$R_w = \tfrac{3}{16}(21)(1.7) = 6.7 \text{ kips/in.}$$

using allowable stresses 1.7 times stated values according to AISC–2.8. In order to satisfy AISC–1.10.4, the maximum force that can be developed over 10 in. by two lines of fillet weld is

$$\text{Force} = 10(6.7)(2) = 134 \text{ kips}$$

This is acceptable since it exceeds the required force of 78.8 kips to be developed over the 10-in. extensions. The minimum length cover plate required is $L_1 + 2a' = 2.9 + 2(0.83) = 4.6$ ft.

Use ₧s 5 ft long, beginning 8'-4" from support as shown in Fig. 10.3.6.

In this example no splicing of member was used; however, splicing will frequently be desired on continuous beams. Splices are discussed in Sec. 10.5.

10.4 DESIGN EXAMPLES—WORKING STRESS DESIGN

For design of ordinary continuous beams having close spacing of lateral supports, plastic design is the preferred method. Working stress design employs some arbitrary adjustments to indirectly account for plastic behavior and redistribution of moments. Such arbitrary adjustments cannot reflect true behavior; consequently a uniform safety factor is not obtained. Occasionally, for designs involving long laterally unsupported lengths where lateral stability instead of plastic strength controls, the working stress method becomes the practical approach.

Example 10.4.1

Redesign the beam of Example 10.3.1 (Fig. 10.3.1) by the working stress method. A36 steel.

SOLUTION

The elastic bending moment diagram (see Fig. 10.4.1) under service loads must first be computed using a method of statically indeterminate analysis.

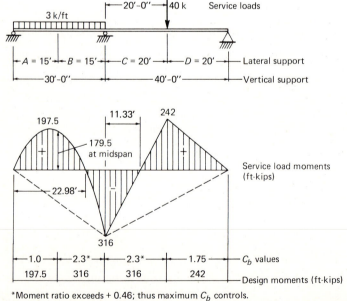

*Moment ratio exceeds + 0.46; thus maximum C_b controls.

Fig. 10.4.1 Example 10.4.1.

The four separate unbraced segments with their C_b values and design moments are shown in Fig. 10.4.1. If lateral bracing is sufficient to permit a plastic moment to form at a given maximum moment location, an allowable stress of $0.66F_y$ is permitted. Further, at the negative moment location if adequate rotation capacity (plastic strain after M_p has been reached) is available, redistribution of moments is accounted for by a 10 percent reduction in maximum negative moment, with the corresponding adjustment of positive moments to maintain equilibrium.

(a) Determine required section assuming reduction in $(-M)$ and using $0.66F_y$.

$$\text{Required } S = \frac{0.9(316)12}{24} = 142 \text{ in.}^3$$

which is the minimum section modulus required assuming the most favorable conditions. Since the laterally unbraced length is 15 ft for segment B, try for $L_c > 15$ ft. Text Appendix Table A3 indicates that selection of a minimum weight beam with $L_c > 10$ ft is not practical.

(b) Whenever the laterally unbraced length exceeds L_u, the suitable approach is to use the AISC Manual curves, "Allowable Moments in Beams." For segment C, enter with

$$M = 316 \text{ ft-kips}$$

$$\frac{L}{C_b} = \frac{20}{2.3} = 8.7 \text{ ft} \quad \text{(if } F_b(1.5\text{--}7) \text{ controls)}$$

$$\frac{L}{\sqrt{C_b}} = \frac{20}{\sqrt{2.3}} = 13.2 \text{ ft} \quad \text{(if } F_b(1.5\text{--}6) \text{ controls)}$$

Noting that $20 \text{ ft} > L_c$, try W24×76; $L_u = 11.8$ ft. From the curves, the capacity is

$$M = 322.5 \text{ ft-kips} \qquad \text{at } L = 8.7 \text{ ft}$$

and $F_b = 0.60F_y$.

(c) Check segment A (Fig. 10.4.1) with W24×76.

$$f_b = \frac{197.5(12)}{176} = 13.5 \text{ ksi}$$

$$\frac{L}{C_b} = \frac{15}{1.0} = 15 \text{ ft} > L_u = 11.8 \text{ ft}; \qquad F_b < 0.60F_y$$

$$F_b(1.5\text{-}7) = \frac{12,000C_b}{Ld/A_f} = \frac{12,000(1.0)}{15(12)3.91} = 17.1 \text{ ksi} > f_b = 13.5 \text{ ksi} \qquad \text{OK}$$

which indicates segment A is OK and no need to check AISC Formula (1.5–6).

(d) Check segment D; W24×76.

$$\frac{L}{C_b} = \frac{20}{1.75} = 11.4 < L_u = 11.8 \text{ ft} \qquad \text{OK}$$

$$F_b = 0.60F_y = 22 \text{ ksi}$$

$$f_b = \frac{242(12)}{176} = 16.5 \text{ ksi} < 22 \text{ ksi} \qquad \text{OK}$$

(e) Show final check for segments B and C; W24×76:

$$F_b(1.5\text{-}7) = \frac{12,000(2.3)}{20(12)(3.91)} = 29.4 \text{ ksi} > 0.60 \, F_y$$

$$F_b = 0.60F_y = 22 \text{ ksi}$$

$$f_b = \frac{316(12)}{176} = 21.5 \text{ ksi} < 22 \text{ ksi} \qquad \text{OK}$$

Use W24×76, $F_y = 36$ ksi.

A comparison with Example 10.3.1 shows that the working stress method produced the lighter section (W24×76 vs W14×109). The reason is that plastic design requires development of the collapse mechanism; which because of the 20-ft laterally unsupported length required a larger r_y to satisfy the severe lateral bracing requirement. The working stress method in effect regarded first yield in the negative zone as the limiting criterion; thus lateral bracing did not have to insure rotation capacity in the plastic range.

Example 10.4.2

Redesign the continuous beam of Example 10.3.2 (Fig. 10.3.3) by the working stress method. $F_y = 60$ ksi.

SOLUTION

The elastic bending moment diagram under service load is shown in Fig. 10.4.2. To get a more appropriate comparison with plastic design, lateral support is provided every 5 ft on the 30-ft span and at the load and vertical support points on the 25-ft span.

(a) Determine the section required using 0.9 of maximum negative moment with $0.66F_y$. L_c must exceed 5 ft.

$$\text{Required } S = \frac{0.9(185.8)(12)}{39.6} = 50.7 \text{ in.}^3$$

Try W16×36; $S = 56.5$ in.^{3}; $L_c = 5.3$ ft; $L_u = 6.2$ ft (from text Appendix Table A3). Check to verify acceptability,

$$L_c = \frac{76b_f}{\sqrt{F_y}} = \frac{76(6.985)}{\sqrt{60}(12)} = 5.7 \text{ ft}$$

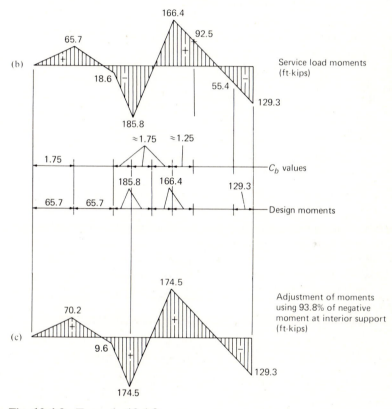

Fig. 10.4.2 Example 10.4.2.

or

$$L_c = \frac{20,000}{(d/A_f)F_y} = \frac{20,000}{5.28(60)12} = 5.3 \text{ ft} \quad \text{(controls)}$$

For local buckling, AISC–1.5.1.4.1(2.),

$$\frac{b_f}{2t_f} = \frac{6.985}{2(0.430)} = 8.1 < \frac{65}{\sqrt{F_y}} = 8.4 \qquad \text{OK}$$

$$\frac{d}{t_w} = \frac{15.86}{0.295} = 53.8 < \frac{640}{\sqrt{F_y}} = 82.6 \qquad \text{OK}$$

Note is made that this local buckling limitation for $b_f/2t_f$ in plastic design was 6.3 and prevented use of this section.

(b) Adjustment of the moment diagram to utilize the reduction in negative moment permitted under AISC–1.5.1.4.1 would make the negative moment $0.9(185.8) = 167$ ft-kips. The resulting increase in positive moment will make the positive moment exceed the adjusted negative moment. Adjust only until the moments in question are equal:

$$x\ 185.8 = \frac{50(10)20}{30} - \frac{129.3}{3} - \frac{2}{3}(185.8\,x)$$

which indicates that using 93.8 percent $(x = 0.938)$ of the negative moment will equalize the positive and negative values at

$$-M = 0.938(185.8) = 174.5 \text{ ft-kips}$$

$$\text{Required } S = \frac{174.5(12)}{39.6} = 52.8 \text{ in.}^3$$

W16×36, $S = 56.5$ in.^{3} The adjusted moments are shown in Fig. 10.4.2c. The decrease in negative moment is 11.3 ft-kips. For equilibrium, the moments in span 1 become

$$-18.6 - 0.8(-11.3) = -9.6 \text{ ft-kips @ 25-kip load}$$

$$+65.7 - 0.4(-11.3) = +70.2 \text{ ft-kips @ 15-kip load}$$

These are only slightly different than the AISC indication of using the average of the change in negative moments [in this case $0.5(11.3)$] to correct positive moments. Further, the authors believe the intention of the AISC method is followed by adjusting less than the full 10 percent and maintaining equilibrium.

(c) Check 10-ft unbraced lengths on 25-ft span for W16×36:

$$F_b(1.5\text{-}7) = \frac{12,000C_b}{Ld/A_f} = \frac{12,000(1.75)}{10(12)(5.28)} = 33.1 \text{ ksi}$$

$$\frac{L}{r_T} = \frac{10(12)}{1.79} = 67.0$$

$$F_b(1.5\text{-}6) = 40.0 - \frac{(L/r_T)^2}{425C_b}$$

$$= 40.0 - \frac{(67.0)^2}{425(1.75)} = 34.0 \text{ ksi} \quad \text{(controls)}$$

$$f_b = \frac{70.2(12)}{56.5} = 14.9 \text{ ksi} < F_b = 34.0 \text{ ksi} \qquad \text{OK}$$

Use W16×36, $F_y = 60$ ksi.

In plastic design, Example 10.3.2 shows that because of more severe local buckling and lateral bracing requirements a W16×36 was not acceptable, even though providing adequate strength.

Example 10.4.3

Redesign the continuous beam of Example 10.3.3 by the working stress method (Fig. 10.4.2). $F_y = 60$ ksi.

SOLUTION

The requirement here is to use a cover plate on a basic section. A study of Fig. 10.4.2c shows that any cover plate would have to extend over the interior support and out into the region under the 50-kip load. The moments under service load conditions (neglecting the change caused by variable moment of inertia) show that the plastic design solution of using cover plates only in the region of the 50-kip load would not be acceptable. It would be inadequate at the support.

(a) Select section for $M = 129.3$ ft-kips, assuming a "compact section" with $F_b = 0.66F_y = 39.6$ ksi.

$$\text{Required } S_x = \frac{129.3(12)}{39.6} = 39.2 \text{ in.}^3$$

Using text Appendix Table A3, try W14×30, W16×31, or W14×34.
Check W14×30, $S_x = 42.0$ in.^{3}, $L_c = 5.2$ ft:

$$\frac{b_f}{2t_f} = \frac{6.730}{2(0.385)} = 8.7 > 8.4 \qquad \text{NG}$$

$$\frac{d}{t_w} = \frac{13.84}{0.270} = 51.3 < 82.6 \qquad \text{OK}$$

Since $b_f/2t_f$ exceeds 8.4 but does not exceed $95/\sqrt{F_y} = 12.3$, the W14×30 may be classified as "partially compact." Thus

$$F_b = F_y \left[0.79 - 0.002 \left(\frac{b_f}{2t_f} \right) \sqrt{F_y} \right] = 39.3 \text{ ksi}$$

This value agrees with that given in text Appendix Table A3.

$$f_b = \frac{129.3(12)}{42.0} = 36.9 \text{ ksi} < F_b = 39.3 \text{ ksi} \qquad \text{OK}$$

There is adequate reserve to add in the stress due to the weight of the beam.

(b) Check W14×30 for 10 ft unbraced length on the 25-ft span. Since

$L = 10$ ft exceeds $L_c = 5.2$ ft, the section may not be considered "compact."

$$F(1.5\text{-}7) = \frac{12{,}000C_b}{Ld/A_f} = \frac{12{,}000(1.75)}{10(12)5.34} = 32.8 \text{ ksi}$$

$$\frac{L}{r_T} = \frac{10(12)}{1.74} = 69.0$$

$$F_b(1.5\text{-}6) = 40.0 - \frac{(L/r_T)^2}{425C_b}$$

$$= 40.0 - \frac{(69.0)^2}{425(1.75)} = 33.6 \text{ ksi} \quad \text{(controls)}$$

$$f_b = \frac{70.2(12)}{42.0} = 20.1 \text{ ksi} < F_b = 33.6 \text{ ksi} \qquad \text{OK}$$

Use W14×30, $F_y = 60$ ksi, as the base section.

(c) Determine plate size required where moment exceeds capacity of W14×30. For lateral bracing at 5-ft intervals in this region,

$$F_b = 39.3 \text{ ksi} \quad \text{[from part (a)]}$$

$$\text{Moment capacity} = 39.3(42.0)\tfrac{1}{12} = 137.6 \text{ ft-kips}$$

The moment capacity to be provided by plates,

$$M \approx 174.5 - 137.6 = 36.9 \text{ ft-kips}$$

$$\text{Required } A = \frac{M}{df_{\text{avg}}} = \frac{36.9(12)}{14.09(38.6)} = 0.81 \text{ sq in.}$$

estimating $\quad d \approx 13.84 + 0.25 = 14.09 \quad$ and $\quad f_{\text{avg}} \approx 39.3(14.09/14.34) = 38.6$ ksi. For a plate 6 in. wide,

$$\text{Thickness } t = \frac{0.81}{6} = 0.14 \text{ in.}$$

Try plates, top and bottom, $\tfrac{1}{4} \times 5$, $A = 1.25$ sq in.
Check local buckling to satisfy "compact section" for a stiffened element,

$$\frac{b}{t} = \frac{5.0}{0.25} = 20 < \frac{190}{\sqrt{F_y}} = 24.5 \qquad \text{OK}$$

Therefore this section with a cover plate continuously welded along its edges may be treated as "compact"; $F_b = 0.66 F_y = 39.6$ ksi.

(d) Recheck flexural capacity.

$$I = 291 \text{ (for W14} \times 30) + 1.25(99.3) = 415 \text{ in.}^4$$

$$S = \frac{I}{d/2} = \frac{415}{14.34/2} = 57.9 \text{ in.}^3$$

$$f_b = \frac{174.5(12)}{57.9} = 36.2 \text{ ksi} < 39.6 \text{ ksi} \qquad \text{OK}$$

Use plates, top and bottom, $\frac{1}{4} \times 5$, $F_y = 60$ ksi.

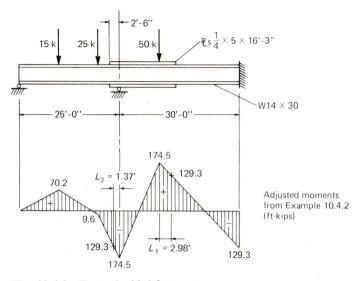

Fig. 10.4.3 Example 10.4.3.

(e) Determine length for plates (see Fig. 10.4.3):
To right of 50-kip load,

$$L_1 = \frac{174.5 - 129.3}{174.4 + 129.3}(20) = \frac{45.2}{303.8}(20) = 2.98 \text{ ft}$$

To left of interior support,

$$L_2 = \frac{174.5 - 129.3}{174.5 - 9.6}(5) = \frac{45.2}{164.9}(5) = 1.37 \text{ ft}$$

$$\text{Force in plate} = \frac{A_{p1}}{A_{f1} + A_{p1}}\left(\frac{M}{d}\right) = \frac{1.25}{3.84}\left[\frac{174.5(12)}{13.45}\right] = 50.7 \text{ kips}$$

For $\frac{3}{16}$ in. fillet weld placed by the submerged arc process using E70 electrodes, the capacity R_w per inch is

$$R_w = \frac{3}{16}(21) = 3.94 \text{ kips/in.}$$

The maximum force that can be developed in the distance a' [2 times width $= 2(5.0) = 10$ in.] is

$$\text{Force} = 10(3.94)2 = 78.8 \text{ kips} > 50.7 \text{ kips} \qquad \text{OK}$$

$$\text{Plate length} = L_1 + 10.0 + L_2 + 2a'$$

$$= 2.98 + 10.0 + 1.37 + 2(10/12) = 16.0 \text{ ft}$$

Use $\text{PL } \frac{1}{4} \times 5$, 16'-3" long, beginning 2'-6" to left of interior support, as shown in Fig. 10.4.3.

10.5 SPLICES

While the design of connections is outside the scope of this chapter, the location of and strength requirements for beam splices are appropriately discussed here. It is obvious that if a splice is designed for the moment and shear capacity for the member spliced, full continuity is maintained and no special precautions are necessary. Some designers prefer to use shear splices at points of contraflexure and thus introduce a real hinge at a point of zero moment. There are two reasons why this should be avoided: (1) the point of contraflexure under service load is not at the same location that it occurs under factored loads (i.e., at its mechanism condition); (2) moments obtained assuming continuity are invalid if real hinges are inserted. Hart and Milek [11] have provided a good discussion of this problem.

According to AISC, connections must be designed for the moments, shears, and axial loads to which they are to be subjected. AISC–1.10.8 refers to service loads, while AISC–2.8 refers to factored loads.

If full continuity is assumed when determining moments, either under service loads or under factored loads, then splices should provide that continuity. A reduced stiffness at a splice may prevent or reduce transmission of moment across that section and significantly change the resulting moment diagram.

Example 10.5.1

Examine the effect of a shear splice at the point of contraflexure in the 40-ft span of the two-span continuous beam of Example 10.3.1. A36 steel. Assume 70 percent of the given load is live load.

SOLUTION

(a) Full dead load plus live load. If full continuity is maintained, partial loading in some spans to account for live load in various locations is unnecessary. Note that the moment $M_p = 453$ ft-kips (see Fig. 10.5.1a) can develop even when the adjacent span has a reduced load. In other words, each span may be treated separately, as long as continuity exists so that the negative moment may be assumed to reach M_p.

(b) Effect of hinge at inflection point. Use of a shear splice in effect

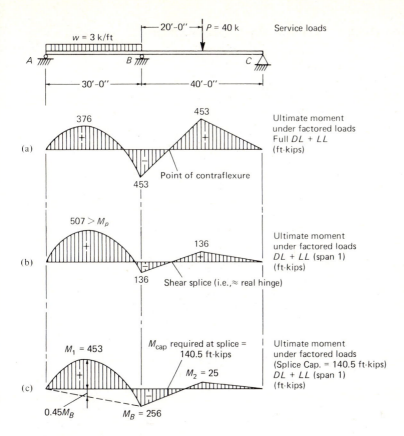

Fig. 10.5.1 Example 10.5.1.

creates a hinge for all stages of loading (i.e., transforms the system into a statically determinate one). The maximum negative moment that can develop is limited to the shear at the hinge times the distance to the support. When load on the span with the hinge is reduced, the negative moment that can develop is reduced. As shown in Fig. 10.5.1b, when only the 30 percent of the factored load that is dead load $[0.3(40)(1.7) = 20.4 \text{ kips}]$ is on the 40-ft span, the maximum negative moment is

$$\text{Shear at splice} = \tfrac{2}{3}(20.4) = 13.6 \text{ kips}$$

$$\text{Moment at support} = 13.6(10) = 136 \text{ ft-kips}$$

With the full factored load in the adjacent span, the maximum positive moment in the 30-ft span is

$$V_{AB} = \frac{3(30)(1.7)}{2} - \frac{136}{30} = 76.5 - 4.53 = 71.97 \text{ kips}$$

$$M = \frac{(71.97)^2}{2(3)(1.7)} = 508 \text{ ft-kips} > 453 \text{ ft-kips} \qquad \text{NG}$$

Since the original design required $M_p = 453$ ft-kips, the 30-ft span is now inadequate as a result of the inability of the negative moment to exceed 136 ft-kips with the reduced load on the 40-ft span.

(c) Minimum strength required for splice. If the splice design is to provide less than full continuity, Fig. 10.5.1c illustrates the minimum capacity needed. If the section being spliced provides $M_p = 453$ ft-kips, the positive moment in the 30-ft span may not exceed that value. Thus, referring to Fig. 10.5.1c, and assuming the maximum occurs at approximately $0.45\,L_1$,

$$M_1 = \frac{3(1.7)(0.45)(0.55)}{2}(30)^2 - 0.45\,M_B = 453 \text{ ft-kips}$$

$$M_B = \frac{568 - 453}{0.45} = \frac{115}{0.45} = 256 \text{ ft-kips}$$

In order to have this moment develop, a moment capacity must be provided at the splice point:

$$M_2 = \frac{40(1.7)(0.30)(30)}{4} - \frac{M_B}{2} = 153 - 128 = 25 \text{ ft-kips}$$

$$M \text{ required at splice} = \frac{256 + 25}{2} = 140.5 \text{ ft-kips}$$

If the splice in this plastic design problem is designed for the shear as determined from Fig. 10.5.1a, plus the 140.5 ft-kip moment, the reduced loading of span BC will not cause overstress in span AB.

Finally, it is noted that reduced load in span AB has no detrimental effect on span BC.

The authors prefer that continuity be maintained by the design of splices for 100 percent of the moment capacity of members spliced; however, reduced capacities providing more economical designs may be used as long as the resulting designs are checked under possible partial loadings. Other examples and a more detailed procedure are given in Ref. 11.

SELECTED REFERENCES

1. Lynn S. Beedle, *Plastic Design of Steel Frames*. New York: John Wiley & Sons, Inc., 1958.
2. C. E. Massonnet and M. A. Save, *Plastic Analysis and Design*, Vol. I, Beams and Frames. Waltham, Mass.; Blaisdell Publishing Company, 1965.
3. Bruce G. Johnston, "Strength as a Basis for Structural Design," *Proc. AISC National Engineering Conf.*, 1956, pp. 7–13.
4. Bruno Thürlimann, "Simple Plastic Theory," *Proc. AISC National Engineering Conf.*, 1956, pp. 13–18.

5. Robert L. Ketter, "Analysis and Design Examples," *Proc. AISC National Engineering Conf.*, 1956, pp. 19–35.

6. Lynn S. Beedle, "Experimental Verification of Plastic Theory," *Proc. AISC National Engineering Conf.*, 1956, pp. 35–49.

7. Bruno Thürlimann, "Modifications to 'Simple Plastic Theory,'" *Proc. AISC National Engineering Conf.*, 1956, pp. 50–57.

8. Frederick S. Merritt, "How to Design Steel by the Plastic Theory," *Engineering News-Record* (Apr. 4, 1957), 38–43.

9. Bruce G. Johnston, C. H. Yang, and Lynn S. Beedle, "An Evaluation of Plastic Analysis as Applied to Structural Design," *Welding Journal Research Suppl.* (May 1953), 1–16.

10. Joint Committee of Welding Research Council and the American Society of Civil Engineers, *Commentary on Plastic Design in Steel*, 2nd Ed., ASCE Manual and Reports on Practice No. 41, New York, 1971.

11. Willard H. Hart and William A. Milek, "Splices in Plastically Designed Continuous Structures," *Engineering Journal*, AISC 2, 2 (April 1965), 33–37.

PROBLEMS

All problems are to be done in accordance with latest AISC Specification, and all given loads are service loads, unless otherwise indicated.

10.1. Determine the maximum value for service load P of the beam in the accompanying figure. Assume adequate lateral support to satisfy any "compact section" requirements. Solve by (a) working stress method and (b) plastic design.

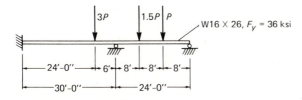

Prob. 10.1

10.2. Determine the uniformly distributed service load a W24×104, $F_y = 50$ ksi, may be permitted to carry as a two-span continuous beam, having equal spans of 40 ft. Assume deflections do not control, and that lateral support is provided at the vertical supports and at 15 and 30 ft from the simply supported ends. Solve by (a) working stress method and (b) plastic design.

10.3. Select the lightest W section for the two-span continuous beam shown in the accompanying figure. Use $F_y = 50$ ksi with (a) working stress method and (b) plastic design. In plastic design, given an alternate design fully utilizing maximum plastic strength and specify the necessary lateral support associated with your design.

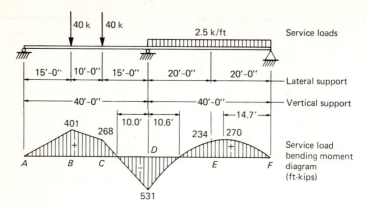

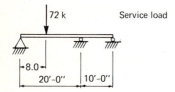

Prob. 10.3

10.4. Determine the lightest W section that is suitable to serve as the two-span continuous beam shown in the accompanying figure. Lateral supports exist at the vertical supports and under the concentrated load. Use $F_y = 60$ ksi with (a) working stress method and (b) plastic design. In the plastic design case, as an alternate if maximum plastic strength cannot be utilized with the given lateral support, specify the required lateral support and select the section for maximum strength.

Prob. 10.4

10.5. Select the lightest W section for the two-span continuous beam of the accompanying figure. The load of 3 kips/ft may be assumed to include the beam weight. Lateral support is provided at the top flange at vertical supports and at 10 ft intervals. Use $F_y = 50$ ksi with (a) working stress method and (b) plastic design. In the plastic design method, if maximum plastic strength cannot be utilized with the given lateral support, specify as an alternate the required lateral support and select a section for maximum strength.

10.6. Select the lightest W section for the two-span continuous beam of the accompanying figure. Assume lateral support is provided every $3\frac{1}{2}$ ft. $F_y = 60$ ksi. Use (a) working stress method and (b) plastic design.

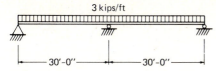

Prob. 10.5

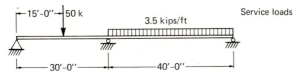

Prob. 10.6

10.7. Select the lightest W section for the three-span continuous beam of the accompanying figure with fixed position service loads. Beam weight is not included. $F_y = 36$ ksi. Use (a) working stress method and (b) plastic design. For plastic design, specify the lateral support locations for economical design instead of using the given locations.

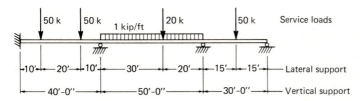

Prob. 10.7

10.8. Select the lightest W section for a three-span (50 ft–65 ft–50 ft) continuous beam to carry a uniform dead load of 2 kips/ft in addition to the beam weight, and a uniform live load of 1.5 kips/ft. Live load deflection (maximum) may not exceed $L/360$. Lateral support is provided every 5 ft. $F_y = 60$ ksi. Use (a) working stress method and (b) plastic design. In the plastic design method, if the 5 ft spacing of lateral supports is too large for economical design, give an alternate design and prescribe the spacing of lateral supports.

10.9. Repeat Prob. 10.8 using steel with $F_y = 50$ ksi and specify lateral bracing for an economical design, instead of using the 5-ft distance of Prob. 10.8. Use (a) working stress method and (b) plastic design.

10.10. Assume one splice is required for the beam selected for Prob. 10.3. Specify its location and the shear and moment for which it should be designed. Assume any loads may be reduced to 30 percent of their value while other loads remain at their maximum values. Use (a) working stress method and (b) plastic design.

10.11. Same as Prob. 10.10. but splice the beam of Prob. 10.7. use (a) working stress method and (b) plastic design.

10.12. For the beam shown in the accompanying figure, (a) select the lightest W section assuming the same section extending over all three spans and (b) redesign, using a smaller base section with welded cover plates where needed. For the cover plate, specify the size, length, and location of the plate. $F_y = 50$ ksi. Specify the necessary lateral support. Use (a) working stress method and (b) plastic design.

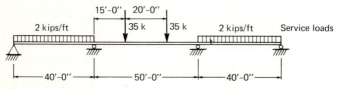

Prob. 10.12

11
Plate Girders

11.1 INTRODUCTION AND HISTORICAL DEVELOPMENT

A plate girder is a beam built up from plate elements to achieve a more efficient arrangement of material than is possible with rolled beams. Plate girders are economical where spans are long enough to permit saving in cost by proportioning for the particular requirements. Plate girders may be of riveted, bolted, or welded construction. Beginning with early railroad bridges during the period 1870–1900, riveted plate girders (Fig. 11.1.1) composed of angles connected to a web plate, with or without cover plates, were extensively used in the United States on spans from about 50 to 150 ft.

Beginning in the 1950s when welding became more widely used (because of improved quality of welding and shop-fabricating economies resulting from increased use of automatic equipment) shop-welded plate girders composed of three plates (Fig. 11.1.2) gradually replaced riveted girders. During this period also, high-tensile-strength bolts were displacing rivets in field construction. In the 1970s plate girders are nearly always shop welded using two flange plates and one web plate to make an I-shaped cross section.

Where practically all *riveted* girders were composed of plate and angle

Plate girders, showing welded stiffeners in place, rocker bearings for vertical supports at the pier, transverse cross bracing between girders, and hinges to provide a simple support for the spans to the right of the hinges. (Photo by C. G. Salmon)

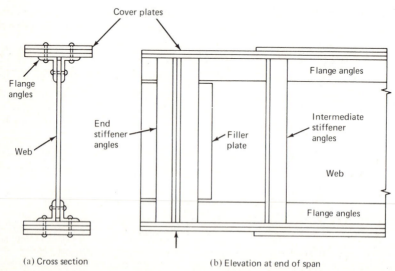

(a) Cross section   (b) Elevation at end of span

Fig. 11.1.1 Typical components of riveted plate girder.

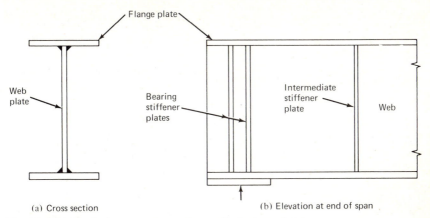

Fig. 11.1.2   Typical components of a welded plate girder.

components having the same material yield point, the tendency now with *welded* girders is to combine materials of different strength. By changing materials at various locations *along the span* so that higher strength materials are available at locations of high moment and/or shear, or by using different strength material for flanges than for web (hybrid girders), more efficient and economical girders can be obtained.

Because few railroad bridges are being built today, discussion of economical spans and other dimensioning comments in this chapter will be limited to highway bridges, where most are continuous over two or more spans; or to buildings where some spans may be assumed as simply supported but more frequently are part of a rigid frame system.

Better understanding of plate-girder behavior, higher-strength steels, and improved welding techniques have combined to make plate girders economical in many situations formerly thought to be ideal for the truss. Generally, simple spans of 70 to 150 ft (20 to 50m) have traditionally been the domain for the plate girder. For bridges, continuous spans frequently using haunches (variable-depth sections) are now the rule for spans 90 ft or more. There are several three-span-continuous plate girders in the United States with center spans exceeding 400 ft, and longer spans are likely to be feasible in the future. The longest plate girder in the world is a three-span continuous structure over the Save River at Belgrade, Yugoslavia, with spans 246–856–246 ft (75–260–75 m). It is a double box girder in cross section varying in depth from 14 ft 9 in. (4.5 m) at midspan to 31 ft 6 in. (9.6 m) at the pier. The structure replaces a suspension bridge destroyed in World War II.

Three types of plate girders whose design is outside the scope of this chapter are shown in Fig. 11.1.3: (a) the box girder, providing improved torsional stiffness for long-span bridges; (b) the hybrid girder, providing variable material strength in accordance with stresses; and (c) the delta

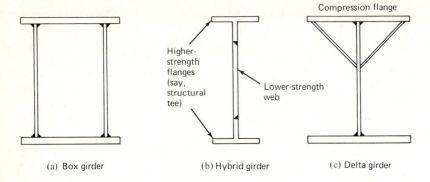

(a) Box girder   (b) Hybrid girder   (c) Delta girder

Fig. 11.1.3 Types of welded plate girders.

girder, providing improved lateral rigidity for long lengths of lateral unsupport.

Prior to studying the theoretical development in this chapter the reader is advised to review Chapter 6, Part II, where the basic elastic stability of plates is treated. If the theoretical development is to be omitted, the reader may proceed directly to Secs. 11.9 through 11.14 which treat the design procedure.

Since the design of riveted girders has been extensively treated in older texts [1, 2] and such girders are rarely used at present, emphasis is placed on welded girders. No example of bolted or riveted girder design is given; however, high-strength bolted splices, commonly found in field connections, are treated in Chapter 13.

General Design Concepts

As with most other aspects of steel design, procedures are increasingly becoming based on ultimate strength. Until adoption of the 1961 AISC Specification the basis for design rules was that elastic buckling should be prevented in plate elements. It was thus assumed that either yielding or elastic instability constituted failure.

The work of Basler and others at Lehigh University has resulted in present AISC provisions which consider post-buckling strength. A plate girder with properly designed regularly spaced stiffeners has the capacity, after instability of the web plate has occurred, to behave much like a truss with its web carrying diagonally the tension forces and the stiffeners carrying the compression forces. This truss-like behavior is referred to as *tension-field action*. Even classical buckling theory recognized that reserve capacity was available by using lower factors of safety against web buckling than for the overall strength of the member.

The classical theory and design procedure are still used by the AREA and AASHTO (railroad bridge and highway bridge) specifications. The ultimate strength concept, including "tension-field action," has formed the basis for the AISC Specification since 1961.

11.2 INSTABILITY RELATING TO LOADS
CARRIED BY THE WEB PLATE

When a designer is free to arrange material to most efficiently carry load, it is apparent that for bending moment which is carried mostly by the flanges, a deep section is desired. A web is needed to make the flanges act as a unit and to carry the shear, but extra web thickness unnecessarily increases the weight. So far as material is concerned, a thin web with stiffeners will give the lightest weight girder. Stability of the thin web plate then is of primary concern.

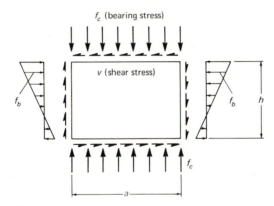

Fig. 11.2.1 Stresses on web plate.

Consider a segment of web plate, as in Fig. 11.2.1, where a represents the length between stiffeners and h is the clear height between longitudinal plate supports (i.e., between flanges, flange and longitudinal stiffener, or between longitudinal stiffeners). In general, such a plate segment is subjected to shear stress v along its edges, linearly varying normal stress f_b over the height h, and compressive stress f_c from loads bearing directly on the girder along the length a. The analysis of such a combined stress situation is complex and not suitable for use in design.

The procedure of this section is to consider separately the three types of stress (shear, v; uniform compression, f_c; and linearly varying compression, such as bending, f_b) on the plate edges. After separate treatment, a combined stress criterion may be established.

The stresses are assumed distributed according to classical beam theory where bending stress is computed $f = Mc/I$ and shear stress is computed as $v = VQ/It$.

In each of the separate treatments the basic approach is the same as illustrated in Secs. 6.14 and 6.15. The problems of inelastic buckling due to residual stress, accidental eccentricity, and initial crookedness apply here to give the strength vs. slenderness function relationship similar to that shown in Fig. 6.15.4.

Elastic Buckling Under Pure Shear

The elastic buckling stress for any plate is given by Eq. 6.14.28 as

$$F_{cr} = k \frac{\pi^2 E}{12(1-\mu^2)(b/t)^2} \qquad [6.14.28]$$

where for the case of pure shear (see Fig. 11.2.2), Eq. 6.14.28 may be written (using τ in place of F for shear stress)

$$\tau_{cr} = k \frac{\pi^2 E}{12(1-\mu^2)\left(\dfrac{\text{short dimension}}{t}\right)^2} \qquad (11.2.1)$$

where for the case of edges simply supported (i.e., displacement prevented but rotation about edges unrestrained)

$$k = 5.34 + 4.0 \left(\frac{\text{short dimension}}{\text{long dimension}}\right)^2 \qquad (11.2.2)$$

the development of which has been given by Timoshenko and Woinowski-Krieger.*

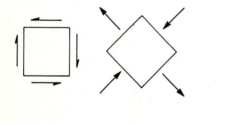
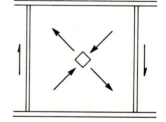

(a) Element in   (b) Principal stress   (c) Principal stress
    pure shear   on element in   on panel in pure
             pure shear   shear

Fig. 11.2.2 Classical shear theory.

For design purposes it may be desirable to put Eqs. 11.2.1 and 11.2.2 in terms of h, the unsupported web height, and a, the stiffener spacing. When this is done two cases must be considered.

    1. If $a/h \leq 1$ (see Fig. 11.2.3a), Eq. 11.2.1 becomes

$$\tau_{cr} = \frac{\pi^2 E[5.34 + 4.0(a/h)^2]}{12(1-\mu^2)(a/t)^2} \frac{(h/a)^2}{(h/a)^2} \qquad (11.2.3)$$

* Reference 40, Chap. 6, pp. 379–385.

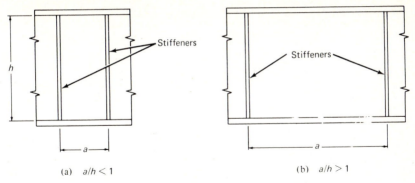

(a) $a/h < 1$ (b) $a/h > 1$

Fig. 11.2.3 Two cases of intermediate stiffener spacing.

2. If $a/h \geq 1$ (see Fig. 11.2.3b), Eq. 11.2.1 becomes

$$\tau_{cr} = \frac{\pi^2 E[5.34 + 4.0(h/a)^2]}{12(1-\mu^2)(h/t)^2} \tag{11.2.4}$$

It is apparent from Eqs. 11.2.3 and 11.2.4 that if one desires to use h/t as the stability ratio in the denominator, then two expressions for k are necessary. For all ranges of a/h, Eqs. 11.2.3 and 11.2.4 may be written

$$\tau_{cr} = \frac{\pi^2 E k}{12(1-\mu^2)(h/t)^2} \tag{11.2.5}$$

where

$$k = 4.0 + 5.34/(a/h)^2 \quad \text{for} \quad a/h \leq 1 \tag{11.2.6}$$

$$k = 4.0/(a/h)^2 + 5.34 \quad \text{for} \quad a/h \geq 1 \tag{11.2.7}$$

For use in AISC–1.10.5, Eq. 11.2.5 has been put in nondimensional form, defining coefficient C_v as the ratio of shear stress at buckling to shear yield stress,

$$C_v = \frac{\tau_{cr}}{\tau_y} = \frac{\pi^2 E k}{\tau_y (12)(1-\mu^2)(h/t)^2} \tag{11.2.8}$$

which is the C_v for *elastic* stability. Substitution of $E = 29{,}000$ ksi, $\mu = 0.3$, and $\tau_y = F_y/\sqrt{3}$ (Eq. 7.5.9),

$$C_v = \frac{\pi^2(29{,}000)\sqrt{3}}{12(1-0.09)} \frac{k}{F_y(h/t)^2}$$

$$C_v = \frac{45{,}000k}{F_y(h/t)^2} \tag{11.2.9}$$

valid for τ_{cr} below the elastic proportional limit, as shown in Fig. 11.2.4.

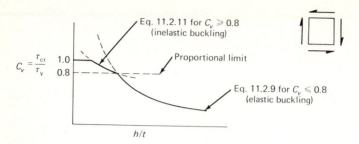

Fig. 11.2.4 Buckling of plates under pure shear.

Inelastic Buckling under Pure Shear

As in all stability situations, residual stresses and imperfections cause inelastic buckling as critical stresses approach yield stress. A transition curve for inelastic buckling was given by Basler [3] based on curve fitting and using test results from Lyse and Godfrey [4]. In the transition zone between elastic buckling and yielding,

$$\tau_{cr} = \sqrt{\tau_{\substack{prop. \\ limit}} \; \tau_{\substack{cr\,ideal \\ elastic}}} \qquad (11.2.10)$$

The proportional limit is taken as $0.8\tau_y$, higher than for compression in flanges, because the effect of residual stress is less. Dividing Eq. 11.2.10 by τ_y to obtain C_v and using Eq. 11.2.9 gives

$$C_v = \frac{\tau_{cr}}{\tau_y} = \sqrt{(0.8)\frac{45,000k}{F_y(h/t)^2}}$$

$$= \frac{190}{h/t}\sqrt{\frac{k}{F_y}} \qquad (11.2.11)$$

which is shown in Fig. 11.2.4.

Bending in the Plane of the Web

As for any plate stability problem, the elastic buckling stress is represented by Eq. 6.14.28,

$$F_{cr} = k\frac{\pi^2 E}{12(1-\mu^2)(b/t)^2} \qquad [6.14.28]$$

where for this case $b = h$.

The theoretical development of the k values for bending in the plane of the plate (Fig. 11.2.5), is given by Timoshenko and Woinowski-Krieger.* For any given type of loading, k varies with the aspect ratio,

* Reference 40, Chap. 6, pp. 373–379.

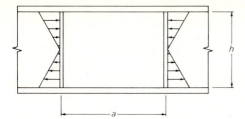

Fig. 11.2.5 Web plate under pure moment.

a/h (see Fig. 11.2.5), and with the support conditions along the edges. If the plate can be considered to have full fixity (full moment resistance against edge rotation) along the edges parallel to the direction of loading (i.e., at the edges joined to the flanges), the minimum value of k is 39.6 for any a/h ratio. If the flanges are assumed to offer no resistance to edge rotation, the minimum k value is 23.9. The variation of k with a/h ratio is given in Fig. 11.2.6.

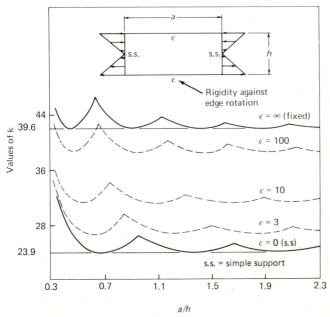

Fig. 11.2.6 Buckling coefficients for plates in pure bending. [From *Handbook of Structural Stability*, Vol. 1 [5] (p. 92)]

Thus the critical stress (using $E = 29,000$ ksi) may be said to lie between

$$F_{cr} = \frac{627,000}{(h/t)^2} \text{ ksi} \qquad \text{for } k = 23.9 \quad \text{(simple support at flanges)}$$

and

$$F_{cr} = \frac{1038,000}{(h/t)^2} \text{ ksi} \qquad \text{for } k = 39.6 \quad \text{(full fixity at flanges)}$$

While each particular girder will have a different degree of flange restraint, fully welded flange to web connections will surely approach the full fixity case. It will be reasonable then to arbitrarily select a k value closer to 39.6, say 80 percent of the difference toward the higher value. One might say that

$$F_{cr} = \frac{954,000}{(h/t)^2} \text{ ksi} \tag{11.2.12}$$

is representative of the stress when elastic buckling is imminent due to bending in the plane of the web. Such "bend buckling" cannot occur if

$$\frac{h}{t} \le \sqrt{\frac{954,000}{F_{cr}, \text{ksi}}} = \frac{975}{\sqrt{F_{cr}}} \tag{11.2.13}$$

Figure 11.2.7 shows the elastic stability relationship as rationalized by the foregoing logic.

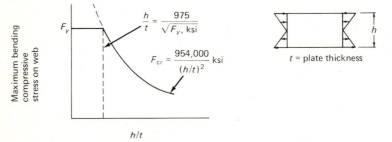

Fig. 11.2.7 Buckling of plates under bending in the plane of the web.

Since the web carries only a small part of the total bending moment to which the girder is subjected, neglect of the transition zone arising from inelastic buckling will not be significant.

Combined Shear and Bending

While there often are regions along a girder span where the cases of bending or shear may be considered separately, usually combined shear and bending stresses are present. Timoshenko and Gere [6] have shown that when $\tau/\tau_{cr} < 0.4$ the effect of shear stress on the critical value of bending stress is small. It is suggested by the *SSRC Guide* [7] that the following approximate interaction formula agrees with theoretical computations for elastic stability,

$$\left[\frac{f_b}{F_{\text{cr}}(\text{bending alone})}\right]^2 + \left[\frac{\tau}{\tau_{\text{cr}}(\text{shear alone})}\right]^2 = 1 \qquad (11.2.14)$$

where f_b and τ are the bending and shear stresses, respectively, which in combination will cause elastic instability.

Elastic Buckling Under Transverse Compression

Heavy transverse loads on plate girders are normally carried by bearing stiffeners so that compressive stress f_c, as shown in Fig. 11.2.1, is usually of low magnitude. However, as the web plate is made thinner, even uniform loading may cause compressive stress high enough to buckle the web vertically. The localized yielding, i.e., web crippling, that may occur at the toe of fillets on beams has already been treated in Sec. 7.6. While the local effect certainly requires checking, thin plate girder webs may also be subject to overall collapse, i.e., vertical buckling due to transverse compression.

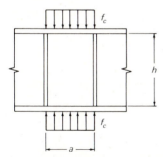

Fig. 11.2.8 Uniformly compressed plate.

Reexamine the situation treated in Sec. 6.14, the uniformly compressed plate, shown in Fig. 11.2.8. The solution obtained in Chapter 6, Eq. 6.14.28, may be rewritten in terms of the plate girder. The width b of the loaded edge, becomes the stiffener spacing a (i.e., the width over which the load may be considered uniformly distributed). The length a in Fig. 6.14.7 becomes the unsupported web height h in the plate girder. Equations 6.14.28 and 6.14.29 become

$$F_{\text{cr}} = k\frac{\pi^2 E}{12(1-\mu^2)(a/t)^2} \qquad (11.2.15)$$

where

$$k = \left[\frac{1}{m(a/h)} + m(a/h)\right]^2 \qquad (11.2.16)$$

with m an integer indicating the number of half-waves that occur over the height h at buckling.

If one desires the stability ratio h/t to be in the denominator of Eq. 11.2.15, multiply the numerator and denominator by $(a/h)^2$, giving

$$F_{cr} = k_c \frac{\pi^2 E}{12(1-\mu^2)(h/t)^2} \qquad (11.2.17)$$

where

$$k_c = \frac{k}{(a/h)^2} = \left[\frac{1}{m(a/h)^2}+m\right]^2 \qquad (11.2.18)$$

Since a single expression for k is desired for all a/h values, consider the actual variation over the practical range. When a/h becomes large, i.e., wide stiffener spacing compared to depth, $k_c \approx m^2$. The minimum value is with $m = 1$, essentially the Euler pinned column. Basler [8] states that since the compression stress actually varies from a maximum at the top of the web plate to near zero at the bottom, Eq. 11.2.18 is overly conservative. Therefore k_c is suggested to have a minimum value of 2 instead of 1.

At the other extreme with closely spaced stiffeners, strength greatly increases. As shown in Fig. 6.15.2, curve C, when the ratio of plate length to width of loaded edge increases, the value of k approaches 4. To satisfy both extremes Basler [8] suggests using

$$k_c = \left[\frac{4}{(a/h)^2}+2\right] \qquad (11.2.19)$$

which is compared with Eq. 11.2.18 for $m = 1$ in Fig. 11.2.9. For small h/a, i.e., widely spaced stiffeners, the decrease of intensity of compression from top to bottom means Eq. 11.2.19 should correctly exceed Eq. 11.2.18. As the stiffener spacing becomes closer, the variation in compressive stress over the depth h has less effect. For h/a greater than about 2, Eqs. 11.2.18 and 11.2.19 become essentially coincident. Thus, while the specific values of Eq. 11.2.19 have not been verified, the result is reasonable.

The preceding discussion and Eq. 11.2.19 apply to situations in which the flange offers negligible rotational restraint at the compressed edge of the web plate. When the loaded edges are fixed, or fully restrained against rotation, the k value is higher as shown in Fig. 6.15.2, curve C (dashed). For wide stiffener spacing (a/b of Fig. 6.15.2 small) the effect of end restraint increases stability, so that k varies from 5.5 upward as the ratio, girder height/stiffener spacing, decreases below 1.5. On the other hand, for closely spaced stiffeners, the minimum k value is about 4, the same as for no rotational restraint at the loaded edge. Basler therefore recommends [8] using

$$k_c = \left[\frac{4}{(a/h)^2}+5.5\right] \qquad (11.2.20)$$

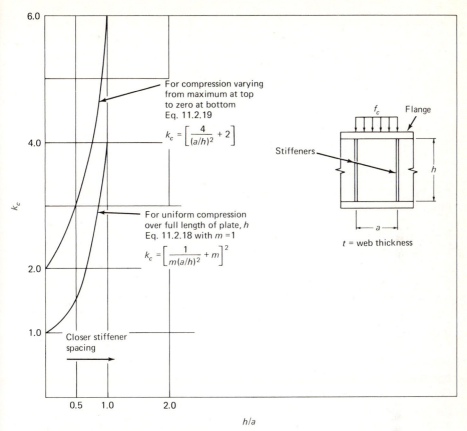

Fig. 11.2.9 Buckling coefficient for vertical buckling under transverse compression, $F_{cr} = k_c \pi^2 E / [12(1 - \mu^2)(h/t)^2]$.

which increases k_c for wide spacing of stiffeners and approaches the same value as Eq. 11.2.19 as a/h becomes small. Again, although the coefficient k_c cannot be rigorously derived, it seems logical.

For uniform loading on the flange of a girder, the unit stress, f_c, is obtained directly by dividing the uniform loading w per inch by the thickness t (Fig. 11.2.10a). Basler [8] recommends distributing concentrated loads over the panel width (Fig. 11.2.10b), or over the depth of the web (Fig. 11.2.10c), whichever gives the greater stress f_c.

11.3 INSTABILITY RELATING TO THE COMPRESSION FLANGE

Flange related instability, *lateral-torsional buckling*, was the subject of Chapter 9. The flange plates in rolled beams are attached together by a relatively thick web which brings both flanges into action (high torsional stiffness) when lateral instability is imminent. As the h/t for the web is

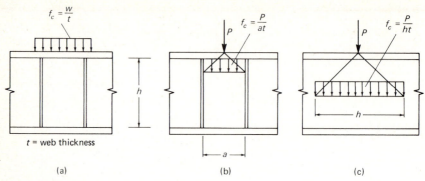

Fig. 11.2.10 Suggestions for distributing loads to evaluate web stability under direct compression.

increased, the effect of the tension flange decreases (column strength of the compression flange based on its lateral bending stiffness predominates). If h/t exceeds the value at which buckling occurs due to bending in the plane of the web, the cross section behaves, for carrying bending stress, as if part of the web were not there. The vertical support the web is able to provide to the compression flange is then significantly reduced and the possibility of *vertical buckling of the flange* must be considered. Furthermore, once this flange support of the web is reduced, depending on how thick is the web and how much of it will perform integrally with the flange plate, the possibility of *torsional buckling* of the tee-shaped flange (flange with segment of web combined) arises. The typical buckling modes of the compression flange behaving as a column are shown in Fig. 11.3.1.

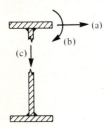

Fig. 11.3.1 (a) Lateral buckling, (b) torsional buckling, (c) vertical buckling.

Lateral-Torsional Buckling

This type of behavior has been theoretically treated in Chapter 9 with the elastic buckling stress computed in accordance with Eq. 9.4.1,

$$F_{cr} = \frac{C_4\sqrt{EI_y GJ}}{S_x L}$$

[9.4.1]

where all terms are defined in Sec. 9.4.

Vertical Buckling

This potential failure mode exists only for cases where buckling can occur due to bending in the plane of the web (so-called "bend-buckling"); i.e., when h/t exceeds $975/\sqrt{F_{cr}}$ (Eq. 11.2.13). For such cases, one may imagine that the flange is a compression member independent of the rest of the girder (see Fig. 11.3.2).

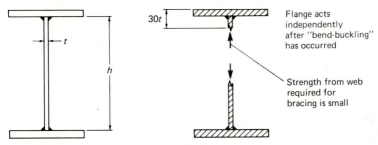

Fig. 11.3.2 Effect of "bend-buckling" of the web.

Neglecting higher-order terms, the curvature of the girder gives rise to flange force components that cause compressive stress on the edges of the web adjacent to the flanges as in Fig. 11.3.3. When the web is stable under the compressive stress produced by the transverse components of

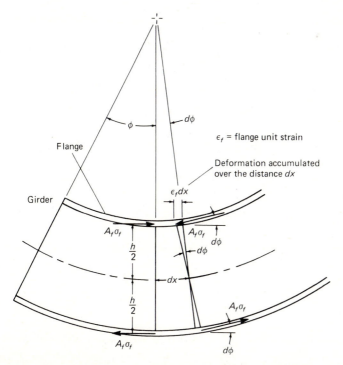

Fig. 11.3.3 Flange forces arising from girder curvature.

the flange forces, the flange cannot buckle vertically. In the following derivation the flange itself is assumed to have *zero* stiffness to resist vertical buckling.

Referring to Fig. 11.3.3, the deformation $\epsilon_f\, dx$ accumulated over the distance dx is

$$\epsilon_f\, dx = d\phi\, \frac{h}{2} \tag{11.3.1}$$

$$d\phi = \frac{2\epsilon_f}{h}\, dx \tag{11.3.2}$$

As shown in Fig. 11.3.4a, the vertical component causing compressive stress is $\sigma_f A_f\, d\phi$. After dividing by the area $t\, dx$ to obtain the compressive unit stress f_c as shown in Fig. 11.3.4b, one may substitute Eq. 11.3.2 for $d\phi$,

$$f_c = \frac{\sigma_f A_f\, d\phi}{t\, dx} = \frac{2\sigma_f A_f \epsilon_f}{th} \tag{11.3.3}$$

Referring again to Eq. 6.4.28, the elastic buckling stress for a plate,

$$F_{cr} = \frac{k\pi^2 E}{12(1-\mu^2)(b/t)^2} \tag{6.4.28}$$

where $b = h$ and $k = 1$ for the case of the Euler plate assumed free along edges parallel to loading, and pinned top and bottom. Thus

$$F_{cr} = \frac{\pi^2 E}{12(1-\mu^2)(h/t)^2} \tag{11.3.4}$$

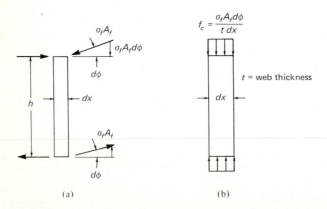

(a) (b)

Fig. 11.3.4 Effect of flange force component normal to flange plate.

Equating applied stress, Eq. 11.3.3, to the critical stress, Eq. 11.3.4, gives

$$\frac{2\sigma_f A_f \epsilon_f}{th} = \frac{\pi^2 E}{12(1-\mu^2)(h/t)^2} \tag{11.3.5}$$

which, letting $th = A_w$, gives

$$\frac{h}{t} = \sqrt{\frac{\pi^2 E}{24(1-\mu^2)} \left(\frac{A_w}{A_f}\right)\left(\frac{1}{\sigma_f \epsilon_f}\right)} \tag{11.3.6}$$

Conservatively assume that σ_f must reach the yield stress F_y for the flange to achieve its ultimate capacity. Furthermore, if residual stress F_r exists in the flange distributed as shown in Fig. 11.3.5, then the total flange strain will be that due to the sum of the residual stress plus the yield stress; therefore

$$\epsilon_f = (F_r + F_y)/E \tag{11.3.7}$$

It is the strain adjacent to the web that is of concern; in Fig. 11.3.5c the change is from F_r in tension (point A) to F_y in compression (point B).

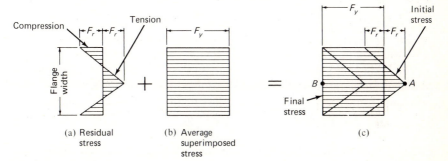

(a) Residual stress
(b) Average superimposed stress
(c)

Fig. 11.3.5 Effect of residual stress.

Substitution of $\sigma_f = F_y$, $\epsilon_f = $ Eq. 11.3.7, $E = 29{,}000$ ksi, and $\mu = 0.3$ into Eq. 11.3.6 gives

$$\frac{h}{t} = \frac{19{,}500 \sqrt{A_w/A_f}}{\sqrt{F_y(F_y + F_r)}} \tag{11.3.8}$$

a conservative estimate of the maximum h/t to prevent vertical buckling. Basler [9] suggests A_w/A_f will rarely be below 0.5 and that $F_r = 16.5$ ksi is realistic. If these substitutions are made,

$$\frac{h}{t} = \frac{13{,}800}{\sqrt{F_y(F_y + 16.5)}} \tag{11.3.9}$$

Equation 11.3.9 has been developed without regard to placement of stiffeners. The effect of stiffeners would certainly be to increase the buckling capacity above the value given by Eq. 11.3.4. Tests reported [11] on hybrid girders with A514 flanges ($F_y = 100$ ksi) indicate that h/t can conservatively be accepted in design as 250 if $a/h \leq 1.0$, and 200 for a/h between 1.0 and 1.5. The limitation of 200 may be expressed for other yield strengths as $2000/\sqrt{F_y}$ for $a/h \leq 1.5$.

Torsional Buckling

If the girder flange is made as wide as feasible to provide good lateral-torsional buckling resistance, it still may be possible that torsional buckling of the flange plate and adjacent portion of the web will occur at a critical stress below the value given by Eq. 9.4.1. Essentially, the torsional buckling is the buckling of a uniformly compressed plate free along one edge and hinged at the other (Case E, Fig. 6.15.2). The width-thickness ($b_f/2t_f$) limitations in design specifications are normally satisfactory to preclude torsional buckling as a potential failure mode.

11.4 POST-BUCKLING CONDITIONS IN THE WEB PLATE

The various types of web plate stability discussed in Sec. 11.2 all were based on small-deflection theory wherein the buckled position is an unstable equilibrium one, close to the flat unbuckled position. Such a small out-of-plane displacement occurs in the web plate at buckling that it is difficult to determine the critical load experimentally. After a small out-of-plane displacement occurs, tensile membrane stresses arise in the mid-thickness plane which tend to stabilize the plate. Any given element of a plate is constrained by all adjacent elements. Because these membrane effects arise gradually during the loading of a plate girder, the sudden buckling as for a compressed bar does not appear. Figure 11.4.1 shows a comparison of load-deflection curves for an isolated bar and a plate such as a plate girder web.

The web of a plate girder cannot acutally collapse without the flanges and stiffeners that surround it also collapsing. The result of web buckling

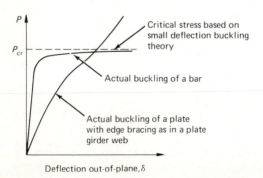

Fig. 11.4.1 Plate buckling compared with buckling of a slender bar.

is to cause a redistribution of stresses. As long as the flanges and stiffeners are *capable* of resisting an increased share of the load, the girder cannot fail.

11.5 ULTIMATE BENDING STRENGTH OF GIRDER— POST-BUCKLING STRENGTH OF WEB IN BENDING

As discussed in Sec. 11.2, the web may buckle due to bending stress unless, according to Eq. 11.2.13,

$$\frac{h}{t} \leq \frac{975}{\sqrt{F_{cr}, \text{ksi}}}$$ [11.2.13]

As described in Sec. 11.4, such buckling does not mean the usefulness of the girder has ended. Figure 11.5.1 shows the ultimate moment capacity of a girder as affected by h/t. It should be remembered that ultimate

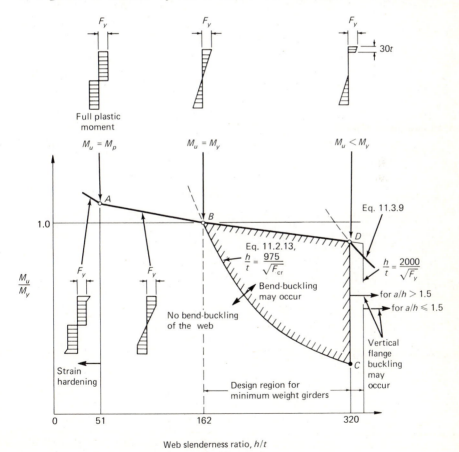

Fig. 11.5.1 Bending strength of girders as affected by bending stress in the web plate: A36 steel.

moment capacity of a deep girder is a function of

$$M_u = f\left(\frac{h}{t}, \frac{L}{r_y}, \frac{b}{t_f}, \frac{A_w}{A_f}\right) \tag{11.5.1}$$

where $\dfrac{h}{t}$ governs web instability (bend-buckling)

$\dfrac{L}{r_y}$ governs flange lateral instability (lateral-torsional buckling)

$\dfrac{b}{t_f}$ governs local buckling (or torsional buckling) of flange

$\dfrac{A_w}{A_f}$ determines the post-web-buckling effect on the flange

Assuming that lateral-torsional buckling and local buckling are prevented as assumed for Fig. 11.5.1, the variables become

$$M_u = f\left(\frac{h}{t}, \frac{A_w}{A_f}\right) \tag{11.5.2}$$

When post-buckling strength of the girder is considered, the capacity is raised from line BC of Fig. 11.5.1 to line BD. The actual position of line BD varies with A_w/A_f.

Example 11.5.1

Using $h/t = 320$, determine the expression for M_u/M_y (point D of Fig. 11.5.1) in terms of A_w/A_f.

SOLUTION

With $h/t = 320$, "bend-buckling" occurs at a low level of flexural stress. Such buckling does not signify the maximum bending moment that can be carried; however, under additional load the flexural stress on the compression side of the neutral axis becomes nonlinear. In order to retain the use of the flexure formula, Mc/I, a reduced effective section must be used. The reduced section entirely neglects much of the web in the region where the buckling (out-of-planeness) has occurred. The effective section shown in Fig. 11.5.2 was proposed by Basler [9].

(a) Determine location of neutral axis. Equating static moments about the neutral axis gives

$$A_f(kh) + \frac{t(kh)^2}{2} = A_f(1-k)h + \frac{3}{32}h\left(\frac{61}{64}h - kh\right)t$$

Divide by $A_f h$:

$$k + k^2\frac{th}{2A_f} = (1-k) + \frac{3}{32}\left(\frac{61}{64} - k\right)\frac{th}{A_f}$$

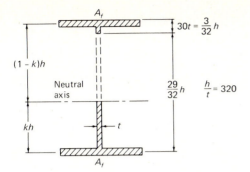

Fig. 11.5.2 Effective section in bending when vertical flange buckling is imminent.

Noting that $th = A_w$ and letting $\rho = A_w/A_f$ gives

$$k^2 + k\left(\frac{4}{\rho} + \frac{3}{16}\right) = \frac{2}{\rho} + \frac{183}{1024}$$

$$k = \sqrt{\frac{192}{1024} + \frac{38}{16\rho} + \frac{4}{\rho^2}} - \left(\frac{3}{32} + \frac{2}{\rho}\right) \tag{a}$$

(b) Determine the effective moment of inertia.

$$I_e = A_f(kh)^2 + \frac{1}{3}t(kh)^3 + A_f(1-k)^2h^2 + \frac{3th}{32}\left(\frac{61}{64}h - kh\right)^2$$

Using $th = A_w$ and $\rho = A_w/A_f$,

$$I_e = A_f h^2\left[\frac{\rho}{3}k^3 + k^2 + (1-k)^2 + \frac{3\rho}{32}\left(\frac{61}{64} - k\right)^2\right] \tag{b}$$

(c) Determine the ultimate moment capacity M_u. Assuming the extreme fiber in compression stressed to the yield stress F_y,

$$M_u = \frac{F_y I_e}{(1-k)h} \tag{c}$$

(d) Determine the moment capacity M_y assuming the entire section elastic (and therefore effective) with the extreme fiber stress equal to F_y. In developing the expression the flange-area concept will be used as shown in Fig. 11.5.3. The moment capacity of the web is approximately (Fig. 11.5.3a)

$$M_{web} = fS_x = f(\tfrac{1}{6}th^2) \tag{d}$$

which assumes web depth, distance between flange centroids, and overall depth are the same. The moment capacity of the equivalent flange area system (Fig. 11.5.3b) is

$$M_{equiv} = fA_f'h \tag{e}$$

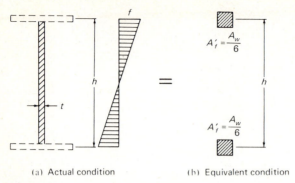

(a) Actual condition   (b) Equivalent condition

Fig. 11.5.3 Equivalent flanges area to replace web.

Equating Eqs. (d) and (e) gives for the equivalent flange area, A'_f,

$$A'_f = \tfrac{1}{6}th = \tfrac{1}{6}A_w \tag{f}$$

The total moment capacity of a girder where the stress $f = F_y$ then becomes

$$M_y = F_y\left[A_f + \frac{A_w}{6}\right]h \tag{g}$$

$$= F_y A_f h\left(1 + \frac{\rho}{6}\right) \tag{h}$$

The vertical ordinate of point D in Fig. 11.5.1 is obtained by dividing Eq. (c) by Eq. (h):

$$\frac{M_u}{M_y} = \frac{\dfrac{\rho}{3}k^3 + k^2 + (1-k)^2 + \dfrac{3\rho}{32}\left(\dfrac{61}{64} - k\right)^2}{(1-k)(1+\rho/6)} \tag{i}$$

which is plotted in Fig. 11.5.4.

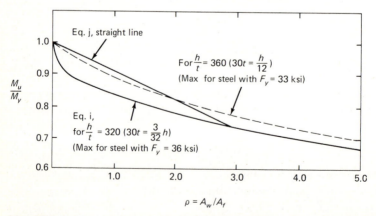

$$\rho = A_w/A_f$$

Fig. 11.5.4 Reduction in moment capacity considering post-buckling strength at maximum h/t for A36 steel.

From Fig. 11.5.4, the variation in M_u/M_y might be approximated by a straight line for A_w/A_f from zero to three with a slope of $-(1.00-0.73)/3.0 = -0.09$.

Thus, at $h/t = 320$,

$$\frac{M_u}{M_y} = 1.0 - 0.09 \frac{A_w}{A_f} \qquad (j)$$

It may be observed that the straight line agrees better for $h/t = 360$, the situation for which this linear equation was originally developed [9], than it does for $h/t = 320$. For higher strength steels, for which the maximum h/t to prevent vertical buckling of the flange is less than 360, more of such a stiffer web participates with the compression flange, causing a greater reduction in M_u/M_y.

The linear reduction based on Eq. (j) does not seem conservative, but is within several percent of the more accurate curve using $30t$ as the effective depth of web participating with the compression flange.

Tests [9] have verified the correctness of this linear reduction method using $h/t = 360$ as its basis.

Reduced Nominal Flange Stress for $M_u < M_y$

By reference to Fig. 11.5.1, it may be reasonably assumed that M_u/M_y varies linearly from point B to D. Thus the reduction in M_u/M_y per A_w/A_f per h/t greater than that at point B is

$$\frac{\text{Slope per } A_w/A_f}{320 - 162} = \frac{0.09}{158} = 0.00057 \quad \text{(say 0.0005)}$$

Thus M_u/M_y for the region from point B to D (Fig. 11.5.1) assuming linear variation, is

$$\frac{M_u}{M_y} = 1.0 - 0.0005 \frac{A_w}{A_f} \left(\frac{h}{t} - \frac{975}{\sqrt{F_y}} \right) \qquad (11.5.3)$$

If stress calculations are to be made using gross properties, then nominal ultimate stress is $M_u/S = F_{ult}$. Since $F_y = M_y/S$, the $M_u/M_y = F_{ult}/F_y$; thus

$$F_{ult} = F_y \left[1.0 - 0.0005 \frac{A_w}{A_f} \left(\frac{h}{t} - \frac{975}{\sqrt{F_y}} \right) \right] \qquad (11.5.4)$$

Equation 11.5.4 assumes no influence of stability relative to the compression flange. If, however, lateral-torsional buckling of the compression flange gives $F_{cr} < F_y$, then F_{cr} should replace F_y in Eq. 11.5.4. In general, then,

$$F_{ult} = F_{cr} \left[1.0 - 0.0005 \frac{A_w}{A_f} \left(\frac{h}{t} - \frac{975}{\sqrt{F_{cr}}} \right) \right] \qquad (11.5.5)$$

In summary, the reader should keep in mind that if $F_{cr} \geq F_y$ and h/t exceeds $975/\sqrt{F_y}$, the extreme fiber stress is actually F_y when $M = M_u$. The section properties used are, however, those of a reduced section, as in Fig. 11.5.2. For the case of stable flanges

$$M_u = F_y S_{\text{reduced}} = F_{\text{ult}} S_{\text{full}} \tag{11.5.6}$$

Using a reduced stress on gross section provides the same capacity as if the real conditions are used. The foregoing treatment for the post-buckled web is the same concept used for stiffened plate elements in Chapter 6, Part II.

11.6 ULTIMATE BENDING STRENGTH—HYBRID GIRDERS

As discussed in the last section, a girder having large h/t may have its web buckle due to flexural stress, thereby increasing the load-carrying requirement of the compression flange. This extra load-carrying requirement for the flanges also may occur when a hybrid girder is used. A hybrid girder is one in which the flanges are of a higher strength steel than the web. The use of a hybrid girder has particular economic advantages in composite construction, as described in Sec. 16.9.

The special behavioral feature of the hybrid girder is the yielding of the lower strength web before the maximum flange strength has been reached. When the flexural strength of the hybrid girder is achieved, the web will have participated to a lesser extent than in a girder using only one grade of steel.

Frost and Schilling [10] studied the hybrid girder under static loads. The state-of-the-art and design recommendations for hybrid girders have been summarized by a Joint ASCE–AASHO Committee under the chairmanship of C. G. Schilling [11]. The following example compares the moment-rotation characteristics of a hybrid girder with those of a girder having only one grade of steel to illustrate some of the special concerns relating to hybrid girders.

Example 11.6.1

For the section given in Fig. 11.6.1 whose flexural properties are $I_x = 13,640$ in.^{4}; $S_x = 910$ in.^{3}, determine the moment-rotation characteristics (a) for the section as a homogeneous one of A514 Grade 100 steel and (b) for the section as a hybrid A514/A36 beam.

SOLUTION
(a) Homogeneous A514 Grade 100 section.

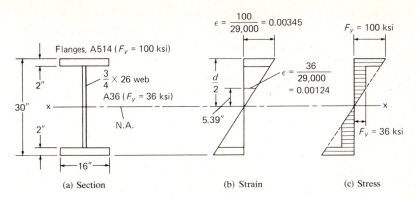

Fig. 11.6.1 Section for Example 11.6.1, showing strain and stress on hybrid section when F_y is reached at extreme fiber of section.

Strain $\epsilon = F_y/E_s$ at extreme fiber of section:

$$M_y = S_x F_y = 910(100)\tfrac{1}{12} = 7580 \text{ ft-kips}$$

$$\epsilon = \frac{d}{2}\phi = \frac{F_y}{E_s} = \frac{100}{29,000} = 0.00345$$

Plastic moment (i.e., ultimate strength):

$$Z_x = 2[16(2)(14) + 13(0.75)(6.5)] = 1020 \text{ in.}^3$$
$$M_p = Z_x F_y = 1020(100)\tfrac{1}{12} = 8520 \text{ ft-kips}$$

(b) Hybrid A514/A36 section.

Strain $\epsilon = F_y/E_s$ at extreme fiber of *web*:

$$f \text{ at extreme fiber} = 36\left(\frac{15}{13}\right) = 41.5 \text{ ksi}$$

$$M_{yw} = S_x(41.5) = 910(41.5)\tfrac{1}{12} = 3150 \text{ ft-kips}$$

$$\epsilon = \frac{d}{2}\phi = \frac{41.5}{29,000} = 0.00143$$

Strain $\epsilon = F_y/E_s$ at extreme fiber of section (Fig. 11.6.1b), c): In this condition the web is partially plastic while the flanges are at initial yielding.

$$M_y = \left\{100\left[\frac{2(32)(14)^2}{15}\right] + 36\left[\frac{1}{6}\left(\frac{3}{4}\right)(10.78)^2\right]\right.$$

$$\left. + 36\left[(13 - 5.39)\left(\frac{3}{4}\right)(2)\left(\frac{13 - 5.39}{2} + 5.39\right)\right]\right\}\frac{1}{12}$$

$$M_y = 6968 + 44 + 315 = 7330 \text{ ft-kips}$$

Fully plastic flanges but partially plastic web on hybrid section: The strain at extreme fiber of the *web* will be $(100/29,000) = 0.00345$. The distance from the neutral axis to the point where the stress on the web is 36 ksi in 4.67 in.

$$M_{pf} = \left\{ 100(32)(15)(2) + 36 \left[\frac{1}{6}\left(\frac{3}{4}\right)(9.34)^2 \right] \right.$$

$$\left. + 36 \left[(13 - 4.67)\left(\frac{3}{4}\right)(2)\left(\frac{13-4.67}{2} + 4.67\right) \right] \right\} \frac{1}{12}$$

$$M_{pf} = 8000 + 33 + 331 = 8360 \text{ ft-kips}$$

Fully plastic hybrid section:

$$M_p = [100(32)(15)(2) + 36(13)(0.75)(6.5)(2)]\frac{1}{12}$$

$$M_p = 8000 + 380 = 8380 \text{ ft-kips}$$

The results are shown in Fig. 11.6.2, comparing the behavior of the hybrid and homogeneous beams.

There are two principal effects apparent from Fig. 11.6.2. First, the onset of yielding of the web at 38 percent of the strength based on

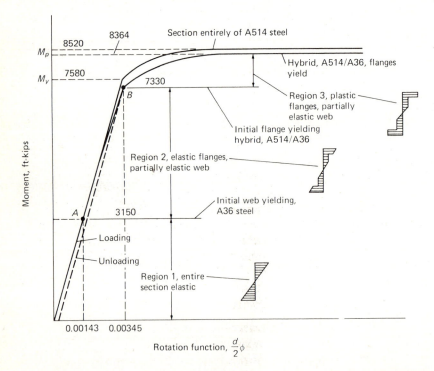

Fig. 11.6.2 Moment-rotation relationships for section of Fig. 11.6.1, assuming zero residual stress.

yielding of the flanges means that even at service load inelastic behavior of the web is to be expected. Second, the strength of the section computed when the flanges have entirely yielded but the web has only partially yielded gives a value not significantly different than obtained for the full plastic strength. In this example, the ratio A_w/A_f, web area to flange area, was only 0.61. For higher ratios the effect of the web increases, but for practical purposes it still does not greatly change the behavior from that of a girder entirely of one grade of steel. Any yielding that occurs in the web is restrained by the elastic flanges.

For design of hybrid girders, Subcommittee 1 of ASCE–AASHO Joint Committee [11] recommends that to account for the small effect on the capacity for a hybrid girder with a low yield strength web, either of the following procedures may be used:

1. An allowable moment be determined by taking the flange-yield moment (point *B* on Fig. 11.6.2) and dividing by a factor of safety; or
2. An allowable moment be calculated as the elastic section modulus of the full section multiplied by a reduced allowable flange extreme fiber stress.

Both AISC–1.10.6 and AASHTO–1.7.50 use the latter approach. The allowable extreme fiber stress for a girder entirely of one grade of steel is determined based on flexural strength (including lateral-torsional buckling) and then is multiplied by a reduction factor to account for yielding of the lower strength web. Thus the reduced allowable stress is given by AISC–1.10.6 as

$$F_b' \le F_b \left[\frac{12 + \beta(3\alpha - \alpha^3)}{12 + 2\beta} \right] \tag{11.6.1}$$

where

$\beta = A_w/A_f$, the ratio of the cross-sectional area of the web to the cross-sectional area of *one* flange

$\alpha = F_y(\text{web})/F_y(\text{flanges})$, the ratio of the yield strength of the web steel to the yield strength of the flange steel

F_b = allowable flexural stress, including lateral-torsional buckling consideration, assuming the entire member is of the grade of steel used in the flanges

F_b' = reduced allowable flexural stress accounting for the lower strength web steel in the hybrid girder.

If the h/t ratio for the web is high, "bend-buckling" may occur as discussed in Sec. 11.5 for a girder entirely of one grade of steel; in which case the reduction in strength relates to web *stability* and Eq. 11.5.5 is applicable. Equation 11.6.1, on the other hand, is a strength reduction based on web *yielding* in a hybrid girder.

The special features of hybrid girders relating to composite construction are contained in Chapter 16.

11.7 ULTIMATE SHEAR STRENGTH—INCLUDING POST-BUCKLING STRENGTH

As discussed in Sec. 11.2, the buckling of plates under pure shear, both elastically and inelastically, gives critical shear stress as illustrated by line *ABCD* in Fig. 11.7.1. A plate stiffened by flanges and stiffeners has considerable post-buckling strength. For efficient use of web plate material in plate girders thin webs are necessary, which results in buckling at low shear stresses.

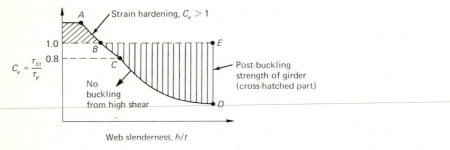

Fig. 11.7.1 Shear capacity available, considering post-buckling strength.

According to Basler [3], the ability of a plate girder to perform in a manner similar to a truss was recognized as early as 1898. As shown in Fig. 11.7.2, the tension forces are carried by membrane action of the web (referred to as *tension-field-action*) while the compression forces are carried by stiffeners. The more recent work of Basler [3] led to a theory that agreed with tests and provides criteria to insure achievement of truss action. The shear strength may then be raised from that based on buckling (*ABCD* on Fig. 11.7.1) to approach a condition corresponding to shear yield in classical beam theory (*ABE* of Fig. 11.7.1).

The ultimate shear strength, in general, may be expressed as the sum of the buckling strength V_{cr} and the post-buckled strength V_{tf} from

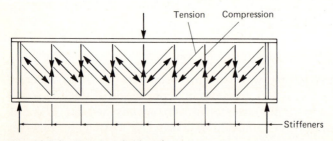

Fig. 11.7.2 Tension-field action.

tension-field action,

$$V_u = V_{cr} + V_{tf} \tag{11.7.1}$$

As discussed in Sec. 11.2, the buckling strength, both elastic and inelastic, may be expressed as

$$V_{cr} = \tau_y ht\, C_v \tag{11.7.2}$$

where $C_v = \tau_{cr}/\tau_y$ and is given by Eq. 11.2.9 and Eq. 11.2.11 for elastic and inelastic buckling, respectively.

The shear strength V_{tf} arising from the tension-field action in the web develops a band of tensile forces that occur after the web has buckled under diagonal compression (principal stresses in ordinary beam theory). Equilibrium is maintained by the transfer of stress to the vertical stiffeners. As the load increases, the angle of the tension field changes to accommodate the greatest carrying capacity. Figure 11.7.3 shows a 50×50-in. (approx. 1.3×1.3 m) panel with a $\frac{1}{4}$-in. (6.4 mm) web which has buckled under diagonal compression when subjected to pure shear. It also illustrates the anchorage requirement wherein the longitudinal component of the tension-field must be transmitted to the flange in the adjacent panel, as shown by the vertical breaks in the whitewash at the

Fig. 11.7.3   Tension-field in test plate girder. (From Ref. 3, Courtesy of Lehigh University)

flange in the corner of the adjacent panel where the tension-field intersects the stiffener and flange.

Tension-Field Action: Optimum Direction

Consider the tensile membrane stress σ_t which develops in the web at the angle ϕ, as shown in Fig. 11.7.4. If such tensile stresses can develop over the full height of the web, then the total diagonal tensile force T would be

$$T = \sigma_t th \cos \phi \qquad (11.7.3)$$

the vertical component of which is the shear force V, given by

$$V = T \sin \phi = \sigma_t th \cos \phi \sin \phi \qquad (11.7.4)$$

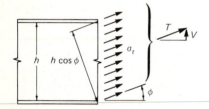

Fig. 11.7.4 Membrane stresses in tension-field action.

If such diagonal tensile stresses could develop *along* the flanges, vertical stiffness of the flanges would be required. Since the flanges have little vertical stiffness and are acting to their capacity in resisting flexure on the girder, the tension-field actually can develop only over a band width such that the vertical component can be transferred at the vertical stiffeners. The stiffeners can be designed to carry the necessary compressive force. It will be assumed that the tension-field (or partial tension-field as some may prefer to call it) may develop over the band width s, shown in Fig. 11.7.5a.

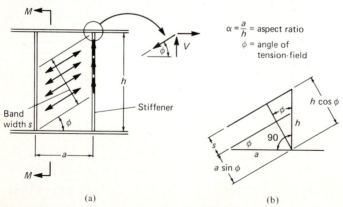

Fig. 11.7.5 Forces arising from tension-field.

The membrane tensile force tributary to one stiffener is $\sigma_t st$, and the partial shear force ΔV_{tf}, developed by compression in the stiffener is

$$\Delta V_{tf} = \sigma_t dt \sin \phi \qquad (11.7.5)$$

and the angle ϕ is the angle providing the maximum shear component from the partial tension-field.

From the geometry shown in Fig. 11.7.5b.

$$s = h \cos \phi - a \sin \phi \qquad (11.7.6)$$

where $a =$ stiffener spacing. Substitution of Eq. 11.7.6 into Eq. 11.7.5 gives

$$\Delta V_{tf} = \sigma_t t(h \cos \phi - a \sin \phi) \sin \phi$$

$$= \sigma_t t\left(\frac{h}{2} \sin 2\phi - a \sin^2 \phi\right) \qquad (11.7.7)$$

For maximum ΔV_{tf}, it is required that $d(\Delta V_{tf})/d\phi = 0$. Thus

$$\frac{d(\Delta V_{tf})}{d\phi} = \sigma_t t\left(\frac{h}{2}(2) \cos 2\phi - 2a \sin \phi \cos \phi\right) = 0 \qquad (11.7.8)$$

$$0 = h \cos 2\phi - a \sin 2\phi$$

or

$$\tan 2\phi = \frac{h}{a} = \frac{1}{a/h} \qquad (11.7.9)$$

From the trigonometry of Eq. 11.7.9,

$$\sin 2\phi = \frac{1}{\sqrt{1+(a/h)^2}} \qquad (11.7.10)$$

also

$$\sin^2 \phi = \frac{1-\cos 2\phi}{2} = \frac{1}{2}\left[1 - \frac{a/h}{\sqrt{1+(a/h)^2}}\right] \qquad (11.7.11)$$

The maximum contribution ΔV_{tf} from tension-field action is then obtained by substituting Eqs. 11.7.10 and 11.7.11 into Eq. 11.7.7 giving

$$\Delta V_{tf} = \sigma_t \frac{ht}{2}[\sqrt{1+(a/h)^2} - a/h] \qquad (11.7.12)$$

It is not practical to use Eq. 11.7.12 directly, since the shear contribution from the part of the section (such as M–M of Fig. 11.7.5) that cuts through the triangles outside the band s must be added. The state of stress in these triangles is unknown, requiring an alternate approach to finding the total shear V_{tf} when the optimum angle ϕ is reached.

An alternate way, as used by Basler [3], is to cut a free body as in

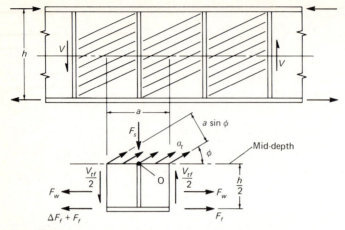

Fig. 11.7.6 Force in stiffener resulting from tension-field action.

Fig. 11.7.6. The section is taken vertically midway between two adjacent stiffeners and horizontally at mid-depth. The mid-depth cut provides access to the tension-field where the state of stress is known, and the shear resultant on each vertical face equals $V_{tf}/2$ from symmetry.

Shear Strength from Tension-Field Action

Using the free body of Fig. 11.7.6, horizontal force equilibrium requires

$$\Delta F_f = (\sigma_t ta \sin \phi) \cos \phi$$

$$= \sigma_t \frac{ta}{2} \sin 2\phi \qquad (11.7.13)$$

Rotational equilibrium, taken about point O, requires

$$\Delta F_f \frac{h}{2} - \frac{V_{tf}a}{2} = 0 \qquad (11.7.14)$$

Solving Eq. 11.7.14 for ΔF_f and substituting into Eq. 11.7.13 gives

$$\frac{V_{tf}a}{h} = \sigma_t \frac{ta}{2} \sin 2\phi \qquad (11.7.15)$$

Solving for V_{tf} and using Eq. 11.7.10 for $\sin 2\phi$ gives

$$V_{tf} = \sigma_t \frac{ht}{2} \left[\frac{1}{\sqrt{1+(a/h)^2}} \right] \qquad (11.7.16)$$

Failure Condition

The actual state of stress in the web involves both τ and σ_t; thus the failure of an element subjected to shear in combination with an inclined

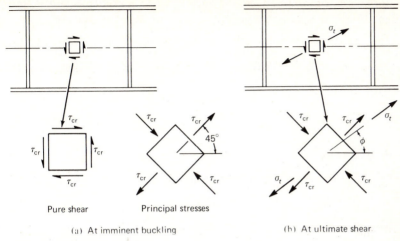

Pure shear Principal stresses

(a) At imminent buckling (b) At ultimate shear

Fig. 11.7.7 State of stress.

tension must be considered, as shown in Fig. 11.7.7. Two basic assumptions are involved: first, τ_{cr} remains at constant value from buckling load to ultimate load and therefore the tension-field stress σ_t acts in addition to the principal stresses τ_{cr}; second, the angle ϕ in Fig. 11.7.7b will be conservatively taken as 45° even though it will always be less than that value.

The generally accepted relationship for failure in plane stress is the "energy of distortion" theory (discussed in Sec. 2.7) shown as the ellipse in Fig. 11.7.8, which may be written

$$\sigma_1^2 + \sigma_2^2 - \sigma_1\sigma_2 = F_y^2 \tag{11.7.17}$$

where σ_1 and σ_2 are principal stresses. Point A represents the case of shear alone and point B represents tension alone. The actual states of

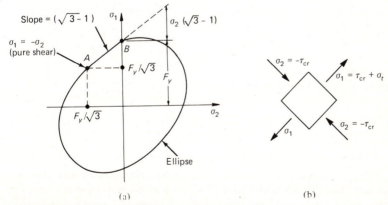

Fig. 11.7.8 Energy of distortion failure criterion.

stress in plate girder webs fall on the ellipse between points A and B, and a straight line is a reasonable approximation of the segment AB,

$$\sigma_1 = F_y + \sigma_2(\sqrt{3} - 1) \tag{11.7.18}$$

and for the stress condition that $\sigma_1 = \tau_{cr} + \sigma_t$ and $\sigma_2 = -\tau_{cr}$, Eq. 11.7.18 becomes

$$\frac{\sigma_t}{F_y} = 1 - \frac{\tau_{cr}}{F_y/\sqrt{3}} = 1 - C_v \tag{11.7.19}$$

Force in Stiffener

Using Fig. 11.7.6, vertical force equilibrium requires

$$F_s = (\sigma_t t a \sin \phi) \sin \phi \tag{11.7.20}$$

and substitution of Eq. 11.7.11 for $\sin^2 \phi$ gives

$$F_s = \sigma_t \left(\frac{at}{2}\right)\left[1 - \frac{a/h}{\sqrt{1+(a/h)^2}}\right] \tag{11.7.21}$$

Substituting Eq. 11.7.19 into Eq. 11.7.21 gives

$$F_s = \frac{F_y(1 - C_v)at}{2}\left[1 - \frac{a/h}{\sqrt{1+(a/h)^2}}\right] \tag{11.7.22}$$

which is the force achieved at ultimate shear strength to accommodate tension-field action.

Ultimate Shear Capacity: Combined Buckling and Post-Buckling Strength

Since thin-web plate girders exhibit some strength in shear before diagonal buckling occurs (V_{cr} from Sec. 11.2) and additional strength in the post-buckling range (V_{tf} from Eq. 11.7.16), their actual capacity is the sum of both components. Substituting Eqs. 11.7.2 and 11.7.16 into 11.7.1 gives

$$V_u = ht\left[\tau_y C_v + \frac{\sigma_t}{2\sqrt{1+(a/h)^2}}\right] \tag{11.7.23}$$

Substituting Eq. 11.7.19 for σ_t and using $\tau_y = F_y/\sqrt{3}$ gives

$$V_u = F_y ht\left[\frac{C_v}{\sqrt{3}} + \frac{1 - C_v}{2\sqrt{1+(a/h)^2}}\right] \tag{11.7.24}$$

In addition to the work of Basler [3] that forms the basis for the AISC Specification use of tension-field action, others, including Rockey and Skaloud [12], Sharp and Clark [13], and Herzog [14], have studied the tension-field mechanism in girder webs. A summary of these mechanisms is available in the *SSRC Guide* [7] (p. 157).

When the plate girder is subjected to cyclic loading fatigue strength is of concern. Though outside the scope of this text, the reader is referred to Yen and Mueller [20], Patterson, Corrado, Huang, and Yen [21], and Hirt, Yen, and Fisher [22]. The static and fatigue strength of unsymmetrical girders has been studied by Dimitri and Ostapenko [23], Schueller and Ostapenko [24], and Parsanejad and Ostapenko [25].

11.8 STRENGTH IN COMBINED BENDING AND SHEAR

In the vast majority of cases the ultimate capacity in bending is not influenced by shear, nor is the ultimate shear capacity influenced by moment. Particularly, in very slender webs where "bend-buckling" may occur, the bending stress is redistributed as discussed in Sec. 11.5, so that the flanges carry an increased share. The shear capacity of the web, however, is not reduced as a result of "bend-buckling" because most of the shear capacity is from tension-field action with only a small contribution from the portion of the web adjacent to the flange. In stockier webs no "bend-buckling" may occur, but high web shear in combination with bending may cause yielding of the web adjacent to the flange; again resulting in a transfer of part of the web's share of the bending moment to the flange. The strength of girders subject to combined bending and shear is the subject of the third major paper by Basler [15].

Since instability is precluded, a plastic analysis may be used. When subjected to high bending moment, the web yields adjacent to the flange and is therefore unable to carry shear. In the mid-depth region of the web the shear causes yielding; thus this part of the web is unable to carry bending moment. Referring to Fig. 11.8.1, the ultimate shear capacity may be expressed as

$$V_u = \tau_y y_0 t \qquad (11.8.1)$$

which if no bending moment were present, $y_0 = h$, giving the maximum shear capacity

$$V_y = \tau_y t h \qquad (11.8.2)$$

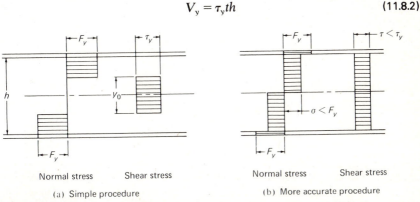

(a) Simple procedure   (b) More accurate procedure

Fig. 11.8.1 Ultimate conditions under combined bending and shear.

Eliminating τ_y from Eqs. 11.8.1 and 11.8.2 gives

$$y_0 = \frac{V}{V_y} h \tag{11.8.3}$$

The moment capacity from Fig. 11.8.1 is

$$M_u = A_f F_y h + F_y t \left(\frac{h}{2}\right)\left(\frac{h}{2}\right) - F_y t \left(\frac{y_0}{2}\right)\left(\frac{y_0}{2}\right) \tag{11.8.4}$$

$$= F_y \left[A_f h + A_w \frac{h}{4} - A_w \frac{h}{4}\left(\frac{V}{V_y}\right)^2 \right]$$

$$= F_y A_f h \left\{ 1 + \frac{1}{4}\left(\frac{A_w}{A_f}\right)\left[1 - \left(\frac{V}{V_y}\right)^2\right]\right\} \tag{11.8.5}$$

The ordinary beam theory moment capacity at first yield with the web fully participating is [see Eq. (g), p. 588]

$$M_y = F_y h A_f \left(1 + \frac{1}{6}\frac{A_w}{A_f}\right) \tag{11.8.6}$$

As the percentage of the maximum shear capacity utilized increases, the available ultimate moment capacity decreases.

In the absence of instability, but in the presence of high shear, the ultimate bending moment capacity may be expressed as

$$M_u = M_y \left[\frac{1 + \frac{1}{4}\rho[1 - (V/V_y)^2]}{1 + \frac{1}{6}\rho}\right] \tag{11.8.7}$$

When $M_u = M_y$, $V/V_y = 0.577$, or approximately 0.6. When more than 60 percent of the maximum shear capacity is used, the available ultimate moment capacity is reduced. Table 11.8.1 gives some values for M_u/M_y for various values of ρ in the practical range, with graphical illustration in Fig. 11.8.2.

Table 11.8.1 Values of M_u/M_y in accordance with Eq. 11.8.7 for $V/V_y \geq 0.6$

$\rho = \dfrac{A_w}{A_f}$	For $\dfrac{V}{V_y} = 0.8$	For $\dfrac{V}{V_y} = 1.0$
0	1.0	1.0
0.5	0.964	0.923
1.0	0.935	0.856
1.5	0.908	0.800
2.0	0.885	0.750

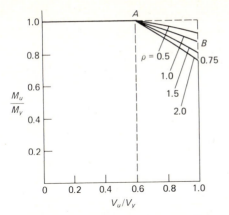

Fig. 11.8.2 Moment-shear strength interaction relationship.

Certainly it would be more accurate to consider both normal stress σ and shear stress τ acting on the web over the entire depth as in Fig. 11.8.1b. Using the Hencky-von Mises failure criterion, Eq. 11.7.17, with σ_1 and σ_2 equal to the principal stresses,

$$\sigma_1 = \frac{\sigma}{2} + \sqrt{\left(\frac{\sigma}{2}\right)^2 + \tau^2}\,; \qquad \sigma_2 = \frac{\sigma}{2} - \sqrt{\left(\frac{\sigma}{2}\right)^2 + \tau^2}$$

which gives for the failure criterion

$$\sigma^2 + 3\tau^2 = F_y^2 \tag{11.8.8}$$

This would provide a more complex relationship for M_u/M_y. Considering that the thin plate girder web typically is subject to a combination of beam action and tension-field action at ultimate condition, the conservative simple approach of Eq. 11.8.7 seems justified.

11.9 AISC PROVISIONS FOR WEB AND FLANGE SELECTION

Specification requirements and limitations are summarized, including a brief statement of the concept involved. The reader is reminded that most provisions relate to ultimate strength behavior, the theory of which has been developed in Secs. 11.2 through 11.8. The present approach was introduced with the adoption of the 1961 AISC Specification.

Flexure and Shear Computations

As discussed in Secs. 7.5 and 7.7, shear stress is computed as the average stress on the gross web area (AISC–1.5.1.2), and flexural stress is computed using the moment of inertia of the gross section except where holes exceed 15 percent of gross flange (AISC–1.10.1).

Flanges

In accordance with AISC–1.10.3, "Flanges of welded plate girders may be varied in thickness or width by splicing a series of plates or by the use of cover plates." In addition, local buckling provisions of AISC–1.9.1.2, must be satisfied.

Provisions for bolted or riveted girders are not discussed in this section since their use is essentially obsolete.

Web Slenderness Ratio Relating to Flexure

A graphical description of behavior was provided by Fig. 11.5.1, showing ultimate moment capacity vs. web slenderness ratio.

Recall from Sec. 7.3 that maximum beam strength is obtained when the full plastic moment develops, and this occurs when the stability conditions of AISC–1.5.1.4.1 are satisfied. In terms of the web slenderness ratio h/t, the fully plastic condition is the region at or to the left of point A, Fig. 11.5.1. Economical plate girders have much higher h/t values, in the region from B to D on Fig. 11.5.1.

As h/t increases, buckling of the web due to bending stress may result. Equation 11.2.13 gives a rational upper limit for h/t to prevent such instability,

$$\frac{h}{t} \leq \frac{975}{\sqrt{F_{cr}}}$$ [11.2.13]

Using the basic AISC factor of safety, the maximum allowable stress in flexure, F_b, would be $0.60F_{cr}$. Substituting this into Eq. 11.2.13 gives

$$\frac{h}{t} \leq \frac{975\sqrt{0.6}}{\sqrt{F_b}} \leq \frac{755}{\sqrt{F_b}}$$

which AISC–1.10.6 prescribes as

$$\frac{h}{t} \leq \frac{760}{\sqrt{F_b, \text{ksi}}}$$ (11.9.1)*

preventing "bend-buckling" of the web

When higher h/t values are used the web will carry less than its share of the moment computed by ordinary flexure theory. To permit the flange to carry part of the moment ordinarily carried by the web. the allowable flange stress must be reduced. This reduction is assumed to be linear, as shown from B to D in Fig. 11.5.1, and may be approximately expressed by Eq. 11.5.5,

$$F_{ult} = F_{cr}\left[1.0 - 0.0005\frac{A_w}{A_f}\left(\frac{h}{t} - \frac{975}{\sqrt{F_{cr}}}\right)\right]$$ [11.5.5]

* For SI units, with F_b in MPa,

$$\frac{h}{t} \leq \frac{2000}{\sqrt{F_b}}$$ (11.9.1)

Again, reducing the expression into the working stress range by letting $F_b' = F_{ult}/1.67$ and $F_b = F_{cr}/1.67$, which gives

$$F_b' = F_b \left[1.0 - 0.0005 \frac{A_w}{A_f} \left(\frac{h}{t} - \frac{760}{\sqrt{F_b}} \right) \right] \qquad (11.9.2)^*$$

which is AISC–1.10.6, Formula (1.10–5). F_b is the allowable stress (in ksi) after considering lateral-torsional buckling. Equation 11.9.2 is shown graphically in Fig. 11.9.1.

Web Slenderness Ratio Relating to Vertical Buckling of Flange

For very large values of h/t the web may be insufficiently stiff to prevent vertical buckling of the flange. Equation 11.3.9 provides a reasonable approximation for the maximum h/t to prevent such instability in the absence of intermediate transverse stiffeners,

$$\frac{h}{t} = \frac{13,800}{\sqrt{F_y(F_y + 16.5)}} \qquad [11.3.9]$$

which when rounded becomes

$$\frac{h}{t} = \frac{14,000}{\sqrt{F_y(F_y + 16.5)}} \qquad (11.9.3)^*$$

which is the AISC–1.10.2 general limitation (F_y in ksi).

In the presence of transverse stiffeners, recent studies indicate higher values may be permitted. Thus AISC in accordance with test results and recommendations of the ASCE–AASHO Joint Committee, Subcommittee 1 on Hybrid Girder Design [11], provides a higher value for maximum h/t,

$$\frac{h}{t} \leq \frac{2000}{\sqrt{F_y, \text{ksi}}} \qquad (11.9.4)^*$$

when transverse stiffeners are provided such that stiffener spacing/web depth, $a/h \leq 1.5$. This is developed using the recommended $h/t = 200$ for $F_y = 100$ ksi‡ and using the typical stability equation format.

---

* For SI units, with F_b', F_b, and F_y in MPa,

$$F_b' = F_b \left[1.0 - 0.0005 \frac{A_w}{A_f} \left(\frac{h}{t} - \frac{2000}{\sqrt{F_b}} \right) \right] \qquad (11.9.2)$$

$$\frac{h}{t} = \frac{96,500}{\sqrt{F_y(F_y + 114)}} \qquad (11.9.3)$$

$$\frac{h}{t} \leq \frac{5250}{\sqrt{F_y}} \qquad (11.9.4)$$

‡ Ref. 11, p. 1412.

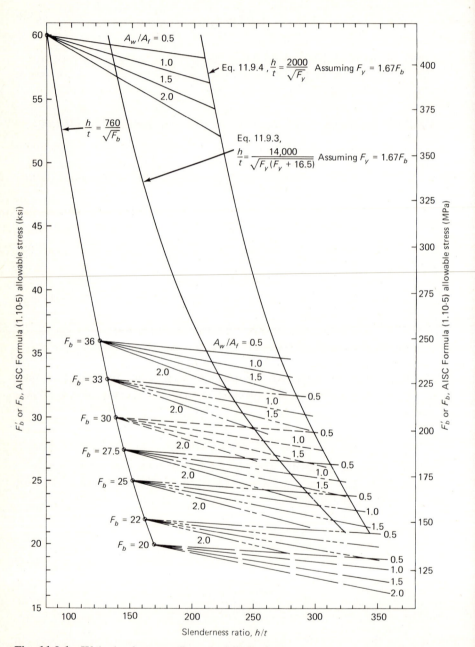

Fig. 11.9.1 Web slenderness effects and limitations.

Table 11.9.1   Maximum h/t Limitations—AISC

F_y (ksi)	h/t for Eq. 11.9.3 for $a/h > 1.5$	h/t for Eq. 11.9.4 for $a/h \leq 1.5$	F_y (MPa)
36	322	333	248
42	282	309	290
45	266	298	310
50	243	283	345
55	223	270	379
60	207	258	414
65	192	248	448
100	130	200	689

Values for Eq. 11.9.3 and 11.9.4 are given in Table 11.9.1.

Shear and Combined Shear and Tension as Affecting Web Selection

In addition to the stability limitations on h/t, the web area must be sufficient to carry the shear. Maximum allowable shear stress is $0.40F_y$; however, when stiffeners are to be used, the allowable value based on stiffener placement, as discussed in the next section, will usually be no greater than about 0.30 to $0.35F_y$.

Furthermore, locations of combined high moment and high shear may limit the nominal shear stress to between $0.25F_y$ to $0.30F_y$.

As discussed in Sec. 11.8, Eq. 11.8.7 and Fig. 11.8.1 show the interaction between shear and bending moment. If one uses a conservative value of $\rho = A_w/A_f = 2.0$ and considers the strength reduction from points A to B of Fig. 11.8.1 as a straight line, the slope of AB would be

$$\text{Slope of } AB \text{ (Fig. 11.8.1)} = \frac{-0.25}{0.40} = -\frac{5}{8}$$

The reduction equation then becomes

$$\frac{M_u}{M_y} = 1 - \frac{5}{8}\left(\frac{V_u}{V_y} - 0.6\right) \leq 1 \tag{11.9.5}$$

$$M_u = M_y\left(\frac{11}{8} - \frac{5}{8}\frac{V_u}{V_y}\right) \leq M_y \tag{11.9.6}$$

Dividing by a 1.67 factor of safety and converting moments and shears to unit stresses gives

$$f_b = 0.6F_y\left(\frac{11}{8} - \frac{5}{8}\frac{f_v}{F_v}\right) \leq 0.60F_y \tag{11.9.7}$$

which gives

$$f_b \leq \left(0.825 - 0.375\,\frac{f_v}{F_v}\right)F_y \leq 0.60F_y \qquad (11.9.8)$$

which is AISC–1.10.7, Formula (1.10–7). Note that f_b and f_v are the maximum flexural and shear stresses in the *web*. The adjacent relatively stiff flange prevents stability from influencing the strength of the web under combined stress. When nondimensionalized, Eq. 11.9.8 may be used in the form shown in Fig. 11.9.2 (note comparison with Fig. 11.8.1).

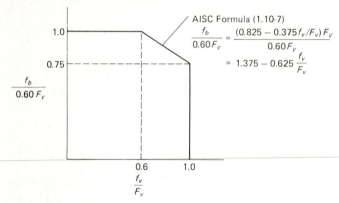

Fig. 11.9.2 AISC relationship for combined shear and tension.

11.10 AISC—INTERMEDIATE TRANSVERSE STIFFENERS

Prior to the 1961 AISC Specification, intermediate stiffeners were required to be spaced sufficiently close together so that buckling of the web due to shear stress could not occur before the bending strength of the section was reached. Post-buckling shear strength was not recognized. Since 1961, the total shear strength consisting of both buckling and post-buckling strength is recognized in design rules.

Requirements to Omit Intermediate Stiffeners

Stiffeners need not be used when the flexural strength of the section can be achieved without diagonal buckling occurring due to shear. Such shear buckling will be avoided if the nominal shear stress $f_v = V/A_w$ does not exceed an allowable value F_v based on the buckling stress τ_{cr} divided by a factor of safety (FS = 1.67). Of course, when τ_{cr} exceeds τ_y, yielding and not stability controls. The allowable stress F_v is

$$F_v = \frac{\tau_{cr}}{FS} \leq \frac{\tau_y}{FS} \qquad (11.10.1)$$

As discussed in Sec. 7.5, the allowable shear stress (τ_y/FS) based on yielding has for many years been taken as $0.40F_y$ and this still serves as the upper limit on F_v. Expressing $\tau_{cr} = C_v\tau_y$, Eq. 11.10.1 becomes

$$F_v = \frac{C_v\tau_y}{FS} \le 0.40F_y \tag{11.10.2}$$

Using $\tau_y = F_y/\sqrt{3}$ and taking FS as 1.67 gives

$$F_v = \frac{F_yC_v}{2.89} \le 0.40F_y \tag{11.10.3}$$

which is AISC Formula (1.10–1). The expressions for C_v are given by Eqs. 11.2.9 and 11.2.11,

$$C_v = \frac{45,000k}{(F_y, \text{ksi})(h/t)^2} \quad \begin{array}{l} \text{for } C_v \le 0.8 \\ \text{(elastic buckling)} \end{array} \qquad [11.2.9]^*$$

$$C_v = \frac{190}{h/t}\sqrt{\frac{k}{F_y, \text{ksi}}} \quad \begin{array}{l} \text{for } C_v > 0.8 \\ \text{(inelastic buckling)} \end{array} \qquad [11.2.11]^*$$

In the C_v expressions k is given by Eqs. 11.2.6 and 11.2.7, as follows:

$$k = 4.0 + 5.34/(a/h)^2 \qquad \text{for } a/h \le 1$$
$$k = 4.0/(a/h)^2 + 5.34 \qquad \text{for } a/h \ge 1$$

In addition to keeping the nominal shear stress below F_v given by Eq. 11.10.3, the maximum h/t may not exceed 260 when stiffeners are not used. This somewhat arbitrary limit was given by Basler [8] as a practical limit. He recommended that fabrication, handling, and erection are facilitated when the smaller panel dimension, a or h, does not exceed $260t_w$. When stiffeners are not used h is less than a.

In summary, intermediate stiffeners are not required when both of the following are satisfied:

1.
$$\frac{h}{t} \le 260 \tag{11.10.4}$$

2.
$$f_v \le \frac{F_yC_v}{2.89} \le 0.40F_y \tag{11.10.5}$$

where $f_v = V/A_w$.

---

* For SI units, with F_y in MPa,

$$C_v = \frac{310,000k}{F_y(h/t)^2} \qquad [11.2.9]$$

$$C_v = \frac{500}{(h/t)}\sqrt{\frac{k}{F_y}} \qquad [11.2.11]$$

Previously, in Table 7.5.1 of Sec. 7.5, maximum h/t values were given such that rolled beams without stiffeners could be designed using $F_v = 0.40F_y$. When no stiffeners are used, a/h is large and k approaches 5.34. From C_v for inelastic buckling (Eq. 11.2.11),

$$C_v = \frac{190}{h/t} \sqrt{\frac{5.34}{F_y, \text{ksi}}} = \frac{439}{(h/t)\sqrt{F_y, \text{ksi}}} \qquad \text{(11.10.6)}$$

Using Eq. 11.10.5,

$$\frac{F_y}{2.89} \frac{439}{(h/t)\sqrt{F_y}} = 0.40F_y$$

$$\frac{h}{t} = \frac{380}{\sqrt{F_y, \text{ksi}}} \qquad \text{(11.10.7)*}$$

Equation 11.10.7 gives the maximum h/t values of Table 7.5.1.

Placement Criteria Including Tension-Field Action

When the nominal shear stress f_v exceeds the allowable stress F_v given by Eq. 11.10.3 with the full span used for the stiffener spacing a, stiffeners will be required. The use of intermediate stiffeners reduces the a/h ratio and increases F_v. Equation 11.10.3 logically applies for situations *with* and *without* intermediate stiffeners when the objective is to prevent buckling due to shear.

Under the current AISC Specification (1978) both buckling strength and post-buckling strength are recognized. The post-buckling behavior, known as *tension-field action*, is similar to truss action as shown in Figs. 11.7.2 and 11.7.3.

The total shear strength when intermediate stiffeners are used is the sum of the buckling strength (Eq. 11.10.3) and the strength contributed by tension-field action, the total of which is given by Eq.11.7.24,

$$V_u = F_y ht \left[\frac{C_v}{\sqrt{3}} + \frac{1-C_v}{2\sqrt{1+(a/h)^2}} \right] \qquad [11.7.24]$$

Conversion into nominal stress on gross web, V_u/ht, and dividing by the factor of safety 1.67, gives for the allowable stress

$$F_v = \frac{F_y}{2.89} \left[C_v + \frac{1-C_v}{1.15\sqrt{1+(a/h)^2}} \right] \qquad \text{(11.10.8)}$$

which is AISC Formula (1.10–2). This equation is to be applied when

* For SI units, with F_y in MPa,

$$\frac{h}{t} = \frac{1000}{\sqrt{F_y}} \qquad \text{(11.10.7)}$$

"tension-field action" can be expected i.e., when stiffeners are used and buckling occurs before shear yielding occurs ($C_v < 1.0$).

While no theoretical upper limits exist for h/t except those of Eqs. 11.9.3 and 11.9.4, practical considerations relating to fabrication, handling, and erection [8] give rise to the restriction of AISC–1.10.5.3, stated as

$$\frac{a}{h} \le \left(\frac{260}{h/t}\right)^2 \le 3.0 \tag{11.10.9}$$

The AISC allowable stresses and limitations, using Eqs. 11.10.3, 11.10.8, and 11.10.9, are shown in Fig. 11.10.1 for $F_y = 50$ ksi (345 MPa).

End Panels

Figure 11.7.5 shows that at the junction of stiffener and flange, equilibrium requires an axial tension to develop in the flange of the adjacent panel. If no such flange is available, as in an end panel, the tension-field cannot adequately develop. AISC–1.10.5.3, therefore, considers that only buckling strength is available. Stiffeners must be provided to insure that buckling will be prevented. Thus for end panels (i.e., panels having *no adjacent panel*), only Eq. 11.10.3 may be used,

$$F_v = \frac{F_y C_v}{2.89} \le 0.40 F_y \tag{11.10.3}$$

Stiffness Requirement

Intermediate stiffeners must be sufficiently rigid to keep the web *at the stiffener* from deflecting out-of-plane when buckling of the web occurs. The stiffener must have a rigidity EI_s that is related to the web plate rigidity $Et^2 a/[12(1-\mu^2)]$.

AISC–1.10.5.4 requires intermediate stiffeners to have

$$I_s \ge \left(\frac{h}{50}\right)^4 \tag{11.10.10}$$

where I_s is the moment of inertia of the stiffener with respect to the axis at the mid-thickness of the web.

Equation 11.10.10 is obviously an oversimplification, since it does not depend on stiffener spacing or web thickness. Various theoretical relationships have been developed for the ratio of the stiffener rigidity to one panel of web-plate rigidity, which may be expressed as

$$\gamma_0 = \frac{EI_s}{Da} = \frac{EI_s[12(1-\mu^2)]}{Et^3 a} \tag{11.10.11}$$

where $I_s =$ optimum stiffener moment of inertia

$D = Et^3/[12(1-\mu^2)] =$ flexural rigidity per unit length of web plate

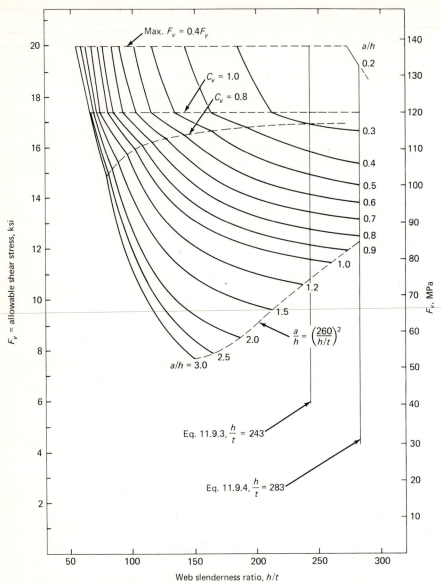

Fig. 11.10.1 Allowable shear stress on plate girders with intermediate transverse stiffeners for $F_y = 50$ ksi. (345 MPa)

The following three expressions for γ_0 are representative of the results of investigations into the stiffness requirement for transverse stiffeners:

$$\gamma_0 = \frac{14}{(a/h)^3} \qquad (11.10.12)$$

as proposed by Moore and reported in *Column Research Council Guide* [16] (p. 136),

$$\gamma_0 = 4\left[7\left(\frac{h}{a}\right)^2 - 5\right]$$

(11.10.13)

as developed by Bleich [17] (p. 417), and

$$\gamma_0 = \frac{3.75}{(a/h)^4}$$

(11.10.14)

as suggested by McGuire [18] (p. 742) to give a stiffener flexural rigidity about twice the theoretical value, in accord with the recommendation of Timoshenko and Gere [6].

According to Eq. 11.10.11 with $\mu = 0.3$,

$$\text{Required } I_s = \frac{\gamma_0 t^3 a}{12(1-\mu^2)} = \frac{\gamma_0 t^3 a}{10.92}$$

(11.10.15)

Equations 11.10.12 through 11.10.14 when substituted into Eq. 11.10.15 give

$$I_s = \frac{1.28 h^4}{(a/t)^2(h/t)}$$

(11.10.16)

$$I_s = \frac{0.366 h^4}{(h/t)^3}[7(h/a) - 5(a/h)]$$

(11.10.17)

$$I_s = \frac{0.343 h^4}{(a/t)^3}$$

(11.10.18)

In order to compare with the AISC requirement, Eq. 11.10.10, let $h/t = 200$ and a/t be 170; consequently $a/h = 0.85$. Equations 11.10.16 through 11.10.18 become $I_s = \left(\frac{h}{46.1}\right)^4$, $\left(\frac{h}{48.4}\right)^4$, and $\left(\frac{h}{61.5}\right)^4$, respectively. The AISC provision seems in the proper order of magnitude.

Strength Requirement

Intermediate stiffeners carry a compression load only after buckling of the web has occurred. As the post-buckling truss-like "tension-field action" increases, the stiffener force increases. The maximum force in the stiffener, reached simultaneously with reaching ultimate shear strength, is given by Eq. 11.7.22,

$$F_s = \frac{F_y(1-C_v)at}{2}\left[1 - \frac{a/h}{\sqrt{1+(a/h)^2}}\right]$$

[11.7.22]

If this force is developed as the stiffener yields, the ultimate capacity of the girder has been achieved. Thus the stiffener area required is

$$A_{st} = \frac{F_s}{F_{y,st}} = \frac{F_{y,w}(1-C_v)at}{F_{y,st}\ 2}\left[1 - \frac{a/h}{\sqrt{1+(a/h)^2}}\right] \quad (11.10.19)$$

where $F_{y,w}$ = yield stress of web material
$F_{y,st}$ = yield stress of stiffener material

Equation 11.10.19 may be rewritten, letting $F_{y,w}/F_{y,st} = Y$ and multiplying and dividing by h, giving

$$A_{st} = \frac{1-C_v}{2}\left[\frac{a}{h} - \frac{(a/h)^2}{\sqrt{1+(a/h)^2}}\right] Yht \quad (11.10.20)$$

which is AISC Formula (1.10–3) when the compressive force from "tension-field action" is axially applied to the stiffener, i.e., the stiffeners are placed in pairs.

Sometimes stiffeners are alternated on each side of the web to gain better economy or they are placed all on one side to improve esthetics. Whenever the stiffeners are not placed in pairs a greater cross-sectional area must be provided to account for eccentric loading. Referring to Fig. 11.10.2a, the symmetrical pair of stiffeners reaches its plastic condition and the ultimate force F_s is

$$F_s = 2wtF_y = A_{st}F_y \quad \text{(for concentric load)} \quad (11.10.21)$$

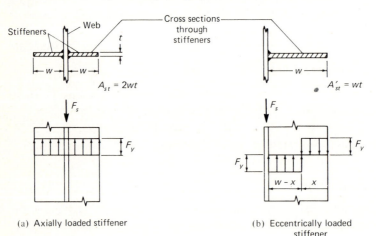

(a) Axially loaded stiffener

(b) Eccentrically loaded stiffener

Fig. 11.10.2 Intermediate stiffeners at ultimate shear condition considering tension-field action.

On the other hand, an eccentrically loaded stiffener becomes plastic with a stress distribution as shown in Fig. 11.10.2b. For this case, force equilibrium requires

$$F_s = (w-x)tF_y - xtF_y \quad (a)$$

and moment equilibrium requires

$$xtF_y\left(w - \frac{x}{2}\right) = F_y(w - x)t\left(\frac{w - x}{2}\right) \tag{b}$$

Solving the quadratic gives $x = 0.293w$. Substitution for x in Eq. (a) gives

$$F_s = [(w - 0.293w)t - 0.293wt]F_y$$
$$= 0.414wtF_y = 0.414A'_{st}F_y \quad \text{(for eccentric load)} \tag{11.10.22}$$

If single plate stiffeners are used on one side only, equating Eqs. 11.10.21 and 11.10.22 shows

$$A'_{st} = \frac{A_{st}}{0.414} = 2.42A_{st} \tag{11.10.23}$$

To correct for eccentric loading of stiffeners, Eq. 11.10.20 is to be multiplied by 2.4 for single plate stiffeners. For a single angle whose center of gravity is closer to the web, the multiplier reduces to 1.8.

The stiffener force used in the foregoing development is that which occurs when the girder is loaded to its ultimate shear capacity. Because unrestricted yielding of the stiffener is assumed to occur prior to stiffener plate buckling, the yield stress is used to determine area required. In other words, Eq. 11.10.20 gives the area required if the panel is *fully stressed* in shear and the local buckling limitations of AISC–1.9.1 are satisfied. For panels stressed to less than their full capacity, the required area of stiffeners may be reduced proportionally.

Connection to Web

Determination of the connection strength presents a difficulty since the exact distribution of shear to be transferred is unknown. Investigation [3] of the shear force expression, Eq. 11.7.22, for various values of a/h and h/t has shown that the maximum force which can occur in the stiffener is

$$\text{Max } F_s = 0.015h^2 \sqrt{\frac{F_y^3}{E}} \tag{11.10.24}$$

when $a/h = 1.18$ and $h/t = 1060/\sqrt{F_y}$, ksi.

Since the shear in the connection is nonuniform, Basler [3] has suggested that a safe procedure would be to consider the force F_s to be transferred over one-third of the girder depth. Thus, as an average for the full height, the shear per unit length f_{vs} to be considered at ultimate load is

$$f_{vs} = \frac{3F_s}{h} \tag{11.10.25}$$

which upon substituting Eq. 11.10.24 and using $E = 29,000$ ksi with a

1.65 safety factor gives

$$f_{vs} = 0.045h \sqrt{\frac{F_y^3}{E} \frac{1}{1.65}} \qquad (11.10.26)$$

$$f_{vs} = h \sqrt{\left(\frac{F_y, \text{ksi}}{340}\right)^3} \qquad (11.10.27)^*$$

which is AISC Formula (1.10–4). The reader is reminded that F_y is the yield stress for the *web* material in ksi, h is in inches, and f_{vs} is in kips/in.

As with the area requirement when adjacent panels are stressed to less than their full capacity, the design shear flow f_{vs} for the connection may also be reduced proportionally.

Connection to Flanges

Intermediate stiffeners are provided to assist the web; to stiffen and create nodal lines during buckling of the web and to accept compression forces transmitted directly from the web. At the compression flange, welding of the stiffener across the flange as shown in Fig. 11.10.3 provides stability to the stiffener and holds it perpendicular to the web; in addition, such welding provides restraint against torsional buckling of the compression flange.

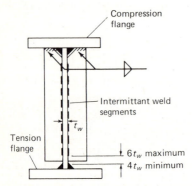

Fig. 11.10.3 Intermediate stiffener connection to flange.

On the tension flange, the effects of stress concentration increase the fatigue or brittle fracture possibilities, i.e., welding in no way helps the tension flange. Since the work of Basler [19] has shown that welding of stiffeners to the tension flange is unnecessary for proper functioning of

* For SI units, with F_y in MPa, h in mm, and f_{vs} in kN/m,

$$f_{vs} = h \sqrt{\left(\frac{F_y}{650}\right)^3} \qquad (11.10.27)$$

stiffeners, AISC–1.10.5.4 permits stopping stiffeners "short of the tension flange provided bearing is not needed to transmit a concentrated load or reaction." The weld by which the stiffener is attached to the web "shall be terminated not closer than 4 times the web thickness nor more than 6 times the web thickness from the near toe of the web-to-flange weld."

For situations where the stiffener serves as the attachment for lateral bracing, the welding to the compression flange should be designed to transmit 1 percent of the compressive force in the flange. For important lateral bracing design in situations involving long unsupported lengths, the strength of lateral bracing connections should be designed using the principles of Sec. 9.11.

11.11 AISC—BEARING STIFFENER DESIGN

Concentrated loads, such as at unframed end reactions, must be carried by stiffeners placed in pairs. Whenever the compressive stress in the vicinity of concentrated loads, as discussed in Sec. 7.6, exceeds the allowable stresses of AISC–1.10.10.1 bearing stiffeners must be provided. Furthermore, interior concentrated loads bearing on or through a flange plate for which web stability as governed by Eqs. 11.2.17, 11.2.19, and 11.2.20 (i.e., AISC–1.10.10.2) is not satisfied must be transmitted through bearing stiffeners.

Bearing stiffeners, unlike intermediate stiffeners, should be close fitting and connected to both tension and compression flanges; furthermore, they should extend approximately to the edge of the flange, whereas economical intermediate stiffeners should not be that wide.

Column Stability Criterion

This provision considers the overall stability of the bearing stiffeners as a column, with the allowable stress given by the ordinary column provisions of AISC–1.5.1.3. A portion of the web is logically assumed to act in combination with the bearing stiffener plates, or angles. The portion of web considered to act with the stiffener according to AISC–1.10.5.1 is shown in Fig. 11.11.1.

The end restraint against column buckling provided by the flanges affords a reduced effective pin-ended length. The AISC Specification states that the effective length may not be taken as less than three-quarters of the actual length; i.e.,

$$\left(\frac{KL}{r}\right)_{\text{eff}} \geq 0.75\,\frac{h}{r} \qquad (11.11.1)$$

where h = web plate depth

r = radius of gyration of the shaded portion shown in Fig. 11.11.1

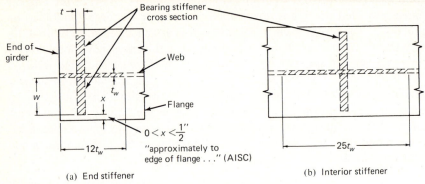

(a) End stiffener

(b) Interior stiffener

Fig. 11.11.1 Bearing stiffener effective cross sections.

Thus

$$\text{Required } A_e = \frac{\text{concentrated load}}{F_a \text{ from AISC--1.5.1.3}} \qquad (11.11.2)$$

where F_a is computed using Eq. 11.11.1 and A_e is the shaded area of Fig. 11.11.1 which includes the stiffener plates plus the tributary web area.

Local Buckling Criterion

Since the width w of the stiffener plates is governed by the flange width (see Fig. 11.11.1), the minimum thickness to prevent local buckling is obtained by applying AISC--1.9.1:

$$\text{Min } t = \frac{w}{95/\sqrt{F_y}} \qquad (11.11.3)$$

unless the special provisions of AISC--Appendix C are applied for cases where lesser thicknesses are desired to be used.

Compression Yield Criterion

Assuming a compression situation in which stability is assured, the stiffener plates are assumed capable of achieving yield stress under the concentrated load. Using a 1.67 factor of safety, the required area based on the yield condition (AISC--1.5.1.3.4) is

$$\text{Required } A_g = \frac{\text{concentrated load}}{0.60F_y} \qquad (11.11.4)$$

where A_g = area of only the stiffener plates.

Bearing Criterion

In order to bring bearing stiffener plates tight against the flanges, some part of the stiffener must be cut off so as to clear the flange-to-web fillet weld. The area of direct bearing is less than the gross area. The allowable stress in direct bearing where this lateral confinement is available is properly greater than $0.60F_y$. AISC–1.5.1.5.1 therefore requires

$$\text{Required contact area} = \frac{\text{concentrated load}}{0.90F_y} \qquad (11.11.5)$$

11.12 LONGITUDINAL WEB STIFFENERS

While longitudinal stiffeners are not as effective as transverse stiffeners, they are frequently desired on highway bridge girders for esthetic reasons. Studies of longitudinal stiffener effectiveness, as related to stiffener size and location, have been made by Cooper [26, 27] and others at Lehigh University. These studies and others are summarized in the *SSRC Guide* [7] (pp. 172–176) and by Bleich [17] (pp. 418–423).

Since the principal use of longitudinal stiffeners is in highway bridge design wherein the buckling strength is assumed to be the maximum strength available, most studies have directly related to stiffness requirements which improve buckling strength. Longitudinal stiffeners effect on, or relationship to, "tension-field" ultimate strength behavior has received little attention.

As discussed in Sec. 11.2, the elastic buckling strength of the web plate in bending (Fig. 11.2.5) may be written

$$F_{cr} = \frac{\pi^2 Ek}{12(1 - \mu^2)(h/t)^2}$$

If the plate is stiffened by a longitudinal stiffener, as shown in Fig. 11.12.1, the value of k will be significantly greater than for the unstiffened case. The stiffener used should be stiff enough so that when buckling occurs a nodal line will be formed along the line of the stiffener.

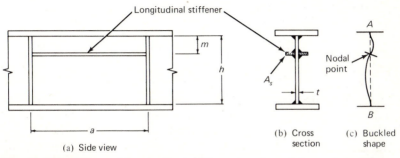

(a) Side view   (b) Cross section   (c) Buckled shape

Fig. 11.12.1   Effect of longitudinal stiffener on plate girder web stability.

Under bending alone, the value of the buckling coefficient k has been found to be as high as 142.6 for the case where the flanges are assumed to provide full restraint to rotation at points A and B of Fig. 11.12.1c and $m = h/5$. For the case where the flanges provide no moment restraint at A and B (simply supported) the stiffener located at $m = h/5$ is also the optimum location. Such stiffener placement in the compression zone serves the purpose of maintaining the full effectiveness of the web in resisting bending stress, which is really the stiffener's principal function.

For webs subjected to shear alone, the longitudinal stiffener should be located at mid-height. For combined shear and bending the stiffener should be located so that $h/5 < m < h/2$; because of its principal function, however, it should preferably be closer to $h/5$.

For design there are two requirements: (1) a moment of inertia to insure adequate stiffness to create a nodal line along the stiffener, and (2) an area adequate to carry axial compression stress while acting integrally with the web.

The design requirement for stiffness can be expressed as a function of the rigidity of the web, using the same approach as discussed for transverse stiffeners. Substituting the web height h for the transverse stiffener spacing a in Eq. 11.10.15 gives

$$\text{Required } I_s = \frac{\gamma_0 t^3 h}{10.92} \tag{11.12.1}$$

Results of theoretical studies [16] to determine γ_0 are shown in Fig. 11.12.2. It becomes apparent from a study of these curves that the selecting of the correct value to be used for γ_0 when the web is subjected to combined bending and shear is not a simple task. The AISC Specification gives no information regarding longitudinal stiffeners. AASHTO–1977 (AASHTO–1.7.43(E)) gives the following expression:

$$\text{Required } I_s = t^3 h \left[2.4 \left(\frac{a}{h} \right)^2 - 0.13 \right] \tag{11.12.2}$$

which from Eq. 11.12.1 indicates

$$\gamma_0 = 10.92 \left[2.4 \left(\frac{a}{h} \right)^2 - 0.13 \right] \tag{11.12.3}$$

which is shown in Fig. 11.12.2.

For design purposes, the AASHTO expression seems reasonable for its requirement that $m = h/5$. For situations not under AASHTO requirement, selection of reasonable γ_0 values can be made from Fig. 11.12.2.

The design requirement for strength is that the computed bending stress in the stiffener not exceed the allowable value in flexure (use same procedure as for flanges).

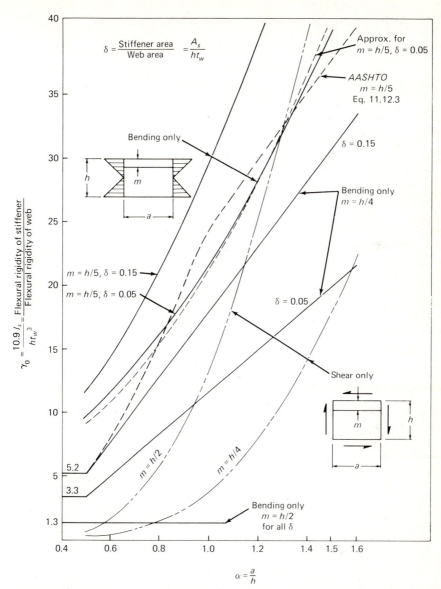

Fig. 11.12.2 Summary of theoretical studies of optimum stiffener relative rigidity, γ_0, for various web conditions. (From equations given by Ref. 16, p. 140)

11.13 PROPORTIONING THE SECTION

The cross section of a girder must be selected such that it adequately performs its functions and requires minimum cost. The function requirements may be summarized as:

1. Strength to carry bending moment (adequate section modulus S_x).

2. Vertical stiffness to satisfy any deflection limitations (adequate moment of inertia I_x).
3. Lateral stiffness to prevent lateral-torsional buckling of compression flange (adequate lateral bracing or low L/r_T).
4. Strength to carry shear (adequate web area).
5. Stiffness to improve buckling or post-buckling strength of the web (related to h/t and a/h ratios).

To satisfy these function requirements at minimum cost, it will be assumed in the discussion that follows that minimum cost is equivalent to minimum weight.

Flange-Area Formula

For simplicity in design it is convenient to replace the real system of Fig. 11.13.1a with a substitute system, Fig. 11.13.1b, which allows the moment to be replaced by a couple with the forces of the couple acting at the flange centroids. The forces can then be treated as direct stress situations. If the distance between flange centroids is approximately $(h+d)/2$, the forces of the couple are

$$C = T = \frac{M}{(h+d)/2} \tag{11.13.1}$$

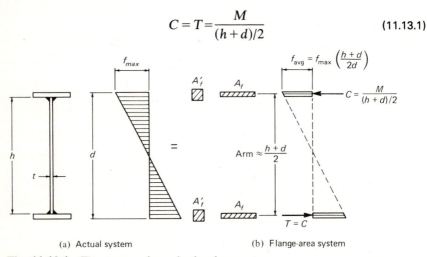

(a) Actual system   (b) Flange-area system

Fig. 11.13.1 Flange-area formula development.

The effective area on which these forces act is equal to the flange plate area A_f plus additional area A_f' to represent the effectiveness of the web in resisting moment.

The average stress on the total effective area is

$$f_{avg} = \frac{\text{Force}}{\text{Area}} = \frac{M}{(h+d)/2}\left(\frac{1}{A_f + A_f'}\right) \tag{11.13.2}$$

The area A'_f must be taken such that the bending moment carried by the web is the same for both the real and substitute systems:

$$M_{\text{real system}} = f_{\text{max}}\left(\frac{h}{d}\right)\frac{th^2}{6} \tag{11.13.3}$$

$$M_{\text{substitute system}} = f_{\text{max}}\left(\frac{h+d}{2d}\right)A'_f\left(\frac{h+d}{2}\right) \tag{11.13.4}$$

Equating Eqs. 11.13.3 and 11.13.4 gives

$$A'_f = \frac{h}{d}\left(\frac{th^2}{6}\right)\left(\frac{2d}{h+d}\right)\left(\frac{2}{h+d}\right) = \frac{th}{6}\left(\frac{2h}{h+d}\right)^2 \tag{11.13.5}$$

which if $A_w = th$, and the squared term is neglected, becomes

$$A'_f = \frac{A_w}{6} \tag{11.13.6}$$

Next, solving Eq. 11.13.2 for A_f gives

$$A_f = \frac{M}{[(h+d)/2]f_{\text{avg}}} - A'_f \tag{11.13.7}$$

which, using Eq. 11.13.5 and $f_{\text{avg}} = f_{\text{max}}(h+d)/2d$, gives

$$A_f = \left[\frac{M}{f_{\text{max}}h}\left(\frac{d}{h}\right) - \frac{A_w}{6}\right]\left(\frac{2h}{h+d}\right)^2 \tag{11.13.8}$$

Letting the squared term equal unity overestimates slightly the value of A_f, while letting $d/h = 1$ underestimates the value. For preliminary design purposes these simplifications are justified to give a simple expression for the required area of one flange plate,

$$A_f = \frac{M}{fh} - \frac{A_w}{6} \tag{11.13.9}$$

In the use of Eq. 11.13.9 if f is taken as the average stress on the flange, the d/h term will be nearly accounted for. When checking a section, of course, the moment of inertia must be obtained and the maximum stress correctly computed.

Optimum Girder Depth

The variation in girder cross-sectional area is to be examined as a function of web depth to determine the depth which will give minimum area. Extended treatment of this subject has been given by Shedd [1], Bresler, Lin, and Scalzi [29] and Blodgett [30]. Schilling [28] has provided an extended treatment of optimum proportioning, including the hybrid girder having a web of lower yield strength than the flanges. The average gross area A_g of the girder for the entire span may be expressed

$$A_g = 2C_1A_f + C_2ht \qquad (11.13.10)$$

where C_1 = factor to account for reducing flange size at regions of lower than maximum moment

C_2 = factor to account for reducing web thickness at regions of reduced shear

Substituting Eq. 11.13.9 into Eq. 11.13.10 gives

$$A_g = 2C_1\left(\frac{M}{fh} - \frac{ht}{6}\right) + C_2ht \qquad (11.13.11)$$

To find the minimum average gross area,

$$\frac{\partial A_g}{\partial h} = 0 \qquad (11.13.12)$$

(a) *Case 1.* No depth restriction; desire large h/t. Assume $K =$ constant $= h/t$; $t = h/K$. Equation 11.13.11 becomes

$$A_g = 2C_1\left(\frac{M}{fh} - \frac{h^2}{6K}\right) + C_2\frac{h^2}{K} \qquad (11.13.13)$$

$$\frac{\partial A_g}{\partial h} = 0 = \frac{-2C_1Mf}{f^2h^2} - \frac{4C_1h}{6K} + \frac{2C_2h}{K} \qquad (11.13.14)$$

$$0 = -6C_1MK - 2C_1h^3f + 6C_2h^3f \qquad (11.13.15)$$

from which

$$h = \sqrt[3]{\frac{3MC_1K}{f(3C_2 - C_1)}} \qquad (11.13.16)$$

and if one neglects the section reduction in regions of lower stress, $C_1 = C_2 = 1$, Eq. 11.13.16 becomes

$$h = \sqrt[3]{\frac{3MK}{2f}} \qquad (11.13.17)$$

Using Eq. 11.13.13 with $C_1 = C_2 = 1$ and substituting for M/f from Eq. 11.13.17 gives

$$A_g = \frac{4h^2}{3K} - \frac{h^2}{3K} + \frac{h^2}{K} = \frac{2h^2}{K} \qquad (11.13.18)$$

from which the girder weight per foot can be estimated using the fact that steel weight is 3.4 lb/sq in./linear ft (0.00784 kg/mm²/linear metre).

$$\text{lb/ft} = 3.4A_g = \frac{6.8h^2}{K} = 8.9\sqrt[3]{\frac{M^2}{f^2K}} \qquad (11.13.19)^*$$

* For SI units, the mass per metre is

$$\text{kg/m} = 1.72\sqrt[3]{\frac{M^2}{f^2K}} \qquad (11.13.19)$$

with mm units for M and f.

using inch units for the variables. Note that stiffeners will generally increase this value by 5 to 10 percent.

(b) *Case 2.* Minimum web thickness; $t =$ const. Differentiating Eq. 11.13.11, $\partial A_g / \partial h = 0$, gives

$$\frac{-2C_1 Mf}{f^2 h^2} - \frac{C_1 t}{3} + C_2 t = 0 \qquad (11.13.20)$$

$$h \doteq \sqrt{\frac{6C_1 M}{ft(3C_2 - C_1)}} \qquad (11.13.21)$$

If $C_1 = C_2 = 1$,

$$h = \sqrt{\frac{3M}{ft}} \qquad (11.13.22)$$

and using A_g from Eq. 11.13.11, the weight per ft is

$$\text{lb/ft} = 3.4 A_g = 4.53 ht = 7.85 \sqrt{\frac{Mt}{f}} \qquad (11.13.23)^*$$

using inch units for the variables. Estimation for the weight of stiffeners should be added to the above value.

(c) *Case 3.* Heavy shear which governs web area; $A_w =$ constant $=$ web area, ht. Equation 11.13.11 becomes

$$A_g = 2C_1 \left(\frac{M}{fh} - \frac{A_w}{6} \right) + C_2 A_w \qquad (11.13.24)$$

from which it is apparent that minimum A_g results from maximum depth, h. This case usually does not govern.

If the same kind of steel is used throughout, the value of C_1 may vary from 0.7 to 0.9 when used with the maximum positive moment: 0.85 to 0.90 is the usual range. The value of C_2 is not as likely to vary except on continuous structures where it might be 1.05 when used with maximum positive moment, or 0.95 when used with maximum negative moment. Because of the complexity of evaluating C_1 and C_2 for continuous structures, it might be well to take them as unity.

Flange Plate Changes in Size

It is usually economical to reduce the size of flange plates in the region of low moment. While no specific rules can be made to help the designer to

---

* For SI units, the mass per metre is

$$\text{kg/m} = 0.0181 \sqrt{\frac{Mt}{f}} \qquad (11.13.23)$$

using mm units for the variables.

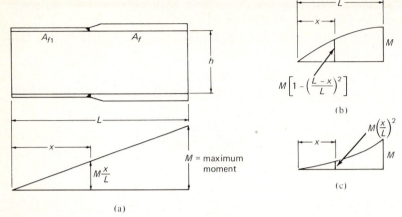

Fig. 11.13.2 Common moment variations for determining changes in flange plate size.

determine when it is desirable to change flange plate size, certain simple relationships are possible if only one change in flange size is desired.

(a) *Case 1.* Linear variation in moment—two flange plate sizes. Consider the situation of Fig. 11.13.2a, and assuming both plates are stressed to their allowable values, the flange-area formula can be used for each plate.

$$A_f = \frac{M}{hf} - \frac{A_w}{6} \tag{11.13.25}$$

$$A_{f1} = \frac{M(x/L)}{hf} - \frac{A_w}{6} \tag{11.13.26}$$

The total flange volume in the length L is

$$\begin{aligned}
\text{Vol} &= A_f(L-x) + A_{f1}x \\
&= \frac{M(L-x)}{hf} - \frac{A_w}{6}(L-x) + \frac{M}{hf}\left(\frac{x^2}{L}\right) - \frac{A_w}{6}x \\
&= \frac{M}{hf}\left(\frac{L^2 - xL + x^2}{L}\right) - \frac{A_w L}{6} \tag{11.13.27}
\end{aligned}$$

For minimum volume,

$$\frac{\partial(\text{Vol})}{\partial x} = 0 = 2x - L; \qquad x = \frac{L}{2}$$

which means

$$\frac{A_{f1}}{A_f} = \frac{\dfrac{M}{2hf} - \dfrac{A_w}{6}}{\dfrac{M}{hf} - \dfrac{A_w}{6}} \approx \frac{1}{2} \tag{11.13.28}$$

(b) *Case 2.* Parabolic variation as for uniformly loaded simple beam (see Fig. 11.13.2b). The total volume in the length L is

$$\text{Vol} = \frac{M}{hf}\left(\frac{L^3 - L^2x + 2Lx^2 - x^3}{L^2}\right) - \frac{A_w L}{6}$$

$$\frac{\partial(\text{Vol})}{\partial x} = 0 = x^2 - \frac{4}{3}Lx + \frac{L^2}{3}; \qquad x = \frac{L}{3}$$

and

$$\frac{A_{f1}}{A_f} \approx \frac{5}{9} \qquad\qquad (11.13.29)$$

(c) *Case 3.* Parabolic variation as for uniformly loaded cantilever (see Fig. 11.13.2c). The total volume in the length L is

$$\text{Vol} = \frac{M}{hf}\left(\frac{L^2 - L^2x + x^3}{L^2}\right) - \frac{A_w L}{6}$$

$$\frac{\partial(\text{Vol})}{\partial x} = 0 = 3x^2 - L^2; \qquad x = \frac{L}{\sqrt{3}}$$

and

$$\frac{A_{f1}}{A_f} \approx \frac{1}{3} \qquad\qquad (11.13.30)$$

The foregoing developments can provide a guide for change of plate sizes. Since making a change involves a groove welded butt joint of the flange plates, enough material must be saved to more than offset the welding cost.

As a rule of thumb, unless 200 to 300 lb of material are saved in a flange plate per added splice the added cost of the butt splice (considering a plate about 2 ft wide and 2 in. thick) is not justified.

Flange Plate Proportions

According to most of the theory developed earlier in this chapter, the flange plate can be any width and thickness as long as it contributes to the girder the properties necessary to satisfy the functional requirements. However, nearly all testing has used flange plate dimensions which were considered reasonable.

In order to assist the engineer who is unsure of what constitutes such reasonable dimensions the following guidelines are suggested.

1. Typically the ratio of girder flange width to girder depth, b/d, varies from about 0.3 for shallow girders to about 0.2 for deep girders.
2. Plate widths should be in 2-in. increments.

3. Plate thickness increments should be as follows:

$\frac{1}{16}$ in. $t \leq \frac{9}{16}$ in.

$\frac{1}{8}$ in. $\frac{5}{8} = t \leq 1\frac{1}{2}$ in.

$\frac{1}{4}$ in.   over $1\frac{1}{2}$ in.

4. Where lateral stability is of concern for the girder, plate width-to-thickness ratios, b/t, should be kept below the limits of AISC–1.9 at the point of maximum moment so that reduced thicknesses can be used to reduce the flange plate size in regions of low moment. For such cases, flange plate area reduction should be made by reducing the thickness.

5. For laterally stable girders, the flange plate area reduction in regions of lower moment may be accomplished by reducing the thickness, reducing the width, or reducing both thickness and width. A slight advantage in fatigue strength accrues by reducing the width rather than the thickness [30]. The transition slope should not exceed 1 in $2\frac{1}{2}$ for either width or thickness, and is usually 1 in 4 to 1 in 12 for the transition in width [30].

6. Excessive thickness of flange plate may require an arbitrarily larger weld size than required for strength in making the flange-to-web connection. AISC–1.17.2 should be considered when deciding on thickness.

11.14 PLATE GIRDER DESIGN EXAMPLE—AISC

Partially design a two-span continuous welded plate girder to support a uniform load of 4 kips/ft plus two fixed concentrated loads of 75 kips in each span as shown in Fig. 11.14.1. Lateral support is provided every 25 ft.

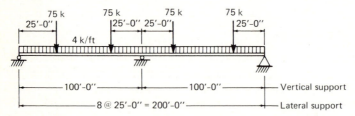

Fig. 11.14.1   Girder loading and support for design example.

Additional specifications and general comments:

1. Assume all loading has fixed position (no moving or partial span loading).

2. Girder should have constant depth of web plate; use A36 steel in positive moment zone and A572 Grade 50 for the negative moment zone.

3. Assume $\frac{5}{16}$ in. is minimum web plate thickness available.
4. Assume no depth restriction; also, that any thin-web deep girder that is selected can be feasibly transported to the construction site without excessive difficulty or cost.
5. Note should be made that laterally unsupported distances as great as 25 ft are uncommon and are used here only for the purpose of illustrating procedure.

SOLUTION

(a) *Preliminary Estimate of Weight.* The maximum moments due to the superimposed loads of Fig. 11.14.1 for the positive and negative regions are computed.

$$+M = 3950 \text{ ft-kips}$$
$$-M = 7100 \text{ ft-kips}$$

Since there is no depth limitation given in this statement, AISC limitations for h/t must be considered. Referring to Table 11.9.1 (based on AISC–1.10.2),

$$\text{Max } h/t = 322 \quad (333 \text{ if } a/h \le 1.5) \quad \text{A36 steel}$$
$$\text{Max } h/t = 243 \quad (283 \text{ if } a/h \le 1.5) \quad \text{A572 Grade 50}$$

Flange stress reduction will be required according to Fig. 11.9.1 (AISC–1.10.6) when

$$h/t > 162 \quad \text{A36 when } F_b = 0.60F_y$$
$$h/t > 139 \quad \text{A572 Grade 50 when } F_b = 0.60F_y$$

For weight estimate, try $K = h/t = 250$ and use Eq. 11.13.19, $M = 3950$ ft-kips, and $f \approx 21$ ksi:

$$\text{Wt/ft} = 8.9 \sqrt[3]{\frac{M^2}{f^2 K}} = 8.9 \sqrt[3]{\frac{[3950(12)]^2}{(21)^2 250}} = 245 \text{ lb/ft}$$

Assume $w = 250$ lb/ft and the positive moment due to girder weight ≈ 175 ft-kips; say total $+M = 4150$ ft-kips:

$$\text{Wt/ft} \approx 250 \text{ lb/ft} \quad \text{(recomputed from formula)}$$

Negative moment will require somewhat heavier flange plates because $-M/+M$ ratio exceeds the ratio of yield stresses for the two materials to be used. Thus the average weight will be somewhat higher than 250 lb/ft. In addition, allowance of about 10 percent should be made for stiffeners. Use a value somewhat larger than 275 lb/ft. Try $w = 290$ lb/ft. The total moment and shear diagrams are then computed and shown in Fig. 11.14.2.

(d) *Determine Web Plate Sizes.* For $+M$ with A36 steel assume $C_1 = C_2 = 1$ and use Eq. 11.13.17. Evaluate optimum value of h for

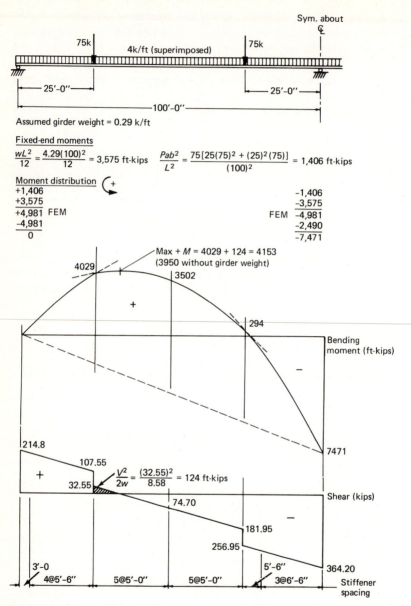

Fig. 11.14.2 Moments and shears for two-span continuous beam of illustrative example.

various h/t values,

$$h = \sqrt[3]{\frac{3MK}{2f}} = \sqrt[3]{\frac{3(4153)(12)320}{2(22)}} = 102.8$$

Assume $0.60F_y$ for f_b

$\dfrac{h}{t} = K$	Formula h (in.)	h (in.)	t (in.)	A_w (sq in.)	$f_v = \dfrac{V}{A_w}$ (ksi)	Actual $\dfrac{h}{t}$
300	100.6	100	$\frac{5}{16}$	31.25	6.9	320
320	102.8	102	$\frac{5}{16}$	31.88	6.7	326
333	104.2	104	$\frac{5}{16}$	32.5	6.6	333

Based on $+M$ the economical depth for bending moment requirements appears to be 100 to 104 in. for near maximum web slenderness ratios. The allowable shear stress for these h/t ratios for $F_y = 36$ ksi is about 10 ksi (see AISC Table 11–36 or compute from Eq. 11.10.8), indicating that the $\frac{5}{16} \times 100$ to 104 plate provides more web area than necessary.

Evaluate optimum h for $(-M)$ requirement,

$$h = \sqrt[3]{\dfrac{3(7470)(12)240}{2(30)}} = 102.5$$

$\Big\uparrow$ assume $0.60F_y$ for f_b

$\dfrac{h}{t} = K$	Formula h (in.)	h (in.)	t (in.)	A_w (sq in.)	$f_v = \dfrac{V}{A_w}$ (ksi)	Actual $\dfrac{h}{t}$
220	99.5	100	7/16	43.75	8.3	229
		100	3/8	37.50	9.7	267
240	102.5	102	7/16	44.63	8.2	233
		102	3/8	38.25	9.5	272
260	105.2	106	3/8	39.75	9.2	283
283	108.3	108	7/16	47.25	7.7	247

On the basis of these preliminary computations, a depth somewhere between 100 and 108 in. seems economical for bending moment. Since it is economical to use intermediate stiffeners, one should note that for large h/t ratios the stiffener spacing requirement is relatively constant for a given shear stress. Thus the deeper the girder the longer must be the stiffener plates, with little advantage to offset this extra weight. Generally, 2-in. increments of depth should be used.

Combined stress on the web (AISC–1.10.7) should also be considered; the unit shear stress at maximum negative moment in this problem should be kept slightly below what would normally be allowed so that the bending capacity of the web may be fully utilized.

Allowing for some flexibility in the design, use $h = 100$ in. which will mean that web slenderness is near but not at its upper limit. Try

$$\tfrac{5}{16} \times 100 \quad (h/t = 320) \text{ for } +M, \qquad F_y = 36 \text{ ksi}$$

$$\tfrac{3}{8} \times 100 \quad (h/t = 267) \text{ for } -M, \qquad F_y = 50 \text{ ksi}$$

(c) *Select Flange Plates for Negative Moment.* $M = 7470$ ft-kips, web plate $= \tfrac{3}{8} \times 96$ ($A_w = 37.5$ sq in.). Allowable stress must be estimated considering the 25-ft distance of lateral unsupport. Normally plate girder stability is controlled by AISC Formula (1.5–6). If a typical ratio of $b/d = 0.25$ is assumed, then $b = 24$ in. Using the radius of gyration of a rectangle about its mid-depth,

$$r = b/\sqrt{12} = 0.288b$$

which gives $r \approx 6.9''$, say $7''$.

$$\text{Estimated } \frac{L}{r_T} = \frac{25(12)}{7} = 42.9$$

Considering the moment gradient in the negative-moment zone,

$$C_b = 1.75 + 1.05 \left(+\frac{294}{7470} \right) + 0.3 \left(\frac{294}{7470} \right)^2$$

$$= 1.79 + \text{or} \approx 1.75 \qquad \text{(AISC–1.5.1.4.5(2.))}$$

Referring to Table 9.5.2 (AISC–1.5.1.4.5(2.)), the allowable stress is $F_b = 0.60F_y$, since $L/r_T < 60$. To obtain the estimated reduction due to high web slenderness, see Fig. 11.9.1 (AISC–1.10.6). For $h/t = 267$,

$$F_b' \approx 28 \text{ ksi}$$

Using the flange-area formula, Eq. 11.13.9, gives the requirement for one flange as

$$A_f = \frac{M}{fh} - \frac{A_w}{6} = \frac{7470(12)}{28(100)} - \frac{37.5}{6} = 25.8 \text{ sq in.}$$

Try **PL** $- 1\tfrac{1}{8} \times 24$: $A_f = 27.0$ sq in.
Check ordinary allowable stress:

$$r_T = \sqrt{\frac{\tfrac{1}{12}(24)^3(1.125)}{27.0 + 37.5/6}} = 6.24 \text{ sq in.}$$

$$\frac{L}{r_T} = \frac{300}{6.24} = 48.1 < 60, \qquad \therefore F_b = 0.60F_y = 30 \text{ ksi}$$

Reduction according to AISC–1.10.6:

$$\frac{A_w}{A_f} = \frac{37.5}{27.0} = 1.39$$

From Fig. 11.9.1, using $h/t = 267$, find $F_b' = 27.3$ ksi. Plate size $1\tfrac{1}{8} \times 24$ still may work even though F_b' is lower than assumed originally.

Check stress using flexure formula. Computing the moment of inertia,

$$1\tfrac{1}{8} \times 24: \qquad 27.0(2)(101.125/2)^2 = 138,000$$
$$\tfrac{3}{8} \times 100: \qquad \tfrac{1}{12}(\tfrac{3}{8})(100)^3 = \quad 31,300$$
$$I = \overline{169,000} \text{ in.}^4$$

$$f = \frac{7470(12)(50+1.125)}{169,000} = 27.1 \text{ ksi}$$

Verifying the allowable stress using AISC Formula (1.10–5),

$$F_b' = 30 \left[1.0 - 0.0005 \frac{A_w}{A_f}\left(\frac{h}{t} - \frac{760}{\sqrt{30}}\right)\right]$$

$$F_b' = 30[1.0 - 0.0005(1.39)(267 - 139)] = 27.3 \text{ ksi}$$

Accept plates $1\tfrac{1}{8} \times 24$.

(d) *Select Flange Plates for Positive Moment.* $M = 4153$ ft-kips; web plate $= \tfrac{5}{16} \times 100$ ($A_w = 31.25$ sq in.). Estimating flange width to be about 24 in., $L/r_T < 48.1$ computed in part (c) above. In this region $C_b = 1$, in which case from Table 9.5.2 for $F_y = 36$ ksi,

$$F_b = 0.60F_y = 22 \text{ ksi}, \qquad \text{since } \frac{L}{r_T} < 53$$

Referring to Fig. 11.9.1, the allowable bending stress resulting from flange stress reduction (AISC–1.10.6) is estimated as

$$F_b' \approx 19.6 \text{ ksi for } \frac{h}{t} = 320$$

$$\text{Required } A_f = \frac{M}{fh} - \frac{A_w}{6} = \frac{4153(12)}{19.6(100)} - \frac{31.25}{6} = 20.2 \text{ sq in.}$$

Try PL $-\tfrac{7}{8} \times 24$: $\qquad A_f = 21.0$ sq in.

Check stress:

$$\tfrac{7}{8} \times 24: \qquad 21.0(2)(100.875/2)^2 = 106,800$$
$$\tfrac{5}{16} \times 100: \qquad \tfrac{1}{12}(\tfrac{5}{16})(100)^3 = \quad 26,000$$
$$I = \overline{132,800} \text{ in.}^4$$

$$f = \frac{4153(12)(50+0.875)}{132,800} = 19.1 \text{ ksi}$$

The allowable stress F_b', with $A_w/A_f = 1.49$ is

$$F_b' = 22.0\,[1.0 - 0.0005(1.49)(320 - 162)] = 19.4 \text{ ksi} > 19.1 \text{ ksi}$$

Accept plates, $\tfrac{7}{8} \times 24$.

(e) *Intermediate Stiffeners—Placement in Positive Moment Zone.* Web plate = $\frac{5}{16} \times 100$, $A_w = 31.25$ sq in., $F_y = 36$ ksi.

Exterior end, $V = 214.8$ kips. Use Eq. 11.10.3, AISC Formula (1.10–1),

$$F_v = \frac{F_y C_v}{2.89} \le 0.40 F_y$$

The nominal shear stress in the end panel is

$$f_v = \frac{V}{A_w} = \frac{214.8}{31.25} = 6.9 \text{ ksi} < 0.40 F_y \qquad \text{OK}$$

Let $F_v = 6.9$ ksi,

$$\text{Required } C_v = \frac{6.9(2.89)}{36} = 0.55$$

Using Eq. 11.2.9 for C_v less than 0.8,

$$\frac{45,000k}{F_y (h/t)^2} = 0.55$$

$$\text{Required } k = \frac{0.55(320)^2 36}{45,000} = 45.4$$

For $a/h < 1$,

$$k = 4.0 + 5.34/(a/h)^2 = 45.4$$

$$\text{Max} \frac{a}{h} = 0.36$$

$$\text{Max } a = 0.36(100) = 36 \text{ in.}$$

Use 3′–0″.

Panel 2:

$$V = 214.8 - 3(4.29) = 201.9 \text{ kips}$$

Equations 11.10.3 and 11.10.8, AISC Formulas (1.10–1) and (1.10–2), considering tension-field action are to be applied here:

$$f_v = \frac{201.9}{31.25} = 6.5 \text{ ksi}, \qquad \frac{h}{t} = \frac{100}{0.3125} = 320$$

Using AISC Appendix Table 11-36, find $a/h \approx 0.7$, computed more exactly as

$$\text{Max} \frac{a}{h} = \left(\frac{260}{h/t}\right)^2 \le 3.0$$

$$= \left(\frac{260}{320}\right)^2 = 0.66$$

$$a = 0.66(100) = 66'' \quad (5.5')$$

Since lateral support occurs every 25 ft and a bearing stiffener is required at concentrated loads, the stiffener spacing is usually arranged to fit these limitations. Considering the first stiffener is at 3'-0", 22'-0" remains. Use 4 spaces @ 5'-6".

For the region from 25 to 50 ft from the end, the a/h limitation of 0.66 still governs. Use 5 spaces @ 5'-0".

For the region from 50 to 75 ft the maximum shear stress is still below the value in panel 2, so that a/h maximum still governs. Use 5 spaces @ 5'-0".

(f) *Intermediate Stiffeners—Placement in Negative Moment Zone.* Web plate $= \frac{3}{8} \times 100$, $A_w = 37.5$ sq in., $F_y = 50$ ksi.

Interior end:

$$V = 364.2 \text{ kips}$$

Use Eqs. 11.10.3 and 11.10.8 for this interior panel (there is an adjacent panel in the other span):

$$f_v = \frac{364.2}{37.5} = 9.7 \text{ ksi}, \qquad \frac{h}{t} = \frac{100}{0.375} = 267$$

Using AISC Appendix Table 11-50, again find the maximum a/h is controlled by the arbitrary limit (see also Fig. 11.10.1).

$$\text{Max} \frac{a}{h} = \left(\frac{260}{267}\right)^2 = 0.95$$

Combined stress on *web* at interior end:

$$f_b(\text{maximum on web}) = 27.1 \frac{50}{51.125} = 26.5 \text{ ksi}$$

$$\frac{f_b}{0.60 F_y} = \frac{26.5}{30} = 0.883 > 0.75$$

In accordance with AISC Formula (1.10–7), Eq. 11.9.8, the usable shear capacity must be reduced from its ordinary allowable value:

$$\text{Max permitted} \frac{f_v}{F_v} = \frac{0.825 - f_b/F_y}{0.375} = \frac{0.825 - 26.5/50}{0.375} = 0.786$$

The required allowable for the panel as computed by AISC Formula (1.10–2) is

$$\text{Required } F_v = \frac{f_v}{0.786} = \frac{9.7}{0.786} = 12.3 \text{ ksi}$$

Therefore from AISC Appendix Table 11-50 the required $a/h \approx 0.85$ for which $F_v = 12.3$ ksi. ~~Try 7'-0" ($a/h = 0.84$).~~ Revised below.

Panel 2:

$$V = 364.2 - 7(4.29) = 334 \text{ kips}$$

$$f_v = \frac{334}{37.5} = 8.9 \text{ ksi}$$

At this location, where $M = 5150$ ft-kips,

$$\frac{f_b \text{(on web)}}{0.60F_y} = \frac{26.5}{30} \left(\frac{5150}{7470}\right) = 0.61 < 0.75$$

Full shear strength may be utilized. Therefore maximum $a/h = 0.95$ controls:

$$a = 0.95(100) = 95''$$

Since two spaces will not work, three spaces must be used. To prevent using a larger spacing in the region of higher shear, reduce the spacing in the first panel. ~~Try 4 spaces @ 6'-3" for the entire 25 ft. section.~~ Revised below. The arrangement of intermediate stiffeners is shown in the design sketch, Fig. 11.14.6.

(g) *Location of Flange and Web Splices.* The location of splices is partially dependent on the type of splices used; for a bolted field splice joining the A36 with the A572 sections both flanges and web usually would be spliced at the same location; for a field welded splice it might be preferable to splice the flanges at a location offset from the web splice by as much as 10 ft. Such an offset reduces stress concentration and probably assists in getting proper alignment at the splice. For this example assume splices for flanges and web are to occur at the same location.

From previous calculations for stiffener spacing it is apparent that splicing could be done at, say, 26 ft from the interior support, just beyond the bearing stiffener. Perhaps, however, it can be made closer to the interior support.

Try 5'-6" from 75-kip load (19'-6" from interior support) which would be the maximum spacing based on the maximum $a/h = 0.66$ for the $\frac{5}{16}$-in. web. See part (e).

$$M = 1210 \text{ ft-kips}; \qquad V = 290 \text{ kips}$$

(scaled from Fig. 11.14.2)

$$f_b \text{(on web)} = \frac{1210(12)50}{132,800} = 5.5 \text{ ksi} < 0.45F_y$$

Combined stress is acceptable.

$$f_v = \frac{290}{31.25} = 9.3 \text{ ksi}; \qquad \frac{h}{t} = 320$$

From AISC Appendix Table 11-36, find $a/h \approx 0.7$ which is acceptable. Splice at 20 ft from interior support. Revise stiffener spacing in end 25-ft section to 3 spaces at 6'-6" and one space at 5'-6", starting from interior support. A desirable 6-in. offset will exist between the splice and the nearest stiffener.

A summary showing complete stiffener spacing appears in Fig. 11.14.6.

(h) *Size of Intermediate Stiffeners.* Frequently, A36 stiffeners are suitable, with higher yield stress material offering little if any saving. Try A36 steel for all stiffeners.

Panel 2 from exterior support: This is the first panel in which tension-field action is presumed to occur.

$$V = 201.9 \text{ kips}; \qquad f_v = 6.5 \text{ ksi}$$

$$\frac{a}{h} = \frac{66}{100} = 0.66, \qquad \frac{h}{t} = 320; \qquad F_v \approx 9.8 \text{ ksi}$$

Using AISC Appendix Table 11–36, find $A_{st}/A_w \approx 0.115$. (Same as using AISC Formula (1.10–3), or Eq. 11.10.20.)

$$\frac{f_v}{F_v} = \frac{6.5}{9.8} = 0.66$$

Required $A_{st} = 0.115 A_w (0.66) = 0.115(31.25)(0.66) = 2.37$ sq in.

This area requirement assumes stiffeners to be used in pairs. Furthermore, local buckling requirements of AISC–1.9.1 must be satisfied; i.e., $w/t = 95/\sqrt{F_y} = 15.8$.

The requirement for stiffness, Eq. 11.10.10, gives

$$\text{Required } I_s = \left(\frac{h}{50}\right)^4 = \left(\frac{100}{50}\right)^4 = 16.0 \text{ in.}^4$$

To find minimum acceptable stiffener width (Fig. 11.14.3),

$$\text{Required } r^2 = \frac{I}{A} = \frac{16.0}{2.37} = 6.8 \text{ in.}^2$$

$$\text{Provided } r^2 = \frac{tW^3}{12tW} = \frac{W^2}{12}$$

$$\text{Required } W = \sqrt{12(6.8)} = 9.0 \text{ in.}$$

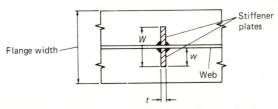

Fig. 11.14.3   Cross section of intermediate stiffener plates.

On this basis, $4\frac{1}{2}$ in. wide plates could be used, which, considering the area requirement, would have to be $\frac{5}{16} \times 4\frac{1}{2}$ in pairs. A designer at this point must decide what stiffener width he considers as minimum. The authors prefer in this case 2 PLs $-\frac{5}{16} \times 5$.

Since the stiffeners for the panel are about the minimum size based on local-buckling provisions, the reduced area required for the more interior panels on the $\frac{5}{16}$-in. web is of little consequence. *Use* $2\,$PLs$-\frac{5}{16}\times5$ for all intermediate stiffeners attached to the $\frac{5}{16}$-in. web.

Examine the end panel adjacent to the interior support:

$$V = 364.2 \text{ kips}; \qquad f_v = 9.7 \text{ ksi}$$

Based on combined stress, however, the required allowable value F_v was 12.3 ksi. Applying AISC Formula (1.10–2), or AISC Appendix Table 11-50,

$$\frac{a}{h} = \frac{78}{100} = 0.78, \qquad \frac{h}{t} = 267; \qquad F_v \approx 12.6 \text{ ksi}$$

Estimate $A_{st}/A_w \approx 0.125$, for stiffeners with $F_y = 50$ ksi.

$$\frac{f_v}{F_v} = \frac{\text{strength utilized}}{\text{full strength}} = \frac{12.3}{12.6} = 0.98$$

Required $A_{st} = 0.125 A_w (0.98) = 0.125(37.5)(0.98) = 4.59$ sq in.

for A572 Grade 50 stiffeners. A36 stiffeners would require $A_{st} = 4.59(50/36) = 6.38$ sq in. A572 stiffeners appear to be economical:

$$2\,\text{PLs} -\tfrac{7}{16}\times5\tfrac{1}{2}, \qquad A_{st} = 4.81 \text{ sq in.}$$
$$2\,\text{PLs} -\tfrac{1}{2}\times5, \qquad A_{st} = 5.00 \text{ sq in.}$$

The maximum w/t limit of 13.4 for $F_y = 50$ ksi must be considered. *Use* $2\,$PLs$-\frac{7}{16}\times5\frac{1}{2}$ of A572 Grade 50 for all intermediate stiffeners connected to the $\frac{3}{8}$-in. web.

   (i) *Connection of Intermediate Stiffeners to Web.* For the $\frac{5}{16}$-in. A36 web; using AISC Formula (1.10–4),

$$f_{vs} = h \sqrt{\left(\frac{F_y}{340}\right)^3} = 100 \sqrt{\left(\frac{36}{340}\right)^3} = 3.44 \text{ kips/in.}$$

Since panels are not fully stressed, this value can be reduced in direct proportion:

$$\text{Required } f_{vs} = 3.44(0.66) = 2.27 \text{ kips/in.}$$
$$\text{Min weld size } a = \tfrac{3}{16} \quad \text{(AISC–1.17.2)}$$

Determine maximum effective weld size (AISC–1.17.3 using shear stress on throat of fillet from AISC–Table 1.5.3 with E70 electrodes) (see also text, Sec. 5.13),

$$a_{\text{max eff.}} = \frac{0.4 F_y t_w}{2(0.707)21.0} = \frac{14.4(0.3125)}{2(0.707)21.0} = 0.151 \text{ in.}$$

Try $\frac{3}{16}$-in. weld of E70 electrodes (basic welding on A36 steel). The safe weld capacity R_w per inch is

$$R_w = 0.151 \, (0.707)21.0 = 2.24 \text{ kips/in.}$$

Since four lines of fillet welds (Fig. 11.14.3) are to provide 2.27 kips/in., then $2.27/4 = 0.57$ kips/in. are required along each line.

$$\% \text{ of continuous } \tfrac{3}{16}\text{-in. weld required} = \frac{0.57}{2.24}(100) = 25.4\%$$

For intermittent welding, minimum segment is $1\frac{1}{2}$ in., according to AISC–1.17.5; the capacity of this segment is

$$L_w R_w = 1.5(2.24) = 3.36 \text{ kips}$$

$$\text{Required pitch } p = \frac{3.36}{0.57} = 5.9 \text{ in.}$$

Use $\tfrac{3}{16}$ in.–$1\frac{1}{2}$ in. segments @ $5\frac{1}{2}''$ pitch, E70 electrodes, for connecting $\tfrac{5}{16} \times 5$ plates to $\tfrac{5}{16}$ in. web.

For the $\tfrac{5}{16}$ in. A572 Grade 50 web:

$$f_{vs} = h \sqrt{\left(\frac{50}{340}\right)^3} = 0.0564h = 0.564(100) = 5.64 \text{ kips/in.}$$

Based on 98% stressed in shear for the most highly stressed panel,

$$\text{Required } f_{vs} = 0.98(5.64) = 5.53 \text{ kips/in.}$$

$$\text{Min weld size } a = \tfrac{3}{16} \text{ in.}$$

$$a_{\max \text{ eff}} = \frac{20.0(0.375)}{2(0.707)21.0} = 0.252 \text{ in.} > \tfrac{3}{16} \qquad \text{try } \tfrac{3}{16} \text{ in.}$$

Using E70 electrodes (text Table 5.12.1), along with the corresponding allowable stress (AISC–Table 1.5.3),

$$R_w \text{ (for } \tfrac{3}{16}\text{-in. weld)} = \tfrac{3}{16}(0.707)21.0 = 2.78 \text{ kips/in.}$$

For $1\frac{1}{2}$-in. segments, the required pitch is

$$p = \frac{1.5(2.78)}{(5.53/4)} = 3.0 \text{ in.}$$

which means nearly continuous $\tfrac{3}{16}$-in. weld. Try $\tfrac{1}{4}$-in. weld (maximum effective $= 0.252$ in.):

$$R_w = \tfrac{1}{4}(0.707)21.0 = 3.71 \text{ kips/in.}$$

$$p = \frac{1.5(3.71)}{(5.53/4)} = 4.03 \text{ in.} \qquad \text{Not sufficient savings.}$$

Use $\tfrac{3}{16}$ in. continuous weld, E70 electrodes, for connecting $\tfrac{7}{16} \times 5\frac{1}{2}$ plates to $\tfrac{3}{8}$-in. web.

(j) *Flange to Web Connection—A36 Steel.* The welding of the flanges to the web must provide for the horizontal shear developed at the joint. The shear flow that must be transmitted may be expressed as

$$\text{Shear flow} = \frac{VQ}{I} \text{ kips/in.}$$

where V = shear at section
Q = statical moment of flange area about neutral axis
I = moment of inertia of section

Welding along both sides of the web provides a shear flow capacity, which, if it exceeds the VQ/I requirement, may be reduced by the use of intermittent welding. Normally flange to web welding is made continuous, primarily because automatic fabricating procedures usually make it more economical to do so. However, for design purposes the minimum percent of continuous weld acceptable in each panel between stiffeners will be prescribed. If it is more economical for the fabricator to use more weld, it is permissible to do so. The following calculations conservatively assume shielded metal arc welding (SMAW) is to be used.

$$\text{Min weld size } a = \tfrac{5}{16} \text{ in.} \qquad (\text{AISC}-1.17.2)$$

$$\text{Max effective size } a = \frac{14.4(0.3125)}{2(0.707)21.0} = 0.151 \text{ in.}$$

for E70 electrodes. For strength, the weld size required is

$$\frac{VQ}{I} = \frac{214.8(21)(50.44)}{132,800} = 1.71 \text{ kips/in.}$$

$$2a(0.707)21.0 = 1.71$$

$$\text{Required } a = \frac{1.71}{2(0.707)21.0} = 0.06 \text{ in.} < 0.151 \text{ in.}$$

Use $\tfrac{5}{16}$-in. weld, E70 electrodes (effective size = 0.151 in.). The safe capacity for continuous weld on both sides of the web is

$$R_w = 2(0.707)(0.151)21.0 = 4.5 \text{ kips/in.}$$

$$\text{Min \% continuous weld} = \frac{1.71}{4.5}(100) = 38\%$$

For each panel along the A36 $\tfrac{5}{16}$-in. web, the minimum percentages of $\tfrac{5}{16}$-in. continuous weld required are similarly computed and given in Fig. 11.14.6. AISC–1.10.5.4. requires at least 23.1% of continuous weld ($1\tfrac{1}{2}$-in. segment @ $6\tfrac{1}{2}$-in. maximum pitch).

(k) *Flange to Web Connection—A572 Steel.*

$$\text{Min weld size } a = \tfrac{5}{16} \text{ in.}(\text{AISC}-1.17.5)$$

$$\text{Max effective size } a = \frac{20.0(0.375)}{2(0.707)21.0} = 0.252 \text{ in.}$$

The maximum strength requirement is

$$\frac{VQ}{I} = \frac{364.2(27.0)50.56}{169,000} = 2.94 \text{ kips/in.}$$

The safe capacity of continuous $\frac{5}{16}$ in. weld on both sides of the web is

$$R_w = 2(0.252)(0.707)21.0 = 7.48 \text{ kips/in.}$$

$$\text{Min \% continuous weld} = \frac{2.94}{7.48}(100) = 39\%$$

Use $\frac{5}{16}$-in. weld, E70 electrodes. The minimum percent of continuous weld required for each panel along the A572 $\frac{3}{8}$ in. web is summarized in Fig. 11.14.6. AISC–1.10.5.4 requires at least 20% of continuous weld ($1\frac{1}{2}$-in. segment of @ $7\frac{1}{2}$-in. maximum pitch).

(1) *Design of Bearing Stiffener—Interior Support.* As discussed in Sec. 11.11, bearing stiffeners are required at concentrated loads. At the interior support,

$$\text{Reaction} = 728.4 \text{ kips}$$

Since bearing stiffeners must extend "approximately to the edge of the flange plates ...," the width must be

$$w = \frac{24 - 0.375}{2} = 11.8 \text{ in.} \qquad (\text{say 11 in.})$$

Column stability criterion (Fig. 11.14.4):

$$r \approx 0.25(22.375) = 5.6 \text{ in.}$$

$$\frac{L}{r} \approx \frac{0.75(96)}{5.6} = 12.9$$

$$F_a = 21 \text{ ksi} \qquad (\text{For A36 steel})$$

$$\text{Required area} = \frac{728.4}{21} = 34.7 \text{ sq in.}$$

$$\text{Required } t = \frac{34.7 - 9.37(0.375)}{2(11)} = 1.42 \text{ in.}$$

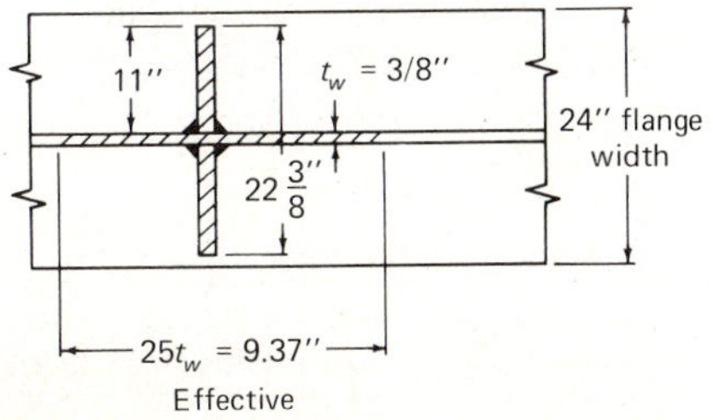

Fig. 11.14.4 Cross section of bearing stiffener at interior support (trial).

Local buckling criterion:

$$\frac{w}{t} = 15.8 \qquad \text{for } F_y = 36 \text{ ksi}$$

$$\text{Required } t = \frac{11}{15.8} = 0.70 \text{ in.}$$

Compression yield criterion:

$$\text{Required } A = \frac{728.4}{0.60F_y} = \frac{728.4}{22} = 33.1 \text{ sq in.}$$

$$\text{Required } t = \frac{33.1}{2(11)} = 1.5 \text{ in.}$$

Bearing criterion:

$$\text{Required contact area} = \frac{728.4}{0.90F_y} = \frac{728.4}{32.4} = 22.5 \text{ sq in.}$$

$$A_c = 2(11 - 0.5)t$$
$$\underset{\text{est. to allow for fillet weld}}{}$$

$$\text{Required } t = \frac{22.5}{2(10.5)} = 1.07 \text{ in.}$$

Compression yield criterion governs. So as to properly distribute this reaction, try $4\,\text{Pls} - \frac{3}{4} \times 11$ to serve as bearing stiffeners. (Alternative, $2\,\text{Pls} - 1\frac{1}{2} \times 11$). Recheck r for column stability:

$$r = \sqrt{\frac{\frac{1}{12}(1.50)(22.375)^3}{22(1.50) + 9(0.375)}} = 6.2 \text{ in.}$$

By inspection, column stability still does not govern. *Use* $4\,\text{Pls} - \frac{3}{4} \times 11$ for $R = 728.4$ kips (A36 steel).

(m) *Connection for Bearing Stiffeners to Web.*

$$\text{Required } f_{vs} = \frac{728.4}{8(100)} = 0.91 \text{ kips/in.}$$

for eight lines of fillet weld (Fig. 11.14.5). Use $\frac{5}{16}$ in., with maximum effective size $= 0.252$, and E70 electrodes, as determined in part (k). The capacity R_w of one line of continuous weld is

$$R_w = 0.252(0.707)21.0 = 3.74 \text{ kips/in.}$$

$$\text{\% of continuous weld required} = \frac{0.91}{3.74}(100) = 24.3\%$$

Use intermittent $\frac{5}{16}$-in. fillet welds, $1\frac{1}{2}$-in. segments @ $5\frac{1}{2}$-in. pitch (27% of continuous weld).

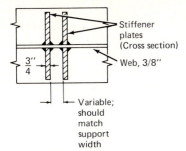

Fig. 11.14.5 Cross section of bearing stiffener at interior support (final).

(n) *Compression Stress Directly on the Web (AISC–1.10.10.2).* This check assures adequate stability under direct compression, as discussed in Sec. 11.2, using Eqs. 11.2.17 with 11.2.19 [AISC Formula (1.10–11)].

The critical region will be the largest stiffener spacing on the thinnest web; therefore the 5'-6" panel on the $\frac{5}{16}$-in. A36 web will be investigated.

$$\frac{a}{h} = \frac{66}{100} = 0.66; \qquad \frac{h}{t} = 320$$

$$\text{Allowable } f_c = \left[2 + \frac{4}{(a/h)^2}\right] \frac{10,000}{(h/t)^2} \text{ ksi}$$

$$= \left[2 + \frac{4}{(0.66)^2}\right] \frac{10,000}{(320)^2} = 1.09 \text{ ksi}$$

$$\text{Actual } f_c = \left[\frac{4}{12(0.3125)}\right] = 1.07 \text{ ksi} < 1.09 \text{ ksi} \qquad \text{OK}$$

(o) *Design Sketch.* Every design must have all decisions summarized in a design sketch such as Fig. 11.14.6. The design of the bearing stiffeners for the exterior support and under the 75-kip load has been omitted because the procedure has been adequately illustrated in items (l) and (m):

Girder weight (for one span):

A572 plates:	$1\frac{1}{8} \times 24 \times 20'$	$= 91.8(40)$	$= 3,672$
	$\frac{3}{8} \times 100 \times 20'$	$= 128(20)$	$= 2,560$
A36 plates:	$\frac{7}{8} \times 24 \times 80'$	$= 71.4(160)$	$= 11,424$
	$\frac{5}{16} \times 100 \times 80'$	$= 106(80)$	$= 8,480$
Stiffeners:	$\frac{7}{16} \times 5\frac{1}{2} \times 8.33 \times 6$	$= 8.18(8.33)6$	$= 409$
	$\frac{5}{16} \times 5 \times 8.33 \times 26$	$= 5.31(8.33)26$	$= 1,150$
	$\frac{3}{4} \times 11 \times 8.33 \times 8$	$= 28.1(8.33)8$	$= 1,872$

$$\text{Total} = 29,567 \text{ lb}$$

Average weight $= \underline{296 \text{ lb/ft.}}$

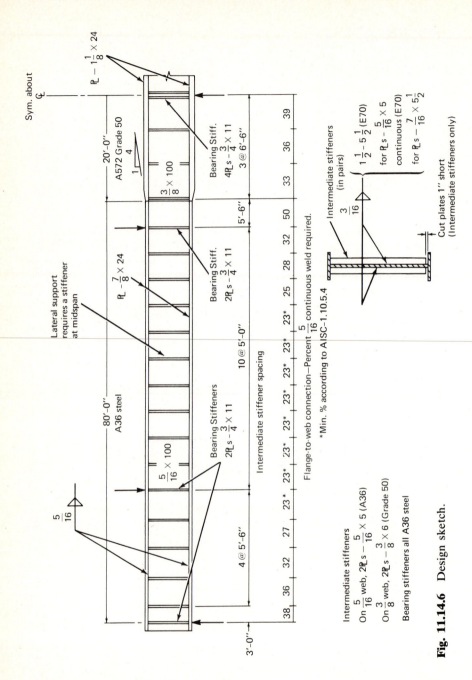

Fig. 11.14.6 Design sketch.

SELECTED REFERENCES

1. Thomas C. Shedd, *Structural Design in Steel.* New York: John Wiley & Sons, Inc., 1934, Chap. 3.
2. Edwin H. Gaylord, Jr. and Charles N. Gaylord, *Design of Steel Structures.* New York: McGraw-Hill Book Company, Inc., 1957, Chap. 8.
3. Konrad Basler, "Strength of Plate Girders in Shear," *Transactions,* ASCE, 128, Part II (1963), 683–719. (Also as Paper No. 2967, *Journal of Structural Division,* ASCE, October 1961).
4. I. Lyse and H. J. Godfrey, "Investigation of Web Buckling in Steel Beams," *Transactions,* ASCE, 100 (1935), 675–706.
5. George Gerard and Herbert C. Becker, *Handbook of Structural Stability:* Vol. I, *Buckling of Flat Plates,* National Advisory Committee for Aeronautics, NACA Tech. Note 3781, July 1957.
6. Stephen P. Timoshenko and James M. Gere, *Theory of Elastic Stability,* 2nd Ed. New York: McGraw-Hill Book Company Inc., 1961, Chap. 8.
7. Bruce G. Johnston, ed., *Structural Stability Research Council, Guide to Stability Design Criteria for Metal Structures,* 3rd Ed. New York: John Wiley & Sons, Inc., 1976, Chap. 7.
8. Konrad Basler, "New Provisions for Plate Girder Design," *Proceedings,* AISC National Engineering Conf., New York, 1961, 65–74.
9. Konrad Basler, "Strength of Plate Girders in Bending," *Transactions,* ASCE, 128, Part II (1963), 655–686. (Also as Paper No. 2913, *Journal of Structural Division,* ASCE, August 1961.)
10. Ronald W. Frost and Charles G. Schilling, "Behavior of Hybrid Beams Subjected to Static Loads," *Journal of Structural Division,* ASCE, 90, ST3 (June 1964), 55–88.
11. C. G. Schilling, Chairman, "Design of Hybrid Steel Beams," Report of Subcommittee 1 on Hybrid Beams and Girders, Joint ASCE–AASHO Committee on Flexural Members, *Journal of Structural Division,* ASCE, 94, ST6 (June 1968), 1397–1426.
12. K. C. Rockey and M. Skaloud, "The Ultimate Load Behavior of Plate Girders Loaded in Shear," *Structural Engineer,* 50, 1(January 1972).
13. M. L. Sharp and J. W. Clark, "Thin Aluminum Shear Webs," *Journal of Structural Division,* ASCE, 97, ST4 (April 1971), 1021–1038.
14. Max A. M. Herzog, "Ultimate Static Strength of Plate Girders from Tests," *Journal of Structural Division,* ASCE, 100, ST5 (May 1974), 849–864.
15. Konrad Basler, "Strength of Plate Girders Under Combined Bending and Shear," *Journal of Structural Division,* ASCE, 87, ST7 (October 1961), 181–197.
16. Bruce G. Johnston, ed. *The Column Research Council Guide to Design Criteria for Metal Compression Members,* 2nd Ed. New York: John Wiley & Sons, Inc., 1966, Chap. 5.
17. Friederich Bleich, *Buckling Strength of Metal Structures.* New York: McGraw-Hill Book Company, Inc., 1952, Chap. 11.
18. William McGuire, *Steel Structures.* Englewood Cliffs, N.J.: Prentice-Hall, Inc., 1968, pp. 734–779.
19. Konrad Basler and Bruno Thürlimann, "Plate Girder Research." *Proceedings,* AISC National Engineering Conference, New York, 1959.

20. B. T. Yen and J. A. Mueller, "Fatigue Tests of Large-Size Welded Plate Girders," *WRC Bulletin 118*, Welding Research Council, New York, November 1966.

21. P. J. Patterson, J. A. Corrado, J. S. Huang, and B. T. Yen, "Fatigue and Static Tests of Two Welded Plate Girders," *WRC Bulletin 155*, Welding Research Council, New York, October 1970.

22. Manfred A. Hirt, Ben T. Yen, and John W. Fisher, "Fatigue Strength of Rolled and Welded Steel Beams," *Journal of Structural Division*, ASCE, 97, ST7 (July 1971), 1897–1911.

23. J. R. Dimitri and A. Ostapenko, "Pilot Tests on the Static Strength of Unsymmetrical Plate Girders," *WRC Bulletin 156*, Welding Research Council, New York, November 1970, 1–22.

24. W. Schueller and A. Ostapenko, "Tests on a Transversely Stiffened and on a Longitudinally Stiffened Unsymmetrical Plate Girder," *WRC Bulletin 156*, Welding Research Council, New York, November 1970, 23–47.

25. S. Parsanejad and A. Ostapenko, "On the Fatigue Strength of Unsymmetrical Steel Plate Girders," *WRC Bulletin 156*, Welding Research Council, New York, November 1970, 48–59.

26. M. A. D'Apice, D. J. Fielding, and P. B. Cooper, "Static Tests on Longitudinally Stiffened Plate Girders," *Welding Research Council Bulletin No. 117*, October 1966 (Includes historical survey and bibliography on longitudinally stiffened plates.)

27. Peter B. Cooper, "Strength of Longitudinally Stiffened Plate Girders," *Journal of Structural Division*, ASCE, 93, ST2 (April 1967), 419–451.

28. Charles G. Schilling, "Optimum Proportions for I-shaped Beams," *Journal of Structural Division*, ASCE, 100, ST12 (December 1974), 2385–2401.

29. Boris Bresler, T. Y. Lin, and John Scalzi, *Design of Steel Structures*, 2nd Ed. New York: John Wiley & Sons, Inc., 1968, pp. 497–554.

30. Omer W. Blodgett, *Design of Welded Structures*. Cleveland, Ohio: James F. Lincoln Arc Welding Foundation, 1966.

PROBLEMS

Analysis

11.1. For the section shown, the shear force acting is 150 kips and the bending moment is 1085 ft-kips.
   - (a) Compute the bending and shear unit stress distribution from top to bottom of the section.
   - (b) Compute total shear carried by the web and that carried by the flanges. What percent of the total is carried by each?
   - (c) Compute the total bending moment carried by the web and that carried by the flanges. What percent of the total is carried by each?

11.2. (This problem forms a basis of Probs. 11.3 to 11.7.) A plate girder supported as shown must carry a live load of 10 kips/ft, not including girder weight, plus the concentrated dead load shown. Compute and draw to scale for later use the moment and shear envelopes for this girder.

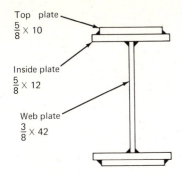

Top plate
$\frac{5}{8} \times 10$

Inside plate
$\frac{5}{8} \times 12$

Web plate
$\frac{3}{8} \times 42$

Prob. 11.1

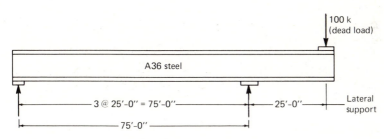

100 k
(dead load)

A36 steel

Lateral
support

3 @ 25'-0" = 75'-0" | 25'-0"

75'-0"

Assume girder weight is 390 lb/ft

Prob. 11.2

11.3. (Uses results of Prob. 11.2.) For the given plate girder designed for the conditions of Prob. 11.2, compute and draw the moment capacity diagram according to AISC Spec. Neglect any reduction that might result from combined shear and moment under AISC–1.10.7. Compare capacity with requirements from Prob. 11.2.

11.4. (Use diagrams of Prob. 11.2.) For the given plate girder, compute and draw shear capacity diagram based on location of intermediate stiffeners. Neglect any limitations of AISC–1.10.7 or 1.10.10. Compare shear capacity with requirements from Prob. 11.2.

11.5. (Use diagrams of Prob. 11.2.) Investigate combined shear and tension stress (AISC–1.10.7) and direct compression on web (AISC–1.10.10.2) by comparing the computed stress with the allowable value at all critical locations.

11.6. (Use diagrams of Prob. 11.2.) Compute and draw the shear capacity diagram for the girder based on flange to web connection. Compare by showing the capacity diagram superimposed on the requirement diagram from Prob. 11.2.

11.7. (Use information from Prob. 11.2.) Check the adequacy of each of the bearing stiffeners, including connection to the web.

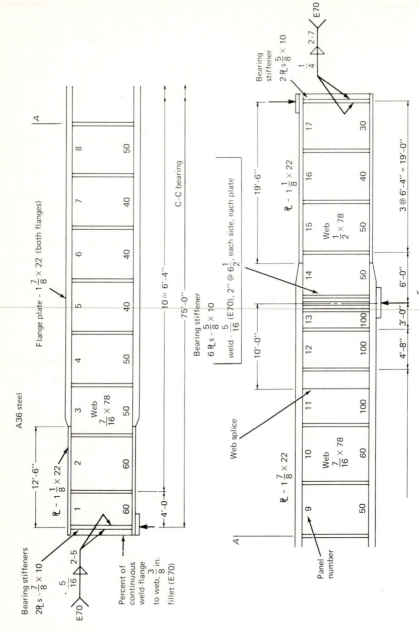

Probs. 11.3 through 11.7

11.8. Given the data for the 50-ft simple span, having lateral support at the ends and at the concentrated loads located 18 ft from each end; AISC Spec.

(a) Investigate the acceptability of the 84-in. spacing for stiffener panel 4.

(b) Investigate combined stress in the web at its most critical location in the girder.

(c) Investigate the adequacy of the size of intermediate stiffeners.

(d) Investigate the adequacy of the size of the bearing stiffener $(2\textbf{PLs} - \frac{1}{2} \times 8)$ at the support.

(e) Specify the flange-to-web connection.

(f) Specify the connection for intermediate stiffeners.

(g) Specify the connection for the support bearing stiffener.

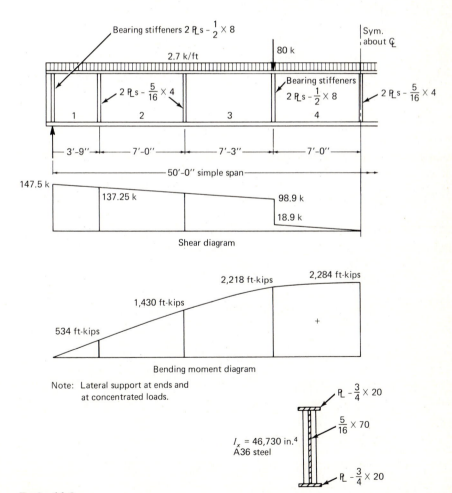

Shear diagram

Bending moment diagram

Note: Lateral support at ends and at concentrated loads.

Prob. 11.8

11.9. Using the given information concerning the portion of the plate girder, determine how close stiffener B must be to stiffener A in order for the design to be acceptable according to the AISC Specification.

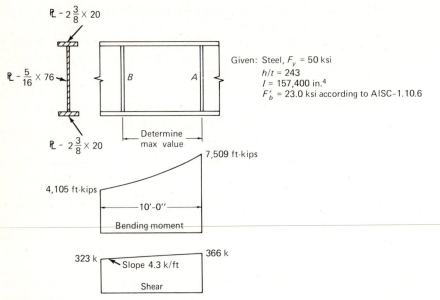

Given: Steel, $F_y = 50$ ksi
$h/t = 243$
$I = 157,400$ in.^{4}
$F'_b = 23.0$ ksi according to AISC–1.10.6

Prob. 11.9

11.10. Using the given information concerning the portion of the plate girder, determine how close stiffener B must be to stiffener A in order for the design to be acceptable according to AISC Specification.

11.11. Given the plate girder interior panel as shown, of A36 material. Using AISC Spec:
(a) Disregarding simultaneous bending stress, determine the shear capacity (kips) of the given panel. What percent of the capacity represents strength prior to elastic buckling (beam action) and what percent comes from "tension-field action"?
(b) If the bending tensile stress is 21 ksi at the extreme fiber of the web, what is the allowable shear capacity (kips)?

11.12. The cross section near the end of a simply supported girder consists of flange plates, $1\frac{1}{2}\times18$, and a web plate, $\frac{3}{8}\times90$ ($I = 135,780$ in.^{4}). The girder is of A36 steel and has lateral support of the compression flange every 12 ft. Use AISC Specification.
(a) Determine the safe moment capacity of this section.
(b) Determine the distance from the reaction to the first intermediate stiffener if the end reaction is 320 kips.

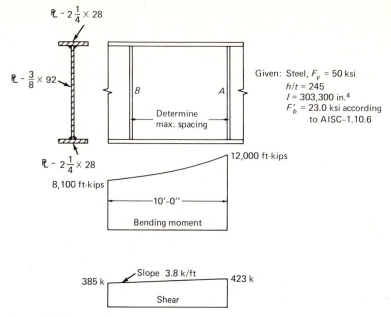

Given: Steel, F_y = 50 ksi
h/t = 245
I = 303,300 in.⁴
F_b' = 23.0 ksi according
to AISC–1.10.6

Prob. 11.10

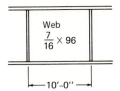

Prob. 11.11

(c) Using E70 electrodes, determine the necessary flange to web welding in the end panel. If intermittent welding can be used, indicate the length of weld segment and the pitch distance.

Design

11.13. Select the main cross section for a plate girder of A36 steel carrying a live load of 3 kips/ft on a simple span of 90 ft. There is no specific clearance limitation and the girder should be as minimum weight as possible. The ratio of flange width to girder depth should be about 0.15 to 0.20. Lateral support of the compression flange is provided every 15 ft.

11.14. Repeat Prob. 11.13 using the following span length, live loading, and material yield stress:

Span	Live Loading	Yield Stress	Lateral Support
(a) 60 ft	3 kips/ft	$F_y = 36$ ksi	Continuous
(b) 70 ft	3 kips/ft	$F_y = 50$ ksi	Every 17.5 ft
(c) 80 ft	3 kips/ft	$F_y = 60$ ksi	Every 10 ft
(d) 60 ft	5 kips/ft	$F_y = 36$ ksi	Every 12 ft
(e) 70 ft	5 kips/ft	$F_y = 50$ ksi	Every 10 ft
(f) 80 ft	5 kips/ft	$F_y = 60$ ksi	Every 10 ft

11.15. Redesign the girder of Probs. 11.2 through 11.7 using a deeper web plate as dictated by the optimum depth equations discussed in Sec. 11.13. Design of connections and bearing stiffeners may be omitted. Compute average weight per foot for the entire girder; for this purpose consider bearing stiffener plates to weigh four times that of intermediate stiffener plates.

11.16. Design a two-span 140 ft–140 ft continuous welded plate girder to support uniform dead load of 1.25 kips/ft in addition to the girder weight, and a uniform live load of 2.25 kips/ft plus two concentrated loads of 60 kips (40 percent live load) located in each span 35 ft from the supports. Consider lateral support every 35 ft (supports, concentrated loads, and midspan). Use AISC Specification.

(a) Consider live load to be applied as necessary to give maximum range of stresses.

(b) Use $F_y = 50$ ksi steel for negative moment zone and A36 steel for positive moment zone.

(c) Consider $\frac{5}{16}$ in. as the minimum web thickness available.

(d) Use A36 steel for stiffeners.

(e) Specify intermittent welding for connections if any material saving results, even though a cost analysis may later dictate continuous welding.

(f) Submit design sketch to approximate scale on $8\frac{1}{2} \times 11$ paper showing all final decisions.

(g) Compute the total average weight per foot of the girder, including all stiffeners. Assume one splice is required in each span because of the lengths involved. An extra butt welded flange splice (two flanges) should be treated as adding 10 lb/ft to the average weight. (This may approximate the added cost effect of such splices.) Web splices in excess of one required per span should be considered as adding 6 lb/ft to the average girder weight. Economical designs should have an adjusted average girder weight under 450 lb/ft.

11.17. Redesign the girder of Prob. 11.16, except consider the spans 125 ft and the concentrated loads located 25 ft from the supports. Lateral support exists every 25 ft. AISC Specification. Other problem requirements are retained, except as indicated below:

(d) Stiffeners may be either A36 or $F_y = 50$ ksi.

(g) Average girder minimum weight suggestion of 450 lb/ft should be disregarded.

11.18. Redesign the girder of Prob. 11.16, except consider the spans 200 ft without any concentrated loads. Lateral support is continuously provided by the floor system (i.e., it is not provided by intermediate stiffener attachments). Use AISC Specification. Other problem requirements are retained, except as indicated below:

(b) Use material with $F_y = 100$ ksi for highest moment region, $F_y = 50$ ksi for medium moment, and $F_y = 36$ ksi for lowest moment regions.

(d) Omit restrictions on stiffener material.

(g) Consider that 85-ft maximum length sections will be spliced with field bolted splices (i.e., minimum of 5 pieces will be required). Extra splices should be considered for comparison purposes by adjusting the average weight as per Prob. 11.16. Disregard the reference to 450 lb/ft.

Review of Theory

11.19. Explain the physical significance of the following C_v values for shear in plate girders. Describe the ultimate behavior based on shear strength in each case.

(a) $C_v \leq 0.8$

(b) $0.8 < C_v \leq 1.0$

(c) $C_v > 1.0$

11.20. What is the specific behavior that is prevented when h/t is kept below $760/\sqrt{F_b}$?

11.21. Show by a diagram of forces in equilibrium why stiffener spacing requirements are different for end panels and panels containing large holes than they are for interior panels of a plate girder.

11.22. In deriving AISC Formula (1.10–3), the force in the stiffener was obtained and then divided by the yield stress to obtain the required area. Why was the yield stress used instead of an allowable stress in the working range, such as $0.60F_y$?

11.23. Why is a reduced allowable flexural stress used for $h/t > 760/\sqrt{F_b}$ in AISC Formula (1.10–5), when actually at ultimate moment the extreme fiber is assumed stressed to F_y for h/t both greater and less than $760/\sqrt{F_b}$? In other words, why does the AISC formula reduce the allowable stress when actually the stress at ultimate does reach F_y?

11.24. Consider the AISC–1.10.2 limitation on the clear distance between flanges:

(a) Show the stress condition on the web plate that gives rise to the limitation equation.

(b) State explicitly what the number 16.5 in the equation means.

(c) If h/t equals the limit value, show the effective girder cross section that might be used to compute moment capacity.

11.25. On a plate girder with $F_y = 50$ ksi, if the web $h/t = 185$, what will happen to the girder before ultimate moment capacity is reached? Particularly describe what happens in a panel between intermediate stiffeners where the moment is high and the shear is low. Be specific and use a sketch.

11.26. Refer to AISC–1.10.5.2.

(a) Explain by diagram the meaning of C_v and state why two different expressions are used for it. Quotations or formulas from the AISC Manual are an inadequate answer.

(b) Why is AISC Formula (1.10–2) restricted to values of C_v less than 1.0?

(c) What does the second term of AISC Formula (1.10–2) represent?

12
Combined Bending
and Axial Load

12.1 INTRODUCTION

Nearly all members in a structure are subjected to both bending moment
and axial load—either tension or compression. When the magnitude of
one or the other is relatively small its effect is usually neglected and the
member is designed as a beam, as an axially loaded column, or as a
tension member. For many situations neither effect can properly be
neglected and the behavior under the combined loading must be considered in design. The member subjected to axial compression and bending
is referred to as a *beam-column*, and is the major element treated in this
chapter. The general subject of strength and stability considerations and
design procedures for beam-columns has been extensively treated by
Massonnet [1] and Austin [2], and a recent summary is given in the *SSRC
Guide* [3].

Since bending is involved, all of the factors considered in Chapters 7
and 9 apply here also; particularly, the stability related factors, such as
lateral-torsion buckling and local buckling of compression elements.
When bending is combined with axial tension, the chance of instability is
reduced and yielding usually governs the design. For bending in combination with axial compression, the possibility of instability is increased with

all of the considerations of Chapter 6 applying. Furthermore, when axial compression is present, a secondary bending moment arises equal to the axial compression force times the deflection.

A number of categories of combined bending and axial load along with the likely mode of failure may be summarized as follows:

1. Axial tension and bending; failure usually by yielding.
2. Axial compression and bending about one axis; failure by instability in the plane of bending, without twisting. (Transversely loaded beam-columns that are stable with regard to lateral-torsional buckling are an example of this category.)
3. Axial compression and bending about the strong axis: failure by lateral-torsional buckling.
4. Axial compression and biaxial bending—torsionally stiff sections; failure by instability in one of the principal directions. (W shapes are usually in this category.)
5. Axial compression and biaxial bending—thin-walled open sections; failure by combined twisting and bending on these torsionally weak sections.
6. Axial compression, biaxial bending, and torsion; failure by combined twisting and bending when plane of bending does not contain the shear center.

Because of the number of failure modes, no simple design procedure is likely to account for such varied behavior. Design procedures generally are in one of three categories: (1) limitation on combined stress; (2) semi-empirical interaction formulas, based on working stress procedures, and (3) semi-empirical interaction procedures based on ultimate strength. Limitations on combined stress ordinarily cannot provide a proper criterion unless instability is prevented, or large safety factors are used. Interaction equations come closer to describing the true behavior since they account for the stability situations commonly encountered. The AISC Specification formulas for beam-columns are of the interaction type.

12.2 DIFFERENTIAL EQUATION FOR AXIAL COMPRESSION AND BENDING

In order to understand the behavior, the basic situation of case 2, Sec. 12.1, will be treated. Failure by instability in the plane of bending is assumed. Consider the general case of Fig. 12.2.1, where the lateral loading $w(z)$ in combination with any end moments, M_1 or M_2, constitute the primary bending moment M_i which is a function of z. The primary moment causes the member to deflect, y, given rise to a secondary moment Py. Stating the moment M_z at the section z of Fig. 12.2.1, gives

$$M_z = M_i + Py = -EI\frac{d^2y}{dz^2}$$

(12.2.1)

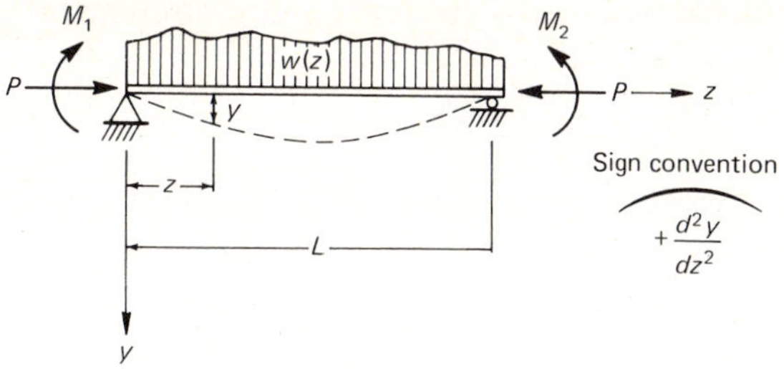

Fig. 12.2.1   General loading of beam-column.

for sections with constant EI, and dividing by EI gives

$$\frac{d^2y}{dz^2} + \frac{P}{EI}\, y = -\frac{M_i}{EI} \tag{12.2.2}$$

For design purposes, the general expression for moment M_z is of greater importance than the deflection y. Differentiating Eq. 12.2.2 twice gives

$$\frac{d^4y}{dz^4} + \frac{P}{EI}\frac{d^2y}{dz^2} = -\frac{1}{EI}\frac{d^2M_i}{dz^2} \tag{12.2.3}$$

From Eq. 12.2.1,

$$\frac{d^2y}{dz^2} = -\frac{M_z}{EI} \quad \text{and} \quad \frac{d^4y}{dz^4} = -\frac{1}{EI}\frac{d^2M_z}{dz^2}$$

Substitution in Eq. 12.2.3 gives

$$-\frac{1}{EI}\frac{d^2M_z}{dz^2} + \frac{P}{EI}\left(\frac{-M_z}{EI}\right) = -\frac{1}{EI}\frac{d^2M_i}{dz^2}$$

or, simplifying and letting $k^2 = P/EI$,

$$\frac{d^2M_z}{dz^2} + k^2M_z = \frac{d^2M_i}{dz^2} \tag{12.2.4}$$

which is of the same form as the deflection differential equation, Eq. 12.2.2.

The homogeneous solution for Eq. 12.2.4 is

$$M_z = A \sin kz + B \cos kz$$

as first discussed in Sec. 6.2. To this must be added the particular solution that will satisfy the right-hand side of the differential equation. Since $M_i = f(z)$, where $f(z)$ is usually a polynomial in z, the particular solution will be of the same form; thus the complete solution may be written

$$M_z = A \sin kz + B \cos kz + f_1(z) \tag{12.2.5}$$

where $f_1(z) = $ value of M_z satisfying Eq. 12.2.4. When M_z is a continuous function, the maximum value of M_z may be found by differentiation:

$$\frac{dM_z}{dz} = 0 = Ak \cos kz - Bk \sin kz + \frac{df_1(z)}{dz} \qquad (12.2.6)$$

For most ordinary loading cases, such as concentrated loads, uniform loads, end moments, or combinations thereof, it may be shown that

$$\frac{df_1(z)}{dz} = 0$$

in which case a general expression for maximum M_z can be established; from Eq. 12.2.6,

$$Ak \cos kz = Bk \sin kz$$

$$\tan kz = \frac{A}{B} \qquad (12.2.7)$$

At maximum M_z,

$$\sin kz = \frac{A}{\sqrt{A^2 + B^2}}, \qquad \cos kz = \frac{B}{\sqrt{A^2 + B^2}} \qquad (12.2.8)$$

and substitution of Eqs. 12.2.8 into Eq. 12.2.5 gives

$$M_{z\,\text{max}} = \frac{A^2}{\sqrt{A^2 + B^2}} + \frac{B^2}{\sqrt{A^2 + B^2}} + f_1(z)$$

$$= \sqrt{A^2 + B^2} + f_1(z) \qquad (12.2.9)$$

It is noted that whenever $df_1(z)/dz \neq 0$, Eq. 12.2.6 must be solved for kz and the result substituted into Eq. 12.2.5.

Case 1—Unequal End Moments Without Transverse Loading

Referring to Fig. 12.2.2, the primary moment M_i may be expressed

$$M_i = M_1 + \frac{M_2 - M_1}{L} z \qquad (12.2.10)$$

Since

$$\frac{d^2 M_i}{dz^2} = 0$$

Eq. 12.2.4 becomes a homogeneous equation, in which case $f_1(z) = 0$ for Eq. 12.2.5. The maximum moment, Eq. 12.2.9, is

$$M_{z\,\text{max}} = \sqrt{A^2 + B^2} \qquad (12.2.11)$$

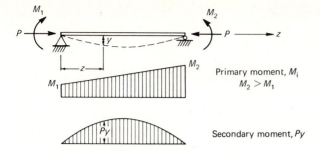

Fig. 12.2.2 Case 1—end moments without transverse loading.

The constants A and B are evaluated by applying the boundary conditions to Eq. 12.2.5. The general equation is

$$M_z = A \sin kz + B \cos kz$$

and the conditions are

(1) at $z = 0$, $M_z = M_1$

$$\therefore B = M_1$$

(2) at $z = L$, $M_z = M_2$

$$M_2 = A \sin kL + M_1 \cos kL$$

$$\therefore A = \frac{M_2 - M_1 \cos kL}{\sin kL}$$

so that

$$M_z = \left(\frac{M_2 - M_1 \cos kL}{\sin kL}\right) \sin kz + M_1 \cos kz \qquad \text{(12.2.12)}$$

and

$$M_{z\,max} = \sqrt{\left(\frac{M_2 - M_1 \cos kL}{\sin kL}\right)^2 + M_1^2}$$

$$= M_2 \sqrt{\frac{1 - 2(M_1/M_2)\cos kL + (M_1/M_2)^2}{\sin^2 kL}} \qquad \text{(12.2.13)}$$

Case 2—Transverse Uniform Loading

Referring to Fig. 12.2.3, the primary moment M_i may be expressed as

$$M_i = \frac{w}{2} z(L - z) \qquad \text{(12.2.14)}$$

Since

$$\frac{d^2 M_i}{dz^2} = -w$$

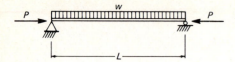

Fig. 12.2.3 Case 2—transverse uniform loading.

$f_1(z) \neq 0$; the particular solution for the differential equation is required. Let $f_1(z) = C_1 + C_2 z$; i.e., any polynomial. Substitute the particular solution into Eq. 12.2.4,

$$\frac{d^2[f_1(z)]}{dz^2} = 0$$

$$0 + k^2(C_1 + C_2 z) = -w$$

Thus

$$C_1 = -w/k^2$$

$$C_2 = 0$$

Equation 12.2.5 then becomes

$$M_z = A \sin kz + B \cos kz - w/k^2 \qquad (12.2.15)$$

Applying the boundary conditions,

(1) at $z = 0$, $M_z = 0$

$$0 = B - w/k^2$$

$$\therefore B = w/k^2$$

(2) at $z = L$, $M_z = 0$

$$0 = A \sin kL + \frac{w}{k^2} \cos kL - \frac{w}{k^2}$$

$$\therefore A = \frac{w}{k^2}\left(\frac{1 - \cos kL}{\sin kL}\right)$$

Since

$$\frac{df_1(z)}{dz} = 0,$$

Eq. 12.2.9 gives maximum moment,

$$M_{z\,\text{max}} = \frac{w}{k^2}\sqrt{\left(\frac{1 - \cos kL}{\sin kL}\right)^2 + 1} - \frac{w}{k^2}$$

$$= \frac{w}{k^2}\left(\sec \frac{kL}{2} - 1\right)$$

$$= \frac{wL^2}{8}\underbrace{\left(\frac{8}{(kL)^2}\right)\left(\sec \frac{kL}{2} - 1\right)}_{\substack{\text{magnification factor due} \\ \text{to axial compression}}} \qquad (12.2.16)$$

Case 3—Equal End Moments Without Transverse Loading (Secant Formula)

Consider that $M_1 = M_2 = M$ in Fig. 12.2.2, in which case Eq. 12.2.13 becomes

$$M_{z\,max} = M\sqrt{\frac{2(1-\cos kL)}{\sin^2 kL}} \qquad (12.2.17)$$

$$= M\sqrt{\frac{2(1-\cos kL)}{1-\cos^2 kL}}$$

$$= M\left(\frac{1}{\cos kL/2}\right)$$

$$= M \sec\frac{kL}{2} \qquad (12.2.18)$$

which was used in Eq. 6.11.3.

Loading a beam-column with constant moment over its length causes a constant compression along one entire flange and constitutes the most severe loading on such a member. In view of this it would always be conservative to multiply the maximum primary moment due to *any* loading by $\sec kL/2$, which is excessively conservative in most cases.

12.3 MOMENT MAGNIFICATION—   SIMPLIFIED TREATMENT FOR MEMBERS IN SINGLE CURVATURE WITHOUT END TRANSLATION

As an alternate to the differential equation approach, a simple approximate procedure is satisfactory for many common situations.

Assume a beam-column is subject to lateral loading $w(z)$ that causes a deflection δ_0 at midspan, as shown in Fig. 12.3.1. The secondary bending moment may be assumed to vary as a sine curve, which is nearly correct for members with no end restraint whose primary bending moment and deflection are maximum at midspan.

The portion of the midspan deflection y_1, due to the secondary bending moment, equals the moment of the M/EI diagram between the support and midspan (shaded portion of Fig. 12.3.1) taken about the support, according to the moment-area principle:

$$y_1 = \frac{P}{EI}(y_1+\delta_0)\left(\frac{L}{2}\right)\frac{2}{\pi}\left(\frac{L}{\pi}\right) = (y_1+\delta_0)\frac{PL^2}{\pi^2 EI} \qquad (12.3.1)$$

or

$$y_1 = (y_1+\delta_0)\frac{P}{P_e} \qquad (12.3.2)$$

where $P_e = \pi^2 EI/L^2$. Solving for y_1,

$$y_1 = \delta_0\left[\frac{P/P_e}{1-\dfrac{P}{P_e}}\right] = \delta_0\left(\frac{\alpha}{1-\alpha}\right) \qquad (12.3.3)$$

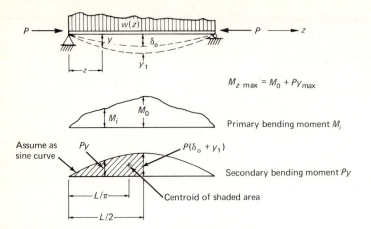

$$M_{z\,max} = M_0 + Py_{max}$$

Fig. 12.3.1 Primary and secondary bending moment.

where $\alpha = P/P_e$. Since y_{max} is the sum of δ_0 and y_1,

$$y_{max} = \delta_0 + y_1 = \delta_0 + \delta_0\left(\frac{\alpha}{1-\alpha}\right) = \frac{\delta_0}{1-\alpha} \qquad (12.3.4)$$

The maximum bending moment including the axial effect becomes

$$M_{z\,max} = M_0 + Py_{max} \qquad (12.3.5)$$

Substituting the expression for y_{max} in Eq. 12.3.5 and setting $P = \alpha P_e = \alpha \pi^2 EI/L^2$, Eq. 12.3.5 may be written as the primary moment M_0 multiplied by a magnification factor A_m; thus

$$M_{z\,max} = M_0 A_m \qquad (12.3.6)$$

where

$$A_m = \frac{C_m}{1-\alpha} \qquad (12.3.7)$$

and

$$C_m = 1 + \left(\frac{\pi^2 EI \delta_0}{M_0 L^2} - 1\right)\alpha = 1 + \psi\alpha \qquad (12.3.8)$$

Example 12.3.1

Compare the differential equation magnification factor for the loading of Fig. 12.2.3, Eq. 12.2.16, with the approximate value, Eq. 12.3.7.

SOLUTION

For the differential equation,

$$A_m = \text{magnification factor} = \frac{2}{(kL/2)^2}\left(\sec\frac{kL}{2} - 1\right) \qquad (a)$$

where

$$\frac{kL}{2} = \frac{L}{2}\sqrt{\frac{P}{EI}} = \frac{\pi}{2}\sqrt{\alpha}$$

For the approximate solution,

$$A_m = \frac{C_m}{1-\alpha}$$

$$\delta_0 = \frac{5wL^4}{384EI}; \qquad M_0 = \frac{wL^2}{8}$$

$$\frac{\delta_0}{M_0} = \frac{5L^2}{48EI}$$

$$C_m = 1 + \left(\frac{\pi^2 EI}{L^2}\frac{5L^2}{48EI} - 1\right)\alpha$$

$$= 1 + 0.028\alpha$$

$$A_m = \frac{1 + 0.028\alpha}{1 - \alpha} \tag{b}$$

α	sec $kL/2$	Eq. (a)	Eq. (b)
0.1	1.137	1.114	1.114
0.2	1.310	1.257	1.257
0.3	1.533	1.441	1.441
0.4	1.832	1.686	1.685
0.5	2.252	2.030	2.028
0.6	2.884	2.546	2.542
0.7	3.941	3.405	3.399
0.8	6.058	5.125	5.112
0.9	12.419	10.284	10.253

Obviously, there is no significant difference between the differential equation solution and the approximate solution for this case.

12.4 MOMENT MAGNIFICATION—   MEMBERS SUBJECTED TO END MOMENTS ONLY; NO JOINT TRANSLATION

For the situation shown in Fig. 12.2.2, which has no transverse loading, the theoretical maximum moment is given by Eq. 12.2.13,

$$M_{z\,max} = M_2\sqrt{\frac{(M_1/M_2)^2 - 2(M_1/M_2)\cos kL + 1}{\sin^2 kL}} \qquad [12.2.13]$$

For this situation the maximum moment may be either (1) the larger end

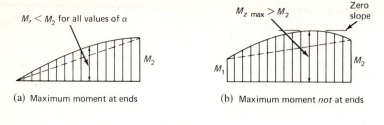

(a) Maximum moment at ends   (b) Maximum moment *not* at ends

(c) Equivalent uniform moment with
maximum magnified moment at midspan

Fig. 12.4.1 Primary plus secondary bending moment for members subjected to end moments only.

moment M_2 at the braced location (Fig. 12.4.1a), or (2) the magnified moment given by Eq. 12.2.13 which occurs at various locations out along the span (Fig. 12.4.1b), depending on the ratio M_1/M_2 and the value of α, since $kL = \pi\sqrt{\alpha}$. In order to make an analysis, one needs to know whether the maximum moment occurs at a location away from the support, and if so, the correct *distance*. To eliminate the need for such information, the concept of equivalent uniform moment (Fig. 12.4.1c) is used. Thus when investigating a member at a location away from the supported point, use of the equivalent moment assumes $M_{z\,max}$ to be at midspan.

To establish the equivalent moment M_E, let the solution for uniform moment, Eq. 12.2.17 with $M = M_E$, be equated with Eq. 12.2.13. One obtains

$$M_E = M_2 \sqrt{\frac{(M_1/M_2)^2 - 2(M_1/M_2)\cos kL + 1}{2(1 - \cos kL)}} \qquad (12.4.1)$$

By the procedure used in Example 12.3.1 it may be shown that for uniform moment, the magnification factor is obtained from Eq. 12.2.18:

$$A_m = \sec\frac{kL}{2} = \frac{1}{1-\alpha} \qquad (12.4.2)$$

and using the equivalent uniform moment, M_E, to replace M_1 and M_2, the full maximum moment may be expressed as

$$M_{z\,max} = M_E\left(\frac{1}{1-\alpha}\right) \qquad (12.4.3)$$

which when compared with Eq. 12.3.6 may be written

$$M_{z\,\text{max}} = C_m M_2 \left(\frac{1}{1-\alpha}\right)$$ (12.4.4)

where

$$
\begin{aligned}
C_m &= M_E/M_2 \\
&= \sqrt{\frac{(M_1/M_2)^2 - 2(M_1/M_2)\cos kL + 1}{2(1 - \cos kL)}}
\end{aligned}
$$ (12.4.5)

Equation 12.4.5 does not consider lateral-torsional buckling, or fully cover the double-curvature cases where M_1/M_2 lies between -0.5 and -1.0. The actual failure of members bent in double curvature with bending-moment ratios -0.5 to -1.0 is generally one of "unwinding" through from double to single curvature in a sudden type of buckling, as discussed by Ketter [4], among others.

Massonnet [1], Horne [5], and the AISC Specification have suggested alternate expressions for practical use. Figure 12.4.2 shows Eq. 12.4.5 compared with the recommendations of Massonnet and the AISC. One should note that for a given value of α, the curve shown terminates when

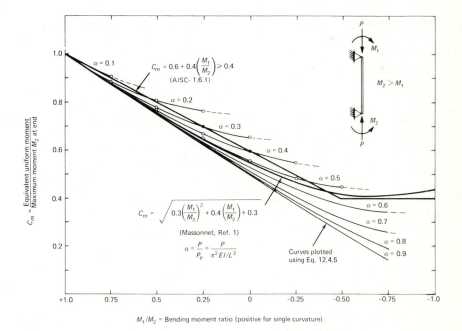

Fig. 12.4.2 Comparison of theoretical C_m with design recommendations for members subject to end moments only, without joint translation.

the moment M_2 at the end of the member exceeds the amplified moment. The most important situations are those where the magnified moment exceeds the moment at the end. The straight line recommended by AISC falls near the upper limit for C_m at any given bending moment ratio, and thus seems to be a realistic and simple approximation. In design, a safety factor is used which overestimates α. By not taking C_m as a function of this increased α, a higher value of C_m will be used; thus it is a conservative procedure.

12.5 MOMENT MAGNIFICATION— MEMBERS SUBJECT TO SIDESWAY

The unbraced frame is dealt with in Chapter 14, where the development is presented to show that the maximum moment M_0 computed in a standard analysis of an elastic frame should be magnified to account for sidesway buckling. One may write

$$M_{\text{max}} = M_0 A_m = M_0 \left(\frac{C_m}{1-\alpha} \right) \tag{12.5.1}$$

in the same manner as for the braced frame cases of Secs. 12.3 and 12.4.

Consider the sidesway situation of Fig. 12.5.1. Whatever the degree of restraint at the top and bottom of the two-story member, the deflection curve, and therefore the secondary bending moment (P times deflection) may be reasonably assumed as a sine curve, in which case the development used when no sidesway occurs (Fig. 12.3.1) is also valid here. Since $2L$ from Fig. 12.5.1 equals L for Fig. 12.3.1, Eq. 12.3.7 becomes

$$C_m = 1 + \left(\frac{\pi^2 EI \delta_0}{4L^2 M_0} - 1 \right) \alpha \tag{12.5.2}$$

The larger effective length ($2L$ instead of L) is also used in the computation of α. Next, referring to Fig. 12.5.1,

$$\delta_0 = \frac{(H/2)L^3}{3EI} \tag{12.5.3}$$

$$M_0 = \frac{HL}{2} \tag{12.5.4}$$

Substitution of Eqs. 12.5.3 and 12.5.4 into Eq. 12.5.2 gives

$$C_m = 1 + \left[\frac{\pi^2 EI}{4L^2} \left(\frac{HL^3}{6EI} \right) \left(\frac{2}{HL} \right) - 1 \right] \alpha$$

or

$$C_m = 1 + \left(\frac{\pi^2}{12} - 1 \right) \alpha = 1 - 0.18\alpha \tag{12.5.5}$$

as suggested in the AISC Commentary–1.6.1.

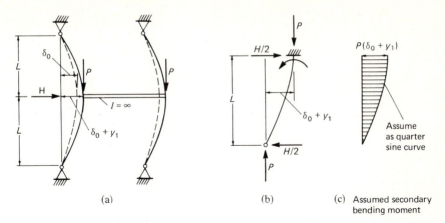

(a)   (b)   (c) Assumed secondary
                                                                          bending moment

Fig. 12.5.1   Beam-column with sidesway.

A comparison of a theoretical analysis with Eq. 12.5.5 is presented in Chapter 14.

The AISC Commentary–1.6.1 recommends but the specification, AISC–1.6.1, *requires* use of

$$C_m = 0.85 \qquad (12.5.6)$$

which seems to be *un*conservative.

12.6* ULTIMATE STRENGTH—INSTABILITY IN THE PLANE OF BENDING

Discussion is limited to the basic beam-column that has moment about one principal axis and which fails by instability in the plane of bending without twisting (case (2) of Sec. 12.1). A study of the differential equation solution will show that the axial effect and the moment effect cannot be computed separately and then combined by superposition. It is a nonlinear relationship. Furthermore, because of bending, as well as due to residual stress, some fibers reach yield before others. The ultimate capacity is reached when the stiffness of the members has been reduced, because of yielding under combined stress, to the state where instability occurs. Similar to the treatment in Secs. 6.4 through 6.6, the ultimate axial load is the critical load on the partially yielded member.

The analysis to determine the actual strength of a beam-column is complicated. First, the moment-curvature-axial compresson $(M{-}\phi{-}P)$ relationship must be developed. This may be done by assuming the yield penetration at various proportions of the depth of the member, measured from the extreme fiber having the highest compressive strain. For a given yield penetration γh, as shown in Fig. 12.6.1, there is a complete range of

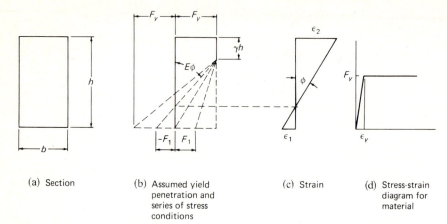

(a) Section   (b) Assumed yield penetration and series of stress conditions   (c) Strain   (d) Stress-strain diagram for material

Fig. 12.6.1 Rectangular member under axial compression and bending, for $\epsilon_1 \leq \epsilon_y$.

M–ϕ–P represented by the dashed line stress distributions. Any chosen position of the dashed line represents a value of ϕ, for which the volume of the stress solid gives the corresponding axial load P. The summation of the moments of the stress solid about a convenient axis gives the corresponding bending moment M. The result is an M–ϕ–P function for each yield penetration depth γh. This series of curves for a rectangular cross section is given in Fig. 12.6.2.

From the auxiliary curves of Fig. 12.6.2, a series of M–ϕ curves can be obtained for constant values of P/P_y. The graphical procedure is shown in Fig. 12.6.2 for $P/P_y = 0.6$. Start at $P/P_y = 0.6$, proceed to the right to intersect a line corresponding to a value of γ, say $\gamma = 0.2$; proceed upward to the $\gamma = 0.2$ line in the M/M_y portion of the curves and read off the value of M/M_y. Repeat for enough values of γ until a complete M/M_y vs ϕ/ϕ_y curve for the given P/P_y can be plotted. The results are shown in Fig. 12.6.3, which gives the beam-column strength for a short member where yielding controls.

For wide-flange sections, curves similar to Figs. 12.6.2 and 12.6.3, as well as the underlying equations, have been presented by Ketter, Kaminsky, and Beedle [6].

Once M–ϕ curves such as Fig. 12.6.3 have been obtained, a specific value of P/P_y and slenderness ratio KL/r is investigated to establish the value of M/M_y at which the member becomes unstable. As an example, Fig. 12.6.4 shows the result of making the analysis for $P/P_y = 0.3$ and $KL/r = 60$. Using θ, the end slope, as the reference deflection-related quantity, various values of M/M_y are applied to the member and the total end slope θ is determined by iteration. The primary moment M/M_y causes a deflection Δ, that in turn gives rise to the secondary bending moment $(P/P_y)\Delta$. A numerical integration method, such as Newmark's

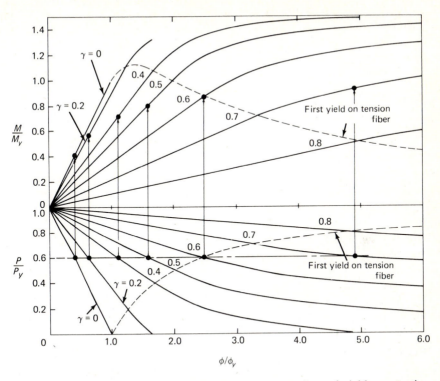

Fig. 12.6.2 Auxiliary curves of M–P–ϕ for constant values of yield penetration at compression face of rectangular beam.

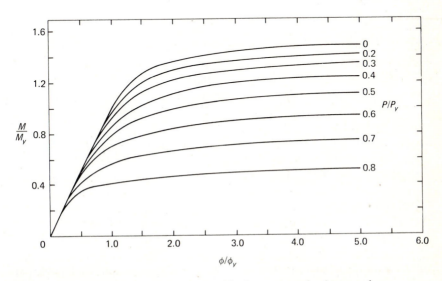

Fig. 12.6.3 Moment-curvature relationship for rectangular beam-column.

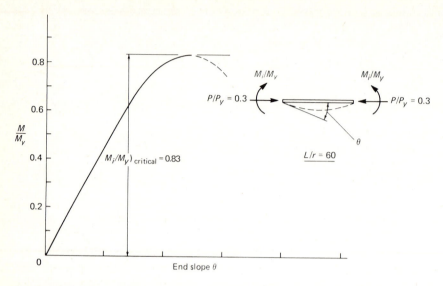

Fig. 12.6.4 Determining critical value for M_i/M_y for constant primary bending moment, with $P/P_y = 0.3$.

Method [7, 8], may be used. In such a method the span is divided into segments, the total moment, $M/M_y + (P/P_y)\Delta$, is evaluated at the ends of each segment. The curvature at each point is obtained from Fig. 12.6.3, or a similar figure. From the curvature the deflection is computed. Since the original Δ was assumed, the process is repeated until the computed deflection agrees with the assumed deflection. From the deflected shape the value of θ may be computed. Each point on Fig. 12.6.4 is obtained by solving for the deflection for different values of M/M_y. The critical value of M/M_y is the one for which the deflection of the member cannot be determined; that is, the value for which the member is unstable. In Fig. 12.6.4, $M/M_y = 0.83$ is the critical value obtained after computing a number of points on the M/M_y vs θ curve.

Although a rectangular section was used to obtain the figures presented here, the same procedure has been used [3] for I-shaped sections, including a linear distribution of residual stress across the flange width (as per Fig. 6.6.6a), with a maximum residual stress of $0.3F_y$. Using a W8×31 section which has reasonably typical cross-sectional properties, ultimate-strength interaction diagrams were developed by Ketter [4] and Ketter and Galambos [9] using the method illustrated herein.

An alternate procedure assuming the flanges to be thin has been used [10–13], which reduces somewhat the graphical or numerical steps. However, residual stress has not been included. Chen and Atsuta [41] have also dealt with the moment-curvature-thrust interaction and have presented interaction equations.

For illustrative purposes, ultimate-strength interaction curves are presented in Fig. 12.6.5 for wide-flange sections with residual stress (as represented by the W8×31) loaded in compression combined with various ratios of end moments. Each point required to establish these curves involved the process discussed in this section.

In studying the interaction curves of Fig. 12.6.5, the reader is reminded that the capacity P when $M = 0$ is based on the slenderness ratio,

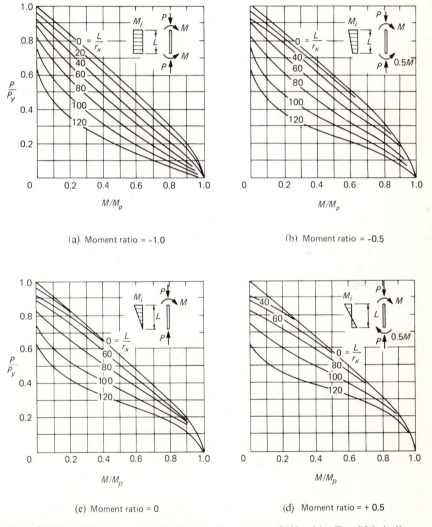

(a) Moment ratio = -1.0

(b) Moment ratio = -0.5

(c) Moment ratio = 0

(d) Moment ratio = + 0.5

Fig. 12.6.5 Ultimate strength interaction curves (W8×31, $F_y = 33$ ksi, linear residual stress $= F_r = 0.3F_y$, strong-axis bending) for braced frame members having various end-moment ratios. *Note*: For $F_y > 33$ ksi, use adjusted $L/r =$ (actual L/r) $\sqrt{F_y/33}$. (Adapted from Figs. 7–10 of Ketter [4])

KL/r_x. As stated at the beginning of the section, the member was assumed to fail by instability in the plane of bending.

Other studies of instability in the plane of bending have included the effects of transverse loading [4, 14–18].

12.7* ULTIMATE STRENGTH—FAILURE BY COMBINED BENDING AND TORSION

The ordinary beam-column unbraced over a finite length involves consideration of instability transverse or oblique to the plane of bending, involving torsional effects. This subject is an extension of lateral-torsional buckling in beams (Chapter 9) and involves both elastic and inelastic considerations. Interaction curves for a number of elastic buckling situations have been developed, including (1) I-section columns with eccentric end loads in the plane of the web [19]; (2) I-columns with unequal end moments but without restraint to rotation about principal axes at the ends of the member [20]; and (3) I-columns with unequal end moments— hinged at ends for rotation about strong axis, but elastically restrained for rotation about the weak axis at the ends [21]. An excellent summary of the topic is presented by Massonnet [1] which includes discussion of plastic effects.

A number of studies are available containing both analytical and experimental treatments of inelastic lateral-torsional buckling [22–25].

In lieu of a rigorous theoretical development which may be found in the literature, including the previously cited references, the authors will attempt to make use of concepts already derived to give a rational basis for explaining failure by combined bending and torsion. A similar discussion is presented by Massonnet [1].

Recall the Euler equation developed in Sec. 6.2:

$$P_{cr} = \frac{\pi^2 EI}{L^2} \qquad [6.2.7]$$

where in the development it was assumed that the member is in a slightly bent position. In other words, failure was assumed to occur by bending only. In Sec. 6.12 it was pointed out that in the slightly bent position the shear equals the buckling load P times the slope of the deflection curve. In Sec. 8.4, it was noted that only when flexural shears act in planes passing through the shear center is there flexure without torsion. Furthermore, in Sec. 7.8, it was shown that only when $I_{xy} = 0$ does the member bend in the same plane that the flexural shears act. Thus the basic Euler expression, Eq. 6.2.7, implies buckling by bending about the weak axis of a member having at least one axis of symmetry ($I_{xy} = 0$); and furthermore, that there is no simultaneous twisting, which means the center of gravity position CG and the shear center SC coincide. Such a condition is true of a rectangle or an I-shaped member (see Fig. 12.7.1a).

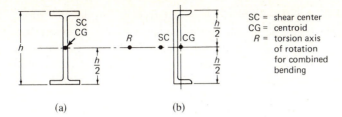

Fig. 12.7.1 Location of geometric axes compared for I-shaped section and channel.

In Sec. 8.11, torsional buckling of an axially loaded compression member was treated. When, as was assumed in that development, the shear center and the centroid coincide, torsional buckling is a possibility, but it is not coupled or linked to the buckling failure by bending. Upon comparison of the homogeneous differential equation for torsion, Eq. 8.5.12, when differentiated once with respect to z,

$$\frac{d^4\phi}{dz^4} - \lambda^2 \frac{d^2\phi}{dz^2} = 0$$

with Eq. 8.11.7 for torsional buckling,

$$\frac{d^4\phi}{dz^4} + p^2 \frac{d^2\phi}{dz^2} = 0 \qquad [8.11.7]$$

it may be noted they are identical if

$$p^2 = -\lambda^2$$

or

$$\left(\frac{\sigma_z I_p - GJ}{EC_w}\right)_{\substack{\text{torsion}\\\text{buckling}}} = \left(-\frac{GJ}{EC_w}\right)_{\text{torsion}} \qquad (12.7.1)$$

Thus one may observe from Eq. 12.7.1,

$$(GJ)_{\substack{\text{torsion}\\\text{without}\\\text{compression}}} = (GJ - \sigma_z I_p)_{\substack{\text{torsion with}\\\text{axial compression}}} \qquad (12.7.2)$$

which indicates that axial compression reduces the effective torsional rigidity. One may conclude torsional buckling is likely when the polar moment of inertia I_p is large compared to torsional rigidity GJ.

When the shear center SC and the center of gravity CG do not coincide, as in Fig. 12.7.1b, buckling occurs by bending (with incremental flexural shears acting through CG) in combination with torsion about SC.

The combined effect is a torsional failure about an axis at R, located more distant than SC from CG and in the same direction. The coupling effect to produce combined bending and torsion is proportional to the distance SC–CG. Figure 12.7.1 gives two extremes, (a) with no coupling effect and (b) with large coupling effect. Locations of SC and CG for some other sections are given in text Appendix Table A2.

It may now be observed that an applied moment combined with axial compression, or an axial compression eccentrically applied along a principal axis produces the same effect as the axial load alone produces on the channel, i.e., a coupled torsional-bending effect.

Figure 12.7.2 illustrates what happens. Note that the behavior is identical to that discussed in Chapter 9. The only difference is that here the torsional moment arises from shear, $P\,dx/dz$, which in turn arises from the axial load; whereas in Chapter 9, the torsional moment was a component, $M_0\,dx/dz$, of the moment applied in the yz plane. The loading involving the beam-column is the more severe one.

Tests by Massonnet [1] and computer studies [22, 23] including inelastic effects indicate that interaction diagrams similar to those in Fig. 12.6.5 will result when including lateral-torsional buckling. The main

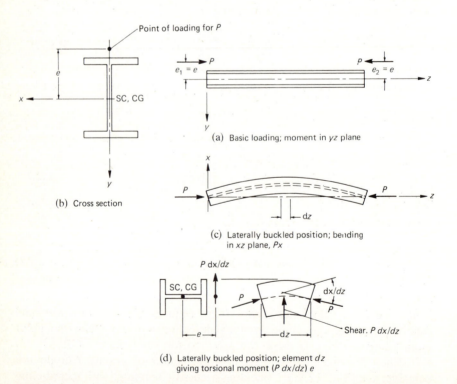

(a) Basic loading; moment in yz plane

(b) Cross section

(c) Laterally buckled position; bending in xz plane, Px

(d) Laterally buckled position; element dz giving torsional moment $(P\,dx/dz)\,e$

Fig. 12.7.2 Combined bending and torsion failure; lateral-torsional failure.

difference is that P/P_y will be lower when based on KL/r_y (weak axis) rather than on KL/r_x (strong axis).

Torsional-Flexural Buckling of Thin-Walled Open Sections

As an extension of the failure mode illustrated by Fig. 12.7.2 where the doubly symmetrical section is subjected to flexure and torsion, consider the singly-symmetric sections such as the channel where the shear center (*SC*) and the center of gravity (*CG*) do not coincide (Fig. 12.7.1b). Instability under combined bending and axial compression on such sections involves the additional complication that there is a torsional moment acting even *without* the member being in the slightly buckled position. The differential equation development has been given by Pekoz and Winter [26], Pekoz and Celebi [27], and the topic has been well summarized by Yu [28]. The AISI Specification [29] provides detailed rules for design to include the possibility of torsional-flexural buckling.

12.8 ULTIMATE STRENGTH—INTERACTION EQUATIONS

The interaction curves discussed and illustrated in Sec. 12.6 may be approximated by simple interaction equations.

Case 1—No Instability

For the braced location where instability cannot occur (i.e., $Kl/r = 0$), the uppermost curves of Fig. 12.6.6 apply and may be approximated by

$$\frac{P}{P_y} + \frac{M}{1.18M_p} = 1.0 \tag{12.8.1}$$

and $M/M_p \leq 1.0$. In the above equation, $P_y = A_g F_y$ and $M_p = $ maximum moment capacity of the member in the absence of axial load (equals the plastic moment for all cases where premature local buckling is prevented). The comparison with the theoretical result in Fig. 12.8.1 shows Eq. 12.8.1 to be a good approximation.

Case 2—Instability in the Plane of Bending

The curves of Fig. 12.6.6 for various combinations of moments and values of L/r_x may be approximated by

$$\frac{P}{P_{cr}} + \frac{M_E}{M_p(1 - P/P_e)} = 1.0 \tag{12.8.2}$$

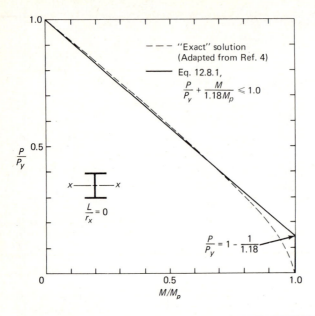

Fig. 12.8.1 "Exact" ultimate strength interaction curve for typical wide-flange sections (including residual stress) compared with interaction equation—Case 1, no instability.

where

P_{cr} = buckling strength under axial load based on slenderness ratio for the axis of bending

M_E = Eq. 12.4.1 (or its alternates from Fig. 12.4.2)

$P_e = \pi^2 EI/L^2$

Massonnet [1] has shown that Eq. 12.8.2 is a good approximation by comparing it with the curves of Galambos and Ketter [9]. A comparison in Fig. 12.8.2 of Eq. 12.8.2 with some curves from Fig. 12.6.5a shows the correlation.

For primary bending moment from transverse loading, Lu and Kamalvand [14] have shown that when M_E is replaced by $C_m M_i$, using C_m as given by Eq. 12.3.7, Eq. 12.8.2 is also a good representation for the "exact" solutions. Thus, in general, the interaction equation may be written

$$\frac{P}{P_{cr}} + \frac{C_m M_i}{M_p(1 - P/P_e)} = 1.0 \qquad (12.8.3)$$

for all cases of instability in the plane of bending.

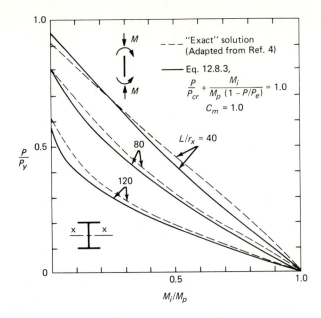

Fig. 12.8.2 "Exact" ultimate strength interaction curve for typical wide-flange sections (including residual stress) compared with interaction equation—Case 2, instability in the plane of bending ("exact" solution from Fig. 7 of Ref. 4). *Note:* For actual use, use adjusted $L/r = $ (actual L/r) $\sqrt{F_y/33}$.

Case 3—Instability by Lateral-Torsional Buckling

Massonnet [1] has shown that with only slight error the form of Eq. 12.8.2 may also be used for this case. P_{cr} for this case may be governed by the slenderness ratio for the axis normal to the axis of bending. Further, since lateral-torsional buckling as a beam may occur for a moment less than M_p, use M_m instead of M_p in Eq. 12.8.3. Using the various definitions of C_m from Secs. 12.3 through 12.5, the general notation $C_m M_i$ should be used. Thus for instability under Cases 2 or 3, the following interaction equation may be considered to apply:

$$\frac{P}{P_{cr}} + \frac{M_i C_m}{M_m(1 - P/P_e)} = 1.0 \qquad (12.8.4)$$

where $P = $ applied axial compression load

$M_i = $ applied primary bending moment

$P_{cr} = A_g F_{cr} = $ ultimate strength for axially loaded compression member (Eq. 6.7.3 for F_{cr} would be a rational value)

$M_m = $ maximum resisting moment in the absence of axial load. For adequately braced members of low slenderness ratios where local buckling is precluded, $M_m = M_p$.

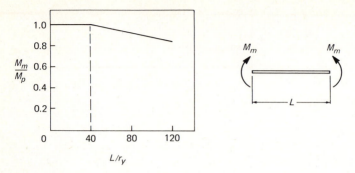

Fig. 12.8.3 Lateral buckling moment for uniform bending. (From Ref. 30; see also Ref. 25)

$$C_m = \text{factors discussed in Secs. 12.3–12.5}$$

$$P_e = \frac{\pi^2 EI}{L^2} = \frac{\pi^2 EP_y}{F_y(L/r)^2}$$

The reduction in ultimate moment capacity M_m when lateral-torsional buckling influences the ultimate strength interaction may be computed as $F_{cr}S_x$ using the SSRC "Basic" design procedure of Sec. 9.4 with no safety factor applied. This will be consistent with working stress procedures involving AISC Formulas (1.5–6) and (1.5–7).

According to Driscoll et al. (Ref. 30, p. 4.25), such a procedure is overly conservative. According to Ref. 30, "The behavior of beam-columns after lateral-torsional buckling has not yet been fully explored and available information is not sufficient to evaluate the effect of lateral-torsional buckling on rotation capacity. The limited results seem to indicate that for columns of low slenderness ratios ($L/r = 40$ or less) the end moment may not be reduced from the in-plane values significantly and drop off suddenly until a large rotation is reached." In other words, in real structures *without* pin-ended beam-columns the lateral-torsional buckling is restrained and the moment capacity does not drop off as drastically as indicated by the theoretical expressions developed in Chapter 9.

Figure 12.8.3 shows an empirical relationship for the reduction in ultimate moment.

12.9 BIAXIAL BENDING

The ultimate strength of members under axial compression and biaxial bending has received little attention until the last few years. Computer studies such as those of Birnstiel and Michalos [31], Culver [32, 33], Harstead, Birnstiel, and Leu [34], Syal and Sharma [35], Santathadaporn and Chen [36], and Chen and Atsuta [37] illustrate that even with a number of simplifying assumptions, the analysis is one of considerable complexity. Some tests have been performed [38] which, though limited, have shown agreement with computer studies. Tebedge and Chen [40] have given interaction surfaces in the form of tables for design. The status of work on biaxial bending of compression members is summarized well by Chen and Santathadaporn [39].

Simple plastic theory becomes inadequate when moments exist about two principal axes. When only one moment exists, plastic behavior is exhibited no matter what the value of axial compression. Figure 12.6.3 showed that under such loading the effective plastic moment reduces as axial compression increases, but plastic behavior does occur.

Upon application of an additional moment about the other principal axis, one might consider an interaction surface relating, P, M_x, and M_y. Even for ideal elastic-plastic material, however, present plastic analysis theorems neglect the influence of deformation on equilibrium. For zero length compression members, the concept of an interaction surface (see Fig. 12.9.1) may be thought of as a first step to the ultimate strength analysis under biaxial bending.

While few designers concern themselves greatly about the sequence of load application, nevertheless loading sequence affects ultimate strength. This is also true for uniaxial bending and compression, but it has less effect on that case than for the biaxial loading.

Figure 12.9.2 illustrates several loading sequences to reach point A, a particular value of P, M_x, and M_y. Point A may be reached by the following paths:

(1) Apply P first, then M_y, then M_x (path 0–1–2–A); (2) apply P first, then apply M_x and M_y proportionally (path 0–1–A); (3) apply P, M_x and

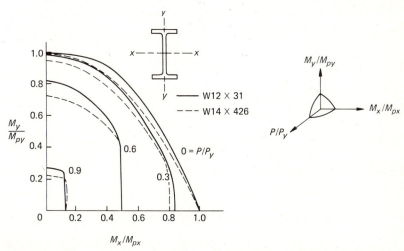

Fig. 12.9.1   Contours on interaction surface for short members where instability does not occur. (Adapted from Birnstiel [38])

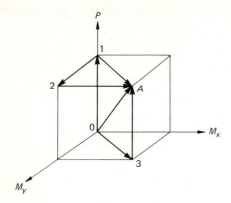

Fig. 12.9.2 Paths of loading for biaxial bending combined with axial force. (Adapted from Chen and Santathadaporn [39])

M_y by increasing magnitude in constant proportion (path $O–A$); (4) apply M_x and M_y in constant proportion, then apply P (path $O–3–A$). Other combinations are possible and in general the loading may become applied via any path through space to get from O to A on Fig. 12.9.2. A given section will exhibit a different strength for each path of loading. Nearly all investigators to date (1979) have used proportional loading (path $O–A$) [38].

To conclude, the ultimate strength of compression members under biaxial bending is not sufficiently well known to make use of it for plastic design of rigid space frames; therefore plastic design should be restricted to planar structures, or ones for which planar behavior represents a reasonable approximation.

For lack of adequate investigation, an interaction formula, such as Eq. 12.8.4, is usually assumed to apply for biaxial bending. Computer studies and some tests indicate that such a procedure is realistic for those cases investigated. Thus the full interaction equation would be

$$\frac{P}{P_{cr}} + \frac{M_x C_{mx}}{M_{mx}(1 - P/P_{ex})} + \frac{M_y C_{my}}{M_{my}(1 - P/P_{ey})} \leq 1 \tag{12.9.1}$$

where all terms are as defined following Eq. 12.8.4, except that now the quantities subscripted x and y must be evaluated for bending about the axis indicated by subscript.

12.10 AISC—WORKING STRESS DESIGN CRITERIA

For working stress design, the strength interaction equations, Eqs. 12.8.1 through 12.8.4, may be converted to unit stresses and a factor of safety (FS) applied to bring them into the service load range.

Stability Interaction Criterion

The ultimate strength interaction equation, Eq. 12.8.4, including lateral-torsional buckling is

$$\frac{P_u}{P_{cr}} + \frac{M_{ui}}{M_m} \frac{C_m}{(1 - P_u/P_e)} = 1 \qquad [12.8.4]$$

where P_u and M_{ui} are the axial force and primary bending moment, respectively, that occur *when failure is imminent*. When both the numerator and denominator are divided by a factor (FS) to bring all terms into the service load range,

$$\frac{P_u/(A_g \text{FS})}{P_{cr}/(A_g \text{FS})} + \frac{M_{ui}/[S(\text{FS})]}{M_m/[S(\text{FS})]} \frac{C_m}{\left[1 - \dfrac{P_u/(A_g \text{FS})}{P_e/(A_g \text{FS})}\right]} = 1.0 \qquad (12.10.1)$$

or

$$\frac{P/A_g}{P_{cr}/(A_g \text{FS})} + \frac{M_i/S}{M_m/[S(\text{FS})]} \frac{C_m}{\left[1 - \dfrac{P/A_g}{P_e/(A_g \text{FS})}\right]} = 1.0 \qquad (12.10.2)$$

which gives as a design requirement,

$$\frac{f_a}{F_a} + \frac{f_b}{F_b} \frac{C_m}{(1 - f_a/F'_e)} \leq 1.0 \qquad (12.10.3)$$

for *uniaxial* bending and compression.

By analogy, for bending about both x- and y-axes, Eq. 12.10.3 would become

$$\frac{f_a}{F_a} + \frac{f_{bx} C_{mx}}{F_{bx}(1 - f_a/F'_{ex})} + \frac{f_{by} C_{my}}{F_{by}(1 - f_a/F'_{ey})} \leq 1.0 \qquad (12.10.4)$$

which is the stability interaction equation, AISC Formula (1.6–1a), where

$f_a = P/A_g = $ nominal axial compression stress at service load

$f_{bx}, f_{by} = $ flexural stresses at service load based on primary bending moment about the x- and y-axes, respectively

$F_a = $ allowable compression stress considering the member as loaded by axial compression only

$F_{bx}, F_{by} = $ allowable flexural stresses for the x- and y-axes, respectively, considering the member loaded in bending only. According to the definition of C_b in AISC–1.5.1.4.5(2.), when the stability equation, Eq. 12.10.4, is used for *braced frames* $C_b = 1.0$, but when Eq. 12.10.4 is used for *unbraced frames*, $C_b = 1.75 + 1.05(M_1/M_2) + 0.3(M_1/M_2)^2 \leq 2.3$

C_m = factors discussed in Secs. 12.3–12.5, to be taken as follows:

(1) For *braced frames* without transverse loading between supports, $C_m = 0.6 - 0.4M_1/M_2 \geq 0.4$, where the moments carry a sign in accordance with end rotational direction; i.e., positive moment ratio for reverse curvature, negative moment ratio for single curvature. In this situation C_m converts the linearly varying (from smaller value M_1 to larger value M_2) primary bending moment into an equivalent uniform moment, $M_E = C_m M_2$. Thus, philosophically the flexural stress due to primary bending moment is $C_m f_b$.

(2) For *braced frames* with transverse lateral loading between supports, C_m is an integral part of the moment magnifier, whose value may be determined by the procedures of Sec. 12.3. Such rational values may be obtained from Eq. 12.3.7 after dividing α by a safety factor (1.67), from Table 12.10.1, or by a direct analysis. In lieu of such an analysis, however, AISC–1.6.1 permits using $C_m = 0.85$ for members with moment restraint at the ends, and $C_m = 1.0$ for members with simply supported ends.

(3) For *unbraced frames*, $C_m = 0.85$. The AISC Commentary suggests that C_m may be taken conservatively as $C_m = 1 - 0.18f_a/F'_e$. The *unbraced frame* is treated in more detail in Chapter 14.

In the application of Eq. 12.10.4 [AISC Formula (1.6–1a)], the term F'_e refers to the *effective pin-end length* in the *plane of bending*:

$$F'_e = \frac{P_e}{A_g \text{FS}} = \frac{\pi^2 EI}{A_g L^2 \text{FS}}$$

$$= \frac{12\pi^2 E}{23(KL_b/r_b)^2} = \frac{149{,}000}{(KL_b/r_b)^2} \text{ ksi} \qquad (12.10.5)^*$$

where K = effective length factor (see Sec. 6.9 and Chapter 14)

  L_b = actual unbraced length in the plane of bending

  r_b = radius of gyration for the axis of bending

It is noted that a nominal safety factor of $23/12 = 1.92$ is used; this is the maximum factor used for long axially loaded members (see Sec. 6.8) and is therefore conservative in the magnification term.

Yielding Interaction Criterion

At support locations in *braced frames* and for low slenderness situations in *unbraced frames*, yielding (plastic strength under the action of P and

* For SI units,

$$F'_e = \frac{1027{,}000}{(KL_b/r_b)^2} \text{ MPa} \qquad (12.10.5)$$

M) under combined axial compression and bending may govern. The strength interaction equation, Eq. 12.8.1, forms the basis for the working stress criterion,

$$\frac{P_u}{P_y} + \frac{M_u}{1.18M_p} = 1.0 \qquad [12.8.1]$$

where P_u and M_u are the axial force and primary bending moment, respectively, that occur when maximum strength is achieved. When the numerator and denominator are divided by a factor (FS) to bring all terms into the service load range,

$$\frac{P_u/(A_g\text{FS})}{P_y/(A_g\text{FS})} + \frac{M_u/[S(\text{FS})]}{1.18M_p/[S(\text{FS})]} = 1.0 \qquad (12.10.6)$$

or

$$\frac{P/A_g}{P_y/(A_g\text{FS})} + \frac{M/S}{1.18M_p/[S(\text{FS})]} = 1.0 \qquad (12.10.7)$$

which gives as a design requirement,

$$\frac{f_a}{0.60F_y} + \frac{f_b}{1.18M_p/[S(\text{FS})]} = 1.0 \qquad (12.10.8)$$

The AISC Specification has used an expression more conservative than Eq. 12.10.8 by using 1.0 instead of 1.18, and instead of $M_p/[S(\text{FS})]$, which for I-shaped sections would correspond to $0.66F_y$ for x-axis bending and $0.75F_y$ for y-axis bending, AISC–1.6.1 has used F_b. For instance, the allowable stress F_{bx} might require reduction below $0.66F_y$ or even below $0.60F_y$ because of the laterally unbraced length adjacent to a braced location.

Thus AISC uses for the yield criterion,

$$\frac{f_a}{0.60F_y} + \frac{f_b}{F_b} \leq 1.0 \qquad (12.10.9)$$

for *uniaxial* bending and compression.

For biaxial bending, the general equation is

$$\frac{f_a}{0.60F_y} + \frac{f_{bx}}{F_{bx}} + \frac{f_{by}}{F_{by}} \leq 1.0 \qquad (12.10.10)$$

which is the yield interaction equation, AISC Formula (1.6-1b), where the quantities are as defined following Eq. 12.10.4, except that in evaluating F_{bx} and F_{by} the moment gradient term C_b is used in exactly the way it is used for bending alone, rather than in the special way it is used for the stability interaction criterion, Eq. 12.10.4.

Table 12.10.1 C_m Values for Braced Frames (from *AISC Commentary*, C1.6.1), Eq. 12.3.8

	Case	ψ	$C_m = 1 + \psi\alpha$
1		0	1.0
2		-0.3	$1 - 0.3f_a/F'_e$
3		-0.4	$1 - 0.4f_a/F'_e$
4		-0.2	$1 - 0.2f_a/F'_e$
5		-0.4	$1 - 0.4f_a/F'_e$
6		-0.6	$1 - 0.6f_a/F'_e$

Simplified Interaction Criterion for Small Axial Compression

When f_a/F_a does not exceed 0.15, AISC–1.6.1 *permits* use of the following instead of the two formulas, AISC Formulas (1.6–1a) and (1.6–1b):

$$\frac{f_a}{F_a} + \frac{f_{bx}}{F_{bx}} + \frac{f_{by}}{F_{by}} \leq 1.0 \qquad (12.10.11)$$

which is AISC Formula (1.6–2).

As is usual with design specifications, current requirements reflect some historical developments. Until 1961, the AISC Specification did not consider the secondary bending moment due to deflection. When axial compression is relatively small, neglect of the secondary effect makes little difference.

Redistribution of Moments to Approximate Plastic Behavior

Regarding bending moment adjustments to account for plastic-behavior moment redistribution, as discussed in Chapter 10, AISC–1.5.1.4.1 allows the 10% reduction in negative moment on a beam to be used in proportioning the column if the *beam or girder* satisfies "compact section" requirements and the unit compressive stress on the column does not exceed $0.15F_a$. The authors further believe the effective laterally unbraced length on the *column*, measured from the end where the adjustment occurs, should also satisfy "compact section" limitations; i.e., it should not exceed L_c.

Example 12.10.1

Investigate the acceptability of a W16×67 used as a beam-column under the loading shown in Fig. 12.10.1. Steel is A572 Grade 60.

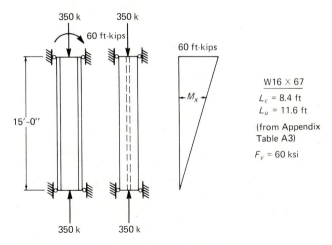

Fig. 12.10.1 Beam-column for Example 12.10.1.

SOLUTION

(a) Column effect.

$$\frac{KL}{r_y} = \frac{15(12)}{2.46} = 73$$

$$\frac{KL/r}{C_c} = \frac{73}{97.7} = 0.747$$

$$F_a = C_a F_y = 0.381(60) = 22.8 \text{ ksi} \quad \text{(AISC Appendix Table 4)}$$

$$f_a = \frac{P}{A} = \frac{350}{19.7} = 17.8 \text{ ksi}$$

$$\frac{f_a}{F_a} = \frac{17.8}{22.8} = 0.78 > 0.15; \qquad \text{use AISC Formulas (1.6–1a) and (1.6–1b)}$$

(b) Beam effect.

$$F_b \ (1.5\text{–}7) = \frac{12{,}000}{Ld/A_f} = \frac{12{,}000}{15(12)(2.40)} = 27.8 \ \text{ksi} < 0.60F_y$$

$$\frac{L}{r_T} = \frac{15(12)}{2.75} = 65.5$$

$$F_b \ (1.5\text{–}6a) = 40.0 - \frac{(L/r_T)^2}{425} = 40.0 - \frac{(65.5)^2}{425} = 29.9 \ \text{ksi}$$

Since $29.9 > 27.8$, F_b (1.5–6) controls. In both formulas C_b is taken as 1.0 (i.e., not used), because C_m in this situation converts the moment diagram into an equivalent uniform moment, for which C_b would be 1.0.

$$C_m = 0.6 - 0.4(M_1/M_2) = 0.60$$

$$f_b = \frac{60(12)}{117} = 6.15 \ \text{ksi}$$

$$\frac{C_m f_b}{F_b} = \frac{0.6(6.15)}{29.9} = 0.12$$

(c) Moment magnification.

$$\frac{KL}{r_x} = \frac{15(12)}{6.96} = 25.9; \qquad F'_e = 223 \ \text{ksi} \quad \text{(AISC Appendix Table 9)}$$

where the x-axis is the axis of bending.

$$\frac{1}{1 - f_a/F'_e} = \frac{1.0}{1 - 17.8/223} = \frac{1.0}{1 - 0.0798} = 1.09$$

(d) Check of AISC Formulas:
For stability, Formula (1.6–1a),

$$\frac{f_a}{F_a} + \frac{C_m f_b}{F_b}\left(\frac{1.0}{1 - f_a/F'_e}\right) = 0.78 + 0.12(1.09) = 0.91 < 1.0$$

For yielding, Formula (1.6–1b), at the braced point,

$$\frac{f_a}{0.60F_y} + \frac{f_b}{F_b} = \frac{17.8}{36} + \frac{6.15}{36} = 0.66 < 1.0$$

For the above equation which does not involve C_m, $F_b(1.5\text{–}7)$ should use $C_b = 1.75$ for this problem. In which case $F_b > 0.60F_y$; use $0.60F_y = 36$ ksi. The W16×67 is acceptable for the given loading.

12.11 DESIGN PROCEDURES AND EXAMPLES—WORKING STRESS METHOD

To aid in selection of a beam-column section, it is usually advantageous to convert, in an approximate way, the resulting bending moment into an equivalent axial compression load and then to make use of column tables. Occasionally, conversion of the axial load into equivalent moment will be helpful.

The stability interaction equation, Eq. 12.10.3, may be written

$$\frac{P}{AF_a} + \frac{M}{F_b S}\left(\frac{C_m}{1 - f_a/F_e'}\right) = 1.0 \tag{12.11.1}$$

Multiplying by AF_a gives

$$P + M\left(\frac{A}{S}\right)\left(\frac{F_a}{F_b}\right)\left(\frac{C_m}{1 - f_a/F_e'}\right) = F_a A = P_{EQ} \tag{12.11.2}$$

Next, examine the magnification term, which may be changed in form, using Eq. 12.10.5 for F_e',

$$\frac{1}{1 - f_a/F_e'} = \frac{F_e'}{F_e' - f_a} = \frac{149{,}000 r^2}{(KL)^2\left(\dfrac{149{,}000 r^2}{(KL)^2} - \dfrac{P}{A}\right)} \tag{12.11.3}$$

$$= \frac{149{,}000 A r^2}{149{,}000 A r^2 - P(KL)^2} \tag{12.11.4}$$

Thus the equivalent column load P_{EQ} may be expressed using Eq. 12.11.2, corresponding to AISC Formula (1.6–1a) for uniaxial bending,

$$P_{EQ} = P + MB\left(\frac{F_a}{F_b}\right)\left(\frac{C_m a}{a - P(KL)^2}\right) \tag{12.11.5}$$

where $B =$ bending factor $= A/S$

$a = 149{,}000 A r^2$ for axis of bending

Note is made that the allowable stress ratio (F_a/F_b) reduces P_{EQ} while the term $[C_m a/(a - P(KL)^2)]$ usually increases P_{EQ}.

When the yield criterion controls, the equivalent column load P_{EQ} may be expressed,

$$P_{EQ} = P + MB\left(\frac{0.60 F_y}{F_b}\right) \tag{12.11.6}$$

corresponding to AISC Formula (1.6–1b) for uniaxial bending.

When $f_a/F_a < 0.15$, P_{EQ} may be computed similarly from Eq. 12.10.11 as

$$P_{EQ} = P + MB\left(\frac{F_a}{F_b}\right) \tag{12.11.7}$$

corresponding to AISC Formula (1.6–2) for uniaxial bending.

Several examples follow which demonstrate application of the interaction formulas, using principles established in Chapters 6, 7, and 9. The following examples use the approaches outlined above; however, alternative approaches and uses of various design aids have been proposed [42–45] including nonconventional cases such as stepped columns.

Example 12.11.1

Select the lightest W14 section to carry an axial compression force of 150 kips in combination with a moment of 500 ft-kips. The member is part of a *braced* system, with support provided in each direction at top and bottom of a 14-ft length. Conservatively assume the moment causes single curvature and varies as shown in Fig. 12.11.1. Use A36 steel and the AISC Specification.

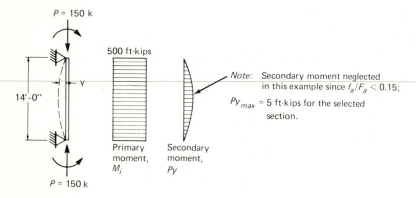

Fig. 12.11.1 Example 12.11.1.

SOLUTION

Conservatively assume the effective length factor $K = 1.0$. If adjacent member stiffness is known, the alignment chart which appears as Fig. 14.3.1a in Chapter 14 may be used to determine $K < 1.0$ for this braced frame.

Since at the start one may have no idea whether $f_a/F_a < 0.15$, use the simplest P_{EQ} expression, Eq. 12.11.7,

$$P_{EQ} = P + MB_x\left(\frac{F_a}{F_b}\right)$$

Referring to AISC Manual, Column Load Tables: for W14 sections, find the average bending factor with respect to the strong axis, $B_x \approx 0.19$. Neglect temporarily F_a/F_b; i.e., assume $F_a \approx F_b$.

$$P_{EQ} \approx 150 + 500(12)(0.19) = 150 + 1140 = 1290 \text{ kips}$$

At this point, since the column part represents less than 15% (150/1290) of the total, assume that AISC Formula (1.6–2) will apply.

Obtain trial sections from AISC Manual using $P = 1290$ kips and $L = 14$ ft, and try one section lighter than indicated,

$$\text{W14} \times 211, \quad B_x = 0.183, \quad r_y = 4.07 \text{ in.}, \quad L_c = 16.7 \text{ ft}$$

$$P = 1183 \text{ kips}$$

Check:

$$\frac{KL}{r_y} = \frac{14(12)}{4.08} = 41.3; \qquad F_a = 19.1 \text{ ksi} \quad \text{(AISC Appendix Table 3-36)}$$

$$L = 14 \text{ ft} < L_c \quad \text{(Sec. 9.5)}; \qquad F_b = 0.66 F_y = 24 \text{ ksi}$$

$$\frac{F_a}{F_b} = \frac{19.1}{24} = 0.80$$

Revise:

$$P_{EQ} = 150 + 500(12)(0.183)(0.80) = 1028 \text{ kips}$$

using closer approximations for the variables.

Select for the revised P_{EQ},

$$\text{W14} \times 193, \quad B_x = 0.183, \quad r_y = 4.05 \text{ in.}, \quad L_c = 16.6 \text{ ft}$$

$$P = 1083 \text{ kips}$$

Little change in properties from preliminary check. Complete check:

$$\frac{KL}{r_y} = \frac{14(12)}{4.05} = 41.5; \qquad F_a = 19.1 \text{ ksi}$$

$$L = 14 \text{ ft} < L_c; \qquad F_b = 24 \text{ ksi}$$

$$f_a = \frac{P}{A} = \frac{150}{56.8} = 2.6 \text{ ksi}$$

$$\frac{f_a}{F_a} = \frac{2.6}{19.1} = 0.14 < 0.15; \qquad \text{Use AISC Formula (1.6–2)}$$

$$f_b = \frac{M}{S_x} = \frac{500(12)}{310} = 19.4 \text{ ksi}$$

$$\frac{f_a}{F_a} + \frac{f_b}{F_b} = 0.14 + 0.81 = 0.95 < 1.0 \qquad \text{OK}$$

Use W14 × 193.

Although the AISC Specification does not require it, check Formula (1.6–1a) to see the effect of the neglected secondary bending moment.

$$C_m = 0.6 - 0.4(M_1/M_2) = 0.6 - 0.4(-1.0) = 1.0$$

$$\frac{KL}{r_x} = \frac{14(12)}{6.50} = 25.8; \qquad F'_e = 225 \text{ ksi}$$

based on the strong axis of bending.

$$\text{Magnification factor} = \frac{C_m}{1 - f_a/F'_e}$$

$$= \frac{1}{1 - 2.6/225} = 1.01$$

Obviously, using AISC Formula (1.6–2) instead of Formula (1.6–1a) involves insignificant error.

Example 12.11.2

Select the lightest W section to carry an axial compression $P = 120$ kips applied with an eccentricity $e = 5$ in. as shown in Fig. 12.11.2. The member is part of a *braced* frame, and is conservatively assumed loaded in single curvature with constant e. Use A36 steel and the AISC Specification.

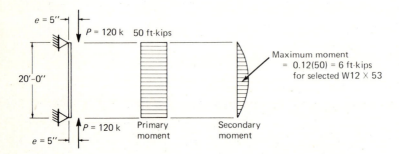

Fig. 12.11.2 Example 12.11.2.

SOLUTION

(a) Compute the equivalent column load. Estimate bending factor and assume other factors to be unity.

$$\text{W10}, \qquad P_{EQ} \approx 120 + 0.26(50)(12) = 276 \text{ kips}$$

$$\text{W12}, \qquad P_{EQ} \approx 120 + 0.22(50)(12) = 252 \text{ kips}$$

$$\text{W14}, \qquad P_{EQ} \approx 120 + 0.20(50)(12) = 240 \text{ kips}$$

It is apparent that the axial compression effect exceeds 15% of the total equivalent load. Since maximum moment occurs at mid–height when stability controls, only AISC Formula (1.6–1a) needs to be checked.

(b) Select trial sections. With a view toward satisfying Eq. 12.11.5, compute

$$P(KL)^2 = 120(240)^2 = 6.92 \times 10^6$$

$$C_m = 1.0$$

$$\left.\begin{array}{l} a_x \approx 60 \text{ for W10} \\ \approx 71 \text{ for W12} \\ \approx 96 \text{ for W14} \end{array}\right\} \begin{array}{l} \text{estimates based on} \\ \text{initial values} \\ \text{of } P_{EQ} \end{array}$$

$$\text{Magnification} = \frac{a_x}{a_x - P(KL)^2} \approx \frac{60}{60-7} = 1.13$$

The reduction from F_a/F_b probably exceeds magnification. Select for $L = 20$ ft,

W10×60	P = 243 kips
W12×58	P = 230 kips
W14×61	P = 237 kips

(c) Check W12×58.

$$\frac{KL}{r_y} = \frac{20(12)}{2.51} = 95.6; \qquad F_a = 13.53 \text{ ksi}$$

$$L_c < L < L_u; \text{ i.e., } 10.6 \text{ ft} < 20 < 24.3 \text{ ft}; \ F_b = 22 \text{ ksi}$$

$$\frac{F_a}{F_b} = \frac{13.53}{22} = 0.615$$

$$P_{EQ} = 120 + 0.22(50)(12)(0.615)(1.13) = 212 \text{ kips}$$

(d) Revise to W12×53 and check AISC Formula (1.6–1a).

$$\frac{KL}{r_y} = \frac{240}{2.48} = 96.8; \qquad F_a = 13.4 \text{ ksi}$$

$$L = 20 \text{ ft} > L_c = 10.6 \text{ ft}$$
$$< L_u = 22.1 \text{ ft}; \qquad F_b = 22 \text{ ksi}$$

$$\frac{KL}{r_x} = \frac{240}{5.23} = 45.9; \qquad F'_e = 70.9 \text{ ksi}$$

$$f_a = \frac{P}{A} = \frac{120}{15.6} = 7.7 \text{ ksi}$$

$$f_b = \frac{M}{S_x} = \frac{50(12)}{70.6} = 8.5 \text{ ksi}$$

AISC Formula (1.6–1a), Eq. 12.10.3:

$$\frac{7.7}{13.4}+\frac{8.5}{22}\left(\frac{1.0}{1-7.7/70.9}\right)=0.57+0.43=1 \quad \text{OK}$$

Use W12×53. A check of W10 and W14 sections will show that no lighter weight section is acceptable. Note that even though AISC Formula (1.6–1b) must also be checked, the check is made by inspection.

Example 12.11.3

Design a beam-column for the service loading conditions shown in Fig. 12.11.3. The member is part of a *braced* system, has support in the weak direction at mid-height, but only at top and bottom for the strong direction. Use AISC Specification and A36 steel.

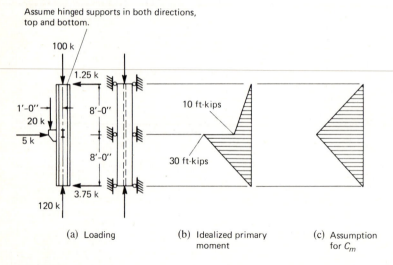

(a) Loading (b) Idealized primary moment (c) Assumption for C_m

Fig. 12.11.3 Example 12.11.3.

SOLUTION

The particular features of this problem are (a) that bracing is not at the same points for both directions; and (b) lateral transverse loading causes the primary bending moment.

(a) Establish effective lengths. The member must be viewed separately as a column, then as a beam, as in Fig. 12.11.4. For column action:

$$KL_x = 16 \text{ ft}; \qquad KL_y = 8 \text{ ft}$$

Required $r_x/r_y \geq 2$ if $L = 8$ ft is valid for entering AISC Manual, Column Load Tables

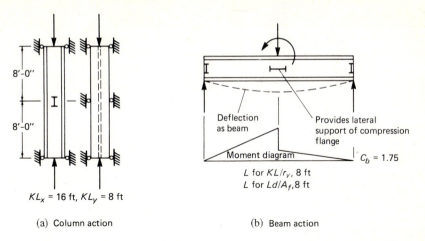

(a) Column action   (b) Beam action

Fig. 12.11.4 Separate beam action and column action from Example 12.11.3.

For beam action:

$$\text{Laterally unbraced length} = 8 \text{ ft}$$

$$L \text{ for moment magnification} = 16 \text{ ft}$$

(b) Estimate equivalent column load. The critical location for checking the interaction formula is at mid-height. Stability controls and AISC Formula (1.6–1a) must be satisfied. Even though bracing in the weak direction is provided at mid-height, that point deflects in the direction of bending. AISC Formula (1.6–1b) is satisfied by inspection.

Using Eq. 12.11.5 for several depths of section, and $KL = 8$ ft,

$$\text{W10} \qquad P_{EQ} \approx 120 + 0.27(30)(12) = 217 \text{ kips}$$
$$\text{W12} \qquad P_{EQ} \approx 120 + 0.23(30)(12) = 203 \text{ kips}$$

Using these values in AISC Column Load Tables suggests W10×39 or W12×40.

(c) Check W10×39:

Column action:

$$f_a = \frac{120}{11.5} = 10.4 \text{ ksi}$$

$$\frac{KL}{r_y} = \frac{8(12)}{1.98} = 48.5, \qquad \frac{r_x}{r_y} = 2.16 > 2.0$$

$$F_a = 18.5 \text{ ksi} \quad \text{(AISC Appendix Table 3-36)}$$

$$\frac{f_a}{F_a} = \frac{10.4}{18.5} = 0.56 > 0.15 \qquad \text{Use Formula (1.6–1a)}$$

Beam action:

$$f_b = \frac{360}{42.1} = 8.55 \text{ ksi}$$

$$L_c = 8.4 \text{ ft} > \text{Actual unbraced length} = 8 \text{ ft}$$

$$F_b = 0.66F_y = 24 \text{ ksi}$$

$$\frac{f_b}{F_b} = \frac{8.55}{24} = 0.36$$

Note: Since an L_c value is given in the tables, the local buckling limitations of AISC–1.5.1.4.1 are automatically satisfied for A36 steel. When no tables are applicable, the following checks must be made in order to use $0.66F_y$:

$$\frac{L}{b_f} \le \frac{76}{\sqrt{F_y}} \quad \text{for lateral-torsional buckling prevention}$$

$$\frac{b_f}{2t_f} \le \frac{65}{\sqrt{F_y}} \quad \text{for flange local buckling prevention}$$

$$\frac{d}{t_w} \le \frac{640}{\sqrt{F_y}}\left(1 - 3.74\frac{f_a}{F_y}\right) \ge \frac{257}{\sqrt{F_y}} \quad \text{for web local buckling prevention}$$

This has been discussed in Secs. 6.17 and 9.5.

Beam-column moment magnification:

$$\frac{KL_b}{r_b} = \frac{KL}{r_x} = \frac{16(12)}{4.27} = 45 \quad \text{(for axis of bending)}$$

$$F_e' = 73.6 \text{ ksi} \quad \text{(AISC Appendix Table 9)}$$

$$\frac{f_a}{F_e'} = \frac{10.4}{73.6} = 0.14, \qquad 1 - \frac{f_a}{F_e'} = 0.86$$

$$C_m \approx 1 - 0.2\frac{f_a}{F_e'} = 0.97$$

(assumes moment variation of Fig. 12.11.3c may approximate that of Fig. 12.11.3b)

$$\text{Magnification factor} = \frac{1}{1 - f_a/F_e'}\frac{0.97}{0.86} = 1.13$$

AISC Formula (1.6–1a):

$$\frac{f_a}{F_a} + \frac{f_b}{F_b}\frac{C_m}{(1 - f_a/F_e')} = 0.56 + 0.36(1.13) = 0.97 < 1.0 \qquad \text{OK}$$

Use W10 × 39.

Example 12.11.4

Select a W14 section for the service loading conditions of Fig. 12.11.5. The member is part of a *braced* system, is to be of A36 steel, and is to be selected according to the AISC Specification. Consider the following two cases of moment variation along the member:

(a) *Case* 1. Assume hinge at bottom, and maximum moment at the top.

(b) *Case* 2. Assume partial fixity at bottom so that an opposite sign bending moment of 40 ft-kips is developed at the bottom.

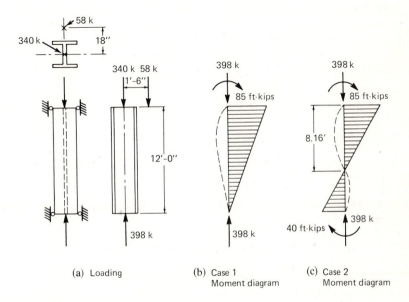

(a) Loading (b) Case 1 Moment diagram (c) Case 2 Moment diagram

Fig. 12.11.5 Example 12.11.4.

SOLUTION

For this problem the design may be governed either by the stability criterion, AISC Formula (1.6–1a), based on stresses out in the span (in the vicinity of a point about 0.4 of the span from the top); or by the yield criterion, Formula (1.6–1b), based on stresses at the braced point (the top).

(a) *Case* 1. Assuming the yield criterion governs,

$$P_{EQ} = P + B_x M = 398 + 0.185(85)(12) = 587 \text{ kips}$$

$$\text{Required } A = \frac{P}{0.60F_y} = \frac{587}{22} = 26.7 \text{ sq in.}$$

Try W14×90: $A = 26.5$ sq in. For this section estimate P_{EQ} for stability criterion:

$$C_m = 0.6 - 0.4M_1/M_2 = 0.6$$

$$a_x = 148.9 \times 10^6$$

$$P(KL)^2 = 398(144)^2 = 8.25 \times 10^6$$

$$\text{Magnification factor} = \frac{a_x}{a_x - P(KL)^2} = \frac{148.9}{(148.9 - 8.25)} = 1.06$$

Because C_m is considerably less than unity, it is unlikely Formula (1.6–1a) will control, even though the allowable column stress F_a is less than $0.60F_y$.

Check Formula (1.6–1b): W14×90

$$f_a = \frac{398}{26.5} = 15.0 \text{ ksi}; \qquad f_b = \frac{85(12)}{143} = 7.1 \text{ ksi}$$

$$L = 12 \text{ ft} < L_c = 15.3 \text{ ft}, \qquad F_b = 0.66F_y = 24 \text{ ksi}$$

$$\frac{f_a}{0.60F_y} + \frac{f_b}{F_b} = \frac{15.0}{22} + \frac{7.1}{24} = 0.68 + 0.30 = 0.98 < 1.0 \qquad \text{OK}$$

Check Formula (1.6–1a):

$$\frac{KL}{r_y} = \frac{144}{3.70} = 38.9; \qquad F_a = 19.3 \text{ ksi}$$

$$\frac{f_a}{F_a} = \frac{15.0}{19.3} = 0.78; \qquad \frac{C_m f_b}{F_b} = \frac{0.6(7.1)}{24} = 0.18$$

Magnification factor $= 1.06$, as computed above.

$$\frac{f_a}{F_a} + \frac{C_m f_b}{F_b}\left(\frac{1}{1 - f_a/F_e'}\right) = 0.78 + 0.18(1.06) = 0.97 < 1 \qquad \text{OK}$$

In this case both Formulas (1.6–1a) and (1.6–1b) were close to the limit. *Use* W14×90.

(b) *Case 2.* Since the stability of the member is increased when moments cause double curvature, obviously the yield criterion will again control. The improved stability is utilized by $C_m = 0.6 - 0.4(40/85) = 0.41$, reduced from 0.6 as used for Case 1. In this case, the section choice will be the same (W14×90) as for Case 1. If the value of L_c for the potential section choice had been less than 12 ft, there is some sentiment among designers to compare L_c with 8.16 ft, the distance from the maximum moment location to the point of inflection. Use of the inflection point as a laterally braced point (see Sec. 9.8) for this case is about 10% on the unsafe side for the laterally unbraced length. For example, if Eq. 9.3.42 is accurate, it would be correct to compare $0.9L_c$ with the 8.16 ft.

Use W14×90—A36.

Example 12.11.5

The W8×24 section, as shown in Fig. 12.11.6, is laterally loaded to cause bending about its weak axis. The steel is A572 Grade 50.

(a) Check the adequacy according to the AISC Specification.
(b) Calculate the maximum deflection and magnified bending moment directly using $M_i + Py_{max}$ (see Sec. 12.3).
(c) Using the result of part (b), show the part of the AISC interaction equation to which it corresponds.

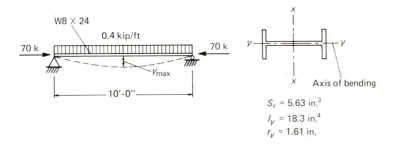

$$S_y = 5.63 \text{ in.}^3$$
$$I_y = 18.3 \text{ in.}^4$$
$$r_y = 1.61 \text{ in.}$$

Fig. 12.11.6 Example 12.11.5.

SOLUTION

(a) AISC Specification check; since maximum moment occurs away from the supports, the stability criterion will govern. Check AISC Formula (1.6–1a).

Note that for this problem buckling as a column occurs in the plane of bending, whereas in the previous four examples column buckling occurred about the axis which was not the axis of bending.

Column action:

$$\frac{KL}{r_y} = \frac{1.0(10)(12)}{1.61} = 74.5; \qquad F_a = 20.1 \text{ ksi} \quad \text{(AISC Appendix Table 3-50)}$$

$$f_a = \frac{P}{A} = \frac{70}{7.08} = 9.9 \text{ ksi}$$

$$\frac{f_a}{F_a} = \frac{9.9}{20.1} = 0.49$$

Beam action: Since bending is about the weak axis, lateral-torsional buckling is not a problem. Assuming the section is "compact" with regard to local buckling, it would be able to develop its plastic moment. Checking AISC–1.5.1.4.1,

$$\frac{b_f}{2t_f} = \frac{6.495}{2(0.400)} = 8.1 < \frac{65}{\sqrt{50}} = 9.2 \qquad \text{OK}$$

$$F_b = 0.75 F_y = 37.5 \text{ ksi}$$

$$M_y = \tfrac{1}{8}(0.4)(10)^2 = 5 \text{ ft-kips}$$

$$f_b = \frac{M_y}{S_y} = \frac{5(12)}{5.63} = 10.7 \text{ ksi}$$

$$\frac{f_b}{F_b} = \frac{10.7}{37.5} = 0.29$$

Moment magnification:

$$F'_e = 26.9 \text{ ksi, using } \frac{KL}{r_y} = 74.5 \quad \text{(AISC Appendix Table 9)}$$

$$\frac{f_a}{F'_e} = \frac{9.9}{26.9} = 0.37; \qquad 1 - \frac{f_a}{F'_e} = 0.63$$

$$C_m = 1.0 \quad \text{(from Table 12.10.1)}$$

$$\text{Magnification factor} = \frac{C_m}{1 - f_a/F'_e} = \frac{1}{0.63} = 1.58$$

AISC Formula (1.6–1a):

$$\frac{f_a}{F_a} + \frac{f_b}{F_b}\left(\frac{C_m}{1 - f_a/F'_e}\right) = 0.49 + 0.29(1.58) = 0.95 < 1 \qquad \text{OK}$$

(b) Compute maximum deflection $y_{\max}$ and magnified stress directly using $M_i + P y_{\max}$.

$$y_{\max} = \frac{\delta_0}{1 - \alpha} \qquad\qquad [12.3.4]$$

$$\delta_0 = \text{primary deflection} = \frac{5 w L^4}{384 EI}$$

$$= \frac{5(0.4)(10)^4 1728}{384(29{,}000)(18.3)} = 0.17 \text{ in.}$$

$$\alpha = \frac{P}{P_e} = \frac{f_a}{1.92 F'_e} = \frac{0.37}{1.92} = 0.19$$

$$y_{\max} = \frac{0.17}{1 - 0.19} = 0.21 \text{ in.}$$

$$M_{\max} = M_i + P y_{\max} = 5.0 + \frac{70(0.21)}{12}$$

$$= 5.0 + 1.2 = 6.2 \text{ ft-kips}$$

which is the *actual* magnified moment for the given loading on a W8×24.

(c) Show how $M_i + Py_{max}$ relates to the AISC Formula (1.6–1a). To compare part (b) with the design criterion, the factor of safety 1.92 must be applied to Eqs. 12.3.4 and 12.3.5, giving

$$M_i + P\frac{1.92\delta_0}{1-1.92\alpha} = \frac{M_i C_m}{1-f_a/F_e'}$$

$$5 + \frac{70}{12}\left(\frac{1.92(0.17)}{1-0.36}\right) = \frac{5(1.0)}{1-0.36}$$

Both sides of the above equation give 7.8 ft-kips, which compares with 6.2 ft-kips (without factor of safety) in part (b). If one divides both sides by the section modulus S, the identity is established,

$$\frac{M_i + P\left(\dfrac{1.92\delta_0}{1-1.92\alpha}\right)}{S} = \frac{f_b C_m}{1-f_a/F_e'}$$

That the factor of safety (1.92) should appear in two places in the above equation may be verified by assuming that both P and M service loads are multiplied by a factor of 1.92, giving

$$\text{Overload moment} = 1.92M_i + 1.92P\left(\frac{1.92\delta_0}{1-1.92\alpha}\right)$$

The safe service load is obtained by dividing the overload moment by 1.92, which gives

$$\text{Service moment} = M_i + P\left(\frac{1.92\delta_0}{1-1.92\alpha}\right)$$

Example 12.11.6

Select a suitable W section for the column member of the frame shown in Fig. 12.11.7. The joints are rigid to give frame action in the plane of bending, but in the transverse direction sway bracing is provided and the attachments may be considered hinged. Use A36 steel and AISC Specification. Solve for the following two cases:
  (a) *Braced* frame in the plane of bending.
  (b) *Unbraced* frame in the plane of bending.

SOLUTION

(a) Frame *braced* in the plane of bending. Estimate equivalent column load:

$$K_x \approx 0.7,$$

or can be obtained from Alignment Chart, Fig. 14.3.1a, as

$$G_A = 1.0 \text{ (fixed)}, \qquad G_B = \frac{I/20}{3I/30} = 0.5, \qquad K_x = 0.73$$

Assume girder moment of inertia three times that of column

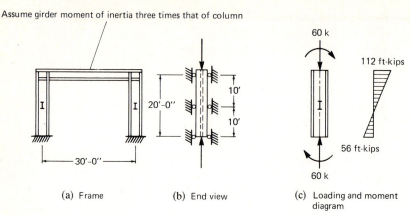

(a) Frame	(b) End view	(c) Loading and moment diagram

Fig. 12.11.7 Example 12.11.6.

$$P(K_xL)^2 = 60(0.7)^2(240)^2 = 1.69 \times 10^6$$

$$P_{EQ} \approx P + B_xM$$

W10, $P_{EQ} = 60 + 0.26(112)(12) = 60 + 350 = 410$ kips

W12, $P_{EQ} = 60 + 0.22(112)(12) = 60 + 296 = 356$ kips

W14, $P_{EQ} = 60 + 0.20(112)(12) = 60 + 269 = 329$ kips

Using the AISC Manual, Column Load Tables, with $K_yL_y = 10$ ft, preliminary sections are selected as

W10×68	$P = 373$ kips @ $L = 10$ ft	$r_x/r_y = 1.71$
W12×65	$P = 367$ kips	$r_x/r_y = 1.75$
W14×61	$P = 330$ kips	$r_x/r_y = 2.44$

For values in the AISC tables to give correct column allowable loads, r_x/r_y of section must exceed $K_xL_x/K_yL_y = 0.7(20)/10 = 1.40$.

Before proceeding, the reader is reminded that due to plastic behavior, moment redistribution may occur (if local buckling and lateral-torsional buckling are prevented in the girder), which according to AISC–1.5.1.4.1 may be accounted for by a 10% reduction in negative moment on the girder. Such a reduced moment may also be used in proportioning the column for combined stress, if f_a/F_a does not exceed 0.15.

Under the conditions when moment redistribution is considered; i.e., plastic moment is assumed to occur at the top of the column, the designer should be wary of using an effective length factor K_x less than 1.0, since end restraint becomes a constant value (M_p), instead of increasing proportionally to the end rotation as it does when the system is elastic.

However, if one is certain the moment gradient is one causing double curvature, as in this case, then it seems rational to do as illustrated in this example.

For this example, the axial compression effect is about $60/329 = 0.18$

from P_{EQ} calculation; therefore the reduction in negative moment at the top of the member may not be used.

Because of the moment gradient indicating double curvature, with $C_m = 0.6 - 0.4(56/112) = 0.4$ min, it is likely that the yield criterion at the braced point will control. Thus AISC Formula (1.6–1b), where both denominator terms are likely to be $0.60F_y$, may be used to obtain required area for a more accurate choice of section. For a W14,

$$\text{Required } A = \frac{P_{EQ}}{0.60F_y} = \frac{329}{22} = 14.9 \text{ sq in.}$$

Try W14×53, W16×50, W18×46, or W21×44. The sections deeper than W14 do not appear in column load tables but may be satisfactory since P_{EQ} decreases as the depth increases. (For W21, $P_{EQ} \approx 60 + 0.16(112)12 = 275$ kips. Required $A = 12.5$ sq in.) Of course, changing depth will affect the relative moments of inertia. For the *braced* frame such changes will have less effect than on the *unbraced* frame.

Check W21×44:
Formula (1.6–1b) based on yielding at the braced point:

$$f_a = \frac{60}{13.0} = 4.6 \text{ ksi}; \qquad f_b = \frac{112(12)}{81.6} = 16.5 \text{ ksi}$$

$$C_b = 1.75 + 1.05(-28/112) + 0.3(28/112)^2 = 1.51$$

$$L_c = 6.6 \text{ ft} < L = 10 \text{ ft}; \qquad \text{Not compact!}$$

$$L_u = 7.0 \text{ ft} < L = 10 \text{ ft}$$

For the yield equation, AISC Formula (1.6–1b), C_b is used in computing F_b.

$$F_b(1.5\text{–}7) = \frac{12,000(1.51)}{10(12)7.06} = 21.4 \text{ ksi (controls)}$$

$$L/r_T = 10(12)/1.57 = 76.4$$

$$F_b(1.5\text{–}6a) = 24.0 - \frac{(76.4)^2}{1181(1.51)} = 20.7 \text{ ksi}$$

$$\frac{f_a}{0.60F_y} + \frac{f_b}{F_b} = \frac{4.6}{22} + \frac{16.5}{21.4} = 0.98 < 1.0 \qquad \text{OK}$$

Next check AISC Formula (1.6–1a) for stability out in the span.

$$\frac{K_y L_y}{r_y} = \frac{1.0(10)(12)}{1.26} = 95.2; \qquad F_a = 13.6 \text{ ksi}$$

For the *braced* frame, C_b is *not used* for computing F_b in the formula where C_m is used.

$$F_b(1.5\text{–}6a) = 24.0 - \frac{(76.4)^2}{1181} = 19.1 \text{ ksi}$$

$$\frac{K_x L_x}{r_x} = \frac{0.7(20)(12)}{8.06} = 20.8; \qquad F_e' = 345 \text{ ksi}$$

$$\frac{f_a}{F_e'} = \frac{4.6}{345} = 0.013; \qquad 1 - \frac{f_a}{F_e'} = 0.987$$

$$\frac{f_a}{F_a} + \frac{f_b}{F_b}\left(\frac{C_m}{1 - f_a/F_e'}\right) = \frac{4.6}{13.6} + \frac{16.5}{19.1}\left(\frac{0.4}{0.987}\right)$$

$$= 0.34 + 0.86(0.41) = 0.69 < 1$$

Use W21×44.

(b) Frame *unbraced* in the plane of bending. The significant difference between this case and previous examples is that here the K_x value in the plane of bending exceeds one. Using the alignment chart from Chapter 14 (Fig. 14.3.1b) for *unbraced* frame (sidesway not prevented) for $G_A = 1.0$ (fixed at bottom) and $G_B = 0.5$ [see case (a)], find $K_x \approx 1.24$.

Equivalent column load estimates are the same as for case (a), but in using column tables, correct allowable loads are obtained for $K_y L_y = 10$ ft only when r_x/r_y exceeds

$$K_x L_x / K_y L_y = 1.24(20)/10 = 2.48$$

Selection of W10 or W12 sections, if the stability criterion governs, will be controlled by column buckling in the plane of bending, while deeper sections will be controlled by weak-axis column buckling.

From AISC Manual, Column Load Tables,

W10, for $L = 1.24(20)/1.71 = 14.5$ ft, $P_{EQ} \approx 410$ kips, find W10×77 with $P \approx 380$ kips.

W12, for $L = 1.24(20)/1.75 = 14.2$ ft, $P_{EQ} \approx 356$ kips, find W12×65 with $P \approx 340$ kips.

W14, for $L = 1.24(20)/2.44 = 10.2$ ft, $P_{EQ} \approx 329$ kips, find W14×61 with $P \approx 330$ kips.

These preliminary sections are likely to be too heavy because F_a/F_b is usually a significant reduction effect, while magnification is normally not more than 10 to 20%. If the yielding equation controls, the section and procedure are as in case (a) where W21×44 was acceptable.

Check W21×44 by AISC Formula (1.6–1a) for stability:

$$K_y L_y/r_y = 95.2 \text{ as from part (a)}$$

$$K_x L_x/r_x = 1.24(240)/8.06 = 36.9$$

$F_a = 13.6$ ksi; weak axis governs

$F_b = 21.4$ ksi; (for $C_b = 1.51$) calculation shown in part (a)

Note that for the *unbraced* frame, the C_b ordinarily used for beams in braced systems is prescribed by AISC to be used in the stability equation.

$$F'_e = 110 \text{ ksi}; \quad \text{based on } KL/r = 36.9$$

$$C_m = 0.85 \quad \text{for unbraced frame}$$

$$\text{Magnification factor} = \frac{C_m}{1 - f_a/F'_e} = \frac{0.85}{1 - 4.6/110} = 0.89$$

$$\frac{f_a}{F_a} + \frac{f_b}{F_b}\left(\frac{C_m}{1 - f_a/F'_e}\right) = 0.34 + 0.77(0.89) = 1.03 > 1$$

For this unbraced frame the stability requirement governs. Thus if a small overstress is acceptable for the unbraced frame, the same W21×44 could be used whether the frame is braced or unbraced. Without exceeding the AISC criterion, the W16×50 would be the choice (W18×46 gives 1.01 for AISC Formula 1.6–1a).

Use W16×50.

Example 12.11.7

Select an economical structural tee for use as the top chord in a roof truss, as shown in Fig. 12.11.8. The chord member is to be designed as continuous over 12-ft spans, with transverse lateral support at 6-ft intervals. Use $F_y = 50$ ksi and the AISC Specification.

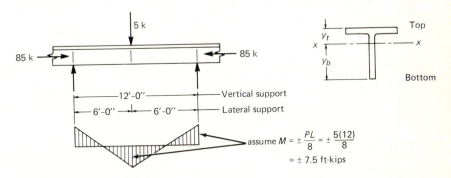

Fig. 12.11.8 Example 12.11.7.

SOLUTION

This design requires satisfying AISC Formula (1.6–1a) at midspan and Formula (1.6–1b) at the supports. Furthermore, because of an unsymmetrical section both top and bottom extreme fibers must be examined.

(a) Estimate equivalent column load. Bending factors are not tabulated in the AISC Manual; therefore Table 12.11.1 may serve as a guide.

Table 12.11.1 Approximate Bending Factors for Structural Tees

WT	$B_{xb} = \dfrac{A}{S_{x \text{ stem}}}$	$B_{xt} = \dfrac{A}{S_{x \text{ flange}}}$
WT9	1–1.2	0.3
WT8	1–1.5	0.3
WT7	1.2–2.0	0.3–0.5
WT6	1.5–2.0	0.5
WT5	2.0–2.5	0.6
WT4	2.5–3.5	0.7

If $(+M)$ governs, combined compression will be in the flange,

$$P_{EQ} \approx P + B_{xt}M = 85 + 0.4(7.5)12 = 121 \text{ kips}$$

If $(-M)$ governs, combined compression will be in the stem,

$$P_{EQ} \approx P + B_{xb}M = 85 + 1.5(7.5)12 = 85 + 135 = 220 \text{ kips}$$

(b) Select trial sections.
It seems likely that the $(-M)$ at the braced point will control:

$$f_a + f_b \leq 0.60F_y = 30 \text{ ksi}$$

$$\text{Required } A = \frac{220}{30} = 7.3 \text{ sq in.}$$

Also, bending is about 60% (135/220) of the total effect; i.e., $f_b/0.60F_y \approx 0.6$, which means

$$\text{Required } S_{xb} \approx \frac{M}{30(0.6)} = \frac{7.5(12)}{30(0.6)} = 5.0 \text{ in.}^3$$

For the section to be fully effective, local buckling must be precluded. If this is desired, AISC–1.9.1.2 requires

$$\frac{d}{t_w} \leq \frac{127}{\sqrt{F_y}} = 18.0$$

Trial sections:

		Computed		
WT6×36	$A = 10.6$	$S_{xb} = 4.54$	$B_{xb} = 2.33$	
WT7×26.5	$A = 7.81$	$S_{xb} = 4.94$	$B_{xb} = 1.58,$	$d/t_w = 18.8$
WT8×25	$A = 7.37$	$S_{xb} = 6.78$	$B_{xb} = 1.09,$	$d/t_w = 21.4$

For WT6,

$$P_{EQ} = 85 + 2.33(90) = 295 \text{ kips}$$

$$\text{Required } A = \frac{295}{30} = 9.8 \text{ sq in.}; \qquad \text{Required } S_{xb} = \frac{90}{30(0.71)} = 4.2 \text{ in.}^3$$

For WT7,

$$P_{EQ} = 85 + 1.58(90) = 227 \text{ kips}$$

$$\text{Required } A = \frac{227}{30} = 7.6 \text{ sq in.;} \qquad \text{Required } S_{xb} = \frac{90}{30(0.63)} = 4.8 \text{ in.}^3$$

Best choices appear to be:

$$\begin{array}{lll} \text{WT6} \times 32.5 & A = 9.54 & S_{xb} = 4.06 \\ \text{WT7} \times 26.5 & A = 7.81 & S_{xb} = 4.94 \end{array}$$

(c) Check WT7$\times$26.5 even though $d/t_w = 18.8$ exceeds the 18.0 basic limit of AISC–1.9.1.

Check AISC Formula (1.6–1b), yielding at braced points:

$$f_a = \frac{85}{7.81} = 10.9 \text{ ksi}$$

$$f_{bt} = \frac{90(1.38)}{27.6} = 4.5 \text{ ksi} \quad \text{(tension at top)}$$

$$f_{bb} = \frac{90}{4.94} = 18.2 \text{ ksi} \quad \text{(compression at bottom)}$$

A structural tee, having r_x and r_y of comparable magnitude, has no tendency to buckle in a direction other than the direction of bending; thus $0.60F_y$ (or $0.66F_y$ if local buckling requirements for "compact section" are satisfied) is the basic value to be used for F_b, modified by Q_s when d/t_w exceeds $127/\sqrt{F_y}$. AISC–1.5.1.4.1 and 1.5.1.4.5(2) would seem to indicate that tee sections must satisfy lateral support requirements for I-shaped sections; however, the AISC Commentary–1.5.1.4.1, first paragraph, confirms the authors' opinion that lateral support is not required. In this example, the web of the tee is in compression and does not satisfy the local buckling requirement of AISC–1.9.1.2; therefore the reduced efficiency must be computed in accordance with AISC Appendix C. Thus AISC Formula (C2–5) applies:

$$Q_s = 1.908 - 0.00715(b/t)\sqrt{F_y}$$

$$= 1.908 - 0.00715(18.8)\sqrt{50} = 0.96$$

$$F_b = Q_s(0.60F_y) = 0.96(30) = 28.8 \text{ ksi*}$$

(bottom) $\qquad f_a + f_b = 10.9 + 18.2 = 29.1 \text{ ksi} \approx 28.8 \text{ ksi} \qquad$ OK

(top) $\qquad f_a + f_b = 10.9 - 4.5 = 6.4 \text{ ksi} \qquad$ OK

---

* For discussion of the Q factor, see Sec. 6.18 and AISC Appendix Secs. C5 and C6.

Check AISC Formula (1.6–1a), stability out in (+M) region:

$$f_{bt} = 4.5 \text{ ksi} \quad \text{(compression at top)}$$
$$f_{bb} = 18.2 \text{ ksi} \quad \text{(tension at bottom)}$$

$K = 1.0$ for truss joints as recommended by the Structural Stability Stability Research Council (see Sec. 6.9)

$$\frac{KL}{r_y} = \frac{72}{1.92} = 37.5 \qquad \frac{KL}{r_x} = \frac{144}{1.88} = 76.6$$

$$C_c' = \sqrt{2\pi^2 E/(Q_a Q_s F_y)} = 109.2$$

$$\frac{KL/r}{C_c'} = \frac{76.6}{109.2} = 0.701$$

$$C_a = 0.400 \quad \text{(AISC Appendix Table 4)}$$
$$F_a = C_a Q_s F_y = 0.400(0.96)50 = 19.2 \text{ ksi}$$
$$F_e' = 25.5 \text{ ksi} \quad \text{(based on } KL_b/r_b = 76.6 \text{ for axis of bending)}$$

$$C_m = 1.0 - 0.6\frac{f_a}{F_e'} \quad \text{(see Table 12.10.1)}$$

$$= 1.0 - 0.6(10.9)/25.5 = 1.0 - 0.6(0.43) = 0.74$$

$$\text{Magnification factor} = \frac{C_m}{1 - f_a/F_e'} = \frac{0.74}{0.57} = 1.30$$

$$\frac{f_a}{F_a} + \frac{f_b}{F_b}\left(\frac{C_m}{1 - f_a/F_e'}\right) = \frac{10.9}{19.2} + \frac{4.5}{28.8}(1.30) = 0.77 < 1.0$$

based on the compression fiber (the top) under flexure.

Use WT7×26.5. (*Note:* WT8×25 also is adequate and may be preferred.)

Example 12.11.8

Select the lightest W12 section to carry an axial compression in addition to biaxial bending, as shown in Fig. 12.11.9. Use A572 Grade 50 steel.

SOLUTION

AISC Formulas (1.6–1a) and (1.6–1b), Eqs. 12.10.4 and 12.10.10, are used in the three term form for biaxial bending.

(a) Determine equivalent column load. Assume that yielding at braced point governs:

$$P_{EQ} \approx P + B_x M_x + B_y M_y$$
$$= 250 + 0.22(25)(12) + 0.8(9)(12)$$
$$= 250 + 66 + 87 = 403 \text{ kips}$$

(b) Select trial section. Using $KL_y = 7.5$ ft, select W12×53 (AISC

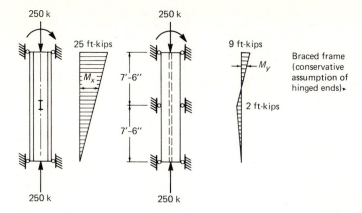

Fig. 12.11.9 Example 12.11.8.

Manual, Column Load Tables):

$$r_x/r_y = 2.11 > KL_x/KL_y = 2.0 \qquad \text{OK}$$

(c) Investigate C_m times magnification factor:

$$P(KL_y)^2 = 250(1)^2(90)^2 = 2.03 \times 10^6$$

$$P(KL_x)^2 = 250(1)^2(180)^2 = 8.1 \times 10^6$$

$$\frac{C_{mx}a_x}{a_x - P(KL_x)^2} = \frac{0.6(63.6)}{63.6 - 8.1} = 0.69$$

$$\frac{C_{my}a_y}{a_y - P(KL_y)^2} = \frac{[0.6 - 0.4(2/9)]14.3}{14.3 - 2.03} = \frac{0.51(14.3)}{12.27} = 0.60$$

(d) Check AISC Formula (1.6–1b), yielding governs:

$$\text{Required } A = \frac{403}{0.60F_y} = \frac{403}{30} = 13.4 \text{ sq in.}$$

W12×50 might satisfy. Check:

$$f_a = \frac{250}{14.7} = 17.0 \text{ ksi}$$

$$L = 7.5 \text{ ft} > L_c = 7.2 \text{ ft}$$

$$F_{bx} = 0.60F_y = 30 \text{ ksi} \quad (L_u = 24.2 \text{ ft})$$

Section "compact" based on local buckling:

$$F_{by} = 0.75F_y = 37.5 \text{ ksi}$$

$$f_{bx} = \frac{25(12)}{64.7} = 4.6 \text{ ksi}; \qquad f_{by} = \frac{9(12)}{13.9} = 7.8 \text{ ksi}$$

$$\frac{f_a}{0.60F_y} + \frac{f_{bx}}{F_{bx}} + \frac{f_{by}}{F_{by}} = \frac{17.0}{30} + \frac{4.6}{30} + \frac{7.8}{37.5} = 0.93 < 1.0 \qquad \text{OK}$$

(e) Check AISC Formula (1.6–1a), stability governs:

$$\frac{KL}{r_y} = \frac{7.5(12)}{1.96} = 45.9; \qquad F_a = 25.0 \text{ ksi}$$

Computing C_m times magnification factor,

$$\text{(for } M_x) = \frac{C_{mx}a_x}{a_x - P(KL_x)^2} = \frac{0.6(58.8)}{58.8 - 8.1} = 0.70$$

$$\text{(for } M_y) = \frac{0.51(8.4)}{8.4 - 2.03} = 0.67$$

$$\frac{f_a}{F_a} + \frac{f_{bx}}{F_{bx}}\left(\frac{C_{mx}}{1 - f_a/F_e'}\right) + \frac{f_{by}}{F_{by}}\left(\frac{C_{my}}{1 - f_a/F_e'}\right)$$

$$= \frac{17.0}{25.0} + \frac{4.6}{30}(0.70) + \frac{7.8}{37.5}(0.67) = 0.93 < 1 \qquad \text{OK}$$

Use W12×50. (W12×45 gives about 3+% overstress.)

12.12 AISC—PLASTIC DESIGN CRITERIA

The provisions for beam-columns are rational for all cases of loading and are essentially the equations discussed in Sec. 12.8.

1. For conditions at a braced location where instability is prevented, AISC–2.4 provides, as per Eq. 12.8.1,

$$\frac{P}{P_y} + \frac{M}{1.18M_p} \le 1.0 \qquad\qquad [12.8.1]$$

and $M \le M_p$. This is AISC Formula (2.4–3). In this equation, P and M are the required strengths (service load times load factor) for which to design; $P_y = F_y A_g$; and M_p is the plastic moment capacity of the section. Equation 12.8.1 is illustrated in Fig. 12.8.1.

2. For locations where instability may govern, AISC–2.4 provides, as per Eq. 12.8.4,

$$\frac{P}{P_{cr}} + \frac{M_i}{M_m}\left(\frac{C_m}{1 - P/P_e}\right) \le 1.0 \qquad\qquad [12.8.4]$$

which is AISC Formula (2.4–2). In the above equation,

P = factored service axial compression load

M_i = factored service primary bending moment

P_{cr} = ultimate strength of an axially loaded compression member, taken as $1.70F_a A_g$

M_m = maximum resisting moment in the absence of axial load, to be taken as

(1) if braced to prevent lateral-torsional buckling,

$$M_m = M_p \tag{12.12.1}$$

(2) if unbraced over the length L,

$$M_m = \left[1.07 - \frac{(L/r_y)\sqrt{F_y, \text{ksi}}}{3160} \right] M_p \le M_p \tag{12.12.2}^*$$

$P_e = 1.92 F_e' A_g$; i.e., $\pi^2 EI/L^2$

C_m = same as for working stress method; see Sec. 12.10 for details

The reduction equation for M_m corresponds to the straight line reduction shown in Fig. 12.8.3 and is less conservative than using AISC Formulas (1.5–6) and (1.5–7) as the basis for ultimate strength. There are two reasons that may explain this. First, Formulas (1.5–6) and (1.5–7) were conservative approximations for the correct stability expression, Eq. 9.4.1, so those formulas indicated excessive reduction in moment capacity; and second, plastic design uses the reduced M_m only for beam-columns, where end restraint provided by the integral rigid frame action increases the buckling and post-buckling strength over that which might be obtained for transversely loaded beams with simple framing. Typical values for M_m are given in Table 12.12.1.

Table 12.12.1 Moment Capacity Under Pure Bending, as Used for Plastic Design of Beam-Columns, Eq. 12.12.2

	M_m as a percent of M_p		
L/r_y	$F_y = 36$ ksi	$F_y = 50$ ksi	$F_y = 60$ ksi
28.6			100
31.3		100	
36.9	100		
40	99	98	97
60	96	93	92
80	92	89	87
100	88	85	82
120	84	80	78

12.13 DESIGN EXAMPLES—PLASTIC DESIGN METHOD

The general procedure for selecting sections for combined axial compression and bending is the same as that for working stress; i.e., compute an

---

* For SI units, with F_y in MPa,

$$M_m = \left[1.07 - \frac{(L/r_y)\sqrt{F_y}}{8300} \right] M_p \le M_p \tag{12.12.2}$$

equivalent axial load and use column tables. The designer may also find that the manual, *Plastic Design of Braced Multistory Steel Frames* [46] provides useful design aids.

Expressed in terms of ultimate loads (factored service loads) the stability criterion, Eq. 12.8.4, converted into equivalent axial compression becomes

$$\text{Equivalent } P_{cr} = P + M_i \frac{P_{cr}}{M_m}\left(\frac{C_m}{1-P/P_e}\right) \qquad (12.13.1)$$

Since $P_{cr} = 1.70F_aA$ and M_m is normally $M_p = ZF_y$,

$$\text{Equivalent } P_{cr} = P + M_i \frac{1.70F_aA}{F_yZ}\left(\frac{C_m}{1-P/P_e}\right) \qquad (12.13.2)$$

The magnification term may be written

$$\frac{C_m}{1-P/P_e} = \frac{C_m}{1-f_a/(1.92F_e')} = \frac{C_m}{1-0.52f_a/F_e'} = \frac{C_m a}{a - 0.52P(KL)^2} \qquad (12.13.3)$$

Also, $Z \approx 1.12S$ and $F_a \approx 0.5F_y$, which gives

$$\frac{1.70F_aA}{F_yZ} \approx \frac{1.7(0.5)F_yA}{1.12F_yZ} \approx 0.75B$$

where $B = $ bending factor, A/S.

Thus the stability criterion, Eq. 12.13.1 [AISC Formula (2.4–2)], becomes approximately

$$\text{Equivalent } P_{cr} = P + M_i(0.75)B\left(\frac{C_m a}{a - 0.52P(KL)^2}\right) \qquad (12.13.4)$$

If stability governs,

$$\text{Equivalent } P_{cr} = 1.70F_aA$$

$$\text{Required } A = \frac{\text{Equivalent } P_{cr}}{1.70F_a} \qquad (12.13.5)$$

On the other hand, if strength (yielding) governs at braced points, Eq. 12.8.1 must form the basis of the Equivalent P; thus

$$\text{Equivalent } P_y = P + M\frac{P_y}{1.18M_p} \qquad (12.13.6)$$

Since $P_y = AF_y$ and $M_p = ZF_y \approx 1.12SF_y$, Eq, 12.13.6 becomes

$$\text{Equivalent } P_y = P + M(0.76)B \qquad (12.13.7)$$

where $B = $ bending factor, A/S. Because the bending factor involves the working stress concept of section modulus, a function of the basic dimensions may be preferred, in which case, for strong-axis bending,

$$S = Ar_x^2/(d/2) = 2Ad(r_x/d)^2 \qquad (12.13.8)$$

$$\frac{P_y}{1.18M_p} \approx \frac{AF_y}{1.18(1.12)SF_y} \approx \frac{1}{1.32(2)d(r_x/d)^2} = \frac{0.37}{d(r_x/d)^2} \qquad (12.13.9)$$

Using $r_x/d \approx 0.43$ (see text Appendix Table A1 or Ref. 46), and substituting Eq. 12.13.9 into Eq. 12.13.6 gives

$$\text{Equivalent } P_y = P + 2M/d \qquad (12.13.10)$$

Reference 46 suggests 2.1 instead of 2. This shows that for strong-axis bending, $B_x \approx 2.6/d$.

Thus for selection purposes Eqs. 12.13.4 and 12.13.10 may be used. In addition, for the selected section, the plastic moment capacity must exceed the factored applied moment.

Example 12.13.1

Repeat Example 12.11.1, except use plastic design (see Fig. 12.11.1). $F_y = 36$ ksi.

SOLUTION
Assuming the plastic analysis of the structure has been made for preliminary sections using a gravity load factor of 1.7,

$$P_u = 255 \text{ kips}; \qquad M_u = 850 \text{ kips}$$

For this braced system, the effective-length factors are $K_x = K_y = 1.0$. Assuming that stability controls, use Eq. 12.13.4 for Equivalent P and as a first approximation assume

$$\text{Equivalent } P_{cr} = P + M_i(0.75)B_x$$

$$= 255 + 850(12)(0.75)(0.19) = 1709 \text{ kips}$$

When AISC Manual, Column Load Tables are available, Eq. 12.13.5 gives

$$\text{Required } AF_a = \frac{\text{Equivalent } P_{cr}}{1.70} = 1005 \text{ kips}; \qquad KL = 14 \text{ ft}$$

A comparison with Example 12.11.1 will show the approach to selecting a section when stability governs is identical for working stress and plastic design. From AISC Manual select as first trial, W14×193, with $B_x = 0.183$, and $a_x = 357.6 \times 10^6$.

Check W14 × 193: Use AISC Formula (2.4–2), Eq. 12.8.4, for stability:

$$\frac{KL}{r_y} = \frac{168}{4.05} = 41.5; \qquad F_a = 19.1 \text{ ksi}$$

$$P_{cr} = 1.70 F_a A = 1.70(19.1)(56.8) = 1840 \text{ kips}$$

$$M_m = \left[1.07 - \frac{\sqrt{F_y}(L/r_y)}{3160} \right] M_p$$

$$= [1.07 - \sqrt{36}(41.5)/3160] M_p = 0.99 M_p$$

$$= 0.99 F_y Z = 0.99(36)(355)\tfrac{1}{12} = 1050 \text{ ft-kips}$$

$$\frac{KL}{r_x} = \frac{168}{6.50} = 25.8; \qquad F'_e = 225 \text{ ksi}$$

$$P_e = (23/12) F'_e A = 1.92(225)56.8 = 24{,}500 \text{ kips}$$

$$\frac{P}{P_{cr}} + \frac{M_i}{M_m} \left(\frac{C_m}{1 - P/P_e} \right) = \frac{255}{1840} + \frac{850}{1050} \left(\frac{1.0}{1 - 255/24{,}500} \right)$$

$$= 0.14 + 0.82 = 0.96 < 1 \qquad \text{OK}$$

Check the strength criterion, AISC Formula (2.4–3):

$$P_y = F_y A = 36(56.8) = 2040 \text{ kips}$$

$$M_p = F_y Z = 36(355)\tfrac{1}{12} = 1070 \text{ ft-kips}$$

$$\frac{P}{P_y} + \frac{M}{1.18 M_p} = \frac{255}{2040} + \frac{850}{1.18(1070)} = 0.80 < 1.0 \qquad \text{OK}$$

Use W14 × 193.

This is the same as obtained by working stress method, as would be expected for this statically determinate system. There would ordinarily be a difference, however, for a real structure where the analysis to obtain the loads would be performed differently—plastic analysis of a rigid frame structure vs elastic analysis with perhaps some arbitrary adjustment of negative moments. For plastic analysis and design of frames, see Chapter 15.

Example 12.13.2

Select the lightest W section to resist the factored (LF = 1.70) gravity loading on the braced frame of Fig. 12.13.1 using plastic design according to AISC Specification. Use A36 steel.

SOLUTION

Use the unbraced length as the equivalent pin-end length. The alignment chart for $K < 1.0$ (Fig. 14.3.1) should _not_ be used for this case since the point of contraflexure is close to the bottom and a plastic condition is

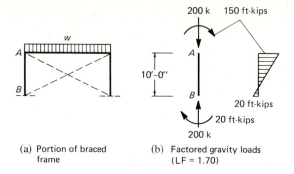

(a) Portion of braced frame

(b) Factored gravity loads (LF = 1.70)

Fig. 12.13.1 Example 12.13.2.

assumed to develop at the top. The failure mode for buckling would probably be one of single curvature, with the moment at the bottom close to zero and the restraint at the top reduced when the plastic condition is reached.

Assume that the strength criterion, AISC Formula (2.4–3), governs. Use Eq. 12.13.10 to obtain Equivalent P_y:

$$\text{Equivalent } P_y = P + \frac{2M}{d}$$

$$= 200 + \frac{2(150)(12)}{d} = 200 + \frac{3600}{d}$$

d	Equiv. P_y	$A = P_y/F_y$	Section
W12	500 kips	13.9 sq in.	W12 × 50, $A = 14.7$ sq in.
W14	457 kips	12.7 sq in.	W14 × 48, $A = 14.1$ sq in.

Check W14 × 48 according to AISC Formula (2.4–3):

$$P_y = AF_y = 14.1(36) = 508 \text{ kips}$$

$$M_p = ZF_y = 78.4(36)\tfrac{1}{12} = 235 \text{ ft-kips}$$

$$\frac{P}{P_y} + \frac{M}{1.18M_p} = \frac{200}{508} + \frac{150}{1.18(235)} = 0.93 < 1 \qquad \text{OK}$$

Check stability according to AISC Formula (2.4–2):

$$\frac{KL}{r_y} = \frac{120}{1.91} = 62.8; \qquad F_a = 17.2 \text{ ksi}$$

$$P_{cr} = 1.70F_aA = 1.70(17.2)14.1 = 412 \text{ kips}$$

$$M_m = \left[1.07 - \frac{\sqrt{F_y}(L/r_y)}{3160}\right]M_p = 0.95M_p = 223 \text{ ft-kips}$$

$$\frac{KL}{r_x} = \frac{120}{5.85} = 20.5; \qquad F'_e = 355 \text{ ksi}$$

$$P_e = (23/12)F'_e A = 1.92(355)14.1 = 9610 \text{ kips}$$

$$C_m = 0.6 - 0.4(20/150) = 0.55$$

$$\frac{P}{P_{cr}} + \frac{M_i}{M_m}\left(\frac{C_m}{1-P/P_e}\right) = \frac{200}{412} + \frac{150}{223}\left(\frac{0.55}{1-200/9610}\right) = 0.86 < 1 \qquad \text{OK}$$

The W14×48 satisfies the local buckling limitations of AISC–2.7:

$$\frac{b_f}{2t_f} = \frac{8.030}{2(0.595)} = 6.7 < 8.5 \qquad \text{OK}$$

For $P/P_y > 0.27$,

$$\frac{d}{t_w} = \frac{13.79}{0.340} = 40.6 < \frac{257}{\sqrt{F_y}} = 42.8 \qquad \text{OK}$$

Use W14×48.

Example 12.13.3

Select a W shape to serve as the beam-column of a gabled frame as shown in Fig. 12.13.2. Use A36 steel. The structure is braced transverse to the plane of the frame, with bracing at top, bottom, and at mid-height on member *AB*.

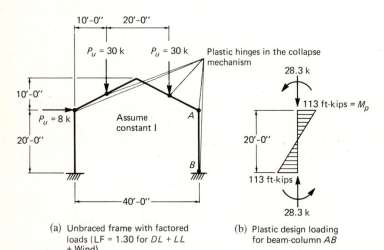

(a) Unbraced frame with factored loads (LF = 1.30 for *DL* + *LL* + Wind)

(b) Plastic design loading for beam-column *AB*

Fig. 12.13.2 Example 12.13.3.

SOLUTION

Assume the plastic analysis has been made for the frame using constant moment of inertia.

(a) Estimate Equivalent *P* based on the strength criterion, AISC

Formula (2.4–3). Using Eq. 12.13.10,

$$\text{Equivalent } P_y = P + \frac{2M}{d} = 28.3 + \frac{2(113)(12)}{8} \quad \text{(for W8)}$$

$$\text{Required } A = \frac{\text{Equivalent } P_y}{F_y} = \frac{367}{36} = 10.2 \text{ sq in.}$$

Try W8×35, $A = 10.3$ sq in.

(b) Check strength criterion, AISC Formula (2.4–3):

$$P_y = AF_y = 10.3(36) = 371 \text{ kips}$$

$$M_p = ZF_y = 34.7(36)\tfrac{1}{12} = 104 \text{ ft-kips}$$

It is immediately noted that the required strength M exceeds the plastic moment capacity of the section; since this cannot be, the section must be increased so that its $M_p \geq 113$ ft-kips. It appears W8×40 or W10×39 will satisfy the strength criterion.

(c) Investigate lateral bracing requirement. Since plastic hinges are expected to occur at the top and bottom of the member, the provisions of AISC–2.9 apply. With lateral support at top, bottom, and mid-height, the moment gradient is $M/M_p = 0$ which is more than (-0.5):

$$L_{cr} = \left[\frac{1375}{F_y} + 25\right] r_y = \left[\frac{1375}{36} + 25\right] r_y = 63.2 r_y$$

If unbraced over a length of 10 ft,

$$\text{Required } r_y = \frac{120}{63.2} = 1.90 \text{ in.}$$

Both the W8×40 and W10×39 will satisfy this requirement.

(d) Complete check of strength requirement for W10×39:

$$P_y = AF_y = 11.5(36) = 414 \text{ kips}$$

$$M_p = ZF_y = 46.8(36)\tfrac{1}{12} = 141 \text{ ft-kips} > 113 \text{ ft-kips} \quad \text{OK}$$

$$\frac{P}{P_y} + \frac{M}{1.18M_p} = \frac{28.3}{414} + \frac{113}{1.18(140)} = 0.75 < 1 \quad \text{OK}$$

(e) Check stability requirement, AISC Formula (2.4–2). *Note*: With large moment restraint at the base of the frame, say $G_B = 1.0$, and with I/L approximately the same for beam and column, $G_A = 1.0$. From Alignment Chart, Fig. 14.3.1b, find $K = 1.3$.

$$\frac{KL}{r_x} = \frac{1.3(240)}{4.27} = 73; \quad F'_e = 28 \text{ ksi}$$

$$P_e = (23/12)F'_e A = 1.92(28)(11.5) = 618 \text{ kips}$$

$$C_m = 0.85 \quad \text{for an unbraced frame}$$

$$\frac{KL}{r_y} = \frac{120}{1.98} = 60.6 < 73; \qquad F_a = 16.1 \text{ ksi}$$

$$P_{cr} = 1.70 F_a A = 1.70(16.1)(11.5) = 315 \text{ kips}$$

$$M_m = \left[1.07 - \frac{\sqrt{F_y}(L/r_y)}{3160}\right] M_p = 0.95 M_p = 133 \text{ ft-kips}$$

$$\frac{P}{P_{cr}} + \frac{M_i}{M_m}\left(\frac{C_m}{1 - P/P_e}\right) = \frac{28.3}{315} + \frac{113}{133}\left(\frac{0.85}{1 - 28.3/618}\right) = 0.85 < 1 \qquad \text{OK}$$

The W10×39 does satisfy the local buckling requirement of AISC–2.7. Though this section is heavier than required, the next lighter W10 does not have adequate plastic moment strength.
Use W10×39.

For practical plastic design of beam-columns, there are a number of design aids and charts available [30, 46].

SELECTED REFERENCES

1. Charles Massonnet, "Stability Considerations in the Design of Steel Columns," _Journal of Structural Division_, ASCE, 85, ST7 (September 1959), 75–111.
2. Walter J. Austin, "Strength and Design of Metal Beam-Columns," _Journal of Structural Division_, ASCE, 87, ST4 (April 1961), 1–32.
3. Bruce G. Johnston, ed., _Structural Stability Research Council, Guide to Stability Design Criteria for Metal Structures_, 3rd Ed. New York: John Wiley & Sons, Inc., 1976, Chap. 8.
4. Robert L. Ketter, "Further Studies of the Strength of Beam-Columns," _Journal of Structural Division_, ASCE, 87, ST6 (August 1961), 135–152.
5. M. R. Horne, "The Stanchion Problem in Frame Structures Designed According to Ultimate Carrying Capacity," _Proc. Inst. Civil Engrs._, 5, 1, Part III (April 1956), 105–146.
6. Robert L. Ketter, Edmund L. Kaminsky, and Lynn S. Beedle, "Plastic Deformation of Wide-Flange Beam-Columns," _Transactions_, ASCE, 120 (1955), 1028–1069.
7. N. M. Newmark, "Numerical Procedure for Computing Deflections, Moments, and Buckling Loads," _Transactions_, ASCE, 108 (1943), 1161–1234.
8. Ping-Chun Wang, _Numerical and Matrix Methods in Structural Mechanics_. New York: John Wiley & Sons, Inc., 1966 (pp. 96–109).
9. Theodore V. Galambos and Robert L. Ketter, "Columns Under Combined Bending and Thrust," _Journal of Engineering Mechanics Division_, ASCE, 85, EM2 (April 1959), 1–30.
10. George F. Hauck and Seng-Lip Lee, "Stability of Elasto-Plastic Wide-Flange Columns," _Journal of Structural Division_, ASCE, 89, ST6 (December 1963), 297–324.
11. S. L. Lee and G. F. Hauck, "Buckling of Steel Columns Under Arbitrary End Loads," _Journal of Structural Division_, ASCE, 90, ST2 (April 1964), 179–200.

12. S. L. Lee and S. C. Anand, "Buckling of Eccentrically Loaded Steel Columns," *Journal of Structural Division*, ASCE, 92, ST2 (April 1966), 351–370.
13. Edwin C. Rossow, George B. Barney, and Seng-Lip Lee, "Eccentrically Loaded Steel Columns with Initial Curvature," *Journal of Structural Division*, ASCE, 93, ST2 (April 1967), 339–358.
14. Le-Wu Lu and Hassan Kamalvand, "Ultimate Strength of Laterally Loaded Columns," *Journal of Structural Division*, ASCE, 94, ST6 (June 1968), 1505–1524.
15. W. F. Chen, "Further Studies of Inelastic Beam-Column Problem," *Journal of Structural Division*, ASCE, 97, ST2 (February 1971), 529–544.
16. Gordon W. English and Peter F. Adams, "Experiments on Laterally Loaded Steel Beam-Columns," *Journal of Structural Division*, ASCE, 99, ST7 (July 1973), 1457–1470.
17. Francois Cheong-Siat-Moy, "Methods of Analysis of Laterally Loaded Columns," *Journal of Structural Division*, ASCE, 100, ST5 (May 1974), 953–970.
18. Francois Cheong-Siat-Moy, "General Analysis of Laterally Loaded Beam-Columns," *Journal of Structural Division*, ASCE, 100, ST6 (June 1974), 1263–1278.
19. B. G. Johnston, "Lateral Buckling of I-Section Columns with Eccentric End Loads in the Plane of the Web," *Transactions*, ASME, 62 (1941), A–176.
20. Mario G. Salvadori, "Lateral Buckling of I-Beams," *Transactions*, ASCE, 120 (1955), 1165–1182.
21. M. Salvadori, "Lateral Buckling of Eccentrically Loaded I-Columns," *Transactions*, ASCE, 121 (1956), 1163–1178.
22. Constancio Miranda and Morris Ojalvo, "Inelastic Lateral-Torsional Buckling of Beam Columns," *Journal of Engineering Mechanics Division*, ASCE, 91, EM6 (December 1965), 21–37.
23. Yushi Fukumoto and T. V. Galambos, "Inelastic Lateral-Torsional Buckling of Beam Columns," *Journal of Structural Division*, ASCE, 92, ST2 (April 1966), 41–61.
24. T. V. Galambos, P. F. Adams, and Y. Fukumoto, "Further Studies on the Lateral-Torsional Buckling of Steel Beam-Columns," *Bulletin No. 115*, Welding Research Council, July 1966.
25. Ralph C. Van Kuren and T. V. Galambos, "Beam Column Experiments," *Journal of Structural Division*, ASCE, 90, ST2 (April 1964), 223–256.
26. T. B. Pekoz and G. Winter, "Torsional-Flexural Buckling of Thin-Walled Sections Under Eccentric Load," *Journal of Structural Division*, ASCE, 95, ST5 (May 1969), 941–963.
27. Teoman B. Pekoz and N. Celebi, "Torsional Flexural Buckling of Thin-Walled Sections Under Eccentric Load," *Cornell Engineering Research Bulletin 69–1*. Ithaca, N.Y.: Cornell University, 1969.
28. Wei-Wen Yu, *Cold-Formed Steel Structures*. New York: McGraw-Hill Book Company, Inc., 1973, Chap. 6.
29. *Specification for the Design of Cold-Formed Steel Structural Members*. Washington, D.C.: American Iron and Steel Institute, 1968, with Addenda (No. 1 dated November 19, 1970, No. 2 dated February 4, 1977).
30. George C. Driscoll, Jr., et al., "Plastic Design of Multistory Frames," Lecture Notes, Fritz Engineering Laboratory Report No. 273.20, 1965.

31. Charles Birnstiel and James Michalos, "Ultimate Load of H-Columns Under Biaxial Bending," *Journal of Structural Division*, ASCE, 89, ST2 (April 1963), 161–197.

32. Charles G. Culver, "Exact Solution of Biaxial Bending Equations," *Journal of Structural Division*, ASCE, 92, ST2 (April 1966), 63–83.

33. Charles G. Culver, "Initial Imperfections in Biaxial Bending," *Journal of Structural Division*, ASCE, 92, ST3 (June 1966), 119–135.

34. Gunnar A. Harstead, Charles Birnstiel, and Keh-Chun Leu, "Inelastic Behavior of H-Columns Under Biaxial Bending," *Journal of Structural Division*, ASCE, 94, ST10 (October 1968), 2371–2398.

35. Ishwar C. Syal and Satya S. Sharma, "Biaxially Loaded Beam-Column Analysis," *Journal of Structural Division*, ASCE, 97, ST9 (September 1971), 2245–2259.

36. Sakda Santathadaporn and Wai F. Chen, "Analysis of Biaxially Loaded Steel H-Columns," *Journal of Structural Division*, ASCE, 99, ST3 (March 1973), 491–509.

37. Wai F. Chen and Toshio Atsuta, "Ultimate Strength of Biaxially Loaded Steel H-Columns," *Journal of Structural Division*, ASCE, 99, ST3 (March 1973), 469–489.

38. Charles Birnstiel, "Experiments on H-Columns Under Biaxial Bending," *Journal of Structural Division*, ASCE, 94, ST10 (October 1968), 2429–2449.

39. Wai F. Chen and Sakda Santathadaporn, "Review of Column Behavior Under Biaxial Loading," *Journal of Structural Division*, ASCE, 94, ST12 (December 1968), 2999–3021.

40. Negussie Tebedge and Wai F. Chen, "Design Criteria for H-Columns Under Biaxial Loading," *Journal of Structural Division*, ASCE, 100, ST3 (March 1974), 579–598.

41. Wai F. Chen and Toshio Atsuta, "Simple Interaction Equations for Beam-Columns," *Journal of Structural Division*, ASCE, 98, ST7 (July 1972), 1413–1426.

42. Ira Hooper, "Design of Beam-Columns," *Engineering Journal*, AISC, 4, 2 (April 1967), 41–61.

43. Moe A. Rubinsky, "Rapid Selection of Beam-Columns," *Engineering Journal*, AISC, 5, 3 (July 1968), 100–122.

44. William Y. Liu, "Steel Column Bending Amplification Factor," *Engineering Journal*, AISC, 2, 2 (April 1965), 50–51.

45. Benjamin Koo, "Amplification Factors for Beam-Columns," *Engineering Journal*, AISC, 5, 2 (April 1968), 66–71.

46. *Plastic Design of Braced Multistory Steel Frames.* New York: American Iron and Steel Institute, 1968 (pp. 8–12, 98–111).

PROBLEMS

For all problems* involving design or safety analysis, use the latest AISC Specification unless otherwise indicated. The requirement of W section is intended to include W, S, and M sections.

---

* Many problems may be solved either as stated in U.S. customary units or in SI units using the numerical data in parenthesis at the end of the statement. The conversions are approximate to avoid having the given data imply accuracy in SI greater than for U.S. customary units.

Problems 12.1 through 12.27 relate to design considerations. Problems 12.28 through 12.38 relate to other theoretical considerations.

12.1. Investigate the adequacy of the section according to the working stress method. No translation of joints can occur, and external lateral support is provided at the ends only. ($w = 12$ kN/m; axial compression $= 70$ kN; span $= 3$ m)

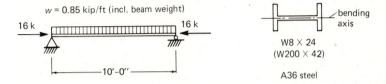

Prob. 12.1

12.2. Investigate the given section with regard to safety if primary bending is in the weak direction. Use working stress method. A572 Grade 50 steel. ($w = 1.5$ kN/m; $P = 90$ kN; span $= 3$ m)

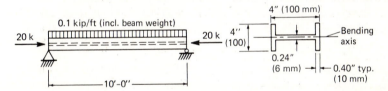

Prob. 12.2

12.3. Determine the allowable load Q (kips) at the mid-height of the beam-column shown. Assume the member is hinged with respect

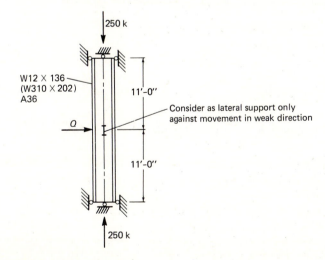

Prob. 12.3

to bending in both x and y directions at the top and bottom. Additionally, lateral support occurs in the weak direction at mid-height. Use working stress method. ($P = 1100$ kN; total column height $= 6.8$ m)

12.4. Investigate the adequacy of the given section, according to the working stress method. No joint translation can occur and external lateral support is provided at the ends only. ($w = 4.4$ kN/m; $P = 220$ kN; $M = 7$ kN $\cdot$ m; $L = 4.5$ m)

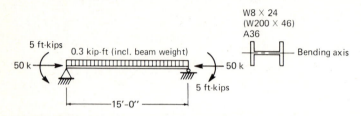

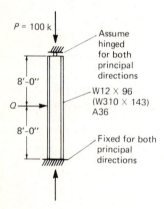

Prob. 12.4

12.5. Determine the safe service load Q permitted according to the working stress method. Braced frame. ($P = 450$ kN; total column height $= 5$ m)

Prob. 12.5

12.6. Determine the allowable axial load P which the W12×45 may be permitted to carry according to the working stress method. La-

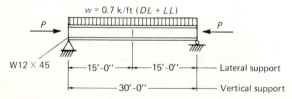

Prob. 12.6

teral support is provided at ends and at midspan. Compare for A36 and A572 Grade 50 steels. (W310×67 section; $w = 10$ kN/m; total span $= 9$ m)

12.7. Select the lightest W14 section to carry a service load $P = 500$ kips with an eccentricity $e = 12$ in. with respect to the strong axis. Assume the member is part of a braced system, and conservatively assume the effective length equals the unbraced height. Use (a) A36 steel; (b) A572 Grade 60 steel. Use working stress method. (W360 section; $P = 2200$ kN; $e = 300$ mm; span $= 4.3$ m)

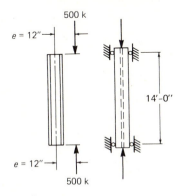

Prob. 12.7

12.8. Select the lightest W14 section to carry an axial compression of 240 kips along with a bending moment of 450 ft-kips, which, for conservative simplicity, is assumed to be constant along the 15-ft equivalent pin-end length of the member in the braced structure. Use (a) A36 steel; (b) A572 Grade 50. Use the working stress method. (W360 section; $P = 1100$ kN; $M = 610$ kN · m; $L = 4.5$ m)

12.9. If the service load P is 125 kips and w is 2 kips/ft for the beam-column span and support conditions of Prob. 12.6, select the lightest W section according to the working stress method. Use (a) A36 steel; (b) A572 Grade 60. ($P = 550$ kN; $w = 30$ kN/m)

12.10. Select the lightest W14 section for the beam-column shown. Use working stress method and assume lateral buckling is adequately prevented such that $L < L_c$. Use A36 steel. (W360 section; $P = 300$ kN; $w = 40$ kN/m; $L = 9.2$ m)

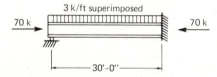

Prob. 12.10

12.11. For the member of a braced system shown, select the lightest W section. Use working stress method with (a) A36 steel and (b) A572 Grade 60 steel. ($P = 440$ kN; $M = 92$ kN · m; total column height = 10 m; 6 m upper segment)

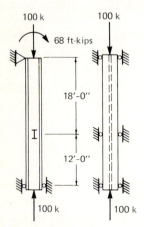

Prob. 12.11

12.12. Redesign the section for Example 12.11.6 for A572 Grade 60 steel. Use the working stress method.

12.13. A frame braced against sidesway has a beam-column loaded as shown resulting from an elastic analysis. The horizontal member has lateral support at its ends and every 9 ft. Select the lightest W section acceptable using working stress method with A36 steel.

12.14. Redesign the member of Prob. 12.13 as part of an unbraced frame. Assume moment of inertia for beam and column to be the same. Use working stress method.

12.15. Design the lightest W12 for column A of the unbraced frame shown. Preliminary design has selected W27 × 94 for all adjacent beams, and it has been decided column A will be approximately the same stiffness as those above and below it. Use working stress method with A572 Grade 50 steel.

12.16. For the vierendeel truss (rigid frame) shown, investigate the safety of members A and B, according to the working stress method. The uniform loading includes the weight of the steel section and the steel is A36. Assume simple cross-bracing between given frame and an adjacent parallel one.

12.17. Redesign member A for A572 Grade 60 steel, using working stress method.

12.18. Design the lightest W section to carry two eccentric 20-kip loads causing primary bending plus an axial compression of 275 kips. Use working stress method and A36 steel. Assume torsional fixity at the vertical supports. Assume lateral bracing at midspan is

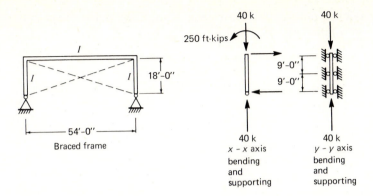

Probs. 12.13 and **12.14**

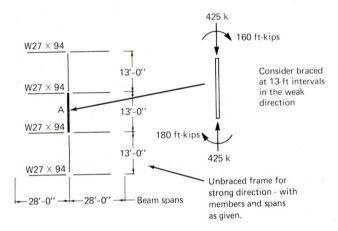

Prob. 12.15

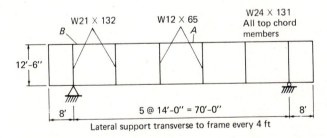

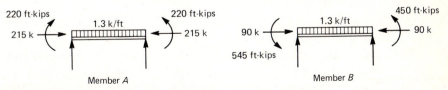

Probs. 12.16 and **12.17**

adequate to prevent lateral-torsional buckling, but does permit enough rotation for torsional moments to develop.

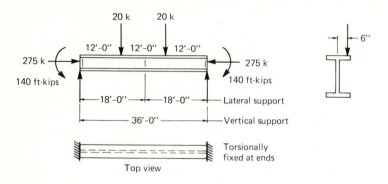

Prob. 12.18

12.19. Determine whether or not the given structural tee is adequate for the loading shown. Assume the uniform loading is delivered through construction which prevents lateral buckling of the member. Assume continuity over the end supports. Use A36 steel and working stress method.

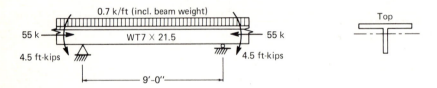

Prob. 12.19

12.20. Select the lightest WT7 structural tee for the loading shown. Use A572 Grade 50 steel with working stress method. Assume vertical and lateral supports at ends only.

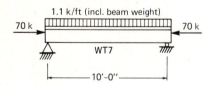

Prob. 12.20

12.21. Select an economical structural tee to serve as a continuous compression chord member of a truss to carry service loads as shown. For design purposes assume fixed ends on the member. Use working stress method. Use the more economical of A36 or

A572 Grade 50 steel, assuming the Grade 50 costs 12% more per lb than A36.

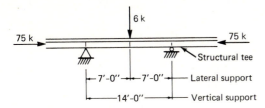

Prob. 12.21

12.22. Select the lightest W section for the member of the unbraced frame using plastic design procedure. Assume simple lateral bracing at 8-ft intervals transverse to the plane of the frame.

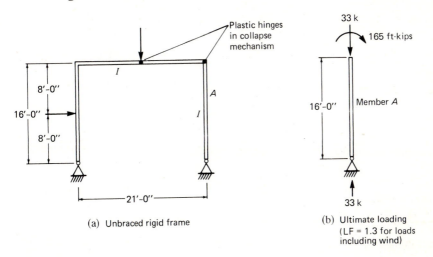

(a) Unbraced rigid frame

(b) Ultimate loading
(LF = 1.3 for loads including wind)

Prob. 12.22

12.23. Redesign the member of Prob. 12.13 using the plastic design method. Assume ultimate axial force and moment (i.e., obtained using factored loads) are $P_u = 66$ kips and $M_u = 380$ ft-kips, and that a plastic hinge in the collapse mechanism occurs at the top of the member.

12.24. Design the lightest W section for the ultimate conditions (M and P obtained using factored loads) using plastic design method. The collapse mechanism does *not* have a plastic hinge in this member.

12.25. Repeat Prob. 12.24 using $P_u = 500$ kips and $M_u = 60$ kips.

12.26. A column in a building has reactions at the top from beams framing into it. Assume the framing beams contribute moments at the top of the column, but the bottom of the column is hinged (no moments). The beams framing into the web are assumed to rest on seats, where reactions are assumed to be 2 in. from the center of the web. The reaction from the other beam is assumed to be acting at the face of the flange. Use A36 steel and select lightest W section. Use working stress method.

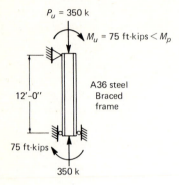

Probs. 12.24 and **12.25**

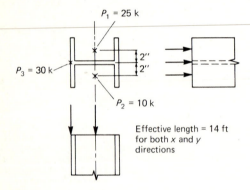

Prob. 12.26

12.27. For the loading shown, select lightest sections for the following conditions using working stress method:
(a) W14 using A36 steel
(b) W in any depth using $F_y = 50$ ksi
(c) W14 using $F_y = 60$ ksi
(d) W14 using $F_y = 70$ ksi

Problems Relating to Theoretical Considerations

12.28 through **12.32.** For the given loading and support conditions, developed the differential equation for M_z, the moment in the plane of bending, and determine the maximum value for M_z.

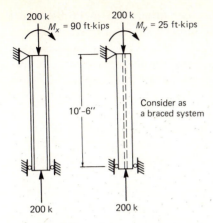

Prob. 12.27

Probs. 12.28 and **12.33**

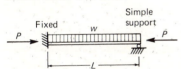

Probs. 12.29 and **12.34**

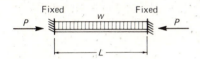

Probs. 12.30 and **12.35**

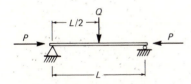

Probs. 12.31 and **12.36**

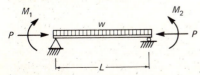

Probs. 12.32 and **12.37**

12.33 through **12.37.** For each loading condition used in Probs. 12.28 through 12.32, compute the maximum total compressive stress on the extreme fiber of a W4×13 bent in the weak direction using (a) the differential equation solution and (b) the approximate procedure of $M_{max} = M_{i\,max} + Py_{max}$. For unknowns use $P = 20$ kips, $w = 0.1$ kip/ft, $Q = 0.5$ kips, $a = 3$ ft, $M_1 = 0.6$ ft-kips, $M_2 = 1.0$ ft-kips, and $L = 10$ ft.

12.38. (a) Develop the differential equation solution for the loading of Prob. 12.10, and determine the expression for maximum bending moment. Assume the fixed-end moment may be approximated as $wL^2/8$.

(b) For the W14 selected in Prob. 12.10 compute the maximum bending moment using both the differential equation and the approximate method, $M_{max} = M_{i\,max} + Py_{max}$.

(c) Develop the differential equation solution assuming zero slope is required at the fixed end. How much effect does moment magnification have on the end moment?

13
Connections

13.1 TYPES OF CONNECTIONS

Under AISC–1.2 for working stress design and AISC–2.1 for plastic design, steel construction is defined in three categories according to the type of connections used. The following three types are indicated.

1. AISC Type 1. *Rigid-frame*, where full continuity is provided at the connection so that the original angles between intersecting members are held virtually constant; i.e., with rotational restraint on the order of 90% or more of that necessary to prevent any angle change. Such connections are used under both the working stress and plastic design methods.
2. AISC Type 2. *Simple framing*, where rotational restraint at the ends of members is as little as practicable. For beams, simple framing provides only shear transfer at the ends. One may consider the framing simple if the original angle between intersecting pieces may change up to 80 percent of the amount it would theoretically change if frictionless hinged connections could be used. Design of simply supported beams under working stress method uses Type 2 connections. Simple framing is not used in

Welded connections for rigid frame construction, showing beam-to-column connections with column web stiffeners. Rural Mutual Insurance Building, Madison, Wis. (Photo by C. G. Salmon)

plastic design, except for connections of members transverse to the plane of the frame in which the plastic strength is to develop. Two or more plastically designed planar systems may be linked together by means of simple framing combined with cross-bracing.

3. AISC Type 3. *Semi-rigid framing*, where rotational restraint is between 20 and 90% of that necessary to prevent any relative angle change. Alternatively, one may consider that with semi-rigid framing the moment transmitted across the joint is neither zero (or a small amount) as in simple framing, nor is it the full continuity moment as assumed in elastic rigid-frame analysis. AISC–1.2 states that design of construction using Type 3 connections may be used when "the connections of beams and girders possess a *dependable and known* moment capacity intermediate in degree between the rigidity of Type 1 and the flexibility of Type 2."

Semi-rigid connections are not used in plastic design and only rarely used in working stress design, primarily because of the difficulty of evaluating the degree of restraint.

Beam Line

In order to better understand the practical distinction between the AISC framing types, the beam line developed by Batho and Rowan [1] and used by Sourochnikoff [2] is a useful graphical device.

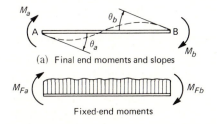

(a) Final end moments and slopes

Fixed-end moments

Fig. 13.1.1 Moments and rotations for slope-deflection equations (shown with positive signs).

As shown in Fig. 13.1.1, consider a beam AB loaded in any manner and subject to end moments M_a and M_b, and with end slopes θ_a and θ_b. The moments necessary to have $\theta_a = \theta_b = 0$ are designated M_{Fa} and M_{Fb}, the fixed-end moments. Writing the slope deflection equations,

$$M_a = M_{Fa} + \frac{4EI}{L}\theta_a + \frac{2EI}{L}\theta_b$$
$$M_b = M_{Fb} + \frac{2EI}{L}\theta_a + \frac{4EI}{L}\theta_b$$

(13.1.1)

Solving Eqs. 13.1.1 for θ_a and θ_b gives

$$\left.\begin{array}{l} \dfrac{6EI}{L}\theta_a = 2(M_a - M_{Fa}) - (M_b - M_{Fb}) \\[2mm] \dfrac{6EI}{L}\theta_b = -(M_a - M_{Fa}) + 2(M_b - M_{Fb}) \end{array}\right\} \qquad (13.1.2)$$

Subtracting the second equation from the first gives

$$\dfrac{6EI}{L}(\theta_a - \theta_b) = 3(M_a - M_b) - 3(M_{Fa} - M_{Fb}) \qquad (13.1.3)$$

If symmetrical loading is considered, then

$$M_b = -M_a, \qquad \theta_b = -\theta_a, \qquad M_{Fb} = -M_{Fa} \qquad (13.1.4)$$

in which case Eq. 13.1.3 becomes

$$\dfrac{2EI}{L}\theta_a = M_a - M_{Fa}$$

or

$$M_a = M_{Fa} + \dfrac{2EI}{L}\theta_a \qquad (13.1.5)$$

which may be called the *beam-line equation*. When $\theta_a = 0$ (a full fixity condition), $M_a = M_{Fa}$; and for a hinged end where $M_a = 0$, the slope becomes $\theta_a = -M_{Fa}/(2EI/L)$.

Figure 13.1.2 shows a diagram of the beam-line equation and also the moment-rotation behavior of typical connections of Types 1, 2, and 3.

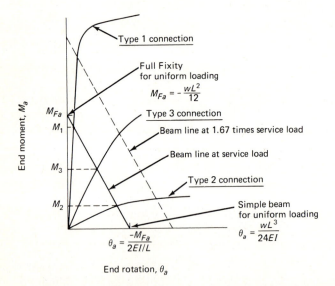

Fig. 13.1.2 Moment-rotation characteristics of AISC connection types.

The typical rigid connection would have to carry an end moment M_1, about 90% or more of M_{Fa}; hence its degree of restraint may be said to be 90%. The simple connection (Type 2) may have to resist only 20% or less of the moment M_{Fa}, as indicated by the moment M_2, while the semi-rigid connection would be expected to resist some intermediate value M_3, at perhaps 50% of the fixed-end moment M_{Fa}.

If one is able to establish the moment-rotation characteristics of a particular connection, then the strength can be designed so that the resulting end rotation θ is compatible with that caused by the loads. A summary and discussion of the moment-rotation characteristics of various connection arrangements is given by Schenker, Salmon, and Johnston [3].

13.2 FRAMED BEAM CONNECTIONS

These simple framing connections, AISC Type 2, are used to connect beams to other beams or to column flanges. For the most part these connections are standardized and given in the AISC Manual under the headings, "Framed Beam Connections" and "Heavy Framed Beam Connections." Typical bolted and welded framed connections are shown in Fig. 13.2.1. It is intended in such connections that the angles be as flexible as possible. The connection to the column (2 rows of 5 fasteners shown in Fig. 13.2.1a) is usually made in the field while the connection to the beam web (one row of 5 fasteners shown in Fig. 13.2.1b) is usually made in the shop. Generally on plans, shop fastener holes are shown as in Fig. 13.2.1b, while field fastener holes are shown as solid black dots.

In today's fabrication practice, the shop connection is usually welded, while the field connection may be either bolted or welded; thus any combination in Fig. 13.2.1 of (a) with (b) or (c); or (d) with (b) or (c), may be used. Recently there has been some use made of an end plate welded to the web of the beam and bolted to the beam or column to which it is attached. Kennedy [4] has presented data on the moment-rotation characteristics of such a connection.

When angles, sometimes known as *clip angles*, are used to attach a beam to a column there is a clearance setback of about $\frac{1}{2}$ in. so that if the beam is too long, within acceptable tolerances, the angles may be relocated without cutting off a piece of the beam. When beams intersect and are attached to other beams so that the flanges of both are at the same elevation, as in Fig. 13.2.1e, the beams framing in have their flanges coped, or cut away. The loss of section is primarily loss of flange that carries little shear anyway, so that normally a cope results in little loss of shear strength. Birkemoe and Gilmor [42] have shown that a coped web subject to high bearing stress in a high-strength bolted beam end connection may fail in a tearing mode along a line through the holes, as shown in Fig. 13.2.2. Particularly, the problem may arise when there are only a few bolts through the beam web and they do not extend uniformly over the

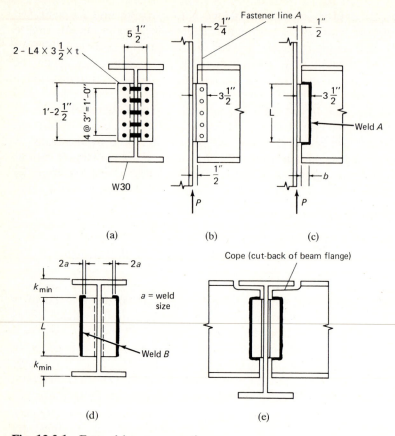

Fig. 13.2.1 Framed beam connections.

entire web depth. AISC–1.5.1.2.2 and AISC–1.16.5.3 contain special provisions relating to such end connections.

The number of high-strength bolts is based on the direct shear, neglecting any eccentricity of loading, while the weld length and size includes the effect of eccentric loading. The fasteners, bolts or welds, are designed in accordance with procedures of Chapters 4 and 5, respectively.

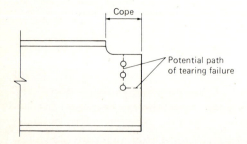

Fig. 13.2.2 Tearing failure at coped ends of framed beam connections.

The thickness of the angles is usually determined so that bearing does not govern. Shear stress on the gross angle section must also be checked. The angles are *expected* to bend in order that only a small amount of end restraint will occur.

Flexural Stress on Connection Angles

Next, examine the stress due to flexure on the clip angles, as shown in Fig. 13.2.3. The tensile force T per inch at the top of the angles is obtained by taking the applied moment (reaction P times eccentricity e of applied load measured to fastener line A or to the centroid of weld A) and computing the flexural stress on the projecting angle legs.

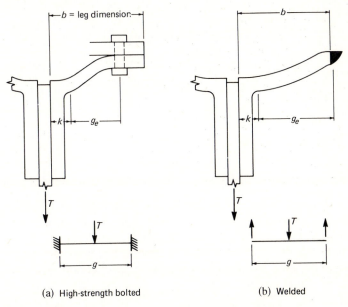

    (a) High-strength bolted   (b) Welded

Fig. 13.2.3 Behavior at tension edge of clip angles.

If the angle thickness is t and the angle length is L, the stress at the top of the angles is

$$f = \frac{Mc}{I} = \frac{Pe(L/2)}{\frac{1}{12}(2t)(L)^3} = \frac{3Pe}{tL^2} \tag{13.2.1}$$

and

$$T = (f)(2)(\text{angle thickness } t) = \frac{6Pe}{L^2} \tag{13.2.2}$$

where T is the load on the top 1 in. length of two connection angles.

The load T bends the connection angles as shown in Fig. 13.2.3. For the high-strength bolted connection the clamping force provides fixity at the bolt and the critical section in flexure is probably at the base of the fillet of the angle, a distance k (a dimension for detailing given by AISC Manual) from the back of angle.

If the bolted connection is considered a fixed-end beam (Fig. 13.2.3a) because of the clamping due to initial tension in the bolts, and the welded connection is considered a simply supported beam (Fig. 13.2.3b) because of the lack of initial tension, the bending stress in the angles may be approximated as:

For bolted angles,

$$f = \frac{M}{S} = \frac{M}{\frac{1}{6}(1)(t^2)} = \frac{6M}{t^2} \tag{13.2.3}$$

$$M \approx \frac{Tg}{8} \tag{13.2.4}$$

$$f = \frac{0.75Tg}{t^2} = \frac{4.5Peg}{t^2L^2} = \frac{9.0Peg_e}{t^2L^2} \tag{13.2.5}$$

For welded angles,

$$M \approx \frac{Tg}{4} \tag{13.2.6}$$

$$f = \frac{1.5Tg}{t^2} = \frac{9.0Peg}{t^2L^2} = \frac{18.0Peg_e}{t^2L^2} \tag{13.2.7}$$

From Eq. 13.2.5 or 13.2.7 one may observe that thick angles will reduce the stress; however, the deformation will also reduce, offsetting the desired flexibility. For some brackets treated later in Secs. 13.4 and 13.5, the rigidity may be desired. For framed connections, adequate deformation of the angles is desired to get close to simple support conditions. The deflections may be approximated as follows:

For bolted angles,

$$\Delta \approx \frac{T(2g_e)^3}{192EI} = \frac{T(2g_e)^3}{192E(\frac{1}{12})t^3} \tag{13.2.8}$$

from Eq. 13.2.5,

$$T = \frac{ft^2}{1.5g_e}$$

$$\Delta = \frac{fg_e^2}{3.0Et} \tag{13.2.9}$$

For welded angles,

$$\Delta \approx \frac{T(2g_e)^3}{48EI} = \frac{T(2g_e)^3}{48E(\frac{1}{12})t^3} \tag{13.2.10}$$

and using T from Eq. 13.2.7 gives

$$\Delta = \frac{fg_e^2}{1.5Et} \qquad (13.2.11)$$

Equations 13.2.9 and 13.2.11 show that for constant deformation Δ the thicker the angle the greater must be the stress. Further, for a constant stress f the thinner the angle the greater the deflection.

Example 13.2.1

Determine the capacity for the 5 row framed beam connection of Fig. 13.2.1 for connecting a W30×99 to a column with a $\frac{3}{4}$-in. flange. Use $\frac{3}{4}$-in.-diam A325 bolts in standard holes with clean mill scale (Class A) surface condition. Compute for a friction-type connection (A325-F) and a bearing-type connection with no threads in the shear plane (A325-X). Use A36 steel.

SOLUTION

(a) Friction-type connection (A325–F). Connection to the web of W30×99, $t_w = 0.520$ in., referring to Table 4.6.1,

$$R_{DS} = A_b F_v = 2(0.4418)17.5 = 15.46 \text{ kips}$$

If the eccentricity e with respect to the fastener line is considered, fastener line A is subject to eccentric shear. It has been common practice to neglect the eccentric shear effect in riveted and bolted framed beam connections. Thus the capacity based on the connection to the W30×99 web is

$$P = 5(15.46) = 77.3 \text{ kips}$$

The single shear connection to the $\frac{3}{4}$-in. flange involves fasteners subject to combined shear and tension as discussed in Sec. 4.10. The basic fastener value is

$$R_{SS} = 0.4418(17.5) = 7.73 \text{ kips}$$

For framed beam connections it is common to neglect the tension resulting from eccentricity of loading (AISC Manual tables of framed beam connection capacity neglect this effect). Neglecting the eccentricity,

$$P = 10(7.73) = 77.3 \text{ kips}$$

Considering eccentricity with respect to fastener line,

$$e = 2\frac{1}{4} \text{ in.}$$

assuming the reaction to be along fastener line A. The nominal forces on the most highly stressed fasteners are

$$Af_v = \frac{P}{10} = 0.10P \quad \text{(direct shear)}$$

$$Af_t = \frac{Pe(6)}{4(3)^2 + 4(6)^2} = \frac{P(2.25)(6)}{180} = 0.075P \quad \text{(tension)}$$

The allowable shear stress F'_v in the presence of tension is

$$F'_v = 17.5(1 - f_t A_b/T_i) = 17.5(1 - 0.075P/28)$$

For fasteners fully stressed in shear, $f_v = 0.10P/A$:

$$\frac{0.10P}{0.4418} = 17.5(1 - 0.075P/28)$$

$$P = 64.1 \text{ kips} < 77.3 \text{ kips}$$

Combined stress, if considered, would reduce capacity about 17 percent for the friction-type connection.

(b) Bearing-type connection (A325-X). For connection to the web ($t_w = 0.520$ in.) of a W30×99, referring to Table 4.6.1,

$$R_{DS} = A_b F_v = 2(0.4418)30.0 = 26.51 \text{ kips} \quad \text{(controls)}$$

$$R_B = 1.5 F_u D t_w = 1.5(58)(\tfrac{3}{4})(0.520) = 33.93 \text{ kips}$$

Neglecting the effect of eccentric shear,

$$P = 5(26.51) = 133 \text{ kips}$$

For the connection to the $\frac{3}{4}$-in. flange,

$$R_{SS} = 0.4418(30.0) = 13.25 \text{ kips}$$

Since the angles will likely be thinner than $\frac{3}{4}$ in., bearing on the angles will probably be more critical than bearing on the $\frac{3}{4}$-in. flange. The minimum thickness for the angles based on bearing so that $R_B = R_{SS}$ is

$$t = \frac{R_{SS}}{1.5 F_u D} = \frac{13.25}{1.5(58)(\tfrac{3}{4})} = 0.20 \text{ in.}$$

When the eccentricity causing combined shear and tension is neglected,

$$P = 10(13.25) = 133 \text{ kips}$$

If the combined shear and tension effect is considered, the forces are $0.10P$ for direct shear and $0.075P$ for tension, as computed in part (a). The allowable tensile stress F'_t in the presence of shear is, for bearing-type connections,

$$F'_t = 55 - 1.4 f_v$$

$$\frac{0.075P}{0.4418} = 55 - 1.4 \frac{0.10P}{0.4418}$$

$$P = 113 \text{ kips} < 133 \text{ kips (neglecting } e)$$

Again, considering combined stress would indicate about 15% reduction in capacity. The AISC Specification does not indicate how the computation is to be made; though usual practice for framed connections has been to neglect any combined shear and tension effect for bolted connections.

(c) Summary of results.

$$P = 77.3 \text{ kips} \quad \text{(A325-F, neglecting } e)$$

$$P = 64.1 \text{ kips} \quad \text{(A325-F, including } e)$$

$$P = 133 \text{ kips} \quad \text{(A325-X, neglecting } e)$$

$$P = 113 \text{ kips} \quad \text{(A325-X, including } e)$$

(d) End and edge distances. Under AISC–1.16.5.2 the minimum angle thickness for a $1\frac{1}{4}$-in. end distance on the angles is

$$\text{Min } t \geq \frac{2P}{F_u(\text{end distance})}$$

$$= \frac{2(133/10)}{58(1.25)} = 0.37 \text{ in.}$$

For the bearing-type connection with capacity 133 kips, $\frac{3}{8}$-in. angles would be required; for the friction-type connection having capacity 77.3 kips, $\frac{1}{4}$-in. angles would be sufficient.

For connections designed for the shear reaction only (i.e., neglecting eccentricity), AISC–1.16.5.3 requires an end distance check for the beam web. In that check the required end distance is

$$\text{Min end distance} \geq \frac{2P_R}{F_u t_w}$$

$$= \frac{2(133/5)}{58(0.520)} = 1.76 \text{ in.}$$

In the above calculation, P_R is the beam reaction divided by the number of bolts in the connection, and the "end distance" is the distance from the center of the nearest hole to the *end of the beam web*. Thus the vertical line of fasteners must be located at least 1.76 in. from the end of the beam.

This requirement may be waived when the bearing stress f_p does not exceed $0.90F_u$. In this case,

$$f_p = \frac{133/5}{(\frac{3}{4})(0.520)} = 68.2 \text{ ksi} > [0.90F_u = 52.2 \text{ ksi}] \quad \text{NG}$$

Thus for the bearing-type connection to carry 133 kips the end distance would have to be at least 1.76 in. For the friction-type connection carrying 77.3 kips, $2P_R/(F_u t_w)$ gives 1.02 in. which is less than the minimum $1\frac{1}{4}$ in. prescribed by AISC–Table 1.16.5.1 and therefore the $1\frac{1}{4}$ in. would control.

(e) Examine the degree of simple support provided by $\frac{3}{8}$-in. angles. The flexural stress in accordance with Eq. 13.2.5 is

$$f = \frac{9.0 Peg_e}{t^2 L^2} = \frac{9.0(133)(2.25)(2.5 - \frac{13}{16})}{(0.375)^2(14.5)^2} = 154 \text{ ksi}$$

using angles $4 \times 3\frac{1}{2} \times \frac{3}{8}$. This indicates that the yield stress is reached and permanent deformation of the angles occurs at service load. Examine the deformation in accordance with Eq. 13.2.9:

$$\Delta = \frac{f(g_e)^2}{3.0Et} = \frac{154(1.75)^2}{3.0(29,000)(0.375)} = 0.0135 \text{ in.}$$

The longest beam span uniformly loaded of a W30×99 having a reaction of 133 kips is

$$M = F_b S = 24(269) = 538 \text{ ft-kips}$$

$$M = \frac{WL}{8} = \frac{133(2)L}{8}$$

$$L = \frac{8(538)}{133(2)} = 16.2 \text{ ft}$$

The end slope is

$$\theta = \frac{WL^2}{24EI} = \frac{ML}{3EI} = \frac{538(12)(16.2)(12)}{3(29,000)3990} = 0.00362 \quad \text{radian}$$

Assuming rotation about the bottom of the angle, the deformation at the top of the angles would be

$$\Delta = 14.5 \ \theta = 14.5(0.00362) = 0.052 \text{ in.} > 0.0135$$

Based on these approximate calculations, the angles must yield to accommodate rotation at the end of the beam, and even if such rotation occurs, the full simple beam end slope is not likely to occur. Some end moment will exist but it is generally not large enough to significantly reduce the midspan positive moment. The conclusion is that simple framed connections should use the thinnest angles consistent with bearing and other practical limitations. Bertwell [5] has provided some additional discussion on the behavior of framed beam connection angles.

Example 13.2.2

Design the connection for a W10×68 having a 46-kip reaction and a W24×104 with a 165-kip reaction to frame into opposite sides of a plate girder having a $\frac{3}{8}$-in. web as shown in Fig. 13.2.4. The connection is to be of $\frac{3}{4}$-in.-diam A325 bolts in a bearing-type connection, and threads are excluded from the shear planes. Base material is A36 steel.

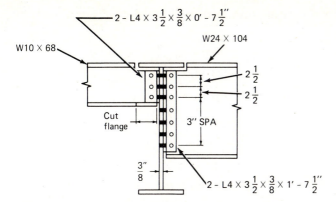

Fig. 13.2.4 Framed beam connection having unequal reactions.

SOLUTION

(a) Connections to webs of W10×68 and W24×104.

$$R_B = 1.5F_u t_w D$$

$$R_B = 1.5(58)(0.470)(0.75) = 30.67 \text{ kips} \quad \text{(W10 web)}$$

$$R_B = 1.5(58)(0.500)(0.75) = 32.63 \text{ kips} \quad \text{(W24 web)}$$

$$R_{DS} = 2(30)(0.4418) = 26.51 \text{ kips} \quad \text{(double shear)}$$

$$\text{Number of bolts} = \frac{46}{26.51} = 1.7, \quad \text{say 2} \quad \text{(W10)}$$

$$\text{Number of bolts} = \frac{165}{26.51} = 6.2 \quad \text{say 7} \quad \text{(W24)}$$

The use of only two bolts to fasten the W10 section may result in a tearing failure along a net section measured vertically downward from the cope to the center of the lowermost fastener and then horizontally to the near end of the beam. Try 3 bolts as in Fig. 13.2.4 and check AISC–1.5.1.2.2.

The net length L_n along the potential tearing line (Fig. 13.2.2) is 5.625 in., computed as $1\frac{1}{4}$ in. (end distance on angles) plus 5 in. (two 2.5-in. spacings) plus 2 in. (gage on 4-in. leg less $\frac{1}{2}$-in. setback) minus $3(\frac{13}{16}+\frac{1}{16})$(i.e., deduction for three standard holes). Thus

$$\frac{\text{Reaction}}{L_n t_w} \leq 0.30F_u = 0.30(58) = 17.4 \text{ ksi}$$

$$\frac{46}{5.625(0.470)} = 17.4 \text{ ksi} = 0.30F_u \qquad \text{OK}$$

Use 3 bolts to connect the W10 section.

(b) Connection to plate girder web. For this connection, the top bolts common to both sides will be governed by double shear or bearing on the

$\frac{3}{8}$-in. plate, while the remainder are governed by single shear or bearing on the $\frac{3}{8}$-in. plate.

$$R_B = 1.5(58)(0.375)(0.75) = 24.46 \text{ kips} \quad (\tfrac{3}{8}\text{-in. web})$$

$$R_{DS} = 2(30)(0.4418) = 26.51 \text{ kips}$$

$$R_{SS} = 30(0.4418) = 13.25 \text{ kips}$$

For the bolts common to both sides, bearing governs:

$$R = 24.46 \text{ kips/bolt}$$

Six bolts in common (two vertical rows of three) are assumed because it makes a convenient pattern with the three needed for the web of the W10×68. Each of the top six bolts then carries 46/6 = 7.7 kips from the W10×68. The remainder is available for W24×104 reaction; i.e., 24.46 − 7.7 = 16.8 kips.

Since 16.8 > 13.25, it will be conservative to assume 13.25 as the capacity per bolt available to the W24×104, assuming all bolts will carry an equal part of the reaction:

$$\text{Number of bolts} = \frac{165}{13.25} = 12.5, \quad \text{say 14}$$

The required angle thickness to allow $1\frac{1}{4}$-in. end distance on angles is (for W24 section)

$$t \geq \frac{2P}{F_u L_e} = \frac{2(165/14)}{58(1.25)} = 0.33 \text{ in.}, \quad \text{say } \tfrac{3}{8} \text{ in.}$$

or (for W10 section)

$$t \geq \frac{2(46/6)}{58(1.25)} = 0.21 \text{ in.}$$

Use 2—L4×$3\frac{1}{2}$×$\frac{3}{8}$×$0'-7\frac{1}{2}''$ for W10×68.

Use 2—L 4×$3\frac{1}{2}$×$\frac{3}{8}$×$1'-7\frac{1}{2}''$ for W24×104.

The angles (see Fig. 13.2.4) are made nonstandard because the length of angle should not exceed the dimension T which is $7\frac{5}{8}$ in. for the W10×68. The girder flange thickness is such that it requires a cope that encroaches on the T dimension; thus the $2\frac{1}{2}$-in. spacing is prescribed so that adequate edge distance will be available on the web of the W10×68.

Weld Capacity in Eccentric Shear on Angle Connections

Since no initial tension is involved with welded connections, the eccentricity of loading, even though small, is considered. The principles of Chapter 5 (Sec. 5.16) are used with the welds treated as lines.

Example 13.2.3

Determine the capacity for weld A on the angle connection shown in Fig. 13.2.1. The beam is a W30×99 and the weld size is $\frac{1}{4}$ in. The angles are $4 \times 3\frac{1}{2} \times \frac{5}{16} \times 1'\text{-}2\frac{1}{2}''$ in length. Use E70 electrodes on A36 steel.

SOLUTION

Using I_p from Table 5.16.1 and referring to Fig. 13.2.1c,

$$I_p = \frac{8(3)^3 + 6(3)(14.5)^2 + (14.5)^3}{12} - \frac{(3)^4}{2(3) + 14.5} = 583.5 \text{ in.}^3$$

When properties of lines are used, one may either consider that stresses are in ksi as computed with a 1-in. effective throat, or one may consider the unit stress has been multiplied by the 1-in. effective throat to give the stresses in kips per inch.

$$f_y' = \frac{P}{2(20.5)} = 0.0244P \quad \text{(direct shear component)} \downarrow$$

$$\bar{x} = \frac{(3)^2}{2(3) + 14.5} = 0.44 \text{ in.}$$

The x and y components of torsional stress are

$$f_y'' = \frac{P(3.50 - 0.44)(3.50 - 0.44 - 0.50)}{2(583.5)} = 0.00671P \downarrow$$

$$f_x'' = \frac{P(3.50 - 0.44)(7.25)}{2(583.5)} = 0.0190P \quad \rightarrow$$

$$f_r = P\sqrt{(0.0244 + 0.0067)^2 + (0.0190)^2} = 0.0364P$$

The capacity per inch of weld is

$$R_w = (\tfrac{1}{4})(0.707)21.0 = 3.71 \text{ kips/in.}$$

$$P = \frac{3.71}{0.0364} = 102 \text{ kips}$$

Tests of welded angle connections by Johnston and Green [6] and Johnston and Diets [7] have demonstrated that performance of web angles agrees generally with assumptions.

Weld Capacity in Tension and Shear on Angle Connections

This is the field-welded connection shown in Fig. 13.2.1d. There is not agreement regarding the strength analysis for this situation. Blodgett [8] considers the strength as an eccentric shear situation in the plane of the welds. With the eccentric load as in Fig. 13.2.5b, the angles bear against themselves for a distance of $L/6$ from the top, and the torsional stress

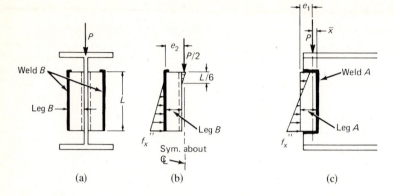

Fig. 13.2.5 Field-welded connection for web framing angles.

over the remaining $\frac{5}{6}$ of the length is resisted by the weld. Neglecting the effects of the returns at the top, moment equilibrium requires

$$\frac{1}{2}\underbrace{f''_x\left(\frac{5}{6}L\right)}_{\text{force/in.}}\underbrace{\frac{2}{3}L}_{\text{arm}}=\frac{P}{2}e_2 \tag{13.2.12}$$

$$f''_x=\frac{9Pe_2}{5L^2}\quad\text{force/unit length} \tag{13.2.13}$$

The direct-shear component is

$$f'_y=\frac{P}{2L}\quad\text{force/unit length} \tag{13.2.14}$$

$$\text{Actual }f_r=\sqrt{\left(\frac{P}{2L}\right)^2+\left(\frac{9}{5}\frac{Pe_2}{L^2}\right)^2}$$

$$=\frac{P}{2L^2}\sqrt{L^2+12.9e_2^2}\quad\text{force/unit length} \tag{13.2.15}$$

which is the equation of Ref. 8 for which a nomograph is available (Nomograph No. 6). This procedure neglects the eccentricity e_1, which tends to cause tension at the top of the weld lines. The authors do not believe the tendency is great for the angle legs to spread at the bottom as indicated by the stress distribution of Fig. 13.2.5b.

Instead of the procedure of Ref. 8, the authors consider the flexural stress distribution of Fig. 13.2.5c to be more appropriate. The flexural component is

$$f''_x=\frac{Mc}{I}=\frac{Pe_1(L/2)}{2L^3/12}=\frac{3Pe_1}{L^2} \tag{13.2.16}$$

where the returns at the top of angles are neglected. The direct shear component is

$$f'_y = \frac{P}{2L} \quad \text{force/unit length} \tag{13.2.17}$$

$$\text{Actual } f_r = \sqrt{\left(\frac{P}{2L}\right)^2 + \left(\frac{3Pe_1}{L^2}\right)^2}$$

$$= \frac{P}{2L^2}\sqrt{L^2 + 36e_1^2} \quad \text{force/unit length} \tag{13.2.18}$$

Or, if returns are considered (distance b of Fig. 13.2.6) the expression becomes complicated. The AISC Manual indicates the returns to be twice the weld size. The returns have the greatest effect when the angle length L is short. It may be reasonable to consider the returns to be $L/12$ (2 times $\frac{1}{4}$ in. weld for $L = 6$ in.).

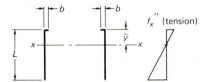

Fig. 13.2.6 Weld configuration for web angles and beam seats.

Using, from Table 5.16.1 (Case 4), $S = I/\bar{y}$ referred to the tension fiber at the top of the configuration,

$$S = 2\left(\frac{4bd + d^2}{6}\right) \tag{13.2.19}$$

which for $d = L$ and $b = L/12$ becomes

$$S = \frac{4L^2}{9} \tag{13.2.20}$$

The flexural-stress component, as shown in Fig. 13.2.6, is

$$f''_x = \frac{M}{S} = \frac{Pe_1}{S} = \frac{Pe_1}{4L^2/9} = \frac{9Pe_1}{4L^2} \tag{13.2.21}$$

Since little of the shear is carried by the returns, they are neglected for the direct-shear component, giving

$$f'_y = \frac{P}{2L} \tag{13.2.22}$$

$$\text{Actual } f_r = \sqrt{\left(\frac{P}{2L}\right)^2 + \left(\frac{9Pe_1}{4L^2}\right)^2}$$

$$= \frac{P}{2L^2}\sqrt{L^2 + 20.25e_1^2} \quad \text{kips/in.} \tag{13.2.23}$$

Example 13.2.4

Determine the capacity of weld B on Fig. 13.2.5 if $\frac{5}{16}$-in. weld is used and $L = 20$ in. E70 electrodes are used in shielded metal arc welding (SMAW). $4 \times 3 \times \frac{3}{8}$ angles are used.

SOLUTION

(a) Best procedure, Eq. 13.2.23

$$R_w = 0.707(\tfrac{5}{16})21 = 4.64 \text{ kips/in.}$$

$$\text{Actual } f_r = \frac{P}{2L^2}\sqrt{L^2 + 20.25e_1^2}$$

$$e_1 = 3.00 - \bar{x} = 3.00 - 0.25 = 2.75 \text{ in.}$$

$$\bar{x} = \frac{2(2.5)(1.25)}{2(2.5) + 20} = 0.25 \text{ in.}$$

$$\text{Actual } f_r = \frac{P}{2(20)^2}\sqrt{(20)^2 + 20.25(2.75)^2} = 0.0294P$$

$$P(\text{capacity}) = \frac{4.64}{0.0294} = 158 \text{ kips}$$

(b) Neglecting returns entirely, Eq. 13.2.18,

$$\text{Actual } f_r = \frac{P}{2(20)^2}\sqrt{(20)^2 + 36(2.75)^2} = 0.0324P$$

$$P(\text{capacity}) = \frac{4.64}{0.0324} = 143 \text{ kips}$$

(c) Using Ref. 8 equation, Eq. 13.2.15,

$$\text{Actual } f_r = \frac{P}{2(20)^2}\sqrt{(20)^2 + 12.9e_2^2}$$

$$e_2 = 4\text{-in. leg}$$

$$\text{Actual } f_r = 0.0308P$$

$$P(\text{capacity}) = \frac{4.64}{0.0308} = 151 \text{ kips}$$

The authors believe method (a) to be appropriate, $P = 158$ kips. The most conservative low estimate is 143 kips. Reference 8 equation gives reasonable results but the rationale of its development seems unrealistic.

13.3 SEATED BEAM CONNECTIONS—UNSTIFFENED

As an alternative to framed beam connections using web angles, or other attachments to the beam web, a beam may be supported on a seat, either unstiffened or stiffened. In this section the unstiffened seat is treated, as shown in Fig. 13.3.1. The unstiffened seat (an angle) is shown in Fig. 13.3.1 and is designed to carry the entire reaction. It must always, however, be used with a top clip angle, whose sole function is to provide lateral support of the compression flange.

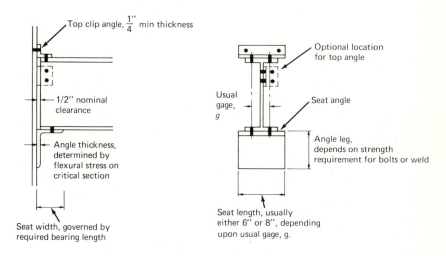

Fig. 13.3.1   Seated beam connections—unstiffened.

As with the case of the framed-beam connection, the seated connection is intended to transfer only the vertical reaction and should not give significant restraining moment on the end of the beam; thus the seat and the top angle should be relatively flexible. The behavior of welded seat angle connections has been studied by Lyse and Schreiner [9].

The thickness of seat angle is determined by the flexural stress on a critical section of the angle, as shown in Fig. 13.3.2. If a bolted connection is used without attachment to the beam (Fig. 13.3.2a), the critical section should probably be taken as the net section through the upper bolt line. When the beam is attached to the seat as in Fig. 13.3.2b, the rotation of the beam at the end creates a force that tends to restrain the pull away from the column. The critical section for flexure will then be at or near the base of the fillet on the outstanding leg. Similarly for the welded seat, the weld completely along the end holds the angle tight against the column, in which case the critical section is as shown in Fig. 13.3.2c, whether or not the beam is attached to the seat. As a practical matter, rarely will the beam be left unattached from the seat, so the design procedures of this section use a critical section as in Figs. 13.3.2b and c, taken at $\frac{3}{8}$ in. from the face of the angle.

The bending moments on the critical section of the angle and on the connection to the column flange are determined by taking the beam reaction times the distances to the critical sections. The beam reaction

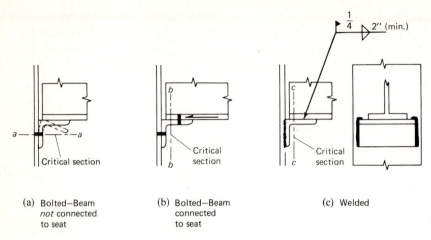

Fig. 13.3.2 Critical section for flexure on seats.

occurs at the centroid of the bearing stress distribution, as shown in Fig. 13.3.3. While the AISC Specification does not state how the computation of this bending moment is to be made, a conservative approach is to assume the reaction at the center of the full contact width (Fig. 13.3.3a). This will lead to excessively thick angles in most cases. The less conservative approach of assuming the reaction at the center of the *required* bearing length N measured from the end of the beam (Fig. 13.3.3b) has been used by Blodgett [8] and has been the approach used for AISC Manual tables. Another rational distribution for a flexible seat angle is the triangular distribution of Fig. 13.3.3c, and if the angle is very stiff the reaction may become heavier on the outer edge, as in Fig. 13.3.3d.

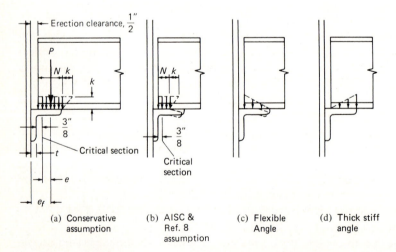

Fig. 13.3.3 Bearing stress assumptions for seated connections.

The design of unstiffened seats involves the following steps:

1. Determine the seat width.
2. Determine the moment arms e and e_f.
3. Determine the length and thickness of the angle.
4. Determine the supporting angle leg dimension, and the weld size; or the number and placement of bolts.

The seat width is based on the bearing length N required in accordance with AISC–1.10.10.1:

$$N = \frac{P}{0.75 F_y t_w} - k \geq k \qquad (13.3.1)$$

where t_w = thickness of web
    k = distance from outer face of flange to web toe of fillet
Generally the seat width should not be less than 3 in., with the AISC tables indicating a standard 4-in. seat width.

The moment arms e and e_f are obtained as follows, referring to Fig. 13.3.3a:

$$e_f = \text{erection clearance} + \frac{N}{2} \qquad (13.3.2)$$

$$e = e_f - t - \tfrac{3}{8} \qquad (13.3.3)$$

The bending moment on the critical section of the angle is

$$M = Pe$$

$$f_b = \frac{M}{S} = \frac{Pe}{\frac{1}{6}bt^2} = \frac{6Pe}{bt^2} \qquad (13.3.4)$$

Using the allowable stress on solid rectangular sections bent about their weak axis, AISC–1.5.1.4.3,

$$F_b = 0.75 F_y$$

in which case Eq. 13.3.4 becomes

$$t^2 = \frac{6Pe}{0.75 F_y b} = \frac{8Pe}{F_y b} \qquad (13.3.5)$$

This length of the seat angle is generally taken as either 6 in. or 8 in. for a beam gage g of $3\tfrac{1}{2}$ in. and $5\tfrac{1}{2}$ in., respectively.

The number of bolts, which are in combined shear and tension, is determined in accordance with the principles of Sec. 4.10.

The weld size and length are obtained using the principles of Sec. 5.16 with Eq. 13.2.23 applicable to this case; direct shear and bending about the $x - x$ axis for the configuration of Fig. 13.2.5 with the returns $b \approx L/12$.

Example 13.3.1

Design the seat angle to support a W12×40 beam on a 25-ft span, assuming the beam has adequate lateral support. Use A36 steel.

SOLUTION

In many cases it will be wise practice to design the seat for the maximum reaction when the beam is fully stressed in flexure.

(a) Determine seat width, length, and thickness:

$$M = 0.66F_yS_x = 24(51.9)\tfrac{1}{12} = 103.8 \text{ ft-kips}$$

$$P = \frac{wL}{2} = \frac{8M}{2L} = \frac{4(103.8)}{25} = 16.6 \text{ kips}$$

Bearing length required is

$$N = \frac{P}{0.75F_yt_w} - k = \frac{16.6}{27(0.295)} - 1.25 = 0.83 \text{ in.} < k$$

Take $N = k = 1.25$ in. Following AISC Manual, "Seated Beam Connections" recommendation, use 4-in. seat width. Using Eq. 13.3.2 with a clearance of $\tfrac{3}{4}$ in. to allow for possible mill underrun,

$$e_f = \frac{1.25}{2} + \frac{3}{4} = 1.375 \text{ in.}$$

Trying $t = \tfrac{1}{2}$ in.,

$$e = e_f - t - \tfrac{3}{8} = 1.375 - 0.50 - 0.375 = 0.50 \text{ in.}$$

Since $g = 5\tfrac{1}{2}$ in. for W12×40, use angle length of 8 in. The angle thickness required is then, by Eq. 13.3.5,

$$t^2 = \frac{8Pe}{F_yb} = \frac{8(16.6)(0.50)}{36(8)} = 0.23; \qquad t = 0.48. \text{ in.}$$

<u>Use</u> seat angle, $\tfrac{1}{2}$ in. thick and 8 in. long.

(b) Determine bolted connection to column, using $\tfrac{3}{4}$-in.-diam A325 bolts in a bearing-type connection with no threads in the shear plane.

$R_{ss} = 13.25$ kips; $\qquad R_B = 32.63$ kips (based on $\tfrac{1}{2}$-in.-thick angle);

$R_t = 19.44$ kips

$$n \approx \sqrt{\frac{6M}{Rp}} = \sqrt{\frac{6(16.6)(1.38)}{13.25(3)(2)}} = 1.3$$

assuming two vertical rows of connectors at a 3-in. pitch.

Try 2 bolts (i.e., $n = 1$) as shown in Fig. 13.3.4. The direct-shear component is

$$Af_v = \frac{P}{n} = \frac{16.6}{2} = 8.3 \text{ kips} < 13.25 \text{ kips} \qquad \text{OK}$$

Since the bolts lie on the center of gravity, no moment of inertia can be computed using $\sum y^2$. However, since initial tension exists, the initial compression, according to Eq. 4.10.16, is

$$f_{ti} = \frac{\sum T_i}{bd} = \frac{2(28)}{8(3)} = 2.33 \text{ ksi}$$

The change in stress due to moment, Eq. 4.10.17, is

$$f_{tb} = \frac{6M}{bd^2} = \frac{6Re}{bd^2} = \frac{6(16.6)1.38}{8(3)^2} = 1.91 \text{ ksi}$$

Since $1.91 < 2.33$, the initial precompression is not eliminated and the connection may be considered safe.

Use 2 bolts, with seat angle, $L\, 4 \times 3\frac{1}{2} \times \frac{1}{2} \times 0' - 8''$.

(c) Determine welded connection to column, using Eq. 13.2.23, with allowable weld value for E70 electrodes with shielded metal arc welding:

Max weld size $= \frac{1}{2} - \frac{1}{16} = \frac{7}{16}$ in.

Min weld size $=$ AISC–Table 1.17.2A based on thickest material being joined.

Try $L = 4$-in. supported leg:

$$f_r = \frac{P}{2L^2} \sqrt{L^2 + 20.25 e_1^2} \qquad\qquad [13.2.23]$$

where $e_1 = e_f$

$$f_r = \frac{16.6}{2(4)^2} \sqrt{(4)^2 + 20.25(1.38)^2} = 3.83 \text{ kips/in.}$$

$$R_w = a(0.707)21.0 = 14.85a$$

Weld size $a = \dfrac{3.8}{14.85} = 0.258$ in., say $\frac{5}{16}$ in.

A more conservative approach is to measure e_f to the center of the contact bearing width of the seat (Fig. 13.3.3a). This traditional AISC method for tables giving weld capacity for seats gives

$$e_f = \frac{N}{2} + \frac{3}{4} = \frac{3.5 - 0.75}{2} + 0.75 = 2.13 \text{ in.}$$

which upon substitution into Eq. 13.2.23 with $R_w = 14.85a = 14.85(0.3125) = 4.64$ kips/in. (for $\frac{5}{16}$ in. weld), gives a capacity of

$$P = \frac{R_w(2L^2)}{\sqrt{L^2 + 20.25 e_f^2}} = \frac{4.64(2)(4)^2}{\sqrt{(4)^2 + 20.25(2.13)^2}} = 14.3 \text{ kips}$$

The weld size required would then be

$$\text{Required } a = \tfrac{5}{16}\left(\frac{16.6}{14.3}\right) = 0.36 \text{ in., say } \tfrac{3}{8} \text{ in.}$$

Use $L4 \times 3\frac{1}{2} \times \frac{1}{2} \times 0' - 8''$ with $\frac{3}{8}$ in. weld. Since the W12 × 40 flange width is 8 in., the beam must be "blocked" or "cut" to have a reduced flange width over the seat so that the necessary welding can be done, or if the column flange permits, the seat may be longer than 8 in. The final designs are shown in Fig. 13.3.4.

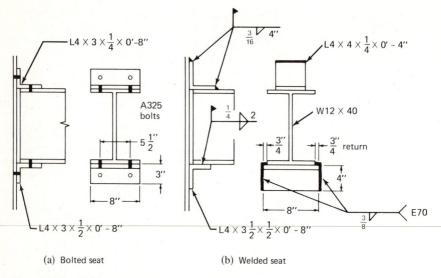

(a) Bolted seat   (b) Welded seat

Fig. 13.3.4 Designs for Example 13.3.1.

13.4 STIFFENED SEAT CONNECTIONS

When reactions become heavier than desirable for unstiffened seats, stiffeners may be used with the seat angle in bolted construction, or a T-shaped stiffened seat may be used in welded construction. The unstiffened seat may become excessively thick when the beam reaction exceeds about 40 kips. There are no AISC restrictions, however, to the maximum load that may be carried by unstiffened seats.

The stiffened seat as discussed herein is not intended to be part of a moment resisting connection, but rather it is only to support vertical loads. Here the stiffened seat is treated as AISC Type 2 construction; i.e., "simple framing." Behavior of welded brackets has been studied by Jensen [10].

There are two basic types of loading used on stiffened seats; the common one where the reaction is carried with the beam web directly in line with the stiffener, as shown in Fig. 13.4.1; the other is with a beam oriented so the plane of the web is at 90 degrees to the plane of the stiffener, as in Fig. 13.4.2. Furthermore, a difference in behavior arises depending on the angle at which the stiffener is cut, as shown in Fig. 13.4.3. If the angle θ is approximately 90 degrees, the stiffener behaves

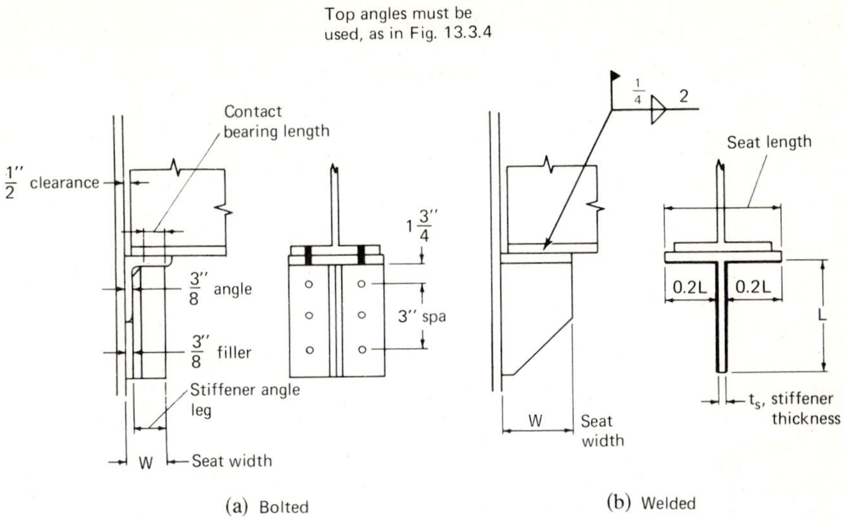

Top angles must be
used, as in Fig. 13.3.4

$\frac{1}{2}''$ clearance

Contact
bearing length

$\frac{1}{4}$ 2

Seat length

$1\frac{3}{4}''$

$\frac{3''}{8}$ angle

$\frac{3''}{8}$ filler

3'' spa

Stiffener angle
leg

W Seat width

0.2L 0.2L

L

t_s, stiffener
thickness

W Seat
width

(a) Bolted

(b) Welded

Fig. 13.4.1 Stiffened seat—beam web in line with stiffener.

similarly to an unstiffened element under uniform compression, and local buckling may be prevented by satisfying AISC–1.9.1.2. When the supporting plate is cut to create a triangular bracket plate a different behavior results, and this case is discussed in Sec. 13.5.

The steps in the design of stiffened seats are as follows:

1. Determine the seat width.
2. Determine the eccentricity of load, e_s.
3. Determine the stiffener thickness, t_s.
4. Determine the angle sizes and arrangement of bolts; or the weld size and length.

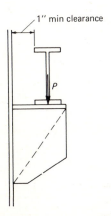

1'' min clearance

P

Fig. 13.4.2 Bracket supporting concentrated load.

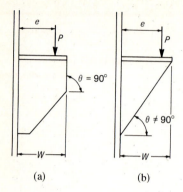

Fig. 13.4.3 Two cases of inclination angle at free edge of stiffener.

The seat width is based on the required bearing length, N, according to AISC–1.10.10.1:

$$N = \frac{P}{0.75F_y t_w} - k \geq k \tag{13.4.1}$$

where t_w and k are defined following Eq. 13.3.1. Because of the rigidity of the stiffener, the most highly stressed portion is at the edge of the seat rather than at the interior side as it was for the unstiffened seat (see Fig. 13.4.4).

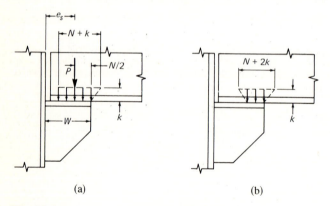

Fig. 13.4.4 Bearing stress on stiffened seats.

For bolted stiffened seats, the AISC Manual tabulates stiffener angle capacities for stiffener outstanding legs of $3\frac{1}{2}$, 4, and 5 in., which will accommodate beam bearing lengths of $3\frac{1}{2}$, $4\frac{1}{4}$, and $5\frac{1}{4}$ in., respectively. For welded stiffened seats the AISC Manual tables give seat widths ranging from 4 to 9 in. The thickness of the seat plate should be comparable to the flange of the supported beam.

Assuming the beam reaction P is located at $N/2$ from the edge of the seat the *thickness of the stiffener t_s* may be established to satisfy several criteria:

1. $$t_s \geq t_w \tag{13.4.2}$$

   in order to be at least as thick as the beam web:

2. $$t_s \geq \frac{W}{95/\sqrt{F_y}} \tag{13.4.3}$$

   according to AISC–1.9.1.2 so as to prevent local buckling:

3. $$t_s \geq \frac{P}{0.90F_y(W-0.5)2} \quad \text{(for stiffener angles)} \tag{13.4.4}$$

   according to AISC–1.5.1.5, bearing on contact area. It is assumed that $\frac{1}{2}$ in. of the stiffener angle is clipped off so as to get close bearing under the seat angle. Equation 13.4.4 assumes no eccentricity of load with respect to center of bearing contact length.

4. $$f_b = \frac{P}{A} + \frac{M}{S}$$

   $$= \frac{P}{Wt_s} + \frac{P(e_s - W/2)}{t_s W^2/6} = \frac{P}{t_s W^2}(6e_s - 2W) \tag{13.4.5}$$

   $$t_s \geq \frac{P(6e_s - 2W)}{0.90F_y W^2} \quad \text{(for welded stiffener)}$$

   where the combined bending and direct stress on the rectangular stiffener plate is taken as a bearing stress. The neutral axis for flexure is taken at $W/2$. While one may argue that the bearing allowable stress should not be used, it seems that $0.60F_y$ is overly conservative. Many designers will prefer to use that value, however.

5. $$t_s \geq \frac{74.2a}{F_y} \tag{13.4.6}$$

   for welding of weld size a with E70 electrodes. Assuming two lines of fillet weld such that the weld may be fully effective and not overstress the stiffener plate in shear, it is required that

   $$2(0.707a)(21.0) = 0.40F_y t_s$$

   which gives Eq. 13.4.6. For some of the common types of steel this means

   $$t_s \geq 2.06a \quad \text{(for A36 steel)}$$
   $$t_s \geq 1.48a \quad \text{(for } F_y = 50 \text{ ksi)}$$

Once the stiffener dimensions have been established, the connection must be designed to transmit the reaction at the moment arm, e_s. For the bolted connection, AISC tables, "Stiffened Seated Beam Connections," consider only direct shear in determining fastener group capacities. One may reason, as in Example 13.3.1, that as long as initial compression between the pieces in contact is not reduced to zero due to flexure, the moment component need not be considered.

For the welded connection suggested by the AISC Manual, as shown in Fig. 13.4.1b, the weld configuration is subject to direct shear and flexure using the combined stress at the top of the weld as the critical one. Thus the configuration is identical to that used for web framing angles (see Fig. 13.2.1d) except the return is longer. Using $d = L$ and $b = 0.2L$ in the S values for Case 4 from Table 5.16.1 gives

$$\bar{y} = \frac{L^2}{2(L+b)} = \frac{L^2}{2(1.2L)} = \frac{L}{2.4}$$

$$S_x = \frac{2(4bL+L^2)}{6} = \frac{4(0.2L)L+L^2}{3} = 0.6L^2$$

Then,

$$f''_x = \frac{M}{S_x} = \frac{Pe_s}{0.6L^2} \quad \text{force/unit length} \rightarrow$$

$$f'_y = \frac{P}{2(L+0.2L)} = \frac{P}{2.4L} \quad \text{force/unit length} \downarrow$$

$$f_r = \sqrt{\left(\frac{Pe_s}{0.6L^2}\right)^2 + \left(\frac{P}{2.4L}\right)^2}$$

$$= \frac{P}{2.4L^2}\sqrt{16e_s^2+L^2} \quad \text{force/unit length} \tag{13.4.7}$$

Equation 13.4.7 is used for obtaining capacities given in the AISC Manual tables, "Stiffened Seated Beam Connections," when e_s is taken as $0.8W$.

Example 13.4.1

Design a welded stiffened seat to support a W30×99 with a reaction of 150 kips. The steel is A572 Grade 50.

SOLUTION
The bearing length required is

$$N = \frac{P}{0.75F_y t_w} - k = \frac{150}{37.5(0.520)} - 1.438 = 6.25 \text{ in.}$$

Required $W = 6.25 + 0.5$ (setback) $= 6.75$ in.   Use 7 in.

Since the W30×99 flange thickness is 0.670 in., use $\frac{5}{8}$-in. seat plate.

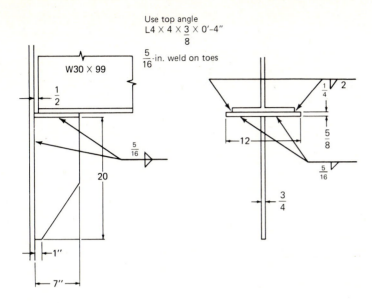

Fig. 13.4.5 Design for Example 13.4.1.

Minimum weld size for welding on $\frac{5}{8}$ in. seat and 0.67-in. flange is $\frac{1}{4}$ in. The stiffener thickness is next to be established:

$$t_s \geq t_w = 0.520 \text{ in.} \qquad\qquad [13.4.2]$$

$$t_s \geq \frac{W}{95/\sqrt{F_y}} = \frac{7}{13.4} = 0.52 \text{ in.} \qquad [13.4.3]$$

$$e_s = W - \frac{N}{2} = 7.0 - \frac{6.25}{2} = 3.88 \text{ in.}$$

$$t_s \geq \frac{P(6e_s - 2W)}{0.90F_y W^2} = \frac{150(23.3 - 14)}{45(7)^2} = 0.63 \text{ in.} \qquad [3.4.5]$$

If the allowable stress is reduced to $0.60F_y$, a 1-in. plate would be indicated. If a $\frac{3}{4}$-in. stiffener plate is used, Eq. 13.4.6 would permit a maximum effective weld size of

$$a_{\max \text{ eff}} = \frac{F_y t_s}{74.2} = \frac{50(0.75)}{74.2} = 0.51 \text{ in.}$$

For weld size and length, use $e_s = 0.8W$ as is used for AISC Manual tables. Using Eq. 13.4.7 and assuming $0.8W$ (5.6 in.) is about $L/4$:

$$f_r = \frac{P}{2.4L^2} \sqrt{16\left(\frac{L^2}{16}\right) + L^2} = 0.59\frac{P}{L}$$

if $\frac{5}{16}$-in. weld is used,

$$R_w = 0.707(\tfrac{5}{16})21 = 4.64 \text{ kips/in.}$$

$$\text{Required } L \approx \frac{0.59(150)}{4.64} = 19.1 \text{ in.}$$

Try 20 in. with $\frac{5}{16}$-in. weld:

$$f_r = \frac{150}{2.4(20)^2} \sqrt{16(5.6)^2 + (20)^2} = 4.69 \text{ kips/in.} \approx 4.64 \text{ kips/in.} \qquad \text{OK}$$
$$\text{(about 1\% high)}$$

Accept the overstress in view of the fact that using $e_s = 0.8\, W$ is believed to overestimate the moment arm.

Use $\frac{5}{16}$*-in. weld with* $L = 20$ *in. Use stiffener plate,* $\frac{9}{16} \times 7 \times 1'$-$8''$*; and seat plate* $\frac{5}{8} \times 7 \times 1'$-$0''$. The seat plate width equals the flange width (10.45 in.) plus enough to easily make the welds (approx. 4 times the weld size is often used). The final design is shown in Fig. 13.4.5.

13.5 TRIANGULAR BRACKET PLATES

When the stiffener for a bracket is cut into a triangular shape, as in Fig. 13.4.3b, the plate behaves in a different manner than when the free edge is parallel to the direction of applied load in the region where the greatest stress occurs, as in Fig. 13.4.5. The triangular bracket plate arrangement and notation are shown in Fig. 13.5.1.

The behavior of triangular bracket plates has been studied analytically by Salmon [11] and experimentally by Salmon, Buettner, and O'Sheridan [12] and design suggestions have been proposed by Beedle et al. [13]. For small stiffened plates to support beam reactions there is little danger of buckling or failure of the stiffener if cut into a triangular shape. In general, it provides a stiffer support when so cut than if left with a rectangular shape.

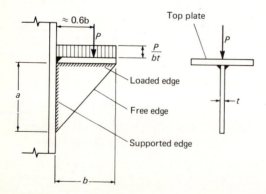

Fig. 13.5.1 Triangular bracket plate.

Most Exact Analysis and Design Recommendations

For many years design of such brackets was either empirical without benefit of theory or tests, or when in doubt, angle or plate stiffeners were used along the diagonal edge. The recommendations presented here are based on certain assumptions: (1) the top plate is solidly attached to the supporting column; (2) the load P is distributed (though not necessarily uniformly) and has its centroid at approximately $0.6b$ from the support; and (3) the ratio b/a, loaded edge to supported edge, lies between 0.50 and 2.0.

The original theoretical analysis was concerned with elastic buckling; however, the experimental work showed that triangular bracket plates have considerable post-buckling strength. Yielding along the free edge frequently occurs prior to buckling, at which point redistribution of stresses occurs. A considerable margin of safety against collapse was observed indicating the ultimate capacity may be expected to be at least 1.6 times the buckling load.

The maximum stress was found to occur at the free edge; however, because of the complex nature of the stress distribution, the stress on the free edge is not obtainable by any simple process. Because of this difficulty, a ratio z was established between the average stress, P/bt, on the loaded edge to the maximum stress f_{max} on the free edge. The original theoretical expression [11] for z was revised as a result of the tests [12] which conformed closely to what one could realistically expect in practice. The relationship is given [12] as

$$z = \frac{P/bt}{f_{max}} = 1.39 - 2.2\left(\frac{b}{a}\right) + 1.27\left(\frac{b}{a}\right)^2 - 0.25\left(\frac{b}{a}\right)^3 \qquad (13.5.1)$$

which for practical purposes may be obtained from Fig. 13.5.2.

If yielding controls strength, and $0.60F_y$ is considered a safe allowable stress, the safety criterion becomes

$$f_{max} = \frac{P/bt}{z} \leq 0.60F_y \qquad (13.5.2)$$

For stability controlling, the plate buckling equation, as discussed in Chapter 6, is

$$f_{cr} = \frac{k\pi^2 E}{12(1-\mu^2)(b/t)^2} \qquad (13.5.3)$$

where f_{cr} is the principal stress along the diagonal free edge. In order to guarantee that yielding is achieved without buckling, let $f_{cr} = F_y$ and solve Eq. 13.5.3 for maximum b/t,

$$\frac{b}{t} \leq \sqrt{\frac{k\pi^2 E}{12(1-\mu^2)F_y}} = 162\sqrt{\frac{k}{F_y, \text{ksi}}} \qquad (13.5.4)$$

where $E = 29{,}000$ ksi and $\mu = 0.3$.

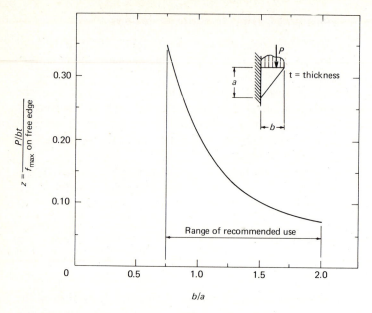

Fig. 13.5.2 Coefficient used to obtain maximum stress on free edge.

Figure 13.5.3 gives the variation in $(b/t)\sqrt{F_y}$ with b/a for the theoretical studies [11] (fixed and simply supported), the welded bracket tests result [12], and the authors' suggested design curve. The design requirement may be expressed (with F_y in ksi) as

$$\text{For } 0.5 \geq b/a \leq 1.0; \qquad \frac{b}{t} \leq \frac{250}{\sqrt{F_y}} \qquad\qquad (13.5.5a)^*$$

$$\text{For } 1.0 \leq b/a \leq 2.0; \qquad \frac{b}{t} \leq \frac{250(b/a)}{\sqrt{F_y}} \qquad\qquad (13.5.5b)^*$$

Satisfying the above limits means that yielding along the *diagonal free edge* will occur prior to buckling.

It is noted that the b/t limits suggested here are higher than those of Ref. 13 (p. 552), which were based solely on the theoretical studies [11]. Reference 13 suggests a coefficient of 180 instead of 250 in Eq. 13.5.5 and reaches a maximum of 300 instead of 500 as indicated by Eq. 13.5.5b

---

* For SI units, with F_y in MPa,

$$\text{For } 0.5 \leq \frac{b}{a} \leq 1.0; \qquad \frac{b}{t} \leq \frac{656}{\sqrt{F_y}} \qquad\qquad (13.5.5a)$$

$$\text{For } 1.0 \leq \frac{b}{a} \leq 2.0; \qquad \frac{b}{t} \leq \frac{656(b/a)}{\sqrt{F_y}} \qquad\qquad (13.5.5b)$$

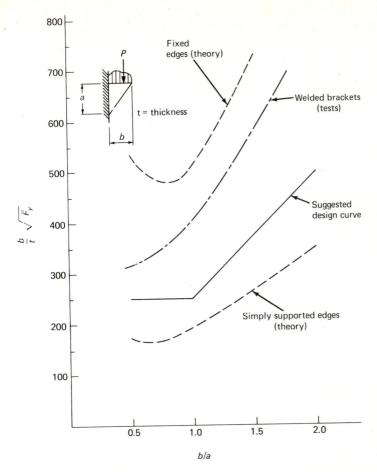

Fig. 13.5.3 Critical b/t values so that yield stress is reached along diagonal free edge without buckling.

for $b/a = 2.0$. The reason for the higher values is found in the test results which showed the principal stress along the diagonal free edge to be lower relative to the stress on the loaded edge than had been established by the theoretical study. In other words, the z value, Eq. 13.5.1, as determined by tests is substantially smaller than assumed for the design suggestion of Ref. 13.

Approximate Beam Analysis for Stress

For many years stress on triangular, as well as other shaped, bracket plates has been based on a beam analysis (see Ref. 10) using the relationships from Fig. 13.5.4.

The combined stress at the free edge on the critical section is

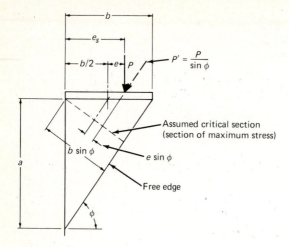

Fig. 13.5.4 Beam analysis for bracket plates.

computed as follows:

$$f_{max} = \frac{P}{A} + \frac{Mc}{I} = \frac{P/\sin \phi}{bt \sin \phi} + \frac{(P/\sin \phi)(e \sin \phi)(b \sin \phi/2)}{(b \sin \phi)^3 t/12}$$

$$= \frac{P}{bt \sin^2 \phi}\left[1 + \frac{6e}{b}\right] \tag{13.5.6}$$

or, in terms of e_s, Eq. 13.5.6 becomes

$$f_{max} = \frac{P}{bt \sin^2 \phi}\left(\frac{6e_s}{b} - 2\right) \tag{13.5.7}$$

If the maximum stress is limited to $0.60F_y$, the required thickness for stress is

$$t \geq \frac{P}{b(0.60F_y)\sin^2 \phi}\left(\frac{6e_s}{b} - 2\right) \tag{13.5.8}$$

When the beam-analysis approach is used, one may imagine, as suggested in Ref. 13, that a strip of width $(b \sin \phi)/4$ acts as a compression element. The problem then arises of what slenderness ratio limitation, if any, should be used. Reference 13 gives no suggestion. The authors suggest that one may imagine a strip of plate acting as a column and use Eq. 13.5.4 with $k = 1.0$ for the pin-end plate; letting $b = L_d$, the length of the diagonal,

$$\frac{L_d}{t} \leq \frac{162}{\sqrt{F_y}, \text{ksi}} \tag{13.5.9*}$$

* For SI units, with F_y in MPa,

$$\frac{L_d}{t} \leq \frac{425}{\sqrt{F_y}} \tag{13.5.9}$$

Plastic Strength of Bracket Plates

Reference 13 suggests that to develop the full plastic strength of brackets the b/t ratios should be restricted to about $\frac{1}{3}$ of those limitations for achieving first yield on the free edge. The test results [12] indicated that ultimate strengths of at least 1.6 times buckling strengths could be achieved due to post-buckling strength. To be certain of developing the plastic capacity of the bracket, it may be realistic to use half of the limitations of Eqs. 13.5.5a and b.

To establish the plastic strength of a bracket plate used in rigid-frame structures, one may follow the approach of Beedle et al. [13] as shown in Fig. 13.5.5. This method assumes that plastic strength develops on the critical section. Taking force equilibrium parallel to the free edge and moment equilibrium about point O gives the Beedle et al. [13] equation for the ultimate load,

$$P_u = F_y t \sin^2\phi (\sqrt{4e^2 + b^2} - 2e) \tag{13.5.10}$$

In addition to the triangular plate being adequate, the top plate must carry the ultimate force $P_u \cot \phi$.

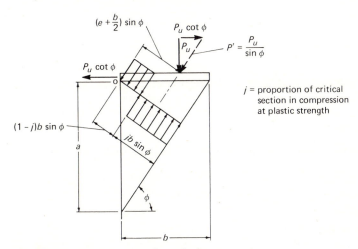

j = proportion of critical section in compression at plastic strength

Fig. 13.5.5 Plastic strength analysis.

Example 13.5.1

Determine the thickness required for a triangular bracket plate 25 in. by 20 in. to carry a load of 40 kips. Assume the load is located 15 in. from the face of support as shown in Fig. 13.5.6, and that A36 material is used.

SOLUTION

(a) Use the more exact method; working stress design. Since the load is approximately at the 0.6 point along the loaded edge, the bracket fits

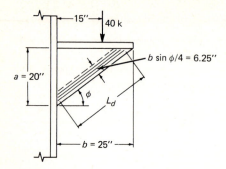

Fig. 13.5.6 Bracket for Example 13.5.1.

the assumption of this method. Using Eq. 13.5.2,

$$f_{avg} = \frac{P}{btz} = 0.60F_y$$

where from Fig. 13.5.2, for $b/a = 25/20 = 1.25$, find $z = 0.135$; for the yield criterion

$$t \geq \frac{P}{bz(0.60F_y)} = \frac{40}{25(0.135)22} = 0.54 \text{ in.}$$

For stability, using Eq. 13.5.5b,

$$t \geq \frac{b\sqrt{F_y}}{250(b/a)} = \frac{25\sqrt{36}}{250(1.25)} = 0.48 \text{ in.}$$

Use $\frac{9}{16}$-in. plate.

(b) Use the approximate beam analysis; working stress method. Using Eq. 13.5.8,

$$t \geq \frac{P}{b(0.60F_y)\sin^2\phi} \left(\frac{6e_s}{b} - 2\right)$$

where $\sin^2\phi = (0.625)^2 = 0.39$

$$t \geq \frac{40}{25(22)0.39} \left(\frac{6(15)}{25} - 2\right) = 0.30 \text{ in.}$$

which by comparison with the previous method is not conservative. Considering stability according to Eq. 13.5.9 gives

$$t \geq \frac{L_d\sqrt{F_y}}{162} = \frac{(\approx 27)(6)}{162} = 1 \text{ in.}$$

where L_d was taken along the center of the edge strip. The authors conclude that the beam analysis does not provide realistic answers; the stress requirement is low and the stability requirement is high, although past practice may well have used either a 1-in. plate, or an edge stiffener on a thinner plate.

Use 1-in. plate.

(c) Plastic strength method. Use a load factor of 1.7. Using Eq. 13.5.10 for the stress requirement,

$$t \geq \frac{P_u}{F_y \sin^2 \phi [\sqrt{4e^2 + b^2} - 2e]}$$

Using $e = 15 - 25/2 = 2.5$ in.,

$$t \geq \frac{1.7(40)}{36(0.39)[\sqrt{4(2.5)^2 + (25)^2} - 2(2.5)]} = 0.24 \text{ in.}$$

For stability, using *one-half* of Eq. 13.5.5b

$$t \geq \frac{b\sqrt{F_y}}{125(b/a)} = \frac{25\sqrt{36}}{125(1.25)} = 0.96 \text{ in.}$$

Use 1-in. plate as a conservative practice to assure deformation well beyond first yield along the free edge.

The authors note that if a 1-in. plate is just stable enough to inhibit buckling until the plastic strength is obtained, that strength would be about 4 times $(1.0/0.24)$ the required P_u.

13.6 CONTINUOUS BEAM-TO-COLUMN CONNECTIONS

In continuous beam-to-column connections it is the design intent to have full transfer of moment and little or no relative rotation of members within the joint (i.e., AISC Type 1—rigid-frame connections). Since the flanges of a beam carry most of the bending moment via tension and compression flange forces acting at a moment arm approximately equal to the beam depth, it is the transfer of these essentially axial forces for which provision must be made. Since the shear is carried primarily by the web of a beam, full continuity requires that it be transferred directly from the web.

Columns being rigidly framed by beams may have attachments to both flanges, as in Figs. 13.6.1a, b, and c, or only to one flange, as in Fig. 13.6.1d, and in Fig. 13.6.2. Alternatively, the rigid attachment of beams may be to the web, from either or both sides, as in Fig. 13.6.3. When the rigid system has rigid attachments *either* to the flanges or the web (but not to both) the system is said to be a two-way, or planar, rigid frame. When the rigid frame system consists of continuous connections to both flange (or flanges) and web (either or both sides), the system becomes a four-way system, or space frame.

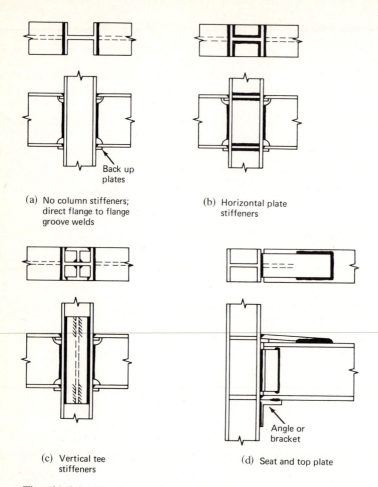

(a) No column stiffeners;
direct flange to flange
groove welds

(b) Horizontal plate
stiffeners

(c) Vertical tee
stiffeners

(d) Seat and top plate

Fig. 13.6.1 Continuous beam-to-column connections: welded attachment to column flange.

The variety of arrangements for a continuous beam-to-column connection is so great as to preclude any complete listing or illustration; however, those shown in Figs. 13.6.1, 13.6.2, and 13.6.3 are believed to be common in current (1979) design use. Most connections are partly shop welded and then completed in the field by either welding or fastening with high-strength bolts.

The principal design concern is with transmission of loads through the joint, and the localized deformations relating thereto. Rigid framing may be used to greatest advantage either (1) in plastically designed structures; or (2) in working stress design when the beams framing to columns are "compact sections" and the 10% reduction permitted under AISC–1.5.1.4.1 may be used. In either case, it is the objective of the connection to be able to develop the full plastic-moment capacity at the joint and in addition be able to undergo plastic-hinge rotation.

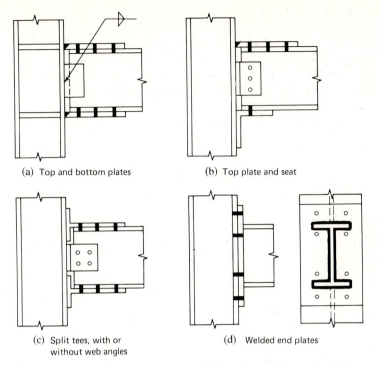

(a) Top and bottom plates   (b) Top plate and seat

(c) Split tees, with or   (d) Welded end plates
without web angles

Fig. 13.6.2 Continuous beam-to-column connections: bolted attachments.

Tests [14–22] have adequately demonstrated the ability of beam-to-column rigid-frame connections to develop the plastic hinge moment and exhibit adequate rotation capacity (ductility). A summary of behavior of bolted beam-to-column connections is given by Fisher and Struik [22].

Horizontal Stiffener in Compression Region of Connection

When the forces in beam flanges are transmitted as compression or tension forces to column flanges, horizontal stiffeners, as in Figs. 13.6.1b and d and Fig. 13.6.2a, may be required. Such stiffeners prevent web crippling in a region where the beam flange causes compression, and prevent distortion of the column flange where the beam flange causes compression. The phenomenon to be protected against is similar to that related to beams as treated in Sec. 7.6, and to plate girders as treated in Sec. 11.2. (See discussion relating to Fig. 11.2.8.)

The following development, and resulting AISC provisions are from research conducted at Lehigh University [14].

Conservative Approach—AISC–1.15.5. Consider a beam compression flange bearing against a column as in Fig. 13.6.4a. When the maximum strength of the column web is reached, the load has been distributed along the base of the fillet (k from the face of flange) on a

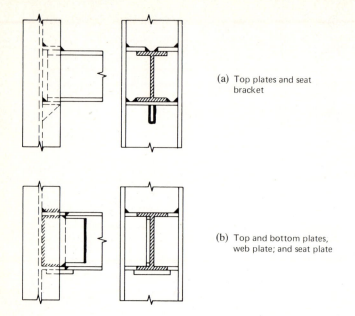

(a) Top plates and seat bracket

(b) Top and bottom plates, web plate; and seat plate

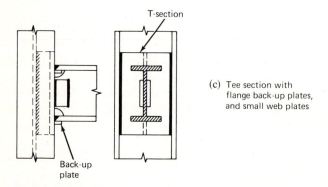

T-section

(c) Tee section with flange back-up plates, and small web plates

Back-up plate

Fig. 13.6.3 Continuous beam-to-column connections: welded attachment to column web.

1:2.5 slope such that for equilibrium

$$P_{bf} = F_{yc}(t_b + 5k)t \qquad (13.6.1)$$

where P_{bf} is the required nominal strength of the column web and F_{yc} is the yield stress of the column web. The strength P_{bf} also corresponds to the load transmitted by the beam flange under *factored loads*.

Under an ultimate strength philosophy, P_{bf} could be as high as $F_{yb}A_f$ if the beam must develop its plastic moment M_p at the connection to the column, or it could be some lesser value based on a factored load moment

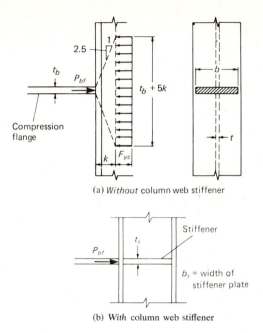

(a) *Without* column web stiffener

(b) *With* column web stiffener

Fig. 13.6.4 Strength of column web in compression region of connection—AISC approach.

less than M_p. When plastic design is used the load factor 1.7 would be used in the analysis of the structure. When working stress method is used it is less obvious how to incorporate a strength concept. AISC–1.15.5.2 requires taking P_{bf} as the service load flange force multiplied by $\frac{5}{3}$ for gravity dead and live load and by $\frac{4}{3}$ when wind or earthquake is included. Such a requirement is consistent with the general use of a 1.67 safety factor for working stress method and is somewhat conservative for plastic design (AISC–2.6). Past practice (such as 1969 AISC Specification) has been conservative in using the maximum possible P_{bf} (i.e., $F_{yb}A_f$) in the development of the design requirements.

Thus the minimum column web thickness t required to prevent web crippling is

$$t \geq \frac{P_{bf}}{(t_b + 5k)F_{yc}} \tag{13.6.2}$$

Not only must web crippling be prevented but overall vertical buckling of the web plate must be avoided. The plate girder provision of AISC–1.10.10.2 as discussed in Sec. 11.2 might seem applicable here. However, the plate girder loading is less severe since the axial compression that *is* present in the column would not be present in the plate girder. Also, in the beam-to-column connection the compression force could be constant over the entire column section depth, whereas the loading on the plate girder is maximum at one flange but zero at the other. As discussed in Sec. 11.2, the elastic buckling of a uniformly compressed plate may be expressed

$$F_{cr} = \frac{k_c \pi^2 E}{12(1 - \mu^2)(h/t)^2}$$ [11.2.17]

where

$$k_c = \left(\frac{1}{m(a/h)^2} + m\right)^2$$ [11.2.18]

When applied to the compression zone of a beam-to-column connection, the depth h in Eq. 11.2.17 becomes d_c and a/h could conservatively be taken as infinity. The buckling strength would be the value with m, the number of half-waves occurring over the depth d_c, having its minimum value of unity. Thus with $k_c = 1.0$ the buckling stress may be approximated as

$$F_{cr} = \frac{\pi^2 E}{12(1 - \mu^2)(d_c/t)^2} = \frac{26{,}200}{(d_c/t)^2}$$ (13.6.3)

or, if $F_{cr} = F_{yc}$, the stability ratio limit would be

$$\frac{d_c}{t} \leq \frac{162}{\sqrt{F_{yc}, \text{ksi}}}$$ (13.6.4)*

Recognizing the conservatism of Eq. 13.6.4, AISC–1969 used a coefficient 180 instead of 162.

Alternatively, and more rational according to Chen and Newlin [23] and Chen and Oppenheim [24], the loading situation may be considered analogous to a plate subject to equal and opposite concentrated loads (rather than uniformly distributed) as shown in Fig. 13.6.5. The elastic buckling strength for such a situation is given by Timoshenko and Gere

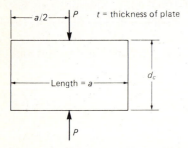

Fig. 13.6.5 Plate subject to equal and opposite concentric concentrated loads.

* For SI units, with F_y in MPa,

$$\frac{d_c}{t} \leq \frac{425}{\sqrt{F_{yc}}}$$ (13.6.4)

[25] (pp. 387–389) for a simply supported plate with a large ratio of a/d_c as the following:

$$P_{cr} = \frac{4\pi E t^3}{12(1-\mu^2)d_c} \tag{13.6.5}$$

which if $P_{cr} = P_{bf}$ gives for the stability ratio limit

$$\frac{d_c}{t} \leq \frac{4\pi E t^2}{12(1-\mu^2)P_{bf}} = \frac{33{,}400t^2}{P_{bf}} \tag{13.6.6}$$

If the rotational restraint provided by the column flanges were the fully fixed condition, the buckling strength would be theoretically twice as great as that given by Eq. 13.6.6. Experimental work [19, 23, 24] has shown that when lower yield stress (such as A36 steel) material is involved, yielding of the web along the junction with the flange occurs at a load level corresponding closely to the simple support case; i.e., Eq. 13.6.6. When 100 ksi (690 MPa) yield stress material was used in tests, the higher yield stress material provided a high degree of rotational restraint along the junction of flange to web, giving a buckling strength about twice that obtained when A36 steel was used. Thus Chen and Newlin [23] suggested that the increase in effective degree of fixity at the loaded edge may be accounted for in practical design by making the strength proportional to the square root of the yield stress. Their stability criterion is

$$\frac{d_c}{t} \leq \frac{33{,}400t^2}{P_{bf}} \sqrt{\frac{F_{yc}}{36}} \tag{13.6.7}$$

which becomes

$$\frac{d_c}{t} \leq \frac{5570t^2\sqrt{F_{yc}}}{P_{bf}} \tag{13.6.8}$$

Adjusting this semirational expression downward to represent a lower bound for all test results gives the AISC–1.15.5.2 requirement that

$$\frac{d_c}{t} \leq \frac{4100t^2\sqrt{F_{yc}}}{P_{bf}} \tag{13.6.9}*$$

with d_c and t in inches, F_{yc} in ksi, and P_{bf} in kips.

When either of Eqs. 13.6.2 or 13.6.9 is not satisfied, a stiffener

---

* For SI units (approximately),

$$\frac{d_c}{t} \leq \frac{10.8t^2\sqrt{F_{yc}}}{P_{bf}} \tag{13.6.9}$$

with d_c and t in mm, F_{yc} in MPa, and P_{bf} in kN.

(area $= A_{st}$) must be provided on the column opposite the compression flange of the beam. When a properly proportioned and connected stiffener is used the force P_{bf} capable of being developed is increased by the amount $A_{st}F_{yst}$; thus

$$P_{bf} = F_{yc}(t_b + 5k)t + A_{st}F_{yst} \qquad (13.6.10)$$

Solving for A_{st} gives AISC Formula (1.15–1),

$$A_{st} \geq \frac{P_{bf} - F_{yc}(t_b + 5k)t}{F_{yst}} \qquad (13.6.11)$$

where F_{yst} is the yield stress of the stiffener material and all other terms are as previously defined.

For proportioning any required compression stiffener, AISC–1.15.5.4 provides empirical rules to use in combination with the area A_{st} requirement from Eq. 13.6.11, or to use alone when Eq. 13.6.11 gives a negative (meaningless) result.

The rules of AISC–1.15.5.4 are:

1. The width b_s of each stiffener plus $\frac{1}{2}$ the thickness t of the column web shall not be less than $\frac{1}{3}$ of the width b of the flange or moment connection plate delivering the concentrated force P_{bf},

$$b_s + \frac{t}{2} \geq \frac{b}{3} \qquad (13.6.12)$$

2. The thickness t_s of a stiffener shall not be less than $t_b/2$ and the local buckling requirements for compression elements must be satisfied. Thus

$$t_s \geq \frac{t_b}{2} \qquad (13.6.13)$$

$$\frac{b_s}{t_s} \leq \begin{array}{l} \text{limits of AISC–1.9 for} \\ \text{working stress design} \\ \text{or AISC–2.7 for plastic} \\ \text{design} \end{array} \qquad (13.6.14)$$

3. When the concentrated force P_{bf} occurs at only one flange of a column, the stiffener length need not exceed $\frac{1}{3}$ of the column depth.

4. The weld joining stiffeners to the column web shall be sized to carry the force in the stiffener caused by unbalanced moments on opposite sides of the column.

Horizontal Stiffener in Tension Region of Connection

At the beam tension flange it is the deformation caused to the column flange which is the main concern, as shown in Fig. 13.6.6. A yield-line

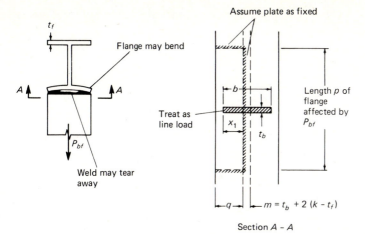

Fig. 13.6.6 Strength of column flange in tension region of connection.

analysis was performed [14] on the portion of the column flange of width q and length p, as in Fig. 13.6.6. Placing a line load on the system, the ultimate capacity was developed as

$$P_u = F_{yc}t_f^2\left[\frac{4/\beta + \beta/\beta_1}{2 - \beta_1/\alpha}\right]$$ (13.6.15)

where t_f = column flange thickness

$\beta = p/q$

$\alpha = x_1/q$

$\beta_1 = \dfrac{\beta}{4}[\sqrt{\beta^2 + 8\alpha} - \beta]$

It is stated [14] that a conservative approximation for Eq. 13.6.15 is

$$P_u = 3.5F_{yc}t_f^2$$ (13.6.16)

Thus the two portions of the column flange of width x_1 may carry $2P_u$. The central portion of width m may be assumed to be stressed to the yield stress:

$$P_{bf} = 7F_{yc}t_f^2 + F_{yc}t_b m$$ (13.6.17)

Since the distribution length, $t + 5k$ or $t + 7k$, is conservative for the compression loading, similar conservatism may be obtained for the tension case by using for P_{bf} only 0.8 of the amount indicated by Eq. 13.6.17. Solving for t_f gives

$$t_f = \sqrt{\frac{P_{bf}}{7F_{yc}}\left(1.25 - \frac{t_b m F_{yc}}{P_{bf}}\right)}$$ (13.6.18)

The minimum value of $t_b m F_{yc}/P_{bf}$ will give the largest t_f requirement. The maximum P_{bf} is $F_{yb}A_f$, and assuming it unlikely that F_{yc} is less than F_{yb}, the negative term may be expressed $t_b m/A_f$, the minimum value for which is 0.15 [12]. Thus Eq. 13.6.18 becomes

$$t_f = \sqrt{\frac{P_{bf}}{F_{yc}} \frac{(1.25-0.15)}{7}} = 0.396\sqrt{\frac{P_{bf}}{F_{yc}}} \tag{13.6.19}$$

AISC–1.15.5.3 requires that a pair of stiffeners shall be provided opposite the tension flange unless the column flange thickness satisfies the following:

$$t_f \geq 0.4\sqrt{\frac{P_{bf}}{F_{yc}}} \tag{13.6.20}$$

The rules of AISC–1.15.5.4 as summarized following Eq. 13.6.11 are the only code-specified requirements for establishing the size of these tension stiffeners.

Example 13.6.1

Design the connection for the rigid framing of two W16×40 beams to the flanges of a W12×65 column using A572 Grade 50 steel, as shown in Fig. 13.6.7. Use A36 steel for stiffeners.

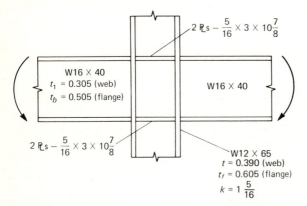

Fig. 13.6.7 Connection with horizontal stiffeners for Example 13.6.1.

SOLUTION

(a) Compression region. Design for the maximum flange force transmitted by the beam. Thus

$$P_{bf} = A_f F_y = 6.995(0.505)50 = 177 \text{ kips}$$

To avoid stiffeners, the column web thickness must satisfy AISC–1.15.5,

Eqs. 13.6.2 and 13.6.9,

$$t \geq \frac{P_{bf}}{(t_b + 5k)F_{yc}} = \frac{177}{[0.505 + 5(1.3125)]50} = 0.50 \text{ in.}$$

and

$$t \geq \sqrt[3]{\frac{P_{bf}d_c}{4100\sqrt{F_{yc}}}} = \sqrt[3]{\frac{177(9.5)}{4100\sqrt{50}}} = 0.39 \text{ in.}$$

Since the column web (W12×65, $t = 0.390$ in.) has less thickness than required to prevent web crippling, compression stiffeners are required.

The area A_{st} of stiffener required according to AISC–1.15.5, Eq. 13.6.11, is

$$A_{st} \geq \frac{P_{bf} - F_{yc}(t_b + 5k)t}{F_{yst}}$$

$$= \frac{177 - 50[0.505 + 5(1.3125)]0.390}{36} = 1.09 \text{ sq in.}$$

Using Eq. 13.6.12, the minimum stiffener width b_s is

$$b_s + \frac{t}{2} \geq \frac{b}{3}$$

$$b_s + \frac{0.390}{2} \geq \frac{6.995}{3}$$

$$\text{Min } b_s = 2.14 \text{ in.,} \qquad \text{say 3 in.}$$

From Eq. 13.6.13, the minimum thickness t_s is

$$\text{Min } t_s = \frac{t_b}{2} = \frac{0.505}{2} = 0.252 \text{ in.,} \qquad \text{say } \tfrac{5}{16} \text{ in.}$$

In addition, local buckling must be prevented; thus one of the following must be satisfied:

1. AISC–1.9.1.2, working stress method, "compact section" *not* utilized at the joint.

$$\frac{\text{width}}{t_s} \leq \frac{95}{\sqrt{F_y}} = 15.8 \qquad \text{for A36 stiffeners}$$

$$t_s \geq \frac{\text{width}}{15.8} = \frac{3}{15.8} = 0.19 \text{ in.}$$

2. AISC–1.5.1.4.1, working stress method, "compact section" requirement.

$$\frac{\text{width}}{t_s} \leq \frac{65}{\sqrt{F_y}} = 10.8 \qquad \text{for A36 stiffeners}$$

$$t_s \geq \frac{\text{width}}{10.8} = \frac{3}{10.8} = 0.28 \text{ in.}$$

3. AISC–2.7, plastic design method.

$$\frac{\text{width}}{t_s} \le 8.5$$

$$t_s \ge \frac{3}{8.5} = 0.35 \text{ in.}$$

Assuming "compact section" is needed, try $\frac{5}{16}$ in.

Provided $A_{st} = 2(3)(0.3125) = 1.88$ sq in. > 1.09 OK

Two plates, $\frac{5}{16} \times 3$, are acceptable for the compression zone.

    (b) Tension region. Check AISC Formula (1.15–3), Eq. 13.6.20:

$$t_f \ge 0.4 \sqrt{\frac{P_{bf}}{F_{yc}}} = 0.4 \sqrt{\frac{177}{50}} = 0.75 \text{ in.}$$

Since the flange thickness of the W12×65 ($t_f = 0.605$) is less than 0.75 in., a stiffener is required. The local buckling requirements do not apply to the tension plates; however, the minimum requirements of AISC–1.15.5.4 still apply, and the minimum $\frac{5}{16} \times 3$ used for the compression region should also be used here.

Use 2 $\text{R}\kern-0.3em\text{s} - \frac{5}{16} \times 3 \times 10\frac{7}{8}$, A36, for both compression and tension sides.

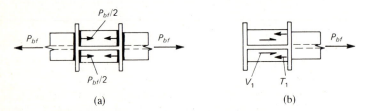

Fig. 13.6.8   Weld requirements for horizontal plates.

    (c) Connection of plates to column. The forces to be considered in the design of welding are shown in Fig. 13.6.8. When beams frame from both sides and contribute equal flange forces P_{bf}, the weld on the ends of the plates must be designed to carry all of the force P_{bf} which is not directly taken by the web. For this example, Fig. 13.6.8a applies, and the proportion of P_{bf} carried by stiffeners is based on the compression region and assumed to be valid also for the tension region; thus

$$\text{Force to stiffeners} = \frac{A_{st}}{t(t_b + 5k) + A_{st}} (P_{bf})$$

$$= \frac{1.88}{(0.39)(7.06) + 1.88} (177) = 72 \text{ kips}$$

Maximum effective weld size, using shielded metal arc welding with E70 electrodes,

$$a_{max\,eff} = \frac{0.60F_y t_s}{2(0.707)21} = \frac{22t_s}{29.7} = 0.74t_s$$

Min weld size $a = \frac{1}{4}$ in.

For fillets top and bottom of plates, capacity required per inch is

$$\frac{72}{2(6)} = 6.0 \text{ kips/in.}$$

$$\text{Required } a = \frac{6.0}{0.707(21)} = 0.40 \text{ in.}, \qquad \text{say } \tfrac{3}{8} \text{ in. is OK}$$

Use $\frac{3}{8}$-in. fillet weld, top and bottom, on both tension and compression plates where they bear against the column flanges. Along the column web, fillet weld is required on only one side of plate. If the $\frac{3}{8}$-in. or larger weld size seems undesirable the stiffener plates could be made wider, say $3\frac{1}{2}$ or 4 in., but they would then have to be thicker to satisfy local buckling requirements.

When a beam frames in from only one side, as in Fig. 13.6.8b, the weld forces T_1 are designed as in the symmetrical case. However, in addition the shear forces V_1 must also be developed (AISC–1.15.5.4, item 4) to take the proportion of the unbalanced flange force which comes to the stiffener plates; in this case $V_1 = T_1$.

Vertical Plate and Tee Stiffeners

Sometimes it may be desirable to use vertical plates, or structural tee sections as shown in Fig. 13.6.1c. Particularly, they may be useful in a four-way system where beams are attached to the tee sections. Research has indicated that a vertical stiffener at the toe of the flange is only one-half as effective as is the web of the column [14].

Thus, assuming two vertical stiffeners (one at each flange toe) each having one-half the web capacity of Eq. 13.6.1, gives the following when vertical stiffeners are used as the counterpart to Eq. 13.6.1,

$$P_{bf} = F_{yc}(t_b + 5k)t + 2\left(\frac{F_{yst}}{2}\right)(t_b + 5k)t_s \qquad \text{(13.6.21)}$$

which upon solving for the stiffener thickness t_s gives

$$t_s \geq \frac{P_{bf}}{(t_b + 5k)F_{yc}} - t\frac{F_{yc}}{F_{yst}} \qquad \text{(13.6.22)}$$

If the vertical stiffener is a plate, then overall buckling must be prevented by satisfying Eq. 13.6.9. When structural tees are used, the web attachment precludes the overall buckling.

For the design of the tee stiffener and connection when a beam is attached to it, as in Fig. 13.6.3c, some special considerations are necessary. The beam flange, if equal in width to the tee, acts on the tee as shown in Fig. 13.6.9a. For analysis, one might consider uniform loading on a two-span beam as shown in Fig. 13.6.9b, wherein the load would be transmitted $\frac{5}{8}$ into the tee web and $\frac{3}{16}$ each to the column flanges. Blodgett [8] suggests when the beam flange extends full width of the tee one may assume the effective flange width b_E (Fig. 13.6.9a) tributary to the tee web to be $\frac{3}{4}$ of the beam flange width.

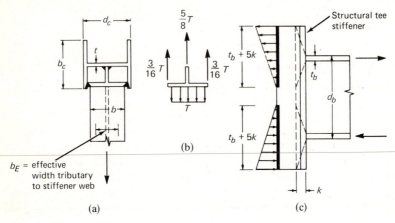

Fig. 13.6.9 Structural tee stiffener.

One may summarize the design requirements when the *beam flange width is approximately the same as the tee flange width:*

1. The tee web thickness t_w must satisfy Eq. 13.6.2:

$$t_w \geq \frac{0.75P_{bf}}{(t_b + 5k)F_{yst}}$$

(13.6.23)

where $0.75P_{bf}$ = the portion of the factored load transmitted by the beam flange tributary to the tee stiffener

k = distance to root of fillet; tee section

t_b = flange thickness of beam

2. The structural tee flange thickness t_s must be able to carry the beam tensile flange force without excessive deformation; hence Eq. 13.6.20 should be satisfied. This will be conservative since the equation was derived for the free-edge condition at the flange toes, whereas here there is a welded connection. Using $0.75P_{bf}$ as in 1,

$$t_s \geq 0.4\sqrt{\frac{0.75P_{bf}}{F_{yst}}} = 0.35\sqrt{\frac{P_{bf}}{F_{yst}}}$$

(13.6.24)

3. The structural tee flange width b_s must extend fully between the column flanges:

$$b_s = d_c - 2t_f \qquad (13.6.25)$$

where d_c = column overall depth
t_f = flange thickness of column

4. The structural tee depth d_s must be adequate to be nearly flush with the outer edges of the column flanges:

$$d_s = \frac{b_c - t}{2} \qquad (13.6.26)$$

where b_c = column flange width
t = column web width

When the beam flange width is significantly less than the tee flange width (say, more than an inch or two), P_{bf} instead of $0.75P_{bf}$ should be used in Eqs. 13.6.23 and 13.6.24.

In making the welded connection when beam flange and tee flange are nearly equal in width, the weld on the tee web (two segments of two fillets) is to resist the moment assuming $0.75P_{bf}$ must be carried [6]. At the toes of the tee, it is suggested [6] to design for $\frac{1}{3}$ of the beam flange effect (somewhat greater than the $\frac{3}{16}$ of Fig. 13.6.9b) to be carried.

Example 13.6.2

Design a vertical tee stiffener connection to frame a W14×61 beam into the web of a W12×65 column. Use A572 Grade 50 steel. Use the type of connection shown in Fig. 13.6.9.

SOLUTION

Since the beam flange width (9.995 in.) is approximately the same as the clear distance between column flanges ($12.12 - 1.21 = 10.91$ in.), Eqs. 13.6.23 and 13.6.24 may be applied.

(a) Determine the stiffener web thickness required to prevent web crippling. The maximum beam flange force P_{bf} is

$$P_{bf} = A_f F_y = 9.995(0.645)50 = 322 \text{ kips}$$

Using Eq. 13.6.23, and estimating $k = 1$ in. for the tee section,

$$\text{Required } t_w = \frac{0.75P_{bf}}{(t_b + 5k)F_{yst}} = \frac{0.75(322)}{[0.645 + 5(1.0)]50} = 0.86 \text{ in.}$$

(b) Determine the stiffener flange thickness to prevent distortion under tension. Using Eq. 13.6.24.

$$\text{Required } t_s = 0.35\sqrt{\frac{P_{bf}}{F_{yst}}} = 0.35\sqrt{\frac{322}{50}} = 0.89 \text{ in.}$$

(c) Select a structural tee section. The maximum flange width permitted is

$$\text{Max } b_s = d_c - 2t_f = 12.12 - 2(0.605) = 10.91 \text{ in.}$$

$$\text{Max depth} = 0.5(12.00 - 0.390) = 5.81 \text{ in.}$$

Try W12×96 cut into tees:

$$t_s = 0.900 \text{ in.} \quad \text{(flange thickness)}$$

$$t_w = 0.550 \text{ in. with } k = 1\tfrac{5}{8} \text{ in.}$$

Rechecking Eq. 13.6.23,

$$\text{Required } t_w = \frac{0.75(322)}{[0.645 + 5(1.625)]50} = 0.55 \text{ in.} = \text{provided } t_w \qquad \text{OK}$$

Try W12×96 cut as shown in Fig. 13.6.10a.

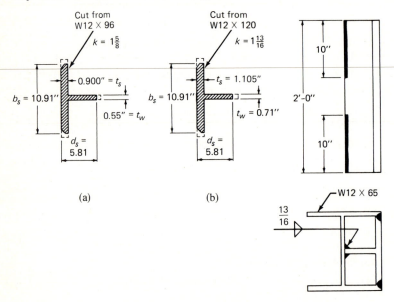

Fig. 13.6.10 Example 13.6.2. Tee selection and web welding.

(d) Welding on stiffener web. Referring to Fig. 13.6.9c, the length of tee stiffener required is

$$\text{Length} = d_b + 5k = 13.89 + 5(1.625) = 22.02 \text{ in.}$$

Try length of tee = 1'-10".
Length of weld, upper, and lower ends:

$$t_b + 5k = 0.645 + 5(1.625) = 8.77 \text{ in.}$$

Try 9-in. weld at each end, connecting web of tee to web of column.

Assume the moment contributed by the center $\frac{3}{4}$ of the flange is tributary to the tee web:

$$0.75M = 0.75F_bS = 0.75(33)(92.2)\tfrac{1}{12} = 190 \text{ ft-kips}$$

where $F_b = 33$ ksi for "compact section." The resisting section modulus of the two 9-in. weld segments treated as lines is

$$S = 2\left(\frac{1}{12}\right)\left[\frac{(22)^3 - (22-18)^3}{11}\right] = 160 \text{ in.}^2$$

$$\text{Actual } R_w = \frac{M}{S} = \frac{190(12)}{160} = 14.3 \text{ kips/in.}$$

$$\text{Allowable } R_w = a(0.707)21 = 14.85a$$

1-in. fillet welds on each side of web would be required.
Check flexural stress on the stiffener web:

$$f = \frac{M}{S} = \frac{190(12)}{0.55(22)^2/6} = 51.3 \text{ ksi;} \qquad \text{too high!}$$

Based on the large weld size required (1 in.) and the high flexural stress which may arise on the web of the stiffener tee, increase the section to W12×120 from which to cut the tee.

$$\text{Length} = d_b + 5k = 13.89 + 5(1\tfrac{13}{16}) = 23.0 \text{ in.,} \qquad \text{say 24 in.}$$

$$\text{Weld segment lengths} = t_b + 5k$$

$$= 0.645 + 5(1.8125) = 9.7 \text{ in.,} \qquad \text{say 10 in.}$$

$$S = 2\left(\frac{1}{12}\right)\left[\frac{(24)^3 - (4)^3}{12}\right] = 191 \text{ in.}^3 \quad \text{(provided by welds)}$$

$$\text{Actual } R_w = \frac{190(12)}{191} = 11.9 \text{ kips/in.}$$

$$\text{Required } a = \frac{11.9}{14.85} = 0.80 \text{ in.,} \qquad \text{say } \tfrac{13}{16} \text{ in.}$$

$$\text{Stress on tee web} = \frac{190(12)}{0.71(24)^2/6} = 33.5 \text{ ksi} \qquad \text{close enough!}$$

Use W12×120 as shown in Fig. 13.6.10b.

(e) Welding on stiffener flanges. Assume conservatively that the tee connection to the column flange (Figs. 13.6.9a and b) may have to carry as much as $\frac{1}{3}$ of the beam flange force. The concentrated forces from the beam flange may be considered as distributed along the column flange over the distance $t_b + 5t_s$, as shown in Fig. 13.6.11.
Since the W14×61 flange may carry an allowable stress, $F_b = 33$ ksi,

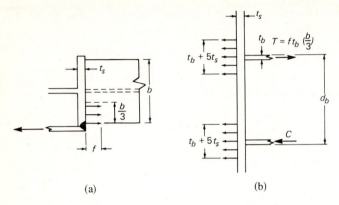

(a) (b)

Fig. 13.6.11 Forces carried by weld along tee stiffener flanges.

the tensile force assumed tributary to the groove weld on one flange is

$$T = ft_b\left(\frac{b}{3}\right) = 33(0.645)\left(\frac{9.995}{3}\right) = 70.9 \text{ kips}$$

$$t_b + 5t_s = 0.645 + 5(1.105) = 6.17 \text{ in.}$$

$$\text{Required weld capacity} = \frac{70.9}{6.17} = 11.5 \text{ kips/in.}$$

Using a partial joint penetration groove weld, E70 electrode material, the required effective throat dimension is

$$\text{Required effective throat} = \frac{11.5}{21} = 0.55 \text{ in.}$$

$$\text{Required groove depth} = 0.55 + \tfrac{1}{8} \text{ in.} = 0.67 \text{ in.} \quad \text{(AISC--1.14.6.1.2)}$$

AISC–1.17.2 gives the minimum effective throat as $\frac{5}{16}$ in.
Use $\frac{11}{16}$-in. single-U-groove (AWS–2.10, BC-P6 prequalified joint) weld along the edges of the flange of the tee.

(f) Effect of beam shear force. Ordinarily the length of weld used is so large that the additional capacity required to carry end shear is negligible. A W14×61 might be expected to carry something on the order of 50 kips shear. In which case it is resisted by

$$\text{Total length of weld} = 4(10) + 2(24) = 88 \text{ in.}$$
$$\text{web} \quad \text{flanges}$$

$$\text{Required extra weld capacity} = \frac{50}{88} = 0.57 \text{ kips/in.}$$

$$\text{Added size for web fillets} = \frac{0.57}{0.707(21)} = 0.04 \text{ in.}$$

$$\text{Added size for flange groove welds} = \frac{0.57}{21} = 0.03 \text{ in.}$$

The shear component and the flexure component actually act at 90 degrees to one another so that adding the requirements algebraically would be overly safe. The authors consider the weld design adequate without an increase for direct shear.

Top Tension Plates

When the beam is connected to the column flange and the column is stiffened by vertical or horizontal plate stiffeners, or when the beam is connected to the column web through a vertical tee stiffener, a simple means of transmitting the moment from the beam is with a tension plate at the top of the beam combined with (1) a bottom compression plate and web plates for shear; (2) a bottom seat angle; or (3) a bottom bracket (stiffened seat). See Figs. 13.6.1d and 13.6.2a and b. The behavior of such top plate connections has been verified by research studies [26, 27].

The design of seat angles and brackets has been treated in Secs. 13.3 and 13.4. Transmission of tension and compression forces into the column has been treated earlier in this section. Emphasis here is on the top plate, a tension member whose design is illustrated in the following example.

Example 13.6.3

Design the top tension plate and its connection by welding or by A325 high-strength bolts for transferring the full end moment from a W14×61 to a column. Use A572 Grade 50 steel. Assume the connection to be of the type shown in Figs. 13.6.1d or 13.6.2a.

SOLUTION

(a) Design plate as a tension member.

Since the W14×61 is a "compact section" for Grade 50 steel, the allowable flexural stress is $0.66F_y$. Design for full capacity. The flange force T is approximately

$$T = \frac{M}{d_b} = \frac{0.66F_yS_x}{d_b} = \frac{33(92.2)}{\approx 14} = 217 \text{ kips}$$

$$\text{Required plate area } A_g = \frac{T}{0.60F_y} \geq \frac{T}{0.50F_u}$$

$$\text{Required } A_g = \frac{217}{0.60(50)} = \underline{7.2 \text{ sq in.}} > \frac{217}{0.50(65)} = 6.7 \text{ sq in.}$$

The W14×61 flange width is 10.0 in., so plate width must be less than 10.0 in., say 9 in.

$$\text{Required plate thickness} = \frac{7.2}{9} = 0.8 \text{ in.}$$

Use ℝ—$\frac{7}{8}$×9, $A_g = 7.9$ sq in., for welded construction.

(b) For welding plate to beam flange, try $\frac{3}{8}$-in. fillet weld, E70 electrodes, assuming shielded metal arc process,

$$R_w = a(0.707)21 = (0.375)(0.707)21 = 5.57 \text{ kips/in.}$$

$$\text{Required } L_w = \frac{217}{5.57} = 39.0 \text{ in.}$$

To reduce the required weld length, try $\frac{1}{2}$-in. weld:

$$\text{Required } L_w = \frac{217}{7.42} = 29.2 \text{ in.}$$

Use $\frac{1}{2}$-in. weld, 9 in. on end and 10 in. on each side for a total of 29 in. The design is summarized in Fig. 13.6.12a.

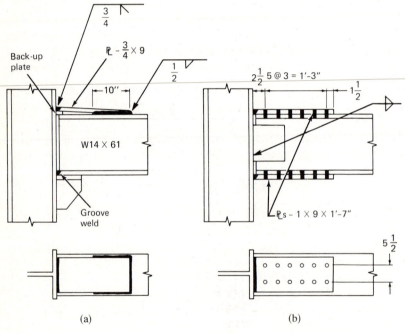

(a)   (b)

Fig. 13.6.12 Example 13.6.3. Moment connection using top plate.

The plate is connected to the column flange by a full penetration single bevel groove weld, using a backup bar. When a seat is used it will serve as backup to make the flange groove weld. Otherwise, a backup plate is needed below the compression flange. Frequently it may be satisfactory to use fillet welds along the top and bottom of the plate to make the connection to the column flange.

(c) Using $\frac{7}{8}$-in.-diam A325 bolts in a bearing-type connection with threads excluded from the shear plane (A325–X),

$$R_{SS} = A_b F_v = 0.6013(30) = 18.04 \text{ kips} \quad \text{(single shear)}$$

$$R_B = \text{does not control}$$

When bolts are used, the holes in the W14×61 flange exceed 15% of the gross flange, so that according to AISC–1.10.1 the excess must be deducted. In this case,

$$\text{Percent} \frac{A_n}{A_g} = \frac{9.995(0.645) - 2(1)(0.645)}{9.995(0.645)} = 80\%$$

Thus 5% (20–15) of the gross flange effect must be deducted and the tensile capacity of the flange becomes 95% of 217 kips.

For the plate as a tension member, the effect of all holes must be deducted in computing net area A_n:

$$\text{Required } A_g = \frac{T}{0.60F_y} = \frac{0.95(217)}{30} = 6.9 \text{ sq in.}$$

or

$$\text{Required } A_n = \frac{T}{0.50F_u} = \frac{0.95(217)}{0.50(65)} = 6.3 \text{ sq in.}$$

Use ℝ —1×9, $A_n = 9.00 - 2(1) = 7.00$ sq in. > 6.3 OK

$$\text{No. of bolts required} = \frac{0.95(217)}{18.04} = 11.4, \qquad \text{Use 12.}$$

Use 12—$\frac{7}{8}$-in.-diam A325 bolts, as shown in Fig. 13.6.12b.

Split-Beam Tee Connections

For bolted moment connections, the split-beam tee as shown in Fig. 13.6.2c is one of the more common types. The design of a connection involving the transfer of a tensile force through a thick-plate bolted connection, such as the tee connected to the flange, requires consideration of "prying action." Prying action was first mentioned in Sec. 4.9.

Consider the deformation of a tee section, as in Fig. 13.6.13, where as the pull on the web deforms the flange and deflects it outward, the edges of the flanges of the tee bear against the connected piece giving rise to the force Q, known as the "prying force." Inclusion of this force is required by AISC–1.5.2.1 wherein it states, "The applied load shall be the sum of the external load and any tension resulting from prying action produced by deformation of the connected parts."

When bolted connections are not subject to distortion such as is evident in Fig. 13.6.13b, the treatment of Sec. 4.9 for bolts in tension is

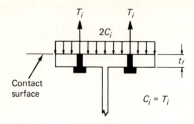

(a) No external load.

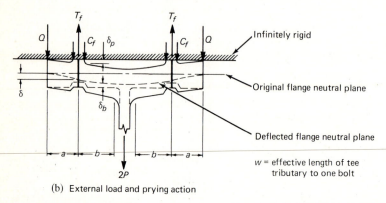

(b) External load and prying action

Fig. 13.6.13 Prying action, including model used by Douty and McGuire [17].

valid. However, when a thick-flanged tee distorts, the flange tips tend to dig in, giving rise to the force Q.

Using the same reasoning as for Eqs. 4.9.2 through 4.9.9, one may obtain the expression for the prying force Q, presented by Douty and McGuire [17]. Before external load is applied,

$$C_i = T_i \tag{13.6.27}$$

After external load is applied, prying action occurs giving for the equilibrium requirement

$$P + C_f + Q = T_f \tag{13.6.28}$$

The increased bolt length is

$$\delta_b = \left(\frac{T_f - T_i}{A_b E}\right) t = \left[\frac{P + Q - (T_i - C_f)}{A_b E}\right] t \tag{13.6.29}$$

The increased plate thickness is

$$\delta_p = \left(\frac{C_i - C_f}{A_p E}\right) t = \left(\frac{T_i - C_f}{A_p E}\right) t \tag{13.6.30}$$

Assume the forces are uniformly distributed along the tee section length w that is tributary to one bolt. The tee flange thickens an amount δ_p and remains in contact with the support, the bolt length increases an amount δ_b, and the neutral plane of the tee flange deflects an amount δ as shown in Fig. 13.6.13. All of these deformations are equal in magnitude.

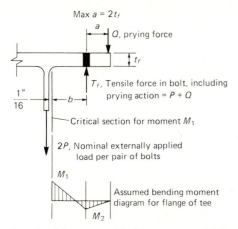

Fig. 13.6.14 Simplified loading model for prying action.

The equation for the prying force Q given by Douty and McGuire is overly complex for design use. Prying action has been studied more recently by Kato and McGuire [18], Nair, Birkemoe, and Munse [28], Agerskov [29, 30], and Krishnamurthy [31]. For the past several years the AISC Manual has recommended the procedure of Nair et al. [28] along with the simple loading model shown in Fig. 13.6.14. The equations for the prying force Q are as follows:

1. For connections using A325 bolts,

$$Q = P\left[\frac{100bD^2 - 18wt_f^2}{70aD^2 + 21wt_f^2}\right] \qquad (13.6.31a)$$

2. For connections using A490 bolts,

$$Q = P\left[\frac{100bD^2 - 14wt_f^2}{62aD^2 - 21wt_f^2}\right] \qquad (13.6.31b)$$

where P = externally applied load per fastener

b and a = distances defined in Fig. 13.6.14

D = the nominal bolt diameter

t_f = the thickness of the flange or plate being attached

w = the length of flange tributary to a pair of bolts

Thus the design requirement satisfying AISC–1.5.2.1 may be stated as

$$P + Q \le R_t \qquad (13.6.32)$$

where $R_t = A_b F_t$ from AISC–Table 1.5.2.1.

Example 13.6.4

Design a split-beam tee connection, such as in Fig. 13.6.2c, to enable a plastic hinge to develop in a W14×53 beam framing to the flange of a W14×159 column. Use A572 Grade 50 steel and $\frac{3}{4}$-in.-diam A325 bolts in a bearing-type connection (A325–N). Use working stress design.

SOLUTION

(a) Determine tensile force to be carried. Since the W14×53 satisfies the local buckling requirements for "compact section," and if the maximum laterally unbraced length is $L_c = 7.2$ ft, the allowable stress is $0.66F_y$.

$$M = 0.66F_y S_x = 33(77.8)\tfrac{1}{12} = 214 \text{ ft-kips}$$

If all moment is transmitted by the tees, the force of the couple is

$$\text{Force} = \frac{M}{d_b} = \frac{214(12)}{13.92} = 184 \text{ kips}$$

(b) Check whether the tensile force can be accommodated by the bolts in tension:

$$\text{Allowable bolt capacity } R_t = 0.4418(44) = 19.4 \text{ kips}$$

Only 8 bolts will fit, as shown in Fig. 13.6.15; therefore the maximum tensile force which may be carried is

$$T = 8(19.4) = 155 \text{ kips} < 184 \text{ kips} \qquad \text{NG}$$

When this difficulty arises one may use a stub beam or a tee stub attached to the bottom of the main beam (Fig. 13.6.15) to increase the moment arm of the couple. Actually, when designing for the support moment, one might have used a beam size required for the midspan moment and then used the stub beam to gain the increased capacity required at the support. The necessary moment arm is

$$\text{Arm} = \frac{214(12)}{155} = 16.6 \text{ in.}$$

$$\text{Extra depth required} = 16.6 - 13.92 = 2.6 \text{ in.}$$

Try as stub beam a WT5×22.5, $t_w = 0.350$ in., $t_f = 0.620$ in., $b_f = 8.020$ in., whose dimensions are comparable to the main W14×53 beam.

$$\text{Force of couple} = \frac{214(12)}{13.92 + 5.050} = 135 \text{ kips}$$

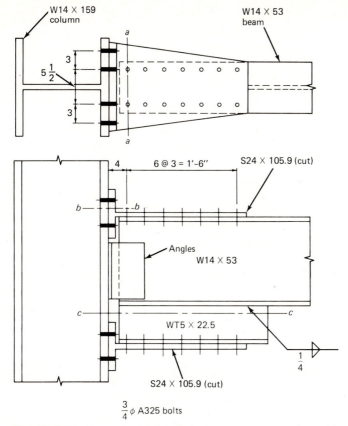

$\frac{3}{4} \phi$ A325 bolts

Fig. 13.6.15 Example 13.6.4. Split-beam tee connection with tee stub.

Using 8 bolts in tension,

$$f_t A = \frac{135}{8} = 16.9 \text{ kips} < 19.4 \text{ kips allowed} \qquad \text{OK}$$

(c) Check shear stress on web (section $c-c$ of Fig. 13.6.15) of WT5×22.5

$$\text{Length of tee required} = \frac{\text{Force}}{t_w(0.40 F_y)} = \frac{135}{0.350(20)} = 19.3 \text{ in.}$$

(d) Determine bolts required to transmit tension and compression forces at the top and bottom of beam:

$$R_{SS} = (0.4418)22 = 9.67 \text{ kips} \quad \text{(controls)}$$
$$R_B = 1.5F_u Dt = 1.5(65)(0.75)t = 73.1t$$

$$\text{Number of bolts} = \frac{135}{9.67} = 14.0 \qquad \text{use 14}$$

Using 3-in. spacing, the minimum length of WT5×22.5 required is 21 in.:
Use WT5×22.5 stub tee, 2'-0" long, welded to the bottom of W14×53, as shown in Fig. 13.6.15.

(e) Determine thickness required to transmit tension on section $a - a$ (Fig. 13.6.15):

$$\text{Required } A_g = \frac{T}{0.60F_y} = \frac{135}{30} = 4.5 \text{ sq in.}$$

and

$$\text{Required } A_n = \frac{T}{0.50F_u} = \frac{135}{0.50(65)} = 4.1 \text{ sq in.}$$

Using the length of section $a - a$ as 13 in. (column flange width $= 15.565$ in.), and deducting two holes, gives

$$t \geq \frac{4.1}{13 - 2(0.875)} = 0.36 \text{ in.}$$

$$t \geq \frac{9.67}{73.1} = 0.13 \text{ in.} \quad \text{(bearing does not control)}$$

(f) Determine flange thickness required for flexure on section $b - b$, Fig. 13.6.15.

$$M_1 = T_f b - Q(b + a)$$

using notation of Fig. 13.6.14. The critical section at M_1 may be taken at $\frac{1}{16}$ in. from the face of the web according to Ref. 28. Thus

$$b = \frac{g}{2} - \frac{t_w}{2} - \frac{1}{16} = \text{estimated 1 in.}$$

Assuming $Q = 0$,

$$M_1 = T_f b = \frac{135}{2}(1) = 63 \text{ in.-kips}$$

assuming the length of tee at the critical section is 14 in.,

$$f_b = \frac{M}{S} = \frac{63}{14t_f^2/6} \leq 0.75F_y \quad \text{(AISC–1.5.1.4.3)}$$

$$\text{Required } t_f = \sqrt{\frac{6(63)}{14(37.5)}} = 0.85 \text{ in.}$$

Try a tee cut from S24×79.9, $t_f = 0.871$ in., $t_w = 0.501$ in.

(g) Check prying force using Eq. 13.6.31a,

$$b = \frac{4}{2} - \frac{0.501}{2} - \frac{1}{16} = 1.69 \text{ in.}$$

$$w = \frac{14}{4} = 3.5 \text{ in.}$$

$$a = \frac{b_f - g}{2} = \frac{7.001 - 4}{2} = 1.5 \text{ in.} < 2t_f = 1.74 \text{ in.} \qquad \text{OK}$$

$$Q = P\left[\frac{100(1.69)(\tfrac{3}{4})^2 - 18(3.5)(0.870)^2}{70(1.5)(\tfrac{3}{4})^2 + 21(3.5)(0.870)^2}\right] = 0.413P$$

$$P + Q = P + 0.413P = 1.413(16.9) = 23.9 \text{ kips}$$

This is not acceptable since it exceeds $R_t = 19.4$ kips allowed. It is best to revise the section to obtain a thicker flange and reduce prying action Try S24×105.9, $t_f = 1.102$ in., $t_w = 0.625$ in.

$$a = \frac{7.875 - 4}{2} = 1.94 \text{ in.} < 2t_f \qquad \text{OK}$$

$$b = \frac{4}{2} - \frac{0.625}{2} - \frac{1}{16} = 1.63 \text{ in.}$$

$$Q = P\left[\frac{100(1.63)(\tfrac{3}{4})^2 - 18(3.5)(1.102)^2}{70(1.94)(\tfrac{3}{4})^2 + 21(3.5)(1.102)^2}\right] = 0.092P$$

$$P + Q = 1.092P = 1.092(16.9) = 18.4 \text{ kips} < R_t = 19.4 \text{ kips} \qquad \text{OK}$$

(h) Recheck flexural stress on flange.

$$M_1 = 1.092Pb - 0.092P(a + b)$$
$$= [1.092(1.63) - 0.092(1.94 + 1.63)]P = 1.45P$$
$$M_1 = 1.45(16.9) = 24.5 \text{ in.-kips/fastener}$$

$$S = \frac{1}{6}\left(\frac{14}{4}\right)(1.102)^2 = 0.708 \text{ in.}^3/\text{fastener}$$

$$f_b = \frac{M}{S} = \frac{24.5}{0.708} = 34.6 \text{ ksi} < 0.75F_y = 37.5 \text{ ksi} \qquad \text{OK}$$

Use tees cut from S24×105.9 to carry tensile and compressive forces.

Not discussed in this example is the development of the shear capacity of the W14×53. A pair of angles may be attached to the beam web for the purpose of providing whatever shear is required. The final design is shown in Fig. 13.6.15.

End-Plate Connections

This simple type of moment resisting connection consists of a plate welded to the end of the beam and then bolted to the column, as shown in Fig. 13.6.2d.

Behavior of end-plate connections has been studied by Beedle and Christopher [16], Onderdonk, Lathrop, and Coel [15], Douty and McGuire [17], Agerskov [29, 30], and Krishnamurthy [31]. Since extensive use of end-plate connections began in the late 1960s, the design approach has changed concurrently with better understanding of their behavior. When properly designed using the minimum number of high-strength bolts, the region near the tension flange of the beam is designed as for the split-tee design discussed previously in this section. The fastener group is designed for shear and tension, including any effect of prying action. The use of Eqs. 13.6.31 for the prying force Q is appropriate for the typical end-plate connection. Recently Krishnamurthy [31] has recommended a modified split-tee method based on an analytical study and correlation with available tests. This modified tee method seems suitable only with adequate design aids, some of which are available in Ref. 31.

Example 13.6.5

Design an end-plate connection for a W14×53 beam section attachment to a W14×176 column; both of A36 steel. Design for full moment capacity of the beam and for 45 kips shear at service load. Use A325 bolts in a bearing-type connection (A325–X). Use working stress design.

SOLUTION

(a) Determine the number of bolts required for the tensile force of the bending moment. The force to be developed in the tension flange is

$$T = 0.66F_y A_f = 24(8.060)(0.660) = 128 \text{ kips}$$

For $\frac{7}{8}$-in.-diam bolts, A325–X,

$$R_t = 0.6013(44) = 26.4 \text{ kips}$$

$$\text{Number of bolts} = \frac{T}{R_t} = \frac{128}{26.4} = 4.8$$

For symmetrical placement above and below the tension flange, either 4 or 8 bolts would be needed. If the bolt size were increased to 1 in. diam, 4 bolts might be satisfactory.

For 1-in.-diam bolts, A325–X,

$$R_t = 0.7854(44) = 34.6 \text{ kips}$$

$$\text{Number of bolts} = \frac{T}{R_t} = \frac{128}{34.6} = 3.7; \qquad \text{Try 4}$$

(b) Establish the plan dimensions of the end-plate. For determining the distance s (Fig. 13.6.16) the fillet weld size (for E70 electrodes) and the bolt installation clearance are needed.

$$\text{Required } R_w = \frac{T}{L_w} = \frac{T}{2b_f - t_w} = \frac{128}{2(8.060) - 0.370} = 8.13 \text{ kips/in.}$$

$$\text{Required weld size} = \frac{8.13}{0.707(21)} = 0.55 \text{ in.,} \qquad \text{say } \frac{9}{16} \text{ in.}$$

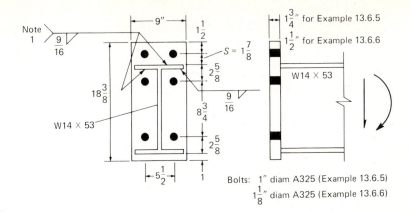

Fig. 13.6.16 End-plate connection for Examples 13.6.5 and 13.6.6.

The minimum erection clearance for 1-in.-diam bolts is $1\frac{5}{16}$ in., as given by AISC Manual table, "Rivets and Threaded Fasteners, Erection Clearances."

$$\text{Distance } s = 1\tfrac{5}{16} + \tfrac{9}{16} = 1\tfrac{7}{8} \text{ in.}$$

Try a plate 9 in. wide and about 18 in. long.

(c) Estimate the plate thickness required. Referring to Fig. 13.6.14, the moment arm b for computing the bending moment on the plate is

$$b = 1\tfrac{7}{8} - \tfrac{1}{16} = 1\tfrac{13}{16} = 1.81 \text{ in.}$$

Assume $Q = 0$; then $P = T/2$, and

$$M_1 = \frac{T}{2}(b) = \frac{128}{2}(1.81) = 116 \text{ in.-kips}$$

$$S = \tfrac{1}{6}(9)t_p^2$$

$$f_b = \frac{M}{S} = \frac{116}{\tfrac{1}{6}(9)t_p^2} = 0.75 F_y = 27 \text{ ksi}$$

$$\text{Required } t_p = \sqrt{\frac{6(116)}{9(27)}} = 1.7 \text{ in.}$$

Try $1\frac{3}{4}$-in.-thick end-plate.

(d) Check prying action. Using Eq. 13.6.31a,

$$a = 1.5 \text{ in.}; \qquad b = 1.81 \text{ in.}; \qquad w = \frac{9}{2} = 4.5 \text{ in.}$$

$$Q = P\left[\frac{100(1.81)(1)^2 - 18(4.5)(1.75)^2}{70(1.5)(1)^2 + 21(4.5)(1.75)^2}\right] = \text{negative}$$

For the $1\frac{3}{4}$-in.-thick plate, the prying force is zero as assumed.

(e) Check combined shear and tension on bolts.

$$f_v = \frac{V}{A_b} = \frac{45}{6(0.7854)} = 9.5 \text{ ksi}$$

Allowable tensile stress F'_t in the presence of shear is (AISC–1.6.3)

$$F'_t = 55 - 1.4 f_v \leq 44 \text{ ksi}$$
$$= 55 - 1.4(9.5) = 41.7 \text{ ksi} < 44 \text{ ksi} \qquad \text{OK}$$

$$f_t = \frac{128}{4(0.7854)} = 40.7 \text{ ksi} < F'_t = 41.7 \text{ ksi} \qquad \text{OK}$$

(f) Length of end-plate. At the end near the compression flange of the beam, it is sometimes desirable to extend the plate outside the beam flange an amount equal to the end-plate thickness t_p. This will increase the length of the critical section used in computing the web crippling stress. If a 45° flow through the end-plate is assumed, the critical section length could be increased by t_p; using $(t_b + 5k + 2t_p)$ instead of $(t_b + 5k + t_p)$.

Check web crippling on the W24×176 without extending the end-plate beyond the compression flange of the beam.

$$f_b = \frac{\text{Force}}{(t_b + 5k + t_p)t} = \frac{128}{[0.660 + 5(2) + 1.75]0.83} = 12.4 \text{ ksi}$$

This is well below the allowable value of $0.75F_y = 27$ ksi.
Use ℝ $1\frac{3}{4} \times 9 \times 1' - 5\frac{3}{8}''$, with 6—1-in.-diam bolts in a bearing-type connection (A325–X), as shown in Fig. 13.6.16.

Example 13.6.6

Repeat Example 13.6.5 except use plastic design.

SOLUTION

(a) Number of bolts required for the tensile force of the bending moment.

$$T_u = F_y A_f = 36(8.060)(0.660) = 192 \text{ kips}$$

Try 4—1-in.-diam A325 bolts,

$$R_t = 1.7 A_b F_t = 1.7(0.7854)44 = 58.7 \text{ kips}$$

$$\text{Number of bolts} = \frac{192}{58.7} = 3.3; \qquad \text{try 4}$$

(b) Establish plan dimensions of the end-plate. The fillet weld size required is

$$\text{Weld size} = \frac{T_u}{L_w(1.7)(0.707)21} = \frac{192}{15.75(1.7)(0.707)21} = 0.48 \text{ in.}$$

For plastic design, $\frac{1}{2}$-in. weld could be used instead of the $\frac{9}{16}$ in. used in the working stress method. Try the same plate size used in Example 13.6.5, as shown in Fig. 13.6.16.

(c) Determine trial plate thickness. Assume the prying force is zero and $b = 1.81$ in. as in Example 13.6.5,

$$M_1 = M_p = \frac{192}{2}(1.81) = 174 \text{ in.-kips}$$

$$Z = \frac{\text{width }(t_p)^2}{4} = \frac{9t_p^2}{4}$$

$$\frac{M_p}{Z} = F_y$$

$$t_p = \sqrt{\frac{4M_p}{9F_y}} = \sqrt{\frac{4(174)}{9(36)}} = 1.46 \text{ in.}$$

Try $1\frac{1}{2}$-in.-thick plate.

(d) Check prying action. Using Eq. 13.6.31a,

$$Q_u = P_u\left[\frac{100(1.81)(1)^2 - 18(4.5)(1.5)^2}{70(1.5)(1)^2 + 21(4.5)(1.5)^2}\right] = \text{negative}$$

Thus the prying force Q_u is to be taken as zero.

(e) Check combined shear and tension on the bolts. Following the provisions of AISC–2.8,

$$f_v = \frac{V_u}{A_b} = \frac{1.7(45)}{6(0.7854)} = 16.2 \text{ ksi}$$

$$F_t' = 1.7(55 - 1.4f_v) = 1.7[55 - 1.4(16.2)] = 54.9 \text{ ksi}$$

$$f_t = \frac{T_u}{4(0.7854)} = \frac{192}{4(0.7854)} = 61.1 \text{ ksi} > F_t' \quad \text{NG}$$

Use larger diameter bolts, say $1\frac{1}{8}$ in. diam.

Use ⅊ $1\frac{1}{2} \times 9 \times 1' - 5\frac{3}{8}''$, with 6—$1\frac{1}{8}$-in.-diam A325 bolts in a bearing-type connection (A325–X).

Beam to Column-Web Direct Connection

Instead of using vertical tee stiffeners (Fig. 13.6.1c) or complicated details such as shown in Fig. 13.6.3, occasionally the designer would prefer to attach a beam directly to the column web, using either a welded or bolted connection. The column web must then resist the moment effect by plate action. Abolitz and Warner [32], Stockwell [33], and Kapp [34] have presented yield line analysis techniques for computing whether or not the column web is satisfactory for making a moment or direct tension connection directly to it. Attaching beams to box sections presents a similar situation. Practical design data is presented by Stockwell [33]; however, a detailed discussion of the yield line analysis is outside the scope of this text.

13.7 CONTINUOUS BEAM-TO-BEAM CONNECTIONS

When beams frame transversely to other beams or girders they may be attached to either or both sides of the girder web using simple framed beam connections (Sec. 13.2) or using simple seats in combination with a framed beam connection (Sec. 13.3). When full continuity of the beam is desired, the connection must develop a higher degree of rigidity than provided by framed connections. For beam-to-beam connections the principal objective is to provide a means of allowing the tensile force developed in one beam flange to be carried across to the adjacent beam framing opposite the girder web. These connections may be divided into two categories; (1) those with intersecting tension flanges not rigidly attached to one another, as in Fig. 13.7.1; and (2) those with rigidly attached intersecting tension flanges, as in Fig. 13.7.2.

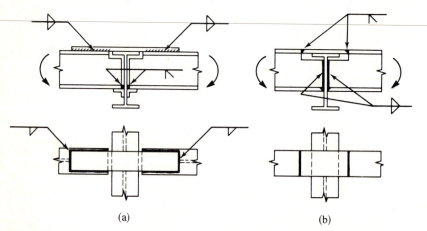

(a)   (b)

Fig. 13.7.1 Intersecting beam connections: tension flanges *not attached* to each other.

When the intersecting tension flanges are not rigidly attached (Fig. 13.7.1) the connection design is essentially a tension member design at the tension flange along with a shear connection.

On the other hand, the designer should be cautious when designing intersecting beams with tension flanges attached since this case becomes one of biaxial instead of uniaxial stress. With biaxial stress the possibility of brittle fracture increases (see Sec. 2.9). In addition, a biaxial stress yield criterion must be used, such as Eq. 2.6.2, to establish safety:

$$\sigma_y^2 = \sigma_1^2 + \sigma_2^2 + \sigma_1\sigma_2 \qquad [2.6.2]$$

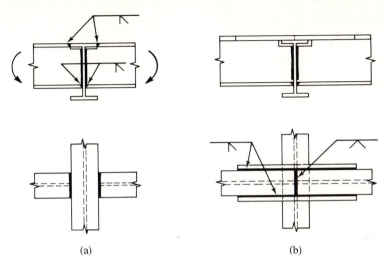

Fig. 13.7.2 Intersecting beam connections: tension flanges *attached* to each other.

where σ_1, σ_2 are the principal stresses acting. When the beams frame at 90 degrees to one another the axial stresses in the flanges are principal stresses.

The designer should make certain the σ_y of Eq. 2.6.2 does not exceed $0.60F_y$ for working stress design, nor F_y for plastic design.

13.8 RIGID-FRAME KNEES

In the design of rigid frames according to either working stress method or plastic design, the safe transmission of stress at the junction of beam and column is of great importance. When members join with their webs lying in the plane of the frame, the junction is frequently referred to as a knee joint. Typical knee joints are (1) the square knee, with or without a diagonal or other stiffener (Fig. 13.8.1a and b); (2) the square knee with a bracket (Fig. 13.8.1c); (3) the straight haunched knee (Fig. 13.8.1d); and (4) the curved haunched knee (Fig. 13.8.1e).

Working stress design ordinarily assumes the span of the members to extend from center-to-center of adjacent knees, and that the moment of inertia of the member varies in accordance with the moment of inertia of a cross section taken at right angles to lines extending center-to-center of knees. Moments and shears may then be determined using statically indeterminate analysis of a frame (involving variable moment of inertia if knees are haunched).

In plastic design using square knees without brackets or haunches, the plastic hinge forms in the member and the knee is proportioned to prevent failure occurring in that region. When haunches are used it is

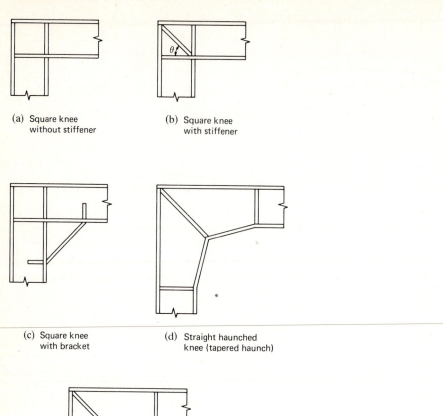

(a) Square knee
without stiffener

(b) Square knee
with stiffener

(c) Square knee
with bracket

(d) Straight haunched
knee (tapered haunch)

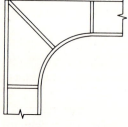

(e) Curved haunched knee

Fig. 13.8.1 Rigid-frame knees.

possible for the plastic hinge to occur either within or outside of the haunch region.

Considerable research has been conducted on rigid-frame knees and the design concepts have been summarized in ASCE Manual No. 41 [35], which forms the basis for what follows.

To be adequately designed, a knee connection must (1) transfer the end moment between the beam and the column, (2) transfer the beam end shear into the column, and (3) transfer the shear at the top of the column into the beam. Furthermore, in performing the three functions relating to strength, the knee must deform in a manner consistent with the analysis by which moments and shears were determined.

If a plastic hinge associated with the failure mechanism is expected to form at or near the knee, adequate rotation capacity must be built into the connection. Square knees have the greatest rotation capacity but are also the most flexible (i.e., deform elastically the most under service load conditions). Curved knees are the stiffest but have the least rotation capacity. Since straight tapered knees provide reasonable stiffness along with adequate rotation capacity, in addition to the fact that they are cheaper than curved haunches to fabricate, the straight haunched knees are commonly used.

Shear Transfer in Square Knees Without Brackets or Haunches

In the design of a rigid frame having square knees, two rolled sections may come together at right angles as shown in Fig. 13.8.1a. A frame analysis, either elastic or plastic, will have established what moments and shears act on the boundaries of the square knee region, as shown in Fig. 13.8.2a. The forces carried by the flanges must be transmitted by shear into the web, as shown in Fig. 13.8.2b.

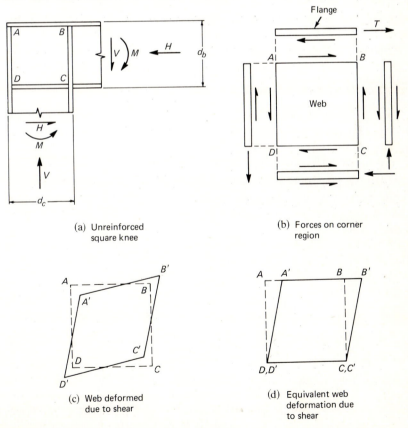

(a) Unreinforced square knee

(b) Forces on corner region

(c) Web deformed due to shear

(d) Equivalent web deformation due to shear

Fig. 13.8.2 Shear transfer in a square knee.

Assuming all bending moment is carried by the flanges, and approximating the distance between flange centroids to be $0.95d_b$, the flange force is

$$T = \frac{M}{0.95d_b} \qquad (13.8.1)$$

The shear capacity of the web across AB is

$$V_{ab} = F_v t_w d_c \qquad (13.8.2)$$

Equating Eqs. 13.8.1 and 13.8.2 and solving for t_w gives

$$t_w = \frac{M}{F_v(0.95d_b)d_c} \qquad (13.8.3)$$

For *plastic design*, the value for F_v is the shear yield stress, taken as $0.55F_y$ by AISC–2.5. Equation 13.8.3 then becomes

$$\text{Required } t_w = \frac{1.91M}{F_y d_b d_c} = \frac{1.91M}{F_y A_{bc}} \qquad (13.8.4)$$

where $A_{bc} = d_b d_c$, the planar area within the knee. In Eq. 13.8.4, M is the factored load moment for which the connection must be designed (maximum is M_p, the plastic moment).

For *working stress method*, the maximum allowable shear F_v is $0.40F_y$, which in Eq. 13.8.3 gives

$$\text{Required } t_w = \frac{2.63M}{F_y d_b d_c} = \frac{2.63M}{F_y A_{bc}} \qquad (13.8.5)$$

where M is the service load moment.

For most W sections the required web thickness will exceed that provided by the section so that reinforcement is usually needed. A doubler plate is sometimes used to lap over the web to obtain the total required thickness, but usually a pair of diagonal stiffeners is the most practical solution.

When diagonal stiffeners are used, the horizontal component of the stiffener force, $T_s \cos \theta$, participates with the web. Equilibrium (Fig. 13.8.3) then requires

$$T = V_{ab} + T_s \cos \theta \qquad (13.8.6)$$

or

$$\frac{M}{0.95d_b} = F_v t_w d_c + A_{st} F_a \cos \theta \qquad (13.8.7)$$

Solving for the stiffener area A_{st} gives

$$\text{Required } A_{st} = \frac{1}{\cos \theta}\left(\frac{M}{0.95d_b F_a} - \frac{F_v t_w d_c}{F_a}\right) \qquad (13.8.8)$$

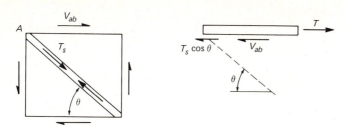

Fig. 13.8.3 Diagonal stiffener effect.

For *plastic design*, $F_a = F_y$ and $F_v = 0.55F_y$, which gives, for Eq. 13.8.8,

$$\text{Required } A_{st} = \frac{1}{\cos\theta}\left[\frac{1.053M}{d_bF_y} - 0.55t_wd_c\right] \tag{13.8.9}$$

For *working stress method*, $F_a = 0.60F_y$ assuming no instability, and $F_v = 0.40F_y$, which gives

$$\text{Required } A_{st} = \frac{1}{\cos\theta}\left[\frac{1.76M}{d_bF_y} - 0.67t_wd_c\right] \tag{13.8.10}$$

Example 13.8.1

Design the square knee connection to join a W27×94 girder to a W14×74 column. Use plastic design with a corner moment of 376 ft-kips under factored load (LF = 1.7). Use A36 steel and E70 electrodes with shielded metal arc welding (SMAW).

SOLUTION

(a) Check whether or not diagonal stiffener is required. Using Eq. 13.8.4, and referring to Fig. 13.8.4,

$$\text{Required } t_w = \frac{1.91M}{A_{bc}F_y} = \frac{1.91(376)(12)}{26.92(14.17)36} = 0.63 \text{ in.}$$

$$\text{Actual } t_w = 0.490 \text{ in. for W27×94} < 0.63 \text{ in.}$$

A diagonal stiffener is required.

(d) Determine stiffener size.

$$\tan\theta = \frac{26.92}{14.17} = 1.900; \quad \cos\theta = 0.466$$

Using Eq. 13.8.9,

$$\text{Required } A_{st} = \frac{1}{0.466}\left[\frac{1.053(376)(12)}{26.92(36)} - 0.55(0.49)14.17\right] = 2.3 \text{ sq in.}$$

Use 2 ℞ s—$\frac{1}{2}$×2$\frac{1}{2}$. The local buckling requirement of AISC–2.7 is satisfied.

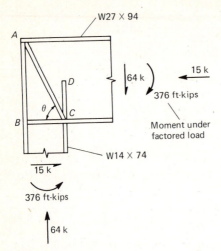

Fig. 13.8.4 Square knee of Example 13.8.1.

(c) Determine fillet weld size along length AB. This weld must be adequate to transmit the plastic moment flange force into the beam web.

$$\text{Flange force} = F_y A_f = 36(10.070)(0.785) = 285 \text{ kips}$$
$$\text{Weld capacity} = a(0.707)(21)(2)(1.7) = 50.5a$$

Note that for plastic design the allowable stress is multiplied by the load factor 1.7.

$$\text{Available length for weld} = 26.92 - 2(0.745) = 25.4 \text{ in.}$$
$$\text{Required } a = \frac{285}{25.4(50.5)} = 0.22 \text{ in.}$$

Use $\frac{1}{4}$-in. E70 fillet weld along length AB (both sides of girder web).

(d) Determine fillet weld size along length BC. The connection of the column web to the beam flange must carry the stress due to flexure and axial load combined with shear. At the most highly stressed location the plastic moment will cause a tensile yield stress over part of the web.

$$\text{Tensile component} = F_y t_w = 36(0.450) = 16.2 \text{ kips/in.}$$
$$\text{Shear component} = \frac{V}{d_c - 2t_f} = \frac{15}{14.17 - 2(0.785)} = 1.2 \text{ kips/in.}$$
$$\text{Resultant loading} = \sqrt{(16.2)^2 + (1.2)^2} = 16.2 \text{ kips/in.}$$
$$\text{Required } a = \frac{16.2}{2(0.707)(21)(1.7)} = 0.32 \text{ in.}$$

Use $\frac{3}{8}$-in. E70 fillet weld along length BC (both sides of girder web).

(e) Determine weld required along stiffener. This weld must develop the stiffener capacity, which in plastic design may be computed using the yield stress:

$$T_s = F_y A_{st} = 36(2.5)(0.5)2 = 90 \text{ kips}$$

$$\text{Required } a = \frac{90/(14.17/\cos \theta)}{4(0.707)(21)1.7} = 0.03 \text{ in.}$$

Use $\frac{3}{16}$ in. E70 fillet weld along diagonal _AC_ (both sides of girder web).

(f) Determine extent of stiffener required from point _C_ vertically into girder. The capacity of the W27×94 web to carry the compression force P_{bf} from the W14×74 flange at _C_ is, using Eq. 13.6.2,

$$\text{Web capacity} = F_y t_w (t_b + 5k)$$

$$= 36(0.490)[0.785 + 5(1.4375)] = 141 \text{ kips}$$

Since the diagonal stiffener is already performing the task of transferring the shear resulting from moment on the W27×94, it may not be included in the resistance of the beam web against the column flange compression force. Thus a vertical stiffener _CD_ must carry

$$\text{Stiffener } CD \text{ force} = 285 - 141 = 144 \text{ kips}$$

If $\frac{3}{16}$-in. weld is used, the stiffener length _CD_ must be

$$\text{Length} = \frac{144}{4(\frac{3}{16})(0.707)(21)1.7} = 7.6 \text{ in., say 8 in.}$$

(g) Establish plate size. The dimensions other than length must satisfy AISC–1.15.5.4, Eqs. 13.6.12 and 13.6.13.

$$\text{Required area} = \frac{144/2}{F_y} = \frac{72}{36} = 2.0 \text{ sq in./plate}$$

$$\text{Max width available} = \frac{10.070}{3} - \frac{0.490}{2} = 3.1 \text{ in.}$$

$$\text{Min thickness} = \frac{0.785}{2} = 0.39 \text{ in.}$$

For local buckling, AISC–2.7 limits width-to-thickness ratio to 8.5 maximum.

Use 2℉ s—$\frac{1}{2}$×4×0′-8″, tapered from full width at _C_ to zero width at _D_ (Fig. 13.8.4).

Straight Haunched Knees

Straight haunched knees, sometimes called _tapered haunches_, may extend over a significant portion of the span; in which case they are not really connections but rather they are an integral part of a variable depth frame.

The design of tapered haunches involves consideration of three factors; (1) bending moment along the tapered region; (2) shear stresses and flange force transmission within and adjacent to the haunched section; and (3) local and lateral-torsional buckling resistance.

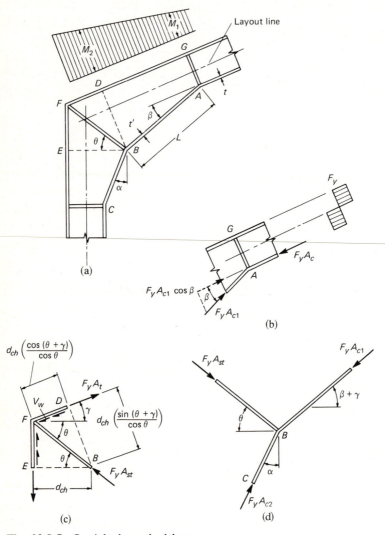

Fig. 13.8.5 Straight haunched knee.

Examine the straight haunched knee of Fig. 13.8.5 showing a moment diagram over the haunched portion. Use of a haunch complicates a plastic-strength frame analysis because it increases the number of possible plastic hinge locations. For working stress method no special disadvantage arises. Even though the moment variation may not be linear along the haunch, a linear assumption satisfying the moment requirements for M_1 and M_2 will be satisfactory in flexure at intermediate points.

The following development relates to plastic design requirements as suggested in ASCE Manual No. 41 [35]. For working stress design, the reader is referred to Griffiths [36]; however, the general philosophy is similar to that of plastic design.

The design for flexure requires that the sections at moments M_1 and M_2 have a plastic moment capacity exceeding those moments. For computation, the resisting sections are taken perpendicular to the layout line; i.e., AG for resisting M_1 and BD for resisting M_2. Since the compressive force in the flange of the haunch acts in the direction AB the moment provided at any section is proportional to the component in the direction parallel to the layout line; thus the effective area is $A_{c1} \cos \beta$. It is usually desirable to increase the flange area A_{c1} in the haunched region to give equal flange forces so that the resisting section may be considered symmetrical. Thus the required area of the flange in the haunch region AB is

$$A_{c1} = \frac{A_c}{\cos \beta} \tag{13.8.11}$$

where A_c = compression flange area outside of haunch region. If the plate width is kept the same (as it usually should be) within and outside the haunch region, then

$$t' = \frac{t}{\cos \beta} \tag{13.8.12}$$

When the angle β is less than $20°$ the increase in area required is less than 6 percent and most practical design would neglect such a difference [35].

The design for shear in the corner is similar to that for a square knee without haunch, except the region to be considered is $EBDF$ and the trigonometry is more complicated. The flange force at D is partly carried as shear on the web over the distance FD. A stiffener should be used along BF, and its size may be determined considering force equilibrium at F (tension in flanges) and at B (compression in flanges). The larger required stiffener force satisfying equilibrium will determine stiffener size.

Consider first the equilibrium of forces at the tension corner (corner at F), as shown in Fig. 13.8.5c. Summation of horizontal forces gives

$$F_y A_{st} \cos \theta + V_w \cos \gamma = F_y A_t \cos \gamma \tag{13.8.13}$$

$$A_{st} = \frac{\cos \gamma}{\cos \theta} \left(A_t - \frac{V_w}{F_y} \right) \tag{13.8.14}$$

Since plastic strength in shear is taken as $0.55F_y$, the shear force V_w carried by the web over the distance FD is

$$V_w = 0.55 F_y t_w d_{ch} \left[\frac{\cos (\theta + \gamma)}{\cos \theta} \right] \tag{13.8.15}$$

which upon substitution into Eq. 13.8.14 gives

$$\text{Required } A_{st} = \frac{\cos \gamma}{\cos \theta}\left[A_t - 0.55t_w d_{ch}\left(\frac{\cos(\theta + \gamma)}{\cos \theta}\right)\right] \qquad \text{(13.8.16)}$$

based on the tension corner.

Consider next the equilibrium of the forces at the compression corner (corner at B), as shown in Fig. 13.8.5d. Summation of horizontal forces gives

$$F_y A_{st}\cos\theta + F_y A_{c2}\sin\alpha = F_y A_{c1}\cos(\beta + \gamma) \qquad \text{(13.8.17)}$$

$$\text{Required } A_{st} = \frac{A_{c1}\cos(\beta + \gamma) - A_{c2}\sin\alpha}{\cos\theta} \qquad \text{(13.8.18)}$$

In compression there is no significant amount of shear transfer into the web, so that the small shear forces along CB and BA are neglected. The stiffener size is based on the larger requirement of Eqs. 13.8.16 and 13.8.18. Because of the web participation near F, and its lack of participation near B, equilibrium at B usually controls—i.e., Eq. 13.8.18.

Regarding lateral bracing, AISC–2.9 indicates that for severe cases where bending moment is nearly constant along an unbraced length and equal in magnitude to the plastic moment (i.e., rotational moments give $-0.5 > M/M_p > -1.0$):

$$L_{cr} = \frac{1375}{F_y} r_y \qquad \text{(13.8.19)}$$

For $F_y = 36$ ksi, and taking $r_y \approx b_f/\sqrt{12}$ for rectangular flanges, gives

$$L_{cr} = \frac{38.2}{\sqrt{12}} b_f = 11 b_f \qquad \text{(13.8.20)}$$

It would seem AISC–2.9 applies for this situation even though ASCE Manual 41 [35] indicates L_{cr} should not exceed $5b_f$ for A36 steel. The AISC Commentary–2.9 does not specifically refer to tapered haunches but indicates AISC–2.9 should be applied to all lateral bracing situations in connections near plastic hinges.

Curved Haunched Knees

For better appearance, curved haunches are sometimes used. Detailed treatment of curved haunches may be found in Blodgett [8] and ASCE Manual 41 [35].

13.9 COLUMN BASES

The design of column bases involves two main considerations. The compressive forces in the column flanges must be distributed by means of a base plate to the supporting medium so that bearing stresses are within permissible values allowed by specifications. The second concern is with the connection, or anchorage, of the base plate and column to the concrete foundation. For frame analysis it may be of importance to evaluate the moment-rotation characteristics of the entire anchorage, including the base plate, anchor bolts, and concrete footing, to determine its stiffness and degree of fixity.

Requirements for Bases Under Axial Load

The method presented and terminology used is that from the AISC Manual, "Column Base Plates." This procedure is not required by the AISC Specification but may be considered as accepted good practice. The dimensions and loads are given in Fig. 13.9.1.

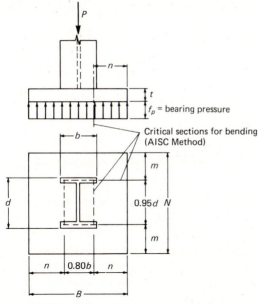

Fig. 13.9.1 Column base plate dimensions.

It is assumed the stress distribution under the base plate is uniform and the plate projections beyond the critical sections act as cantilever beams.

The bending moments on cantilevers of spans m and n are respectively

$$M = \frac{f_p N n^2}{2} \quad \text{(on section parallel to column web)} \quad \text{(13.9.1)}$$

$$= \frac{f_p B m^2}{2} \quad \text{(on section parallel to column flanges)}$$

$$\text{(13.9.2)}$$

The unit stress is

$$f = \frac{M}{S} = \frac{f_p N n^2/2}{N t^2/6} = \frac{3 f_p n^2}{t^2} \qquad (13.9.3)$$

or

$$f = \frac{3 f_p m^2}{t^2} \qquad (13.9.4)$$

The larger of Eqs. 13.9.3 or 13.9.4 governs.

AISC–1.5.1.4.3 allows a stress of $0.75 F_y$ for bending on solid rectangular sections. Thus the required plate thickness can be found:

$$f = \frac{3 f_p m^2}{t^2} \le 0.75 F_y \qquad (13.9.5)$$

$$\text{Required } t = \sqrt{\frac{3 f_p m^2}{0.75 F_y}} \quad \text{or} \quad \sqrt{\frac{3 f_p n^2}{0.75 F_y}} \qquad (13.9.6)$$

Recent tests by DeWolf [37] have indicated the AISC suggested procedure is conservative.

Example 13.9.1

Design a base plate for a W14×145 column of A36 steel to carry 775 kips axial compression resulting from dead load, live load, and wind. Assume allowable unit pressure F_p under plate is 1.0 ksi under gravity loading.

SOLUTION

Once the column has been designed it is frequently best to design the base plate for the column capacity rather than the applied load, in order to avoid having the connection as the weakest element. The column capacity, of course, depends on the effective (equivalent pinned-end) length. AISC tables, "Column Base Plates," provide base plate sizes assuming the shortest practical effective length, in most cases 6 ft.

(a) Determine plan dimensions for plate:

$$\text{Required } A = \frac{P}{1.33 F_p} = \frac{775}{1.33(1.0)} = 582 \text{ sq in.}$$

Note the allowable stress is increased $33\frac{1}{3}\%$ for loading that includes wind in accordance with AISC–1.5.6. The plate should be approximately square, with slight differences in dimensions B and N to give nearly equal values for m and n.

$$0.80b = 0.80(15.50) = 12.40 \text{ in.}$$
$$0.95d = 0.95(14.78) = 14.04 \text{ in.}$$

Try $B = 24$ in. and $N = 25$ in.

$$A = 25(24) = 600 \text{ sq in.} > 582 \text{ sq in.} \qquad \text{OK}$$

(b) Determine plate thickness:

$$n = 0.5(24 - 12.40) = 5.8 \text{ in.} \quad \text{(controls)}$$
$$m = 0.5(25 - 14.04) = 5.5 \text{ in.}$$
$$f_p = \frac{775}{600} = 1.29 \text{ ksi}$$
$$\text{Required } t = \sqrt{\frac{3f_p n^2}{0.75 F_y}} = \sqrt{\frac{3(1.29)(5.8)^2}{27(1.33)}} = 1.90 \text{ in.}$$

Use ℝ —2×24×2'-1". Note AISC–1.21.3 which states that plates over 2 in. but not over 4-in. thickness may be straightened by rolling or pressing. When over 4 in. thick, extra thickness must be specified so that the bearing surface can be planed. Design aids for selection of base plates have been presented by Sandhu [38], Stockwell [39], and Bird [40].

Column Bases to Resist Moment

Column bases frequently must resist moment in addition to axial compression. The situation has some similarities to the behavior of bolted connections discussed in Chapter 4, and in many respects is analogous to the situation of reinforcing bars in concrete construction. The axial force causes a precompression between the base plate and the contact surface (frequently a concrete wall or footing). When the moment is applied the precompression on the tension side in flexure is reduced, often to zero, leaving only the anchor bolt to provide the tensile force resistance. On the compression side, the contact area remains in compression. The anchorage will have an ability to undergo rotational deformation depending primarily on the length of anchor bolt available to deform elastically. Also, the behavior is influenced by whether or not the anchor bolts are given an initial pretension (similar to the installation of high-strength bolts as discussed in Chapter 4). The moment-rotation characteristics of column anchorages are treated in detail by Salmon, Schenker, and Johnston [41].

A number of elaborate methods are available for designing moment resisting bases, with variations depending on the magnitude of the eccentricity of loading and the specific details of the anchorage. Some simple details are shown in Fig. 13.9.2.

When the eccentricity of loading, $e = M/P$, is small such that it does not exceed $\frac{1}{6}$ of the plate dimension N in the direction of bending (i.e., within the middle third of the plate dimension), the ordinary combined stress formula applies. Thus for small e

$$f = \frac{P}{A} \pm \frac{M}{S} \tag{13.9.7}$$

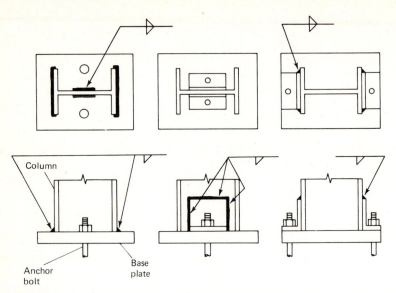

Fig. 13.9.2 Column bases to resist moment.

where $M = Pe$

$$S = Ar^2/(N/2) = AN/6$$
$$r^2 = N^2/12$$
$$f = \frac{P}{A} \pm \frac{6Pe}{AN} = \frac{P}{A}\left[1 \pm \frac{6e}{N}\right] \tag{13.9.8}$$

where N = plate dimension in the direction of bending. Equation 13.9.8 is correct for $e \le N/6$ when there is no bolt pretension, and is considered satisfactory for practical purposes at least up to $e = N/2$ without serious error.

When the eccentricity e exceeds $N/6$, part of the base plate at the tension face becomes inactive and the stress distribution becomes as shown in Fig. 13.9.3. A simple practical assumption to use for such a situation is that the resultant of the triangular distribution is located directly under the compression flange of the column. When large moment is designed for, generally the attachment arrangement to the base of the column becomes more complex than those of Fig. 13.9.2. For a more detailed presentation on designing column bases for moment, the reader is referred to Ref. 8 (Sec. 3.3, pp. 1–32).

Example 13.9.2

Design a column base of the type shown in Fig. 13.9.3 to carry an axial compression of 150 kips and a moment of 190 ft-kips on a W14×82 column section. Use A36 steel with E70 electrodes in shielded metal arc welding, and the working stress method. The base rests on a large

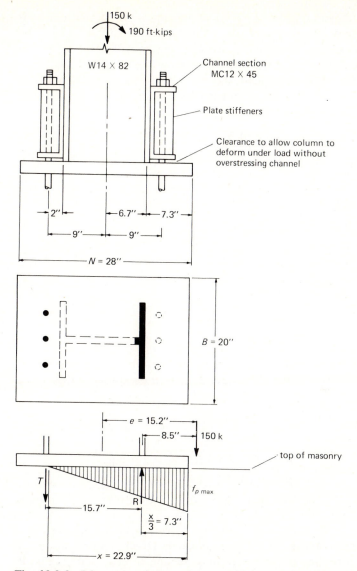

Fig. 13.9.3 Moment-resisting column anchorage of Example 13.9.2.

concrete footing having $f'_c = 3000$ psi such that the allowable bearing stress F_p is $0.7f'_c$ (AISC–1.5.5).

SOLUTION

(a) Estimate the base plate plan dimensions. The eccentricity of load is

$$e = \frac{M}{P} = \frac{190(12)}{150} = 15.2 \text{ in.}$$

Allowing 3 in. beyond the flanges for attachments, the dimension N is estimated at 20 in., making $e/N > 0.5$; therefore Eq. 13.9.8 will not be practical to use since a triangular stress distribution will occur under the base plate. Referring to Fig. 13.9.3, assume the compressive stress resultant force lies directly beneath the compression flange for bending. If the tension anchor bolts can be centered at 2 in. from the column face, the internal couple moment arm will be

$$\text{Arm} = d_c - t_f + 2 = 14.31 - 0.855 + 2 = 15.46 \text{ in.}$$

If the compressive stress distribution is assumed to be as in Fig. 13.9.3, the length x is

$$x = \tfrac{3}{2}(\text{arm}) = 1.5(15.46) = 23.2 \text{ in.} \quad \text{(actually used 22.9 in.)}$$

$$\frac{x}{3} = 7.73 \text{ in.} \quad \text{(actually used 7.3 in.)}$$

The distance from the compression edge of plate to the nearest anchor bolt is

$$\text{Edge distance} = 7.73 - 0.5(0.855) - 2 = 5.3 \text{ in.}$$

$$N = x + 5.3 = 23.2 + 5.3 = 28.5 \text{ in.}$$

Use $N = 28$ in. since the foregoing procedure is not precise.

The width B required will depend on the allowable bearing pressure. Moment equilibrium about R (Fig. 13.9.3) gives the required tensile force T,

$$T = \frac{150(8.5)}{15.7} = 81.2 \text{ kips}, \quad \text{say } 81$$

$$R = 150 + T = 150 + 81 = 231 \text{ kips}$$

$$R = \tfrac{1}{2}f_p xB = \tfrac{1}{2}(2.1)(22.9)B = 24.0B$$

$$\text{Required } B = \frac{231}{24.0} = 9.6 \text{ in.}$$

Since the flange width is 10.13 in., B must be larger than required for bearing stress. To allow for connection of section to base plate, 12 in. is probably minimum. To minimize deformation of the plate, try something larger, say 15 in. Try a plate with 15×28 in. plan dimensions.

(b) Determine plate thickness:

$$f_{p\max} = \frac{2R}{xB} = \frac{2(231)}{22.9(15)} = 1.34 \text{ ksi}$$

Referring to Fig. 13.9.1, the m dimension is

$$m = \frac{28 - 0.95(14.31)}{2} = 7.2 \text{ in.}$$

The bearing pressure at the critical section is

$$f_p = \frac{f_{p\,max}}{x}(22.9 - m) = \frac{1.34}{22.9}(22.9 - 7.2) = 0.92 \text{ ksi}$$

At the critical section,

$$M = \left[0.92\frac{(7.2)^2}{2} + (1.34 - 0.92)\frac{(7.2)^2}{3}\right]15 = 467 \text{ in.-kips}$$

$$0.75F_y = \frac{M}{Bt^2/6}$$

$$\text{Required } t = \sqrt{\frac{6(467)}{27(15)}} = 2.63 \text{ in.}$$

If $2\frac{5}{8}$ in. is considered too thick, the width B may be increased. (For $B = 20$ in., $t = 2\frac{1}{4}$ in.)

Use ⅊—$2\frac{1}{4} \times 20 \times 2' - 4''$ as the base plate.

 (c) Select anchor bolts. Assume A307 bolts; $F_u = 60$ ksi.

$$\text{Required } A_g = \frac{T}{0.33F_u} = \frac{81}{0.33(60)} = 4.1 \text{ sq in.}$$

From AISC table, "Threaded Fasteners," select 3—$1\frac{1}{2}$-in.-diam anchor bolts at each flange of column, $A = 3(1.77) = 5.31$ sq in.

Use 6—$1\frac{1}{2}$-in.-diam A307 threaded rods for anchor bolts.

 (d) Channel stiffeners and their connections. The bolt load is transmitted through the channel (MC12×45), which is stiffened by four plate stiffeners inside the channel, as shown in Fig. 13.9.4. Each interior plate stiffener may be assumed to carry the full bolt load.

$$\text{Load to stiffener} = \frac{81.2}{3} = 27.1 \text{ kips}$$

$$\text{Required } A_{st} = \frac{27.1}{0.60F_y} = \frac{27.1}{22} = 1.23 \text{ sq in.}$$

Use ⅊ s—$\frac{7}{16} \times 3\frac{1}{4}$ as stiffeners.

 As shown in Fig. 13.9.4b, horizontal weld forces F_1 and a vertical weld force F_2 combine to resist the 27.1 kips acting with an eccentricity of 2 in. from channel web. One may assume the forces F_1 carry the moment and the force F_2 carries the shear. Such an approach will give

$$F_2 = 27.1 \text{ kips;} \qquad F_1 = \frac{27.1(2)}{10.6} = 5.12 \text{ kips}$$

$$f'_y = \frac{27.1}{10.6} = 2.55 \text{ kips/in.}$$

$$f''_x = \frac{5.12}{3.25} = 1.58 \text{ kips/in.}$$

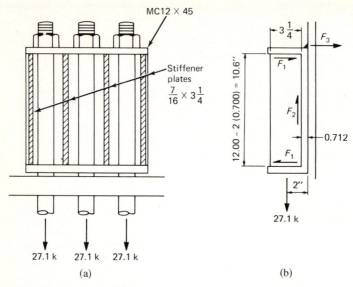

MC12 × 45

Stiffener plates $\frac{7}{16} \times 3\frac{1}{4}$

$12.00 - 2\,(0.700) = 10.6''$

$3\frac{1}{4}$

F_3

F_1

F_2

F_1

0.712

2''

27.1 k

27.1 k   27.1 k   27.1 k

(a)   (b)

Fig. 13.9.4   Details for Example 13.9.2.

Using $\frac{1}{4}$-in. fillet on both sides of stiffener, as controlled by AISC–1.17.2,

$$R_w = \tfrac{1}{4}(0.707)(21)2 = 7.4 \text{ kips/in.}$$

The strength requirement is low so that intermittent weld (say several $1\frac{1}{2}$-in. segments) would be acceptable. Weld should be made on both sides of stiffener so as to avoid eccentricity with respect to the plane of the stiffener.

(e) Connection to the column flange. The sum of the forces F_1 must be carried by a fillet weld connecting the channel flange to the column flange, indicated in Fig. 13.9.4 by the force F_3.

$$F_3 = \frac{\text{Moment}}{\text{Channel depth}} = \frac{81.2(2)}{12} = 13.5 \text{ kips}$$

The length available for weld is the flange width of the W14×84; i.e., about 12 in. Using $\frac{1}{4}$-in. weld, the length required to be welded is

$$L_w = \frac{13.5}{0.25(0.707)21} = \frac{13.5}{3.71} = 3.6 \text{ in.}$$

Use 4 in. of $\frac{1}{4}$ in. fillet weld to carry force F_3.

The shear to be carried by weld along the channel web is 81.2 kips. Using $\frac{1}{4}$-in. fillet, the length of weld required is

$$L_w = \frac{81.2}{2(3.71)} = 10.95 \text{ in.} \quad (12 \text{ in. available})$$

Use $\frac{1}{4}$-in. continuous fillet weld along the channel web to connect to the column flange.

This example has been presented to illustrate some of the reasoning that may be used to design the welds for a column base. The stiffened channel arrangement represents but one of many possible and commonly used means of transmitting the bolt forces into the column. When only one or two anchor bolts are required on each side, often angles are used as anchor bolt sleeves with their toes welded directly to the column flange. Frequently wing plates are used in various positions to better distribute the loads over the base plate.

13.10 BEAM SPLICES

There are several reasons why a rolled beam or plate girder may be spliced, such as: (1) the full length may not be available from the mill; (2) the fabricator may find it economical to splice even when the full length could be obtained; (3) the designer may desire to use splice points as an aid to cambering; and (4) the designer may desire a change in section to fit the variation in strength required along the span.

One each side of a joint, one must provide for the transverse shear and bending moment. For welded plate girders, and frequently for rolled beams, the splice may be accomplished by a full-penetration groove weld. Splices of material made entirely in the shop are nearly always groove welded. Field splices are becoming more frequently all welded, though usually using lapping of pieces and fillet-welded connections, instead of groove welding where dimensional control is a critical factor. Since full-penetration groove-welded splices are as strong as the base material, they require no further comment here.

Many field welded splices for beams and welded plate girders use lapping splice plates and high-strength bolts as connectors. In this section the four- and eight-plate beam splices are teated.

Prior to the almost universal use of welding to fabricate plate girders, riveted or bolted girders had splices made of many plates through which a complex load distribution had to be considered. Treatment of such splices is adequately presented in older texts* and, because of their rare current use, is omitted here.

Splices are designed for the moment M and the shear V occurring at the spliced point, or they are designed for some arbitrary or specification prescribed higher values. AISC–1.10.8 requires groove-welded splices to "develop the full strength of the smaller spliced section." "Other types of splices in cross sections of plate girders and in beams shall develop the strength required by the stresses, at the point of splice." Earlier AISC

---

* See Thomas C. Shedd, *Structural Design in Steel* (John Wiley & Son, Inc., New York, 1934), pp. 142–166.

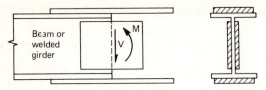

Fig. 13.10.1 Four-plate beam splice.

Specifications required splices to develop not less than 50% of the effective strength of the material spliced. AASHO–1.7.16 (1977) requires designing for "not less than 75% of the strength of the member."

For beam splices, each element of the splice is designed to do the work the material underlying the splice plates could do if uncut. Plates on the flange do the work of the flange, while plates over the web do the work of the web. One typical arrangement of a four-plate splice is shown in Fig. 13.10.1. When the beam flange is thick and there are large forces to be transmitted, plates on both inside and outside of the flanges may be desired, so the eight-plate splice as shown in Fig. 13.10.2 may be used. In the eight-plate splice the flange splices can be designed so that the centroid of the splice area coincides with the flange centroid.

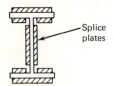

Fig. 13.10.2 Cross section of eight-plate splice.

When determining the forces for which to design, one should recognize that the splice actually covers a finite distance along the span (perhaps 1 or 2 ft). Along this distance the moment and shear vary. In accordance with the principles used for designing bolted connections, the forces designed for should be those acting on the center of gravity of the bolt group. From Fig. 13.10.3 it may be noted that theoretically the moment at the centroid of the connectors on one side of the splice is different from that at the centroid on the other. Some designers, therefore, proportion the connection on each side of the splice for a moment; $M_1 = M + Ve$. Such a procedure rarely seems justified. In those unusual situations when a splice is located where both shear and bending moment are high, such a procedure might seem desirable. Most splices are located in regions where either the shear or the bending moment are low so that often the design is for a specification required minimum strength. As discussed in Sec. 10.6 for plastic design, one must be wary of designing a splice for a low moment just because the splice is near an inflection point. *If moments for the structure were computed using theory of statically*

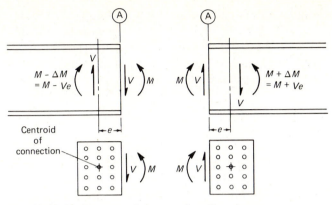

Fig. 13.10.3 Forces acting on web splice plates.

indeterminate structures without a hinge within the span, one should not later design a splice which has low stiffness to act as a hinge.

For most situations the authors recommend designing for the actual or specification required forces *at the splice* and neglecting any eccentricity effect. Furthermore, if one considers the effective eccentricity as discussed in Sec. 4.8, the effective value will be nearly zero for most situations.

Example 13.10.1

Design a rolled-beam splice (four-plate type of Fig. 13.10.1) for a W24×84 beam of A36 steel to be located where the bending moment is 270 ft-kips and the shear is 70 kips. Use A36 splice material with $\frac{3}{4}$ in. diam. A325 bolts in a bearing-type connection (A325–X). In addition to AISC requirements, assume the splice must have not less than 50% of the capacity of the material spliced.

SOLUTION

(a) Total capacity of W24×84.

$$M = F_b S_x = 0.66 F_y S_x = 24(196)\tfrac{1}{12} = 392 \text{ ft-kips}$$
$$V = F_v dt_w = 0.40 F_y dt_w = 14.5(24.10)(0.470) = 164 \text{ kips}$$

(b) Design conditions.

Actual $M = 270$ ft-kips $> 50\%$ of 392 ft-kips

Design $M = 270$ ft-kips

Actual $V = 70$ kips $< 50\%$ of 164 kips

Design $V = 82$ kips

Even though the design is to be made for the combination of 270 ft-kips moment and 82 kips vertical shear acting simultaneously, it is probable that these two values (or even 270 ft-kips and 70 kips) do not occur under the same loading condition. In an actual structure the designer should evaluate the chances of the loads acting simultaneously and consider them accordingly.

(c) Web plates. The web plates are selected to carry the shear on the gross section (AISC–1.10.1).

$$\text{Required } A_g = \frac{V}{0.40F_y} = \frac{82}{14.5} = 5.65 \text{ sq in.}$$

$$\text{Available depth} = T \text{ dimension} = 21 \text{ in.}$$

$$\text{Required } t = \frac{A_g}{2(\text{depth})} = \frac{5.65}{2(21)} = 0.134 \text{ in.}$$

A thickness of $\frac{1}{4}$ in. should be considered a practical minimum.
Use 2 ℞ s—$\frac{1}{4}$×21.

(d) Flange plates. In this approach, flange splice area is provided which will do the work the unspliced flange was required to do.

$$A_f(\text{W24} \times 84) = 9.020(0.770) = 6.95 \text{ sq in.}$$

$$\text{Percent of full capacity required} = \frac{270}{392}(100) = 68.9\%$$

$$\text{Required plate area } A_g = 6.95(0.689) = 4.80 \text{ sq in.}$$

AISC–1.10.1 requires deduction for area of holes exceeding 15% of gross flange.
Try plate $\frac{9}{16} \times 9$, $A_g = 5.06$ sq in. Use 2 lines of fasteners.

$$\text{Ordinary deduction for 2 holes} = 2(\tfrac{13}{16}+\tfrac{1}{16})(\tfrac{9}{16})$$
$$\text{(principles of Chapter 3)} \qquad\qquad\qquad = 0.985 \text{ sq in.}$$
$$15\% \text{ of gross flange splice plate} = 0.15(5.06) = 0.76$$
$$\text{Excess that must be deducted} \qquad\qquad = 0.225 \text{ sq in.}$$

Effective area provided $= 5.06 - 0.23 = 4.83$ sq in. > 4.80 sq in.   OK
Use 2 ℞ s—$\frac{9}{16} \times 9$.

It would also be acceptable to consider the increased effectiveness resulting from the splice plate being centered farther from the neutral axis than the flange. In such a treatment,

$$\text{Required effective area} = 4.80\left(\frac{24.10-0.770}{24.10+0.5625}\right) = 4.54 \text{ sq in.}$$

assuming a $\frac{9}{16}$-in. plate. It does not appear that this would justify use of a $\frac{1}{2} \times 9$ plate.

(e) Check plate $\frac{9}{16}\times 9$ using $f = Mc/I$.

$$
\begin{aligned}
I\text{(web plates)} &= 386 \\
I\text{(flange plates) } 2(4.83)(12.326)^2 &= 1\,469 \\
\hline
I &= 1\,855 \text{ in.}^4
\end{aligned}
$$

$$f = \frac{270(12)(12.61)}{1855} = 22.0 \text{ ksi} < 24 \text{ ksi} \qquad \text{Understressed.}$$

Use ℞ s—$\frac{9}{16}\times 9$, since $\frac{1}{2}\times 9$ plates will be overstressed.

(g) Flange bolts. A325 bearing-type connection (A325–X).

$$R_{SS} = 13.25 \text{ kips}; \qquad R_B > 13.25 \text{ kips}$$

In order to carry the full capacity of the splice material,

$$P = f_{avg}A_n = 24\left(\frac{12.33}{12.61}\right)(4.83) = 113 \text{ kips}$$

$$\text{Number of bolts} = \frac{P}{R_{SS}} = \frac{113}{13.25} = 8.5 \qquad \text{say 10}$$

Use 2 rows of 5 bolts, each side of splice.

(h) Web bolts. The web plates must carry $V = 82$ kips and $M = 61.3$ ft-kips. Using Eq. 4.8.27 to gain an estimate of the connectors required, try 2 rows of 4 bolts each (Fig. 13.10.4); 3 spaces at approximately 6 in.

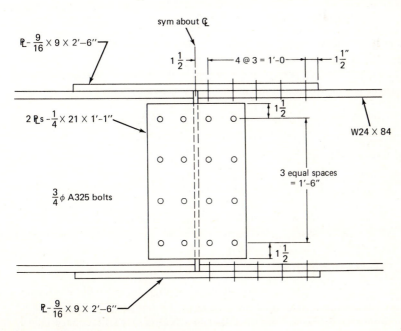

sym about ℄

℞-$\frac{9}{16}$ X 9 X 2'–6"

$1\frac{1}{2}$ ← → ←——4 @ 3 = 1'-0——→ ← $1\frac{1}{2}''$

$1\frac{1}{2}$

2 ℞s -$\frac{1}{4}$ X 21 X 1'–1"

W24 X 84

3 equal spaces = 1'–6"

$\frac{3}{4}\phi$ A325 bolts

$1\frac{1}{2}$

$1\frac{1}{2}$

℞-$\frac{9}{16}$ X 9 X 2'–6"

Fig. 13.10.4 Design for Example 13.10.1.

vertically:

$$f_s A = \frac{82}{8} = 10.3 \text{ kips } \downarrow$$

$$f_x A = \frac{My}{\Sigma x^2 + \Sigma y^2} = \frac{61.3(12)9}{378} = 17.5 \text{ kips } \rightarrow$$

where

$$\Sigma x^2 + \Sigma y^2 = 8(1.5)^2 + 4(9)^2 + 4(3.0)^2 = 378 \text{ in.}^2$$

$$f_y A = \frac{Mx}{\Sigma x^2 + \Sigma y^2} = \frac{61.3(12)(1.5)}{378} = 2.9 \text{ kips } \downarrow$$

$$\text{Actual } R = \sqrt{(10.3 + 2.9)^2 + (17.5)^2} = 21.9 \text{ kips}$$

For allowable bolt values,

$$R_{DS} = 26.5 \text{ kips}$$

$$R_B = 1.5(58)(0.470)(0.75) = 30.7 \text{ kips} > 26.5 \text{ kips} \qquad \text{OK}$$

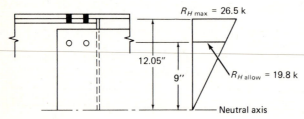

Fig. 13.10.5 Check for adequacy of bolts at top of web plates, Example 13.10.1.

However, since there are bolts in the flange plates at this same section, the allowable horizontal component in the web bolts must be restricted so as not to overstress the flange bolts. The two flange fasteners at a section acting in single shear can carry a total load of

$$R_{H \text{ max}} = 2(13.25) = 26.5 \text{ kips}$$

The allowable value on the web plate fastener most distant from the neutral axis is (see Fig. 13.10.5)

$$R_{H \text{ allow}} = 26.5 \left(\frac{9}{12.05} \right) = 19.8 \text{ kips} > f_x A = 17.5 \qquad \text{OK}$$

<u>Use</u> 2 vertical rows of 4 bolts each, each side of joint.
The final design is shown in Fig. 13.10.4.

SELECTED REFERENCES

1. C. Batho and H. C. Rowan, "Investigations of Beam and Stanchion Connections," 2nd Report, Steel Structures Research Committee, Dept. of Scientific and Industrial Research of Great Britain, His Majesty's Stationery Office, London, 1934.

2. Basil Sourochnikoff, "Wind Stresses in Semi-Rigid Connections of Steel Framework," *Transactions*, ASCE, 115 (1950), 382–402.
3. Leo Schenker, Charles G. Salmon, and Bruce G. Johnston, *Structural Steel Connections*, Armed Forces Special Weapons Project, Report No. 352, Engineering Research Institute, University of Michigan, June 1954.
4. D. J. L. Kennedy, "Moment-Rotation Characteristics of Shear Connections," *Engineering Journal*, AISC, 6, 4 (October 1969), 105–115.
5. W. Bertwell, Discussion of "Design of Bolts or Rivets Subject to Combined Shear and Tension," by Alfred Zweig, *Engineering Journal*, AISC, 3, 4 (October 1966), 165–167.
6. Bruce Johnston and Lloyd F. Green, "Flexible Welded Angle Connections," *Welding Journal*, 19, 10 (October 1940), 402s–408s.
7. Bruce G. Johnston, and Gordon R. Diets, "Tests of Miscellaneous Welded Building Connections," *Welding Journal*, 21, 1 (January 1942), 5s–27s.
8. Omer W. Blodgett, *Design of Welded Structures*. Cleveland, Ohio: James F. Lincoln Arc Welding Foundation, 1966.
9. Inge Lyse and Norman G. Schreiner, "An Investigation of Welded Seat Angle Connections," *Welding Journal*, 14, 2 (February 1935), Research Supplement, 1–15.
10. Cyril D. Jensen, "Welded Structural Brackets," *Welding Journal*, 15, 10 (October 1936), Research Supplement, 9–15.
11. Charles G. Salmon, "Analysis of Triangular Bracket-Type Plates," *Journal of Engineering Mechanics Division*, ASCE, 88, EM6 (December 1962), 41–87.
12. Charles G. Salmon, Donald R. Buettner, and Thomas C. O'Sheridan, "Laboratory Investigation of Unstiffened Triangular Bracket Plates," *Journal of Structural Division*, ASCE, 90, ST2 (April 1964), 257–278.
13. Lynn S. Beedle et al., *Structural Steel Design*. New York: The Ronald Press Company, 1964 (pp. 550–555).
14. J. D. Graham, A. N. Sherbourne, R. N. Khabbaz, and C. D. Jensen, *Welded Interior Beam-to-Column Connections*. New York: American Institute of Steel Construction, Inc., 1959.
15. A. B. Onderdonk, R. P. Lathrop, and Joseph Coel, "End Plate Connections in Plastically Designed Structures," *Engineering Journal*, AISC, 1, 1 (January 1964), 24–27.
16. Lynn S. Beedle and Richard Christopher, "Tests of Steel Moment Connections," *Engineering Journal*, AISC, 1, 4 (October 1964), 116–125.
17. Richard T. Douty and William McGuire, "High Strength Bolted Moment Connections," *Journal of Structural Division*, ASCE, 91, ST2 (April 1965), 101–128.
18. Ben Kato and William McGuire, "Analysis of T-Stub Flange-to-Column Connections," *Journal of Structural Division*, ASCE, 99, ST5(May 1973), 865–888.
19. J. S. Huang, W. F. Chen, and L. S. Beedle, "Behavior and Design of Steel Beam-to-Column Moment Connections," *WRC Bulletin 188*, Welding Research Council, New York, October 1973, 1–23.
20. J. E. Regec, J. S. Huang, and W. F. Chen, "Test of a Fully-Welded Beam-to-Column Connection," *WRC Bulletin 188*, Welding Research Council, New York, October 1973, 24–35.
21. John Parfitt, Jr. and Wai F. Chen, "Tests of Welded Steel Beam-to-Column

Connections," *Journal of Structural Division*, ASCE, 102, ST1 (January 1976), 189–202.

22. John W. Fisher and John H. A. Struik, *Guide to Design Criteria for Bolted and Riveted Joints*. New York: Wiley-Interscience, 1974.

23. Wai F. Chen and David E. Newlin, "Column Web Strength in Beam-to-Column Connections," *Journal of Structural Division*, ASCE, 99, ST9 (September 1973), 1978–1984.

24. Wai F. Chen and Irving J. Oppenheim, "Web Buckling Strength of Beam-to-Column Connections," *Journal of Structural Division*, ASCE, 100, ST1 (January 1974), 279–285.

25. Stephen P. Timoshenko and James M. Gere, *Theory of Elastic Stability*, 2nd Ed. New York: McGraw-Hill Book Company, Inc., 1961 (pp. 367–369).

26. R. Ford Pray and Cyril Jensen, "Welded Top Plate Beam-Column Connections," *Welding Journal*, 35, 7 (July 1956), 338s–347s.

27. J. L. Brandes and R. M. Mains, "Report of Tests of Welded Top-Plate and Seat Building Connections," *Welding Journal*, 24, 3 (March 1944), 146s–165s.

28. Rajasekharan S. Nair, Peter C. Birkemoe, and William H. Munse, "High Strength Bolts Subject to Tension and Prying," *Journal of Structural Division*, ASCE, 100, ST2 (February 1974), 351–372.

29. Henning Agerskov, "High-Strength Bolted Connections Subject to Prying," *Journal of Structural Division*, ASCE, 102, ST1 (January 1976), 161–175.

30. Henning Agerskov, "Analysis of Bolted Connections Subject to Prying," *Journal of Structural Division*, ASCE, 103, ST11 (November 1977), 2145–2163.

31. N. Krishnamurthy, "A Fresh Look at Bolted End-Plate Behavior and Design," *Engineering Journal*, AISC, 15, 2 (2nd Quarter 1978), 39–49.

32. A. Leon Abolitz and Marvin E. Warner, "Bending Under Seated Connections," *Engineering Journal*, AISC, 2, 1 (January 1965), 1–5.

33. Frank W. Stockwell, Jr., "Yield Line Analysis of Column Webs with Welded Beam Connections," *Engineering Journal*, AISC, 11, 1 (First Quarter 1974), 12–17.

34. Richard H. Kapp, "Yield Line Analysis of a Web Connection in Direct Tension," *Engineering Journal*, AISC, 11, 2 (Second Quarter 1974), 38–41.

35. Joint Committee of the Welding Research Council and the American Society of Civil Engineers, *Plastic Design in Steel, A Guide and Commentary*, 2nd Ed., ASCE, Manuals and Reports on Engineering Practice No. 41, 1971, pp. 167–186.

36. John D. Griffiths, *Single Span Rigid Frames in Steel*. New York: American Institute of Steel Construction, Inc., 1948.

37. John T. DeWolf, "Axially Loaded Column Base Plates," *Journal of Structural Division*, ASCE, 104, ST5 (May 1978), 781–794.

38. Balbir S. Sandhu, "Steel Column Base Plate Design," *Engineering Journal*, AISC, 10, 4 (Fourth Quarter 1973), 131–133.

39. Frank W. Stockwell, Jr., "Preliminary Base Plate Selection," *Engineering Journal*, AISC, 12, 3 (Third Quarter 1975), 92–99.

40. William R. Bird, "Rapid Selection of Column Base Plates," *Engineering Journal*, AISC, 13, 2 (Second Quarter 1976), 43–47.

41. Charles G. Salmon, Leo Schenker, and Bruce G. Johnston, "Moment-Rotation Characteristics of Column Anchorages," *Transactions*, ASCE, 122 (1957), 132–154.
42. Peter C. Birkemoe and Michael I. Gilmor, "Behavior of Bearing Critical Double-Angle Beam Connections," *Engineering Journal*, AISC, 15, 4 (4th Quarter 1978), 109–115.

PROBLEMS

All problems are to be done in accordance with the working stress method provisions of the latest AISC Specification and all given loads are service loads, unless otherwise indicated.

13.1. Determine the capacity for the framed beam connection shown. First, neglect eccentricity as is customary on framed beam connections. Then consider eccentricity of loading; using the ultimate strength approach (Sec. 4.8) for eccentric shear, and one of the combined shear and tension approaches (Sec. 4.10) for the connection to the column flange. Show all calculations and compare with tabulated values from the AISC Manual.
(a) Friction-type connection.
(b) Bearing-type connection, with threads excluded from shear plane.

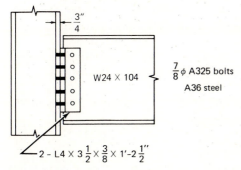

Prob. 13.1

13.2. Design the framed beam connection for a W16×50 beam with a reaction of 50 kips to connect to the column flange of a W8×67. Determine the angles to be used, and show the number and placement of fasteners. Use A36 steel and A325 bolts in a bearing-type connection, with threads excluded from the shear plane. Use an economical bolt size.

13.3. Redesign the framed beam connection of Prob. 13.2 as a friction-type connection.

13.4. Design the welded framed beam connection for a W33×118 connecting to a column flange which is $\frac{3}{4}$ in. thick. The beam reaction is 125 kips. E70 electrodes are to be used with shielded metal arc welding (SMAW), and the base steel is A572 Grade 50. Design using basic principles and then compare with applicable AISC Manual tables.

13.5. What is the maximum allowable reaction the W14×30 may safely transmit to the seat angle *B*? What size should be used for angle *A* and what is its function? Assume a bearing-type connection with threads excluded from the shear plane.

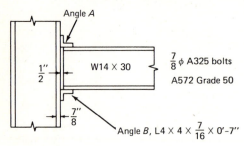

Prob. 13.5

13.6. For the unstiffened seat of Prob. 13.5, if E70 electrodes are used with shielded metal arc welding (SMAW), what is the maximum capacity that could be achieved? Would the 7-in. length of seat angle be satisfactory?

13.7. Assume a W16×57 is to have the maximum allowable reaction for a 15-ft simple span, uniformly loaded. Design the unstiffened seat to be used and specify its welded connection. Show all calculations and compare the result with any applicable AISC Manual tables. Use A36 steel, and E70 electrodes with shielded metal arc welding (SMAW).

13.8. Redesign the unstiffened seat of Prob. 13.7 using $\frac{3}{4}$-in.-diam A325 bolts in a friction-type connection.

13.9. Design an unstiffened seat angle to support a W10×22 beam on the web of a W12×65 column. The beam reaction is 22 kips. There is no beam on the opposite side of the web of the column. Use $\frac{3}{4}$-in.-diam A325 bolts (bearing-type connection with threads excluded from shear planes) and A36 steel. Design by calculations, and do not merely select from AISC Manual. Specify a size for the top clip angle.

13.10. Design a welded stiffened seat similar to that of Fig. 13.4.4, to support a W21×83 with a reaction of 130 kips. Use E70 electrodes with shielded metal arc welding (SMAW) and A572 Grade 60 for base material. The stiffened seat is attached to a W14×74 column flange.

13.11. Design a welded stiffened seat to support a W18×65 with a reaction of 90 kips. The beam lies at right angles to the stiffened seat as in Fig. 13.4.2. Use E70 electrodes, shielded metal arc welding (SMAW), and A572 Grade 60 steel.

13.12. Specify the weld size and the bracket plate thickness for the given situation. Use $P = 33$ kips, $e = 6$ in., and $L_w = 10$ in. Use E70 electrodes with shielded metal arc welding (SMAW), and A36 steel for base material.

13.13. Repeat Prob. 13.12 using $P = 45$ kips, $e = 3$ in., and $L_w = 12$ in.

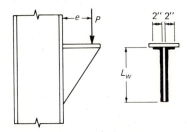

Probs. 13.12 and **13.13**

13.14. Design a bolted bracket plate for the loading shown. As part of the design, determine (a) size and length of angles; (b) number and location of fasteners in angles; (c) thickness and dimensions of bracket plate. Neglect concern about dimensions of the seat plate. Connection is bearing-type with threads excluded from the shear plane.

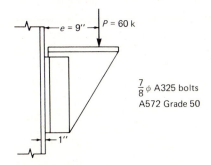

Prob. 13.14

13.15. Repeat the design of Prob. 13.14 using a welded connection with E70 electrodes and shielded metal arc welding (SMAW).

13.16. Check the connector number and determine the bracket thickness to support a load of 20 kips to be applied $12\frac{1}{2}$ in. from centerline of a W8×31 column. Use $\frac{7}{8}$-in.-diam A325 bolts (friction-type connection) and A36 steel. Use a rational analysis for plate thickness.

13.17. Design the continuous beam-to-column connection for W21×55 beams to connect to both flanges of a W14×84 column. Use the type of connection shown in Fig. 13.6.1b if stiffeners are required. Use the plastic design method assuming a plastic hinge must be

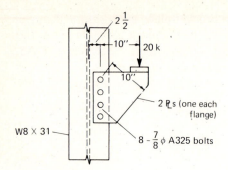

Prob. 13.16

able to form in the beams at this connection. Use A36 steel, and E70 electrodes with shielded metal arc welding. The shear under factored (LF = 1.7) loads is 50 kips.

13.18. Design for the conditions of Prob. 13.17 except use vertical tee stiffeners as shown in Fig. 13.6.1c if stiffeners are required.

13.19. Design for the conditions of Prob. 13.17 except consider the beam to frame in from one side only, and use the type of connection shown in Fig. 13.6.1d.

13.20. The split-tee beam connection shown is subject to a moment of 80 ft-kips and an end reaction of 30 kips. (a) Determine whether the structural tees are adequate, and if not, increase their size. (b) Determine the number of $\frac{7}{8}$-in.-diam A325 bolts to connect the tees to the column flanges, the tees to the beam flanges, the clip angles to the column flange, and the clip angles to the beam web. Assume connections are bearing-type and threads are excluded from shear planes.

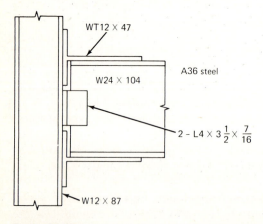

Prob. 13.20

13.21. For the split-tee beam connection with stub beam shown, determine the allowable moment and shear capacity at the face of the
column when $\frac{7}{8}$-in.-diam A325 bolts are used in a bearing-
type connection (no threads in shear planes) along with A36
steel.

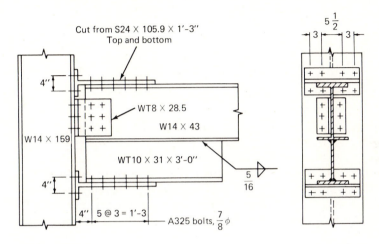

Cut from S24 × 105.9 × 1'–3''
Top and bottom

WT8 × 28.5

W14 × 43

W14 × 159

WT10 × 31 × 3'–0''

4''   5 @ 3 = 1'–3

A325 bolts, $\frac{7}{8}\phi$

$5\frac{1}{2}$

Prob. 13.21

13.22. Design a split-tee connection as shown in Fig. 13.6.2c to connect a
W16×50 beam to a W14×132 column. Use A572 Grade 50 steel
and assume the beam's full flexural capacity is to be developed.
The shear is 20 kips under service load. Use $\frac{3}{4}$-in.-diam A325
bolts, or if it seems advantageous use A490 bolts. Use a bearing-
type connection with no threads in the shear planes.

13.23. Redesign the connection of Prob. 13.22 using an end-plate type of
connection as shown in Fig. 13.6.2d.

13.24. Design a vertical tee stiffener (as in Figs. 13.6.3 and 13.6.9)
connection to frame a W16×67 beam into the web of a W12×96
column. Use A572 Grade 60 steel, and shielded metal arc welding
with E80 electrodes. The beam is required to develop maximum
flexural capacity and service load shear of 30 kips.

13.25. Redesign the connection of Prob. 13.22 using (a) top plate and
seat connection similar to that of Fig. 13.6.12a, and (b) top and
bottom plate connection similar to that of Fig. 13.6.12b. Use
shielded metal arc welding (SMAW) with E70 electrodes.

13.26. Specify the plate size, weld size using E70 electrodes, and length
of weld to develop the full continuity of the W10×39 beam.
Consider only plate A and not other aspects of the connection.
Use A36 steel and shielded metal arc welding (SMAW).

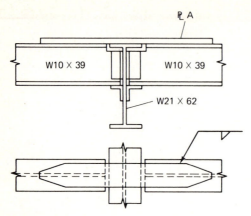

$\mathcal{R}$ A

W10 × 39   W10 × 39

W21 × 62

Prob. 13.26

13.27. Design a square knee connection (Fig. 13.8.1a or b) to join a W24×84 girder to a W30×108 column. Assume the plastic moment develops in the girder at collapse condition. Use A36 steel and E70 electrodes (shielded metal arc welding). Use plastic design.

13.28. A W14×211 of A572 Grade 50 steel has a concentric reaction of 850 kips resulting from dead load plus wind load. Using the AISC Manual design procedure, select a column base plate. The base plate is to be located in the center of a 6-ft square concrete footing having $f'_c = 3000$ psi.

13.29. Design the smallest size base plate for a W12×65 column to carry an axial compression load of 220 kips. The load is due to dead load plus live load. The base plate is to be centered on a 5-ft square footing with concrete having $f'_c = 3000$ psi.

13.30. A W14×120 column base connection of the type in Fig. 13.9.3 is subject to a moment of 320 ft-kips and a direct compression of 210 kips. The base plate is $3\frac{1}{2}$ in. ×28 in. ×30 in., with $N = 30$ in. Determine the total force to be resisted by the anchor bolts, the maximum stress in the concrete, and the maximum bending stress in the base plate.

13.31. Design a four-plate beam splice for a W14×53 beam of A36 steel. Design for full capacity using $\frac{3}{4}$-in.-diam A325 bolts in a bearing-type connection with no threads in shear planes.

13.32. Design an eight-plate beam splice as shown in Fig. 13.10.2 for a W27×94 section of A36 steel. Use $\frac{3}{4}$-in.-diam A325 bolts (friction-type connection), and design for full shear capacity and 80% of full moment capacity.

13.33. Design an eight-plate beam splice for a welded girder consisting of $1\frac{7}{8}$×22 flanges and a $\frac{7}{16}$×78 web. Use a design moment of 4000 ft-kips and design shear of 300 kips. Use A36 steel and $\frac{7}{8}$-in.-diam A325 bolts in a friction-type connection.

13.34. Design an eight-plate splice for a welded girder consisting of $\frac{7}{8} \times 24$ flanges and a $\frac{5}{16} \times 96$ web of A36 steel. Design for 270 kips shear and 3000 ft-kips moment. Use $\frac{3}{4}$-in.-diam A490 bolts in a bearing-type connection with no threads in shear plane.

14
Frames—Braced and Unbraced

14.1 GENERAL

As discussed in Sec. 6.9, the effective length of the column members in frames is dependent on whether the frame is *braced* or *unbraced*. For the *braced* frame the effective length KL is equal to or less than the actual length. For the *unbraced* frame, the effective length KL always is greater than the actual length.

In order to understand frame behavior, consider in Fig. 14.1.1 the forces that arise in a column member of a frame as a result of lateral deflection due to a force such as wind. The moments M_Δ and shears Q_Δ are those portions of the moments and shears required to balance the moment $P\Delta$. In addition, there will be moments and shears due to gravity loads at the particular floor level. Equilibrium in Fig. 14.1.1a requires

$$P\Delta = Q_\Delta h + 2M_\Delta \qquad (14.1.1)$$

The lateral deflection Δ is commonly called *drift* [1] when it results from wind loading in multistory frames, as shown in Fig. 14.1.2. Drift consists of two parts; that resulting directly from horizontal load, and that arising from vertical load times the drift.

Welded unbraced rigid frame (Vierendeel truss). (Courtesy Bethlehem Steel Corporation)

A frame will deflect under lateral loading such as wind regardless of the pattern of its component members. However, the manner in which equilibrium is maintained against the moment $P\Delta$ differs depending on the restraint conditions. If the building were a vertical pin-jointed truss under lateral loading there would be no continuity at the joints to allow

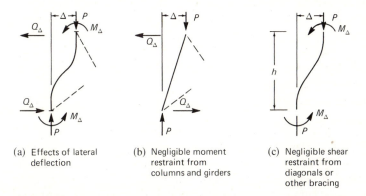

(a) Effects of lateral deflection

(b) Negligible moment restraint from columns and girders

(c) Negligible shear restraint from diagonals or other bracing

Fig. 14.1.1 Secondary bending moment due to $P\Delta$ in frames.

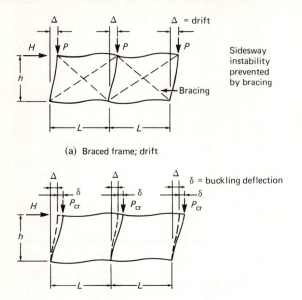

(a) Braced frame; drift

(b) Unbraced frame; drift and sideway buckling

Fig. 14.1.2 Comparison of braced and unbraced frames.

the moment M_Δ to develop. In which case, as in Fig. 14.1.1b,

$$Q_\Delta = \frac{P\Delta}{h} \tag{14.1.2}$$

Diagonal and horizontal members (web members of the truss) would have to carry the entire shear Q_Δ.

On the other hand, if the members are rigidly joined together but without diagonal members, there would be little shear resistance. Neglecting the shear resistance entirely would result in

$$M_\Delta = \frac{P\Delta}{2} \tag{14.1.3}$$

as in Fig. 14.1.1c. In this case, the girders and columns would have to accommodate the moment M_Δ.

Braced Frame

A braced frame has relatively small moment resistance from columns and girders to counterbalance $P\Delta$, in comparison with the actual restraint from diagonals or other bracing. In other words, Eq. 14.1.2 is assumed to represent the braced frame in simplified design procedures. As is shown later in this chapter, there is both flexural resistance and shear resistance developed in the braced and the unbraced frame. It is the relative magnitudes of these resistances that make the difference between the braced and unbraced frame.

Basically, a braced frame is more appropriately defined as one in which *sidesway buckling is prevented* by bracing elements of the structure other than the structural frame itself. As will be seen in the next section, the theoretical elastic stability analysis of a braced frame assumes no relative joint displacements, which obviously could occur only with infinitely stiff bracing. However, it is practical for design and reasonably correct to assume negligible moment resistance as implied by Eq. 14.1.2, and to assume that for stability purposes the sidesway mode is thereby also prevented.

The term "sidesway" is used to refer to stable elastic lateral movement of a frame, usually due to lateral loads, such as wind. Sidesway buckling is the sudden lateral movement, such as δ of Fig. 14.1.2b, caused by axial loads reaching a certain critical value.

In conclusion, the braced frame accommodates the $P\Delta$ moments by developing shears Q_Δ in the bracing system.

Unbraced Frame

In the unbraced frame, as Fig. 14.1.2b, if the horizontal load H is maintained constant and the compressive loads P are increased sufficiently to cause failure, such failure will occur with a side lurch known as sidesway buckling. The lateral deflection will *suddenly* become greater than the drift as shown in Fig. 14.1.2b. For cases where there is no lateral loading H and, therefore, no initial deflection, the sudden sidesway will still occur when the vertical load reaches a critical value.

The practical design treatment of the unbraced frame assumes that, referring to Fig. 14.1.1c, no shears Q_Δ are capable of developing and Eq. 14.1.3 applies. Any $P\Delta$ effects are balanced by column and girder moments in the unbraced frame.

Elastic buckling of frames has received considerable attention of researchers so that classical analysis methods are widely available [2]. Detailed study of one-story frames has been presented by Goldberg [3], and Zweig and Kahn [4]. Galambos [5] has studied the effect of partial base fixity on one- and two-story frames. Multistory frames have been treated by Switzky and Wang [6].

Since the use of digital computers has become widespread, the matrix formulation for the solution of elastic buckling loads using flexibility and/or stiffness coefficients seems to be the most efficient approach [7, 8]. The stiffness and flexibility coefficients are developed in Sec. 14.2 and used with the well-known slope deflection method in explaining frame behavior in the remainder of this chapter.

Inelastic Buckling

Since some fibers of a cross section usually yield prior to buckling, inelastic buckling probably governs the actual strength of a frame. Many studies of inealastic buckling have been conducted, including for braced frames the work of Ojalvo and Levi [9], Levi, Driscoll, and Lu [10], and for unbraced frames the work of Merchant [11], Yura and Galambos [12], Lu [13], Levi, Driscoll, and Lu [14], Korn and Galambos [15], Springfield and Adams [16], Daniels and Lu [17], Liapunov [18], Cheong-Siat-Moy [19, 20, 21], Lu, Ozer, Daniels, and Okten [22], and Haris [23]. Inelastic buckling studies have been extended to the hybrid frame by Arnold, Adams, and Lu [24], and to space frames by McVinnie and Gaylord [25].

14.2 ELASTIC BUCKLING OF FRAMES

The distinction between the braced and unbraced frame has been made in Sec. 14.1. In addition, there are two kinds of loading that may contribute to instability. For the *braced* frame of Fig. 14.2.1a, there are no primary bending moments; the only loading is axial compression. The critical load for such a situation is usually defined as a buckling load. No bending of members occurs until the buckling load is reached.

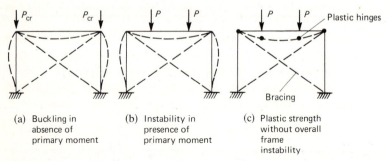

(a) Buckling in absence of primary moment

(b) Instability in presence of primary moment

(c) Plastic strength without overall frame instability

Fig. 14.2.1 Strength of braced frames.

For Fig. 14.2.1b, bending moments exist when the structure is entirely stable. Due to the rigid joints, moments are transmitted into column members. In addition to the primary bending moments in the column members, the axial compression induces secondary moments equal to P times the deflection. This was discussed in Chapter 12. Under certain combinations of axial compression and moment, the lateral deflection of the column increases without achieving equilibrium; this is usually referred to as instability, or instability in the presence of primary bending moment.

When primary bending moments are present, enough plastic hinges may develop prior to achieving frame instability so that a mechanism forms, in which case for the braced frame the ultimate strength is the plastic strength (Fig. 14.2.1c).

The strength of unbraced frames, as shown in Fig. 14.2.2, also may be separated into three categories; buckling in the absence of primary moment (Fig. 14.2.2a), instability in the presence of primary moment (Fig. 14.2.2b), and plastic strength (Fig. 14.2.2c). For the unbraced frame, achieving plastic strength frequently (though not always) means achieving a mechanism associated with overall geometric instability.

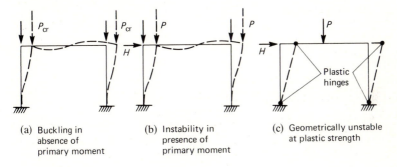

(a) Buckling in absence of primary moment

(b) Instability in presence of primary moment

(c) Geometrically unstable at plastic strength

Fig. 14.2.2 Strength of unbraced frames.

The remainder of this section treats the case of elastic buckling in the absence of primary bending moment, with the purpose of having the reader understand the difference in behavior of braced and unbraced frames.

In order to investigate the elastic stability of a rigid frame, it is first necessary to establish the relationships between end moments and end slopes for the individual frame member and than apply the compatibility of deformations requirement for rigid joints.

General Flexibility and Stiffness Coefficients for Beam-Columns

The reader is presumed to have some familiarity with the slope-deflection equations used in frame analysis when axial effect is not considered. For a prismatic section without axial load and without transverse load, as in Fig. 14.2.3a,

$$M_a = \theta_a\left(\frac{4EI}{L}\right) + \theta_b\left(\frac{2EI}{L}\right) \tag{14.2.1a}$$

$$M_b = \theta_a\left(\frac{2EI}{L}\right) + \theta_b\left(\frac{4EI}{L}\right) \tag{14.2.1b}$$

The following treatment, though similar to that for beam-columns in Sec. 12.2, is a more general approach that begins by expressing the

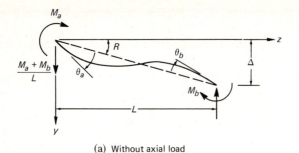

(a) Without axial load

(b) Including axial compression

Fig. 14.2.3 Definition of terms and sign convention for slope-deflection equations.

moment at any section z of Fig. 14.2.3b:

$$-EI\frac{d^2y}{dz^2} = M_z = M_a + Py - \left(\frac{M_a + M_b + PRL}{L}\right)z \qquad (14.2.2)$$

$$\frac{d^2y}{dz^2} + \frac{P}{EI}y = -\frac{M_a}{EI} + \frac{M_a + M_b}{EIL}z + \frac{P}{EI}Rz \qquad (14.2.3)$$

Letting $k^2 = P/EI$, the solution for Eq. 14.2.3 is

$$y = A\sin kz + B\cos kz - \frac{M_a}{P} + \frac{M_a + M_b}{PL}z + Rz \qquad (14.2.4)$$

Applying the boundary conditions of zero deflection at $z = 0$ and $y = RL$ at $z = L$ gives

$$B = \frac{M_a}{P}$$

and

$$A = \frac{1}{P\sin kL}(M_a\cos kL + M_b)$$

Then

$$y = \frac{M_a}{P}\left[\frac{\sin k(L-z)}{\sin kL} - \frac{(L-z)}{L}\right] - \frac{M_b}{P}\left(\frac{\sin kz}{\sin kL} - \frac{z}{L}\right) + Rz \qquad (14.2.5)$$

Differentiating once to obtain the slope,

$$\frac{dy}{dz} = \frac{M_a}{P}\left[\frac{-k\cos k(L-z)}{\sin kL} + \frac{1}{L}\right] - \frac{M_b}{P}\left(\frac{k\cos kz}{\sin kL} - \frac{1}{L}\right) + R \qquad (14.2.6)$$

when $z = 0$, $\dfrac{dy}{dz} = \theta_a + R$ and when $z = L$, $\dfrac{dy}{dz} = \theta_b + R$.

Letting $\phi^2/L^2 = k^2 = P/EI$, θ_a and θ_b after some manipulation of terms may be expressed as

$$\theta_a = \frac{M_a L}{EI}\left(\frac{\sin\phi - \phi\cos\phi}{\phi^2\sin\phi}\right) + \frac{M_b L}{EI}\left(\frac{\sin\phi - \phi}{\phi^2\sin\phi}\right) \qquad (14.2.7a)$$

$$\theta_b = \frac{M_a L}{EI}\left(\frac{\sin\phi - \phi}{\phi^2\sin\phi}\right) + \frac{M_b L}{EI}\left(\frac{\sin\phi - \phi\cos\phi}{\phi^2\sin\phi}\right) \qquad (14.2.7b)$$

where the ϕ functions are known as *flexibility coefficients*, f_{ii}, f_{ij}, f_{ji}, and f_{jj}. To obtain the beam-column counterparts to Eq. 14.2.1, solve Eqs. 14.2.7 (i.e., invert the matrix of coefficients) to obtain

$$M_a = \theta_a\frac{EI}{L}\left(\frac{\phi\sin\phi - \phi^2\cos\phi}{2 - 2\cos\phi - \phi\sin\phi}\right) + \theta_b\frac{EI}{L}\left(\frac{\phi^2 - \phi\sin\phi}{2 - 2\cos\phi - \phi\sin\phi}\right)$$

$$(14.2.8a)$$

$$M_b = \theta_a\frac{EI}{L}\left(\frac{\phi^2 - \phi\sin\phi}{2 - 2\cos\phi - \phi\sin\phi}\right) + \theta_b\frac{EI}{L}\left(\frac{\phi\sin\phi - \phi^2\cos\phi}{2 - 2\cos\phi - \phi\sin\phi}\right)$$

$$(14.2.8b)$$

where the ϕ functions are known as *stiffness coefficients*. The θ values are the end slopes measured with reference to the axis of the member.

Note that since $\phi^2 = PL^2/EI$, $\phi = 0$ means no axial compression and Eq. 14.2.8 should become Eq. 14.2.1. To verify the coefficient 4 when $\phi = 0$ in Eq. 14.2.8a, the numerator and denominator of the bracketed term must be differentiated four times in accordance with L'Hospital's Rule and then apply the $\phi = 0$ limit.

In order to simplify the use of Eqs. 14.2.8 for the slope-deflection solution of frame buckling problems, let the stiffness coefficients be referred to as S_{ii}, S_{ij}, S_{ji}, and S_{jj}. Because they are symmetrical $S_{ji} = S_{ij}$ and $S_{jj} = S_{ii}$. Thus Eqs. 14.2.8 become

$$M_a = \theta_a\frac{EI}{L}S_{ii} + \theta_b\frac{EI}{L}S_{ij} \qquad (14.2.9a)$$

$$M_b = \theta_a\frac{EI}{L}S_{ij} + \theta_b\frac{EI}{L}S_{ii} \qquad (14.2.9b)$$

Braced Frame—Slope-Deflection Method

The analysis of the rigid frame of Fig. 14.2.4 is presented using the slope-deflection method. Using clockwise rotations and rotational end moments as positive, the slope-deflection equations are as follows, using Eqs. 14.2.9:

$$M_{12} = \theta_1 \frac{EI_c}{h} S_{ii} + \theta_2 \frac{EI_c}{h} S_{ij} \qquad (14.2.10)$$

$$M_{21} = \theta_1 \frac{EI_c}{h} S_{ij} + \theta_2 \frac{EI_c}{h} S_{ii} \qquad (14.2.11)$$

$$M_{23} = \theta_2 \frac{EI_g}{L} S_{ii} + \theta_3 \frac{EI_g}{L} S_{ij} = \frac{2EI_g}{L} \theta_2 \qquad (14.2.12)$$

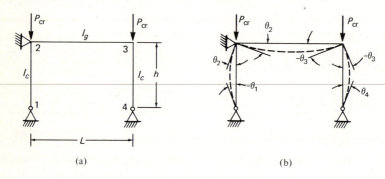

(a)   (b)

Fig. 14.2.4 Braced frame—hinged base.

Because no axial compression acts on member 2–3, $S_{ii} = 4$ and $S_{ij} = 2$ for Eq. 14.2.12. Use of summetry reduces the number of moment equations from six to three, and $\theta_3 = -\theta_2$. The equilibrium equations for the joints are

$$M_{12} = 0 \qquad (14.2.13)$$

$$M_{21} + M_{23} = 0 \qquad (14.2.14)$$

Substitution of Eqs. 14.2.10 through 14.2.12 into Eqs. 14.2.13 and 14.2.14 gives

$$\theta_1 \frac{EI_c}{h} S_{ii} + \theta_2 \frac{EI_c}{h} S_{ij} = 0$$

$$\theta_1 \frac{EI_c}{h} S_{ij} + \theta_2 \left(\frac{EI_c}{h} S_{ii} + \frac{2EI_g}{L} \right) = 0$$

$$(14.2.15)$$

Since the θ values cannot be zero when buckling occurs, the determinant of the coefficients of the θs must be zero. Thus the determinant, which is the stability equation, is

$$\left(\frac{EI_c}{h}\right)^2\left(S_{ii}^2+\frac{2I_gh}{I_cL}S_{ii}-S_{ij}^2\right)=0 \tag{14.2.16}$$

Since EI_c/h cannot be zero the other term must be zero. Thus

$$S_{ii}-\frac{S_{ij}^2}{S_{ii}}=-\frac{2I_gh}{I_cL} \tag{14.2.17}$$

which in terms of ϕ becomes

$$\frac{\phi^2\sin\phi}{\sin\phi-\phi\cos\phi}=-\frac{2I_gh}{I_cL} \tag{14.2.18}$$

Example 14.2.1

Determine the buckling load P_{cr} and effective length KL for a braced rigid frame, as in Fig. 14.2.4, which has $I_g=2100$ in.[4] (W24×76); $I_c=796$ in.[4] (W14×74); $L=36$ ft; and $h=14$ ft.

SOLUTION

The stability equation to be satisfied is Eq. 14.2.18, which inverted is

$$\frac{\sin\phi-\phi\cos\phi}{\phi^2\sin\phi}=-\frac{I_cL}{2I_gh}=\frac{-796(36)}{2(2100)(14)}=-0.487$$

The smallest value of ϕ satisfying the buckling equation is the critical value; i.e., the one that governs. As I_g approaches zero, an isolated pinned column is indicated, with $\phi^2=\pi^2$. As girder stiffness increases, ϕ should become greater than π. For an infinitely stiff girder, according to Fig. 6.9.1 which indicates $K=0.7$, one would expect $\phi^2=(\pi/0.7)^2=(4.49)^2$.

The solution for $\phi=3.60$ obtained by trial using values greater than π but less than 4.49 is shown in Fig. 14.2.5.

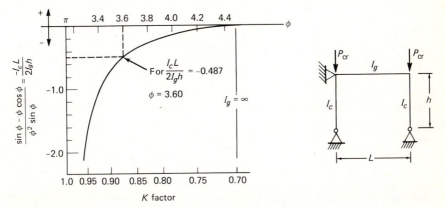

Fig. 14.2.5   Braced frame—hinged base, Example 14.2.1.

Comparing with the isolated pinned column,

$$\frac{\phi^2 EI}{h^2} = \frac{\pi^2 EI}{(Kh)^2} \qquad (14.2.19)$$

it may be noted that the effective length factor K may be expressed

$$K = \frac{\pi}{\phi} \qquad (14.2.20)$$

which for this problem means $K = \pi/3.60 = 0.87$. In other words, to account for frame buckling the column member could be designed using $0.87h$ as the pinned length of the isolated member. The K factor axis is also shown on Fig. 14.2.5 so that for various frame properties one could obtain the K factor directly. Note that a large increase in girder stiffness achieves only a small reduction in K.

The elastic buckling load is

$$P_{cr} = \frac{(3.60)^2 EI_c}{h^2} = \frac{(3.60)^2 29,000(796)}{(14)^2(12)^2} = 10,600 \text{ kips}$$

Unbraced Frame—Slope-Deflection Method

Next, the same frame that was treated as braced against sidesway at joint 2 is now to be analyzed as unbraced; i.e., with the horizontal support at joint 2 removed, as in Fig. 14.2.6. Even for the unbraced frame it is conceivable to consider a symmetrical buckling mode (Fig. 14.2.1a) exactly as when it was braced. It will be demonstrated that the sidesway buckling mode (Fig. 14.2.6b) will occur at a smaller load than the value obtained for the symmetrical case.

Equations 14.2.9 are applied with the added factor that the axes of some members are tilted due to sidesway, so that letting θ represent *total*

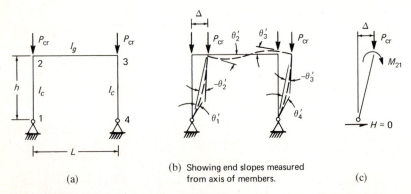

(a)

(b) Showing end slopes measured from axis of members.

(c)

Fig. 14.2.6 Unbraced frame—hinged base. ($\theta' = \theta$ from Eqs. 14.2.9 and Fig. 14.2.3.)

rotation, Δ/h must be subtracted from it to get the end slope θ' measured with respect to the member axis.

$$M_{12} = \left(\theta_1 - \frac{\Delta}{h}\right)\frac{EI_c}{h}S_{ii} + \left(\theta_2 - \frac{\Delta}{h}\right)\frac{EI_c}{h}S_{ij} \qquad (14.2.21)$$

$$M_{21} = \left(\theta_1 - \frac{\Delta}{h}\right)\frac{EI_c}{h}S_{ij} + \left(\theta_2 - \frac{\Delta}{h}\right)\frac{EI_c}{h}S_{ii} \qquad (14.2.22)$$

$$M_{23} = \theta_2\frac{EI_g}{L}S_{ii} + \theta_3\frac{EI_g}{L}S_{ij} \qquad (14.2.23)$$

With no axial compression considered on member 2–3, $S_{ii} = 4$ and $S_{ij} = 2$ for Eq. 14.2.23. This time if the structure is symmetrical, the sidesway buckling gives an antisymmetrical deflected curve; thus $\theta_3 = \theta_2$ and only three end moment equations are needed instead of six. The equilibrium equations are

$$M_{12} = 0 \qquad (14.2.24)$$

$$M_{21} + M_{23} = 0 \qquad (14.2.25)$$

which are the same as for the braced case. In addition, the sum of the base shears must be zero since no external horizontal force is acting. Because of antisymmetry $H_1 = -H_4$ so that only one column member needs to be considered. Referring to Fig. 14.2.6c,

$$H = \frac{M_{21} + P_{cr}\Delta}{h} = 0 \qquad (14.2.26)$$

Equations 14.2.24, 14.2.25, and 14.2.26 then become

$$\left.\begin{array}{l} \theta_1\dfrac{EI_c}{h}S_{ii} + \theta_2\dfrac{EI_c}{h}S_{ij} \qquad\qquad + \Delta\dfrac{EI_c}{h^2}(-S_{ii} - S_{ij}) \quad = 0 \\[3mm] \theta_1\dfrac{EI_c}{h}S_{ij} + \theta_2\left(\dfrac{EI_c}{h}S_{ii} + \dfrac{6EI_g}{L}\right) + \Delta\dfrac{EI_c}{h^2}(-S_{ii} - S_{ij}) \quad = 0 \\[3mm] \theta_1\dfrac{EI_c}{h}S_{ij} + \theta_2\dfrac{EI_c}{h}S_{ii} \qquad\qquad + \Delta\dfrac{EI_c}{h^2}(\phi^2 - S_{ii} - S_{ij}) = 0 \end{array}\right\} \quad (14.2.27)$$

Since θ_1, θ_2, and Δ cannot be zero (if buckling occurs), the determinate of the coefficients must be zero. Thus algebraic elimination of θ_1 and computation of the remaining four element determinant gives

$$\frac{EI_c}{h^2}(S_{ii}^2 - S_{ij}^2)\left[\frac{6EI_g}{L}\phi^2\frac{S_{ii}}{(S_{ii}^2 - S_{ij}^2)} + \frac{\phi^2 EI_c}{h} - \frac{6EI_g}{L}\right] = 0 \qquad (14.2.28)$$

Since $S_{ii} \neq S_{ij}$ the bracketed term must be zero. Substitution of the S_{ii} and

S_{ij} expressions in terms of ϕ from Eqs. 14.2.8 gives the stability equation,

$$\frac{1}{\phi^2} - \frac{\sin\phi - \phi\cos\phi}{\phi^2\sin\phi} = \frac{I_cL}{6I_gh} \tag{14.2.29}$$

or

$$\phi\tan\phi = \frac{6I_gh}{I_cL} \tag{14.2.30}$$

Example 14.2.2

Determine the buckling load P_{cr} and the effective length KL for the unbraced frame consisting of the same members and span as in Example 14.2.1.

SOLUTION

$$\phi\tan\phi = \frac{6I_gh}{I_cL} = \frac{6(2100)(14)}{796(36)} = +6.156$$

which is solved by trial to obtain $\phi = 1.354$, the smallest value satisfying the equation (see Fig. 14.2.7).

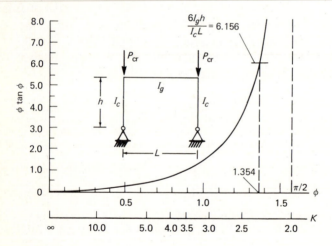

Fig. 14.2.7 Unbraced frame—hinged base, Example 14.2.2.

$$K = \frac{\pi}{\phi} = \frac{\pi}{1.354} = 2.32$$

Thus if the frame is unbraced it may be designed as an isolated member of effective length $2.32h$, whereas if it were braced the effective length would be $0.87h$. The range of K values may be studied for this frame in Fig. 14.2.7. With hinged bases, even with an infinitely rigid beam the effective length factor K is never less than 2 for an unbraced frame.

The buckling load is

$$P_{cr} = \frac{(1.354)^2 EI_c}{h^2} = \frac{(1.354)^2 29,000(796)}{(14)^2(12)^2} = 1500 \text{ kips}$$

or about $\frac{1}{7}$ of what it was for the same frame when braced.

14.3 GENERAL EQUATION FOR EFFECTIVE LENGTH

For ordinary design, it is entirely impractical to analyze an entire frame to determine its buckling strength and its effective length (equivalent pinned-end length). Thus it is desirable to have some general way of obtaining the K factor without the full analysis.

A number of investigators have provided charts to permit easy determination of frame buckling loads and effective lengths for commonly encountered situations. Effective length factors K are given by Hassan [26] for one-story, one-bay frames, with vertical loads applied to the columns at an intermediate point in addition to the load at the top. Galambos [5] has presented them for one- and two-story, one-bay-wide frames, and Gurfinkel and Robinson [27] have given K values for the general case of an elastic rotationally restrained column (both with and without sidesway elastic restraint). Switsky and Wang [6] have summarized buckling load data from which effective lengths can be obtained for frames one bay wide and up to five stories high.

The most commonly used procedure for obtaining effective length is to use the alignment charts from the SSRC Guide [28] originally developed by O. J. Julian and L. S. Lawrence, and presented in detail by T. C. Kavanagh [29]. The alignment chart method using Fig. 14.3.1 is also suggested by the AISC Commentary as satisfying the "rational method" requirement of AISC–1.8.3. Chu and Chow [30] have presented some modifications to use the alignment chart method for unsymmetrical frames.

An extension of the use of the alignment charts to account for *inelastic* column buckling that governs strength to some extent in nearly all compression members has been presented by Yura [31]; provoking additional discussion by Adams [32], Johnston [33], Disque [34], Smith [35], Matz [36], and Stockwell [37]. The AISC Commentary endorses Yura's procedure, which is discussed later in this chapter.

Alignment Chart Equation—Braced Frame (No Sidesway Buckling)

The following assumptions are used in the development of the *elastic* stability equation:

1. All behavior is elastic.
2. The members are prismatic.

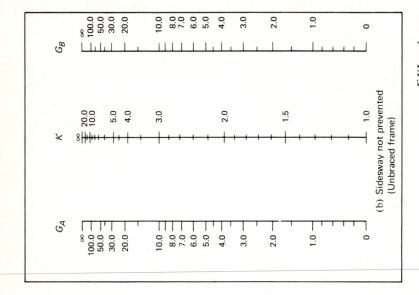

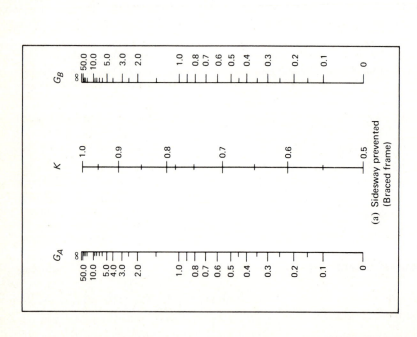

Fig. 14.3.1 Alignment Charts for effective length of columns in continuous frames (Ref. 29), where $G = \dfrac{\Sigma I/L, \text{columns}}{\Sigma I/L, \text{girders}}$

(a) Sidesway prevented (Braced frame)

(b) Sidesway not prevented (Unbraced frame)

3. All columns reach their respective buckling loads simultaneously.
4. The structure consists of symmetrical rectangular frames.
5. At a joint, the restraining moment provided by the girders is distributed among the columns in proportion to their stiffnesses.
6. The girders are elastically restrained at their ends by the columns, and at the onset of buckling the rotations of the girder at its ends are equal and opposite.
7. The girders carry no axial loads.

Consider a framework with no sidesway possible, consisting of two columns and two beams rigidly joined at a but having rotation elastically restrained at their opposite ends, as shown in Fig. 14.3.2. No translation of joints is possible.

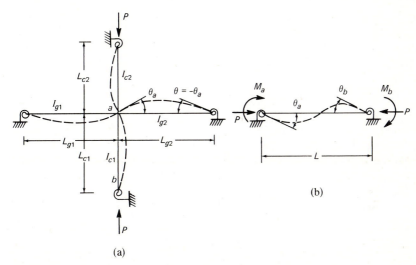

(a)

(b)

Fig. 14.3.2 Position of braced frame with elastically restrained ends—no sidesway.

Next, consider a beam-column with elastically restrained ends, as in Fig. 14.3.2b. If the elastic restraints are α and β, then

$$\theta_a = -\frac{M_a}{\alpha} \quad \text{and} \quad \theta_b = -\frac{M_b}{\beta} \tag{14.3.1}$$

i.e., the restraint moment is opposite to the positive directions of M_a and M_b.

The stability relationship may be developed using either stiffness coefficients, Eqs. 14.2.8 or 14.2.9 (slope-deflection equations), or flexibility coefficients, Eqs. 14.2.7. Because they have a single term denominator the flexibility coefficients are easier for solving a problem having a closed form solution. Therefore, substituting Eqs. 14.3.1 into Eqs. 14.2.7 gives

$$
\left.\begin{array}{l}
0 = \dfrac{M_a L}{EI}\left(\dfrac{\sin\phi - \phi\cos\phi}{\phi^2 \sin\phi} + \dfrac{EI}{\alpha L}\right) + \dfrac{M_b L}{EI}\left(\dfrac{\sin\phi - \phi}{\phi^2 \sin\phi}\right) \\[3mm]
0 = \dfrac{M_a L}{EI}\left(\dfrac{\sin\phi - \phi}{\phi^2 \sin\phi}\right) \qquad + \dfrac{M_b L}{EI}\left(\dfrac{\sin\phi - \phi\cos\phi}{\phi^2 \sin\phi} + \dfrac{EI}{\beta L}\right)
\end{array}\right\} \quad (14.3.2)
$$

Since no applied moments are assumed acting, M_a and M_b exist only after buckling occurs and their magnitudes cannot be determined. Either M_a and M_b are both zero (i.e., no buckling) or the determinant of their coefficients must equal zero.

Thus it is required that

$$
\dfrac{1}{\alpha\beta}\left(\dfrac{EI}{L}\right)^2 + \left(\dfrac{EI}{L}\right)\left(\dfrac{1}{\alpha} + \dfrac{1}{\beta}\right)\left(\dfrac{\sin\phi - \phi\cos\phi}{\phi^2 \sin\phi}\right)
$$
$$
+ \left(\dfrac{\sin\phi - \phi\cos\phi}{\phi^2 \sin\phi}\right)^2 - \left(\dfrac{\sin\phi - \phi}{\phi^2 \sin\phi}\right)^2 = 0 \quad (14.3.3)
$$

which may be simplified to become

$$
\dfrac{\phi^2}{\alpha\beta}\left(\dfrac{EI}{L}\right)^2 + \left(\dfrac{1}{\alpha} + \dfrac{1}{\beta}\right)\left(\dfrac{EI}{L}\right)\left(1 - \dfrac{\phi}{\tan\phi}\right) + \dfrac{2}{\phi}\tan\dfrac{\phi}{2} - 1 = 0 \quad (14.3.4)
$$

Next, the elastic restraint factors α and β must be established. Consider the frame of Fig. 14.3.2a; if on the girders which have no axial compression the θ at the far end from joint a equals $-\theta_a$, the moments on the girders at end a become, using Eqs. 14.2.1,

$$
\left.\begin{array}{l}
M_a \text{ (for girder 1)} = \theta_a\left(\dfrac{4EI_{g1}}{L_{g1}}\right) - \theta_a\left(\dfrac{2EI_{g1}}{L_{g1}}\right) = \dfrac{2EI_{g1}}{L_{g1}}\theta_a \\[3mm]
M_a \text{ (for girder 2)} = \theta_a\left(\dfrac{4EI_{g2}}{L_{g2}}\right) - \theta_a\left(\dfrac{2EI_{g2}}{L_{g2}}\right) = \dfrac{2EI_{g2}}{L_{g2}}\theta_a
\end{array}\right\} \quad (14.3.5)
$$

or the sum of the reactive moments M_{ag} developed due to girder stiffness may be expressed

$$
M_{ag} = \sum \dfrac{2EI_g}{L_g}\theta_a \quad (14.3.6)
$$

The moments at a on the column members may be expressed, using Eq. 14.2.9a,

$$
\left.\begin{array}{l}
M_a \text{ (for col. 1)} = \theta_a\left(\dfrac{EI_{c1}}{L_{c1}}\right)S_{ii} - \theta_a\left(\dfrac{EI_{c1}}{L_{c1}}\right)S_{ij} \\[3mm]
M_a \text{ (for col. 2)} = \theta_a\left(\dfrac{EI_{c2}}{L_{c2}}\right)S_{ii} - \theta_a\left(\dfrac{EI_{c2}}{L_{c2}}\right)S_{ij}
\end{array}\right\} \quad (14.3.7)
$$

The sum of the moments at joint a on the column members M_{ac} may be written

$$M_{ac} = \sum \frac{EI_c}{L_c}(S_{ii} - S_{ij})\theta_a \qquad (14.3.8)$$

where it is assumed that $(S_{ii} - S_{ij})$ is the same for all column members framing at joint a.

Solving for θ_a from Eq. 14.3.8 and substituting into Eq. 14.3.7a gives

$$M_a \text{ (col. 1)} = \frac{(EI_c/L_c)(S_{ii} - S_{ij})M_{ac}}{\sum \frac{EI_c}{L_c}(S_{ii} - S_{ij})} \qquad (14.3.9)$$

The term $(S_{ii} - S_{ij})$ cancels since it is assumed identical for all column members. Further, with no external joint moment acting, $M_{ac} = -M_{ag}$; thus using Eq. 14.3.6 in 14.3.9 gives

$$M_a \text{(col. 1)} = -\frac{2EI_c}{L_c}\frac{\sum \frac{EI_g}{L_g}}{\sum \frac{EI_c}{L_c}}\theta_a \qquad (14.3.10)$$

From Eqs. 14.3.1, M_a (col. 1) $= -\alpha\theta_a$; then, from Eq. 14.3.10,

$$\alpha = \frac{2EI_c}{L_c}\frac{\sum \frac{EI_g}{L_g}}{\sum \frac{EI_c}{L_c}} \quad \text{(for joint } a) \qquad (14.3.11)$$

and similarly,

$$\beta = \frac{2EI_c}{L_c}\frac{\sum \frac{EI_g}{L_g}}{\sum \frac{EI_c}{L_c}} \quad \text{(for joint } b) \qquad (14.3.12)$$

Defining, as in the AISC Commentary Fig. C1.8.2,

$$G = \frac{\sum \frac{I_c}{L_c}}{\sum \frac{I_g}{L_g}} \qquad (14.3.13)$$

in which the case the elastic restraint factors become

$$\alpha = \frac{2EI}{L}\left(\frac{1}{G_A}\right); \qquad \beta = \frac{2EI}{L}\left(\frac{1}{G_B}\right) \qquad (14.3.14)$$

with the subscripts A and B referring to the two ends of the column member ab. The subscripts on EI/L have been dropped; the EI/L in Eqs. 14.3.14 as well as in the stability equation, Eq. 14.3.4 refer to the column ab.

Substitution of Eqs. 14.3.14 into Eq. 14.3.4 and using $\phi = \pi/K$, where K = effective length factor, gives

$$\frac{G_A G_B}{4}\left(\frac{\pi^2}{K^2}\right) + \left(\frac{G_A + G_B}{2}\right)\left(1 - \frac{\pi/K}{\tan \pi/K}\right) + \frac{2}{\pi/K}\tan\frac{\pi}{2K} = 1$$

$$(14.3.15)$$

which is the equation used for Fig. 14.3.1a, the alignment chart for the *braced frame*.

If the far ends of the girders do not have rotations equal and opposite to θ_a (i.e., symmetrical single curvature) but are instead either fixed or hinged, adjustments on G may be made.

For girder far ends fixed, θ_b in Eq. 14.2.1 becomes zero, and Eq. 14.3.6 becomes

$$M_{ag} = \sum \frac{4EI_g}{L_g}\theta_a \qquad (14.3.16)$$

and Eq. 14.3.14 will be

$$\alpha = \frac{4EI_c}{L_c}\left(\frac{1}{G_A}\right); \qquad \beta = \frac{4EI_c}{L_c}\left(\frac{1}{G_B}\right)$$

or

$$\alpha = \frac{2EI_c}{L_c}\left(\frac{2}{G_A}\right); \qquad \beta = \frac{2EI_c}{L_c}\left(\frac{2}{G_B}\right) \qquad (14.3.17)$$

For girder far ends hinged, $M_b = 0$ and $\theta_b = -\theta_a/2$; thus Eqs 14.3.14 are

$$\alpha = \frac{3EI_c}{L_c}\left(\frac{1}{G_A}\right); \qquad \beta = \frac{3EI_c}{L_c}\left(\frac{1}{G_B}\right)$$

or

$$\alpha = \frac{2EI_c}{L_c}\left(\frac{1.5}{G_A}\right); \qquad \beta = \frac{2EI_c}{L_c}\left(\frac{1.5}{G_B}\right) \qquad (14.3.18)$$

In other words, to adjust for the far ends of girders fixed, divide G by 2.0; and for the far ends hinged, divide G by 1.5.

Alignment Chart Equation—Unbraced Frame (Sideway Buckling Possible)

The unbraced member *ab*, as shown in Fig. 14.3.3, having elastically restrained ends, is shown in Fig. 14.3.3.

The assumptions for the unbraced frame are the same as for the braced frame, except assumption 4 which for the braced frame assumed the girders to be in single curvature. For the unbraced frame, the girder is assumed to be in double curvature (Fig. 14.3.3a) with the rotation at both ends equal in magnitude and direction.

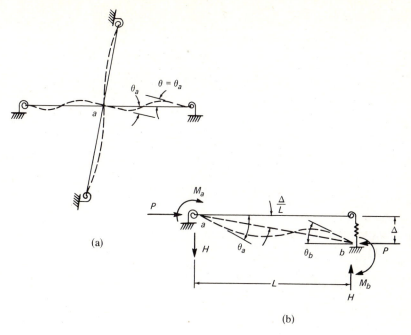

(a)

(b)

Fig. 14.3.3 Portion of unbraced frame with elastically restrained ends—with sidesway.

The elastic restraints α and β are given by Eqs. 14.3.1, the same as for the braced frame.

For the same reason as given in the braced frame development, the stability relationship is developed here using the flexibility coefficients, Eqs. 14.2.7, where the original θ of Eqs. 14.2.9, which was measured from the axis connecting the ends of the member, is replaced by θ which represents the total rotation.

Thus, using $(-M/\alpha - \Delta/L)$ for θ in Eqs. 14.2.7 gives

$$
\left.
\begin{aligned}
0 &= \frac{M_a L}{EI}\left(\frac{\sin\phi - \phi\cos\phi}{\phi^2\sin\phi} + \frac{EI}{\alpha L}\right) + \frac{M_b L}{EI}\left(\frac{\sin\phi - \phi}{\phi^2\sin\phi}\right) + \frac{\Delta}{L} \\
0 &= \frac{M_a L}{EI}\left(\frac{\sin\phi - \phi}{\phi^2\sin\phi}\right) + \frac{M_b L}{EI}\left(\frac{\sin\phi - \phi\cos\phi}{\phi^2\sin\phi} + \frac{EI}{\beta L}\right) + \frac{\Delta}{L}
\end{aligned}
\right\} \quad (14.3.19)
$$

Since three unknowns (M_a, M_b, and Δ) are involved a third equation is required—to satisfy rotational equilibrium of the structure,

$$M_a + M_b - HL + P\Delta = 0 \qquad (14.3.20)$$

where the net horizontal force H must be zero in the absence of any external horizontal force.

Rearranging the terms for Eqs. 14.3.19 and 14.3.20, recognizing that

$P = \phi^2 EI/L^2$, gives

$$
\left.
\begin{aligned}
0 &= \frac{M_a L}{EI}\left(\frac{\sin\phi - \phi\cos\phi}{\phi^2\sin\phi} + \frac{EI}{\alpha L}\right) + \frac{M_b L}{EI}\left(\frac{\sin\phi - \phi}{\phi^2\sin\phi}\right) &&+ \frac{\Delta}{L} \\
0 &= \frac{M_a L}{EI}\left(\frac{\sin\phi - \phi}{\phi^2\sin\phi}\right) &&+ \frac{M_b L}{EI}\left(\frac{\sin\phi - \phi\cos\phi}{\phi^2\sin\phi} + \frac{EI}{\beta L}\right) + \frac{\Delta}{L} \\
0 &= \frac{M_a L}{EI}\left(\frac{EI}{L^2}\right) &&+ \frac{M_b L}{EI}\left(\frac{EI}{L^2}\right) + \frac{\Delta}{L}\left(\frac{\phi^2 EI}{L^2}\right)
\end{aligned}
\right\}
$$

$$(14.3.21)$$

Because there are no applied moments, M_a, M_b, and Δ can exist only after buckling occurs. Equations 14.3.21 may be satisfied when M_a, M_b, and Δ are zero (i.e., no buckling) or the determinant of the coefficients must be equal to zero. The determinant is

$$
\phi^2\left(\frac{\cancel{\sin^2\phi} - 2\phi\cos\phi\sin\phi + \phi^2\cos^2\phi - \cancel{\sin^2\phi} + 2\phi\sin\phi - \phi^2}{\phi^4\sin^2\phi}\right)
$$

$$
+ \left(\frac{1}{\alpha} + \frac{1}{\beta}\right)\left(\frac{EI}{L}\right)\left(\frac{\phi^2(\cancel{\sin\phi} - \phi\cos\phi) - \phi^2\cancel{\sin\phi}}{\phi^2\sin\phi}\right)
$$

$$
+ \frac{\phi^2}{\alpha\beta}\left(\frac{EI}{L}\right)^2 - 2\left(\frac{\phi - \phi\cos\phi}{\phi^2\sin\phi}\right) = 0 \qquad (14.3.22)
$$

Combining the first and last terms and multiplying by $\tan\phi$ gives the stability equation as

$$
\left[\frac{\phi^2}{\alpha\beta}\left(\frac{EI}{L}\right)^2 - 1\right]\tan\phi - \left(\frac{1}{\alpha} + \frac{1}{\beta}\right)\left(\frac{EI}{L}\right)^2 = 0 \qquad (14.3.23)
$$

The development of the expression for the elastic restraint factors α and β is similar to that illustrated for the braced frame. The only difference is that the assumed deformation for the girders differs; in this case θ at the far end of the girder from joint a equals θ_a as shown in Fig. 14.3.3a.

Using the slope-deflection equations for no axial load, Eq. 14.2.1, the moment at end a for each girder is

$$
M_a \text{ (one girder)} = \theta_a\left(\frac{4EI_g}{L_g}\right) + \theta_a\left(\frac{2EI_g}{L_g}\right) = \frac{6EI_g}{L_g}\theta_a \qquad (14.3.24)
$$

or the sum of reactive moments developed due to girder stiffness M_{ag} may be expressed

$$
M_{ag} = \sum \frac{6EI_g}{L_g}\theta_a \qquad (14.3.25)
$$

For the column, the assumptions behind Eq. 14.3.9 are still valid; and in the absence of any external joint moment, $M_{ac} = -M_{ag}$. Thus using Eq.

14.3.25 in Eq. 14.3.9 gives

$$M_a \text{ (col. 1)} = -\frac{6EI_c}{L_c} \frac{\sum \frac{EI_g}{L_g}}{\sum \frac{EI_c}{L_c}} \theta_a \tag{14.3.26}$$

and by comparison with Eqs. 14.3.11 through 14.3.14, α and β may be written

$$\alpha = \frac{6EI_c}{L_c}\left(\frac{1}{G_A}\right); \qquad \beta = \frac{6EI_c}{L_c}\left(\frac{1}{G_B}\right) \tag{14.3.27}$$

Substitution of Eqs. 14.3.27 into Eq. 14.3.23 and using $\phi = \pi/K$ gives

$$\left[\frac{(\pi/K)^2 G_A G_B}{36} - 1\right] \tan\frac{\pi}{K} - \left(\frac{G_A + G_B}{6}\right)\frac{\pi}{K} = 0$$

or

$$\frac{G_A G_B (\pi/K)^2 - 36}{6(G_A + G_B)} = \frac{\pi/K}{\tan(\pi/K)} \tag{14.3.28}$$

which is the equation used for Fig. 14.3.1b, the alignment chart for the unbraced frame.

For the case of the far ends of the girder hinged and fixed, the girder stiffnesses are $3EI_g/L_g$ and $4EI_g/L_g$ respectively. In which case, to adjust for far ends of girders hinged, multiply G by 2.0; and for the far ends of girders fixed, multiply G by 1.5.

Adjustment of Alignment Chart K Factors for Inelastic Column Behavior

As discussed in Chapter 6, most columns have slenderness ratios KL/r less than C_c (see Sec. 6.8); therefore their strength is based on inelastic buckling (including residual stress) and is approximated by the SSRC parabola for buckling stress above $0.5F_y$. The inelastic buckling strength is given by Eq. 6.3.1 as

$$F_{cr} = \frac{\pi^2 E_t}{(KL/r)^2} \tag{6.3.1}$$

The restraint factors α and β (see Eqs. 14.3.11 and 14.3.12) used in development of the alignment chart expressions involve the modulus of elasticity E for both columns and beams. For elastic behavior this cancels from the equation. If the elastic E applies for the beam members but the inelastic E_t applies for the columns, this can be accounted for by an adjustment of the G values (Eq. 14.3.13); thus

$$G_{\text{inelastic}} = \frac{\sum (E_t I/L)_{\text{col}}}{\sum (EI/L)_{\text{beam}}} = G_{\text{elastic}}\left(\frac{E_t}{E}\right) \tag{14.3.29}$$

The ratio of inelastic buckling strength to elastic buckling strength is approximately

$$\frac{F_{cr}(\text{inelastic})}{F_{cr}(\text{elastic})} \approx \frac{\text{SSRC parabola, Eq. 6.7.3}}{\text{Euler formula, Eq. 6.3.1 with } E_t = E} \tag{14.3.30}$$

For the working stress method, the basic column allowable stress F_a, Eq. 6.8.2, could represent the numerator. The long column allowable stress F_a, Eq. 6.8.6, could represent the denominator. This will be conservative since the FS used for short columns ranges from 1.67 to 1.92, whereas the long column equation uses FS = 1.92. Thus

$$\frac{E_t}{E} \approx \frac{F_a(\text{AISC Formula 1.5-1})}{F_a(\text{AISC Formula 1.5-2})} \tag{14.3.31}$$

Yura [31] suggests using F'_e from AISC–1.6.1 for the denominator because F'_e is readily available for all KL/r ratios in AISC Appendix Table 9. Disque [34] proposes using $0.60F_y$ for the numerator along with F'_e in the denominator. That approach is conservative and practical. While conceptually the same, the use of F'_e is better than AISC Formula (1.5-2) because the latter would have to be used for low KL/r values; that is, outside its usual range of applicability. Further, Disque suggests that for members carrying little bending moment, the nominal stress $f_a = P/A$ could be used for the numerator. Thus the Disque [34] approach is

$$\frac{E_t}{E} \approx \frac{0.60F_y}{F'_e} \quad \text{or} \quad \frac{E_t}{E} \approx \frac{f_a}{F'_e} \tag{14.3.32}$$

Smith [35] suggests using the basic equation, Eq. 14.3.30, and Matz [36] has provided a table of values of E_t/E for various KL/r values.

For examples of the use of the alignment charts (Fig. 14.3.1) and this inelastic behavior modification, see Examples 15.4.1 and 15.4.2 of Chapter 15.

14.4 STABILITY OF FRAMES
UNDER PRIMARY BENDING MOMENTS

Referring to Figs. 14.2.1 and 14.2.2, the distinction should be noted between (1) buckling in the absence of primary bending moments; i.e., *no* moments exist until buckling occurs; and (2) magnification of primary bending moments (which exist even without the presence of compressive loads) due to axial compression P times deflection Δ. The buckling of frames, which involved solving for the compressive loads P that make a determinant equal to zero, was treated in the preceding section.

This section considers frame behavior when primary moments also exist. Consider a simple rectangular fixed base frame subject to primary moments as in Fig. 14.4.1a. If simple bending theory is used, and the stiffness of the girder is *not* considered reduced due to axial compression, any ordinary procedure of statically indeterminate analysis method will give the moments.

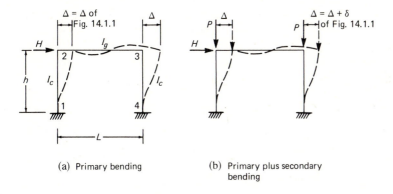

(a) Primary bending

(b) Primary plus secondary bending

Fig. 14.4.1 Frame with fixed bases.

If compressive loads P are applied, there will be additional moments $P\Delta$, as in Fig. 14.4.1. The total effect may be treated as a magnification factor times the primary bending moments. This has been discussed in detail in Chapter 12 for members with no translation of joints (braced frames such as in Fig. 14.2.1b), and has been approximately presented for the unbraced frame beam-column in Sec. 12.5.

The following treatment is intended to provide an understanding of the mathematics and compare the actual magnification factor with the simple expression, Eq. 12.5.5, suggested by AISC Commentary–1.6.1.

Primary Bending of Fixed-Base Frame

In order to determine magnification of primary moments, the primary moments must first be determined. While design practice could use any method of elastic frame analysis, the slope-deflection method is used here so that the reader may compare what follows with the modifications required to include the $P\Delta$ effect.

For the frame of Fig. 14.4.1a, use Eqs. 14.2.1 noting that θ is the angle measured from the axis of the member, and that for θ to represent the total rotation, Δ/h must be subtracted from it. The end moment equations are

$$M_{12} = \left(\theta_1 - \frac{\Delta}{h}\right)\frac{4EI_c}{h} + \left(\theta_2 - \frac{\Delta}{h}\right)\frac{2EI_c}{h}$$

$$M_{21} = \left(\theta_1 - \frac{\Delta}{h}\right)\frac{2EI_c}{h} + \left(\theta_2 - \frac{\Delta}{h}\right)\frac{EI_c}{h} \qquad (14.4.1)$$

$$M_{23} = \theta_2\frac{4EL_g}{L} + \theta_3\frac{2EI_g}{L}$$

Because of symmetry in the structure,

$$\left.\begin{array}{c} M_{43} = M_{12} \\ M_{34} = M_{21} \\ \theta_2 = \theta_3 \end{array}\right\} \qquad (14.4.2)$$

Also, because of the fixed base, $\theta_1 = 0$.

For equilibrium, it is required that

$$M_{21} + M_{23} = 0 \qquad (14.4.3)$$

and also the shears on the columns must equal H:

$$\frac{M_{12} + M_{21}}{h} + \frac{M_{43} + M_{34}}{h} + H = 0 \qquad (14.4.4)$$

Substitution of Eqs. 14.4.1 into Eqs. 14.4.3 and 14.4.4 gives

$$\left.\begin{array}{c} \theta_2\left(\dfrac{4EI_c}{h} + \dfrac{6EI_g}{L}\right) - \dfrac{\Delta}{h}\left(\dfrac{6EI_c}{h}\right) = 0 \\[3mm] \theta_2\left(\dfrac{12EI_c}{h}\right) \quad - \dfrac{\Delta}{h}\left(\dfrac{24EI_c}{h}\right) = -Hh \end{array}\right\} \qquad (14.4.5)$$

Solving for θ_2 and Δ/h gives

$$\theta_2 = \frac{Hh^2}{4EI_c}\left(\frac{I_cL/I_gh}{I_cL/I_gh + 6}\right) \qquad (14.4.6)$$

and

$$\frac{\Delta}{h} = \frac{Hh^2}{12EI_c}\left(\frac{2I_cL/I_gh + 3}{I_cL/I_gh + 6}\right) \qquad (14.4.7)$$

Substitution into the M_{12} and M_{21}, Eqs. 14.4.1, gives

$$M_{12} = \frac{-Hh}{2}\left(\frac{I_cL/I_gh + 3}{I_cL/I_gh + 6}\right) \qquad (14.4.8)$$

and

$$M_{21} = \frac{-Hh}{2}\left(\frac{3}{I_cL/I_gh + 6}\right) \qquad (14.4.9)$$

M_{12} and M_{21} are the primary moments which are magnified when the axial loads P are applied.

Magnification Factor for Fixed-Base Frame

The frame of Fig. 14.4.1b is investigated next to determine the magnified moments M_{12} and M_{21}. Use the slope-deflection equations, Eqs. 14.2.9, that include effect of axial compression on stiffness. Again the rotation angle Δ/h must be subtracted from the full angle to obtain the value measured from the member axis:

$$
\left.\begin{aligned}
M_{12} &= \left(\theta_1 - \frac{\Delta}{h}\right)\frac{EI_c}{h}S_{ii} + \left(\theta_2 - \frac{\Delta}{h}\right)\frac{EI_c}{h}S_{ij} \\
M_{21} &= \left(\theta_1 - \frac{\Delta}{h}\right)\frac{EI_c}{h}S_{ij} + \left(\theta_2 - \frac{\Delta}{h}\right)\frac{EI_c}{h}S_{ii} \\
M_{23} &= \theta_2 \frac{4EI_g}{L} + \theta_3 \frac{2EI_g}{L}
\end{aligned}\right\}
\tag{14.4.10}
$$

whereas for primary moment determination, the compression effect on the girder stiffness is not considered. The same symmetry and fixed-base condition ($\theta_1 = 0$) as in Eqs. 14.4.2 apply.

The equilibrium conditions are

$$
M_{21} + M_{23} = 0 \tag{14.4.11}
$$

the same as when the axial force P was not present; and

$$
\frac{M_{12} + M_{21}}{h} + \frac{M_{43} + M_{34}}{h} + \frac{2P\Delta}{h} + H = 0 \tag{14.4.12}
$$

where the term $P\Delta/h$ is added to the previous condition of Eq. 14.4.4. Let $P = \phi^2 EI_c/h^2$.

Substitution of Eqs. 14.4.10 into Eqs. 14.4.11 and 14.4.12, and recalling $M_{43} = M_{12}$ and $M_{34} = M_{21}$, gives

$$
\left.\begin{aligned}
\theta_2\left(S_{ii} + \frac{6I_g h}{I_c L}\right) - \frac{\Delta}{h}(S_{ii} + S_{ij}) &= 0 \\
\theta_2[2(S_{ii} + S_{ij})] - \frac{\Delta}{h}[4(S_{ii} + S_{ij}) - 2\phi^2] &= \frac{-Hh^2}{EI_c}
\end{aligned}\right\}
\tag{14.4.13}
$$

Equations 14.4.13 compare with Eqs. 14.4.5 when Eqs. 14.4.5 are divided by EI_c/h.

Solving Eqs. 14.4.13 for θ_2 and Δ/h gives

$$
\frac{\Delta}{h} = \left(\frac{S_{ii} + 6I_g h/I_c L}{S_{ii} + S_{ij}}\right)\theta_2 \tag{14.4.14}
$$

and

$$\theta_2 = \frac{Hh^2}{2EI_c} \left\{ \frac{(S_{ii} + S_{ij})(I_cL/I_gh)}{(S_{ii}^2 - S_{ij}^2 - \phi^2 S_{ii})(I_cL/I_gh) + 6[2(S_{ii} + S_{ij}) - \phi^2]} \right\}$$

(14.4.15)

Note that Eq. 14.4.15 becomes Eq. 14.4.6 when $P = 0$. When $P = 0$, $\phi^2 = Ph^2/EI_c = 0$, $S_{ii} = 4$, and $S_{ij} = 2$.

Substituting Eqs. 14.4.14 and 14.4.15 into Eqs. 14.4.10 gives

$$M_{12} = \theta_2 \frac{EI_c}{h} (S_{ij} - S_{ii} - 6I_gh/I_cL)$$

$$= \frac{-Hh}{2} \left\{ \frac{(S_{ii} + S_{ij})(S_{ii} - S_{ij} + 6I_gh/I_cL)(I_cL/I_gh)}{(S_{ii}^2 - S_{ij}^2 - \phi^2 S_{ii})I_cL/I_gh + 6[2(S_{ii} + S_{ij}) - \phi^2]} \right\}$$

(14.4.16)

$$M_{12} = \frac{-Hh}{2} \left\{ \frac{(S_{ii}^2 - S_{ij}^2)I_cL/I_gh + 6(S_{ii} + S_{ij})}{(S_{ii}^2 - S_{ij}^2 - \phi^2 S_{ii})I_cL/I_gh + 6[2(S_{ii} + S_{ij}) - \phi^2]} \right\}$$

and

$$M_{21} = \frac{-Hh}{2} \left\{ \frac{6(S_{ii} + S_{ij})}{(S_{ii}^2 - S_{ij}^2 - \phi^2 S_{ii})I_cL/I_gh + 6[2(S_{ii} + S_{ij}) - \phi^2]} \right\}$$ (14.4.17)

which become the same as Eqs. 14.4.8 and 14.4.9 when $\phi = 0$; i.e., no axial compression.

The magnification factor is the ratio of the moment including the $P\Delta$ effect to the primary moment without $P\Delta$. Thus, dividing Eq. 14.4.16 by Eq. 14.4.8 and Eq. 14.4.17 by Eq. 14.4.9 gives the magnification factor for the moment at the bottom of the column

$$A_{m12} = \frac{\text{Eq. } 14.4.16}{\text{Eq. } 14.4.8}$$

$$= \frac{(I_cL/I_gh + 6)[(S_{ii}^2 - S_{ij}^2)I_cL/I_gh + 6(S_{ii} + S_{ij})]}{(I_cL/I_gh + 3)\{(S_{ii}^2 - S_{ij}^2 - \phi^2 S_{ii})I_cL/I_gh + 6[2(S_{ii} + S_{ij}) - \phi^2]\}}$$

(14.4.18)

and the magnification factor for the moment at the top of the column,

$$A_{m21} = \frac{\text{Eq. } 14.4.17}{\text{Eq. } 14.4.9}$$

$$= \frac{6(I_cL/I_gh + 6)(S_{ii} + S_{ij})}{3\{(S_{ii}^2 - S_{ij}^2 - \phi^2 S_{ii})I_cL/I_gh + 6[2(S_{ii} + S_{ij}) - \phi^2]\}}$$ (14.4.19)

The determination of the actual magnification factor is excessively complicated to be used in design practice. The AISC Commentary suggests the magnification factor may be expressed by Eq. 12.5.5 for C_m

as follows:

$$A_m = \frac{C_m}{1-\alpha} = \frac{1-0.18\alpha}{1-\alpha} \tag{14.4.20}$$

where $\alpha = P/P_{cr}$.

The term P_{cr} is the buckling load occurring in the absence of primary moment. From the discussion in Sec. 14.3, P_{cr} may be determined for the present problem by letting $H=0$, and setting the determinant of the coefficients in Eqs. 14.4.13 equal to zero. This is equivalent to setting the denominator of Eqs. 14.4.15 through 14.4.19 equal to zero. Setting the determinant equal to zero and substituting the ϕ expressions of Eqs. 14.2.8 for S_{ii} and S_{ij}, one obtains after considerable manipulation

$$\phi \sin \phi \cos \phi \left(\frac{\cot \phi}{\phi} - \frac{I_c L}{6 I_g h} \right) = 1 \tag{14.4.21}$$

which for example, if $I_c L / I_g h = 1$, one obtains $\phi_{cr} = 0.865\pi$. For the fixed-base frame, the buckling load would be

$$P_{cr} = \frac{(0.865\pi)^2 E I_c}{h^2}$$

and

$$\alpha = \frac{P}{P_{cr}} = \left(\frac{\phi}{0.865\pi} \right)^2, \qquad \text{for } \frac{I_c L}{I_g h} = 1.0$$

The effective length factor K is

$$K = \frac{\pi}{\phi_{cr}} = \frac{1.0}{0.865} = 1.16$$

Table 14.4.1 provides a comparison of the theoretical magnification factor with $(1-0.18\alpha)/(1-\alpha)$, the suggested practical design equation. In design, when computing α the effective length KL is used and also a

Table 14.4.1 Comparison of Theoretical Magnification Factor A_{m21} with $A_m = \dfrac{1-0.18\alpha}{1-\alpha}$ and $A_m = \dfrac{0.85}{1-\alpha}$

ϕ	$\dfrac{I_c L}{I_g h} = 0.5$			1.0			5.0		
	Exact*	$\dfrac{1-0.18\alpha}{1-\alpha}$	$\dfrac{0.85}{1-\alpha}$	Exact*	$\dfrac{1-0.18\alpha}{1-\alpha}$	$\dfrac{0.85}{1-\alpha}$	Exact*	$\dfrac{1-0.18\alpha}{1-\alpha}$	$\dfrac{0.85}{1-\alpha}$
0.60	1.041	1.037	0.888	1.050	1.042	0.894	1.092	1.073	0.926
1.00	1.124	1.110	0.964	1.151	1.128	0.983	1.305	1.243	1.102
1.40	1.278	1.249	1.108	1.350	1.296	1.157	1.836	1.655	1.539
1.80	1.572	1.512	1.381	1.756	1.641	1.514	3.949	3.335	3.271
2.20	2.233	2.107	1.997	2.842	2.560	2.467	—	—	—
2.60	4.691	4.326	4.298	11.558	9.874	10.048	—	—	—

* Eq. 14.4.19.

safety factor is applied. The K in design would usually be determined using the alignment chart, Fig. 14.3.1, instead of from a theoretical elastic buckling solution.

The magnification factor using C_m suggested by the AISC Commentary, Eq. 14.4.20, seems to be in line with the theoretical value, though somewhat lower. The use of $0.85/(1-\alpha)$ as required by AISC–1.6.1 seems even less conservative. However, when actual structures are built, attachments which are designed as nonload supporting do actually contribute some bracing effect. In other words, a real building can never be as flexible as the skeleton elastic frame. In addition, exterior walls, partitions, stairways, etc., all tend to add to the overall stiffness. The amount of additional stiffness will depend on the actual structures and in many cases can be large.

For further study of the instability of frames in the presence of primary moment, the reader is referred to the work of Bleich [2], Lu [38], McGuire [39], Galambos [40], and the *SSRC Guide* [28].

14.5 BRACING REQUIREMENTS—BRACED FRAME

One of the decisions faced by the designer is the determination of whether a frame is braced or unbraced. Most efficient use of material in compression members is obtained when the frame is braced so that sidesway buckling or instability cannot occur. Some design guidelines are provided by Galambos [41]. AISC–1.8.2 indicates that lateral stability of frames is provided by attachment "to diagonal bracing, shear walls, to an adjacent structure having adequate lateral stability, or to floor slabs or roof decks secured horizontally by walls or by bracing systems parallel to the plane of the frame."

No indication is provided by the AISC Specification of the amount of stiffness required to prevent sidesway buckling. It has been suggested [41] that "in many instances nonstructural building elements, curtain walls for example, can also provide the necessary stiffness against sidesway buckling."

The following presentation from Galambos [41] may provide assistance in making the necessary engineering judgment regarding the strength required to create a braced frame. This discussion complements that in Sec. 9.11, where the emphasis was beam and column bracing.

Stiffness Required from Bracing

It is the objective to use bracing to convert Fig. 14.2.2a into Fig. 14.2.1a. A simple and conservative procedure is to idealize the braced frame, as in Fig. 14.5.1. The following assumptions are made:

1. Column does not participate in resisting sidesway.

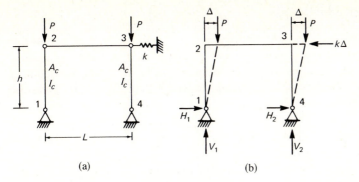

Fig. 14.5.1 Idealized bracing arrangement.

2. Columns are hinged at ends.
3. Bracing acts independently as a spring at the top of columns.

Equilibrium using summation of moments about point 1 of Fig. 14.5.1 gives

$$PA + P(\Delta + L) - k\Delta h - V_2 L = 0$$

$$V_2 = P + 2P\left(\frac{\Delta}{L}\right) - k\Delta\frac{h}{L} \tag{14.5.1}$$

Similarly, moments about point 4 gives

$$V_1 = P - 2P\left(\frac{\Delta}{L}\right) + k\Delta\frac{h}{L} \tag{14.5.2}$$

Moments about the hinges at 2 and 3 give

$$H_1 = V_1\frac{\Delta}{h} \tag{14.5.3}$$

$$H_2 = V_2\frac{\Delta}{h} \tag{14.5.4}$$

Summation of horizontal forces gives

$$H_1 + H_2 - k\Delta = 0 \tag{14.5.5}$$

Substitution of Eqs. 14.5.1 and 14.5.2 into Eqs. 14.5.3 and 14.5.4, and then into Eq. 14.5.5, gives

$$\frac{\Delta}{h}(2P - kh) = 0 \tag{14.5.6}$$

where Eq. 14.5.6 is the approximate buckling equation. For a simple rectangular frame, as in Fig. 14.2.4, the exact buckling equation is given by Eq. 14.2.18.

From Eq. 14.5.6,

$$k = \frac{2P_{cr}}{h} \qquad (14.5.7)$$

In design, if the buckling load P_{cr} is assumed to be twice the service load P (i.e., factor of safety of 2.0 against buckling), then Eq. 14.5.7 gives the required k for a given P:

$$k_{reqd} = \frac{2P_{cr}}{h} = \frac{2(2P)}{h} = \frac{4P}{h} \qquad (14.5.8)$$

For a frame of several bays, Eq. 14.5.6 becomes

$$\frac{\Delta}{h}\left(\sum P - kh\right) = 0$$

where $\sum P =$ summation of loads causing buckling. When the factor of safety (FS) is applied and service loads P are used, Eq. 14.5.8 becomes

$$k_{reqd} = \frac{(FS)\sum P}{h} \qquad (14.5.9)$$

Equation 14.5.9 assumes that only one of the bays in a multibay structure is braced or that k_{reqd} is the combined stiffness of all bracing. The $\sum P$ includes the design loads in all columns for which the particular brace is to provide support.

Bracing Provided by Stiffer Members in an Overall Unbraced System

Overall sidesway buckling can occur only if the total lateral or sidesway resistance to horizontal movement is overcome. When the loading on individual members is less than their strength, the reserve strength can be utilized to provide a bracing force for other members. Yura [31] has made an excellent presentation of this concept. Examine the unbraced frame of Fig. 14.5.2a. Assuming the columns hinged at the junction with the beam, $K = 2.0$ for this cantilever-type situation. Members A, B, and C were proportioned for the axial loads 100, 300, and 400 kips, respectively. When sidesway occurs as in Fig. 14.5.2b, moments $P\Delta$ are produced at the base and the total load is the 800 kips.

If the system of Fig. 14.5.2 were a *braced* one, the effective length factor K would be 1.0 instead of 2.0 and the strengths of the members would be four times as great; 400, 1200, and 1600 kips, respectively, at columns A, B, and C.

Suppose the loads applied to B and C are only 200 and 300 kips instead of the capacity values 300 and 400 kips. As in Fig. 14.5.2c, columns B and C will not sidesway buckle until the moments developed

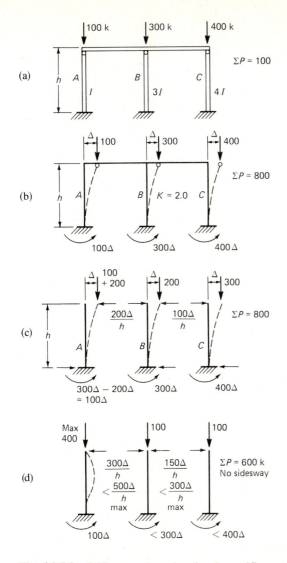

Fig. 14.5.2 Stiffer members bracing less stiff member in frame.

at the bottom reach 300Δ and 400Δ, respectively. Those columns can therefore sustain horizontal forces at their tops sufficient to cause sidesway buckling. Column C can develop 100Δ/h as a shear and column B can develop 100Δ/h as a shear. These resistances are additive and act as horizontal restraint at the top of column A. Thus column A can carry the 100 kips original load plus 200 kips as a result of the extra bracing from columns B and C. The total frame load is still 800 kips.

The maximum increase in strength that horizontal bracing can provide is the strength the member will have when its ends cannot translate with respect to one another. In other words, the shear developed at the top to prevent horizontal movement is the maximum resistance from other members that is usable. For instance, in Fig. 14.5.2d, the horizontal shear that can be developed based on columns B and C is $500\Delta/h$; however, when the shear at the top of column A is $300\Delta/h$, no movement occurs and the member is fully braced. The maximum capacity for member A in a braced system is 400 kips.

As stated by Yura [31], "the total gravity loads which produce sidesway can be distributed among the columns in a story in any manner. Sidesway will not occur until the total frame load on a story reaches the sum of the *potential* individual column loads for the *unbraced* frame." There still is the limitation that an individual column can carry no more than it could carry in a *braced* frame; i.e., with $K = 1$. Salem [42] has shown theoretically this procedure is valid regardless of the type of framing and ratio of member sizes.

This concept may be used to advantage when an overall rigid unbraced frame contains an internal portion consisting of nonrigid connections. In effect the *unbraced* part provides the bracing for the portion that must be braced to be stable. This fits the previously quoted statement in AISC–1.8.2 wherein reference is made to "adequate attachment . . . " for obtaining bracing.

An example design employing this concept is given in Chapter 15 on the design of rigid frames.

Diagonal Bracing

When cross-bracing is used it is generally assumed that it can only act in tension; i.e., the diagonal which would be in compression buckles slightly and becomes inactive. Under a horizontal force F the diagonal brace in Fig. 14.5.3 must carry the force

$$\text{Brace force} = \frac{F}{\cos \alpha} \tag{14.5.10}$$

and the elongation of the brace, $\Delta \cos \alpha$, is

$$\text{Elongation} = \frac{(\text{brace force})(\text{brace length})}{(\text{area of brace})\,E} \tag{14.5.11}$$

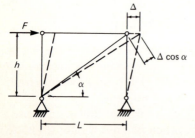

Fig. 14.5.3 Deformation of the bracing member.

or

$$\Delta \cos \alpha = \frac{(F/\cos \alpha)\sqrt{h^2+L^2}}{A_b E} \qquad (14.5.12)$$

Solving for F, using $\cos \alpha = L/\sqrt{h^2+L^2}$, gives

$$F = \frac{A_b E L^2}{(h^2+L^2)^{3/2}} \Delta \qquad (14.5.13)$$

Since $F = k\Delta$ according to Fig. 14.5.1b,

$$k = \frac{A_b E L^2}{(h^2+L^2)^{3/2}} \qquad (14.5.14)$$

Using the required k in terms of the service loads, Eq. 14.5.9, the required brace area is determined:

$$A_b = \frac{(FS)\left[1+\left(\frac{L}{h}\right)^2\right]^{3/2} \sum P}{\left(\frac{L}{h}\right)^2 E} \qquad (14.5.15)$$

Example 14.5.1

Determine the area required to convert the unbraced frame of Example 14.2.2, Fig. 14.2.6a, into a braced frame.

SOLUTION
For a frame with $I_g = 2100$ in.^{4}, $I_c = 796$ in^4, $L = 36$ ft, and $h = 14$ ft, the *braced* frame buckling load from Example 14.2.1 is

$$P_{cr} = \frac{(3.60)^2\, EI_c}{h^2}$$

Assuming $P_{cr}/2$ equals the service load and there are two loads, one on each column, and FS = 2,

$$\sum P = 2\left(\frac{P_{cr}}{2}\right) = \frac{(3.60)^2 EI_c}{h^2}$$

The required area of the bracing, using Eq. 14.5.15, is

$$A_b = \frac{2\left[1+\left(\frac{36}{14}\right)^2\right]^{3/2}(3.60)^2 E(796)}{\left(\frac{36}{14}\right)^2 E(14)^2(144)}$$

$$= \frac{2(7.6)^{3/2}(12.95)(796)}{6.6(196)(144)} = 2.32 \text{ sq in.}$$

The brace area required is about 10% of the column cross-sectional area (21.8 in. for W14×74). It is noted that the example unbraced frame was very flexible with a higher than usual effective length factor K of 2.32; thus a somewhat heavier than usual brace was required.

Usually bracing is designed to carry about 5% of the vertical column loads since horizontal forces applied at panel points will provide adequate stiffness to create a braced frame.

14.6 OVERALL STABILITY WHEN PLASTIC HINGES FORM

General

This chapter has focused primarily on the basic concepts of frame behavior, using the single-story frame as the example. In fact, for plastic design of one- and two-story frames (braced or unbraced), AISC–2.3 states, "the maximum strength may be determined by a routine plastic analysis procedure and the frame instability effect ($P\Delta$) may be ignored." For multistory *braced* frames, the design of the bracing system must include the $P\Delta$ effect. For multistory *unbraced* frames the $P\Delta$ effect must be included directly in the maximum frame strength calculation.

Braced Frames

Such frames are usually designed to cause the plastic hinges associated with the collapse mechanism to form in the girders. For one- and two-story frames, the $P\Delta$ effect may be ignored according to AISC–2.3. The columns are then designed as beam-columns in accordance with concepts treated in Chapter 12. Design procedures for the one- and two-story braced frames are presented in Chapter 15.

Multistory braced frames may be designed in accordance with *Plastic Design of Braced Multi-Story Steel Frames* [43].

Unbraced Frames

AISC–2.3.2 states "The strength of an unbraced *multistory* frame shall be determined by a rational analysis which includes the effect of frame instability and column axial deformation." Such a frame must be stable under factored gravity loads as well as factored horizontal loads (such as wind or earthquake). Further, the axial force in column when factored loads are applied must not exceed $0.75P_y$.

A "rational analysis" as required in AISC–2.3.2 may be performed in a manner similar to the procedure illustrated in Secs. 14.2 and 14.3 for one-story frames. Such a process would require a step-by-step loading, with a load increment ending when a plastic hinge forms at some location in the structure. For the next higher increment of load, the stiffness will change on the member (or members) where the plastic hinge formed. Use of a digital computer with a large memory core is generally required for such an analysis. The behavior of multistory frames is generally outside the scope of this text; however, the reader is referred to the work of Cheong-Siat-Moy, Ozer, and Lu [20], Springfield and Adams [16], Liapunov [18], Daniels and Lu [17], and LeMessurier [44, 45] for techniques of evaluating the strength of multistory unbraced frames.

SELECTED REFERENCES

1. T. R. Higgins, "Effective Column Length—Tier Buildings," *Engineering Journal*, AISC, 1, 1 (January 1964), 12–15.
2. Frederich Bleich, *Buckling Strength of Metal Structures*, New York: McGraw-Hill Book Company, Inc., 1952, Chaps. 6–7.
3. John E. Goldberg, "Buckling of One-Story Frames and Buildings," *Journal of Structural Division*, ASCE, 86, ST10 (October 1960), 53–85.
4. Alfred Zweig and Albert Kahn, "Buckling Analysis of One-Story Frames," *Journal of Structural Division*, ASCE, 94, ST9 (September 1968), 2107–2134.
5. Theodore V. Galambos, "Influence of Partial Base Fixity on Frame Stability," *Journal of Structural Division*, ASCE, 86, ST5 (May 1960), 85–108.
6. Harold Switsky and Ping Chun Wang, "Design and Analysis of Frames for Stability," *Journal of Structural Division*, ASCE, 95, ST4 (April 1969), 695–713.
7. Chu-Kia Wang, "Stability of Rigid Frames with Nonuniform Members," *Journal of Structural Division*, ASCE, 93, ST1 (February 1967), 275–294.
8. Ottar P. Halldorsson and Chu-Kia Wang, "Stability Analysis of Frameworks by Matrix Methods," *Journal of Structural Division*, ASCE, 94, ST7 (July 1968), 1745–1760.
9. M. Ojalvo and V. Levi, "Columns in Planar Continuous Structures," *Journal of Structural Division*, ASCE, 89, ST1 (February 1963), 1–23.
10. Victor Levi, George C. Driscoll, Jr., and Le-Wu Lu, "Structural Subasemblages Prevented from Sway," *Journal of Structural Division*, ASCE, 91, ST5 (October 1965), 103–127.
11. W. Merchant, "The Failure Load of Rigid Jointed Frameworks as Influenced by Stability," *Structural Engineer*, 32 (July 1954), 185–190.
12. Joseph A. Yura and Theodore V. Galambos, "Strength of Single Story Steel Frames," *Journal of Structural Division*, ASCE, 91, ST5 (October 1965), 81–101.
13. Le-Wu Lu, "Inelastic Buckling of Steel Frames," *Journal of Structural Division*, ASCE, 91, ST6 (December 1965), 185–214.
14. Victor Levi, George C. Driscoll, Jr., and Le-Wu Lu, "Analysis of Restrained Columns Permitted to Sway," *Journal of Structural Division*, ASCE, 93, ST1 (February 1967), 87–107.
15. Alfred Korn and Theodore V. Galambos, "Behavior of Elastic-Plastic Frames," *Journal of Structural Division*, ASCE, 94, ST5 (May 1968), 1119–1142.
16. John Springfield and Peter F. Adams, "Aspects of Column Design in Tall Steel Buildings," *Journal of Structural Division*, ASCE, 98, ST5 (May 1972), 1069–1083.

17. J. Hartley Daniels and Le-Wu Lu, "Plastic Subassemblage Analysis for Unbraced Frames," *Journal of Structural Division*, ASCE, 98, ST8 (August 1972), 1769–1788.
18. Sviatoslav Liapunov, "Ultimate Strength of Multistory Steel Rigid Frames," *Journal of Structural Division*, ASCE, 100, ST8 (August 1974), 1643–1655.
19. Francois Cheong-Siat-Moy, "Inelastic Sway Buckling of Multistory Frames," *Journal of Structural Division*, ASCE, 102, ST1 (January 1976), 65–75.
20. Francois Cheong-Siat-Moy, Erkan Ozer, and Le-Wu Lu, "Strength of Steel Frames under Gravity Loads," *Journal of Structural Division*, ASCE, 103, ST6 (June 1977), 1223–1235.
21. Francois Cheong-Siat-Moy, "Consideration of Secondary Effects in Frame Design," *Journal of Structural Division*, ASCE, 103, ST10 (October 1977), 2005–2019.
22. Le-Wu Lu, Erkan Ozer, J. Hartley Daniels, and Omer S. Okten, "Strength and Drift Characteristics of Steel Frames," *Journal of Structural Division*, ASCE, 103, ST11 (November 1977), 2225–2241.
23. Ali A. K. Haris, "Approximate Stiffness Analysis of High-Rise Buildings," *Journal of Structural Division*, ASCE, 104, ST4 (April 1978), 681–696.
24. Peter Arnold, Peter F. Adams, and Le-Wu Lu, "Strength and Behavior of an Inelastic Hybrid Frame," *Journal of Structural Division*, ASCE, 94, ST1 (January 1968), 243–266.
25. William W. McVinnie and Edwin H. Gaylord, "Inelastic Buckling of Unbraced Space Frames," *Journal of Structural Division*, ASCE, 94, ST8 (August 1968), 1863–1885.
26. Kamal Hassan, "On the Determination of Buckling Length of Frame Columns," *Publications*, International Association for Bridge and Structural Engineering, 28-I, 1968, 91–101 (in German).
27. German Gurfinkel and Arthur R. Robinson, "Buckling of Elastically Restrained Columns," *Journal of Structural Division*, ASCE, 91, ST6 (December 1965), 159–183.
28. Bruce G. Johnston, ed., *Structural Stability Research Council, Guide to Stability Design Criteria for Metal Structures*, 3rd Ed. New York: John Wiley & Sons, Inc., 1976, Chap. 15.
29. Thomas C. Kavanagh, "Effective Length of Framed Columns," *Transactions*, ASCE, 127, Part II (1962), 81–101.
30. Kuang-Han Chu and Hsueh-Lien Chow, "Effective Column Length in Unsymmetrical Frames," *Publications*, International Association for Bridge and Structural Engineering, 29-I, 1969, 1–15.
31. Joseph A. Yura, "The Effective Length of Columns in Unbraced Frames," *Engineering Journal*, AISC, 8, 2 (April 1971), 37–42. Discussion, 9, 3 (October 1972), 167–168.
32. Peter F. Adams, Discussion of "The Effective Length of Columns in Unbraced Frames," by J. A. Yura, *Engineering Journal*, AISC, 9, 1 (January 1972), 40–41.
33. Bruce G. Johnston, Discussion of "The Effective Length of Columns in Unbraced Frames," by J. A. Yura, *Engineering Journal*, AISC, 9, 1 (January 1972), 46.
34. Robert O. Disque, "Inelastic *K*-factor for Column Design," *Engineering Journal*, AISC, 10, 2 (Second Quarter 1973), 33–35.

35. C. V. Smith, Jr., "On Inelastic Column Buckling, " *Engineering Journal*, AISC, 13, 3 (Third Quarter 1976), 86–88; Discussion, 14, 1 (First Quarter 1977), 47–48.

36. Charles A. Matz, Discussion of "On Inelastic Column Buckling," by C. V. Smith, Jr., *Engineering Journal*, AISC, 14, 1 (First Quarter 1977), 47–48.

37. Frank W. Stockwell, Jr., "Girder Stiffness Distribution for Unbraced Columns," *Engineering Journal*, AISC, 13, 3 (Third Quarter 1976), 82–85.

38. Le-Wu Lu, "Stability of Frames Under Primary Bending Moments," *Journal of Structural Division*, ASCE, 89, ST3 (June 1963), 35–62.

39. William McGuire, *Steel Structures*. Englewood Cliffs, N.J.: Prentice-Hall, Inc., 1968 (pp. 448–474).

40. Theodore V. Galambos, *Structural Members and Frames*. Englewood Cliffs, N.J.: Prentice-Hall, Inc., 1968 (pp. 176–189).

41. Theodore V. Galambos, "Lateral Support for Tier Building Frames," *Engineering Journal*, AISC, 1, 1 (January 1964), 16–19, Discussion by Ira Hooper, 1, 4 (October 1964), 141.

42. Adel Helmy Salem, Discussion of "Buckling Analysis of One-story Frames," by Alfred Zweig, *Journal of Structural Division*, ASCE, 95, ST5 (May 1969), 1017–1029.

43. *Plastic Design of Braced Multi-Story Steel Frames*. New York: American Iron and Steel Institute, 1968.

44. Wm. J. LeMessurier, "A Practical Method of Second Order Analysis, Part 1—Pin Jointed Systems," *Engineering Journal*, AISC, 13, 4 (Fourth Quarter 1976), 89–96.

45. Wm. J. LeMessurier, "A Practical Method of Second Order Analysis, Part 2—Rigid Frames," *Engineering Journal*, AISC, 14, 2 (Second Quarter 1977), 49–67.

15
Design of Rigid Frames

15.1 INTRODUCTION

Many concepts and procedures developed in previous chapters are combined in this chapter by the use of illustrative examples. Plastic analysis and design concepts are extended from the continuous beam treatment in Chapter 10 to one-story rigid frames and the working stress method is compared with plastic design. However, no development is included regarding methods of statically indeterminate analysis of frames. The reader is assumed to be familiar with elastic methods such as moment distribution and slope deflection.

15.2 PLASTIC STRENGTH ANALYSIS OF ONE-STORY FRAMES

As discussed in Chapter 10 for continuous beams, the plastic strength of a structure may be obtained either by using the equilibrium method, or the energy method. In braced frames, where joints cannot displace (i.e., no sidesway), the plastic strength may be obtained exactly as for continuous beams. For unbraced frames the sidesway mechanism creates a somewhat more complicated analysis than that for continuous beams. The following examples are intended to illustrate principles of plastic analysis for

Welded multi-story rigid frame. (Courtesy Bethlehem Steel Corporation)

one-story unbraced frames. For a more extended treatment than is included here, the reader is referred to the AISC plastic design manual [1], textbooks devoted entirely to plastic design [2, 3], and published papers on strength and analysis of frames [4, 5].

Equilibrium Method

As discussed in Sec. 10.2, equilibrium must be satisfied at every stage of loading from a small load until the collapse mechanism has been achieved. When a sufficient number of plastic hinges have been developed to allow instantaneous hinge rotations without developing increased resistance, a mechanism is said to have occurred.

Example 15.2.1

Determine the plastic strength for the frame of Fig. 15.2.1, using the equilibrium method.

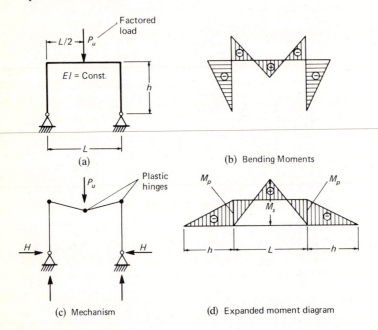

(a)

(b) Bending Moments

(c) Mechanism

(d) Expanded moment diagram

Fig. 15.2.1 Example 15.2.1.

SOLUTION

The elastic moment diagram shape is shown in Fig. 15.2.1b. For ease in solution it is often desirable to show the diagram with a horizontal base line as shown in Fig. 15.2.1d. Assuming overall frame stability is adequate, the collapse mechanism is that shown in Fig. 15.2.1c.

The simple beam moment for the girder is

$$M_s = \frac{P_u L}{4} \tag{a}$$

which when superimposed on the moments at the ends of the girder

$(H \times h)$ gives for equilibrium when the mechanism occurs,

$$M_p = M_s - M_p = \frac{P_u L}{4} - M_p \tag{b}$$

$$P_u = \frac{8M_p}{L} \tag{c}$$

From the concepts of plastic design it might have been expected that only two plastic hinges should be required for a collapse mechanism, since the structure is statically indeterminate to the first degree. In this case both corner hinges form simultaneously, since the horizontal base reactions are equal and opposite. Depending on the ratio of h to L either the midspan plastic hinge, or the two corner plastic hinges, will occur first. For small values of h/L the positive moment plastic hinge forms first and the structure remains stable until the corner plastic hinges form. If, however, h/L is large, the corner plastic hinges occur first and the structure has reached its collapse condition. If the beam and columns are of different cross section the structure can be designed so the positive moment plastic hinge occurs first. This should be the objective. The occurrence first of the corner plastic hinges creates overall frame instability prior to utilizing the flexural strength of the girder; such a result should be avoided.

Example 15.2.2

Determine the plastic strength of the same frame as in Example 15.2.1 except in addition apply a horizontal load $0.5P_u$ at the top of the column (see Fig. 15.2.2a). Use the equilibrium method.

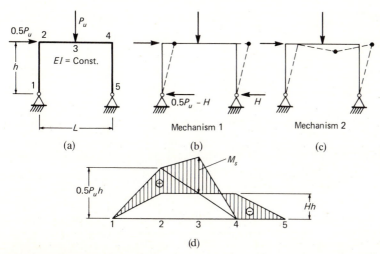

Fig. 15.2.2 Example 15.2.2.

SOLUTION

Again, two plastic hinges should provide the collapse mechanism. This time, however, additional mechanisms are possible, as shown in Fig. 15.2.2. It is possible that (a) plastic hinges form at points 2 and 4 (Fig. 15.2.2b); (b) plastic hinges form at points 3 and 4 (Fig. 15.2.2c); and also (c) plastic hinges form at points 2, 3, and 4. The last situation would occur only if two of the hinges formed simultaneously.

(a) Plastic hinges at points 2 and 4,

$$M_s = \frac{P_u L}{4} \tag{a}$$

Equilibrium requires the positive moment M_p at point 2 $(0.5 P_u h - Hh)$ to equal the negative moment M_p at point $4 (Hh)$:

$$M_p = 0.5 P_u h - M_p \tag{b}$$

$$M_p = \frac{P_u h}{4}; \qquad P_u = \frac{4 M_p}{h} \tag{c}$$

Further, in order that Eq. (c) is valid, the resulting moment at point 3 cannot exceed M_p:

$$M_3 = M_s + \tfrac{1}{2}(0.5 P_u h) - M_p \tag{d}$$

$$= \frac{P_u L}{4} + \frac{P_u h}{4} - \frac{P_u h}{4} = \frac{P_u L}{4} \leq M_p \tag{e}$$

Equation (e) requires $P_u L/4 < P_u h/4$, which means if $L < h$ plastic hinges form at points 2 and 4. If $L > h$ the plastic hinges will form at points 3 and 4.

(b) Plastic hinges at points 3 and 4. For this, Eq. (d) is to be used letting $M_3 = M_p$,

$$M_p = M_s + \tfrac{1}{2}(0.5 P_u h) - M_p \tag{f}$$

$$M_p = \frac{P_u}{8}(L + h) \tag{g}$$

$$P_u = \frac{8 M_p}{L + h} \tag{h}$$

The above analysis has assumed constant moment of inertia (constant M_p) for both girder and column. If the girder and column are different a combined mechanism with plastic hinges at points 2, 3, and 4 could be achieved.

Example 15.2.3

Determine the plastic capacity P_u for the gabled frame of Fig. 15.2.3

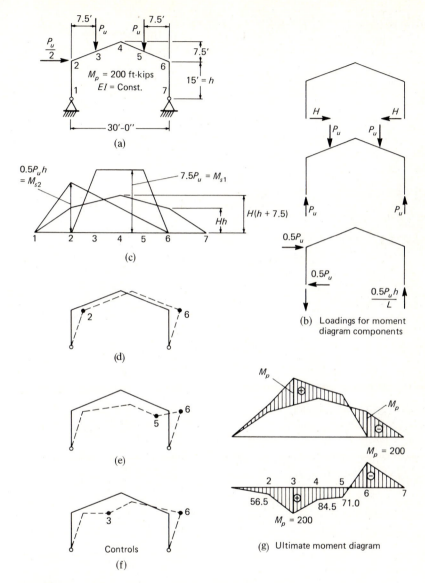

Fig. 15.2.3 Gabled frame analysis—Example 15.2.3.

using the equilibrium method. All elements of the frame are identical, having a plastic moment capacity of 200 ft-kips.

SOLUTION

Equilibrium requires a compatibility between the moment diagrams of the component loadings as shown in Fig. 15.2.3b. The moment due to H may be thought of as causing negative bending, while the other two components are causing positive bending. The moment diagrams for the components are shown on a horizontal baseline in Fig. 15.2.3c.

The maximum so-called positive moments, M_{s1} and M_{s2} are

$$M_{s1} = 7.5P_u$$

$$M_{s2} = 0.5P_u h = 7.5P_u$$

Several possible collapse mechanisms must be considered.

(a) Consider a sidesway mechanism (Fig. 15.2.3d) with plastic hinges at points 2 and 6. Equilibrium requires at point 2:

$$M_2 = M_p = M_{s2} - Hh \qquad \text{(a)}$$

and at point 6:

$$M_6 = M_p = Hh = 15H \qquad \text{(b)}$$

Thus

$$M_p = M_{s2} - M_p = 7.5P_u - M_p$$

$$M_p = 3.75P_u \qquad \text{(c)}$$

(b) Consider a combination mechanism with plastic hinges at points 5 and 6 (Fig. 15.2.3e). At point 5,

$$M_5 = M_p = M_{s1} + \tfrac{1}{4}M_{s2} - H(h + 3.75)$$

$$M_p = 7.5P_u + \frac{7.5}{4}P_u - \frac{M_p}{15}(15 + 3.75)$$

$$M_p = 4.17P_u$$

Actually, the mechanism consisting of plastic hinges at points 5 and 6 could have been eliminated as a possibility by comparing the total positive moment at points 3 and 5 (Fig. 15.2.3c) which will indicate that point 3 will achieve a plastic hinge before point 5 can attain it.

(c) Consider the mechanism with plastic hinges at points 3 and 6:

$$M_3 = M_p = M_{s1} + \tfrac{3}{4}M_{s2} - H(h + 3.75)$$

$$M_p = 7.5P_u + \frac{3}{4}(7.5P_u) - \frac{M_p}{15}(15 + 3.75)$$

$$M_p = 5.83P_u \qquad \text{Governs}$$

The largest M_p, or the smallest P_u, for a mechanism to occur indicates the governing one. A check may be made by determining the ultimate moment diagram assuming $M_p = 200$ ft-kips. Thus

$$P_u = \frac{200}{5.83} = 34.2 \text{ kips}$$

$$H = \frac{M_p}{15} = \frac{200}{15} = 13.33 \text{ kips}$$

$$M_{s1} = 7.5P_u = 7.5(34.2) = 256.5 \text{ ft-kips}$$

$$M_{s2} = 256.2 \text{ ft-kips}$$

At the critical locations, the ultimate moment is

$$M_2 = 256.5 - 200 = 56.5 \text{ ft-kips} < M_p \qquad \text{OK}$$

$$M_3 = 256.5 + \tfrac{3}{4}(256.5) - 13.33(18.75) = 200 \text{ ft-kips} = M_p$$

$$M_4 = 256.5 + \tfrac{1}{2}(256.5) - 13.33(22.5) = 84.5 \text{ ft-kips} < M_p \qquad \text{OK}$$

$$M_5 = 256.5 + \tfrac{1}{4}(256.5) - 13.33(18.75) = 71.0 \text{ ft-kips} < M_p \qquad \text{OK}$$

$$M_6 = 13.33(15) = 200 \text{ ft-kips} = M_p$$

The final ultimate moment diagram is shown in Fig. 15.2.3g, both as it would be from graphical superposition and as it would be if the net moment is plotted on a horizontal baseline.

Example 15.2.4

Determine the plastic moment required for the frame of Fig. 15.2.4. Consider (a) uniform gravity loading w alone; and (b) uniform gravity loading in combination with uniform lateral loading.

SOLUTION

(a) Gravity loading only. The simple beam moment is

$$M_s = \frac{w_u L^2}{8} = \frac{w_u (75)^2}{8} = 704 w_u \tag{a}$$

For a properly designed frame, the plastic hinges will occur at the ends and midspan of the girder:

$$M_p = M_s - M_p$$

$$M_p = \frac{1}{2} M_s = \frac{w_u L^2}{16} = 352 w_u \tag{b}$$

and the superposition of the bending moments due to the loading components is shown in Fig. 15.2.4d.

(b) Combined vertical and lateral loading. There is an additional component of loading as seen in Fig. 15.2.4e which contributes an unsymmetrical effect of the same bending moment sign as caused by the gravity uniform loading. It may be observed from Fig. 15.2.4f that the maximum moments will occur slightly to the left of point 3 and at point 4. One could mathematically solve for the exact location and magnitude of the plastic moment. However, for design purposes one may use a graphical construction and divide the maximum value of the parabolic curve by two to establish the horizontal line Hh. In other words, equalize the positive and negative moments. The moment at point 2 is obviously less than that at point 4 so that no plastic hinge will form at point 2.

By scaling, the maximum ordinate of the parabola near point 3 is

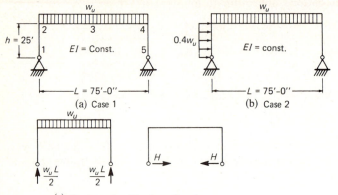

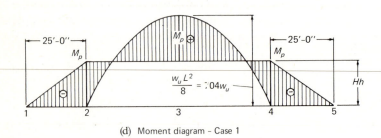

(c) Components of loading – Case 1

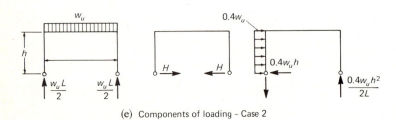

(d) Moment diagram – Case 1

(e) Components of loading – Case 2

(f) Moment diagram – Case 2

Fig. 15.2.4 Example 15.2.4.

$768w_u$. In which case

$$M_p = \frac{768w_u}{2} = 384w_u \tag{c}$$

The uniform lateral loading causes about $9\frac{1}{2}\%$ reduction in vertical carrying capacity, or alternatively it would require a section having about $9\frac{1}{2}\%$ greater moment capacity.

Example 15.2.5

Re-solve Example 15.2.2 if the bases are fixed instead of hinged, as shown in Fig. 15.2.5. Use $L = 80$ ft and $h = 20$ ft.

SOLUTION

Using the equilibrium method, separate the loading components (Fig. 15.2.5b) and then superimpose the moment diagrams for each component of load, as shown in Fig. 15.2.5c. Since the structure is statically indeterminate to the third degree, four plastic hinges are required for the collapse mechanism.

A study of the moment-diagram components indicates maximum moments at points 1 and 5 and probable maximums at points 3 and 4. Assume this combination and check the resulting moment at point 2.

At point 1

$$M_1 = M_p \tag{a}$$

At point 5

$$M_5 = M_p \tag{b}$$

At point 4

$$M_4 = Hh - M_5 \tag{c}$$
$$M_p = 20H - M_p$$
$$H = M_p/10$$

At point 3

$$M_3 = \frac{M_5}{2} + 20P_u + 5P_u - \frac{M_1}{2} - Hh \tag{d}$$

$$M_p = \frac{\cancel{M_p}}{\cancel{2}} + 25P_u - \frac{\cancel{M_p}}{\cancel{2}} - 2M_p$$

$$M_p = 8.33P_u \tag{e}$$

Check point 2

$$M_2 = Hh + M_1 - 10P_u \tag{f}$$
$$= 2M_p + M_p - 10P_u$$
$$= 25P_u - 10P_u = 15P_u > M_p$$

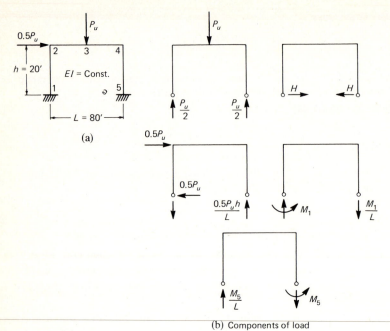

(a)

(b) Components of load

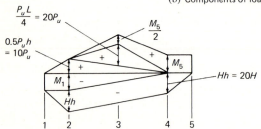

(c) Components of moment diagram

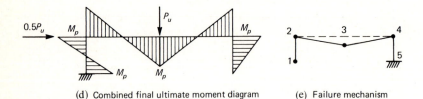

(d) Combined final ultimate moment diagram   (e) Failure mechanism

Fig. 15.2.5 Example 15.2.5.

Thus a plastic hinge occurs at point 2. A beam mechanism therefore is a probability. Furthermore, if a negative moment plastic hinge occurs at point 2, a plastic moment must also exist at point 1. The equilibrium conditions with plastic hinges at 1, 2, 3, and 4 are

$$M_1 = M_p \tag{g}$$

At point 2

$$M_2 = Hh + M_1 - 10P_u \tag{h}$$

$$M_p = 20H + M_p - 10P_u$$

$$20H = 10P_u \tag{i}$$

At point 3

$$M_3 = 0.5M_5 + 20P_u + 5P_u - 0.5M_1 - Hh$$

$$M_p = 0.5M_5 + 25P_u - 0.5M_p - 10P_u$$

$$1.5M_p = 0.5M_5 + 15P_u \tag{j}$$

At point 4

$$M_4 = Hh - M_5$$

$$M_p = 10P_u - M_5 \tag{k}$$

Multiply Eq. (k) by 0.5 and add to Eq. (j), which gives

$$2M_p = 20P_u$$

$$M_p = 10P_u \tag{l}$$

The resulting moment at point 5, using Eq. (k), is

$$M_5 = M_p - 10P_u = 10P_u - 10P_u = 0$$

which is less than M_p, as it should be. The correct mechanism, containing plastic hinges at points 1, 2, 3, and 4 and the ultimate moment diagram appear in Fig. 15.2.5.

Energy Method

Just as discussed for beams in Chapter 10, the energy method may also be used for the plastic analysis of frames. For certain frame arrangements, the energy method will be found easier. The following examples show how the energy method may be applied to the same structures previously analyzed by the equilibrium method.

Example 15.2.6

Repeat Example 15.2.1 except use the energy method.

SOLUTION
The mechanism is shown in Fig. 15.2.6. The plastic hinge locations are assumed and from the geometry the angles θ are established. The external work done by the load equals the internal strain energy of the plastic

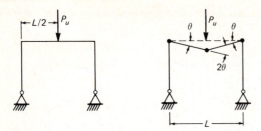

Fig. 15.2.6 Example 15.2.6.

moments moving through their angles of rotation:

$$W_e = W_i$$

$$P_u \frac{\theta L}{2} = M_p(\theta + 2\theta + \theta)$$

$$P_u = \frac{8M_p}{L}$$

exactly as obtained in Example 15.2.1.

Example 15.2.7

Repeat Example 15.2.2 using the energy method.

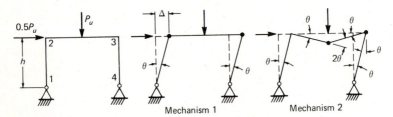

Mechanism 1   Mechanism 2

Fig. 15.2.7 Example 15.2.7.

SOLUTION
The possible mechanisms are shown in Fig. 15.2.7.

(a) Mechanism 1
$$0.5P_u\theta h = M_p(\theta + \theta)$$

$$P_u = \frac{4M_p}{h}$$

(b) Mechanism 2
$$0.5P_u\theta h + P_u\theta\frac{L}{2} = M_p(2\theta + 2\theta)$$

$$P_u = \frac{8M_p}{L+h}$$

The results are identical to those of Example 15.2.2.

Example 15.2.8

Repeat Example 15.2.3 using the energy method.

SOLUTION
The possible mechanisms are shown in Fig. 15.2.8.

(a) Mechanism 1

$$0.5P_u(15\theta) = M_p(2\theta)$$

$$M_p = 3.75P_u$$

(b) Mechanism 2. In order to treat this more complex mechanism, the concept of *instantaneous center* is used. When plastic hinges form at points 5 and 6, three rigid bodies remain which rotate as the structure moves. Segment 1–2–3–4–5 rotates about point 1; segment 6–7 rotates about point 7; segment 5–6 rotates and translates an amount which is controlled by the movement of points 5 and 6 on the adjacent rigid segments. If the body is rigid, point 5′ is perpendicular to line 1–5, and point 6′ is perpendicular to line 6–7. Thus the points 5 and 6 may be thought of as rotating about point 0, the intersection of line 1–5 and line 6–7; i.e., the instantaneous center.

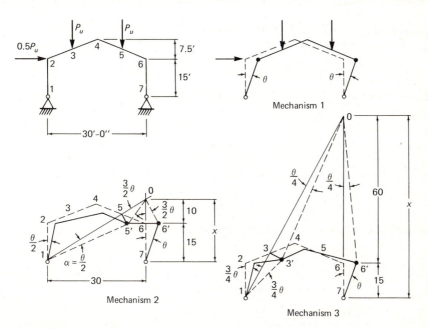

Fig. 15.2.8 Example 15.2.8.

The first step in the energy method using the instantaneous center is to determine its location; since point 5 is 22.5 ft horizontally and 18.75 ft vertically from point 1, the vertical distance to point 0 from point 7 is

$$\frac{x}{30} = \frac{18.75}{22.5}, \qquad x = 25 \text{ ft}$$

Next, a reference angle θ is established arbitrarily as shown in Fig. 15.2.8. By proportion, the angle of rotation with respect to point 0 is $3\theta/2$. The rigid-body segment 5–6 rotates through this angle $3\theta/2$. By inverse proportion as distance 0–5 is to 1–5, the rotation of rigid-body 1–2–3–4–5 about point 1 is

$$\frac{\frac{1}{4}}{\frac{3}{4}} = \frac{\alpha}{3\theta/2}, \qquad \alpha = \frac{\theta}{2}$$

The relative plastic hinge rotation at point 5 is

$$\frac{\theta}{2} + \frac{3\theta}{2} = 2\theta$$

The relative plastic hinge rotation at point 6 is

$$\theta + \frac{3\theta}{2} = 2.5\theta$$

In order to compute external work done by the applied loads, the vertical distances moved due to rotation of points 3 and 5 and the horizontal distance moved at point 2 are required.

The vertical displacement of point 3 equals the angle of rotation times the horizontal projection from points 2 to 3. The load at point 3 moves vertically through a distance

$$\frac{\theta}{2}(7.5) = 3.75\theta$$

The load at point 5 moves vertically through the distance

$$\frac{\theta}{2}(22.5) = 11.25\theta$$

The load at point 2 moves horizontally through the distance

$$\frac{\theta}{2}(15) = 7.5\theta$$

The complete energy equation then becomes

$$W_e = W_i$$

$$0.5P_u(7.5\theta) + P_u(3.75\theta) + P_u(11.25\theta) = M_p(2\theta + 2.5\theta)$$

$$M_p = \frac{18.75}{4.5} P_u = 4.17 P_u$$

the same as from the equilibrium method.

(c) Mechanism 3. The instantaneous center is found by intersecting the line 1–3 with line 6–7:

$$\frac{x}{30} = \frac{18.75}{7.5}, \qquad x = 75 \text{ ft}$$

If θ is defined by Fig. 15.2.8, then the angle of rotation with respect to 0 is $\theta/4$, since the distance 0–6 is four times the distance 6–7. Since the distance 0–3 is three times the distance 3–1, the angle 3–1–3' is $3\theta/4$ (3 times the rotation angle about 0).

The external work done by the various loads is

Load at 2, $\qquad 0.5P_u\left(\dfrac{3\theta}{4}\right)(15) = \dfrac{22.5}{4} P_u\theta$

Load at 3, $\qquad P_u\left(\dfrac{3\theta}{4}\right)(7.5) = \dfrac{22.5}{4} P_u\theta$

Load at 5, $\qquad P_u\left(\dfrac{\theta}{4}\right)(7.5) = \dfrac{7.5}{4} P_u\theta$

The internal strain energy is

Moment at 3, $\qquad M_p\left(\dfrac{3\theta}{4} + \dfrac{\theta}{4}\right) = M_p\theta$

Moment at 6, $\qquad M_p\left(\theta + \dfrac{\theta}{4}\right) = M_p\dfrac{5\theta}{4}$

$$W_e = W_i$$

$$P_u\theta\left(\frac{22.5}{4} + \frac{22.5}{4} + \frac{7.5}{4}\right) = M_p\theta\left(1 + \frac{5}{4}\right)$$

$$P_u\left(\frac{52.5}{4}\right) = M_p\left(\frac{9}{4}\right)$$

$$M_p = \frac{52.5}{9} P_u = 5.83 P_u \qquad \text{Governs}$$

The result is identical with Example 15.2.3.

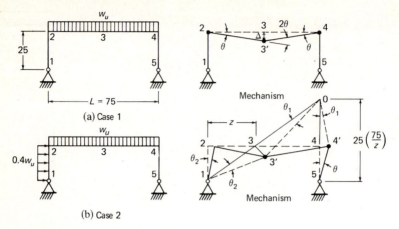

Fig. 15.2.9 Example 15.2.9.

Example 15.2.9

Repeat Example 15.2.4 using the energy method.

SOLUTION

(a) For Case 1 with no lateral loading, the mechanism is shown in Fig. 15.2.9a. The external work done by uniformly distributed load equals the intensity of load w_u times the area displaced as the mechanism forms. In Fig. 15.2.9a the area displaced is the triangular area 2–3–4–3'. Thus

$$W_e = w_u \frac{1}{2}(\Delta)L = w_u \frac{1}{2}\left(\theta \frac{L}{2}\right)L = w_u \theta \frac{L^2}{4}$$

Finally, equating

$$W_e = W_i$$

$$w_u \theta \frac{L^2}{4} = M_p(\theta + 2\theta + \theta) = 4M_p \theta$$

$$M_p = \frac{w_u L^2}{16} = \frac{w_u(75)^2}{16} = 352 w_u$$

(b) Case 2—combined vertical and lateral loading. For this situation, the position of the plastic hinge at point 3 is an unknown distance z from point 2, as shown in Fig. 15.2.9b. One could assume several positions of point 3 and determine the maximum M_p (or the minimum w_u), or the problem can be set up mathematically to determine minimum w_u.

Try $z = L/2 = 37.5$ ft:

$$\theta_1 = \theta$$

$$\theta_2 = \theta$$

$$W_e \text{ (vertical load)} = w_u\theta\left(\frac{L^2}{4}\right) = 1408 w_u\theta$$

$$W_e \text{ (horizontal load)} = 0.4w_u(\tfrac{1}{2})\theta h^2 = 125 w_u\theta$$

$$W_e = W_i$$

$$w_u\theta(1408 + 125) = M_p(2\theta + 2\theta)$$

$$M_p = 382.8 w_u$$

Try $z = 35$ ft:

$$\theta_1 = \theta\left(\frac{z}{75-z}\right) = \frac{35}{40}\theta = 0.875\theta$$

$$\theta_2 = \theta_1\left(\frac{75-z}{z}\right) = \frac{40}{35}\theta_1 = \theta$$

Vertical Δ at point 3,

$$\Delta = z\theta$$

$$W_e = W_i$$

$$w_u(\tfrac{1}{2})35\theta(75) + 125 w_u\theta = M_p(1.875\theta + 1.875\theta)$$

$$w_u(1312.5 + 125) = 3.75 M_p$$

$$M_p = 383.0 w_u$$

This result is little different from that obtained for $z = L/2$. The answer, in other words, is not too sensitive to the exact position of the plastic hinge, and the approximate procedure illustrated is considered adequate for design purposes.

If one wishes, the "exact" solution may be obtained, as follows:

$$W_e \text{ (vertical load)} = w_u(\tfrac{1}{2})z\theta(75) = 37.5 w_u\theta z$$

$$W_e \text{ (horizontal load)} = 125 w_u\theta$$

$$W_i \text{ (at point 3)} = M_p(\theta + \theta_1)$$

$$= M_p\left(\theta + \theta\frac{z}{75-z}\right)$$

$$= M_p\theta\left(\frac{75}{75-z}\right)$$

$$W_i \text{ (at point 4)} = M_p\theta\left(\frac{75}{75-z}\right)$$

$$W_e = W_i$$

$$w_u(125+37.5z) = M_p\left(\frac{150}{75-z}\right)$$

$$M_p = w_u\left[\frac{(125+37.5z)(75-z)}{150}\right] \tag{a}$$

$$M_p = w_u(-z^2+71.67z+150)$$

$$\frac{dM_p}{dz} = 0 = -2z+71.67$$

$$z = 35.83 \text{ ft}$$

Substituting z into Eq. (a) above gives

$$M_p = w_u\left[\frac{125+37.5(35.83)}{150}\right](75-35.83) = 383.5w_u$$

This result is obviously little different than the approximate values.

Example 15.2.10

Repeat Example 15.2.5 using the energy method.

SOLUTION
The possible mechanisms are shown in Fig. 15.2.10.
 (a) Mechanism 1. The beam mechanism shown is always a good possibility when the vertical load is larger than the horizontal load.

$$W_e = W_i$$

$$P_u(40\theta) = M_p(\theta+2\theta+\theta)$$

$$M_p = 10P_u$$

 (b) Mechanism 2. The sidesway mechanism is not expected to control since the larger load is the vertical one:

$$0.5P_u(20\theta) + P_u(0) = M_p(\theta+\theta+\theta+\theta)$$

$$M_p = 2.5P_u$$

 (c) Mechanism 3. This mechanism is a combination of mechanisms 1 and 2. Note that the angle change at point 2 is equal and opposite for mechanisms 1 and 2; thus the combination would not expect a plastic hinge at point 2:

$$W_e = W_i$$

$$0.5P_u(20\theta) + P_u(40\theta) = M_p(\theta+2\theta+2\theta+\theta)$$

$$50P_u = 6M_p$$

$$M_p = 8.33P_u$$

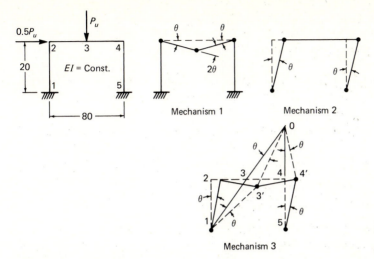

Fig. 15.2.10 Example 15.2.10.

The largest value of M_p is $10P_u$, so Mechanism 1, the beam mechanism, governs. It is noted that in the energy method it was not established that a plastic hinge develops at point 1 as determined in the equilibrium method; unnecessary here since no rotation occurs at point 1 when the mechanism is achieved.

In conclusion, either the equilibrium method or the energy method may be used for the plastic analysis of frames. It is suggested that one method be used to check the other.

15.3 PLASTIC DESIGN EXAMPLES—ONE-STORY FRAMES

One-story frames are most practicably designed by the plastic design method. Working stress design, as illustrated in the next section, requires preliminary assumptions of relative stiffness of members and a tedious statically indeterminate structural analysis. The plastic design method is simpler and more efficiently utilizes steel.

The following examples incorporate the plastic analysis of Sec. 15.2 with all the design concepts from earlier chapters.

Example 15.3.1

Design a rectangular frame of 75-ft span and 25-ft height to carry a gravity uniform load of 1.0 kip/ft when no lateral load is acting; or a gravity uniform load of 1.10 kips/ft combined with a horizontal uniform wind load of 0.44 kips/ft acting on either side of frame. Lateral bracing from purlins occurs every 6 ft on the girder, and bracing is provided by wall girts on the column every 5 ft. Use A36 steel.

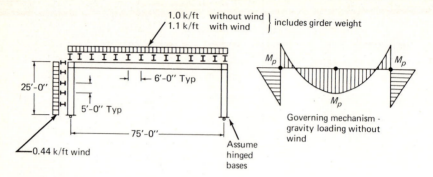

Fig. 15.3.1 Example 15.3.1.

SOLUTION

The frame is shown in Fig. 15.3.1.

(a) Factored loads. Apply load factors to the given service loads. Two cases must be considered, loading with and without wind. AISC–2.1 prescribes factors of 1.7 and 1.3 for loads without wind and with wind, respectively. The factored loads are therefore

$$\text{Case 1: } w_u = 1.0(1.7) = 1.7 \text{ kips/ft}$$

$$\text{Case 2: } w_u = 1.0(1.3) = 1.43 \text{ kips/ft} \quad \text{(vertical)}$$

$$0.4w_u = 0.572 \text{ kip/ft} \quad \text{(horizontal)}$$

(b) Consider gravity loading without wind; Case 1. Using the plastic analysis as illustrated in Example 15.2.4,

$$\text{Required } M_p = \frac{w_u L^2}{16} = \frac{1.7(75)^2}{16} = 598 \text{ ft-kips}$$

$$\text{Required } Z = \frac{M_p}{F_y} = \frac{598(12)}{36} = 199 \text{ in.}^3$$

Try W24×76, $M_p = 36(200)/12 = 600$ ft-kips, or a deeper heavier section if deflection is a critical factor. Check local buckling requirements of AISC–2.7.

$$\frac{b}{2t_f} = \frac{8.990}{2(0.680)} = 6.6 < 8.5 \quad \text{OK}$$

$$\frac{d}{t} = \frac{23.92}{0.440} = 54.4 < \left[\frac{412}{\sqrt{F_y}} = 68.7\right] \quad \text{OK}$$

(c) Check whether Case 2 including wind may govern. Using plastic analysis from Example 15.2.4,

$$\text{Required } M_p = 384w_u = 384(1.43) = 549 \text{ ft-kips}$$

Since this is less than required for vertical load only, Case 1 controls. The reader is reminded that use of the 1.3 load factor for wind instead of 1.7 is the same as in the working stress method (AISC–1.5.6) allowing a $33\frac{1}{3}\%$ overstress due to the temporary nature of wind.

(d) Check W24×76 for action as a beam-column. Determine the effective length factor K (see AISC Commentary–2.4)

$$G_A \approx 10.0 \quad \text{(Estimated for hinge at bottom)}$$

$$G_B \approx \frac{I/L \ \text{(column)}}{I/L \ \text{(girder)}} = \frac{I/25}{I/75} = 3.0$$

From Fig. 14.3.1b, unbraced frame, find $K_x = 2.25$.

$$\frac{K_x L}{r_x} = \frac{2.25(25)(12)}{9.69} = 69.7; \quad F'_e = 30.7 \ \text{ksi} \quad \text{(AISC Appendix Table 9)}$$

At this point the G factor may be adjusted to account for inelastic column behavior as discussed in Sec. 14.3.

$$\text{Inelastic } G_B = \text{Elastic } G_B\left(\frac{0.60F_y}{F'_e}\right) = 3.0\left(\frac{22}{30.7}\right) = 2.1$$

$$\text{Revised } K_x = 2.1; \quad \frac{K_x L}{r_x} = 65; \quad F_a = 16.9 \ \text{ksi}; \quad F'_e = 35.3 \ \text{ksi}$$

$$\frac{K_y L}{r_y} = \frac{1.0(5)(12)}{1.92} = 31.2$$

$$P_e = (23/12)AF'_e = (23/12)(22.4)(35.3) = 1520 \ \text{kips}$$

$$C_m = 0.85, \quad \text{for an unbraced frame}$$

$$P_{cr} = 1.70F_a A = 1.70(16.9)(22.4) = 644 \ \text{kips}$$

$$M_m = \left[1.07 - \frac{\sqrt{F_y}(L/r_y)}{3160}\right]M_p > M_p, \quad \text{use } M_m = M_p$$

$$P = 0.5w_u L = 0.5(1.7)(75) = 63.8 \ \text{kips}$$

Check column stability using AISC Formula (2.4–2):

$$\frac{P}{P_{cr}} + \frac{M_i}{M_m}\left(\frac{C_m}{1 - P/P_e}\right) = \frac{63.8}{644} + \frac{598}{600}\left(\frac{0.85}{1 - 63.8/1520}\right)$$
$$= 0.10 + 0.89 = 0.99 < 1.0 \quad \text{OK}$$

Check strength criterion, AISC Formula (2.4–3):

$$P_y = F_y A = 36(22.4) = 806 \ \text{kips}$$

$$\frac{P}{P_y} + \frac{M}{1.18M_p} = \frac{63.8}{806} + \frac{598}{1.18(600)} = 0.08 + 0.84 = 0.92 < 1.0 \quad \text{OK}$$

Since $P/P_y < 0.15$, the strength criterion is automatically satisfied and the full plastic moment may be utilized. Check local buckling on beam-column, AISC Formula (2.7–1a),

$$\frac{d}{t_w} = \frac{412}{\sqrt{F_y}}\left(1 - 1.4\frac{P}{P_y}\right) = 68.7 - 96.1\frac{P}{P_y}$$

$$= 68.7 - 96.1(0.08) = 61.0 > 54.3 \text{ actual} \qquad \text{OK}$$

Use W24×76 for beam and column members.

(e) Check shear (AISC–2.5). Outside the corner connection,

$$V_u = \tfrac{1}{2}(1.7)(75) = 63.8 \text{ kips}$$

Strength $V_u = 0.55F_y td = 0.55(36)(0.440)(23.92) = 208$ kips

This is acceptable since the strength exceeds the factored shear caused by loading.

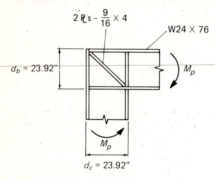

Fig. 15.3.2 Corner connection for Example 15.3.1.

Inside the corner connection, referring to Fig. 15.3.2, the shear produced may be expressed, according to Eq. 13.8.4, as

$$\text{Required } t_w = \frac{1.91M}{A_{bc}F_y} = \frac{1.91(600)(12)}{(23.92)^2 36} = 0.67 \text{ in.}$$

Since the web thickness is only 0.44 in., a diagonal web stiffener is required.

(f) Determine size of diagonal stiffener plate. Force to be carried by diagonal stiffeners,

$$\text{Shear force} = \frac{M}{d_b}\left(\frac{0.67 - 0.44}{0.67}\right) = \frac{600(12)}{23.92}\left(\frac{0.23}{0.67}\right) = 103 \text{ kips}$$

$$\text{Stiffener force} = (\text{shear force})\sqrt{2}$$

$$= 103(1.414) = 146 \text{ kips}$$

$$\text{Required } A_{st} = \frac{\text{stiffener force}}{F_y} = \frac{146}{36} = 4.1 \text{ sq in.}$$

Since the flange width of W24×76 is 8.990 in. and the web thickness is 0.440 in., the width available for a stiffener plate is $(8.990-0.440)/2 = 4.28$ in.

Try 4-in. wide plate. For local buckling,

$$\frac{b}{t} \leq 8.5; \qquad t_{min} = \frac{4}{8.5} = 0.47 \text{ in.}$$

Use 2 $\text{Rs}—\frac{9}{16} \times 4$, $A = 4.50$ sq in.

(g) Check lateral support, AISC–2.9. Little moment variation occurs over the 6-ft laterally unbraced length. Thus the end moment ratio on the segment ≈ -1, requiring use of AISC Formula (2.9–1b),

$$\frac{L_{cr}}{r_y} = \frac{1375}{F_y} = 38.2$$

$$L_{cr} = 38.2r_y = 38.2(1.92)\tfrac{1}{12} = 6.1 \text{ ft} > 6 \text{ ft} \qquad \text{OK}$$

Consider lateral bracing acceptable at 6 ft on the girder and 5 ft on the column.

(h) Overall frame stability. Use of AISC Formula (2.4–2) using $K > 1.0$ is considered to automatically satisfy the overall stability necessary to achieve plastic strength for one- and two-story unbraced frames.

Example 15.3.2

Redesign the frame of Example 15.3.1 (Fig. 15.3.1) if the column cannot be deeper than 14.50 in. actual depth.

SOLUTION
For this situation the column and the beam will have different plastic moment capacities.

(a) Select section for plastic strength. There are many possible solutions using various sizes of column. Try a W14×82 section with $Z = 139$ in.^{3} Assuming gravity loading governs, the positive plastic moment required at girder midspan is

$$M_p = \frac{w_u L^2}{8} - M_p \text{ (column)} = \frac{1.7(75)^2}{8} - \frac{139(36)}{12}$$

$$= 1195 - 417 = 778 \text{ ft-kips}$$

$$\text{Required } Z = \frac{778(12)}{36} = 259 \text{ in.}^3$$

A W27×94 might be selected to satisfy strength. For practicality of making a corner connection the two members should have approximately the same flange width. Revise sections.

Try W14×74 column, $b_f = 10.070$ in.

$$M_p \text{ (column)} = 126(36)/12 = 378 \text{ ft-kips}$$

$$\text{Required } M_p \text{ (girder)} = 1195 - 378 = 817 \text{ ft-kips}$$

$$\text{Required } Z = \frac{817(12)}{36} = 272 \text{ in.}^3$$

Try W27×94 girder, $Z = 278$ in.^{3}, $b_f = 9.99$ in.

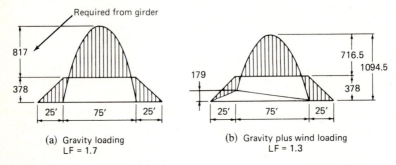

(a) Gravity loading
LF = 1.7

(b) Gravity plus wind loading
LF = 1.3

Fig. 15.3.3 Example 15.3.2.

(b) Check whether loading that includes wind may govern. Refer to Fig. 15.2.4 for plastic strength analysis using the same section throughout and Fig. 15.3.3 for the result using different column and girder sections,

$$M_s \text{ (at knee)} = 0.2 w_u h^2 = 0.2(1.43)(25)^2 = 179 \text{ ft-kips}$$

$$M_s \text{ (at girder midspan)} = \tfrac{1}{8} w_u L^2 = \tfrac{1}{8}(1.43)(75)^2 = 1005 \text{ ft-kips}$$

The required plastic moment from the girder assuming the plastic hinge to occur approximately at midspan is

$$M_p = 1005 + \tfrac{1}{2}(179) - M_p \text{ (column)}$$
$$= 1005 + 89.5 - 378 = 716.5 \text{ ft-kips}$$

which is less than the 819 ft-kips required from gravity load alone. Wind loading does not control.

(c) Check beam-column capacity. If axial compression were large, plastic moment might not govern selection of column. In which case, using Eq. 12.13.10 one should compute equivalent P_y,

$$\text{Equivalent } P_y = P + \frac{2M}{d} \qquad \text{[12.13.10]}$$

$$= 63.8 + \frac{2(378)12}{14} = 712 \text{ kips}$$

$$\text{Required } A = \frac{\text{Equivalent } P_y}{F_y} = \frac{712}{36} = 19.8 \text{ sq in.}$$

The selected W14×74 has $A = 21.8$ sq in., and so is likely to be acceptable or closely so. With the axial effect 15% or less, the girder may be selected solely for M_p; then checked for stability as follows:

$$G_A \approx 10.0 \quad \text{(hinged at bottom)}$$

$$G_B = \frac{I/L \text{ (column)}}{I/L \text{ (girder)}} = \frac{796/25}{3270/75} = 0.73$$

From Fig. 14.3.1b, find $K_x = 1.8$.

$$\frac{K_x L_x}{r_x} = \frac{1.8(25)(12)}{6.04} = 89.4; \qquad F'_e = 18.7 \text{ ksi}$$

Investigate a possible revision of K_x for inelastic column behavior as discussed in Sec. 14.3. In this case there is no reduction since F'_e is less than $0.60F_y$, indicating elastic behavior.

$$\frac{K_y L_y}{r_y} = \frac{1.0(5)12}{2.48} = 24.2$$

$$F_a = 14.3 \text{ ksi} \quad \text{(based on } KL/r = 89.4)$$

$$P_e = (23/12)AF'_e = (23/12)(21.8)(18.7) = 781 \text{ kips}$$

$$P_{cr} = 1.70F_a A = 1.70(14.3)(21.8) = 530 \text{ kips}$$

$$M_m = M_p = 378 \text{ ft-kips}$$

AISC Formula (2.4–2),

$$\frac{P}{P_{cr}} + \frac{M_i}{M_m}\left(\frac{C_m}{1 - P/P_e}\right) = \frac{63.8}{530} + \frac{378}{378}\left(\frac{0.85}{1 - 63.8/781}\right)$$

$$= 0.12 + 0.93 = 1.05$$

Accept the section even though the criterion indicates an overstress of about $4\frac{1}{2}\%$, since the hinged-base assumption $(G_A = 10.0)$ underestimates the moment restraint. AISC Formula (2.4–3) does not control, since $P/P_y < 0.15$,

$$P_y = F_y A = 36(21.8) = 785 \text{ kips}$$

$$\frac{P}{P_y} = \frac{63.8}{785} = 0.081 < 0.15 \qquad \text{OK}$$

Check local buckling (AISC-2.7):

$$\frac{d}{t} = \frac{412}{\sqrt{F_y}}\left(1 - 1.4\frac{P}{P_y}\right) = 68.7 - 96.1(0.081) = 60.9$$

$$\text{Actual } \frac{d}{t} = \frac{14.17}{0.450} = 31.5 < 60.9 \qquad \text{OK}$$

Use W14×74 for the column.

(d) Check the W27×94 beam for local buckling under AISC–2.7:

$$\frac{b_f}{2t_f} = \frac{9.990}{2(0.745)} = 6.7 < 8.5 \qquad \text{OK}$$

$$\frac{d}{t} = \frac{26.92}{0.490} = 54.9 < 68.7 \qquad \text{OK}$$

Use W27×94 *for the girder.*

(e) Check shear in the corner region. This is illustrated in Example 13.8.1.

(f) Check lateral support, AISC 2–9. For the girder, W27×94,

$$L_{cr} = 38.2r_y = 38.2(2.12)\tfrac{1}{12} = 6.8 \text{ ft} > 6.0 \text{ ft} \qquad \text{OK}$$

For the column, W14×74,

$$L_{cr} = 38.2r_y = 38.2(2.48)\tfrac{1}{12} > 5.0 \text{ ft} \qquad \text{OK}$$

Use W14×74 *column,* W27×94 *girder with a corner joint containing diagonal stiffeners, similar to Fig. 15.3.2.*

Example 15.3.3

Design a gabled frame as shown in Fig. 15.3.4a to carry a uniform gravity load of 1.1 kips/ft, and a horizontal wind load of 0.6 kips/ft. Use A572 Grade 50 steel.

SOLUTION

(a) Establish plastic strength requirement for gravity loading alone. Apply the 1.7 load factor and compute simple beam moment at point 3,

$$M_{s3} = \frac{w_u L^2}{8} = \frac{1.7(1.1)(55)^2}{8} = 707 \text{ ft-kips}$$

plotted in Fig. 15.3.4b. Then superimpose the effect of the equal horizontal reactions H at points 1 and 5 until the positive and negative moments are equal. Referring back to Example 15.2.3 for the general procedure, the moment due to H is

$$M \text{ (top of gable)} = 28H$$

$$M \text{ (top of column)} = 20H$$

$$\text{Distance to rotation point} = 27.5\left(\frac{28H}{8H}\right) = 96.3 \text{ ft}$$

from midspan of the frame

$$M_p = 295 \text{ ft-kips} \quad \text{(scaled from Fig. 15.3.4b)}$$

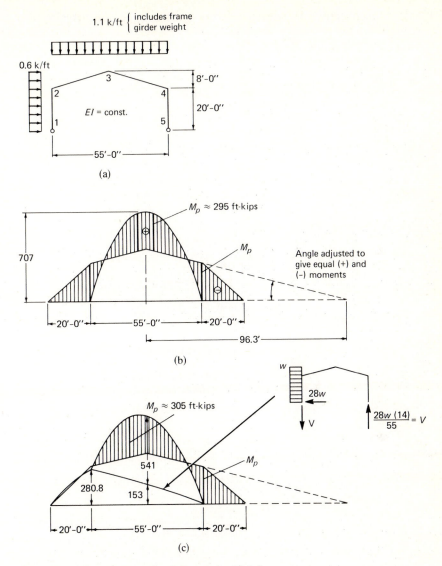

Fig. 15.3.4 Gabled frame of Example 15.3.3.

(b) Establish plastic strength requirement for loading with wind. Apply 1.3 load factor.

$$M_{s3} = \tfrac{1}{8}(1.3)(1.1)(55)^2 = 541 \text{ ft-kips}$$

$$M_3 \text{ (wind)} = 1.3(0.6)(28)(14)\tfrac{1}{2} = 153 \text{ ft-kips}$$

$$M_{s2} \text{ (wind)} = 1.3(0.6)(20)(18) = 280.8 \text{ ft-kips}$$

$$M_p = 305 \text{ ft-kips} \quad \text{(scaled from Fig. 15.3.4c)}$$

Thus for this example the wind condition governs maximum strength by a slim margin.

(c) Select section:

$$\text{Required } Z = \frac{M_p}{F_y} = \frac{305(12)}{50} = 73.2 \text{ in.}^3$$

Try W16×45 which satisfies AISC–2.7 for local buckling. A W18×40 will work for strength but will require lateral bracing every 2.9 ft. The W16×45 permits lateral bracing every 3.6 ft.

$$M_p = ZF_y = 82.3(50)\tfrac{1}{12} = 342 \text{ ft-kips}$$

(d) Check requirements for beam-column. Without wind,

$$P = 1.7w_u\frac{L}{2} = 1.7(1.1)\frac{55}{2} = 51.5 \text{ kips}$$

$$\text{Equivalent } P_y = P + 2M/d$$

$$= 51.5 + 2(295)(12)/16 = 494 \text{ kips}$$

$$\text{Required } A = \frac{\text{Equivalent } P_y}{F_y} = \frac{494}{50} = 9.9 \text{ sq in.}$$

Try W16×45, $A = 13.3$ sq in. Since P/P_y appears to be less than 0.15, check stability first.

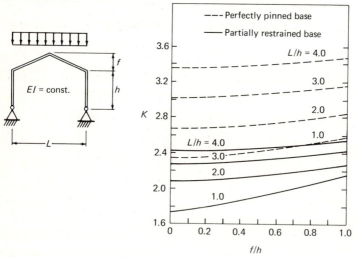

Fig. 15.3.5 Effective length factors for gabled frames. (From Ref. 6)

The use of G_A and G_B along with the alignment chart, Fig. 14.3.1, may not be appropriate for a gabled frame. A chart for such effective length factors given by Lu [6] appears as Fig. 15.3.5. This is applicable for the single bay frame with girder and column sections the same. It is further suggested in Ref. 6 that if $I_g \neq I_c$ one may obtain an approximate solution by using the Fig. 14.3.1 alignment chart with

$$G_{top} = \frac{I_c/h(\text{column})}{I_g/2q(\text{girder})}$$

where h = column height

q = length of girder to ridge

In this example (see Fig. 15.3.5),

$$\frac{f}{h} = \frac{8}{20} = 0.4 \quad \text{and} \quad \frac{L}{h} = \frac{55}{20} = 2.75$$

find $K_x \approx 2.25$ for hinged base $(G = 10.0)$.

$$\frac{K_x L}{r_x} = \frac{2.25(20)12}{6.65} = 81.2; \qquad F'_e = 22.7 \text{ ksi}; \qquad F_a = 18.8 \text{ ksi}$$

Assume lateral bracing will be adequate so that KL/r_y will not control.

$$P_e = (23/12)AF'_e = (23/12)(13.3)(22.7) = 577 \text{ kips}$$

$$P_{cr} = 1.70F_a A = 1.70(18.8)(13.3) = 424 \text{ kips}$$

$$M_m = M_p = 342 \text{ ft-kips}$$

AISC Formula (2.4–2) for stability (without wind):

$$\frac{P}{P_{cr}} + \frac{M_i}{M_m}\left(\frac{C_m}{1 - P/P_e}\right) = \frac{51.5}{424} + \frac{295}{342}\left(\frac{0.85}{1 - 51.5/577}\right)$$

$$= 0.12 + 0.81 = 0.93 < 1.0 \qquad \text{OK}$$

AISC Formula (2.4–3) for strength:

$$P_y = 50(13.3) = 665 \text{ kips}$$

$$\frac{P}{P_y} = \frac{51.5}{665} = 0.077 < 0.15 \qquad \text{OK}$$

(e) Check shear in corner region. Use M_p for wind condition since it is the largest,

$$\text{Required } t_w = \frac{1.91M}{A_{bc}F_y} = \frac{1.91(305)(12)}{(16.13)^2(50)} = 0.54 \text{ in.} > t_w = 0.345$$

Diagonal stiffeners required.

(f) Lateral support requirement:

$$L_{cr} = \left(\frac{1375}{F_y}\right)r_y = \left(\frac{1375}{50}\right)r_y = 27.5r_y = 27.5(1.57)\tfrac{1}{12} = 3.6 \text{ ft}$$

Lateral support from purlins or joists is required every $3\frac{1}{2}$ ft.

Use W16×45, $F_y = 50$ for column and girder.

More detail concerning plastic design of gabled frames may be found in *Plastic Design in Steel* [1], and design tables are available in *Steel Gables and Arches* [7].

15.4 WORKING STRESS DESIGN—ONE-STORY FRAMES

While the authors believe the plastic design method is the method of choice for single-story rigid frames, a comparison of the methods seems desirable. The principal difficulty with the working stress method is the necessity of doing an elastic analysis of a rigid frame without knowing the final sizes of the members. The elastic analysis, using the matrix displacement method, ordinary slope deflection, moment distribution, or some other method, is outside the scope of this text; the many available standard references on statically indeterminate structures should be referred to as needed.*

For practical purposes, various design aids [8] are available giving elastic solutions for standard frame dimensions and loadings.

For design, an elastic analysis is needed to select member sizes; however, one cannot determine moments, shears, and axial forces without knowing the relative moments of inertia. This problem does not arise in plastic design. However, there is one advantage in the working stress method; if lateral bracing is not satisfactory to develop the plastic strength mechanism, the working stress method allows for design using reduced strengths based on lateral stability. Plastic design requires the mechanism to develop.

Example 15.4.1

Design the frame of Example 15.3.1 using the working stress method. A36 steel.

SOLUTION

In the working stress method the elastic moments under service load are required. For this design the girder and column are assumed to have constant moment of inertia. Figure 15.4.1 shows the two loading cases and the corresponding bending moment diagrams. The relatively lengthy process of doing the statically indeterminate analysis has not been illustrated here. When wind loading is included, AISC–1.5.6 allows a $33\frac{1}{3}\%$ overstress. The use of 0.75 of the applied loading has the same effect as using the full loading and 1.33 of the allowable stresses.

(a) Select the W section. For the loading without wind, the maximum moment occurring at the corner (381 ft-kips) may be reduced according to AISC–1.5.1.4.1.

---

* See, for example, C. K. Wang, *Statically Indeterminate Structures*. New York. (McGraw-Hill Book Company, 1953); or C. K. Wang, *Matrix Methods of Structural Analysis*, 2d ed. (American Book Co., Madison, Wisc., 1970).

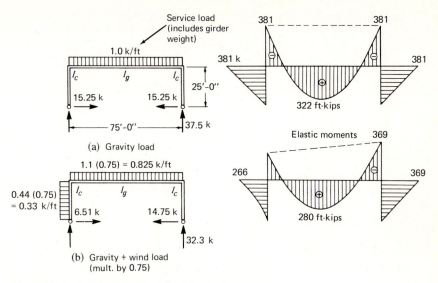

Fig. 15.4.1 Frame of Example 15.4.1: $I_g = I_c$.

Assume $F_b = 0.66F_y = 24$ ksi

$$\text{Reduced } (-M) = 0.9(381) = 343 \text{ ft-kips}$$

which would require for equilibrium an increase in the positive moment,

$$\text{Increased } (+M) = 322 + (381 - 343) = 360 \text{ ft-kips}$$

Thus the full reduction in negative moment should not be made. Adjust until positive and negative moments are equal:

$$+M = -M = 351.5 \text{ ft-kips}$$

$$\text{Required } S = \frac{351.5(12)}{24} = 175.8 \text{ in.}^3$$

Try W24×84, $S = 196$ in.^{3} (The W24×76 will be slightly overstressed on the girder and also may not satisfy beam-column requirements.)

$$f_b = \frac{351.5(12)}{196} = 21.5 \text{ ksi}$$

Note that "compact section" requirements must be satisfied even though stress is less than $0.60F_y$ because if the section is noncompact, the unadjusted moment of 381 ft-kips would have to be used.

Check AISC–1.5.1.4.1 for "compact section,"

$$\frac{b_f}{2t_f} = \frac{9.020}{2(0.770)} = 5.9 < 10.8 \qquad \text{OK}$$

$$\frac{d}{t} = \frac{24.10}{0.470} = 51.3 < 106.7 \qquad \text{OK}$$

(b) Check beam-column. When axial compression is low as in this case, selection as a beam is the correct approach. Exactly as for plastic design, determine G_A and G_B and find K_x from Fig. 14.3.1b correcting for inelastic column behavior if desired. $K_x = 2.1$ as in Example 15.3.1,

$$\frac{K_x L}{r_x} = \frac{2.1(25)12}{9.79} = 64.4; \qquad \frac{K_y L}{r_y} = \frac{1.0(5)12}{1.95} = 30.8$$

$$F_a = 17.0 \text{ ksi}; \qquad F'_e = 36.0 \text{ ksi}$$

Check inelastic G_B (see Sec. 14.3),

$$\text{Inelastic } G_B = \text{Elastic } G_B \left(\frac{0.60F_y}{F'_e}\right) = 3.00 \left(\frac{22}{36}\right) = 1.8$$

$$\text{Revised } K_x = 2.05 \quad \text{(close to previous value)}$$

Using previously computed values for F_a and F'_e,

$$P = 1.0(37.5) = 37.5 \text{ kips}$$

$$\frac{f_a}{F_a} = \frac{37.5/24.7}{17.0} = 0.09 < 0.15$$

AISC Formula (1.6.2) may be used:

$$\frac{f_a}{F_a} + \frac{f_b}{F_b} = 0.09 + \frac{21.5}{24} = 0.09 + 0.90 = 0.99 < 1.0 \qquad \text{OK}$$

Local buckling, AISC–1.5.1.4.1(4.):

$$\frac{d}{t} = \frac{640}{\sqrt{F_y}}\left(1 - \frac{3.74f_a}{F_y}\right) = \frac{640}{\sqrt{36}}[1 - 3.74(0.09)] = 70.8 > 51.3 \qquad \text{OK}$$

(c) Lateral bracing, W24×84:

$$L_c = \frac{76b_f}{\sqrt{F_y}} = 12.7b_f = 12.7(9.020)\tfrac{1}{12} = 9.5 \text{ ft}$$

or

$$\frac{20,000}{(d/A_f)F_y} = \frac{20,000}{3.47(36)12} > 9.5 \text{ ft}; \qquad L_c = 9.5 \text{ ft}$$

Since lateral support is provided at 5-ft intervals on the column and at 6-ft intervals on the girder, lateral-torsional buckling is adequately prevented.

(d) Check shear at the corner connection. Using Eq. 13.8.5, as discussed in AISC Commentary–1.5.1.2,

$$\text{Required } t_w = \frac{2.63M}{A_{bc}F_y} = \frac{2.63(351.5)12}{(24.10)^2(36)} = 0.53 \text{ in.} > t_w = 0.47 \text{ in.}$$

Diagonal stiffeners are required.

(e) Since $f_a/F_a < 0.15$, there is no reason to question the overall stability of the frame. Use of AISC Formula (1.6–1a) involving the magnification factor $C_m/(1-f_a/F'_e)$ would be the appropriate check if one is desired.

(f) Conclusions. <u>Use W24×84 section</u> which compares with W24×76 used in plastic design method. The greater design time required in doing an elastic frame analysis (with and without wind) and the somewhat arbitrary provision for reducing negative moment 10%, leave the conclusion that plastic design is the preferred method for designing one-story (and probably two-story) frames.

Example 15.4.2

Redesign the rectangular frame of Example 15.4.1 (see Fig. 15.4.1a and b) by the working stress method using a column no deeper than 14.5 in. overall (compares with plastic design Example 15.3.2). A36 steel.

SOLUTION

(a) Determination of moments. There is no direct process for this except to estimate from experience the relative moments of inertia of the column and girder. For this problem one might examine a W14 section which has a depth not exceeding 14.5 in., say the W14×120 ($I_c = 1380$), and compare it to the section used when $I_c = I_g$, W24×76 ($I_g = 2100$). Of course, one would not have known that value without doing an elastic analysis with $I_c = I_g$, as in Example 15.4.1. From the above, it seems the ratio I_g/I_c will be at least 2.0.

Assume $I_g/I_c = 4$ so as to compare results with plastic design; the elastic moments are given in Fig. 15.4.2.

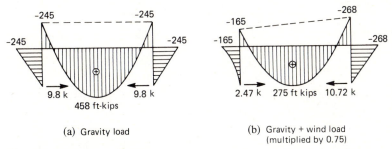

(a) Gravity load   (b) Gravity + wind load (multiplied by 0.75)

Fig. 15.4.2 Frame of Example 15.4.2: $I_g = 4I_c$.

(b) Selection of section for girder. Clearly, gravity load without wind governs. No negative moment reduction may be made even if a "compact section" is used, because the positive moment is already the larger one.

$$\text{Assume } F_b = 0.66F_y = 24 \text{ ksi}$$

$$\text{Required } S = \frac{458(12)}{24} = 229.0 \text{ in.}^3$$

Try W27×94, $S = 243$ in.^{3} From text Appendix Table A3, this section is compact for local buckling; also

$$L_c = 10.5 \text{ ft} > \text{unbraced length} = 6 \text{ ft} \qquad \text{OK}$$

(c) Selection of beam-column section. The two loading cases are

$$P = 37.5 \text{ kips}, \qquad M = 245 \text{ ft-kips} \quad \text{(without wind)}$$
$$P = 32.3 \text{ kips} \qquad M = 268 \text{ ft-kips} \quad \text{(with wind)}$$

Using Eq. 12.11.7, neglecting F_a/F_b for preliminary selection,

$$P_{EQ} \approx P + B_x M$$
$$P_{EQ} = 37.5 + 0.194(245)12 = 37.5 + 570 = 608 \text{ kips}$$

or

$$P_{EQ} = 32.3 + 0.194(268)12 = 32.8 + 622 = 657 \text{ kips}$$

Thus wind loading governs. Select preliminary section from AISC Manual Column Load Tables, using effective weak-axis unbraced length,

$$\text{Assume } K_x \approx 1.8^*$$
$$\text{Equivalent } K_y L_y = \frac{K_x L_x}{r_x/r_y} = \frac{1.8(25.0)}{2.45} = 18.4 \text{ ft}$$

Estimate W14×109:

$$G_A = 10.0 \text{ (hinged)}; \qquad G_B = \frac{1240/25}{3270/75} = 1.14$$
$$\text{Find } K_x = 1.9$$

$$\frac{K_x L}{r_x} = \frac{1.9(25)12}{6.22} = 91.6; \qquad F_a = 14.0 \text{ ksi}$$

if "compact section," $F_b = 0.66 F_y = 24$ ksi

$$\frac{F_a}{F_b} = \frac{14.03}{24} = 0.585$$

$$\text{Revised } P_{EQ} = 32.3 + 0.585(624) = 32.3 + 365 = 397 \text{ kips}$$

For Equivalent $K_y L_y \approx 18$ ft, try W14×90:

$$G_A = 10.0 \text{ (hinged)}; \qquad G_B = \frac{999/25}{3270/75} = 0.92; \qquad K_x = 1.85$$

$$\frac{K_x L}{r_x} = \frac{1.85(25)12}{6.14} = 90.4 > \frac{K_y L_y}{r_y} = \frac{1.0(5)12}{r_y}$$

$$F_a = 14.2 \text{ ksi}; \qquad F_e' = 18.3 \text{ ksi}$$

---

* Theoretically should be ≥ 2.0 if frictionless hinge exists at base ($G = \infty$). Using $G = 10.0$ assumes some restraint, in which case K may be less than 2.0.

Since $F'_e < 0.60F_y$, there is no adjustment of G_B for inelastic behavior (see Sec. 14.3).

$$f_a = \frac{P}{A} = \frac{32.3}{26.5} = 1.22 \text{ ksi}$$

$$\frac{f_a}{F_a} = \frac{1.22}{14.2} = 0.086 < 0.15 \qquad \begin{array}{l} \text{AISC Formula (1.6–2)} \\ \text{may be used} \end{array}$$

$$f_b = \frac{268(12)}{143} = 22.5 \text{ ksi}$$

AISC Formula (1.6–2):

$$\frac{f_a}{F_a} + \frac{f_b}{F_b} = \frac{1.22}{14.2} + \frac{22.5}{24} = 0.09 + 0.94 = 1.03 > 1$$

The W14×90 is within 3% of required strength criterion and may be considered acceptable. If not, use the next heavier section, W14×99, which is also a "compact section." AISC Formula (1.6–2) gives 0.93 < 1.0.

Use W14×99. This compares with W14×74 using plastic design (Example 15.3.2).

The check for required diagonal stiffeners should be made as in other examples. The lateral support spacing requirement for elastic design is much less severe than required in plastic design. Maximum unbraced length $L_c = 10.5$ ft on the girder and $L_c = 15.4$ ft on the column. In plastic design, lateral bracing was necessary at 6.5 ft on the girder and at 7.9 ft on the column.

Example 15.4.3

Redesign the gabled frame of Example 15.3.3 by the working stress method. A573 Grade 50 steel.

SOLUTION
Assuming the moment of inertia of all frame elements to be the same, the elastic analysis is performed and is shown in Fig. 15.4.3. It is seen that the wind loading case governs.

(a) Select the section to satisfy beam-column requirements.

$$P_{EQ} \approx P + B_x M$$

$$= 25.9 + 0.2(227)12 = 25.9 + 545 = 570 \text{ kips}$$

Loading is about 95% flexure; select as a beam. If the section is "compact," the negative moment may be reduced to 90% of its elastic

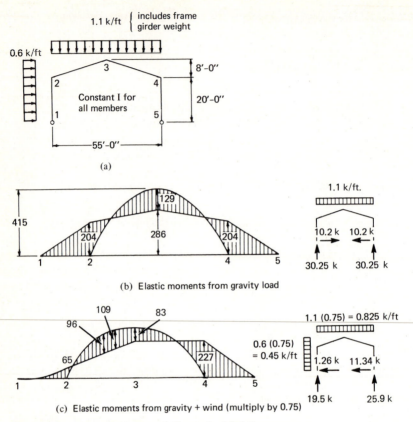

(a)

(b) Elastic moments from gravity load

(c) Elastic moments from gravity + wind (multiply by 0.75)

Fig. 15.4.3 Gabled frame of Example 15.4.3.

value:

$$-M = 0.9(227) = 204 \text{ ft-kips}$$

$$\text{Required } S = \frac{(204)12}{33(0.95)} = 78 \text{ in.}^3$$

where $F_b = 0.66F_y = 33$ ksi and 0.95 accounts for the percentage of capacity required for flexure. The increased positive moment is about 130, well below the negative moment of 204 used for design, and thus may be ignored.

Try W16×50, which is a "compact section" for $F_y = 50$ ksi according to text Appendix Table A3, and $L_c = 6.3$ ft.

$$f_b = \frac{204(12)}{81.0} = 30.2 \text{ ksi} < 33 \text{ ksi} \qquad \text{OK}$$

Using Fig. 15.3.5 with $f/h = 0.4$ and $L/h = 2.75$, find $K_x \approx 2.25$ for partially restrained bases.

$$\frac{K_x L}{r_x} = \frac{2.25(20)(12)}{6.68} = 81; \qquad F_a = 18.8 \text{ ksi}$$

$$f_a = \frac{25.9}{14.7} = 1.76 \text{ ksi}$$

$$\frac{f_a}{F_a} = \frac{1.76}{18.8} = 0.094 < 0.15 \qquad \text{May use AISC Formula (1.6--2)}$$

$$\frac{f_a}{F_a} + \frac{f_b}{F_b} = 0.094 + \frac{30.2}{33} = 0.09 + 0.92 = 1.01 \approx 1 \qquad \text{OK}$$

Again, shear must be checked in corner regions to ascertain whether diagonal stiffeners are needed. Lateral support is necessary at the top of the column and at intervals of 6'-4" ($L_c = 6.3$ ft).

Use W16×50, $F_y = 50$ ksi.

Plastic design was able to use a W16×45, but lateral bracing was required every $3\frac{1}{2}$ ft.

Example 15.4.4

Select W sections for the column of the frame of Fig. 15.4.4a, containing rigid joints at the tops of columns *A* and *D*. The tops of columns *B* and *C* are to be considered pinned. The bases of columns *A* and *D* are restrained such that $G = (\Sigma I/L)_{\text{col}}/(\Sigma I/L)_{\text{beam}} = 3.5$. The structure is braced in the direction perpendicular to the given frame, with each column braced at the top and bottom. Use A36 steel. (This example illustrates the unbraced frame concept discussed in the latter part of Sec. 14.5.)

SOLUTION

(a) Design of columns *B* and *C*. For these columns, $K = 1.0$ since they are treated as pinned-end within a braced system. The bracing is to be provided by the rigid framing and the design of columns *A* and *D*.

$$K_y L_y = 15 \text{ ft}; \qquad P = 85 \text{ kips}$$

Select from AISC Manual Column Load Tables,

$$\text{W8×28} \qquad P = 95 \text{ kips} > 85 \text{ kips required} \qquad \text{OK}$$

Check:

$$\frac{K_y L_y}{r_y} = \frac{1.0(15)(12)}{1.62} = 111; \qquad F_a = 11.5 \text{ ksi}$$

$$f_a = \frac{85}{8.25} = 10.3 \text{ ksi} < F_a = 11.5 \text{ ksi} \qquad \text{OK}$$

Use W8×28 for interior columns *B* and *C*.

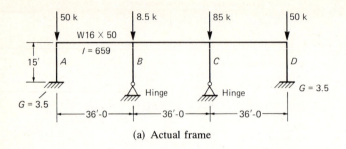

(a) Actual frame

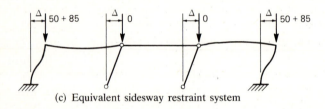

(b) Sidesway $P\Delta$ effects

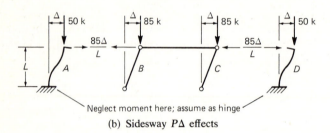

(c) Equivalent sidesway restraint system

Fig. 15.4.4 Frame of Example 15.4.4.

(b) Design exterior columns A and D. These columns provide bracing for the entire assemblage. Due to sidesway the interior columns B and C develop combined $P\Delta$ moments of $2(85)\Delta = 170\Delta$, as shown in Fig. 15.4.4b. To stabilize the system requires development of a horizontal bracing force at the top of the frame of $170\Delta/L$. Since *two* identical exterior columns are intended to provide the bracing, each exterior column must contribute $85\Delta/L$. Thus on columns A and D the total sidesway moment is $(85\Delta/L)(L)+50\Delta = 135\Delta$, neglecting end moment that may be developed by the partial restraint at the column bases. In other words, the extra bracing force required is the same as if the member itself were carrying 135 kips. Thus when the exterior columns are designed for 135 kips they will be capable of bracing the entire system.

Assume $K_x = 2$.

$$K_x L_x = 2(15) = 30 \text{ ft}$$

For use of AISC Manual Column Load Tables, since they are based on

$K_y L_y$,

$$\text{Equivalent } K_y L_y \text{ for W10} = \frac{30}{2.16} = 13.9 \text{ ft}; \qquad \text{Try W10} \times 33$$

$$\text{Equivalent } K_y L_y \text{ for W12} = \frac{30}{3.4} = 8.8 \text{ ft}; \qquad \text{Try W12} \times 30$$

Check the W12×30.

For the y–y axis (weak-axis buckling), the actual load of 50 kips must be carried.

$$\frac{K_y L_y}{r_y} = \frac{1.0(15)(12)}{1.52} = 118; \qquad F_a = 10.5 \text{ ksi}$$

$$f_a = \frac{50}{8.79} = 5.7 \text{ ksi} < F_a \qquad \text{OK}$$

For the x–x axis, the end restraint must be evaluated:

$$G_{\text{top}} = \frac{\Sigma(I/L)_{\text{col}}}{\Sigma(I/L)_{\text{beam}}} = \frac{238/15}{659/36} = 0.87$$

To account for the far end of the girder hinged (see Fig. 15.4.4c), multiply G by 2.0, as discussed in Sec. 14.3.

$$G_{\text{top}} = 0.87(2.0) = 1.74$$

$$G_{\text{bottom}} = 3.5 \quad \text{(given)}$$

From the alignment chart, Fig. 14.3.1, find $K_x = 1.71$.

$$\frac{K_x L_x}{r_x} = \frac{1.71(15)12}{5.21} = 59; \qquad F_a = 17.5 \text{ ksi}$$

$$f_a = \frac{P}{A} = \frac{135}{8.79} = 15.4 \text{ ksi} < F_a \qquad \text{OK}$$

The section is larger than necessary but the next lighter section, W12×26, is slightly overstressed.

Use W12×30 for exterior column.

15.5 MULTISTORY FRAMES

The design of multistory frames, either braced or unbraced, is outside the scope of this text. _Braced_ multistory frames have been designed by the working stress method since the earliest design of steel structures. Plastic design of multistory braced frames was first accepted by AISC with the 1969 AISC Specification. The theoretical and experimental background, developed primarily at Lehigh University, was first presented to the engineering profession in 1965 [9]. Additional background information may be found in the work of Driscoll and Beedle [10], Lu [11], Williams and Galambos [12], Goldberg [13], and Yura and Lu [14]. The design procedure is well-described in *Plastic Design of Braced Multistory Steel Frames* [15].

Regarding the *unbraced* multistory frame, design procedures are continuously being developed. The 1978 AISC Specification (AISC–2.3) for plastic design prescribes no maximum limit for number of stories. It does say that "for unbraced multistory frames, the frame stability effect should be included directly in the calculations for maximum strength." The *SSRC Guide* [16, Chap. 15] contains excellent treatment of methods to evaluate the frame stability of multistory frames. Early background on unbraced frame stability is available in the Lehigh Lecture Notes [9], Driscoll and Beedle [10], and Daniels [17]. Recent developments in evaluating the stability of unbraced frames is available from Cheong-Siat-Moy [18, 19, 21] and Cheong-Siat-Moy and Lu [20].

SELECTED REFERENCES

1. *Plastic Design in Steel.* New York: American Institute of Steel Construction, Inc., 1959.
2. Lynn S. Beedle, *Plastic Design of Steel Frames.* New York: John Wiley & Sons, Inc., 1958.
3. C. E. Massonnet and M. A. Save, *Plastic Analysis and Design, Vol. I, Beams and Frames.* Waltham, Mass.: Blaisdell Publishing Company, 1965.
4. Lynn S. Beedle, "Plastic Strength of Steel Frames," *Proceedings ASCE*, 81, Paper No. 764 (August 1955).
5. Edward R. Estes, Jr., "Design of Multi-span Rigid Frames by Plastic Analysis," *Proceedings of National Engineering Conference*, AISC, 1955, 27–39.
6. Le-Wu Lu, "Effective Length of Columns in Gabled Frames," *Engineering Journal*, AISC, 2, 1 (January 1965), 6–7.
7. *Steel Gables and Arches.* New York: American Institute of Steel Construction, Inc., 1963.
8. John D. Griffiths, *Single Span Rigid Frames in Steel.* New York: American Institute of Steel Construction, Inc., 1948.
9. George C. Driscoll, Jr., et al., *Plastic Design of Multi-Story Frames*, Lecture Notes, Fritz Engineering Laboratory Report No. 273.20, Lehigh University, Bethlehem, Pa., 1965.
10. George C. Driscoll and Lynn S. Beedle, "Research in Plastic Design of Multistory Frames," *Engineering Journal*, AISC, 1, 3 (July 1964), 92–100.
11. Le-Wu Lu, "Design of Braced Multi-Story Frames by the Plastic Method," *Engineering Journal*, AISC, 4, 1 (January 1967), 1–9.
12. James B. Williams and Theodore V. Galambos, "Economic Study of a Braced Multi-Story Steel Frame," *Engineering Journal*, AISC, 5, 1 (January 1968), 2–11.
13. John E. Goldberg, "Lateral Buckling of Braced Multistory Frames," *Journal of Structural Division*, ASCE, 94, ST12 (December 1968), 2963–2983.

14. Joseph A. Yura and Le-Wu Lu, "Ultimate Load Tests of Braced Multistory Frames," *Journal of Structural Division*, ASCE, 95, ST10 (October 1969), 2243–2263.

15. *Plastic Design of Braced Multistory Steel Frames.* New York: American Iron and Steel Institute, 1968.

16. Bruce G. Johnston, ed., *Structural Stability Research Council, Guide to Stability Design Criteria for Metal Structures*, 3rd Ed. New York: John Wiley & Sons, Inc., 1976, Chap. 15.

17. J. Hartley Daniels, "A Plastic Method for Unbraced Frame Design," *Engineering Journal*, AISC, 3, 4 (October 1966), 141–149.

18. F. Cheong-Siat-Moy, "Stiffness Design of Unbraced Steel Frames," *Engineering Journal*, AISC, 13, 1 (First Quarter 1976), 8–10.

19. Francois Cheong-Siat-Moy, "Multistory Frame Design Using Story Stiffness Concept," *Journal of Structural Division*, ASCE, 102, ST6 (June 1976), 1197–1212.

20. F. Cheong-Siat-Moy and Le-Wu Lu, "Stiffness and Strength Design of Multistory Frames," *Publications*, International Association for Bridge and Structural Engineering, 36-II, 1976, 31–47.

21. Francois Cheong-Siat-Moy, "Frame Design Without Using Effective Column Length," *Journal of Structural Division*, ASCE, 104, ST1 (January 1978), 23–33.

PROBLEMS

All problems are to be done in accordance with latest AISC Specification, unless otherwise indicated, and all loads given are service loads.

15.1. Design an unbraced rectangular frame of 60-ft span as shown in the accompanying figure, to carry uniform gravity dead and live loads of 1.5 and 2.5 kips/ft, respectively, not including the weight of the girder. Assume the wind may be approximated by 0.8 kips/ft acting uniformly over the 18-ft height. Lateral bracing occurs every 4 ft on the girder and 4.5 ft on the columns. Include design of the knee region at the top of the column. Use A36 steel. (a) Use plastic design; (b) use working stress method.

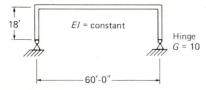

Prob. 15.1

15.2. Design for the conditions of Prob. 15.1 except assume the moment of inertia for the beam to be three times that of the column.

15.3. Design the gabled frame of the accompanying figure, to carry uniform gravity dead and live loads of 0.7 and 0.8 kips/ft, respectively. Assume the loading is per foot of horizontal projection. The wind load is approximated by 0.85 kips/ft acting uniformly on the vertical projection over the 24-ft height. Assume lateral bracing at the knees (points 2 and 4), at the eave (point 3), and approximately every 4 ft otherwise. Use A36 steel.

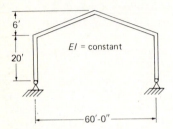

Prob. 15.3

15.4. Design the unbraced rectangular frame of the accompanying figure to carry uniform gravity dead and live loads of 1.4 and 2.2 kips/ft, respectively. Assume the wind may be approximated by 0.8 kips/ft acting uniformly over the height of the column. Assume the structure is braced perpendicular to the plane of the frame so that the columns are braced top and bottom in their weak (y–y axis) bending direction. Use A36 steel.

(a) Design for all joints at the tops of columns to be rigid (i.e., moment resisting).

(b) Design as in Example 15.4.4 using simple connections (assumed as pinned) at the tops of columns BG, CH, and DI, with rigid moment resisting connections only at A and E.

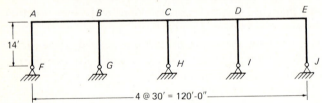

Prob. 15.4

16

Composite Steel-Concrete Construction

16.1 HISTORICAL BACKGROUND

Steel framing supporting cast-in-place reinforced concrete slab construction was historically designed on the assumption that the concrete slab acts independently of the steel in resisting loads. No consideration was given to the composite effect of the steel and concrete acting together. This neglect was justified on the basis that the bond between the concrete floor or deck and the top of the steel beam could not be depended upon. However, with the advent of welding, it became practical to provide mechanical shear connectors to resist the horizontal shear which develops during bending.

Steel beams encased in concrete were widely used from the early 1900s until the development of lightweight materials for fire protection in the past 25 years. Some such beams were designed compositely and some were not. In the early 1930s bridge construction began to use composite sections. Not until the early 1960s was it economical to use composite construction for buildings. However, current practice (1979) utilizes composite action in nearly all situations where concrete and steel are in contact, both on bridges and buildings.

An early study of composite structural action by MacKay et al. [1]

Stud shear connectors on flanges of bridge girders to be embedded in the concrete slab in order to make steel section and concrete slab act as a unit (i.e., compositely). (Courtesy Gregory Industries, Inc., Lorain, Ohio)

involved steel beams encased in concrete. It was noted that such encased beams supporting a monolithically cast reinforced concrete slab had good interaction between the steel beam and the slab. In these early studies of encased beams, bond was depended upon to provide the interaction between steel and concrete. Composite beams, including encased beams as well as slabs on I-shaped beams, exhibited sufficient reserve strength so that Caughey [2] in 1929 recommended that they be designed on the basis of a homogeneous section wherein the concrete area is transformed into an equivalent area of steel.

Since stresses in a wide slab supported on a steel beam are not uniform across the slab width, the ordinary flexure formula, $f = Mc/I$, does not apply. Similarly to T-sections entirely of reinforced concrete, an equivalent width of the wide slab is used such that the flexure formula may be applied to obtain the correct moment capacity. Theoretical studies to determine the correct effective width have been carried out by von Karman [3] and Reissner [4], and an excellent summary of the subject has been presented by Brendel [5].

Viest [6], in his 1960 review of research, notes that the important factor in composite action is that the bond between concrete and steel remain unbroken. As designers began to place slabs on top of supporting steel beams, investigators began to study the behavior of mechanical shear connectors. The shear connectors provided the interaction necessary for the composite action between slab and steel I-shaped beams which had previously been supplied by bond for the fully encased beams.

Studies of mechanical shear connectors began in the 1930s with the work of Voellmy reported by Ros [7]. Since that time numerous studies have been conducted on a wide variety of mechanical shear connectors, with the work summarized by Viest [6] and in the State-of-the-Art Survey [8]. The latter contains an excellent Bibliography.

For overall study of composite construction, the reader is referred to the text by Cook [9] and the *Handbook of Composite Construction Engineering* [10].

16.2 COMPOSITE ACTION

Composite action is developed when two load-carrying structural members such as a concrete floor system and the supporting steel beams are integrally connected and deflect as a single unit. Typical examples of composite cross sections are shown in Fig. 16.2.1. The extent to which composite action is developed depends on the provisions made to insure a single linear strain from the top of the concrete slab to the bottom of the steel section.

In developing the concept of composite behavior, consider first the noncomposite beam of Fig. 16.2.2a, wherein if friction between the slab and beam is neglected the beam and slab each carry separately a part of the load. This is further shown in Fig. 16.2.3a. When the slab deforms under vertical load, its lower surface is in tension and elongates; while the upper surface of the beam is in compression and shortens. Thus a discontinuity will occur at the plane of contact. Since friction is neglected, only vertical internal forces act between the slab and beam.

When a system acts compositely (Fig. 16.2.2b and 16.2.3c) no relative slippage occurs between the slab and beam. Horizontal forces (shears) are developed that act on the lower surface of the slab to compress and shorten it, while simultaneously they act on the upper surface of the beam to elongate it.

By an examination of the strain distribution that occurs when there is no interaction between the concrete slab and the steel beam (Fig. 16.2.3a), it is seen that the total resisting moment is equal to

$$\sum M = M_{\text{slab}} + M_{\text{beam}} \tag{16.2.1}$$

It is noted that for this case there are two neutral axes; one at the center of gravity of the slab and the other at the center of gravity of the beam. The horizontal slippage resulting from the bottom of the slab in tension and the top of the beam in compression is also indicated.

Consider next the case where only partial interaction is present, Fig. 16.2.3b. The neutral axis of the slab is closer to the beam and that of the beam closer to the slab. Due to the partial interaction, the horizontal slippage has now decreased. The result of the partial interaction is the partial development of the compressive and tensile forces C' and T', the maximum capacities of the concrete slab and steel beam, respectively.

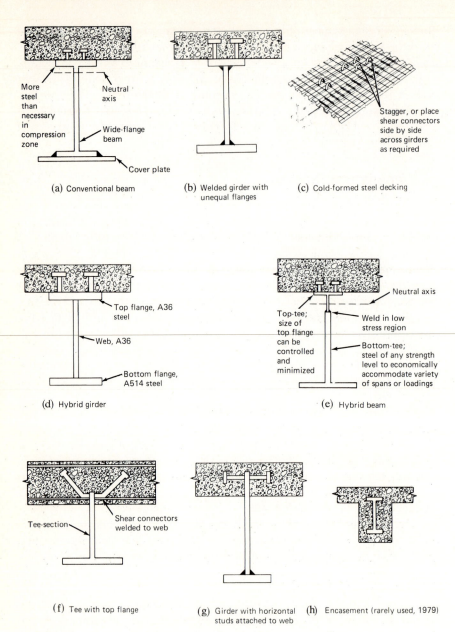

Fig. 16.2.1   Various types of composite steel-concrete sections.

The resisting moment of the section is now increased by the amount $T'e'$ or $C'e'$.

When complete interaction between the slab and the beam is developed, no slippage occurs and the resulting strain diagram is shown in Fig. 16.2.3c. Under this condition a single neutral axis occurs which lies

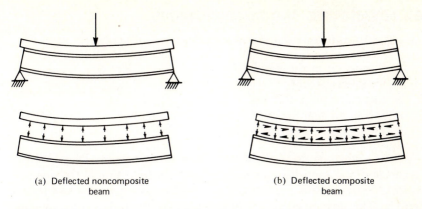

(a) Deflected noncomposite
beam

(b) Deflected composite
beam

Fig. 16.2.2   Comparison of deflected beams with and without composite action.

below that of the slab and above that of the beam. In addition, the compressive and tensile forces C'' and T'' are larger than the C' and T' existing with partial interaction. The resisting moment of the fully developed composite section then becomes

$$\sum M = T''e'' \quad \text{or} \quad C''e'' \qquad (16.2.2)$$

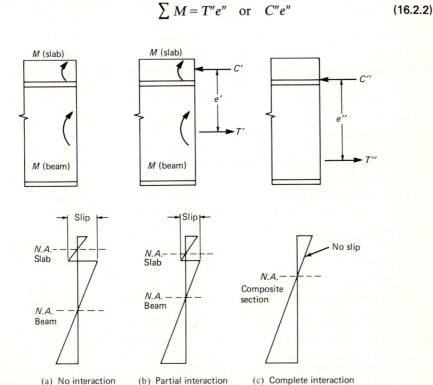

(a) No interaction

(b) Partial interaction

(c) Complete interaction

Fig. 16.2.3   Strain variation in composite beams.

16.3 ADVANTAGES AND DISADVANTAGES

The basic advantages resulting from composite design are

1. Reduction in the weight of steel
2. Shallower steel beams
3. Increased floor stiffness
4. Increased span length for a given member
5. Increased overload capacity

A weight savings in steel of 20 to 30% is often possible by taking full advantage of a composite system. Such a weight reduction in the supporting steel beams usually permits the use of a shallower as well as a lighter member. This advantage may reduce the height of a multistoried building significantly so as to provide savings in other building materials such as outside walls and stairways.

The stiffness of a composite floor is substantially greater than that of a concrete floor with its supporting beams acting independently. Normally the concrete slab acts as a one-way plate spanning between the supporting beams. In composite design, an additional use is made of the slab by its action in a direction parallel to and in combination with the supporting steel beams. The net effect is to greatly increase the moment of inertia of the floor system in the direction of the steel beams. The increased stiffness considerably reduces the live-load deflections and, if shoring is provided during construction, also reduces dead-load deflections. Assuming full composite action, the ultimate strength of the section greatly exceeds the sum of the strengths of the slab and the beam considered separately, providing high overload capacity.

The overall economy of using composite construction when considering total building costs appears to be good and is steadily improving. The development of new combinations of floor systems is continually expanding and the use of high-strength steels and hybrid members (see Fig. 16.2.1d, e) can be expected to produce further economies. In addition, composite wall systems and composite columns are beginning to be used in buildings.

While there are no major disadvantages in composite construction, there are some limitations which should be recognized. These are:

1. Effect of continuity
2. Long-term deflections

At the present time (1979), only the portion of the concrete slab acting in compression is considered to be effective. In the case of continuous beams, the advantage of composite action is reduced in the area of negative bending moment, with only bar reinforcement to provide continuity of composite action.

The matter of long-term deflections might be important if the composite section is resisting a substantial portion of the dead loads or if the live loads are of a long duration. This problem, however, can be minimized by assuming a reduced effective width of the slab or by assuming an increased modulus of elasticity ratio n. This is discussed more fully in Sec. 16.12.

16.4 EFFECTIVE WIDTH

In order to compute in a practical way the section properties of a composite section it is necessary to utilize the concept of effective width. Referring to Fig. 16.4.1, consider the composite section under stress in which the slab is infinitely wide. The intensity of the extreme fiber compressive stress, σ_x, maximum over the steel beam, decreases non-linearly as the distance from the supporting beam increases.

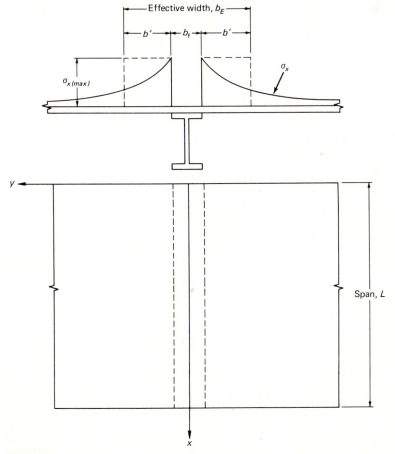

Fig. 16.4.1 Nonuniform distribution of compressive stress σ_x and effective width b_E.

The effective width of a flange for a composite member may be taken as

$$b_E = b_f + 2b' \tag{16.4.1}$$

where $2b'$ times the maximum stress, $\sigma_{x\,max}$ is equal to the area under the curves for σ_x. Various investigators, including Timoshenko [11] and von Kármán [3] have derived expressions for the effective width of homogeneous beams with wide flanges; and Johnson and Lewis [12] have shown the expressions to be valid also for beams in which the flange and stem are of different materials.

The analysis for effective width involves theory of elasticity applied to plates, using an infinitely long continuous beam on equidistant supports, with an infinitely wide flange having a small thickness compared to the beam depth. The total compression carried by the equivalent system is the same as that carried by the actual system. The value of b' depends on the span length and type of loading. Johnson and Lewis [12] have shown that for loading which produces bending moment having a half-sine wave shape, the effective width is

$$b_E = b_f + \frac{2L}{\pi(3 + 2\mu - \mu^2)} \tag{16.4.2}$$

where L = span length of beam
$\quad b_f$ = steel beam flange width
$\quad \mu$ = Poisson's ratio for the slab
Assuming $\mu = 0.2$ for concrete,

$$b_E = b_f + \frac{2L}{\pi(3 + 2(0.2) - (0.2)^2)} = b_f + 0.196L \tag{16.4.3}$$

As a simplification for design purposes, AISC–1.11.1 has adopted the same method of computing effective flange widths as the ACI Code [13] does for reinforced concrete beams. Referring to Fig. 16.4.2, the maximum value of the effective width b_E permitted by the AISC–1.11.1 is

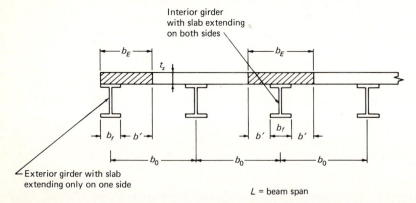

Fig. 16.4.2 Dimensions governing effective width b_E on composite steel-concrete beams.

the least value computed by the following relations:

1. For an interior girder with slab extending on both sides of girder:

$$b_E \leq L/4 \tag{16.4.4a}$$

$$b_E \leq b_0 \quad \text{(for equal beam spacing)} \tag{16.4.4b}$$

$$b_E \leq b_f + 16t_s \tag{16.4.4c}$$

2. For an exterior girder with slab extending only on one side:

$$b_E \leq L/12 + b_f \tag{16.4.5a}$$

$$b_E \leq \tfrac{1}{2}(b_0 + b_f) \tag{16.4.5b}$$

$$b_E \leq b_f + 6t_s \tag{16.4.5c}$$

Similarly, for highway bridge design [14], the effective width according to AASHTO–1.7.48 is identical with AISC except Eq. 16.4.4c for an interior girder is replaced by

$$b_E \leq 12t_s \tag{16.4.6}$$

and for an exterior girder, Eqs. 16.4.5a and c are replaced by

$$b_E \leq L/12 \tag{16.4.7a}$$

$$b_E \leq 6t_s \tag{16.4.7b}$$

16.5 COMPUTATION OF SECTION PROPERTIES

The section properties of a composite section can be computed by the transformed area method. In contrast to reinforced concrete design, where the steel area is transformed into an equivalent concrete area, the concrete is transformed into equivalent steel. As a result, the concrete area is reduced by using a slab width equal to b_E/n, where n is the ratio of steel modulus of elasticity, E_s, to concrete modulus of elasticity, E_c.

Modular Ratio, n

The modulus of elasticity of concrete in psi may be taken [13] as

$$E_c = w^{1.5}33\sqrt{f_c'} \tag{16.5.1*}$$

where w is weight of concrete in lb/cu ft (pcf) and f_c' is in psi. For normal weight concrete, weighing approximately 145 pcf, the value may be taken according to the ACI Code [13], as

$$E_c = 57,000\sqrt{f_c'} \tag{16.5.2*}$$

* For SI units, with E_c in MPa,

$$E_c = w^{1.5}(0.043)\sqrt{f_c'} \tag{16.5.1}$$

where w is in kg/m³ and f_c' is in MPa.

For normal weight (mass) concrete, the value may be taken as

$$E_c = 4730\sqrt{f_c'} \quad \text{with } f_c' \text{ in MPa} \tag{16.5.2}$$

Table 16.5.1 Values of Modulus of Elasticity for Concrete (using $E_c = w^{1.5}(33)\sqrt{f_c'}$ for normal weight concrete weighing 145 pcf)

f_c' (psi)	E_c (psi)	f_c' (MPa)	E_c (MPa)
3000	3,150,000	21*	21 700
3500	3,400,000	24	23 200
4000	3,640,000	28	25 000
4500	3,860,000	31	26 300
5000	4,070,000	35	28 000

* These SI values are rounded values approximating strengths in U.S. customary units. For normal weight (mass) concrete,

$$E_c = 4730\sqrt{f_c'} \qquad \text{for } f_c' \text{ and } E_c \text{ in MPa}$$

Values of modulus of elasticity for various concrete strengths appear in Table 16.5.1.

Example 16.5.1

Compute the modular ratio, n, for normal weight concrete (145 pcf) with a 28-day compressive strength f_c' of 3000 psi.

SOLUTION
From Eq. 16.5.1,

$$E_c = \left[(145)^{1.5}33\sqrt{3000}\right]\tfrac{1}{1000} = 3170 \text{ ksi}$$

which gives

$$n = \frac{E_s}{E_c} = \frac{29,000}{3170} = 9.15 \approx 9$$

The minimum value for n permitted by the ACI Code and the AASHTO Specifications is 6. For practical design purposes, the value of n from Table 16.5.2 is suggested, although the ACI Code states the modular ratio may be taken as the nearest whole number.

Table 16.5.2 Practical Design Values for Modular Ratio n

f_c' (psi)	Modular Ratio $n = E_s/E_c$	f_c' (MPa)
3000	9	21
3500	$8\frac{1}{2}$	24
4000	8	28
4500	$7\frac{1}{2}$	31
5000	7	35
6000	$6\frac{1}{2}$	42

Effective Section Modulus

A complete beam may be considered as a steel member to which has been added a cover plate on the top flange. This "cover plate" being concrete is considered to be effective only when the top flange is in compression. In continuous beams, the concrete slab is usually ignored in regions of negative moment. If the neutral axis falls within the concrete slab, present practice is to consider only that portion of the concrete slab which is in compression.

AISC–1.11.2.2 permits reinforcement parallel to the steel beam and lying within the effective slab width to be included in computing properties of composite sections. These reinforcing bars usually make little difference to the composite section modulus and are frequently neglected.

Example 16.5.2

Compute the section properties of the composite section shown in Fig. 16.5.1 according to the AISC Specification assuming $f'_c = 3000$ psi and $n = 9$.

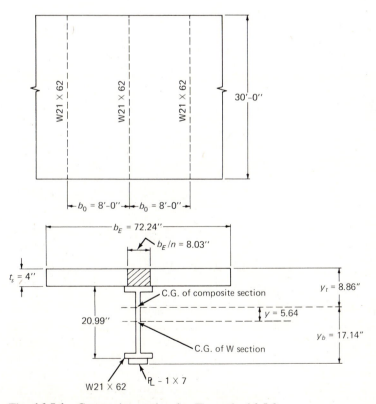

Fig. 16.5.1 Composite section for Example 16.5.2.

SOLUTION

First, determine effective width:

$$b_E = (0.25)(\text{span length}) = 0.25(30)12 = 90 \text{ in.}$$

$$b_E = b_0 = 8(12) = 96 \text{ in.}$$

$$b_E = b_f + 2(8)t_s = 8.24 + 64 = 72.24 \text{ in.} \quad \text{Controls}$$

The width of equivalent steel is $b_E/n = 8.03$ in. The computation of the moment of inertia I_x about the center of gravity of the W21×62 is shown, as follows:

Element	Transformed Area A (sq in.)	Moment Arm from Centroid y (in.)	Ay (in.³)	Ay^2 (in.⁴)	I_0 (in.⁴)
Slab	32.1	+12.495	+401	5012	43
W21×62	18.3	0	0	0	1330
Cover plate	7.0	−10.995	−77	846	
	57.4		+324	5858	1373

$$I_x = I_0 + Ay^2 = 1373 + 5858 = 7231 \text{ in.}^4$$

$$\bar{y} = \frac{+324}{57.4} = +5.64 \text{ in.}$$

$$I = I_x - A\bar{y}^2 = 7231 - 57.4(5.64)^2 = 5402 \text{ in.}^4$$

$$y_t = 10.50 - 5.64 + 4.0 = 8.86 \text{ in.}$$

$$y_b = 10.50 + 5.64 + 1.0 = 17.14 \text{ in.}$$

The section modulus with respect to the top fiber, S_t, is

$$S_t = I/y_t = 5402/8.86 = 610 \text{ in.}^3$$

and the section modulus with respect to the bottom fiber, S_b, is

$$S_b = I/y_b = 5402/17.14 = 315 \text{ in.}^3$$

16.6 SERVICE LOAD STRESSES WITH AND WITHOUT SHORING

The actual stresses that result due to a given loading on a composite member are dependent upon the manner of construction.

The simplest construction occurs when the steel beams are placed first and used to support the concrete slab formwork. In this case the steel beam acting noncompositely (i.e., by itself) supports the weight of the forms, the wet concrete, and its own weight. Once forms are removed and concrete has cured, the section will act compositely to resist all dead and live loads placed after the curing of concrete. Such construction is said to be *without temporary shoring* (i.e., unshored).

Alternatively, to reduce the service load stresses, the steel beams may be supported on temporary shoring; in which case, the steel beam, forms, and wet concrete, are carried by the shores. After curing of the concrete, the shores are removed and the section acts compositely to resist all loads. This system is called *shored* construction.

The following example illustrates the difference in service load stresses under the two systems of construction.

Example 16.6.1

For the steel W21×62 with the 1 by 7-in. plate of Fig. 16.5.1, determine the service load stresses considering that (a) construction is without temporary shoring, and (b) construction uses temporary shores. The dead- and live-load moment M_L to be superimposed on the system after the concrete has cured is 560 ft-kips.

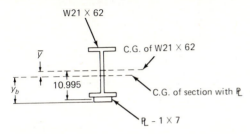

Fig. 16.6.1 Steel section for Example 16.6.1.

SOLUTION
The composite section properties as computed in Example 16.5.2 are

$$S_{top} = 610 \text{ in.}^3 \quad \text{(top of concrete)}$$

$$S_{bottom} = S_{tr}(\text{AISC–1978}) = 315 \text{ in.}^3 \quad \text{(bottom of steel)}$$

The noncomposite properties for the steel section alone (see Fig. 16.6.1) are computed as follows:

$$\bar{y} = \frac{7.0(10.995)}{7.0 + 18.3} = 3.04 \text{ in.}$$

$$y_b = 10.495 - 3.04 + 1.00 = 8.45 \text{ in.}$$

$$I = I_0(\text{W21} \times 62) + A_p y^2 - A\bar{y}^2$$
$$= 1330 + 7.0(10.995)^2 - 25.3(3.04)^2$$
$$= 1330 + 846 - 234 = 1942 \text{ in.}^4$$

$$S_{st} = \frac{1942}{13.55} = 143 \text{ in.}^3 \quad \text{(top)}$$

$$S_{sb} = \frac{1942}{8.45} = 230 \text{ in.}^3 \quad \text{(bottom)}$$

(a) *Without Temporary Shores.* Weight due to the concrete slab and steel beam,

$$w \text{ (concrete slab)}, \quad \frac{4}{12}(8)0.15 = 0.40$$

$$w \text{ (steel beam)}, \quad\quad\quad\quad = 0.06$$
$$\overline{}$$
$$0.46 \text{ kips/ft}$$

$$M_D \text{ (DL on noncomposite)} = \tfrac{1}{8}(0.46)(30)^2 = 51.8 \text{ ft-kips}$$

$$f_{\text{top}} = \frac{M_D}{S_{st} \text{ (steel section)}} = \frac{51.8(12)}{143} = 4.3 \text{ ksi}$$

$$f_{\text{bottom}} = \frac{M_D}{S_{sb} \text{ (steel section)}} = \frac{51.8(12)}{230} = 2.7 \text{ ksi}$$

The additional stresses after the concrete has cured are

$$f_{\text{top}} = \frac{M_L}{S_{\text{top}} \text{ (composite)}} = \frac{560(12)}{610(9)} = 1.22 \text{ ksi} \quad \text{(concrete stress)}$$

where the stress in the concrete is $1/n$ times the stress on equivalent steel (transformed section).

$$f_{\text{bottom}} = \frac{M_L}{S_{tr}} = \frac{560(12)}{315} = 21.3 \text{ ksi}$$

The total maximum tensile stress in the steel is

$$f = f(\text{noncomposite}) + f(\text{composite}) = 2.7 + 21.3 = 24.0 \text{ ksi}$$

(b) *With Temporary Shores.* Under this condition all loads are resisted by the composite section.

$$f_{\text{top}} = \frac{M_D + M_L}{S_{\text{top}} \text{ (composite)}} = \frac{(560 + 51.8)12}{610(9)} = 1.34 \text{ ksi on concrete}$$

$$f_{\text{bottom}} = \frac{M_D + M_L}{S_{tr}} = \frac{(560 + 51.8)12}{315} = 23.3 \text{ ksi}$$

Stress distributions for both with and without shores are given in Fig. 16.6.2. Since the dead load was small in this example use of shores gave insignificant reduction in service load stress. Where thicker slabs are used the dead load stresses may become as high as 30%, in which case using or not using shores will make a significant difference.

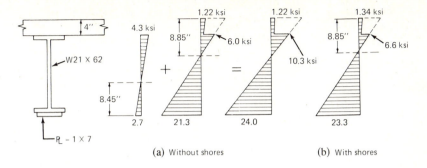

Fig. 16.6.2 Service load stresses for Example 16.6.1.

16.7 ULTIMATE STRENGTH OF FULLY COMPOSITE SECTIONS

The ultimate strength of a composite section is dependent upon the yield strength and section properties of the steel beam, the concrete slab strength, and the interaction capacity of the shear connectors joining the slab to the beam.

The provisions of AISC–1.11 are nearly all based on ultimate strength behavior, even though all relationships are adjusted to be in the service load range. These ultimate strength concepts were applied to design practice as recommended by the ASCE-ACI Joint Committee on Composite Construction [15], and further modified as a result of the work of Slutter and Driscoll [16].

The ultimate strength in terms of an ultimate moment capacity gives a clearer understanding of composite behavior as well as provides a more accurate measure of the true factor of safety. The true factor of safety is the ratio of the ultimate moment capacity to the actually applied moment. In both cases, whether the *slab* is termed "adequate" or "inadequate" as compared to the tensile yield capacity of the beam, the *connection* between the slab and beam is considered adequate in the following development. Complete shear transfer at the steel-concrete interface is assumed.

In determining the ultimate moment capacity, the concrete is assumed to take only compressive stress. Although concrete is able to sustain a limited amount of tensile stress, the tensile strength at the strains occurring during the development of the ultimate moment capacity is negligible.

The procedure for determining the ultimate moment capacity depends on whether the neutral axis occurs within the concrete slab or within the steel beam. If the neutral axis occurs within the slab, the slab is said to be adequate; i.e., the slab is capable of resisting the total compressive force. If the neutral axis falls within the steel beam, the slab

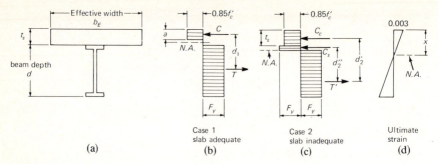

Fig. 16.7.1 Stress distribution at ultimate moment capacity.

is considered inadequate, i.e., the slab is able to resist only a portion of the compressive force, the remainder being taken by the steel beam. Figure 16.7.1 shows the stress distribution for these two cases.

Case 1—Slab Adequate

Referring to Fig. 16.7.1b and assuming the Whitney rectangular stress block* (uniform stress of $0.85f_c'$ acting over a depth a), the ultimate compressive force C is

$$C = 0.85f_c'ab_E \tag{16.7.1}$$

The ultimate tensile force T is the yield strength of the beam times its area:

$$T = A_s F_y \tag{16.7.2}$$

Equating the ultimate compressive force C to the ultimate tensile force T gives

$$a = \frac{A_s F_y}{0.85f_c'b_E} \tag{16.7.3}$$

According to the ACI-accepted [13, Section 10.2.7] rectangular stress block approach, the neutral axis distance x, as shown in Fig. 16.7.1d, equals $a/0.85$ for $f_c' = 4000$ psi. The ultimate moment capacity M_u becomes

$$M_u = Cd_1 \quad \text{or} \quad Td_1 \tag{16.7.4}$$

Since the slab is assumed adequate, it is capable of developing a compressive force equal to the full yield capacity of the steel beam. Expressing the

---

* For the development of the concept of replacing the true distribution of compressive stress by a rectangular stress distribution, see for example, Chu-Kia Wang and Charles G. Salmon, *Reinforced Concrete Design*, 3rd ed. (Harper & Row, New York, 1979, Chap. 3).

ultimate moment in terms of the steel force gives

$$M_u = A_s F_y \left(\frac{d}{2} + t_s - \frac{a}{2} \right)$$
(16.7.5)

The usual procedure is to determine the depth of the stress block a by Eq. 16.7.3, and if a is less than the slab thickness t_s, determine the ultimate moment capacity by Eq. 16.7.5.

Case 2—Slab Inadequate

If the depth a of the stress block as determined in Eq. 16.7.3 exceeds the slab thickness, the stress distribution will be as shown in Fig. 16.7.1c. The ultimate compressive force C_c in the slab is

$$C_c = 0.85 f'_c b_E t_s$$
(16.7.6)

The compressive force in the steel beam resulting from the portion of the beam above the neutral axis is shown in Fig. 16.7.1c as C_s.

The ultimate tensile force T' which is now less than $A_s f_y$ must equal the sum of the compressive forces:

$$T' = C_c + C_s$$
(16.7.7)

Also,

$$T' = A_s f_y - C_s$$
(16.7.8)

Equating Eqs. 16.7.7 and 16.7.8, C_s becomes

$$C_s = \frac{A_s F_y - C_c}{2}$$

or

$$C_s = \frac{A_s F_y - 0.85 f'_c b_E t_s}{2}$$
(16.7.9)

Considering the compressive forces C_c and C_s, the ultimate moment capacity M_u for Case 2 is

$$M_u = C_c d'_2 + C_s d''_2$$
(16.7.10)

the moment arms d'_2 and d''_2 are as shown in Fig. 16.7.1c.

Whenever the Case 2 situation occurs, the steel beam is assumed to accommodate plastic strain in both tension and compression at ultimate strength. Certainly, it is implied that such a steel section satisfy the requirements of a "compact section"; that is, it should have proportions that insure its ability to develop its plastic moment capacity. Little research has been performed on Case 2 situations because they rarely occur in practice.

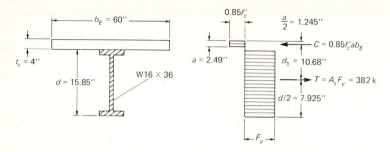

Fig. 16.7.2 Example 16.7.1.

Example 16.7.1

Determine the ultimate moment capacity of the composite section shown in Fig. 16.7.2. Assume A36 steel, $f'_c = 3000$ psi and $n = 9$.

SOLUTION

Referring to Fig. 16.7.2, determine the adequacy of the concrete slab. Assuming the slab is fully adequate; i.e., Case 1,

$$a = \frac{A_s F_y}{0.85 f'_c b_E} = \frac{10.6(36)}{0.85(3)60} = 2.49 \text{ in.} < t_s \qquad \text{OK}$$

$$C = 0.85 f'_c a b_E = 0.85(3)(2.49)(60) = 382 \text{ kips}$$

$$T = A_s F_y = 10.6(36) = 382 \text{ kips} \quad \text{(checks)}$$

$$\text{Arm } d_1 = \frac{d}{2} + t - \frac{a}{2} = 7.925 + 4.0 - 1.245 = 10.68 \text{ in.}$$

Ultimate composite moment capacity,

$$M_u = C d_1 = T d_1$$
$$= 382(10.68)\tfrac{1}{12} = 340 \text{ ft-kips}$$

Example 16.7.2

Determine the ultimate moment capacity of the composite section shown in Fig. 16.7.3. Assume A36 steel is used with $f'_c = 3000$ psi, and $n = 9$.

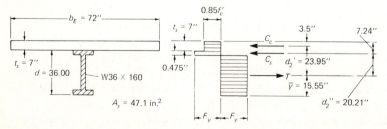

Fig. 16.7.3 Example 16.7.2.

SOLUTION

Referring to Fig. 16.7.3 determine whether or not the slab is adequate. Assuming the slab is adequate to balance tensile capacity of the steel section (i.e., Case 1),

$$a = \frac{A_s F_y}{0.85 f'_c b_E} = \frac{47.1(36)}{0.85(3)(72)} = 9.24 \text{ in.} > t_s = 7 \text{ in.} \qquad \text{NG}$$

Since the concrete slab is only 7 in. thick, the slab is inadequate to carry a compressive force equal to the tensile force that can be developed by the W36×160. Thus, Case 2 applies. Using Eq. 16.7.6,

$$C_c = 0.85 f'_c b_E t_s = 0.85(3)72(7) = 1285 \text{ kips}$$

Using Eq. 16.7.9,

$$C_s = \frac{A_s F_y - 0.85 f'_c b_E t_s}{2} = \frac{47.1(36) - 1285}{2} = 205 \text{ kips}$$

Assuming only the flange of the W36×160 ($b_f = 12.00$ in.) is in compression, the portion of the flange d_f to the neutral axis is

$$d_f = \frac{205}{36(12.00)} = 0.475 \text{ in.}$$

The location of the centroid of the tension portion of the steel beam from the bottom is

$$\bar{y} = \frac{47.1(18) - 0.475(12)35.76}{47.1 - 0.475(12)} = 15.55 \text{ in.}$$

Referring to Fig. 16.7.3, the ultimate composite moment capacity from Eq. 16.7.10 is

$$M_u = C_c d'_2 + C_s d''_2$$
$$= [1285(23.95) + 205(20.21)]/12 = 2910 \text{ ft-kips}$$

In the development of the ultimate strength of composite sections, it has been implicitly assumed that sufficient interaction between the concrete slab and the steel beam existed. Such interaction is usually assured by providing a sufficient number of "shear connectors." This matter is treated in Sec. 16.8.

The foregoing development and examples have emphasized the computation of ultimate moment capacity. Tests have verified that such capacities are achieved. Furthermore, whether during construction the steel beam is *shored* or *unshored*, the ultimate strength is identical. Even though service load stresses, as discussed in Sec. 16.6, are lower when temporary shoring is used to support the beam during construction than when such shoring is omitted, the same ultimate moment capacity is achieved. Current (1979) AISC design practice, as discussed in Sec. 16.8, uses this ultimate strength concept as the basis for its working stress method.

16.8 SHEAR CONNECTORS

The horizontal shear that develops between the concrete slab and the steel beam during loading must be resisted so the composite section acts monolithically. Although the bond developed between the slab and the steel beam may be significant, it cannot be depended upon to provide the required interaction. Neither can the frictional force developed between the slab and the steel beam.

Instead, mechanical shear connectors attached to the top of the beam must be provided. Typical shear connectors are shown in Fig. 16.8.1.

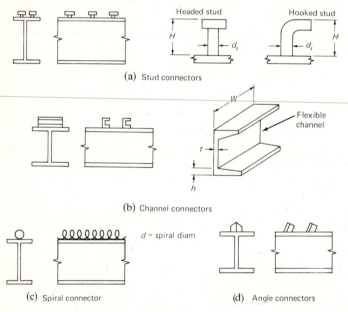

(a) Stud connectors

(b) Channel connectors

(c) Spiral connector   (d) Angle connectors

Fig. 16.8.1 Typical shear connectors.

Ideally, the shear connectors should be stiff enough to provide the complete interaction shown in Fig. 16.2.3c. This, however, would require that the stiffeners be infinitely rigid. Also, by referring to the shear diagram of a uniformly loaded beam as shown in Fig. 16.8.2, it would be inferred, theoretically at least, that more shear connectors would be required near the ends of the span than at the midspan. Consider the shear stress distribution of Fig. 16.8.2b wherein the stress v_1 must be developed by the connection between the slab and beam. Under service load the stress on the beam of Fig. 16.8.2 varies from zero at midspan to

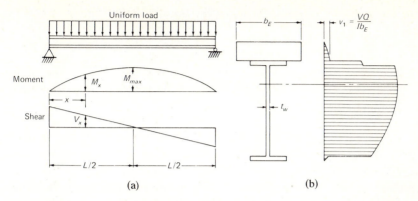

Fig. 16.8.2 Shear variation for uniform loading and distribution across a steel-concrete composite section.

a maximum at the support. Next, examine the equilibrium of an elemental slice of the beam, as in Fig. 16.8.3. The shear force per unit distance along the span is $dC/dx = v_1 b_E = VQ/I$. Thus if a given connector has an allowable capacity of q kips, the maximum spacing p to provide the required capacity is

$$p = \frac{q}{VQ/I} \qquad (16.8.1)$$

Composite design until recent years has used Eq. 16.8.1 to space connectors. AASHTO–1.7.48(E) requires using Eq. 16.8.1 to design for fatigue; then a check is required for ultimate strength [14].

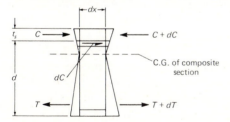

Fig. 16.8.3 Force required of shear connectors at service loads.

If one uses an ultimate strength concept, the shear connectors under ultimate bending moment share equally in carrying the total maximum compressive force developed in the concrete slab. This would mean, referring to Fig. 16.8.2a, that shear connection is required to transfer the compressive force developed in the slab at midspan to the steel beam in the distance $L/2$, since no compressive force can exist in the slab at the end of the span where zero moment exists. The ultimate compressive force to be accommodated could not exceed that which the concrete can carry:

$$C_{\max} = 0.85 f'_c b_E t_s \qquad (16.8.2)$$

or if the ultimate tensile force below the bottom of the slab is less than C_{max},

$$T_{max} = A_s f_y \qquad (16.8.3)$$

Thus if a given connector has an ultimate capacity q_{ult} the total number of connectors N required between the points of maximum and zero bending moment is

$$N = \frac{C_{max}}{q_{ult}} \quad \text{or} \quad \frac{T_{max}}{q_{ult}} \qquad (16.8.4)$$

whichever is smaller. Under the ultimate strength approach, the total required number of shear connectors are distributed uniformly over the region of the beam between maximum and zero bending moment.

The determination of the connector capacity analytically is complex, since the shear connector deforms under load and the concrete which surrounds it is also a deformable material. Moreover, the amount of deformation a shear connector undergoes is dependent upon factors such as its own shape and size, its location along the beam, the location of the maximum moment, and the manner in which it is attached to the top flange of the steel beam. In addition, any particular shear connector may yield sufficiently to cause a slipping between the beam and the slab. In the latter case the adjacent shear connectors pick up the additional shear.

As a result of the extremely complex behavior of shear connectors, their capacities are not based solely on a theoretical analysis. In order to develop a rational approach, a number of research programs, summarized by Viest [6, 8], were undertaken to develop the strengths of the various types of shear connectors.

Investigators determined that shear connectors will not fail if the average load per connector is below that causing 0.003 in. (0.076 mm) residual slip between concrete and steel. The amount of slip is also a function of the strength of the concrete that surrounds the shear connector. Relating connector capacity to a specified slip may be realistic for bridge design where fatigue strength is important, but it is overly conservative with respect to failure loads. So-called "ultimate" capacities used prior to 1965 [16] were based on slip limitation, giving values about one-third of the ultimate strengths obtained when actual failure of a connector is the criterion.

When ultimate flexural strength of the composite section is the basis for design, the connectors must be adequate to satisfy equilibrium of the concrete slab between the points of maximum and zero moment, as discussed in the development of Eqs. 16.8.2, 16.8.3, and 16.8.4. Slip is not a criterion for this equilibrium requirement. As stated by Slutter and Driscoll [16], "the magnitude of slip will not reduce the ultimate moment provided that (1) the equilibrium condition is satisfied, and (2) the magnitude of slip is no greater than the lowest value of slip at which an individual connector might fail." More recent studies by Ollgaard, Slutter, and Fisher [17] and McGarraugh and Baldwin [18] included the effect of lightweight concrete on stud connector capacity.

Two currently accepted [14] expressions for ultimate connector capacity are as follows:

1. Hooked or headed stud connectors welded to flange (Fig. 16.8.1a). The 1977 AASHTO Specifications [14] give essentially the expression developed at Lehigh [17],

$$q_{ult} = 0.4d_s^2\sqrt{f_c'E_c} \qquad \text{for } H/d_s \geq 4 \qquad (16.8.5)^*$$

where H is the height of the stud (in.); d_s is the stud diameter (in.); q_{ult} is the connector capacity (lb) for one stud; f_c' is the 28-day compressive strength (psi) of the concrete; and E_c is the modulus of elasticity (psi) of concrete. ($E_c = w^{1.5}33\sqrt{f_c'}$, where w is the unit weight (pcf) of concrete; for normal weight concrete of 145 pcf, $E_c = 57,000\sqrt{f_c'}$.)

2. Channel connectors (Fig. 16.8.1b). The AASHTO Specification [14] gives

$$q_{ult} = 550(h + 0.5t)W\sqrt{f_c'} \qquad (16.8.6)^*$$

where h is the average thickness (in.) of the channel flange; t is the thickness (in.) of the channel web; and W is the length (in.) of the channel shear connector.

Connector Design—Ultimate Strength Concept

It may be noted that the connection and the beam must resist the same ultimate load. However, under service loads the *beam* resists dead and live loads, but unless shores are used the *connectors* resist essentially only the live load. Working stress method might design the connection only for live load; however, an increased factor of safety should be used, since the ultimate capacity would otherwise be inadequate.

AISC–1.11 uses an ultimate strength concept but converts both the forces to be designed for and the connector capacities into the service load range by dividing them by a factor. The loads to be carried, either Eq. 16.8.2 or Eq. 16.8.3, are divided by a nominal factor of 2. Thus, for design under service loads,

$$V_h = \frac{C_{max}}{2} = \frac{0.85f_c'A_c}{2} \qquad (16.8.7)$$

* For SI units,

$$q_{ult} = 0.0004d_s^2\sqrt{f_c'E_c} \qquad \text{(for stud)} \qquad (16.8.5)$$

with d_s, mm; f_c' and E_c, MPa; and q_{ult}, kN. For this, $E_c = w^{1.5}(0.043)\sqrt{f_c'}$ with w in kg/m³.

$$q_{ult} = 0.588(h + 0.5t)W\sqrt{f_c'} \qquad \text{(for channel)} \qquad (16.8.6)$$

with h, t, and W, mm; f_c', MPa; and q_{ult}, kN.

which is AISC Formula (1.11–3), and where $A_c = b_E t_s$, the effective concrete area. Equation 16.8.3 divided by 2 becomes

$$V_h = \frac{T_{max}}{2} = \frac{A_s F_y}{2} \qquad (16.8.8)$$

which is AISC Formula (1.11–4), In Eqs. 16.8.7 and 16.8.8,

$V_h =$ horizontal shear to be resisted between the points of maximum positive moment and points of zero moment, the smaller of Eq. 16.8.7 or Eq. 16.8.8 to be used

$f'_c =$ 28-day compressive strength of concrete

$A_c = b_E t_s =$ effective concrete area

$A_s =$ area of steel beam

$F_y =$ yield point stress for steel beam

The connector ultimate capacities must also be divided by factors to give "allowable values" for working stress method. The AISC allowable values are obtained by dividing ultimate capacities from Eqs. 16.8.5 and 16.8.6 by a factor of safety of approximately 2.0. AISC allowable values are given in Table 16.8.1 for hooked or headed studs and channels. Since the ultimate capacity expressions for hooked or headed studs are valid for $H/d_s \geq 4$, the values in Table 16.8.1 are also applicable to studs longer than the lengths indicated in the table.

When lightweight aggregate concrete is used, the connector values in Table 16.8.1 are to be multiplied by the coefficients given in Table 16.8.2.

The number N_1 of connectors required is obtained by dividing the

Table 16.8.1 (from AISC–1.11.4) AISC Allowable Horizontal Shear Load for One Connector

Connector	Allowable Shear Load q (kips) (Applicable only to normal-weight concrete)		
	Concrete strength, f'_c (psi)		
	3000	3500	4000
1/2″ diam × 2″ hooked or headed stud	5.1	5.5	5.9
5/8″ diam × 2 1/2″ hooked or headed stud	8.0	8.6	9.2
3/4″ diam × 3″ hooked or headed stud	11.5	12.5	13.3
7/8″ diam × 3 1/2″ hooked or headed stud	15.6	16.8	18.0
Channel C3 × 4.1	4.3W*	4.7W	5.0W
Channel C4 × 5.4	4.6W	5.0W	5.3W
Channel C5 × 6.7	4.9W	5.3W	5.6W

* $W =$ length of channel, in.

Table 16.8.2 (from AISC–1.11.4) Reduction Factors for Connector Capacities when Using Lightweight Aggregate Concretes*

Air Dry Unit Weight, pcf	90	95	100	105	110	115	120
Coefficient, for $f'_c \leq 4.0$ ksi	0.73	0.76	0.78	0.81	0.83	0.86	0.88
Coefficient, for $f'_c \geq 5.0$ ksi	0.82	0.85	0.87	0.91	0.93	0.96	0.99

* ASTM C330 aggregates.

smaller value of V_h by the allowable shear per connector:

$$N_1 = \frac{\text{smaller } V_h}{q} \qquad (16.8.9)$$

where q is the allowable load from Table 16.8.1.

The smaller value of V_h, determined by Eqs. 16.8.7 or 16.8.8, is used since it represents the maximum force to give equilibrium at ultimate strength, as discussed in developing Eqs. 16.8.2 and 16.8.3. To provide more shear resistance than either the concrete slab or the steel beam could develop would be needless. Also, it is doubtful that an excessive number of shear connectors would perceptibly reduce deflection.

AASHTO–1.7.48(E) uses the ultimate strength concept directly (i.e., without dividing by a factor to convert the computation into the nominal service load comparison). However, the strength calculation is not used as the sole design procedure but rather as an additional check after determining the shear connectors required for the fatigue criterion. The fatigue requirement is an elastic procedure based on a slip limitation.

Example 16.8.1

Determine the number of $\frac{3}{4}$ in. diam 3 in. shear stud connectors required according to the AISC Specification for the composite section shown in Fig. 16.8.4. Assume a uniform loading and simple beam supports. Use $F_y = 36$ ksi and $f'_c = 3$ ksi.

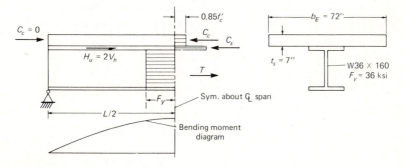

Fig. 16.8.4 Example 16.8.1.

SOLUTION
Using Eqs. 16.8.7 and 16.8.8,

$$V_h = \frac{0.85 f'_c A_c}{2} = \frac{0.85(3.0)72(7)}{2} = 643 \text{ kips}$$

or

$$V_h = \frac{A_s F_y}{2} = \frac{47.1(36)}{2} = 848 \text{ kips}$$

From Table 16.8.1, the allowable shear per connector is 11.5 kips. Taking the smaller value of V_h, the number of shear connectors, N, required for each half of the span is therefore

$$N = \frac{643}{11.5} = 56$$

Use 56—$\frac{3}{4} \times 3$ in. studs per half span.

Ordinarily, AISC design would not involve an ultimate strength analysis which would permit direct computation of H_u (Fig. 16.8.4) equal to C_c as was done in Example 16.7.2. In lieu of such computation the two-formula procedure is necessary.

In the case of continuous beams, the longitudinal reinforcing bar steel within the effective width of the concrete slab may be assumed to act compositely with the steel beam in the areas of negative moment. The total horizontal shear to be resisted by the shear connectors between the interior support and each adjacent point of contraflexure equals the maximum tensile force which can be developed in the reinforced concrete slab; i.e., neglecting tensile capacity of the concrete.

$$T_{slab} = A_{sr} F_{yr} \tag{16.8.10}$$

where A_{sr} = total area of longitudinal reinforcing bar steel at the interior support located within the effective flange width.

F_{yr} = specified minimum yield strength of the longitudinal reinforcing bar steel

For working stress design, the ultimate shear force developed between maximum negative moment and point of contraflexure, T_{slab}, is divided by 2 to bring it into the service load range,

$$V_h = \frac{T_{slab}}{2} = \frac{A_{sr} F_{yr}}{2} \tag{16.8.11}$$

It is logically presumed that the tensile capacity of the reinforced concrete slab will be less than the tensile capacity of the steel beam, so that for negative moment, only Eq. 16.8.11 is used.

Connector Design—Elastic Concept for Static Loads

In the elastic approach to shear connector design, the connectors are distributed according to the variation in horizontal shear between the slab and the steel beam. The connector capacities would be based on a limitation on slip. Formerly, design rules for both buildings and bridges used this approach with Eq. 16.8.1. When this elastic method was used where slip was limited, fatigue was not a controlling factor. The number of connectors required was conservatively large, indicating more than enough to develop the ultimate flexural strength of the composite member. If the shear connector requirements are reduced to the number required to just develop the ultimate flexural strength, fatigue failure may then become the governing factor. Present bridge design considers both fatigue strength as well as ultimate strength.

Connector Design—Elastic Concept for Fatigue Strength

The 1977 AASHTO Specification requirements for fatigue are based largely on the work of Slutter and Fisher [19]. For fatigue, the *range* of stress rather than the magnitude is the major variable. Fatigue strength may be expressed

$$\log N = A + BS_r \qquad (16.8.12)$$

where S_r is the range of horizontal shear stress; N is the number of cycles to failure; and A and B are empirical constants. The equation used for design is shown in Fig. 16.8.5.

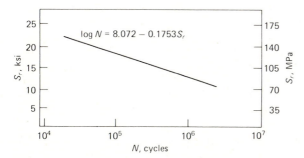

Fig. 16.8.5 Fatigue strength of stud shear connectors. (From Ref. 19)

Since the magnitude of shear force transmitted by individual connectors when service loads act agrees well with prediction by elastic theory, the horizontal shear is calculated by the elastic relation VQ/I. Fatigue is critical under repeated application of service load; thus it is reasonable to determine variation in shear stress using elastic theory. Using Eq. 16.8.1, the equation for *static load* is

$$\frac{VQ}{I} = \frac{\text{Allowable load } q}{p} \tag{16.8.13}$$

For *cyclical load*, Eq. 16.8.1 gives

$$\frac{(V_{max} - V_{min})Q}{I} = \frac{\text{Allowable range } Z_r}{p} \tag{16.8.14}$$

where p is the connector spacing. AASHTO–1.7.48(E) gives Eq. 16.8.14 as

$$S_r = \frac{V_r Q}{I} \le Z_r \tag{16.8.15}$$

where $V_r = V_{max} - V_{min}$
  $Z_r = \alpha d_s^2$ for welded studs
  $\alpha = 13{,}000$ for 100,000 cycles
    $10{,}600$ for 500,000 cycles
    7850 for 2 million cycles

Example 16.8.2

Redesign the shear connectors for the beam of Example 16.8.1 (Fig. 16.8.4) using the working stress fatigue requirement of AASHTO with $\frac{3}{4}$ in. diam $\times 3$ in. stud connectors. Assume 500,000 cycles of loading of live load is to be designed for. Whether or not the beam is shored, only the live load is the cyclical load. Use uniform live load of 3.5 kips/ft, a spacing of 7 ft for beams, a beam span of 45 ft, $F_y = 36$ ksi, and $f_c' = 3$ ksi.

SOLUTION
  (a) Loads and shears. For the fatigue requirement in AASHTO–1.7.48(E)(1) only the *range* of live load is needed. At the support with full span loaded,

$$V = \tfrac{1}{2}wL = 0.5(3.5)45 = 78.8 \text{ kips}$$

Using partial span loading of live load,

$$\text{Max } V(\text{at } \tfrac{1}{4} \text{ point}) = 3.5(45)(0.75)(0.375) = 44.3 \text{ kips}$$

$$\text{Max } V(\text{at midspan}) = \tfrac{1}{8}wL = \tfrac{1}{8}(3.5)45 = 19.7 \text{ kips}$$

The envelope showing the *range* of live load shear is given in Fig. 16.8.6. Inclusion of dead load shear would change both V_{max} and V_{min} by the same amount at any section along the beam; however, $(V_{max} - V_{min})$, that is, the range V_r would not be affected.

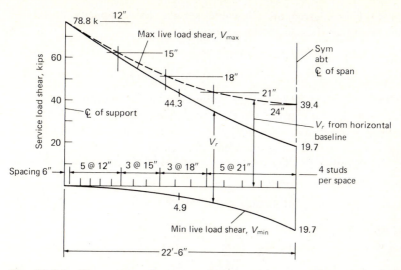

Fig. 16.8.6 Shear range diagram and stud spacing according to elastic fatigue theory used by AASHTO.

(b) Compute composite section properties ($n = 9$) (see Fig. 16.8.4).

Element	Effective Area, A (sq in.)	Arm from C.G. of steel beam, y (in.)	Ay (sq in.)	Ay^2 (in.³)	I_0 (in.⁴)
Slab, 72(7)/9	56.0	21.5	1204	25,900	230
W36×160	47.1	—	—	—	9760
	103.1		1204	25,900	9990

$$I_x = Ay^2 + I_0 = 25,900 + 9990 = 35,900 \text{ in.}^4$$

$$\bar{y} = \frac{1204}{103.1} = 11.68 \text{ in.}$$

$$I = 35,900 - 103.1(11.68)^2 = 21,800 \text{ in.}^4$$

$$y_t = 18.0 + 7.0 - 11.68 = 13.32 \text{ in.}$$

$$y_b = 18.0 + 11.68 = 29.68 \text{ in.}$$

$$S_t = \frac{21,800}{13.32} = 1640 \text{ in.}^3 \qquad S_b = \frac{21,800}{29.68} = 735 \text{ in.}^3$$

Determine the static moment of the effective concrete area about the centroid of the composite section,

$$Q = 56.0(y_t - 3.5) = 56.0(9.82) = 550 \text{ in.}^3$$

(c) Determine the allowable load for $\frac{3}{4}$ in. diam × 3 in. stud connectors.

AASHTO–1.7.48 gives an allowable service load capacity based on fatigue for 500,000 cycles of loading as

$$\text{Allowable } S_r = 10.6d_s^2$$
$$= 10.6(0.75)^2 = 5.96 \text{ kips}$$

The 1977 AASHTO allowable values are higher than those of earlier editions but are still lower than AISC values because the AASHTO values are related to a slip limitation and a range of stress. This is appropriate whenever fatigue loading may occur.

(d) Determine spacing of connectors. Use 4 studs across the beam flange width at each location:

$$S_r \text{ for 4 studs} = 4(5.96) = 23.8 \text{ kips}$$

Using Eq. 16.8.1,

$$p = \frac{S_r}{V_r Q / I} = \frac{S_r I}{V_r Q}$$

where $I/Q = 21,800/550 = 39.6$ in.

$$p = \frac{23.8(39.6)}{V_r} = \frac{943}{V_r (\text{kips})}$$

The values are computed in the table below and the spacing is determined graphically on the shear diagram of Fig. 16.8.6.

p (in.)	V_r (kips)
12	79
15	63
18	52
21	45
24	39

The fatigue service load criterion requires nearly 18% more connectors (66 vs 56 per half span) than the procedure based on ultimate strength concept.

16.9 HYBRID COMPOSITE GIRDERS

The general discussion of the hybrid plate girder appears in Sec. 11.6. The hybrid girder is one that has either the tension flange or both flanges of the steel section made with a higher strength grade of steel than used for the web (see Fig. 16.2.1d, e). There are particular economic advantages to the hybrid girder in composite construction where the concrete slab provides a large compression capacity. The neutral axis will lie near the compression face of the composite section, causing the higher stressed tension flange to become the element controlling the strength. Use of a stronger grade of steel for the tension flange will avoid the use of a large plate.

The design and behavior of hybrid girders has been presented in detail in a Joint ASCE–AASHO Committee Report [20]. The theoretical aspects of the bending behavior of composite members has been given by Schilling [21]. The primary behavioral feature of practical concern is the yielding of the web prior to reaching maximum strength in the flanges. As discussed in Sec. 11.6, the flange is designed to have extra strength to make up for the reduced moment strength of the web. This is accomplished in a practical way by reducing the stress permitted at the extreme fiber when the bending moment is considered resisted by the full cross section. Thus Eq. 11.6.1 is used for noncomposite girders according to AISC–1.10.6.

When extending the hybrid concept to composite steel-concrete members, the behavior is essentially the same [20, 21]. Additional complicating factors are (1) the neutral axis of the composite section is not at mid-depth and therefore requires an evaluation of an unsymmetrical hybrid member; and (2) the relative stiffness of the concrete deck and steel section are continuously changing as the yielding progresses. In the composite hybrid section the principal concern is with the tension flange (bottom flange in positive moment zones). Since a greater percentage of the depth of the web is located below the neutral axis (tension side) the early yielding of the lower-strength web means a greater reduction in strength for a hybrid beam in composite construction than for a noncomposite symmetrical beam.

To account for a variable distance from the tension flange to the neutral axis the Joint ASCE–AASHO Committee [20] recommended the following equation as applicable "to hybrid beams that support the dead weight of the slab without composite action but act compositely with the slab in support of live load:"

$$F_b' = F_b \left[1 - \frac{\beta\psi(1-\alpha)^2(3-\psi+\alpha\psi)}{6+\beta\psi(3-\psi)} \right] \tag{16.9.1}$$

where ψ is the ratio of the distance from the bottom of the beam to the neutral axis of the transformed section (composite section) to the overall depth of the steel section. All other variables are as indicated in regard to Eq. 11.6.1. *Equation 16.9.1 is not to be used if the top flange has a higher yield strength or larger area than the bottom flange.*

AASHTO–1.7.50 or 1.7.67 uses Eq. 16.9.1 for both symmetrical noncomposite and composite hybrid girders. AISC–1.11 makes no reference to hybrid composite construction but presumably it is permitted. The allowable stress reduction factor may be obtained using Eq. 11.6.1. When the section is symmetrical and $\psi = 0.5$ for Eq. 16.9.1, the value of the reduction factor is about the same whichever formula is used.

Table 16.9.1 Multipliers to Reduce Allowable Bending Stress for Hybrid Composite and Noncomposite Girders

α \ β	0.50	1.0	2.0	3.0	4.0	
$\psi = 0.5$ *(Neutral axis at mid-depth)*						
0.36	0.963	0.931	0.879	0.839	0.807	AISC Formula
0.50	0.976	0.955	0.922	0.896	0.875	(1.10-6),
0.72	0.992	0.985	0.973	0.964	0.957	Eq. 11.6.1
0.36	0.959	0.924	0.871	0.831	0.800	AASHTO
0.50	0.974	0.953	0.919	0.894	0.875	Eq. 16.9.1
0.72	0.992	0.985	0.974	0.965	0.959	
$\psi = 0.75$						
0.36	0.943	0.899	0.835	0.790	0.757	
0.50	0.964	0.936	0.895	0.866	0.846	AASHTO
0.72	0.988	0.979	0.965	0.956	0.949	
$\psi = 1.00$						
0.36	0.931	0.879	0.807	0.758	0.724	
0.50	0.955	0.922	0.875	0.844	0.821	AASHTO
0.72	0.985	0.973	0.957	0.947	0.939	

The reduction factors given by Eqs. 11.6.1 and 16.9.1 are in Table 16.9.1.

16.10 AISC DESIGN FOR FLEXURE

As discussed in Sec. 16.6, the actual stresses that occur under service load in a given composite member depend on the manner of construction. The slab formwork must be supported by either the steel beam acting alone or by temporary shoring which also would support the beam. When temporary shoring is used service load stresses will be lower than when such shoring is not used, since *all* loads will be supported by the composite section. If the system is built without temporary shoring, the steel beam alone must support itself and the slab without benefit of composite action.

For economical construction, it is desired to avoid use of shoring wherever possible. In Sec. 16.7 it was shown that no matter which construction system is used, the ultimate moment capacity is identical. It is a simple procedure, therefore, to design as if the entire load is to be carried compositely (i.e. assume shores are used) *even when shores are not to be used*. Strength is assured; however, it is necessary to insure the stress in the steel beam does not approach too closely the yield stress under service load conditions.

In order to resist loads compositely, the concrete strength must be adequately developed. AISC–1.11.2.2 requires that 75% of the compressive strength f'_c of the concrete must be developed before composite action may be assumed.

The AISC design procedure for flexure may be summarized by the following steps:

1. Select section as if shores are to be used; the required composite section modulus, S_{tr}, with reference to the tension fiber is

$$\text{Required } S_{tr} = \frac{M_D + M_L}{F_b} \qquad (16.10.1)$$

where M_D = the service load moment caused by loads applied *prior* to the time the concrete has achieved 75% of its required strength

M_L = the service load moment caused by loads applied *after* the concrete has achieved 75% of its required strength

F_b = allowable service load stress, $0.66F_y$ for positive moment region (where sections are exempt from the "compact section" requirements of AISC–1.5.1.4.1)

Lateral support is adequately provided by the concrete slab and its shear connector attachments.

2. Check AISC Formula (1.11–2). When shores are actually not to be used, service load stress on the steel section must be assured of being less than yield stress. AISC–1.11.2.2 uses an indirect procedure for checking this. This section modulus of the composite section, S_{tr}, may not exceed (or be considered more effective than) the following:

$$S_{tr}(\text{effective}) \le \left(1.35 + 0.35 \frac{M_L}{M_D}\right) S_s \qquad (16.10.2)$$

which is AISC Formula (1.11–2).

To understand the development of Eq. 16.10.2, the reader is referred back to Sec. 16.6 where service load stresses are computed for construction with and without shores. Service load tension stresses on the steel beam may be expressed in general as

$$f_b = \frac{M_D}{S_s} + \frac{M_L}{S_{tr}} \le k_1 F_y \qquad \text{without shores} \qquad (a)$$

$$f_b = \frac{M_D + M_L}{S_{tr}} \le k_2 F_y \qquad \text{with shores} \qquad (b)$$

where S_s = section modulus of the steel beam referred to its bottom flange (tension flange)

S_{tr} = section modulus of composite section referred to its bottom flange (tension flange)

k_1, k_2 = constants to obtain the allowable stresses in tension without shores and with shores, respectively

Divide Eq. (a) by Eq. (b), letting $kS_s = S_{tr}$:

$$\frac{k_1}{k_2} \geq \frac{\dfrac{M_D}{S_s} + \dfrac{M_L}{kS_s}}{\dfrac{M_D + M_L}{kS_s}} = \frac{kM_D + M_L}{M_D + M_L} \qquad \text{(c)}$$

$$\frac{k_1}{k_2}(M_D + M_L) - M_L \geq kM_D \qquad \text{(d)}$$

Divide by M_D,

$$k \leq \frac{k_1}{k_2}\left(1 + \frac{M_L}{M_D}\right) - \frac{M_L}{M_D} \qquad \text{(e)}$$

Replacing k by S_{tr}/S_s gives the AISC formula in general terms.

$$S_{tr} \leq \left[\frac{k_1}{k_2} + \frac{M_L}{M_D}\left(\frac{k_1}{k_2} - 1\right)\right]S_s \qquad \text{(f)}$$

The AISC value of $k_1/k_2 = 1.35$ is obtained if a compact section ($F_b = 0.66F_y$) is allowed to reach a service load stress of $0.89F_y$ ($0.89/0.66 = 1.35$). As is seen from Eq. (f), this limitation of stress is valid no matter what ratio of M_L to M_D is used.

3. Check stress on steel beam supporting the loads acting before concrete has hardened.

$$\text{Required } S_s = \frac{M_D}{F_b} \qquad \text{(16.10.3)}$$

where F_b may be $0.66F_y$, $0.60F_y$, or some lower value if adequate lateral support is not provided. It is to be noted that Eq. 16.10.3 is frequently controlling on the compression fiber (top in positive moment zone), particularly if a steel cover plate is used on the bottom.

4. Partial composite action. When fewer connectors are used than necessary to develop full composite action, an effective section modulus may be obtained by linear interpolation. AISC–1.11.2.2 allows

$$S_{\text{eff}} = S_s + \frac{V_h'}{V_h}(S_{tr} - S_s) \qquad \text{(16.10.4)}$$

where V_h = design horizontal shear for full composite action
V_h' = actual capacity of connectors used; less than V_h
S_s and S_{tr} as defined previously in this section

For this case, S_{eff} is used in design calculations in place of that computed from beam dimensions, and is the quantity that may not exceed the value given by Eq. 16.10.2.

16.11 EXAMPLES—SIMPLY SUPPORTED BEAMS

Example 16.11.1

Design an interior member of the floor shown in Fig. 16.11.1 assuming it is constructed without temporary shoring. Assume $F_y = 36$ ksi, $n = 9$, $f_c' = 3000$ psi, $f_c = 1350$ psi, and a 4-in. slab. Use AISC Spec.

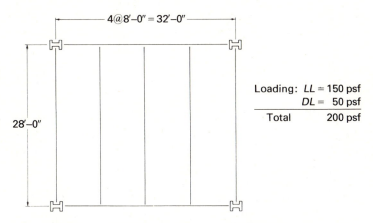

Loading: $LL = 150$ psf
$DL = 50$ psf
Total 200 psf

Fig. 16.11.1 Beam framing plan for Example 16.11.1.

SOLUTION

(a) Loads and bending moments.

Load carried on steel beam:

$$\text{concrete slab,} \quad \tfrac{4}{12}(0.15)(8) = 0.40 \text{ kip/ft}$$
$$\text{beam weight (estimated)} = 0.04 \text{ kip/ft}$$
$$\overline{\phantom{\text{beam weight (estimated) }} 0.44 \text{ kip/ft}}$$
$$M_D = \tfrac{1}{8}(0.44)(28)^2 = 43.1 \text{ ft-kips}$$

Load carried by composite section:

$$\text{live load,} \quad 0.15(8) \quad = 1.20 \text{ kip/ft}$$
$$M_L = \tfrac{1}{8}(1.20)(28)^2 = 118 \text{ ft-kips}$$

(b) Select beam as if shores were to be used. For $M_D + M_L$ the allowable stress is $0.66F_y$ on the composite section.

$$\text{Required } S_{tr} = \frac{(43 + 118)12}{24} = 80.5 \text{ in.}^3$$

For M_D acting on the steel section alone, the allowable stress is at least $0.60F_y$ if adequate lateral support is provided,

$$\text{Required } S_s = \frac{M_D}{0.60F_y} = \frac{43(12)}{22} = 23.4 \text{ in.}^3$$

Enter AISC Manual, "Composite Beam Selection Table" and select W16×36 with no cover plate.

Try W16×36: Properties of the steel section alone are:

$$A = 10.6 \text{ sq in.} \quad I_x = 448 \text{ in.}^4 \quad S_x = 56.5 \text{ in.}^3 \quad b_f = 6.985 \text{ in.}$$

Next, compute properties of the composite section.

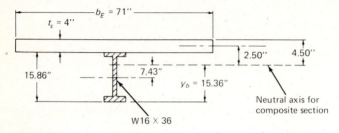

Fig. 16.11.2 Beam cross section for Example 16.11.1.

Determine effective width (see Fig. 16.11.2):

$$b_E = \tfrac{1}{4} \text{ of span} = 0.25(28)(12) = 84 \text{ in.}$$
$$b_E = \text{beam spacing} = 96 \text{ in.}$$
$$b_E = 16 \text{ (thickness of slab)} + b_f$$
$$= 16(4) + 6.985 \qquad = 71 \text{ in.} \quad \text{Controls}$$

$$y_b = \frac{10.6\left(\dfrac{15.86}{2}\right) + \left(\dfrac{71(4)}{9}\right)17.86}{10.6 + \dfrac{(71)4}{9}} = 15.36 \text{ in.}$$

$$I_{\text{comp}} = 448 + 10.6(7.43)^2 + \frac{1}{12}\left(\frac{71}{9}\right)(4)^3 + \frac{71(4)}{9}(2.50)^2 = 1270 \text{ in.}^4$$

$$S_{tr} = \frac{1270}{15.36} = 82.6 \text{ in.}^3 \quad \text{(for bottom of steel beam)}$$

$$S_{\text{top}} = \frac{1270}{4.50} = 282 \text{ in.}^3 \quad \text{(for top of concrete)}$$

Recomputing moments, $w = 0.05(8) + 0.036 = 0.436$ kips/ft

$$M_D = \tfrac{1}{8}(0.436)(28)^2 = 42.8 \text{ ft-kips}$$
$$M_L = \tfrac{1}{8}(0.150)(8)(28)^2 = \underline{117.7 \text{ ft-kips}}$$

$$\text{Total} \qquad 160.5 \text{ ft-kips}$$

Check stresses:

At top of concrete slab; allowable $f_c = 0.45f'_c = 1.35$ ksi

$$f_c = \frac{160.5(12)}{9(282)} = 0.76 \text{ ksi} < 1.35 \text{ ksi} \qquad \text{OK}$$

At bottom of steel beam; $F_b = 0.66F_y = 24$ ksi

$$f_b = \frac{160.5(12)}{82.6} = 23.3 \text{ ksi} < 24.0 \text{ ksi} \qquad \text{OK}$$

(c) Check Eq. 16.10.2 to determine maximum transformed section modulus S_{tr} that can be used.

$$S_{tr} = \left(1.35 + 0.35\frac{117.7}{42.8}\right)56.5 = 130.0 \text{ in.}^3 > 82.6 \text{ in.}^3 \qquad \text{OK}$$

Thus shores need not be used.

(d) Check steel stress for loads carried noncompositely, using Eq. 16.10.3,

$$f_b = \frac{M_D}{S_s} = \frac{42.8(12)}{56.5} = 9.1 \text{ ksi} < 0.60F_y$$

The above assumes adequate lateral support during construction so that the laterally unbraced length is less than L_u based on either $\sqrt{102,000C_b r_T^2/F_y}$ or $20,000C_b/[(d/A_f)F_y]$, as given in AISC–1.5.1.4.5(2.). Stress on the steel section resulting from noncomposite loads is more likely to govern when a cover plate is used on the bottom than when such a plate is not used. The beam section selected is therefore satisfactory. *Use* W16×36.

(e) Design shear connectors:

$$\text{From Eq. 16.8.7,} \qquad V_h = \frac{0.85(3)71(4)}{2} = 362 \text{ kips}$$

$$\text{From Eq. 16.8.8,} \qquad V_h = \frac{10.6(36)}{2} = 191 \text{ kips}$$

From Table 16.8.1, $\frac{5}{8}$ in. diam × $2\frac{1}{2}$ in. headed stud, $q = 8.0$ kips/stud

$$N = \frac{V_h}{q} = \frac{191}{8.0} = 23.8, \quad \text{say 24}$$

Use 24 shear connectors on each side of the centerline at midspan. Use a uniform spacing with 2 studs at a section across the beam width:

$$p = \frac{L/2}{N/2} = \frac{28(12)}{24} = 14.1 \text{ in.}$$

Use a 14-in. spacing for the pairs of stud connectors, starting at the support.

Example 16.11.2

Design a composite section, without shores, for use as an interior floor beam of an office building. $f'_c = 3000$ psi; $n = 9$; $F_y = 36$ ksi. Use AISC Spec.

Span	= 30 ft	Live load	= 150 psf
Beam spacing =	8 ft	Partitions =	25 psf
Slab thickness =	5 in.	Ceiling	= 7 psf

SOLUTION

(a) Determine moments:

$$\text{5-in. slab, } \tfrac{5}{12}(8)0.15 = 0.50 \text{ kips/ft}$$
$$\underline{\text{Steel beam (assumed)} = 0.03}$$
$$0.53 \text{ kips/ft}$$

$$M_D = \tfrac{1}{8}(0.53)(30)^2 = 60 \text{ ft-kips}$$

$$\text{Live load } 0.15(8) = 1.2 \text{ kips/ft}$$
$$\text{Partitions } 0.025(8) = 0.2$$
$$\underline{\text{Ceiling } 0.007(8) = 0.05}$$
$$1.45 \text{ kips/ft}$$

$$M_L = \tfrac{1}{8}(1.45)(30)^2 = 163 \text{ ft-kips}$$

(b) Select section.

$$\text{Required } S_{tr} = \frac{M_D + M_L}{0.66F_y} = \frac{223(12)}{24} = 111.5 \text{ in.}^3$$

If shores are not used, stress on the steel section prior to developing composite action must not be excessive. Assuming adequate lateral bracing such that unbraced length $L < L_u$,

$$\text{Required } S_s = \frac{M_D}{0.60F_y} = \frac{60(12)}{22} = 32.7 \text{ in.}^3$$

If the unbraced length $L < L_c$ and the section is "compact" for local buckling, $0.66F_y$ could be used for the allowable stress. Use AISC Manual "Composite Beam Selection Table" to select this member. Find W14×22 with 4 sq in. cover plate ($S_{tr} = 113$ in.^{3}).

(c) Determine effective width and compute properties.

Effective width:

$$b_E \le L/4 = 90 \text{ in.}$$
$$\le 16t_s + b_f = 16(5) + 5.0 = 85.0 \text{ in.} \quad \text{Controls}$$
$$\le \text{spacing of beams} = 96 \text{ in.}$$

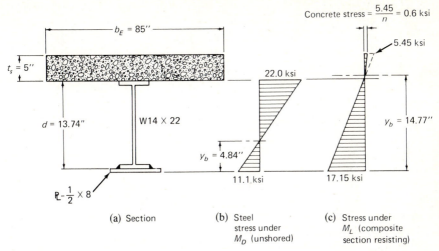

Fig. 16.11.3 Solution for Example 16.11.2, showing stresses under service loads.

Properties are computed as shown in previous examples.

Composite section properties: $S_{top} = 362.0$ in.^{3}

$S_{bottom} = 113.0$ in.^{3}

Steel section alone: $S_{top} = 34.4$ in.^{3}

$S_{bottom} = 65.8$ in.^{3}

(d) Check stress under $M_D + M_L$ (ultimate strength concept; nominal working stress procedure). Since girder weight is 32 lb/ft $\approx$ 30 as assumed, original moments are used.

$$f_b = \frac{M_D + M_L}{S_{tr}} = \frac{223(12)}{113} = 23.7 \text{ ksi} < 0.66 F_y = 24 \text{ ksi} \qquad \text{OK}$$

(e) Check AISC Formula (1.11–2).

$$S_{tr} = \left(1.35 + 0.35 \frac{M_L}{M_D}\right) S_s$$

$$= \left(1.35 + 0.35 \frac{163}{60}\right) 65.8 = 2.30(65.8) = 151 \text{ in.}^3$$

Since actual $S_{tr} = S_{bottom} = 113$ in.^{3} does not exceed the upper limit of 151, shores are not required to prevent service load stresses from becoming too close to the yield stress. As discussed earlier, the same conclusion could be reached if actual service load stresses were computed, as follows, for no shoring,

$$f_b = \frac{M_D}{S_s} + \frac{M_L}{S_{tr}} = \frac{60(12)}{65.8} + \frac{163(12)}{113} = 10.9 + 17.3 = 28.2 \text{ ksi}$$

which is acceptable, since it does not exceed $0.89\,F_y = 32$ ksi at service load. AISC requires the formula check, instead of the stress check, because without the understanding of ultimate strength, the $0.89F_y$ might appear as an unsafe value.

(f) Check stress on steel beam prior to development of composite action for the unshored system. The maximum stress occurs in compression at the top of the beam:

$$f_b = \frac{M_D}{S_s} = \frac{60(12)}{34.4} = 20.9 \text{ ksi} < F_b = 0.60F_y$$

Maximum laterally unsupported length during construction is $L = L_u = 5.7$ ft.

Use W14×22 with $\frac{1}{2}×8$ cover plate. The section with service load stresses is given in Fig. 16.11.3. Design of connectors is not illustrated since no new principles are involved.

Example 16.11.3

Design a composite section, without shores, for the same loading conditions as in Example 16.11.2, except use A572 Grade 50 steel.

1. Select a section using no cover plate and with the minimum number of $\frac{3}{4}$ in. diam × 3 in. stud shear connectors.

2. Compare with a section having a cover plate. Determine the length and connection for the cover plate.

SOLUTION

(a) Loads and moments (from Example 16.11.2).

$$M_D = 60 \text{ ft-kips} \quad \text{(assumes 30 lb/ft beam)}$$

$$M_L = 163 \text{ ft-kips}$$

(b) Select section. $F_b = 0.66F_y = 33$ ksi

$$\text{Required } S_{tr} = \frac{M_D + M_L}{0.66F_y} = \frac{223(12)}{33} = 81.1 \text{ in.}^3$$

For the steel section alone,

$$\text{Required } S_s = \frac{M_D}{0.60F_y} = \frac{60(12)}{30} = 24 \text{ in.}^3$$

Select W18×35 ($S_{tr} = 93.6$ in.^{3}) (from "Composite Design Selection Table"—AISC Manual). For the controlling b_E of $16t_s + b_f = 16(5)+6.0 = 86.0$ in. the properties are:

$$\text{Composite section properties:} \qquad S_{top} = 379 \text{ in.}^3$$

$$S_{bottom} = 93.6 \text{ in.}^3$$

Steel section alone: $\qquad S_{top} = 57.6$ in.^{3}

$$S_{bottom} = 57.6 \text{ in.}^3$$

(c) Check AISC Formula (1.11–2).

$$\text{Max } S_{tr} = \left(1.35 + 0.35 \frac{M_L}{M_D}\right) S_s$$

$$= \left(1.35 + 0.35 \frac{163}{60}\right) 57.6 = 133 \text{ in.}^3 > 93.6 \text{ in.}^3 \qquad \text{OK}$$

Use W18×35.

(d) Determine minimum number of $\frac{3}{4}$ in. diam×3 in. shear studs required.

Sometimes savings can be made by utilizing less than full shear transfer between the concrete slab and the steel beam. For partial composite action, the section modulus used is obtained from Eq. 16.10.4. From Eqs. 16.8.7 and 16.8.8,

$$V_h = \frac{A_s F_y}{2} = \frac{(10.3)50}{2} = 258 \text{ kips} \quad \text{Controls}$$

or

$$V_h = \frac{0.85 f'_c A_c}{2} = \frac{0.85(3)(86)5}{2} = 548 \text{ kips}$$

Solving Eq. 16.10.4 for V'_h,

$$V'_h = \frac{(S_{eff} - S_s)}{(S_{tr} - S_s)} V_h = \frac{(81.1 - 57.6)}{(93.6 - 57.6)}(258) = 168 \text{ kips}$$

The number of connectors required between midspan (point of maximum moment) and the end of the beam (point of zero moment) is

$$N_1 = \frac{V'_h}{q} = \frac{168}{11.5} = 14.6, \qquad \text{say 16} \quad \text{(32 per span)}$$

If full composite action were developed, the effective S would be 93.6 in.^{3} and the number of connectors needed would be

$$N_1 = \frac{V_h}{q} = \frac{258}{11.5} = 22.4, \qquad \text{say 24} \quad \text{(48 per span)}$$

The spacing required for 32 studs per span (16 pairs) is

$$\text{Spacing} = \frac{30(12)}{16} = 22.5 \text{ in.}$$

Max spacing (AISC–1.11.4) $= 8t = 8(5) = 40$ in. > 22.5 in. $\qquad$ OK

Use 32—$\frac{3}{4}$ in. diam×3 in. studs per beam.

(e) Alternate design with cover plate. If a cover plate had been used, a W12×19 with 1×3 cover plate would be selected. The S_{tr} provided (83.3) would have been only slightly higher than that required (81.1); thus nearly full composite action would have to be developed.

$$V_h = \frac{(5.59+3.0)50}{2} = 215 \text{ kips}$$

$$V_h' = \left(\frac{81.1-44.9}{83.3-44.9}\right)215 = 203 \text{ kips}$$

$$N_1 = \frac{203}{11.5} = 17.7, \quad \text{say 18} \quad (36 \text{ per span})$$

It is problematical whether the W18×35 without cover plate with 32 studs, or the W12×19 with a 1×3 cover plate with 36 studs, is the better choice.

(f) Determine the cover plate length and specify its connection.

$$\text{provided } S_x \text{ (with } 1\times3 \text{ cover plate)} = 83.3 \text{ in.}^3$$

$$\text{provided } S_x \text{ (without cover plate)} = 41.1 \text{ in.}^3$$

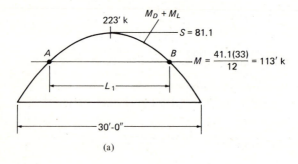

(a)

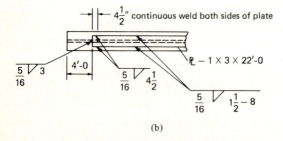

(b)

Fig. 16.11.4 Cover plate for Example 16.11.3.

From Fig. 16.11.4 the cover plate is required over the distance L_1 between points A and B.

$$\frac{\left(\dfrac{L_1}{2}\right)^2}{\left(\dfrac{L}{2}\right)^2} = \frac{223 - 113}{223} = 0.493$$

$$L_1 = 0.702L$$

Under provisions of AISC–1.10.4 the cover plate must develop the cover plate's proportion of the flexural stresses in the beam at the theoretical cutoff point. The stress at mid-thickness of cover plate is

$$f = \frac{(M_D + M_L)}{I_{tr}}(y_b - 0.5)$$

$$= \frac{113(12)(14.03 - 0.5)}{1170} = 15.7 \text{ ksi}$$

The force F in the cover plate is

$$F = fA = 15.7(3.0) = 47.0 \text{ kips}$$

The length of plate required beyond the theoretical termination point is next determined:

Min weld size $= \frac{5}{16}$ in.   (AISC–Table 1.17.2 based on the thicker of the cover plate or the flange of the W12×19)

Weld capacity per inch, $R_w = \frac{5}{16}(0.707)21 = 4.64 \text{ kips/in.}$
(for E70 electrodes)

$$\text{Weld length} = \frac{\text{Force}}{R_w} = \frac{47}{4.64} = 10.1 \text{ in.}$$

Try using weld along the end (3 in.) and use 4 in. along each side. Check AISC–1.10.4, Case 2:

Min length beyond theoretical cutoff $= 1\frac{1}{2}$ times plate width
$= 1.5(3) = 4.5 \text{ in.}$

Cover plate length $= 0.702(30) + 2(4.5/12) = 21.8 \text{ ft}$

Use bottom cover plate $1 \times 3 \times 22'\text{-}0''$ welded and positioned as shown in Fig. 16.11.4b.

Except for the first 4.5 in. on each side of the cover plate, the remainder of the connection can be welded with intermittent welds. According to AISC–1.17.5,

Min weld segment $= 4(\frac{5}{16}) = 1.25 \text{ in.} < 1.5 \text{ in.}$ (controls)

Segment capacity $= 1.5(4.64) = 6.96 \text{ kips}$

The maximum horizontal shear to be transferred occurs at the cutoff location. Neglecting partial span loading the shear is

$$V = \frac{w_{D+L}L}{2} - w_{D+L}(4.0)$$

$$= \frac{(0.53 + 1.45)(30)}{2} - 1.98(4) = 21.8 \text{ kips}$$

$$\frac{VQ}{I} = 21.8\left(\frac{0.42}{12}\right) = 0.76 \text{ kips/in.}$$

$$\text{Required spacing} = \frac{2(6.96)}{VQ/I} = \frac{2(6.96)}{0.76} = 18.4 \text{ in.}$$

$$\text{Max spacing permitted} = 24t = 24(0.350) = 8.4 \text{ in.} \quad \text{(controls)}$$

and not more than 24 in. in any case (AISC–1.18.3.1).
Use—Intermittent $\frac{5}{16}$ in. fillet welds, $1\frac{1}{2}$ in. segments @ 8 in. pitch, except for the first 4.5 in. at each end where continuous weld is to be used. See Fig. 16.11.4b.

Comparison:
1. W18×35 with 32 studs,   35   lb/ft
2. W12×19 with 1×3×22′-0″ plate with 36 studs, 26.3 lb/ft

It is likely that the economical choice is to use the cover plated beam, although it is a borderline decision. If less than 7 or 8 lb/ft is saved by using a cover plate, the plate should not be used.

Example 16.11.4

Design a composite hybrid beam to carry $M_D = 90$ ft-kips and $M_L = 220$ ft-kips. Use an A36 web and A514 ($F_y = 100$ ksi) steel for either the tension flange only or both flanges. The span is 30 ft, beam spacing is 8 ft, and a 5-in. slab ($f_c' = 3000$ psi, $n = 9$) is used.

SOLUTION
   (a) Moments and required section modulus values.

$$M_D = 90 \text{ ft-kips}; \qquad M_L = 210 \text{ ft-kips}$$

$$\text{Required } S_{tr} = \frac{M_D + M_L}{0.60F_y} = \frac{300(12)}{60} = 60 \text{ in.}^3$$

Hybrid girders are not permitted to be treated as "compact sections" under AISC–1.5.1.4.1; thus the maximum allowable stress is $0.60F_y$. The use of an A36 web will reduce the allowable stress below $0.60F_y$ in accordance with Eq. 11.6.1(AISC–1.10.6).

For the steel section alone,

$$\text{Required } S_s = \frac{M_D}{0.60F_y} = \frac{90(12)}{60} = 18 \text{ in.}^3$$

The use of $0.60F_y$ for the steel section acting noncompositely assumes lateral support at spacings no greater than $32r_T\sqrt{C_b}$ (AISC–1.5.1.4.5(2.)). Note that AISC Formula (1.5-7) does *not* apply to hybrid plate girders.

(b) Select trial section. As a guideline for establishing depth, use L/d about 20 for the steel section alone for situations where deflection control is an important consideration. (See Sec. 16.12 for some details regarding deflection.)

$$d \approx \frac{L}{20} = \frac{30(12)}{20} = 18 \text{ in.}$$

In this case, the slab is relatively stiff and the steel beam will represent a smaller than usual proportion of the total effective areas; thus a shallower than 18 in. section may be acceptable. The AISC Manual "Composite Beam Selection Table" for 5-in. slab indicates a very light steel section in the range of 14 to 16 in. depth. Try a 14 in. deep and minimum $\frac{1}{4}$ in. thick web plate. Assume a symmetrical section,

$$S_s = \frac{I}{d/2} \approx \frac{2A_f(d/2)^2 + t_w d^3/12}{d/2}$$

$$= A_f d + \tfrac{1}{6}t_w d^2 = A_f d + \tfrac{1}{6}A_w d$$

$$\text{Required } A_f = \frac{\text{required } S_s - A_w d/6}{d} = \frac{18}{14} - \frac{0.25(14)}{6} = 0.70 \text{ sq in.}$$

for the steel section alone. An unsymmetrical section will likely give the most economical arrangement but with the small flange area required in this case, minimum size plates will be necessary; thus little advantage will accrue to an unsymmetrical section for this problem.

Try flanges $\frac{1}{4}\times 3$, $A_f = 0.75$ sq in. with $\frac{1}{4}\times 14$ web. Properties of the section:

$$\begin{aligned}
\text{Flanges, } 2(0.75)(7.125)^2 &= 76.1 \text{ in.}^4 \\
\text{Web, } 0.25(14)^3\tfrac{1}{12} &= \underline{57.2} \text{ in.}^4 \\
I &= 133.3 \text{ in.}^4
\end{aligned}$$

$$\text{Area} = 2(0.75) + 0.25(14) = 5.0 \text{ sq in.}$$

$$S_s = \frac{I}{d/2} = \frac{133.3}{7.25} = 18.4 \text{ in.}^3$$

Properties of the composite section:

$$b_E = 16t + b_f = 16(5) + 3 = 83 \text{ in.}$$

$$S_{tr} = 39 \text{ in.}^3 < 60 \text{ in.}^3 \text{ required} \qquad \text{NG}$$

Composite section modulus governs! Increase section to $\frac{1}{4} \times 16$ web, $\frac{1}{4} \times 4$ top flange, $\frac{3}{8} \times 6$ bottom flange.

Properties of the steel section:

Element	Area, A (sq in.)	Moment arm from top of slab, y (in.)	Ay (in.^{3})	Ay^2 (in.^{4})	I_0 (in.^{4})
Top flange	1.0	0.5	0.50	0.25	—
Web	4.0	8.25	33.0	272.25	85.3
Bottom flange	2.25	16.44	37.0	608.12	—
	7.25		70.5	881	85
				85	
				$I_{top} = 966$	

$$\bar{y}_{top} = \frac{70.5}{7.25} = 9.72 \text{ in.}$$

$$I = 966 - 7.25(9.72)^2 = 280 \text{ in.}^4$$

$$S_s(\text{bottom}) = \frac{280}{16.63 - 9.72} = 40.6 \text{ in.}^3$$

$$S_s(\text{top}) = \frac{280}{9.72} = 28.8 \text{ in.}^3$$

Properties of the composite section:

	A (sq in.)	y (in.)	Ay (in.^{3})	Ay^2 (in.^{4})	I_0 (in.^{4})
Slab	46.1	2.5	115.3	288	96
Steel section	7.25	14.72	106.7	1571	280
	53.35		222.0	1859	376
				376	
				$I_{top} = 2235$	

$$\bar{y}_{top} = \frac{222}{53.35} = 4.16 \text{ in.}$$

$$I = 2235 - 53.35(4.16)^2 = 1312 \text{ in.}^4$$

Note that 0.84 in. of concrete slab near the neutral axis is in tension (and presumably cracked) but was considered effective in computing the properties. AISC–1.11.2.2 states that concrete tension stresses shall be neglected. It makes little difference here and simplifies computing properties ($S = 74.7$ "exact" vs 75.1 as computed below).

$$S_{tr}(\text{tension flange}) = \frac{1312}{21.63 - 4.16} = 75.1 \text{ in.}^3$$

(c) Check stress on section. Because of the hybrid section, the A36 web will yield before the strength of the A514 flanges has been developed. The flange allowable stress is reduced according to AISC-1.10.6 to account for this (see Sec. 11.6). Using Eq. 11.6.1 (see also Table 16.9.1),

$$F_b' = F_b \left[\frac{12 + \beta(3\alpha - \alpha^3)}{12 + 2\beta} \right]$$

$$\beta = \frac{A_w}{A_f} = \frac{4.0}{2.25} = 1.78 \quad \text{for tension flange}$$

$$\alpha = \frac{\text{web } F_y}{\text{flange } F_y} = \frac{36}{100} = 0.36$$

$$F_b' = 60 \left[\frac{12 + 1.78[3(0.36) - (0.36)^3]}{12 + 2(1.78)} \right] = 53.4 \text{ ksi}$$

$$f_b = \frac{M_D + M_L}{S_{tr}} = \frac{300(12)}{75.1} = 47.9 \text{ ksi} < 53.4 \text{ ksi} \quad \text{OK}$$

For the noncomposite loading the compression stress on the steel beam alone controls. Since the compression flange has less area than the tension flange the allowable stress F_b' will be different.

$$\beta = \frac{A_w}{A_f} = \frac{4.0}{1.0} = 4.0 \quad \text{for compression flange}$$

$$F_b' = 60(0.807) = 48.4 \text{ ksi}$$

(from Table 16.9.1 for $\psi = 0.5$ and $\beta = 4$)

It is noted that the AISC formula does not account for the unsymmetrical section.

$$f_b = \frac{M_D}{S_s(\text{top})} = \frac{90(12)}{28.8} = 37.5 \text{ ksi} < 48.4 \text{ ksi} \quad \text{OK}$$

The stresses are somewhat low. Try reducing the tension flange to $\frac{5}{16} \times 6$. Retain $\frac{1}{4} \times 16$ web and $\frac{1}{4} \times 4$ top flange.

Computed steel section properties:

$$A = 6.88 \text{ sq in.} \quad \bar{y}_{\text{top}} = 9.35 \text{ in.} \quad I = 262 \text{ in.}^4 \quad S_{\text{top}} = 28.0 \text{ in.}^3$$

Computed composite section properties based on $b_E = 16(5) + 4 = 84$ in.:

$$A = 52.98 \text{ sq in.} \quad \bar{y}_{\text{top}} = 4.04 \text{ in.} \quad I = 1198 \text{ in.}^4 \quad S_{tr} = 68.4 \text{ in.}^3$$

For tension flange, $\beta = A_w / A_f = 2.13$,

$$F_b' = 52.4 \text{ ksi}$$

$$f_b = \frac{M_D + M_L}{S_{tr}} = \frac{300(12)}{68.4} = 52.6 \text{ ksi} \approx 52.4 \text{ ksi} \quad \text{OK}$$

On steel section at top flange,

$$f_b = \frac{M_D}{S_s} = \frac{90(12)}{28.0} = 28.6 \text{ ksi} < F_b' = 48.4 \text{ ksi} \qquad \text{OK}$$

Use ℞—$\frac{1}{4}$×16 (A36) for web; ℞—$\frac{1}{4}$×4 (A514) for top flange; and ℞—$\frac{5}{16}$×6 (A514) for bottom flange. See Fig. 16.11.5.

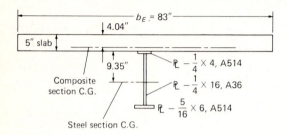

Fig. 16.11.5 Hybrid composite section of Example 16.11.4.

The design of shear connectors for this girder requires no new concepts and is therefore not illustrated.

16.12 DEFLECTIONS

In order to accurately determine the deflections of composite members, a number of factors must be taken into account which are not normally considered. These are: the method of construction, the separation of the live-load and dead-load moments, and the effect of creep and shrinkage in the concrete slab.

The method of construction determines the manner in which the composite cross section carries the dead-load stresses. If the steel is shored from below during the hardening of the concrete slab, the composite section will assume both the dead-load *and* the live-load stresses. However, if the steel beam is *not* shored, the steel beam will assume the dead-load stresses and the composite section will take only the live-load stresses.

If the construction is *without* shoring, the total deflection will be the sum of the dead-load deflection of the steel beam and the live-load deflection of the composite section.

Example 16.12.1

Determine the total deflection of the composite section in Example 16.9.1, and check against maximum deflection permitted by AISC if no shoring is used (see Fig. 16.12.1).

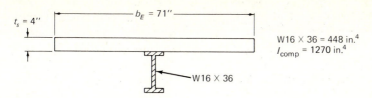

Fig. 16.12.1   Example 16.12.1.

SOLUTION
Dead-load deflection:

$$\Delta_{DL} = \frac{5wL^4}{384EI} = \frac{5\left[\dfrac{8(50)+36}{12,000}\right](28)^4(12)^4}{384(29,000)(447)} = 0.46 \text{ in.}$$

Live-load deflection:

$$\Delta_{LL} = \frac{5wL^4}{384EI_{comp}} = \frac{5\left[\dfrac{8(150)}{12,000}\right](28)^4(12)^4}{384(29,000)(1270)} = 0.45 \text{ in.}$$

Total deflection:

$$\Delta = \Delta_{DL} + \Delta_{LL} = 0.46 + 0.45 = 0.91 \text{ in.}$$

Check maximum deflection permitted,

$$\text{Max } \Delta = \frac{L}{360} = \frac{28(12)}{360} = 0.93 \text{ in.}$$

Since $\Delta < 0.93$ in., deflection criterion is satisfied.

If shoring provides support during the hardening of the concrete, the total deflection will be a function of the total composite section. Account must be taken of the fact that concrete is subject to creep under long time loadings and that shrinkage will occur. This inelastic behavior may be approximated by multiplying the modular ratio n by a factor to reduce the net effective width. The result is a reduced moment of inertia for the composite section which is used in computing the dead-load deflection. The live-load deflection is then usually computed on the basis of the elastic composite moment of inertia. Occasionally, the conservative approach is to use the reduced composite moment of inertia when the live loads are expected to remain for extended periods of time.

Because the concrete slab in building construction is normally not too thick (say $t_s \leq 5$ in.) creep deflection is not considered to be a problem. AISC gives no indication that one need be concerned with anything but live-load short-time deflection. The ACI–ASCE Joint Committee [15] recommends using one-half the concrete modulus of elasticity, $E_c/2$ instead of E_c when computing sustained load creep deflection.

AASHTO [14] uses $E_c/3$ instead of E_c. Such arbitrary procedures can at best give an estimate of creep effects, probably not better than ±30%. The steel section, exhibiting no creep, and representing the principal carrying element, insures that creep problems will usually be minimal.

More accurate procedures for computing deflections to account for creep and shrinkage on composite steel-concrete beams are available in a paper by Roll [22], and particularly in the work of Branson [23].

16.13 CONTINUOUS BEAMS

It has been traditional to design the positive moment region on continuous beams as a composite section and the negative moment region as a noncomposite section. However, some composite action has been known to exist in the negative moment region. Significant contribution to the knowledge about the strength of continuous composite beams has been made by Barnard and Johnson [24], Johnson, Van Dalen, and Kemp [25], Park [26], Daniels and Fisher [27], and more recently by Hamada and Longworth [28, 29].

According to both AISC–1.11.2.2 and AASHTO–1.7.48 and 1.7.63, the steel reinforcement that extends parallel to the beam span and that is contained within the concrete slab effective width b_E *may be used* as part of the effective composite section. This is true for both positive and negative bending regions. The inclusion of such steel reinforcement has little effect in positive moment regions but can help in negative moment regions. In the negative moment region the concrete is ordinarily all in tension and is therefore not considered effective (AISC–1.11.2.2 and AASHTO–1.7.48 and 1.7.63).

When the reinforcing steel in the concrete slab is utilized as part of the composite section, the force developed by it must be transferred in shear by mechanical shear connectors. The ultimate force developed would be

$$T(\text{for } -M \text{ region}) = A_{sr}F_{yr} \tag{16.13.1}$$

$$C(\text{for } +M \text{ region}) = A'_s F_{yr} \tag{16.13.2}$$

where A_{sr} = total area of longitudinal reinforcing steel at the interior support located within the effective flange width b_E

A'_s = total area of longitudinal compression steel acting with the concrete slab at the location of maximum positive moment and lying within the effective width b_E

F_{yr} = specified minimum yield stress of the longitudinal reinforcing steel

In order to use a working stress method, division by a factor of 2 is used to reduce these forces into the service load range (see discussion relating to Eqs. 16.8.7 and 16.8.8). Thus the service load horizontal shear force to be designed for in the *negative moment zone* is

$$V_h = \frac{A_{sr}F_{yr}}{2}$$
(16.13.3)

In the *positive moment zone*, when the compression steel is included when computing the composite section properties, Eq. 16.13.2 divided by 2 is added to Eq. 16.8.7; thus

$$V_h = \frac{0.85f'_cA_c}{2} + \frac{A'_sF_{yr}}{2}$$
(16.13.4)

Equations 16.13.3 and 16.13.4 are as prescribed by AISC–1.11.4. AASHTO–1.7.48 uses Eq. 16.13.3 but makes no reference to utilizing the compression steel in positive moment regions; however, the AASHTO load factor provisions of AASHTO–1.7.62 do use the concept of Eq. 16.13.4.

Under both AISC and AASHTO the inclusion of the longitudinal reinforcing bars A_{sr} in the negative moment zone appears optional. If A_{sr} is included in computing properties the horizontal shear V_h produced by such bars must be developed by shear connectors. Under AASHTO–1.7.48(E)(c), there are additional shear connectors required at points of contraflexure when A_{sr} is *not* utilized in computing section properties. The minimum number N_e of added connectors for the fatigue-related requirement is

$$N_e = \frac{A_{sr}f_r}{Z_r}$$
(16.13.5)

where f_r = range of stress due to live load plus impact, in the slab reinforcement over the support (in lieu of more accurate computations, f_r may be taken as equal to 10,000 psi)

Z_r = allowable range of horizontal shear on an individual shear connector (see Eq. 16.8.15)

As discussed in Sec. 16.7, the usual failure mode in the positive moment region is crushing of the concrete slab. This assumes no shear connector failure and no longitudinal splitting or shear failure in the concrete slab. In the negative moment region, the usual failure mode is local buckling [29].

Under AISC and AASHTO current provisions, the usual lateral buckling provisions for noncomposite steel sections apply to the negative moment regions of continuous composite beams. In the use of the lateral-torsional buckling formulas of AISC–1.5.1.4.5(2.) and AASHTO–1.7.1 or 1.7.59(D) the point of contraflexure may generally be treated as a brace point if the top flange in the negative moment region is braced. Local buckling limitations for the flange and web also apply (AISC–1.5.1.4.1 and 1.9; AASHTO–1.7.43 or 1.7.59).

Hamada and Longworth [28] indicate that the negative moment region of composite continuous beams has measurably greater resistance to lateral buckling than a noncomposite steel section having no concrete slab attached to its top flange. They conclude that "The ultimate moment capacity of composite beams in negative bending is affected by local flange buckling unless the compression flange is stiffened by a cover plate." The cover plate increases the torsional rigidity and may cause lateral buckling to be more critical than local buckling. The most recent local buckling recommendations [29] are summarized as follows for cases where the compression flange is a single plate element:

For $A_{sr}/A_w \leq 1.0$,

$$\frac{b_f}{2t_f} \leq \frac{54}{\sqrt{F_y}} \qquad (16.13.6)^*$$

For $1.0 < A_{sr}/A_w \leq 2.0$,

$$\frac{b_f}{2t_f} \leq \frac{49}{\sqrt{F_y}} \qquad (16.13.7)^*$$

These local buckling limitations for the compression flange are more conservative than required under AISC or AASHTO, even for "compact sections." It would appear that present design procedures are conservative with regard to lateral-torsional buckling in the negative moment regions of continuous beams but may not be conservative with regard to local buckling.

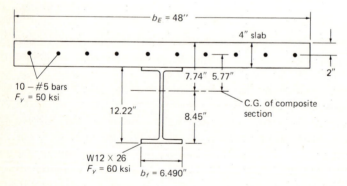

Fig. 16.13.1 Composite section for negative bending of Example 16.13.1.

---

* For SI, with F_y in MPa,

$$\frac{b_f}{2t_f} \leq \frac{142}{\sqrt{F_y}} \qquad \text{for } \frac{A_{sr}}{A_w} \leq 1.0 \qquad (16.13.6)$$

$$\frac{b_f}{2t_f} \leq \frac{129}{\sqrt{F_y}} \qquad \text{for } 1.0 < \frac{A_{sr}}{A_w} \leq 2.0 \qquad (16.13.7)$$

Example 16.13.1

Investigate the section of Fig. 16.13.1 subject to a negative bending moment of 135 ft-kips acting as a composite section under the AISC Spec. The reinforcing steel in the concrete slab has $F_y = 50$ ksi and the W12×26 section has $F_y = 60$ ksi.

SOLUTION

(a) Compute section properties under negative bending. The concrete slab is not considered to participate since it is on the tension side of the neutral axis.

	Area, A (sq in.)	Moment arm from top, y (in.)	Ay (in.^{3})	Ay^2 (in.^{4})	I_0 (in.^{4})
#5 bars ($A_s = 0.31$)	3.10	2.0	6.2	12	—
W12×26	7.65	10.11	77.3	782	204
	10.75		83.5	794	204
				204	
				$I_{top} = 998$	

$$\bar{y}_{top} = \frac{83.5}{10.75} = 7.77 \text{ in.}$$

$$I = 998 - 10.75(7.77)^2 = 349 \text{ in.}^4$$

$$S_{tr}(\text{bottom}) = \frac{349}{8.45} = 41.3 \text{ in.}^3$$

$$S_{tr}(\text{at } \#5 \text{ bars}) = \frac{349}{5.77} = 60.5 \text{ in.}^3$$

(b) Check stresses. On the composite section,

$$f_b = \frac{M_{D+L}}{S_{tr}} = \frac{135(12)}{41.3} = 39.2 \text{ ksi}$$

Assuming that the distance of lateral unsupport does not exceed $L_c = 5.3$ ft (AISC–1.5.1.4.1 or text Appendix Table A3), the section is "partly compact" with an allowable stress F_b equal to 39.5 ksi according to AISC Formula (1.5-6a) (see text Appendix Table A3). The W12×26 satisfies the "compact section" local buckling requirement for the web ($d/t_w < 82.6$) but exceeds the limit for the flange ($b_f/(2t_f) > 8.4$). Thus

$$f_b = 39.2 \text{ ksi} < F_b = 39.5 \text{ ksi} \qquad \text{OK}$$

If the section had been required to satisfy the more conservative requirements of Eqs. 16.13.6 or 16.13.7 for local flange buckling, the section would not have qualified. Thus

$$W12 \times 26: \qquad \frac{b_f}{2t_f} = 8.5 > \left[\frac{54}{\sqrt{F_y}} = 7.0 \right] \qquad \text{NG}$$

(c) Shear connectors. Use $\frac{5}{8}$ in. diam $\times 2\frac{1}{2}$ in. long studs: $q = 8.0$ kips for $f'_c = 3000$ psi concrete.

$$V_h = \frac{A_{sr}F_{yr}}{2} = \frac{10(0.31)50}{2} = 77.5 \text{ kips}$$

The number N_1 of studs required between the maximum negative moment location (the support) and the point of contraflexure is

$$N_1 = \frac{V_h}{q} = \frac{77.5}{8.0} = 9.7, \qquad \text{say } 10$$

Use 10 shear connectors for the negative moment region.

SELECTED REFERENCES

1. H. M. MacKay, P. Gillespie, and C. Leluau, "Report on the Strength of Steel I-Beams Haunched with Concrete." *Engineering Journal*, Engineering Institute of Canada, 6, 8 (1923), 365–369.
2. R. A. Caughey, "Composite Beams of Concrete and Structural Steel," Proceedings, 41st Annual Meeting, Iowa Engineering Society, 1929.
3. Theodore von Kármán, "Die Mittragende Breitte," *August-Föppel-Festschrift*, 1924. (See also Collected Works of Theodore von Kármán, Volume II, p. 176).
4. Eric Reissner, "Über die Berechnung von Plattenbalkan," *Der Stahlbau*, 26, December 1934.
5. Gottfried Brendel, "Strength of the Compression Slab of T-Beams Subject to Simple Bending," *ACI Journal, Proceedings*, 61, January 1964, 57–76.
6. Ivan M. Viest, "Review of Research on Composite Steel-Concrete Beams," *Journal of Structural Division*, ASCE, 86, ST6 (June 1960), 1–21.
7. M. Ros, Les constructions acier-béton, system alpha," *L'Ossature Metallique* (Bruxelle), 3, 4 (1934), 195–208.
8. Ivan M. Viest, Chairman, "Composite Steel-Concrete Construction," Report of the Subcommittee on the State-of-the-Art Survey of the Task Committee on Composite Construction of the Committee on Metals of the Structural Division, *Journal of the Structural Division*, ASCE, 100, ST5 (May 1974), 1085–1139.
9. John P. Cook, *Composite Construction Methods*. New York; John Wiley & Sons, Inc., 1977.
10. Charles G. Salmon and James M. Fisher, "Composite Steel-Concrete Construction," *Handbook of Composite Construction Engineering*, ed. by Gajanan Sabnis. New York: D. Van Nostrand, 1979, Chap. 2.
11. S. Timoshenko and J. Goodier, *Theory of Elasticity*. New York: McGraw-Hill Book Company, Inc., 1959, Chap. 6.
12. John E. Johnson and Albert D. M. Lewis, "Structural Behavior in a Gypsum

Roof-Deck System," *Journal of Structural Division*, ASCE, 92, ST2 (April 1966), 283–296.

13. ACI Committee 318, *Building Code Requirements for Reinforced Concrete.* Detroit, Mich.: American Concrete Institute, 1977.

14. *Standard Specifications for Highway Bridges*, 12th Edition, American Association of State Highway and Transportation Officials, Washington, D.C., 1977.

15. Joint ASCE-ACI Committee on Composite Construction, "Tentative Recommendations for the Design and Construction of Composite Beams and Girders for Buildings," *Journal of Structural Division*, ASCE, 86, ST12 (December 1960), 73–92.

16. Roger G. Slutter and George C. Driscoll, "Flexural Strength of Steel-Concrete Composite Beams," *Journal of Structural Division*, ASCE, 91, ST2 (April 1965), 71–99.

17. Jorgen G. Ollgaard, Roger G. Slutter, and John W. Fisher, "Shear Strength of Stud Connectors in Lightweight and Normal-Weight Concrete," *Engineering Journal*, AISC, 8, 2 (April 1971), 55–64.

18. Jay B. McGarraugh and J. W. Baldwin, Jr., "Lightweight Concrete-on-Steel Composite Beams," *Engineering Journal*, AISC, 8, 3 (July 1971), 90–98.

19. Roger G. Slutter and John W. Fisher, "Fatigue of Shear Connectors," *Highway Research Record No. 147*, Highway Research Board, 1966, pp. 65–88.

20. C. G. Schilling, Chairman, "Design of Hybrid Steel Beams," Report of Subcommittee 1 on Hybrid Beam and Girders, Joint ASCE–AASHO Committee on Flexural Members, *Journal of Structural Division*, ASCE, 94, ST6 (June 1968), 1397–1426.

21. Charles G. Schilling, "Bending Behavior of Composite Hybrid Beams," *Journal of Structural Division*, ASCE, 94, ST8 (August 1968), 1945–1964.

22. Frederic Roll, "Effects of Differential Shrinkage and Creep on a Composite Steel-Concrete Structure," *Designing for Effects of Creep, Shrinkage, Temperature in Concrete Structures*, SP-27. Detroit, Mich.: American Concrete Institute, 1971 (pp. 187–214).

23. Dan E. Branson, *Deformation of Concrete Structures.* New York: McGraw-Hill Book Company, Inc., 1977.

24. P. R. Barnard and R. P. Johnson, "Plastic Behavior of Continuous Composite Beams," *Proceedings*, Institute of Civil Engineers, October 1965.

25. R. P. Johnson, K. Van Dalen, and A. R. Kemp, "Ultimate Strength of Continuous Composite Beams," *Proceedings of the Conference on Structural Steelwork*, British Constructional Steelwork Association, November 1967.

26. Robert Park, "The Ultimate Strength of Continuous Composite Beams," *Civil Engineering Transactions*, Australia, CE9, October 1967.

27. J. H. Daniels and J. W. Fisher, "Static Behavior of Continuous Composite Beams," *Fritz Engineering Laboratory Report* No. 324.2, Lehigh University, Bethlehem, Pa., March 1967.

28. Sumio Hamada and Jack Longworth, "Buckling of Composite Beams in Negative Bending," *Journal of Structural Division*, ASCE, 100, ST11 (November 1974), 2205–2222.

29. Sumio Hamada and Jack Longworth, "Ultimate Strength of Continuous Composite Beams," *Journal of Structural Division*, ASCE, 102, ST7 (July 1976), 1463–1478.

PROBLEMS

16.1. Determine the composite section properties for the given section, using AISC procedures.

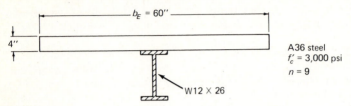

Prob. 16.1

16.2. Determine the composite section properties for the member of Prob. 16.1 according to AISC procedures if a W18×50 section of A572 Grade 50 steel is used.

16.3. Determine the ultimate moment capacity of the composite section in Prob. 16.1 using ultimate strength concepts.

16.4. Determine the ultimate moment capacity of the composite section in Prob. 16.2 using ultimate strength concepts.

16.5. Determine the number of shear connectors for the beam of Prob. 16.1 required by AISC if $\frac{1}{2}$ in. diam×2 in. studs are used.

16.6. Repeat Prob. 16.5 using A572 Grade 60 steel instead of A36.

16.7. Repeat Prob. 16.5 using a W18×50 and A572 Grade 50 steel instead of a W12×26 and A36 steel.

16.8. Using AISC procedures, select a W section for span BD without using a bottom cover plate on the basis of composite design, and design the required shear connectors. Use A36 steel, a 4 in. thick slab with $f_c' = 3000$ psi and $n = 9$. Assume a live load of 100 psf and no temporary shoring is used. Limit the live-load deflection to $L/360$.

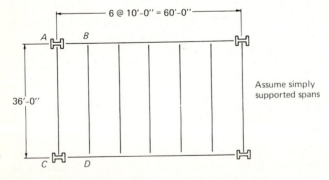

Prob. 16.8. Framing plan

16.9. Repeat Prob. 16.8 using a cover plate on the bottom flange.

16.10. Repeat Prob. 16.8 using temporary shoring.

16.11. Repeat Prob. 16.8 selecting a W section for the exterior span *AB*.

16.12. Repeat Prob. 16.8 using A572 Grade 60 steel and a W12 section.

16.13. Design the lightest weight composite section for use as an interior floor beam of an office building. The dead-load deflection is not to exceed $\frac{7}{8}$ in. and the live-load deflection is limited to $L/360$. Temporary shoring will not be used. If tables are used to select composite section, verify all properties by showing computations. How much saving in steel weight is achieved by using composite construction compared to noncomposite?

Data:

Simple span, 28 ft	Live load, 125 psf
Beam spacing, 9 ft	Partitions, 25 psf
Slab thickness, $4\frac{1}{2}$ in.	Ceiling, 7 psf
$f'_c = 3000$ psi, A36 steel, $n = 9$	

16.14. Repeat Prob. 16.13 using minimum feasible depth and shores.

16.15. Repeat Prob. 16.13 using A572 grade 50 steel.

16.16. Repeat Prob. 16.14 using A572 Grade 50 steel.

16.17. Repeat Prob. 16.13 using ultimate strength method (not AISC method) and a factor of safety of 2.5.

Appendix

Table A1 Approximate Radius of Gyration

Section	Section	Section
$r_x = 0.29h$ $r_y = 0.29b$	$r_x = 0.42h$ $r_y = 0.42b$	$r_x = 0.31h$ $r_y = 0.48b$
$r_x = 0.40h$ $h = \text{mean } h$	$r_y = \text{same as for 2 } L$	$r_x = 0.37h$ $r_y = 0.28b$
$r_x = 0.25h$	$r_x = 0.42h$ $r_y = \text{same as for 2 } L$	$r_x = 0.31h$
$r = \sqrt{\dfrac{H^2 + h^2}{16}}$ $r = 0.35H_m$	$r_x = 0.39h$ $r_y = 0.21b$	$r_x = 0.31h$
$r_x = 0.31h$ $r_y = 0.31h$ $r_z = 0.197h$	$r_x = 0.45h$ $r_y = 0.235b$	$r_x = 0.40h$ $r_y = 0.21b$
$r_x = 0.29h$ $r_y = 0.32b$ $r_z = 0.18\dfrac{h+b}{2}$	$r_x = 0.36h$ $r_y = 0.45b$	$r_x = 0.38h$ $r_y = 0.22b$
$r_x = 0.31h$ $r_y = 0.215b$ $= b(0.21+0.02s)$	$r_x = 0.36h$ $r_y = 0.60b$	$r_x = 0.39h$
$r_x = 0.32h$ $r_y = 0.21b$ $= b(0.19+0.02s)$	$r_x = 0.36h$ $r_y = 0.53b$	$r_x = 0.35h$
$r_x = 0.29h$ $r_y = 0.24b$ $= b(0.23+0.02s)$	$r_x = 0.39h$ $r_y = 0.55b$	$r_x = 0.435h$ $r_y = 0.25b$
$r_x = 0.30h$ $r_y = 0.17b$	$r_x = 0.42h$ $r_y = 0.32b$	$r_x = 0.42h$
$r_x = 0.25h$ $r_y = 0.21b$	$r_x = 0.44h$ $r_y = 0.28b$	$r_x = 0.42h$
$r_x = 0.21h$ $r_y = 0.21b$ $r_z = 0.19h$	$r_x = 0.50h$ $r_y = 0.28b$	$r_x = 0.285h$ $r_y = 0.37b$
$r_x = 0.38h$ $r_y = 0.19b$	$r_x = 0.39h$ $r_y = 0.21b$	$r_x = 0.42h$ $r_y = 0.23b$

* J.A.L. Waddell, "Bridge Engineering," John Wiley & Sons, Inc., New York, 1916. Reproduced by permission.

Table A2   Torsional Properties

O = shear center	J = torsion constant, C_w = warping constant
G = centroid	I_p = polar moment of inertia about shear center

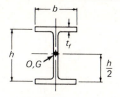

$$J = \tfrac{1}{3}(2bt_f^3 + ht_w^3)$$

$$C_w = \frac{I_f h^2}{2} = \frac{t_f b^3 h^2}{24} = \frac{h^2 I_y}{4}$$

$$I_p = I_x + I_y$$

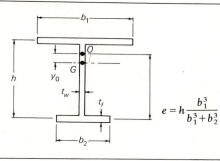

$$J = \tfrac{1}{3}(b_1 t_f^3 + b_2 t_f^3 + ht_w^3)$$

$$C_w = \frac{t_f h^2}{12}\left(\frac{b_1^3 b_2^3}{b_1^3 + b_2^3}\right)$$

$$e = h\frac{b_1^3}{b_1^3 + b_2^3}$$

$$I_p = I_y + I_x + Ay_0^2$$

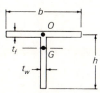

$$J = \tfrac{1}{3}(bt_f^3 + ht_w^3)$$

$$C_w = \frac{1}{36}\left(\frac{b^3 t_f^3}{4} + h^3 t_w^3\right)$$

$$\approx \text{zero for small } t$$

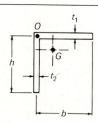

$$J = \tfrac{1}{3}(bt_1^3 + ht_2^3)$$

$$C_w = \frac{1}{36}(b^3 t_1^3 + h^3 t_2^3)$$

$$\approx \text{zero for small } t$$

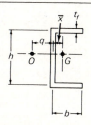

$$J = \tfrac{1}{3}(2bt_f^3 + ht_w^3)$$

$$C_w = \frac{t_f b^3 h^2}{12}\left(\frac{3bt_f + 2ht_w}{6bt_f + ht_w}\right) = \frac{h^2}{4}(I_y + A\bar{x}^2 - q\bar{x}A)$$

$$q = \frac{th^2 b^2}{4I_x}$$

Table A3 Elastic Section Modulus S_x Including Limiting Values of Unbraced Length (L_c and L_u) and Maximum Allowable Flexural Stress*

SX	SHAPE	COMPACT SECTION-SATISFIES AISC 1.5.1.4.1 (YES OR NO)						
		FY= 36.	42.	45.	50.	55.	60.	65.
1280.	W14X730	YES	YES	YES	YES	YES	YES	YES
		LC= 18.9	17.5	16.9	16.0	15.3	14.6	14.1
		LU= 181.4	155.5	145.1	130.6	118.7	108.8	100.5
		FB= 24.0	28.0	29.7	33.0	36.3	39.6	42.9
1150.	W14X665	YES	YES	YES	YES	YES	YES	YES
		LC= 18.6	17.2	16.7	15.8	15.1	14.4	13.9
		LU= 170.7	146.3	136.5	122.9	111.7	102.4	94.5
		FB= 24.0	28.0	29.7	33.0	36.3	39.6	42.9
1110.	W36X300	YES	YES	YES	YES	YES	YES	YES
		LC= 17.6	16.3	15.7	14.9	14.2	13.6	13.1
		LU= 35.3	30.2	28.2	25.4	23.1	21.2	19.5
		FB= 24.0	28.0	29.7	33.0	36.3	39.6	42.9
1040.	W14X605	YES	YES	YES	YES	YES	YES	YES
		LC= 18.4	17.0	16.4	15.6	14.9	14.2	13.7
		LU= 160.3	137.4	128.3	115.4	104.9	96.2	88.8
		FB= 24.0	28.0	29.7	33.0	36.3	39.6	42.9
1030.	W36X280	YES	YES	YES	YES	YES	YES	YES
		LC= 17.5	16.2	15.7	14.9	14.2	13.6	13.0
		LU= 33.0	28.3	26.4	23.8	21.6	19.8	18.3
		FB= 24.0	28.0	29.7	33.0	36.3	39.6	42.9
953.	W36X260	YES	YES	YES	YES	YES	YES	YES
		LC= 17.5	16.2	15.6	14.8	14.1	13.5	13.0
		LU= 30.4	26.1	24.3	21.9	19.9	18.3	16.9
		FB= 24.0	28.0	29.7	33.0	36.3	39.6	42.9
931.	W14X550	YES	YES	YES	YES	YES	YES	YES
		LC= 18.2	16.8	16.2	15.4	14.7	14.1	13.5
		LU= 129.0	110.6	103.2	92.9	84.5	77.4	71.5
		FB= 24.0	28.0	29.7	33.0	36.3	39.6	42.9
895.	W36X245	YES	YES	YES	YES	YES	YES	YES
		LC= 17.4	16.1	15.6	14.8	14.1	13.5	13.0
		LU= 28.6	24.5	22.9	20.6	18.7	17.2	15.8
		FB= 24.0	28.0	29.7	33.0	36.3	39.6	42.9
838.	W14X500	YES	YES	YES	YES	YES	YES	YES
		LC= 18.0	16.6	16.1	15.2	14.5	13.9	13.4
		LU= 140.6	120.5	112.5	101.2	92.0	84.4	77.9
		FB= 24.0	28.0	29.7	33.0	36.3	39.6	42.9
837.	W36X230	YES	YES	YES	YES	YES	YES	YES
		LC= 17.4	16.1	15.5	14.8	14.1	13.5	12.9
		LU= 26.8	22.9	21.4	19.3	17.5	16.1	14.8
		FB= 24.0	28.0	29.7	33.0	36.3	39.6	42.9
829.	W33X241	YES	YES	YES	YES	YES	YES	YES
		LC= 16.7	15.5	15.0	14.2	13.5	13.0	12.5
		LU= 30.1	25.8	24.1	21.7	19.7	18.0	16.7
		FB= 24.0	28.0	29.7	33.0	36.3	39.6	42.9
757.	W33X221	YES	YES	YES	YES	YES	YES	YES
		LC= 16.7	15.4	14.9	14.2	13.5	12.9	12.4
		LU= 27.5	23.6	22.0	19.8	18.0	16.5	15.2
		FB= 24.0	28.0	29.7	33.0	36.3	39.6	42.9
756.	W14X455	YES	YES	YES	YES	YES	YES	YES
		LC= 17.8	16.5	15.9	15.1	14.4	13.8	13.2
		LU= 131.5	112.7	105.2	94.7	86.1	78.9	72.9
		FB= 24.0	28.0	29.7	33.0	36.3	39.6	42.9
719.	W36X210	YES	YES	YES	YES	YES	YES	YES
		LC= 12.9	11.9	11.5	10.9	10.4	10.0	9.6
		LU= 20.9	17.9	16.7	15.0	13.7	12.5	11.6
		FB= 24.0	28.0	29.7	33.0	36.3	39.6	42.9

* Satisfying local buckling requirements of AISC–1.5.1.4.1 indicated by YES, NO, or PTY (partly). PTY indicates F_b computed from AISC Formula (1.5–5a).

Table A3 *(Continued)*

		FY=	36.	42.	45.	50.	55.	60.	65.
707.	W14X426		YES	YES	YES	YES	YES	YES	YES
		LC=	17.6	16.3	15.8	15.0	14.3	13.7	13.1
		LU=	125.6	107.7	100.5	90.5	82.2	75.4	69.6
		FB=	24.0	28.0	29.7	33.0	36.3	39.6	42.9
684.	W33X201		YES	YES	YES	YES	YES	YES	YES
		LC=	16.6	15.4	14.9	14.1	13.4	12.9	12.4
		LU=	24.9	21.3	19.9	17.9	16.3	14.9	13.8
		FB=	24.0	28.0	29.7	33.0	36.3	39.6	42.9
664.	W36X194		YES	YES	YES	YES	YES	YES	YES
		LC=	12.8	11.8	11.4	10.9	10.3	9.9	9.5
		LU=	19.4	16.6	15.5	13.9	12.7	11.6	10.7
		FB=	24.0	28.0	29.7	33.0	36.3	39.6	42.9
663.	W30X211		YES	YES	YES	YES	YES	YES	YES
		LC=	15.9	14.8	14.3	13.5	12.9	12.4	11.9
		LU=	29.7	25.5	23.8	21.4	19.5	17.8	16.5
		FB=	24.0	28.0	29.7	33.0	36.3	39.6	42.9
656.	W14X398		YES	YES	YES	YES	YES	YES	YES
		LC=	17.5	16.2	15.7	14.9	14.2	13.6	13.0
		LU=	119.5	102.4	95.6	86.0	78.2	71.7	66.2
		FB=	24.0	28.0	29.7	33.0	36.3	39.6	42.9
623.	W36X182		YES	YES	YES	YES	YES	YES	YES
		LC=	12.7	11.8	11.4	10.8	10.3	9.9	9.5
		LU=	18.2	15.6	14.5	13.1	11.9	10.9	10.1
		FB=	24.0	28.0	29.7	33.0	36.3	39.6	42.9
607.	W14X370		YES	YES	YES	YES	YES	YES	YES
		LC=	17.4	16.1	15.6	14.8	14.1	13.5	12.9
		LU=	113.2	97.0	90.6	81.5	74.1	67.9	62.7
		FB=	24.0	28.0	29.7	33.0	36.3	39.6	42.9
598.	W30X191		YES	YES	YES	YES	YES	YES	YES
		LC=	15.9	14.7	14.2	13.5	12.8	12.3	11.8
		LU=	26.9	23.1	21.5	19.4	17.6	16.1	14.9
		FB=	24.0	28.0	29.7	33.0	36.3	39.6	42.9
580.	W36X170		YES	YES	YES	YES	YES	YES	YES
		LC=	12.7	11.8	11.4	10.8	10.3	9.8	9.4
		LU=	16.9	14.5	13.6	12.2	11.1	10.4	10.0
		FB=	24.0	28.0	29.7	33.0	36.3	39.6	42.9
559.	W14X342		YES	YES	YES	YES	YES	YES	YES
		LC=	17.3	16.0	15.4	14.7	14.0	13.4	12.9
		LU=	106.7	91.4	85.3	76.8	69.8	64.0	59.1
		FB=	24.0	28.0	29.7	33.0	36.3	39.6	42.9
542.	W36X160		YES	YES	YES	YES	YES	YES	YES
		LC=	12.7	11.7	11.3	10.7	10.2	.9.4	8.7
		LU=	15.7	13.5	12.6	11.4	10.8	10.4	10.0
		FB=	24.0	28.0	29.7	33.0	36.3	39.6	42.9
539.	W30X173		YES	YES	YES	YES	YES	YES	YES
		LC=	15.8	14.6	14.1	13.4	12.8	12.3	11.8
		LU=	24.3	20.8	19.4	17.5	15.9	14.6	13.4
		FB=	24.0	28.0	29.7	33.0	36.3	39.6	42.9
506.	W14X311		YES	YES	YES	YES	YES	YES	YES
		LC=	17.1	15.9	15.3	14.5	13.9	13.3	12.7
		LU=	99.2	85.0	79.4	71.4	64.9	59.5	54.9
		FB=	24.0	28.0	29.7	33.0	36.3	39.6	42.9
504.	W36X150		YES	YES	YES	YES	YES	YES	YES
		LC=	12.6	11.7	11.3	10.5	9.5	8.7	8.1
		LU=	14.5	12.5	11.9	11.3	10.7	10.3	9.9
		FB=	24.0	28.0	29.7	33.0	36.3	39.6	42.9

Table A3 (*Continued*)

			FY=	36.	42.	45.	50.	55.	60.	65.
502.	W27X178			YES	YES	YES	YES	YES	YES	YES
		LC=		14.9	13.8	13.3	12.6	12.0	11.5	11.1
		LU=		27.9	23.9	22.3	20.1	18.3	16.7	15.5
		FB=		24.0	28.0	29.7	33.0	36.3	39.6	42.9
487.	W33X152			YES	YES	YES	YES	YES	YES	YES
		LC=		12.2	11.3	10.9	10.4	9.9	9.5	9.1
		LU=		16.9	14.5	13.5	12.1	11.0	10.1	9.7
		FB=		24.0	28.0	29.7	33.0	36.3	39.6	42.9
459.	W14X283			YES	YES	YES	YES	YES	YES	YES
		LC=		17.0	15.7	15.2	14.4	13.8	13.2	12.7
		LU=		92.2	79.1	73.8	66.4	60.4	55.3	51.1
		FB=		24.0	28.0	29.7	33.0	36.3	39.6	42.9
455.	W27X161			YES	YES	YES	YES	YES	YES	YES
		LC=		14.8	13.7	13.2	12.6	12.0	11.5	11.0
		LU=		25.4	21.8	20.3	18.3	16.6	15.2	14.1
		FB=		24.0	28.0	29.7	33.0	36.3	39.6	42.9
448.	W33X141			YES	YES	YES	YES	YES	YES	YES
		LC=		12.2	11.3	10.9	10.3	9.9	9.2	8.5
		LU=		15.4	13.2	12.3	11.1	10.5	10.0	9.6
		FB=		24.0	28.0	29.7	33.0	36.3	39.6	42.9
439.	W36X135			YES	YES	YES	YES	YES	YES	YES
		LC=		12.3	10.5	9.8	8.9	8.0	7.4	6.8
		LU=		13.0	12.0	11.6	11.0	10.5	10.1	9.7
		FB=		24.0	28.0	29.7	33.0	36.3	39.6	42.9
415.	W14X257			YES	YES	YES	YES	YES	YES	YES
		LC=		16.9	15.6	15.1	14.3	13.7	13.1	12.6
		LU=		85.4	73.2	68.4	61.5	55.9	51.3	47.3
		FB=		24.0	28.0	29.7	33.0	36.3	39.6	42.9
414.	W24X162			YES	YES	YES	YES	YES	YES	YES
		LC=		13.7	12.7	12.2	11.6	11.1	10.6	10.2
		LU=		29.3	25.1	23.4	21.1	19.2	17.6	16.2
		FB=		24.0	28.0	29.7	33.0	36.3	39.6	42.9
411.	W27X146			YES	YES	YES	YES	YES	YES	YES
		LC=		14.7	13.6	13.2	12.5	11.9	11.4	11.0
		LU=		23.0	19.7	18.4	16.6	15.1	13.8	12.8
		FB=		24.0	28.0	29.7	33.0	36.3	39.6	42.9
406.	W33X130			YES	YES	YES	YES	YES	YES	YES
		LC=		12.1	11.2	10.9	9.9	9.0	8.3	7.6
		LU=		13.8	11.8	11.4	10.8	10.3	9.9	9.5
		FB=		24.0	28.0	29.7	33.0	36.3	39.6	42.9
380.	W30X132			YES	YES	YES	YES	YES	YES	YES
		LC=		11.1	10.3	10.0	9.4	9.0	8.6	8.3
		LU=		16.1	13.8	12.9	11.6	10.5	9.7	8.9
		FB=		24.0	28.0	29.7	33.0	36.3	39.6	42.9
375.	W14X233			YES	YES	YES	YES	YES	YES	YES
		LC=		16.8	15.5	15.0	14.2	13.6	13.0	12.5
		LU=		78.9	67.6	63.1	56.8	51.6	47.3	43.7
		FB=		24.0	28.0	29.7	33.0	36.3	39.6	42.9
371.	W24X146			YES	YES	YES	YES	YES	YES	YES
		LC=		13.6	12.6	12.2	11.6	11.0	10.5	10.1
		LU=		26.3	22.6	21.1	18.9	17.2	15.8	14.6
		FB=		24.0	28.0	29.7	33.0	36.3	39.6	42.9
359.	W33X118			YES	YES	YES	YES	YES	YES	YES
		LC=		12.0	10.3	9.6	8.6	7.8	7.2	6.6
		LU=		12.6	11.7	11.3	10.7	10.2	9.8	9.4
		FB=		24.0	28.0	29.7	33.0	36.3	39.6	42.9

Table A3 (*Continued*)

	FY=	36.	42.	45.	50.	55.	60.	65.
355.	W30X124	YES	YES	YES	YES	YES	YES	YES
	LC=	11.1	10.3	9.9	9.4	9.0	8.6	8.3
	LU=	15.0	12.9	12.0	10.8	9.8	9.1	8.8
	FB=	24.0	28.0	29.7	33.0	36.3	39.6	42.9
338.	W14X211	YES	YES	YES	YES	YES	YES	YES
	LC=	16.7	15.4	14.9	14.2	13.5	12.9	12.4
	LU=	72.6	62.2	58.1	52.3	47.5	43.6	40.2
	FB=	24.0	28.0	29.7	33.0	36.3	39.6	42.9
329.	W30X116	YES	YES	YES	YES	YES	YES	YES
	LC=	11.1	10.3	9.9	9.4	9.0	8.3	7.6
	LU=	13.8	11.8	11.0	9.9	9.5	9.1	8.7
	FB=	24.0	28.0	29.7	33.0	36.5	39.6	42.9
329.	W21X147	YES	YES	YES	YES	YES	YES	YES
	LC=	13.2	12.2	11.8	11.2	10.7	10.2	9.8
	LU=	30.2	25.9	24.2	21.7	19.8	18.1	16.7
	FB=	24.0	28.0	29.7	33.0	36.3	39.6	42.9
329.	W24X131	YES	YES	YES	YES	YES	YES	YES
	LC=	13.6	12.6	12.1	11.5	11.0	10.5	10.1
	LU=	23.3	20.0	18.7	16.8	15.3	14.0	12.9
	FB=	24.0	28.0	29.7	33.0	36.3	39.6	42.9
310.	W14X193	YES	YES	YES	YES	YES	YES	YES
	LC=	16.6	15.4	14.8	14.1	13.4	12.8	12.3
	LU=	67.7	58.0	54.1	48.7	44.3	40.6	37.5
	FB=	24.0	28.0	29.7	33.0	36.3	39.6	42.9
299.	W27X114	YES	YES	YES	YES	YES	YES	YES
	LC=	10.6	9.8	9.5	9.0	8.6	8.2	7.9
	LU=	15.9	13.6	12.7	11.4	10.4	9.5	8.8
	FB=	24.0	28.0	29.7	33.0	36.3	39.6	42.9
299.	W30X108	YES	YES	YES	YES	YES	YES	YES
	LC=	11.1	10.2	9.9	8.9	8.1	7.4	6.8
	LU=	12.4	10.7	10.4	9.8	9.4	9.0	8.6
	FB=	24.0	28.0	29.7	33.0	36.3	39.6	42.9
295.	W21X132	YES	YES	YES	YES	YES	YES	YES
	LC=	13.1	12.2	11.7	11.1	10.6	10.2	9.8
	LU=	27.3	23.4	21.8	19.7	17.9	16.4	15.1
	FB=	24.0	28.0	29.7	33.0	36.3	39.6	42.9
291.	W24X117	YES	YES	YES	YES	YES	YES	YES
	LC=	13.5	12.5	12.1	11.5	10.9	10.5	10.1
	LU=	20.8	17.8	16.6	14.9	13.6	12.5	11.5
	FB=	24.0	28.0	29.7	33.0	36.3	39.6	42.9
281.	W14X176	YES	YES	YES	YES	YES	YES	YES
	LC=	16.5	15.3	14.8	14.0	13.4	12.8	12.3
	LU=	62.4	53.5	49.9	44.9	40.8	37.4	34.5
	FB=	24.0	28.0	29.7	33.0	36.3	39.6	42.9
273.	W21X122	YES	YES	YES	YES	YES	YES	YES
	LC=	13.1	12.1	11.7	11.1	10.6	10.1	9.7
	LU=	25.4	21.8	20.3	18.3	16.6	15.2	14.1
	FB=	24.0	28.0	29.7	33.0	36.3	39.6	42.9
269.	W30X 99	YES	YES	YES	YES	YES	YES	YES
	LC=	10.9	9.4	8.7	7.9	7.2	6.6	6.1
	LU=	11.4	10.6	10.2	9.7	9.2	8.8	8.5
	FB=	24.0	28.0	29.7	33.0	36.3	39.6	42.9
267.	W27X102	YES	YES	YES	YES	YES	YES	YES
	LC=	10.6	9.8	9.5	9.0	8.6	8.2	7.9
	LU=	14.2	12.2	11.4	10.2	9.3	8.8	8.5
	FB=	24.0	28.0	29.7	33.0	36.3	39.6	42.9

Table A3 *(Continued)*

		FY=	36.	42.	45.	50.	55.	60.	65.
263.	W12X190		YES	YES	YES	YES	YES	YES	YES
		LC=	13.4	12.4	12.0	11.3	10.8	10.4	10.0
		LU=	70.8	60.7	56.6	51.0	46.3	42.5	39.2
		FB=	24.0	28.0	29.7	33.0	36.3	39.6	42.9
258.	W24X104		YES	YES	YES	YES	YES	PTY	PTY
		LC=	13.5	12.5	12.0	11.4	10.9	10.4	10.0
		LU=	18.4	15.8	14.7	13.2	12.0	11.5	11.1
		FB=	24.0	28.0	29.7	33.0	36.3	39.5	42.4
254.	W14X159		YES	YES	YES	YES	YES	YES	YES
		LC=	16.4	15.2	14.7	13.9	13.3	12.7	12.2
		LU=	57.2	49.1	45.8	41.2	37.5	34.3	31.7
		FB=	24.0	28.0	29.7	33.0	36.3	39.6	42.9
252.	S24X120		YES	YES	YES	YES	YES	YES	YES
		LC=	8.5	7.9	7.6	7.2	6.9	6.6	6.3
		LU=	17.1	14.7	13.7	12.3	11.2	10.3	9.5
		FB=	24.0	28.0	29.7	33.0	36.3	39.6	42.9
249.	W21X111		YES	YES	YES	YES	YES	YES	YES
		LC=	13.0	12.1	11.7	11.1	10.5	10.1	9.7
		LU=	23.2	19.9	18.6	16.7	15.2	13.9	12.9
		FB=	24.0	28.0	29.7	33.0	36.3	39.6	42.9
243.	W27X 94		YES	YES	YES	YES	YES	YES	YES
		LC=	10.5	9.8	9.4	8.9	8.4	7.7	7.1
		LU=	12.8	11.0	10.2	9.5	9.1	8.7	8.4
		FB=	24.0	28.0	29.7	33.0	36.3	39.6	42.9
236.	S24X105.9		YES	YES	YES	YES	YES	YES	YES
		LC=	8.3	7.7	7.4	7.1	6.7	6.4	6.2
		LU=	16.7	14.3	13.4	12.1	11.0	10.0	9.3
		FB=	24.0	28.0	29.7	33.0	36.3	39.6	42.9
235.	W12X170		YES	YES	YES	YES	YES	YES	YES
		LC=	13.3	12.3	11.9	11.3	10.7	10.3	9.9
		LU=	64.7	55.5	51.8	46.6	42.4	38.8	35.8
		FB=	24.0	28.0	29.7	33.0	36.3	39.6	42.9
232.	W14X145		YES	YES	YES	YES	YES	YES	YES
		LC=	16.4	15.1	14.6	13.9	13.2	12.7	12.2
		LU=	52.9	45.4	42.3	38.1	34.6	31.8	29.3
		FB=	24.0	28.0	29.7	33.0	36.3	39.6	42.9
231.	W18X119		YES	YES	YES	YES	YES	YES	YES
		LC=	11.9	11.0	10.6	10.1	9.6	9.2	8.8
		LU=	29.1	25.0	23.3	21.0	19.1	17.5	16.1
		FB=	24.0	28.0	29.7	33.0	36.3	39.6	42.9
227.	W21X101		YES	YES	YES	YES	YES	YES	YES
		LC=	13.0	12.0	11.6	11.0	10.5	10.0	9.7
		LU=	21.3	18.3	17.0	15.3	13.9	12.8	11.8
		FB=	24.0	28.0	29.7	33.0	36.3	39.6	42.9
222.	W24X 94		YES	YES	YES	YES	YES	YES	YES
		LC=	9.6	8.9	8.6	8.1	7.7	7.4	7.1
		LU=	15.1	12.9	12.1	10.9	9.9	9.1	8.4
		FB=	24.0	28.0	29.7	33.0	36.3	39.6	42.9
213.	W27X 84		YES	YES	YES	YES	YES	YES	YES
		LC=	10.5	9.5	8.8	8.0	7.2	6.6	6.1
		LU=	11.0	10.2	9.9	9.4	8.9	8.6	8.2
		FB=	24.0	28.0	29.7	33.0	36.3	39.6	42.9
209.	W12X152		YES	YES	YES	YES	YES	YES	YES
		LC=	13.2	12.2	11.8	11.2	10.7	10.2	9.8
		LU=	59.0	50.6	47.2	42.5	38.6	35.4	32.7
		FB=	24.0	28.0	29.7	33.0	36.3	39.6	42.9

Table A3 (*Continued*)

		FY=	36.	42.	45.	50.	55.	60.	65.
209.	W14X132		YES	YES	YES	YES	YES	YES	YES
		LC=	15.5	14.4	13.9	13.2	12.6	12.0	11.6
		LU=	47.9	41.1	38.3	34.5	31.4	28.7	26.5
		FB=	24.0	28.0	29.7	33.0	36.3	39.6	42.9
204.	W18X106		YES	YES	YES	YES	YES	YES	YES
		LC=	11.8	10.9	10.6	10.0	9.6	9.2	8.8
		LU=	26.0	22.3	20.8	18.7	17.0	15.6	14.4
		FB=	24.0	28.0	29.7	33.0	36.3	39.6	42.9
199.	S24X100		YES	YES	YES	YES	YES	YES	YES
		LC=	7.6	7.1	6.8	6.5	6.2	5.9	5.7
		LU=	12.2	10.4	9.7	8.8	8.0	7.3	6.7
		FB=	24.0	28.0	29.7	33.0	36.3	39.6	42.9
196.	W24X 84		YES	YES	YES	YES	YES	YES	YES
		LC=	9.5	8.8	8.5	8.1	7.7	7.4	7.1
		LU=	13.3	11.4	10.7	9.6	8.7	8.0	7.6
		FB=	24.0	28.0	29.7	33.0	36.3	39.6	42.9
192.	W21X 93		YES	YES	YES	YES	YES	YES	YES
		LC=	8.9	8.2	7.9	7.5	7.2	6.9	6.6
		LU=	16.8	14.4	13.4	12.1	11.0	10.1	9.3
		FB=	24.0	28.0	29.7	33.0	36.3	39.6	42.9
190.	W14X120		YES	YES	YES	YES	YES	YES	YES
		LC=	15.5	14.3	13.9	13.1	12.5	12.0	11.5
		LU=	44.1	37.8	35.3	31.7	28.9	26.5	24.4
		FB=	24.0	28.0	29.7	33.0	36.3	39.6	42.9
188.	W18X 97		YES	YES	YES	YES	YES	YES	YES
		LC=	11.8	10.9	10.5	10.0	9.5	9.1	8.8
		LU=	24.1	20.7	19.3	17.4	15.8	14.5	13.4
		FB=	24.0	28.0	29.7	33.0	36.3	39.6	42.9
187.	S24X 90		YES	YES	YES	YES	YES	YES	YES
		LC=	7.5	7.0	6.7	6.4	6.1	5.8	5.6
		LU=	12.0	10.3	9.6	8.6	7.8	7.2	6.6
		FB=	24.0	28.0	29.7	33.0	36.3	39.6	42.9
186.	W12X136		YES	YES	YES	YES	YES	YES	YES
		LC=	13.1	12.1	11.7	11.1	10.6	10.1	9.7
		LU=	53.5	45.9	42.8	38.5	35.0	32.1	29.6
		FB=	24.0	28.0	29.7	33.0	36.3	39.6	42.9
176.	W24X 76		YES	YES	YES	YES	YES	YES	YES
		LC=	9.5	8.8	8.5	8.1	7.7	7.1	6.6
		LU=	11.8	10.1	9.5	8.6	8.2	7.9	7.6
		FB=	24.0	28.0	29.7	33.0	36.3	39.6	42.9
175.	W16X100		YES	YES	YES	YES	YES	YES	YES
		LC=	11.0	10.2	9.8	9.3	8.9	8.5	8.2
		LU=	29.8	25.5	23.8	21.4	19.5	17.9	16.5
		FB=	24.0	28.0	29.7	33.0	36.3	39.6	42.9
175.	S24X 79.9		YES	YES	YES	YES	YES	YES	YES
		LC=	7.4	6.8	6.6	6.3	6.0	5.7	5.5
		LU=	11.8	10.1	9.4	8.5	7.7	7.1	6.5
		FB=	24.0	28.0	29.7	33.0	36.3	39.6	42.9
173.	W14X109		YES	YES	YES	YES	YES	PTY	PTY
		LC=	15.4	14.3	13.8	13.1	12.5	11.9	11.5
		LU=	40.6	34.8	32.5	29.2	26.6	24.4	22.5
		FB=	24.0	28.0	29.7	33.0	36.3	39.5	42.5
171.	W21X 83		YES	YES	YES	YES	YES	YES	YES
		LC=	8.8	8.2	7.9	7.5	7.1	6.8	6.6
		LU=	15.1	12.9	12.1	10.9	9.9	9.0	8.3
		FB=	24.0	28.0	29.7	33.0	36.3	39.6	42.9

Table A3 (*Continued*)

		FY=	36.	42.	45.	50.	55.	60.	65.
166.	W18X 86		YES	YES	YES	YES	YES	YES	YES
		LC=	11.7	10.8	10.5	9.9	9.5	9.1	8.7
		LU=	21.5	18.4	17.2	15.5	14.1	12.9	11.9
		FB=	24.0	28.0	29.7	33.0	36.3	39.6	42.9
163.	W12X120		YES	YES	YES	YES	YES	YES	YES
		LC=	13.0	12.0	11.6	11.0	10.5	10.1	9.7
		LU=	48.0	41.2	38.4	34.6	31.4	28.8	26.6
		FB=	24.0	28.0	29.7	33.0	36.3	39.6	42.9
161.	S20X 95		YES	YES	YES	YES	YES	YES	YES
		LC=	7.6	7.0	6.8	6.4	6.1	5.9	5.7
		LU=	15.3	13.1	12.2	11.0	10.0	9.2	8.5
		FB=	24.0	28.0	29.7	33.0	36.3	39.6	42.9
157.	W14X 99		YES	YES	YES	PTY	PTY	PTY	PTY
		LC=	15.4	14.2	13.8	13.0	12.4	11.9	11.4
		LU=	37.1	31.8	29.7	26.7	24.3	22.3	20.6
		FB=	24.0	28.0	29.7	32.9	35.8	38.7	41.6
155.	W16X 89		YES	YES	YES	YES	YES	YES	YES
		LC=	10.9	10.1	9.8	9.3	8.9	8.5	8.1
		LU=	25.1	21.5	20.1	18.0	16.4	15.0	13.9
		FB=	24.0	28.0	29.7	33.0	36.3	39.6	42.9
154.	W24X 68		YES	YES	YES	YES	YES	YES	YES
		LC=	9.5	8.8	8.2	7.4	6.7	6.1	5.7
		LU=	10.2	9.3	9.0	8.5	8.1	7.8	7.5
		FB=	24.0	28.0	29.7	33.0	36.3	39.6	42.9
152.	S20X 85		YES	YES	YES	YES	YES	YES	YES
		LC=	7.4	6.9	6.7	6.3	6.0	5.8	5.5
		LU=	15.0	12.8	12.0	10.8	9.8	9.0	8.3
		FB=	24.0	28.0	29.7	33.0	36.3	39.6	42.9
151.	W21X 73		YES	YES	YES	YES	YES	YES	YES
		LC=	8.8	8.1	7.8	7.4	7.1	6.8	6.5
		LU=	13.4	11.5	10.7	9.6	8.8	8.0	7.4
		FB=	24.0	28.0	29.7	33.0	36.3	39.6	42.9
146.	W18X 76		YES	YES	YES	YES	YES	YES	PTY
		LC=	11.6	10.8	10.4	9.9	9.4	9.0	8.7
		LU=	19.1	16.4	15.3	13.7	12.5	11.4	10.6
		FB=	24.0	28.0	29.7	33.0	36.3	39.6	42.8
145.	W12X106		YES	YES	YES	YES	YES	YES	YES
		LC=	12.9	11.9	11.5	10.9	10.4	10.0	9.6
		LU=	43.5	37.2	34.8	31.3	28.4	26.1	24.1
		FB=	24.0	28.0	29.7	33.0	36.3	39.6	42.9
143.	W14X 90		YES	PTY	PTY	PTY	PTY	PTY	PTY
		LC=	15.3	14.2	13.7	13.0	12.4	11.9	11.4
		LU=	34.0	29.2	27.2	24.5	22.3	20.4	18.9
		FB=	24.0	27.6	29.4	32.3	35.1	37.9	40.6
140.	W21X 68		YES	YES	YES	YES	YES	YES	YES
		LC=	8.7	8.1	7.8	7.4	7.1	6.8	6.5
		LU=	12.4	10.6	9.9	8.9	8.1	7.4	7.0
		FB=	24.0	28.0	29.7	33.0	36.3	39.6	42.9
134.	W16X 77		YES	YES	YES	YES	YES	YES	YES
		LC=	10.9	10.1	9.7	9.2	8.8	8.4	8.1
		LU=	21.9	18.8	17.5	15.8	14.4	13.2	12.1
		FB=	24.0	28.0	29.7	33.0	36.3	39.6	42.9
131.	W12X 96		YES	YES	YES	YES	YES	YES	YES
		LC=	12.8	11.9	11.5	10.9	10.4	9.9	9.6
		LU=	39.9	34.2	31.9	28.7	26.1	23.9	22.1
		FB=	24.0	28.0	29.7	33.0	36.3	39.6	42.9

Table A3 (*Continued*)

		FY=	36.	42.	45.	50.	55.	60.	65.
131.	W24X 62		YES	YES	YES	YES	YES	YES	YES
		LC=	7.4	6.9	6.5	5.8	5.3	4.9	4.5
		LU=	8.1	7.0	6.8	6.4	6.1	5.9	5.6
		FB=	24.0	28.0	29.7	33.0	36.3	39.6	42.9
128.	S20X 75		YES	YES	YES	YES	YES	YES	YES
		LC=	6.7	6.2	6.0	5.7	5.5	5.2	5.0
		LU=	11.7	10.0	9.3	8.4	7.6	7.0	6.5
		FB=	24.0	28.0	29.7	33.0	36.3	39.6	42.9
127.	W18X 71		YES	YES	YES	YES	YES	YES	YES
		LC=	8.1	7.5	7.2	6.8	6.5	6.2	6.0
		LU=	15.5	13.3	12.4	11.2	10.1	9.3	8.6
		FB=	24.0	28.0	29.7	33.0	36.3	39.6	42.9
127.	W21X 62		YES	YES	YES	YES	YES	YES	YES
		LC=	8.7	8.1	7.8	7.4	7.0	6.7	6.2
		LU=	11.2	9.6	8.9	8.0	7.5	7.2	6.9
		FB=	24.0	28.0	29.7	33.0	36.3	39.6	42.9
126.	W10X112		YES	YES	YES	YES	YES	YES	YES
		LC=	11.0	10.2	9.8	9.3	8.9	8.5	8.2
		LU=	53.1	45.5	42.4	38.2	34.7	31.8	29.4
		FB=	24.0	28.0	29.7	33.0	36.3	39.6	42.9
123.	W14X 82		YES	YES	YES	YES	YES	YES	YES
		LC=	10.7	9.9	9.6	9.1	8.7	8.3	8.0
		LU=	28.0	24.0	22.4	20.2	18.3	16.8	15.5
		FB=	24.0	28.0	29.7	33.0	36.3	39.6	42.9
118.	S20X 65.4		YES	YES	YES	YES	YES	YES	YES
		LC=	6.6	6.1	5.9	5.6	5.3	5.1	4.9
		LU=	11.4	9.8	9.1	8.2	7.5	6.8	6.3
		FB=	24.0	28.0	29.7	33.0	36.3	39.6	42.9
118.	W12X 87		YES	YES	YES	YES	YES	YES	YES
		LC=	12.8	11.8	11.4	10.9	10.4	9.9	9.5
		LU=	36.3	31.1	29.0	26.1	23.8	21.8	20.1
		FB=	24.0	28.0	29.7	33.0	36.3	39.6	42.9
117.	W18X 65		YES	YES	YES	YES	YES	YES	YES
		LC=	8.0	7.4	7.2	6.8	6.5	6.2	6.0
		LU=	14.4	12.3	11.5	10.3	9.4	8.6	8.0
		FB=	24.0	28.0	29.7	33.0	36.3	39.6	42.9
117.	W16X 67		YES	YES	YES	YES	YES	YES	YES
		LC=	10.8	10.0	9.7	9.2	8.7	8.4	8.0
		LU=	19.3	16.5	15.4	13.9	12.6	11.6	10.7
		FB=	24.0	28.0	29.7	33.0	36.3	39.6	42.9
114.	W24X 55		YES	YES	YES	YES	YES	YES	YES
		LC=	6.9	6.0	5.6	5.0	4.5	4.2	3.8
		LU=	7.5	6.9	6.7	6.3	6.0	5.8	5.5
		FB=	24.0	28.0	29.7	33.0	36.3	39.6	42.9
112.	W10X100		YES	YES	YES	YES	YES	YES	YES
		LC=	10.9	10.1	9.8	9.3	8.8	8.5	8.1
		LU=	48.3	41.4	38.6	34.8	31.6	29.0	26.8
		FB=	24.0	28.0	29.7	33.0	36.3	39.6	42.9
112.	W14X 74		YES	YES	YES	YES	YES	YES	YES
		LC=	10.6	9.8	9.5	9.0	8.6	8.2	7.9
		LU=	25.8	22.1	20.7	18.6	16.9	15.5	14.3
		FB=	24.0	28.0	29.7	33.0	36.3	39.6	42.9
111.	W21X 57		YES	YES	YES	YES	YES	YES	YES
		LC=	6.9	6.4	6.2	5.9	5.6	5.4	5.1
		LU=	9.4	8.0	7.5	6.7	6.1	5.6	5.4
		FB=	24.0	28.0	29.7	33.0	36.3	39.6	42.9

Table A3 (*Continued*)

			FY=	36.	42.	45.	50.	55.	60.	65.
108.	W18X 60			YES	YES	YES	YES	YES	YES	YES
		LC=		8.0	7.4	7.1	6.8	6.5	6.2	5.9
		LU=		13.3	11.4	10.7	9.6	8.7	8.0	7.4
		FB=		24.0	28.0	29.7	33.0	36.3	39.6	42.9
107.	W12X 79			YES	YES	YES	YES	YES	YES	PTY
		LC=		12.8	11.8	11.4	10.8	10.3	9.9	9.5
		LU=		33.2	28.5	26.6	23.9	21.7	19.9	18.4
		FB=		24.0	28.0	29.7	33.0	36.3	39.6	42.7
103.	W14X 68			YES	YES	YES	YES	YES	YES	YES
		LC=		10.6	9.8	9.5	9.0	8.6	8.2	7.9
		LU=		23.8	20.4	19.1	17.2	15.6	14.3	13.2
		FB=		24.0	28.0	29.7	33.0	36.3	39.6	42.9
103.	S18X 70			YES	YES	YES	YES	YES	YES	YES
		LC=		6.6	6.1	5.9	5.6	5.3	5.1	4.9
		LU=		11.1	9.5	8.9	8.0	7.3	6.7	6.2
		FB=		24.0	28.0	29.7	33.0	36.3	39.6	42.9
98.5	W10X 88			YES	YES	YES	YES	YES	YES	YES
		LC=		10.8	10.0	9.7	9.2	8.8	8.4	8.1
		LU=		43.4	37.2	34.7	31.2	28.4	26.0	24.0
		FB=		24.0	28.0	29.7	33.0	36.3	39.6	42.9
98.3	W18X 55			YES	YES	YES	YES	YES	YES	YES
		LC=		7.9	7.4	7.1	6.7	6.4	6.2	5.9
		LU=		12.1	10.4	9.7	8.7	7.9	7.3	6.7
		FB=		24.0	28.0	29.7	33.0	36.3	39.6	42.9
97.4	W12X 72			YES	YES	YES	YES	PTY	PTY	PTY
		LC=		12.7	11.8	11.4	10.8	10.3	9.8	9.5
		LU=		30.5	26.1	24.4	22.0	20.0	18.3	16.9
		FB=		24.0	28.0	29.7	33.0	36.1	39.0	41.9
94.5	W21X 50			YES	YES	YES	YES	YES	YES	YES
		LC=		6.9	6.4	6.2	5.6	5.1	4.7	4.3
		LU=		7.8	6.7	6.3	6.0	5.7	5.5	5.3
		FB=		24.0	28.0	29.7	33.0	36.3	39.6	42.9
92.2	W14X 61			YES	YES	YES	YES	YES	YES	YES
		LC=		10.6	9.8	9.4	9.0	8.5	8.2	7.9
		LU=		21.5	18.4	17.2	15.5	14.1	12.9	11.9
		FB=		24.0	28.0	29.7	33.0	36.3	39.6	42.9
92.2	W16X 57			YES	YES	YES	YES	YES	YES	YES
		LC=		7.5	7.0	6.7	6.4	6.1	5.8	5.6
		LU=		14.3	12.3	11.5	10.3	9.4	8.6	7.9
		FB=		24.0	28.0	29.7	33.0	36.3	39.6	42.9
89.4	S18X 54.7			YES	YES	YES	YES	YES	YES	YES
		LC=		6.3	5.9	5.7	5.4	5.1	4.9	4.7
		LU=		10.7	9.1	8.5	7.7	7.0	6.4	5.9
		FB=		24.0	28.0	29.7	33.0	36.3	39.6	42.9
88.9	W18X 50			YES	YES	YES	YES	YES	YES	YES
		LC=		7.9	7.3	7.1	6.7	6.4	6.1	5.9
		LU=		11.0	9.4	8.8	7.9	7.2	6.7	6.4
		FB=		24.0	28.0	29.7	33.0	36.3	39.6	42.9
87.9	W12X 65			YES	YES	PTY	PTY	PTY	PTY	PTY
		LC=		12.7	11.7	11.3	10.7	10.2	9.8	9.4
		LU=		27.7	23.8	22.2	20.0	18.2	16.6	15.4
		FB=		24.0	28.0	29.6	32.5	35.4	38.2	41.0
85.9	W10X 77			YES	YES	YES	YES	YES	YES	YES
		LC=		10.8	10.0	9.6	9.1	8.7	8.3	8.0
		LU=		38.7	33.2	31.0	27.9	25.3	23.2	21.4
		FB=		24.0	28.0	29.7	33.0	36.3	39.6	42.9

Table A3 (*Continued*)

			FY=	36.	42.	45.	50.	55.	60.	65.
81.6	W21X 44			YES	YES	YES	YES	YES	YES	YES
		LC=		6.6	5.6	5.2	4.7	4.3	3.9	3.6
		LU=		7.0	6.4	6.2	5.9	5.6	5.4	5.2
		FB=		24.0	28.0	29.7	33.0	36.3	39.6	42.9
81.0	W16X 50			YES	YES	YES	YES	YES	YES	YES
		LC=		7.5	6.9	6.7	6.3	6.0	5.8	5.6
		LU=		12.7	10.9	10.1	9.1	8.3	7.6	7.0
		FB=		24.0	28.0	29.7	33.0	36.3	39.6	42.9
78.8	W18X 46			YES	YES	YES	YES	YES	YES	YES
		LC=		6.4	5.9	5.7	5.4	5.2	5.0	4.8
		LU=		9.4	8.1	7.5	6.8	6.2	5.6	5.2
		FB=		24.0	28.0	29.7	33.0	36.3	39.6	42.9
78.0	W12X 58			YES	YES	YES	YES	YES	YES	YES
		LC=		10.6	9.8	9.5	9.0	8.5	8.2	7.9
		LU=		24.3	20.9	19.5	17.5	15.9	14.6	13.5
		FB=		24.0	28.0	29.7	33.0	36.3	39.6	42.9
77.8	W14X 53			YES	YES	YES	YES	YES	YES	YES
		LC=		8.5	7.9	7.6	7.2	6.9	6.6	6.3
		LU=		17.7	15.2	14.2	12.7	11.6	10.6	9.8
		FB=		24.0	28.0	29.7	33.0	36.3	39.6	42.9
75.7	W10X 68			YES	YES	YES	YES	YES	YES	YES
		LC=		10.7	9.9	9.6	9.1	8.7	8.3	8.0
		LU=		34.7	29.8	27.8	25.0	22.7	20.8	19.2
		FB=		24.0	28.0	29.7	33.0	36.3	39.6	42.9
72.7	W16X 45			YES	YES	YES	YES	YES	YES	YES
		LC=		7.4	6.9	6.6	6.3	6.0	5.8	5.5
		LU=		11.4	9.8	9.1	8.2	7.5	6.8	6.3
		FB=		24.0	28.0	29.7	33.0	36.3	39.6	42.9
70.6	W12X 53			YES	YES	YES	YES	YES	PTY	PTY
		LC=		10.6	9.8	9.4	9.0	8.5	8.2	7.9
		LU=		22.1	18.9	17.6	15.9	14.4	13.2	12.2
		FB=		24.0	28.0	29.7	33.0	36.3	39.3	42.2
70.3	W14X 48			YES	YES	YES	YES	YES	YES	YES
		LC=		8.5	7.8	7.6	7.2	6.9	6.6	6.3
		LU=		16.0	13.7	12.8	11.5	10.5	9.6	8.9
		FB=		24.0	28.0	29.7	33.0	36.3	39.6	42.9
68.4	W18X 40			YES	YES	YES	YES	YES	YES	YES
		LC=		6.3	5.9	5.7	5.4	5.1	4.9	4.5
		LU=		8.2	7.0	6.5	5.9	5.5	5.2	5.0
		FB=		24.0	28.0	29.7	33.0	36.3	39.6	42.9
66.7	W10X 60			YES	YES	YES	YES	YES	YES	YES
		LC=		10.6	9.9	9.5	9.0	8.6	8.2	7.9
		LU=		31.1	26.6	24.8	22.4	20.3	18.6	17.2
		FB=		24.0	28.0	29.7	33.0	36.3	39.6	42.9
64.8	S15X 50			YES	YES	YES	YES	YES	YES	YES
		LC=		6.0	5.5	5.3	5.1	4.8	4.6	4.4
		LU=		10.8	9.3	8.7	7.8	7.1	6.5	6.0
		FB=		24.0	28.0	29.7	33.0	36.3	39.6	42.9
64.7	W12X 50			YES	YES	YES	YES	YES	YES	YES
		LC=		8.5	7.9	7.6	7.2	6.9	6.6	6.3
		LU=		19.6	16.8	15.7	14.1	12.9	11.8	10.9
		FB=		24.0	28.0	29.7	33.0	36.3	39.6	42.9
64.7	W16X 40			YES	YES	YES	YES	YES	YES	YES
		LC=		7.4	6.8	6.6	6.3	6.0	5.7	5.5
		LU=		10.2	8.8	8.2	7.4	6.7	6.3	6.0
		FB=		24.0	28.0	29.7	33.0	36.3	39.6	42.9

Table A3 (Continued)

			FY=	36.	42.	45.	50.	55.	60.	65.
62.7	W14X 43			YES	YES	YES	YES	YES	YES	YES
			LC=	8.4	7.8	7.5	7.2	6.8	6.5	6.3
			LU=	14.4	12.3	11.5	10.3	9.4	8.6	8.0
			FB=	24.0	28.0	29.7	33.0	36.3	39.6	42.9
60.4	W 8X 67			YES	YES	YES	YES	YES	YES	YES
			LC=	8.7	8.1	7.8	7.4	7.1	6.8	6.5
			LU=	39.8	34.1	31.9	28.7	26.1	23.9	22.1
			FB=	24.0	28.0	29.7	33.0	36.3	39.6	42.9
60.0	W10X 54			YES	YES	YES	YES	YES	YES	PTY
			LC=	10.6	9.8	9.5	9.0	8.6	8.2	7.9
			LU=	28.3	24.3	22.6	20.4	18.5	17.0	15.7
			FB=	24.0	28.0	29.7	33.0	36.3	39.6	42.8
59.6	S15X 42.9			YES	YES	YES	YES	YES	YES	YES
			LC=	5.8	5.4	5.2	4.9	4.7	4.5	4.3
			LU=	10.6	9.1	8.4	7.6	6.9	6.3	5.8
			FB=	24.0	28.0	29.7	33.0	36.3	39.6	42.9
58.1	W12X 45			YES	YES	YES	YES	YES	YES	YES
			LC=	8.5	7.9	7.6	7.2	6.9	6.6	6.3
			LU=	17.8	15.2	14.2	12.8	11.6	10.7	9.8
			FB=	24.0	28.0	29.7	33.0	36.3	39.6	42.9
57.6	W18X 35			YES	YES	YES	YES	YES	YES	YES
			LC=	6.3	5.7	5.3	4.8	4.4	4.0	3.7
			LU=	6.7	6.1	5.9	5.6	5.3	5.1	4.9
			FB=	24.0	28.0	29.7	33.0	36.3	39.6	42.9
56.5	W16X 36			YES	YES	YES	YES	YES	YES	PTY
			LĈ=	7.4	6.8	6.6	6.3	5.7	5.3	4.9
			LU=	8.8	7.5	7.1	6.7	6.4	6.2	5.9
			FB=	24.0	28.0	29.7	33.0	36.3	39.6	42.8
54.6	W14X 38			YES	YES	YES	YES	YES	YES	YES
			LC=	7.1	6.6	6.4	6.1	5.8	5.5	5.3
			LU=	11.4	9.8	9.2	8.2	7.5	6.9	6.3
			FB=	24.0	28.0	29.7	33.0	36.3	39.6	42.9
54.6	W10X 49			YES	YES	YES	YES	PTY	PTY	PTY
			LC=	10.6	9.8	9.4	9.0	8.5	8.2	7.9
			LU=	26.0	22.3	20.8	18.7	17.0	15.6	14.4
			FB=	24.0	28.0	29.7	33.0	36.2	39.1	42.0
52.0	W 8X 58			YES	YES	YES	YES	YES	YES	YES
			LC=	8.7	8.0	7.8	7.4	7.0	6.7	6.5
			LU=	35.2	30.2	28.2	25.4	23.1	21.1	19.5
			FB=	24.0	28.0	29.7	33.0	36.3	39.6	42.9
51.9	W12X 40			YES	YES	YES	YES	YES	YES	YES
			LC=	8.4	7.8	7.6	7.2	6.8	6.5	6.3
			LU=	16.0	13.7	12.8	11.5	10.5	9.6	8.9
			FB=	24.0	28.0	29.7	33.0	36.3	39.6	42.9
50.8	S12X 50			YES	YES	YES	YES	YES	YES	YES
			LC=	5.8	5.4	5.2	4.9	4.7	4.5	4.3
			LU=	13.9	11.9	11.1	10.0	9.1	8.4	7.7
			FB=	24.0	28.0	29.7	33.0	36.3	39.6	42.9
49.1	W10X 45			YES	YES	YES	YES	YES	YES	YES
			LC=	8.5	7.8	7.6	7.2	6.8	6.6	6.3
			LU=	22.8	19.5	18.2	16.4	14.9	13.7	12.6
			FB=	24.0	28.0	29.7	33.0	36.3	39.6	42.9
48.6	W14X 34			YES	YES	YES	YES	YES	YES	YES
			LC=	7.1	6.6	6.4	6.0	5.8	5.5	5.3
			LU=	10.2	8.7	8.1	7.3	6.7	6.1	5.8
			FB=	24.0	28.0	29.7	33.0	36.3	39.6	42.9

Table A3 (*Continued*)

			FY=	36.	42.	45.	50.	55.	60.	65.
47.2	W16X 31			YES	YES	YES	YES	YES	YES	YES
		LC=		5.8	5.4	5.2	4.9	4.6	4.3	3.9
		LU=		7.1	6.1	5.7	5.2	5.0	4.8	4.6
		FB=		24.0	28.0	29.7	33.0	36.3	39.6	42.9
45.6	W12X 35			YES	YES	YES	YES	YES	YES	YES
		LC=		6.9	6.4	6.2	5.9	5.6	5.4	5.2
		LU=		12.6	10.8	10.1	9.1	8.3	7.6	7.0
		FB=		24.0	28.0	29.7	33.0	36.3	39.6	42.9
45.4	S12X 40.8			YES	YES	YES	YES	YES	YES	YES
		LC=		5.5	5.1	5.0	4.7	4.5	4.3	4.1
		LU=		13.4	11.4	10.7	9.6	8.7	8.0	7.4
		FB=		24.0	28.0	29.7	33.0	36.3	39.6	42.9
43.3	W 8X 48			YES	YES	YES	YES	YES	YES	YES
		LC=		8.6	7.9	7.7	7.3	6.9	6.6	6.4
		LU=		30.3	25.9	24.2	21.8	19.8	18.2	16.8
		FB=		24.0	28.0	29.7	33.0	36.3	39.6	42.9
42.1	W10X 39			YES	YES	YES	YES	YES	YES	YES
		LC=		8.4	7.8	7.5	7.2	6.8	6.5	6.3
		LU=		19.8	16.9	15.8	14.2	12.9	11.9	10.9
		FB=		24.0	28.0	29.7	33.0	36.3	39.6	42.9
42.0	W14X 30			YES	YES	YES	YES	YES	PTY	PTY
		LC=		7.1	6.6	6.4	6.0	5.7	5.2	4.8
		LU=		8.7	7.4	6.9	6.5	6.2	6.0	5.7
		FB=		24.0	28.0	29.7	33.0	36.3	39.3	42.2
38.6	W12X 30			YES	YES	YES	YES	YES	YES	YES
		LC=		6.9	6.4	6.2	5.8	5.6	5.3	5.1
		LU=		10.8	9.2	8.6	7.7	7.0	6.5	6.0
		FB=		24.0	28.0	29.7	33.0	36.3	39.6	42.9
38.4	W16X 26			YES	YES	YES	YES	YES	YES	YES
		LC=		5.6	4.8	4.5	4.0	3.7	3.4	3.1
		LU=		6.0	5.6	5.4	5.1	4.9	4.7	4.5
		FB=		24.0	28.0	29.7	33.0	36.3	39.6	42.9
38.2	S12X 35			YES	YES	YES	YES	YES	YES	YES
		LC=		5.4	5.0	4.8	4.5	4.3	4.2	4.0
		LU=		10.7	9.1	8.5	7.7	7.0	6.4	5.9
		FB=		24.0	28.0	29.7	33.0	36.3	39.6	42.9
36.4	S12X 31.8			YES	YES	YES	YES	YES	YES	YES
		LC=		5.3	4.9	4.7	4.5	4.3	4.1	3.9
		LU=		10.5	9.0	8.4	7.6	6.9	6.3	5.8
		FB=		24.0	28.0	29.7	33.0	36.3	39.6	42.9
35.5	W 8X 40			YES	YES	YES	YES	YES	YES	YES
		LC=		8.5	7.9	7.6	7.2	6.9	6.6	6.3
		LU=		25.4	21.7	20.3	18.3	16.6	15.2	14.0
		FB=		24.0	28.0	29.7	33.0	36.3	39.6	42.9
35.3	W14X 26			YES	YES	YES	YES	YES	YES	YES
		LC=		5.3	4.9	4.7	4.5	4.3	4.1	3.9
		LU=		7.0	6.0	5.6	5.1	4.6	4.4	4.2
		FB=		24.0	28.0	29.7	33.0	36.3	39.6	42.9
35.0	W10X 33			YES	YES	YES	YES	PTY	PTY	PTY
		LC=		8.4	7.8	7.5	7.1	6.8	6.5	6.3
		LU=		16.5	14.1	13.2	11.9	10.8	9.9	9.1
		FB=		24.0	28.0	29.7	33.0	36.0	38.9	41.8
33.4	W12X 26			YES	YES	YES	YES	YES	PTY	PTY
		LC=		6.9	6.3	6.1	5.8	5.5	5.3	5.1
		LU=		9.3	8.0	7.5	6.7	6.2	5.9	5.7
		FB=		24.0	28.0	29.7	33.0	36.3	39.5	42.4

Table A3 *(Continued)*

			FY= 36.	42.	45.	50.	55.	60.	65.
32.4	W10X 30		YES	YES	YES	YES	YES	YES	YES
		LC=	6.1	5.7	5.5	5.2	5.0	4.8	4.6
		LU=	13.1	11.2	10.5	9.4	8.6	7.9	7.3
		FB=	24.0	28.0	29.7	33.0	36.3	39.6	42.9
31.2	W 8X 35		YES	YES	YES	YES	YES	YES	PTY
		LC=	8.5	7.8	7.6	7.2	6.8	6.6	6.3
		LU=	22.6	19.4	18.1	16.3	14.8	13.6	12.5
		FB=	24.0	28.0	29.7	33.0	36.3	39.6	42.9
29.4	S10X 35		YES	YES	YES	YES	YES	YES	YES
		LC=	5.2	4.8	4.7	4.4	4.2	4.0	3.9
		LU=	11.2	9.6	9.0	8.1	7.4	6.7	6.2
		FB=	24.0	28.0	29.7	33.0	36.3	39.6	42.9
29.1	M 8X 34.3		YES	YES	YES	YES	YES	PTY	PTY
		LC=	8.4	7.8	7.6	7.2	6.8	6.5	6.3
		LU=	21.3	18.2	17.0	15.3	13.9	12.8	11.8
		FB=	24.0	28.0	29.7	33.0	36.3	39.3	42.2
29.0	W14X 22		YES	YES	YES	YES	YES	YES	YES
		LC=	5.3	4.8	4.5	4.1	3.7	3.4	3.1
		LU=	5.6	5.1	5.0	4.7	4.5	4.3	4.1
		FB=	24.0	28.0	29.7	33.0	36.3	39.6	42.9
28.4	M 8X 32.6		YES	YES	YES	YES	YES	PTY	PTY
		LC=	8.4	7.8	7.5	7.1	6.8	6.5	6.2
		LU=	21.1	18.1	16.9	15.2	13.8	12.7	11.7
		FB=	24.0	28.0	29.7	33.0	36.3	39.4	42.3
27.9	W10X 26		YES	YES	YES	YES	YES	YES	YES
		LC=	6.1	5.6	5.4	5.2	4.9	4.7	4.5
		LU=	11.4	9.8	9.1	8.2	7.4	6.8	6.3
		FB=	24.0	28.0	29.7	33.0	36.3	39.6	42.9
27.5	W 8X 31		YES	YES	YES	YES	PTY	PTY	PTY
		LC=	8.4	7.8	7.5	7.2	6.8	6.5	6.3
		LU=	20.1	17.3	16.1	14.5	13.2	12.1	11.1
		FB=	24.0	28.0	29.7	33.0	36.0	38.9	41.7
26.6	M10X 29.1		YES	YES	YES	YES	YES	YES	YES
		LC=	6.3	5.8	5.6	5.3	5.1	4.9	4.7
		LU=	10.8	9.3	8.7	7.8	7.1	6.5	6.0
		FB=	24.0	28.0	29.7	33.0	36.3	39.6	42.9
25.4	W12X 22		YES	YES	YES	YES	YES	YES	YES
		LC=	4.3	3.9	3.8	3.6	3.4	3.3	3.2
		LU=	6.4	5.5	5.2	4.6	4.2	3.9	3.6
		FB=	24.0	28.0	29.7	33.0	36.3	39.6	42.9
24.7	S10X 25.4		YES	YES	YES	YES	YES	YES	YES
		LC=	4.9	4.6	4.4	4.2	4.0	3.8	3.7
		LU=	10.6	9.1	8.5	7.6	6.9	6.4	5.9
		FB=	24.0	28.0	29.7	33.0	36.3	39.6	42.9
24.3	W 8X 28		YES	YES	YES	YES	YES	YES	YES
		LC=	6.9	6.4	6.2	5.9	5.6	5.3	5.1
		LU=	17.5	15.0	14.0	12.6	11.4	10.5	9.7
		FB=	24.0	28.0	29.7	33.0	36.3	39.6	42.9
23.6	M10X 22.9		YES	YES	YES	YES	YES	YES	YES
		LC=	6.1	5.6	5.4	5.2	4.9	4.7	4.5
		LU=	10.5	9.0	8.4	7.5	6.9	6.3	5.8
		FB=	24.0	28.0	29.7	33.0	36.3	39.6	42.9
23.2	W10X 22		YES	YES	YES	YES	YES	YES	YES
		LC=	6.1	5.6	5.4	5.2	4.9	4.7	4.5
		LU=	9.4	8.1	7.5	6.8	6.2	5.7	5.2
		FB=	24.0	28.0	29.7	33.0	36.3	39.6	42.9

Table A3 (*Continued*)

		FY=	36.	42.	45.	50.	55.	60.	65.
21.3	W12X 19		YES	YES	YES	YES	YES	YES	YES
		LC=	4.2	3.9	3.8	3.6	3.4	3.2	3.0
		LU=	5.3	4.6	4.3	3.8	3.6	3.4	3.3
		FB=	24.0	28.0	29.7	33.0	36.3	39.6	42.9
21.1	M14X 17.2		YES	YES	YES	YES	YES	YES	YES
		LC=	3.6	3.1	2.9	2.6	2.4	2.2	2.0
		LU=	4.1	3.8	3.7	3.5	3.3	3.2	3.1
		FB=	24.0	28.0	29.7	33.0	36.3	39.6	42.9
20.9	W 8X 24		YES	YES	YES	YES	YES	YES	PTY
		LC=	6.9	6.3	6.1	5.8	5.5	5.3	5.1
		LU=	15.2	13.0	12.1	10.9	9.9	9.1	8.4
		FB=	24.0	28.0	29.7	33.0	36.3	39.6	42.8
18.8	W10X 19		YES	YES	YES	YES	YES	YES	YES
		LC=	4.2	3.9	3.8	3.6	3.4	3.3	3.2
		LU=	7.2	6.2	5.7	5.2	4.7	4.3	4.0
		FB=	24.0	28.0	29.7	33.0	36.3	39.6	42.9
18.2	W 8X 21		YES	YES	YES	YES	YES	YES	YES
		LC=	5.6	5.2	5.0	4.7	4.5	4.3	4.1
		LU=	11.8	10.1	9.4	8.5	7.7	7.1	6.5
		FB=	24.0	28.0	29.7	33.0	36.3	39.6	42.9
17.1	W12X 16		YES	YES	YES	YES	YES	YES	YES
		LC=	4.1	3.5	3.3	2.9	2.7	2.4	2.3
		LU=	4.3	4.0	3.8	3.6	3.5	3.3	3.2
		FB=	24.0	28.0	29.7	33.0	36.3	39.6	42.9
17.1	M 8X 22.5		YES	YES	YES	YES	YES	YES	YES
		LC=	5.7	5.3	5.1	4.8	4.6	4.4	4.2
		LU=	11.0	9.4	8.8	7.9	7.2	6.6	6.1
		FB=	24.0	28.0	29.7	33.0	36.3	39.6	42.9
16.7	W 6X 25		YES	YES	YES	YES	YES	YES	YES
		LC=	6.4	5.9	5.7	5.4	5.2	5.0	4.8
		LU=	20.1	17.2	16.1	14.5	13.1	12.0	11.1
		FB=	24.0	28.0	29.7	33.0	36.3	39.6	42.9
16.2	W10X 17		YES	YES	YES	YES	YES	YES	YES
		LC=	4.2	3.9	3.8	3.6	3.4	3.3	3.2
		LU=	6.1	5.2	4.8	4.4	4.0	3.6	3.4
		FB=	24.0	28.0	29.7	33.0	36.3	39.6	42.9
16.2	S 8X 23		YES	YES	YES	YES	YES	YES	YES
		LC=	4.4	4.1	3.9	3.7	3.6	3.4	3.3
		LU=	10.3	8.8	8.2	7.4	6.7	6.2	5.7
		FB=	24.0	28.0	29.7	33.0	36.3	39.6	42.9
15.5	M 8X 18.5		YES	YES	YES	YES	YES	YES	YES
		LC=	5.5	5.1	5.0	4.7	4.5	4.3	4.1
		LU=	10.7	9.2	8.6	7.7	7.0	6.4	5.9
		FB=	24.0	28.0	29.7	33.0	36.3	39.6	42.9
15.2	W 8X 18		YES	YES	YES	YES	YES	YES	YES
		LC=	5.5	5.1	5.0	4.7	4.5	4.3	4.1
		LU=	9.9	8.4	7.9	7.1	6.4	5.9	5.5
		FB=	24.0	28.0	29.7	33.0	36.3	39.6	42.9
14.9	W12X 14		YES	YES	YES	YES	PTY	PTY	PTY
		LC=	3.5	3.0	2.8	2.5	2.3	2.1	1.9
		LU=	4.2	3.9	3.8	3.6	3.4	3.3	3.1
		FB=	24.0	28.0	29.7	33.0	36.3	39.2	42.1
14.4	S 8X 18.4		YES	YES	YES	YES	YES	YES	YES
		LC=	4.2	3.9	3.8	3.6	3.4	3.3	3.1
		LU=	9.8	8.4	7.9	7.1	6.4	5.9	5.5
		FB=	24.0	28.0	29.7	33.0	36.3	39.6	42.9

Table A3 (Continued)

		FY=	36.	42.	45.	50.	55.	60.	65.
13.8	W10X 15		YES	YES	YES	YES	YES	YES	YES
		LC=	4.2	3.9	3.8	3.6	3.3	3.0	2.8
		LU=	5.0	4.3	4.0	3.7	3.5	3.4	3.3
		FB=	24.0	28.0	29.7	33.0	36.3	39.6	42.9
13.7	M 6X 22.5		YES	YES	YES	YES	YES	YES	YES
		LC=	6.4	5.9	5.7	5.4	5.2	5.0	4.8
		LU=	17.7	15.2	14.2	12.8	11.6	10.6	9.8
		FB=	24.0	28.0	29.7	33.0	36.3	39.6	42.9
13.4	W 6X 20		YES	YES	YES	YES	YES	YES	PTY
		LC=	6.4	5.9	5.7	5.4	5.1	4.9	4.7
		LU=	16.4	14.1	13.1	11.8	10.7	9.8	9.1
		FB=	24.0	28.0	29.7	33.0	36.3	39.6	42.7
13.0	M 6X 20		YES	YES	YES	YES	YES	YES	YES
		LC=	6.3	5.8	5.6	5.3	5.1	4.9	4.7
		LU=	17.4	14.9	13.9	12.5	11.4	10.4	9.6
		FB=	24.0	28.0	29.7	33.0	36.3	39.6	42.9
12.1	S 7X 20		YES	YES	YES	YES	YES	YES	YES
		LC=	4.1	3.8	3.6	3.5	3.3	3.2	3.0
		LU=	10.0	8.6	8.0	7.2	6.6	6.0	5.5
		FB=	24.0	28.0	29.7	33.0	36.3	39.6	42.9
12.0	M12X 11.8		YES	YES	YES	YES	YES	YES	YES
		LC=	2.7	2.3	2.1	1.9	1.7	1.6	1.5
		LU=	3.1	2.8	2.7	2.6	2.5	2.4	2.3
		FB=	24.0	28.0	29.7	33.0	36.3	39.6	42.9
11.8	W 8X 15		YES	YES	YES	YES	YES	YES	YES
		LC=	4.2	3.9	3.8	3.6	3.4	3.3	3.2
		LU=	7.2	6.2	5.8	5.2	4.7	4.3	4.0
		FB=	24.0	28.0	29.7	33.0	36.3	39.6	42.9
10.9	W10X 12		YES	YES	YES	PTY	PTY	PTY	PTY
		LC=	3.9	3.3	3.1	2.8	2.6	2.3	2.2
		LU=	4.3	4.0	3.8	3.6	3.5	3.3	3.2
		FB=	24.0	28.0	29.7	32.8	35.8	38.6	41.5
10.5	S 7X 15.3		YES	YES	YES	YES	YES	YES	YES
		LC=	3.9	3.6	3.5	3.3	3.1	3.0	2.9
		LU=	9.5	8.1	7.6	6.8	6.2	5.7	5.3
		FB=	24.0	28.0	29.7	33.0	36.3	39.6	42.9
10.2	W 6X 16		YES	YES	YES	YES	YES	YES	YES
		LC=	4.3	3.9	3.8	3.6	3.4	3.3	3.2
		LU=	12.0	10.3	9.6	8.7	7.9	7.2	6.7
		FB=	24.0	28.0	29.7	33.0	36.3	39.6	42.9
10.2	W 5X 19		YES	YES	YES	YES	YES	YES	YES
		LC=	5.3	4.9	4.7	4.5	4.3	4.1	4.0
		LU=	19.4	16.7	15.6	14.0	12.7	11.7	10.8
		FB=	24.0	28.0	29.7	33.0	36.3	39.6	42.9
9.91	W 8X 13		YES	YES	YES	YES	YES	YES	YES
		LC=	4.2	3.9	3.8	3.6	3.4	3.3	3.1
		LU=	5.9	5.1	4.7	4.3	3.9	3.5	3.3
		FB=	24.0	28.0	29.7	33.0	36.3	39.6	42.9
9.72	W 6X 15		PTY	PTY	PTY	PTY	PTY	PTY	PTY
		LC=	6.3	5.9	5.7	5.4	5.1	4.9	4.7
		LU=	12.0	10.3	9.6	8.7	7.9	7.2	6.7
		FB=	23.5	26.9	28.6	31.4	34.1	36.7	39.3
9.63	M 5X 18.9		YES	YES	YES	YES	YES	YES	YES
		LC=	5.3	4.9	4.7	4.5	4.3	4.1	3.9
		LU=	19.3	16.5	15.4	13.9	12.6	11.6	10.7
		FB=	24.0	28.0	29.7	33.0	36.3	39.6	42.9

Table A3 (*Continued*)

		FY=	36.	42.	45.	50.	55.	60.	65.
8.77	S 6X 17.25		YES	YES	YES	YES	YES	YES	YES
		LC=	3.8	3.5	3.4	3.2	3.0	2.9	2.8
		LU=	9.9	8.5	7.9	7.1	6.5	5.9	5.5
		FB=	24.0	28.0	29.7	33.0	36.3	39.6	42.9
8.55	W 5X 16		YES	YES	YES	YES	YES	YES	YES
		LC=	5.3	4.9	4.7	4.5	4.3	4.1	3.9
		LU=	16.6	14.3	13.3	12.0	10.9	10.0	9.2
		FB=	24.0	28.0	29.7	33.0	36.3	39.6	42.9
7.81	W 8X 10		YES	YES	YES	PTY	PTY	PTY	PTY
		LC=	4.2	3.9	3.7	3.4	3.1	2.8	2.6
		LU=	4.7	4.1	3.9	3.7	3.6	3.4	3.3
		FB=	24.0	28.0	29.7	32.7	35.6	38.5	41.3
7.76	M10X 9		YES	YES	YES	YES	YES	YES	YES
		LC=	2.6	2.2	2.1	1.8	1.7	1.5	1.4
		LU=	2.7	2.5	2.4	2.3	2.2	2.1	2.0
		FB=	24.0	28.0	29.7	33.0	36.3	39.6	42.9
7.37	S 6X 12.5		YES	YES	YES	YES	YES	YES	YES
		LC=	3.5	3.3	3.1	3.0	2.8	2.7	2.6
		LU=	9.2	7.9	7.4	6.6	6.0	5.5	5.1
		FB=	24.0	28.0	29.7	33.0	36.3	39.6	42.9
7.31	W 6X 12		YES	YES	YES	YES	YES	YES	YES
		LC=	4.2	3.9	3.8	3.6	3.4	3.3	3.1
		LU=	8.6	7.4	6.9	6.2	5.6	5.2	4.8
		FB=	24.0	28.0	29.7	33.0	36.3	39.6	42.9
6.09	S 5X 14.75		YES	YES	YES	YES	YES	YES	YES
		LC=	3.5	3.2	3.1	2.9	2.8	2.7	2.6
		LU=	9.9	8.5	7.9	7.1	6.5	5.9	5.5
		FB=	24.0	28.0	29.7	33.0	36.3	39.6	42.9
5.56	W 6X 9		YES	YES	YES	YES	PTY	PTY	PTY
		LC=	4.2	3.9	3.7	3.5	3.4	3.2	3.1
		LU=	6.6	5.7	5.3	4.8	4.4	4.0	3.7
		FB=	24.0	28.0	29.7	33.0	36.0	38.9	41.7
5.46	W 4X 13		YES	YES	YES	YES	YES	YES	YES
		LC=	4.3	4.0	3.8	3.6	3.5	3.3	3.2
		LU=	15.6	13.4	12.5	11.2	10.2	9.4	8.6
		FB=	24.0	28.0	29.7	33.0	36.3	39.6	42.9
5.42	M 4X 13.8		YES	YES	YES	YES	YES	YES	YES
		LC=	4.2	3.9	3.8	3.6	3.4	3.3	3.1
		LU=	17.2	14.7	13.7	12.4	11.2	10.3	9.5
		FB=	24.0	28.0	29.7	33.0	36.3	39.6	42.9
5.24	M 4X 13		YES	YES	YES	YES	YES	YES	YES
		LC=	4.2	3.9	3.7	3.5	3.4	3.2	3.1
		LU=	16.9	14.5	13.5	12.2	11.1	10.2	9.4
		FB=	24.0	28.0	29.7	33.0	36.3	39.6	42.9
4.92	S 5X 10		YES	YES	YES	YES	YES	YES	YES
		LC=	3.2	2.9	2.8	2.7	2.6	2.5	2.4
		LU=	9.1	7.8	7.3	6.5	5.9	5.4	5.0
		FB=	24.0	28.0	29.7	33.0	36.3	39.6	42.9
4.62	M 8X 6.5		YES	YES	YES	YES	YES	YES	YES
		LC=	2.4	2.1	2.0	1.8	1.6	1.5	1.4
		LU=	2.5	2.2	2.1	2.0	1.9	1.8	1.8
		FB=	24.0	28.0	29.7	33.0	36.3	39.6	42.9
3.44	M 7X 5.5		YES	YES	YES	YES	YES	YES	YES
		LC=	2.2	2.0	2.0	1.8	1.6	1.5	1.4
		LU=	2.5	2.1	2.0	1.9	1.8	1.7	1.6
		FB=	24.0	28.0	29.7	33.0	36.3	39.6	42.9

Table A3 (*Continued*)

			FY=	36.	42.	45.	50.	55.	60.	65.
3.39	S 4X	9.5		YES	YES	YES	YES	YES	YES	YES
			LC=	3.0	2.7	2.6	2.5	2.4	2.3	2.2
			LU=	9.5	8.1	7.6	6.8	6.2	5.7	5.3
			FB=	24.0	28.0	29.7	33.0	36.3	39.6	42.9
3.04	S 4X	7.7		YES	YES	YES	YES	YES	YES	YES
			LC=	2.8	2.6	2.5	2.4	2.3	2.2	2.1
			LU=	9.0	7.7	7.2	6.5	5.9	5.4	5.0
			FB=	24.0	28.0	29.7	33.0	36.3	39.6	42.9
2.40	M 6X	4.4		YES	YES	YES	YES	YES	YES	YES
			LC=	1.9	1.8	1.7	1.7	1.6	1.5	1.3
			LU=	2.4	2.1	1.9	1.8	1.6	1.5	1.5
			FB=	24.0	28.0	29.7	33.0	36.3	39.6	42.9
1.95	S 3X	7.5		YES	YES	YES	YES	YES	YES	YES
			LC=	2.6	2.5	2.4	2.2	2.1	2.1	2.0
			LU=	10.1	8.6	8.1	7.2	6.6	6.0	5.6
			FB=	24.0	28.0	29.7	33.0	36.3	39.6	42.9
1.68	S 3X	5.7		YES	YES	YES	YES	YES	YES	YES
			LC=	2.5	2.3	2.2	2.1	2.0	1.9	1.8
			LU=	9.3	8.0	7.5	6.7	6.1	5.6	5.2
			FB=	24.0	28.0	29.7	33.0	36.3	39.6	42.9

Table A4 Plastic Section Modulus Z_x

ZX	SHAPE	RY	SATISFIES AISC-2.7 (YES OR NO)						
		FY= 36.	42.	45.	50.	55.	60.	65.
1660.	W14X730	4.69	YES	YES	YES	YES	YES	YES	YES
1480.	W14X665	4.61	YES	YES	YES	YES	YES	YES	YES
1320.	W14X605	4.55	YES	YES	YES	YES	YES	YES	YES
1260.	W36X300	3.84	YES	YES	YES	YES	YES	YES	YES
1180.	W14X550	4.48	YES	YES	YES	YES	YES	YES	YES
1170.	W36X280	3.82	YES	YES	YES	YES	YES	YES	YES
1080.	W36X260	3.77	YES	YES	YES	YES	YES	YES	YES
1050.	W14X500	4.43	YES	YES	YES	YES	YES	YES	YES
1010.	W36X245	3.74	YES	YES	YES	YES	YES	YES	NO!
943.	W36X230	3.73	YES	YES	YES	YES	YES	NO!	NO!
939.	W33X241	3.63	YES	YES	YES	YES	YES	YES	YES
936.	W14X455	4.37	YES	YES	YES	YES	YES	YES	YES
869.	W14X426	4.35	YES	YES	YES	YES	YES	YES	YES
855.	W33X221	3.59	YES	YES	YES	YES	YES	YES	NO!
833.	W36X210	2.58	YES	YES	YES	YES	YES	YES	YES
801.	W14X398	4.31	YES	YES	YES	YES	YES	YES	YES
772.	W33X201	3.56	YES	YES	YES	YES	NO!	NO!	NO!
767.	W36X194	2.56	YES	YES	YES	YES	YES	YES	YES
749.	W30X211	3.49	YES	YES	YES	YES	YES	YES	YES
736.	W14X370	4.27	YES	YES	YES	YES	YES	YES	YES
718.	W36X182	2.54	YES	YES	YES	YES	YES	YES	YES
673.	W30X191	3.46	YES	YES	YES	YES	YES	NO!	NO!
672.	W14X342	4.23	YES	YES	YES	YES	YES	YES	YES
668.	W36X170	2.53	YES	YES	YES	YES	YES	NO!	NO!
624.	W36X160	2.51	YES	YES	YES	YES	YES	NO!	NO!
605.	W30X173	3.43	YES	YES	YES	NO!	NO!	NO!	NO!
603.	W14X311	4.20	YES	YES	YES	YES	YES	YES	YES
581.	W36X150	2.47	YES	YES	YES	YES	NO!	NO!	NO!
567.	W27X178	3.26	YES	YES	YES	YES	YES	YES	YES
559.	W33X152	2.47	YES	YES	YES	YES	YES	YES	NO!
542.	W14X283	4.16	YES	YES	YES	YES	YES	YES	YES
514.	W33X141	2.43	YES	YES	YES	YES	YES	NO!	NO!
512.	W27X161	3.24	YES	YES	YES	YES	YES	NO!	NO!
509.	W36X135	2.38	YES	YES	NO!	NO!	NO!	NO!	NO!
487.	W14X257	4.13	YES	YES	YES	YES	YES	YES	YES
468.	W24X162	3.05	YES	YES	YES	YES	YES	YES	YES
467.	W33X130	2.39	YES	YES	YES	YES	NO!	NO!	NO!
461.	W27X146	3.21	YES	YES	YES	NO!	NO!	NO!	NO!
437.	W30X132	2.24	YES	YES	YES	YES	YES	YES	YES
436.	W14X233	4.10	YES	YES	YES	YES	YES	YES	YES
418.	W24X146	3.02	YES	YES	YES	YES	YES	YES	YES
415.	W33X118	2.32	YES	YES	NO!	NO!	NO!	NO!	NO!
408.	W30X124	2.23	YES	YES	YES	YES	YES	YES	NO!
390.	W14X211	4.08	YES	YES	YES	YES	YES	YES	YES
378.	W30X116	2.19	YES	YES	YES	YES	YES	YES	NO!
373.	W21X147	2.95	YES	YES	YES	YES	YES	YES	YES
370.	W24X131	2.97	YES	YES	YES	YES	NO!	NO!	NO!
355.	W14X193	4.05	YES	YES	YES	YES	YES	YES	YES
346.	W30X108	2.15	YES	YES	YES	YES	YES	NO!	NO!
343.	W27X114	2.18	YES	YES	YES	YES	YES	YES	YES
333.	W21X132	2.93	YES	YES	YES	YES	YES	YES	NO!
327.	W24X117	2.94	YES	YES	NO!	NO!	NO!	NO!	NO!
320.	W14X176	4.02	YES	YES	YES	YES	YES	YES	YES
312.	W30X 99	2.10	YES	YES	NO!	NO!	NO!	NO!	NO!
311.	W12X190	3.25	YES	YES	YES	YES	YES	YES	YES
307.	W21X122	2.91	YES	YES	YES	YES	YES	NO!	NO!

Table A4 (*Continued*)

			FY= 36.	42.	45.	50.	55.	60.	65.
305.	W27X102	2.15	YES	YES	YES	YES	YES	YES	NO!
299.	S24X120	1.54	YES	YES	YES	YES	YES	YES	YES
289.	W24X104	2.91	YES	NO!	NO!	NO!	NO!	NO!	NO!
287.	W14X159	4.00	YES	YES	YES	YES	YES	NO!	NO!
279.	W21X111	2.89	YES	YES	YES	NO!	NO!	NO!	NO!
278.	W27X 94	2.12	YES	YES	YES	YES	NO!	NO!	NO!
275.	W12X170	3.22	YES	YES	YES	YES	YES	YES	YES
274.	S24X105.9	1.59	YES	YES	YES	YES	YES	YES	YES
261.	W18X119	2.68	YES	YES	YES	YES	YES	YES	YES
260.	W14X145	3.98	YES	YES	YES	NO!	NO!	NO!	NO!
254.	W24X 94	1.98	YES	YES	YES	YES	YES	YES	YES
253.	W21X101	2.88	YES	YES	NO!	NO!	NO!	NO!	NO!
244.	W27X 84	2.07	YES	YES	NO!	NO!	NO!	NO!	NO!
243.	W12X152	3.19	YES	YES	YES	YES	YES	YES	YES
240.	S24X100	1.28	YES	YES	YES	YES	YES	YES	YES
234.	W14X132	3.76	YES	YES	YES	NO!	NO!	NO!	NO!
230.	W18X106	2.66	YES	YES	YES	YES	YES	YES	YES
224.	W24X 84	1.95	YES	YES	YES	YES	YES	YES	NO!
222.	S24X 90	1.30	YES	YES	YES	YES	YES	YES	YES
221.	W21X 93	1.84	YES	YES	YES	YES	YES	YES	YES
214.	W12X136	3.16	YES	YES	YES	YES	YES	YES	YES
212.	W14X120	3.74	YES	YES	NO!	NO!	NO!	NO!	NO!
211.	W18X 97	2.66	YES	YES	YES	YES	YES	NO!	NO!
205.	S24X 79.9	1.34	YES	YES	YES	YES	YES	YES	YES
200.	W24X 76	1.92	YES	YES	YES	YES	NO!	NO!	NO!
198.	W16X100	2.52	YES	YES	YES	YES	YES	YES	YES
196.	W21X 83	1.83	YES	YES	YES	YES	YES	YES	YES
194.	S20X 95	1.33	YES	YES	YES	YES	YES	YES	YES
192.	W14X109	3.74	YES	NO!	NO!	NO!	NO!	NO!	NO!
186.	W18X 86	2.63	YES	YES	YES	NO!	NO!	NO!	NO!
186.	W12X120	3.13	YES	YES	YES	YES	YES	YES	YES
179.	S20X 85	1.36	YES	YES	YES	YES	YES	YES	YES
177.	W24X 68	1.87	YES	YES	NO!	NO!	NO!	NO!	NO!
175.	W16X 89	2.49	YES	YES	YES	YES	YES	YES	YES
173.	W14X 99	3.72	NO!	NO!	NO!	NO!	NO!	NO!	NO!
172.	W21X 73	1.81	YES	YES	YES	YES	YES	YES	YES
164.	W12X106	3.11	YES	YES	YES	YES	YES	YES	NO!
163.	W18X 76	2.61	YES	NO!	NO!	NO!	NO!	NO!	NO!
160.	W21X 68	1.80	YES	YES	YES	YES	YES	YES	NO!
157.	W14X 90	3.70	NO!	NO!	NO!	NO!	NO!	NO!	NO!
153.	S20X 75	1.16	YES	YES	YES	YES	YES	YES	YES
153.	W24X 62	1.38	YES	YES	YES	YES	YES	NO!	NO!
150.	W16X 77	2.47	YES	YES	YES	YES	NO!	NO!	NO!
147.	W12X 96	3.09	YES	YES	YES	YES	NO!	NO!	NO!
147.	W10X112	2.68	YES	YES	YES	YES	YES	YES	YES
145.	W18X 71	1.70	YES	YES	YES	YES	YES	YES	YES
144.	W21X 62	1.77	YES	YES	YES	YES	NO!	NO!	NO!
139.	W14X 82	2.48	YES	YES	YES	YES	YES	YES	YES
138.	S20X 65.4	1.19	YES	YES	YES	YES	YES	YES	YES
134.	W24X 55	1.34	YES	YES	YES	NO!	NO!	NO!	NO!
133.	W18X 65	1.69	YES	YES	YES	YES	YES	YES	YES
132.	W12X 87	3.07	YES	YES	NO!	NO!	NO!	NO!	NO!
130.	W10X100	2.65	YES	YES	YES	YES	YES	YES	YES
130.	W16X 67	2.46	YES	YES	NO!	NO!	NO!	NO!	NO!
129.	W21X 57	1.35	YES	YES	YES	YES	YES	YES	NO!
126.	W14X 74	2.48	YES	YES	YES	YES	YES	NO!	NO!

Table A4 (*Continued*)

		FY= 36.	42.	45.	50.	55.	60.	65.
125.	S18X 70	1.08	YES	YES	YES	YES	YES	YES	YES
123.	W18X 60	1.69	YES	YES	YES	YES	YES	YES	YES
119.	W12X 79	3.05	YES	NO!	NO!	NO!	NO!	NO!	NO!
115.	W14X 68	2.46	YES	YES	YES	YES	NO!	NO!	NO!
113.	W10X 88	2.63	YES	YES	YES	YES	YES	YES	YES
112.	W18X 55	1.66	YES	YES	YES	YES	YES	YES	YES
110.	W21X 50	1.30	YES	YES	YES	YES	YES	NO!	NO!
108.	W12X 72	3.04	NO!	NO!	NO!	NO!	NO!	NO!	NO!
105.	S18X 54.7	1.14	YES	YES	YES	YES	YES	YES	YES
105.	W16X 57	1.60	YES	YES	YES	YES	YES	YES	YES
102.	W14X 61	2.44	YES	YES	NO!	NO!	NO!	NO!	NO!
101.	W18X 50	1.65	YES	YES	YES	YES	YES	NO!	NO!
97.6	W10X 77	2.61	YES	YES	YES	YES	YES	YES	YES
96.8	W12X 65	3.02	NO!	NO!	NO!	NO!	NO!	NO!	NO!
95.4	W21X 44	1.26	YES	YES	YES	NO!	NO!	NO!	NO!
92.0	W16X 50	1.59	YES	YES	YES	YES	YES	YES	YES
90.7	W18X 46	1.29	YES	YES	YES	YES	YES	YES	YES
87.1	W14X 53	1.92	YES	YES	YES	YES	YES	YES	NO!
86.4	W12X 58	2.51	YES	YES	NO!	NO!	NO!	NO!	NO!
85.3	W10X 68	2.59	YES	YES	YES	YES	YES	NO!	NO!
82.3	W16X 45	1.57	YES	YES	YES	YES	YES	YES	NO!
78.4	W18X 40	1.27	YES	YES	YES	YES	NO!	NO!	NO!
78.4	W14X 48	1.91	YES	YES	YES	YES	NO!	NO!	NO!
77.9	W12X 53	2.48	NO!	NO!	NO!	NO!	NO!	NO!	NO!
77.1	S15X 50	1.03	YES	YES	YES	YES	YES	YES	YES
74.6	W10X 60	2.57	YES	YES	NO!	NO!	NO!	NO!	NO!
72.9	W16X 40	1.56	YES	YES	YES	YES	NO!	NO!	NO!
72.4	W12X 50	1.96	YES	YES	YES	YES	YES	NO!	NO!
70.2	W 8X 67	2.12	YES	YES	YES	YES	YES	YES	YES
69.6	W14X 43	1.89	YES	YES	NO!	NO!	NO!	NO!	NO!
69.3	S15X 42.9	1.07	YES	YES	YES	YES	YES	YES	YES
66.6	W10X 54	2.55	YES	NO!	NO!	NO!	NO!	NO!	NO!
66.5	W18X 35	1.22	YES	YES	YES	NO!	NO!	NO!	NO!
64.7	W12X 45	1.95	YES	YES	YES	YES	NO!	NO!	NO!
64.0	W16X 36	1.52	YES	NO!	NO!	NO!	NO!	NO!	NO!
61.5	W14X 38	1.54	YES	YES	YES	YES	YES	NO!	NO!
61.2	S12X 50	1.03	YES	YES	YES	YES	YES	YES	YES
60.4	W10X 49	2.55	NO!	NO!	NO!	NO!	NO!	NO!	NO!
59.8	W 8X 58	2.10	YES	YES	YES	YES	YES	YES	YES
57.5	W12X 40	1.93	YES	YES	NO!	NO!	NO!	NO!	NO!
54.9	W10X 45	2.00	YES	YES	YES	YES	YES	NO!	NO!
54.6	W14X 34	1.53	YES	YES	NO!	NO!	NO!	NO!	NO!
54.0	W16X 31	1.17	YES	YES	YES	YES	NO!	NO!	NO!
53.1	S12X 40.8	1.06	YES	YES	YES	YES	YES	YES	YES
51.2	W12X 35	1.54	YES	YES	YES	YES	YES	NO!	NO!
49.0	W 8X 48	2.08	YES	YES	YES	YES	YES	YES	YES
47.3	W14X 30	1.49	NO!	NO!	NO!	NO!	NO!	NO!	NO!
46.8	W10X 39	1.98	YES	YES	NO!	NO!	NO!	NO!	NO!
44.8	S12X 35	.98	YES	YES	YES	YES	YES	YES	YES
44.2	W16X 26	1.12	YES	YES	NO!	NO!	NO!	NO!	NO!
43.1	W12X 30	1.52	YES	YES	NO!	NO!	NO!	NO!	NO!
42.0	S12X 31.8	1.00	YES	YES	YES	YES	YES	YES	YES
40.2	W14X 26	1.08	YES	YES	YES	YES	YES	NO!	NO!
39.8	W 8X 40	2.05	YES	YES	YES	NO!	NO!	NO!	NO!
38.8	W10X 33	1.94	NO!	NO!	NO!	NO!	NO!	NO!	NO!
37.2	W12X 26	1.50	NO!	NO!	NO!	NO!	NO!	NO!	NO!

Table A4 (Continued)

		FY= 36.	42.	45.	50.	55.	60.	65.
36.6	W10X 30	1.37	YES	YES	YES	YES	YES	YES	YES
35.4	S10X 35	.90	YES	YES	YES	YES	YES	YES	YES
34.7	W 8X 35	2.03	YES	NO!	NO!	NO!	NO!	NO!	NO!
33.2	W14X 22	1.04	YES	YES	NO!	NO!	NO!	NO!	NO!
32.6	M 8X 34.3	1.86	NO!	NO!	NO!	NO!	NO!	NO!	NO!
31.6	M 8X 32.6	1.89	NO!	NO!	NO!	NO!	NO!	NO!	NO!
31.3	W10X 26	1.36	YES	YES	YES	YES	YES	NO!	NO!
30.9	M10X 29.1	1.14	YES	YES	NO!	NO!	NO!	NO!	NO!
30.4	W 8X 31	2.02	NO!	NO!	NO!	NO!	NO!	NO!	NO!
29.3	W12X 22	.85	YES	YES	YES	YES	YES	YES	YES
28.4	S10X 25.4	.95	YES	YES	YES	YES	YES	YES	YES
27.2	W 8X 28	1.62	YES	YES	YES	NO!	NO!	NO!	NO!
26.4	M10X 22.9	1.22	YES	YES	YES	NO!	NO!	NO!	NO!
26.0	W10X 22	1.33	YES	YES	NO!	NO!	NO!	NO!	NO!
24.8	M14X 17.2	.72	YES	NO!	NO!	NO!	NO!	NO!	NO!
24.7	W12X 19	.82	YES	YES	YES	YES	YES	YES	NO!
23.2	W 8X 24	1.61	YES	NO!	NO!	NO!	NO!	NO!	NO!
21.6	W10X 19	.87	YES	YES	YES	YES	YES	YES	YES
20.4	W 8X 21	1.26	YES	YES	YES	YES	YES	NO!	NO!
20.1	W12X 16	.77	YES	YES	NO!	NO!	NO!	NO!	NO!
19.7	M 8X 22.5	1.06	YES	YES	NO!	NO!	NO!	NO!	NO!
19.3	S 8X 23	.80	YES	YES	YES	YES	YES	YES	YES
18.9	W 6X 25	1.53	YES	YES	YES	YES	NO!	NO!	NO!
18.7	W10X 17	.84	YES	YES	YES	YES	YES	YES	NO!
17.4	W12X 14	.75	NO!	NO!	NO!	NO!	NO!	NO!	NO!
17.4	M 8X 18.5	1.12	YES	YES	NO!	NO!	NO!	NO!	NO!
17.0	W 8X 18	1.23	YES	YES	NO!	NO!	NO!	NO!	NO!
16.5	S 8X 18.4	.83	YES	YES	YES	YES	YES	YES	YES
16.0	W10X 15	.81	YES	YES	NO!	NO!	NO!	NO!	NO!
15.6	M 6X 22.5	1.37	YES	YES	NO!	NO!	NO!	NO!	NO!
14.9	W 6X 20	1.51	YES	NO!	NO!	NO!	NO!	NO!	NO!
14.5	S 7X 20	.73	YES	YES	YES	YES	YES	YES	YES
14.5	M 6X 20	1.40	YES	YES	NO!	NO!	NO!	NO!	NO!
14.3	M12X 11.8	.53	YES	NO!	NO!	NO!	NO!	NO!	NO!
13.6	W 8X 15	.88	YES	YES	YES	YES	YES	NO!	NO!
12.6	W10X 12	.78	NO!	NO!	NO!	NO!	NO!	NO!	NO!
12.1	S 7X 15.3	.77	YES	YES	YES	YES	YES	YES	YES
11.7	W 6X 16	.97	YES	YES	YES	YES	YES	YES	YES
11.6	W 5X 19	1.28	YES	YES	YES	YES	YES	YES	YES
11.4	W 8X 13	.84	YES	YES	NO!	NO!	NO!	NO!	NO!
11.0	M 5X 18.9	1.19	YES	YES	YES	YES	YES	YES	NO!
10.8	W 6X 15	1.45	NO!	NO!	NO!	NO!	NO!	NO!	NO!
10.6	S 6X 17.25	.67	YES	YES	YES	YES	YES	YES	YES
9.63	W 5X 16	1.26	YES	YES	YES	YES	NO!	NO!	NO!
9.19	M10X 9	.48	YES	NO!	NO!	NO!	NO!	NO!	NO!
8.87	W 8X 10	.84	NO!	NO!	NO!	NO!	NO!	NO!	NO!
8.47	S 6X 12.5	.70	YES	YES	YES	YES	YES	YES	YES
8.30	W 6X 12	.92	YES	YES	YES	NO!	NO!	NO!	NO!
7.42	S 5X 14.75	.62	YES	YES	YES	YES	YES	YES	YES
6.31	M 4X 13.8	.94	YES	YES	YES	YES	YES	YES	YES
6.28	W 4X 13	1.00	YES	YES	YES	YES	YES	YES	YES
6.23	W 6X 9	.91	NO!	NO!	NO!	NO!	NO!	NO!	NO!
6.06	M 4X 13	.94	YES	YES	YES	YES	YES	YES	YES
5.67	S 5X 10	.64	YES	YES	YES	YES	YES	YES	YES
5.42	M 8X 6.5	.42	YES	YES	YES	NO!	NO!	NO!	NO!
4.04	S 4X 9.5	.57	YES	YES	YES	YES	YES	YES	YES

Table A4 (*Continued*)

				FY= 36.	42.	45.	50.	55.	60.	65.
4.03	M 7X	5.5	.39	YES	YES	YES	YES	YES	NO!	NO!
3.51	S 4X	7.7	.58	YES	YES	YES	YES	YES	YES	YES
2.80	M 6X	4.4	.36	YES	YES	YES	YES	YES	YES	NO!
2.36	S 3X	7.5	.51	YES	YES	YES	YES	YES	YES	YES
1.95	S 3X	5.7	.52	YES	YES	YES	YES	YES	YES	YES

Index